ADSORPTION REFRIGERATION TECHNOLOGY

ADSORPTION REFRIGERATION TECHNOLOGY

THEORY AND APPLICATION

Ruzhu Wang, Liwei Wang and Jingyi Wu

Shanghai Jiao Tong University, China

WILEY

This edition first published 2014

Registered office
John Wiley & Sons Singapore Pte. Ltd., 1 Fusionopolis Walk, #07-01 Solaris South Tower, Singapore 138628.

For details of our global editorial offices, for customer services and for information about how to apply for permission to reuse the copyright material in this book please see our website at www.wiley.com.

Library of Congress Cataloging-in-Publication Data

Wang, Ruzhu.
Adsorption refrigeration technology : theory and application / Ruzhu Z. Wang, Liwei Wang, Jingyi Wu.
1 online resource.
Includes bibliographical references and index.
Description based on print version record and CIP data provided by publisher; resource not viewed.
ISBN 978-1-118-19746-2 (Adobe PDF) – ISBN 978-1-118-19747-9 (ePub) – ISBN 978-1-118-19743-1 (hardback) 1. Refrigeration and refrigerating machinery – Research. 2. Refrigeration and refrigerating machinery – Technological innovations. 3. Refrigeration and refrigerating machinery – Environmental aspects. 4. Adsorption. I. Wang, Liwei (Professor) II. Wu, Jingyi, Ph.D. III. Title.
TP492.5
621.5′7 – dc23

2014003757

Typeset in 10/12pt Times by Laserwords Private Limited, Chennai, India

ISBN: 978-1-118-19743-1

1 2014

Contents

About the Authors

Ruzhu Wang (R.Z. Wang) is a Professor of Institute of Refrigeration and Cryogenics at Shanghai Jiao Tong University. His major contributions are adsorption refrigeration, heat transfer of superfluid helium, heat pumps, CCHPs (cogeneration systems for cooling, heat, and power), and solar energy systems. He has published about 300 journal papers; about 200 of them are in international journals. He has written five books regarding Refrigeration Technologies. He was elected as CheungKong Chaired Professor in 2000 by the Ministry of Education (MOE) of China. Currently he is the vice president of the Chinese Association of Refrigeration, the vice chairman of the Chinese Society of Heat Transfer. Professor Wang was elected as one of the top 100 outstanding professors in Chinese universities in 2007. He was awarded as the model teacher of China in 2009. Professor Wang won second prize for the National Invention Award in 2010 on "Solar air conditioning and efficient heating units and their application," and also received the second prize for the National Award for Education in 2009 for his ideas and successful practices on "Innovative, Globalization, and Research Learning" for talents education in the field of refrigeration.

Liwei Wang (L.W. Wang) is Professor of the Institute of Refrigeration and Cryogenics at Shanghai Jiao Tong University. Her research experience focuses on the conversion of low grade waste heat using the technology of adsorption, such as the adsorption refrigeration cycle, intensification of the heat and mass transfer performance of adsorbents, and adsorption cogeneration cycle for refrigeration and power generation. For her research work she received awards such as the National 100 Outstanding PhD Theses, IIR Young Researchers Award, Royal Society International Incoming Fellowship in the UK, and the EU Marie Curie International Incoming Fellowship.

Jingyi Wu (J.Y. Wu) is a Professor of the Institute of Refrigeration and Cryogenics at Shanghai Jiao Tong University. Her achievements are mainly in the utilization of low grade heat and cryogenics for aerospace. She has published over 130 papers and has led various research projects funded by National Natural Science Foundation of China (NSFC), Hi-Tech Research and Development Program, Aerospace Research Funding, and so on. As a main member, she won second prize at the National Invention Award (second prizes) in 2010 and the second prize in the National Award for Education in 2009.

Preface

The supply and demand of energy determine the course of global development in every sphere of human activity. Finding sufficient supplies of energy to satisfy the world's growing demand is one of society's foremost challenges. Sorption refrigeration, which is driven by the low grade heat and provides the air conditioning and refrigeration effect, is paid more and more attention as one of the energy conversion technologies.

Sorption technology includes absorption and adsorption technology. The main differences between two types of technologies are the sorbents. The absorbents generally are liquid such as LiBr and NH_3, and the adsorbents are granular or compact solids, such as silica gel, zeolite, and chlorides. Compared with the absorption technology, the adsorption technology has the advantages of the wide choices of adsorbents for the wide scopes of driven temperatures for different heat sources, which generally ranges from 50 to 400 °C. The feature of solid adsorbents also makes it more feasible under the conditions with serious vibration. It doesn't need the rectifying equipments, nor does it have the problems of crystallization that can easily occur in absorption systems.

Adsorption refrigeration has two working processes. The first process is adsorption and refrigeration. In this process the adsorption heat releases cooling water or air to the heat sink and the pressure inside the adsorber decreases to a level lower than the evaporating pressure. The refrigerant evaporates and is adsorbed by the adsorbent under the function of pressure difference, and the evaporation process provides the refrigeration output. The second process is desorption and condensation. In this process the endothermic process of desorption is driven by the low grade heat. The desorbed refrigerant vapor is cooled by the heat sink and condensed in the condenser.

The earliest record of the phenomena of adsorption refrigeration was that AgCl adsorbed NH_3, which was discovered by Faraday in 1848. After that several refrigerators were developed for storing food and air conditioning. In the 1930s, the compression refrigeration technology was accelerated by technology innovations such as the discovery of Freon, the manufacture of a fully closed compressor, the application of compound refrigerants, and so on, and adsorption refrigeration could not compete with the CFCs (chlorofluorocarbons) system because of its low efficiency.

Since the late twentieth century, more and more research concentrated on sustainable development and the technology of adsorption refrigeration began to develop. There were two reasons for the fast development of sorption technologies: one is the need to solve the problems of energy shortage, which became more and more important since the worldwide energy crisis after the Middle East War during 1973. It takes about 7 million years to form petroleum and current supplies have almost been used up after more than 200 years' of exploitation. The stock of coal is greater than petroleum, but it is also consumed quickly especially with increasing demand as people all over the world desire comfortable living standards. The recovery of the low grade heat is one of the main technologies that may overcome the increasing constraints related to energy utilization. Another reason is related to climate change caused by ozonosphere depletion. There is a common recognition by international academics that depletion of the ozonosphere is caused by CFCs, which are found in refrigerators, air conditioners, and heat pumps. The green refrigerants, which are common in sorption technologies, are now being focused on as a replacement for traditional compression refrigeration technology.

The main technologies on adsorption refrigeration which are being researched by academics are mainly advanced adsorbents, advanced cycles, and advanced design for refrigeration systems. For example, Professor Critoph in the UK has worked on adsorption refrigeration for over 20 years. He and his research team have developed the consolidated activated carbon needed for the refrigeration and thermal wave cycle for the high coefficient of performance. The research team in France, such as Spinner, Meunier, and Mauran have worked on chemisorption thermodynamics and developed IMPEX for refrigeration. The research team of Kashiwagi and Saha developed the silica gel–water adsorption chiller and proposed the multi-stage cycle; Lebrun studied the heat and mass transfer of adsorbers; Vasiliev developed the heat pipe type adsorbers; Aristov studied the composite adsorbents of silica gel and the thermodynamics of composite adsorbents; and the academics in Korea studied the heat and mass transfer performances of solidified adsorbent, and so on. But there are no books which have systematically summarized the technology of adsorption refrigeration although it has now been developed for over 150 years.

As researchers in Shanghai Jiao Tong University, P.R. China, we have researched adsorption refrigeration for over 20 years. The research aspects include adsorbents, adsorption working pairs, adsorption refrigeration cycles, and adsorption applications. In order to share our research experience with international academics we have summarized our achievements as well as other researchers' outcomes. In this book the history of the development of adsorption refrigeration, development of adsorbents, thermodynamic theories, design of adsorption systems, adsorption refrigeration cycles have been discussed step by step. The main objective of the book is to give the readers a comprehensive guide to the research on adsorption refrigeration.

Ruzhu Wang, Liwei Wang, Jingyi Wu
2014

Acknowledgments

We are grateful for the contributions from academics and students in our research team. They are: Dr. Z.Z. Xia and Dr. Z.S. Lu who contributed to the design and development of adsorption refrigeration systems, which were cited in the book; Prof. Y.J. Dai and Dr. X.Q. Zhai who contributed to the work on solar powered adsorption air conditioning; Dr. T.X. Li who contributed to the adsorption refrigeration cycles. Some of the contents of this book are from the theses of the Ph.D. students in the research team, and they are M. Li, T.F. Qu, Y.Z. Lu, S.G. Wang, Y.L. Liu, X.Q. Kong, X.Q. Zhai, H.L. Luo, K. Daou, D.C. Wang, K. Wang, Z.S. Lu, Y. Teng, and T.X. Li, et al. The research work of post doctors also was cited in the book, that is, the research work of Prof. W. Wang, S. Jiangzhou, Y.J. Dai, and R.G. Oliveira.

We also appreciate the support from the National Key Fundamental Research Program, National Natural Science Foundation of China (NSFC) for Distinguished and Excellent Young Scholars, NSFC Key Projects for Young Academics, and the Foundation from Science and Technology Commission of Shanghai Municipality, P.R. China.

Nomenclature

a	Coefficient for the equilibrium reaction, coefficient in the van der Waals equation
a_p	The surface area per unit mass of adsorbent, m^2/kg
a_v	The surface area per unit volume of the adsorbent m^2/m^3
A	Coefficient in Clausius-Clapeyron equation
A_0	Dynamic coefficient
A_{0b}	The area of two back plates, m^2
A_a	Adsorbent cross-sectional area in the unit, m^2
A_{adb}	The heat transfer area of adsorber, m^2
A_c	The heat transfer area at the cooling side of the heat exchanger, m^2
A_{eff}, $A_{a,eff}$	Heat transfer area of heat exchanger at the solid adsorbent side, m^2
A_{evf}	The area at the fluid side of the heat pipe type evaporator, m^2
A_f	Heat transfer area of heat exchanger at the fluid side, m^2
A_{fa}	Internal surface area of the fin tube, m^2
A_{fe}	Anterior factor
A_{fin}	The area for the cross section of the fin, m^2
A_{fm}	The surface area of condensation pipe, m^2
A_g	Gas flow cross-sectional area in the unit, m^2
A_m	Heat transfer area of the metal wall at the adsorbent side, m^2
A_{mr}	Cross-sectional area of mass recovery channel, m^2
A_{rx},A_{ry}	Constants in Mazet reaction models
A_s	The area of solar collector, m^2
A_{seff}	Effective collector area, m^2
b	Coefficient in the van der Waals equation
B	Parameter for the pore structure of the adsorbent
c	Concentration of adsorbate, kg/m^3
c^*	Equilibrium concentration corresponding to the adsorption capacity x, kg/m^3
c_i	Concentration of the adsorbate on the surface of the adsorbent, kg/m^3
C	Constant in the Clausius-Clapeyron equation, specific heat, (J/(kg °C))
$C_{0\sim3}$	Coefficients in Tykodi models
C_a, C_{pa}	Specific heat of adsorbent, J/(mol K), J/(kg °C)
C_{ca}	Specific heat of composite adsorbent, J/(mol K), J/(kg °C)
C_{Ha}	Adsorbent heat capacity in the high-temperature adsorbent bed, J/(mol K), J/(kg °C)
C_{hb}	Specific heat of the liquid in the boiler, J/(mol K), J/(kg °C)

C_{Lc}	Specific heat of liquid refrigerant, J/(mol K), J/(kg °C)
C_{Lv}, C_{vg}	Specific heat of refrigerant vapor, J/(mol K), J/(kg °C)
C_m, C_{pm}	Specific heat of metal materials, J/(mol K), J/(kg °C)
C_{mal}	Specific heat of the aluminum, J/(mol K), J/(kg °C)
C_{mcu}	Specific heat of the copper, J/(mol K), J/(kg °C)
C_{mh}	Metal heat capacity of the heating boiler, J/(kg °C)
C_p	Isobaric specific heat, J/(mol K), J/(kg °C)
C_{pb}	The total thermal capacity, J/(mol K) or J/(kg °C)
C_{pc}, C_{pg}	Isobaric specific heat of refrigerant vapor, J/(mol K), J/(kg °C)
C_{pf}	The thermal capacity of the fluid, J/(mol K), J/(kg °C)
C_{pr}, C_{pl}	Isobaric specific heat of liquid refrigerant, J/(mol K) or J/(kg °C)
C_{ps}	The isobaric specific heat of solid material, J/(mol K), J/(kg °C)
C_{pw}	Thermal capacity of the metal walls, J/(mol K) or J/(kg °C)
C_{ra}	Proportional coefficient determined by evaporator type
C_{vf}	Specific heat at constant volume of the liquid refrigerant, J/(kg K)
COP	Coefficient of performance for refrigeration
COP_{AC}	COP for activated carbon adsorber
COP_{carnot}	COP for Carnot cycle
COP_{hp}	COP of heat pump
COP_i	Ideal COP
COP_{int}	COP for intermittent cycle
COP_Z	COP for zeolite adsorber
d	Distance, distance between molecules, diameter, m
d_a	The diameter of the adsorbent particles, m
d_{ave}	Average pore diameter, m
d_e	Equivalent diameter, m
d_p	Equivalent diameter of the solid particles, m
d_{pi}	Inlet diameter of the tube, Inner diameter of the pipe, m
d_{po}	Outer diameter of the pipe, m
d_v	Equivalent diameter for the flowing process of the vapor, m
d_w	The channel width, m
D'	The coefficient in D-A equation
D_e	Diffusion coefficient in the micropore, effective diffusion coefficient
D_{go}	Diameter of the outer glass tube, m
D_i	Effective diffusion coefficient, m^2/s
D_k	Knudsen diffusion coefficient
D_{ms}	Mass diffusion coefficient of the fluid, m^2/s
D_s, D_{so}	Surface diffusion coefficient, m^2/s
e_{eff}	Effective thickness of adsorbent, m
e_{so}	The internal energy for the solid adsorbent skeleton, kJ/kg
E	Specific adsorption power, J/mol
E_a	Activated energy for adsorption, J/mol
E_d	Activated energy for desorption, J/mol
E_{ij}	Thermal dispersion coefficient
E_p	Pseudo-activated energy, J/mol
f	The fugacity under the pressure of p, Pa

f_0	The fugacity under the pressure of p_s, Pa
f_S	The ratio between the area of airflow area and area of the cross-section area of wheel, m^2/m^2
f_V	Surface area of unit volume of adsorbent, m^2/m^3
g	Acceleration of gravity, m/s^2
G	Free enthalpy, J
h	Specific enthalpy, J/kg
h_a,h_d	Adsorption heat, desorption heat, kJ/kg
h_{ev}	The height for the evaporating section of the heat pipe, m
h_f	Specific enthalpy of the refrigerant liquid, J/kg
h_r	Specific enthalpy of the ammonia liquid at the condensation temperature, J/kg
h_w	The depth of the channels, m
H	Enthalpy, J
H_a,H_d	Adsorption heat, desorption heat, kJ
H_{adb}	The thickness (i.e., height) of the adsorbent bed, m
$\overline{H}_2$	Partial molar enthalpy, J/mol
$\overline{H}_g$	Molar enthalpy, J/mol
$H_{\max}$	Maximum capillary height, m
H_r	Chemical reaction heat, J
H_{st}	Isobaric adsorption/desorption heat, kJ/kg
I	The solar radiation intensity, W/m^2
I_0	Direct sunlight intensity, W/m^2
I_{ref}	Reflected sunlight intensity from back plate, W/m^2
J	Heat flux, W/m^2
k	Coefficient in D-R equation
k_1,k_2,k_3	Stability constants
k_F	Mass transfer coefficient, kg/(m^2 s)
k_{ij}	The component of permeability tensor, m^2
k_p	Permeability of porous medium, m^2
k_s	Mass transfer coefficient inside the solid phase film, kg/(m^2 s)
k_y	Convection mass transfer coefficient, kg/(m^2 s)
K	The coefficient for D-R equation, equilibrium constant of the reaction, permeability (m^2)
K_a	Coefficient for the reaction rate in adsorption process, 1/(m^2 s)
K_d	Coefficient for the reaction rate in desorption process, 1/(m^2 s)
K_F	Mass transfer coefficient of the fluid side, m/s
K_i	The dynamic coefficient
K_{ms}	Coefficient of the mass transfer
K_{md}	Coefficient for the influence of chemical kinetics on the reaction
K_n	Knudsen diffusion rate
K_r	Reaction kinetic constant
K_s	The total mass transfer coefficient (kg/(m^2 s)), permeability (m^2/s)
$K_s a_p$	Surface diffusion rate coefficient 1/s
K_v	Net adsorption rate, (kg/kg)/s
K_x	Reaction coefficient in Iloeje's equation, °C/s
l	Length, mass transfer scale, m

l_{ah}	Heat pipe height in the adsorbent bed, m
l_{fin}	The perimeter of the cross section, m
L	Latent heat of vaporization of refrigerant, kJ/kg
L_a	Adsorbent thickness along the direction of L_y, m
L_{ad}	The length of adsorbent bed, m
L_b	The width of the adsorbent bed along the direction of L_y, m
L_{bw}	Thickness of the wall, m
L_B	Unit lateral equivalent width, m
L_c	The condensation heat of the refrigerant in the condenser, kJ/kg
L_e	The evaporating heat of the refrigerant in the evaporator, kJ/kg
L_{ev}	The length of the evaporation section of the heat pipe type evaporator (m); the latent evaporation heat of the refrigerant (kJ/kg)
L_{fin}	The half distance between fins in the adsorption bed, m
L_{hp}	Evaporation latent heat of the fluid inside the heat pipe, kJ
L_m	Height of the heat medium along the direction of L_z, m
L_{pi}	The length of the pipe, m
L_{sat}	Evaporation latent heat of the refrigerant at the temperature of T_s, J/kg
L_x, L_y, L_z	Three coordinates, m
L_{xt}	The total length along the direction of L_x, m
$\dot{\mathrm{m}}$	Gas flow rate from a unit to the next unit, kg/s
m	Flow rate (kg/s, g/s)
m_{am}	Mass flow rate of ammonia, kg/s
m_{air}	The airflow rate, kg/s
m_C	The molar mass of $CaCl_2$, 110.99
m_e	The mass flow rate of the vapor, kg/s
m_f	Volume flow rate of the fluid, m^3/s
m_i	Air flow through the unit cross-sectional area of wheel, $kg/(m^2 s)$
m_{mr}	The mass flow rate for the vapor in mass recovery phase, kg/s
m_N	Molar mass of NH_3, 17
m_{oil}	Fuel quantity, kg/h
m_{uA}	Mass flow rate per unit area, $kg/(m^2\ s)$
m_w	Mass flow rate of heating/cooling fluid, kg/s
m_{water}	Flow rate of the water, kg/s
m_x, m_y	Reaction order
m_y	Flow rate of the exhaust gas, kg/s
M	Mass, kg
M_a	The mass of adsorbent, kg
M_{av}	The adsorbent mass in unit volume, kg/m^3
M_C	The mass of $CaCl_2$, kg
M_{ca}	The mass of composite adsorbent, kg
M_{e0}	Mass of the refrigerant in the evaporator under equilibrium conditions, kg
M_{eqh}	The mass of the working fluid in the liquid pumping boiler, kg
M_{ev}	The mass of the refrigerant in the evaporator, kg
M_{ew}	Mass of the refrigerant liquid inside the evaporator, kg
M_g	The mass of graphite, kg

M_{ha}	The mass of the working fluid in the fin tube of the adsorbent bed and in the liquid chamber, kg
M_{hb}	Total mass of the working fluid in the boiler, kg
M_{hp}	The initial mass of the working fluid in the liquid pumping boiler, kg
M_{Ha}	Adsorbent mass in the high temperature adsorbent bed, kg
M_m	The mass of support body in the unit volume, kg/m^3
M_{madb}	Metal mass of the adsorbent bed, kg
M_{mal}	The mass of aluminum inside the adsorber, kg
M_{mcu}	The mass of the copper material inside the adsorber, kg
$M_{m,con}$	The metal mass of the condenser, kg
$M_{m,eva}$	The metal mass of the evaporator, kg
M_{me}	Mass of methanol desorbed from adsorber, kg
$M_{meva,cond}$	The metal mass of evaporator and condenser, kg
M_{mh}	Metal mass of the heating boiler, kg
M_{pbf}	Mass of the liquid in the liquid pumping boiler that cannot be pumped into the adsorbent bed, kg
M_r	Reaction kinetic constant
M_{Re}	The function of the Reynolds number
M_z	Total mass of the working fluid filled into the heat pipe system, kg
Ma	Reaction dynamic coefficient for adsorption
Md	Reaction dynamic coefficient for desorption
$\dot{n}$	The total molar flow rate, mol/s
n	Coefficient in D-A equation, coefficient for reaction equilibrium, reaction order
n_2^s	Molar adsorption quantity on the surface of solid adsorbent, mol/mol
n_s	Number of flow channels
N	Molar mass (mol), layer numbers of the glass cover
N_g	Molar adsorption quantity, mol/mol
p	Pressure, Pa
p'	Pressure on the metal chloride's surface, Pa
p_{ae}, p_{ads}	The pressure inside the adsorber at the end of the adsorption phase, Pa
p_c	Constrained pressure, Pa
p_{de}, p_{des}	The pressure of the adsorber at the end of the desorption phase, Pa
p_{ea}	Equilibrium pressure of adsorption state, Pa
p_{ed}	Equilibrium pressure of desorption state, Pa
p_h	Pressure of reaction interface, Pa
p_i	Pressure of the vapor reactant interface, Pa
p_m	The pressure of the system after the mass recovery, Pa
P_{el}	The electricity generation of the cogeneration system, W
p_i	Pressure inside pore, Pa
PER	Primary energy ratio
Pr	Prandtl number
Pr_s	Prandtl number of the media under the saturated temperature
Pr_w	Prandtl number of the media under the plate surface temperature of the heat exchanger
q	Heat flux density, W/m^2

q_{ads}	Average differential adsorption heat, J/kg
q_c	Heat adsorbed by the adsorbent, J/kg
$q_{c,st}$	The cold storage quantity per unit mass of adsorbent, kJ/kg
$q_{h,st}$	The heat storage quantity per unit mass of adsorbent, kJ/kg
q_{in}	Endothermic heat, J/kg
q_r	The sum of the radiation, W/m^2
q_{reg}	Required heat of the adsorbent bed without heat recovery process, J
$q_{reg}{}^*$	Heat recovered in a heat recovery process of the adsorbent bed, J
q_{st}	Isosteric heat, J/mol, J/kg
Q	Heat, J or kJ
Q_{bind}	The difference between the heat required for desorption Q_{des} and the condensation heat Q_{cond}, J or kJ
Q_{cc}	The sensible heat of the liquid refrigerant, J or kJ
Q_{char}	Charging heat, J or kJ
Q_{chill}	The heat at the refrigeration section of the heat pipe type evaporator, J or kJ
Q_{eref}	Cooling power generated by the evaporation of the refrigerant in evaporator, J or kJ
Q_{evas}	The sensible heat of liquid refrigerant in evaporator, J or kJ
Q_{ew}	The heat at the condensation section of the heat pipe type evaporator, J or kJ
Q_{hg}	Heat from the heat source, kJ
Q_{hs}	Heat quantity for convective heat transfer process, J or kJ
$Q_{h,st}$	The heat stored, J or kJ
Q_{reg}	Regenerative heat, J or kJ
Q_{sens}	Prerequisite energy to heat up the reactor to a required desorption temperature, J or kJ
Q_{Hd}	The desorption heat of the high temperature adsorber, J or kJ
Q_{Hs}	The synthetization heat of high temperature adsorber, J or kJ
Q_{seff}	Heat transformed from the actual solar radiation, J or kJ
Q_{st}	Isobaric adsorption heat, J or kJ
Q_{sen}	Sensible heat of the adsorber, J or kJ
Q_{solar}	Solar radiation, J or kJ
r	Radius, m
r_{as}	Ratio between expansion space and volume of adsorbent
r_c	Diameter of reaction surface, m
r_g	Radius of grain, m
r_{hc}	Heat recovery coefficient
r_{sh}	Shape factor of isothermal adsorption process of ideal adsorbent material
R	The universal gas constant, J/(mol K)
R_0	Thermal resistance of tube, $(m^2\ °C)/W$
R_f	The thermal resistance of the fouling between the fluid and the metal wall, °C/W
R_{go}	The radius of the outer glass tube, m
R_H	Relative humidity, %
R_i	Thermal resistance of dirt, $(m^2\ °C)/W$
R_m	The radius of metal tube, m
R_p	Average diameter of the adsorbent granules, m

R_t	Thermal contact resistance between the metal wall and the adsorbent particles, °C/W
R_T	The temperature change rate for adsorption/desorption, °C/s
$R_{\Delta x}$	Adsorption/desorption rate, (kg/kg)/min
$R_{\Delta xT}$	Non-equilibrium adsorption/desorption rate, (kg/kg)/°C
Re	Reynolds number
s	The constant in Arhenius law
S	Entropy, J/K
S_A	The ratio between the outside area of pipe and the inside area of the pipe
S_h	Heat exchange rate in the unit volume by the solid adsorbent side W/m^3
S_{solar}	The effective irradiation, W/m^2
$\underline{S}_2$	Partial molar entropy, J/(mol K)
$\overline{S}_g$	Molar entropy, J/(mol K)
Sc	Schmidt number
SCP	Specific cooling power per kg adsorbent, W/kg
SHP	Specific power of heat pump per kg adsorbent, W/kg
t	Variation of time, s or min
t_c	Cycle time, s or min
t_{hc}	Half cycle time, s or min
t_m	Mass recovery time, s or min
T	Temperature, K or °C
T_{adb}	Temperature of adsorbent bed, K or °C
T_b	Temperature of the space inside the adsorbent bed, K or °C
T_c	Constrained temperature, K or °C
T_{chill}	The temperature at the refrigeration section of the heat pipe type evaporator, K or °C
T_{cm}	The temperature at the condensation side of the heat exchanger, K or °C
T_{ew}	The temperature of the condensation section of the heat pipe type evaporator, K or °C
T_f	The temperature of the fluid, K or °C
T_{go}	Temperature of the outer glass tube, K or °C
T_{hb}	Temperature of the working fluid in the boiler, K or °C
T_L	Lowest temperature
T_m	The average temperature of the collector, temperature of the metal tube, K or °C
T_{me}	Ambient temperature, K or °C
T_{mi}	Temperature for the wall of the tube, K or °C
T_{reg}	Regenerative temperature, K or °C
T_{mo}	The temperature of the metal tube wall connected with the adsorbent, K or °C
T_p	The temperature of the heat absorbing plate, K or °C
T_{pa}	Temperature of the working fluid for the heat pipe working fluid after liquid pumping process and the liquid return process, K or °C
T_{pb}	Temperature of the working fluid in the liquid pumping boiler, K or °C
T_s	Saturation temperature, K or °C
T_{sa}	Temperature of adsorbent surface, the saturation temperature of the working fluid in the fin tubes of the bed after adsorption, K or °C
T_{sk}	The sky temperature, K or °C

T_v	The temperature of the vapor, K or °C
T_w	Temperature of the wall, K or °C
T_{web}	Wet bulb temperature, K or °C
u, u_f	The velocity of the fluid, m/s
$u_{f,aver}$	Generic variable of the skeleton of the porous adsorbent
u_l	The coefficient of heat loss
u_{lo}	The locomotive speed, km/h
U_V, U_s	Reference volume of vapor or solid, m^3
U_b	The heat loss coefficient at the bottom of the collector
U_t	The heat loss coefficient at the surface of the collector
v_{wv}	Specific volume of the water vapor, m^3/kg
V_0	Pore volume, maximum pore volume, m^3
V_c	Volume occupied by the refrigerant, m^3
V_C	The volume of $CaCl_2$ solid, m^3
V_m	The molar volume of ammoniate chlorides, m^3/mol
V_p	Internal porosity volume of the IMPEX, m^3
$\overline{V}_2$	Partial molar volume, m^3/mol
$\overline{V}_g$	Molar volume, m^3/mol
w	The mass ratio of ENG in IMPEX, %
W	Power, W or kW
W_b	The heat loss at the bottom of the collector, W or kW
W_{in}	The heat input of the system, W or kW
W_{rgb}	The radiation between the evacuated tubes collector and the back plate, W or kW
W_{sref}	The cooling power of single bed system, W or kW
W_t	The facial heat loss, W or kW
W_{thref}	Cooling power of the triple-bed system, W or kW
x	Adsorption quantity, kg/kg
x^*	The local equilibrium adsorption quantity, kg/kg
x_0	Maximum adsorption rate
x_{am}	Adsorption quantity of the bed after desorption before the mass recovery, kg/kg
x_{dm}	Adsorption quantity of the adsorbent bed before the mass recovery, kg/kg
x_i	Equilibrium adsorption capacity corresponding to the concentration c_i, kg/kg
x_V	Volume adsorption amount, kg/m^3
Y	Moisture content of the air, kg water/kg dry air
Y_W	Moisture content of the air on the surface of the adsorbent, kg water/kg dry air
Z	Compression factor of gas
Z_c	The volume ratio between the hex-ammoniate chlorides and binary ammoniate chlorides

Greek Symbols

α	Heat transfer coefficient, convection heat transfer coefficient, W/(m^2 °C)
α_{ab}	The natural convection heat transfer coefficient, W/(m^2 °C)
α_{ac}	Convective heat transfer coefficient between the activated carbon fiber and ammonia flow, W/(m^2K)
α_{am}	Heat transfer coefficient of the outer glass tube to the air, W/(m^2 °C)
$\alpha_{a,eff}$	Effective heat transfer coefficient inside the adsorbent, W/(m^2 °C)

α_b	Heat transfer coefficient between the wall and the adsorbent bed, W/(m^2 °C)
α_c	The heat transfer coefficient at the condensation side, W/(m^2 °C)
α_f	Heat transfer coefficient of the heat exchanger by the fluid side, W/(m^2 °C)
α_{fc}	Heat transfer coefficient of inner fin tube in the adsorbent bed, W/(m^2 °C)
α_{fi}	The heat transfer coefficient of the cooling water and the surface of the tube, W/(m^2 °C)
α_m	Equivalent heat transfer coefficient of the adsorbent bed metal, W/(m^2 °C)
α_{mi}	The heat transfer coefficient between the adsorbent and cooling water tube, W/(m^2 °C)
α_{pme}	The evaporating heat transfer coefficient of the methanol, W/(m^2 °C)
α_{pwater}	Heat transfer coefficient of the water, W/(m^2 °C)
α_{rg}	Radiation heat transfer coefficient of the outer glass tube to metal tube, W/(m^2 °C)
α_{rs}	Radiation heat transfer coefficient of the outer glass tube to the sky, W/(m^2 °C)
α_t	The total heat transfer coefficient, W/(m^2 °C)
α_w	Heat transfer coefficient of the heat exchanger by the solid adsorbent side W/(m^2 °C)
$\alpha_{we,vap}$	Evaporative heat transfer coefficient outside of tube, W/(m^2 °C)
β	Affinity coefficient, the angle of the collector
β_p	Porosity of the solid adsorbent
ε	The adsorption potential of reference adsorbate (benzene), J/mol
ε_a	Porosity of adsorbent, kg/m^3
ε_b	Porosity of adsorbent bed, kg/m^3
ε_{ev}	Evaporative cooling efficiency
ε_g	The emissivity of the glass cover of the solar collector
ε_{IMPEX}	Porosity of IMPEX
ε_l	Adsorption potential of non-reference adsorbates, J/mol
ε_p	The emissivity of the heat adsorbing plate of the solar collector
ε_r	Adsorption potential per mole real gas, J/mol
ε_μ	The energy consumption rate of fluid caused by the fluid viscosity, W/m^3
ξ_b	The total thermal diffusion coefficient
ξ_f	Thermal diffusivity of the fluid
ξ_w	Thermal diffusivity of the metal walls
ς_{go}	Sunlight absorption rate of outer glass tube
ς_m	Absorption rate of the metal pipe
ς_{solar}	The absorbing rate of sunshine by the collector of adsorber
τ	Time, s
τ_{go}	Sunlight transmittance of the glass tube
τ_{solar}	The sunshine transmittance through the glass cover
μ	Chemical potential, dynamic viscosity (kg/(ms))
μ_f	Surface chemical potential
μ_g	Chemical potential of the adsorbed gas
μ_v	Dynamic viscosity of the vapor, kg/(ms)
σ	Tension force at the liquid surface, N/m
σ_b	Boltzmann constant
ΔG	Variation of the free enthalpy, J or kJ
ΔG^0	Standard reaction free enthalpy change, J

Δh	Change of specific adsorption/desorption heat, J/kg
ΔH	Variation of the enthalpy, J or J/mol
$\Delta H, \Delta H_r$	Change of the chemical reaction heat, J/mol
ΔH^0	Change of the standard enthalpy, J
ΔH_r	Reaction enthalpy, adsorption heat, J or J/mol
ΔM_a	Adsorption/desorption mass of ammonia, kg
ΔS	Variation of the entropy, J/K or J/(mol K)
ΔS^0	Change of the standard entropy, J/K
ΔT	Temperature difference, °C
ΔT_{ah}	Temperature difference between the adsorbent and vapor inside the fin tube, °C
ΔT_{ev}	Fluctuating value of evaporation temperature, °C
ΔT_{wc}	Temperature difference between the water inlet and outlet of the coil cooler, °C
Δx	Cycle adsorption quantity, kg/kg
Δx_{md}	Desorption quantity during the mass recovery process, kg/kg
Δx_{ma}	Adsorption quantity during the mass recovery process, kg/kg
ρ	Density, kg/m^3
ρ_a, ρ_{ad}	Density of the adsorbent, kg/m^3
ρ_b	Volume density of the graphite, kg/m^3
ρ_{bt}	The total density, kg/m^3
ρ_f	Density of liquid membrane, density of fluid, kg/m^3
ρ_g	Density of gas flow, kg/m^3
ρ_i	Density of air, kg/m^3
ρ_L, ρ_l	Density of liquid, kg/m^3
$\rho_{Q\text{-}m}$	Energy density by mass, J/kg or kJ/kg
$\rho_{Q\text{-}V}$	Energy density by volume, J/m^3 or kJ/m^3
ρ_{refg}	Density for the adsorbate gas, kg/m^3
ρ_s	Apparent density of adsorbent, kg/m^3
ρ_v	The density of the vapor, kg/m^3
ρ_w	Density of the metal walls, kg/m^3
λ, λ_a, λ_{ad}	Thermal conductivity of adsorbent, W/(m °C)
λ_{eff}	Effective thermal conductivity, W/(m °C)
λ_f	Thermal conductivity of the fluid, W/(m °C)
λ_{go}	Thermal conductivity of outer glass tube, W/(m °C)
λ_L, λ_l	Thermal conductivity of the liquid, W/(m °C)
λ_m	Thermal conductivity of the metal, W/(m °C)
δ	Thickness, thickness of the falling film, m
δ_{eff}	Equivalent thickness of the liquid film, m
δ_{go}	The thickness of the outer glass tube, m
δ_m	The thickness of the outer metal tube, m
δ_{mi}	The thickness of the cooling water tube, m
δx	The change of the adsorption quantity, kg/kg
υ	Adsorption rate, (kg/kg)/s
ψ_a	The ratio of airflow area to the cross-section area per unit mass of adsorbent
η	Collector efficiency
η_{boiler_conv}	Thermal efficiency of the boiler in the conventional distributed energy system
η_{el_conv}	The power generation efficiency of the conventional distributed energy system

ν	Kinematic viscosity, m^2/s
θ	Degree of coverage, solar elevation angle, the heat load friction
γ	Filling density, kg/m^3, air coefficient
ω	The speed of the wheel, rad/s
Γ	Adsorption quantity per unit area of solid surface

General Subscripts

a, ad, ads	Adsorption, adsorbent
adb, bed	Adsorber
am	Ammonia
c	Condensation, cooling
C	Refrigeration, cooling
ca	Composite adsorbent
cal	Calculation
cond	Condensation
Cool	Refrigeration, cooling
d, des	Desorption
dil	Dilution
e, eva	Evaporation, refrigeration
ea	Equilibrium adsorption
ed	Equilibrium desorption
eff	Effective
eq	Equilibrium
exp	Experiment
f	Fluid
g	Generation, gas
h, H	Heating, highest, heat pump
hp	Heat pipe
Heat	Heating, heat pump
in	Inlet
l	Liquid
m	Cooling media, metal
max	Maximum
mb	The metal back plate
mi	The metal cooling pipe
ref	Refrigerant, refrigeration
reg	Regeneration, heat recovery
s, syn	Synthesization
s, sat	Saturation
so	Solid
w	Wall
wv	Water vapor

1

Introduction

Sustainable development is a common pursuit for people worldwide and energy utilization is a key element. Generally, energy will be consumed in large amounts as the economy of society develops rapidly, and a careful eye needs to be kept on environmental pollution. How to coordinate the balance between energy utilization, economy development, and environmental protection is one of the most important strategies for sustainable development.

With regards to environmental protection, the ozonosphere depletion by chlorofluorocarbons (CFCs), which causes the ultraviolet rays of the sun to be insufficiently blocked and thus threatens life on the earth, has been commonly recognized worldwide. CFCs are very important substances in compression refrigeration. As a type of substitute substance, HCFCs can only be temporally utilized because they also have a negative influence on the ozonosphere. Meanwhile, with regards to central heating systems, the combustion of gases and coal releases CO_2 into the environment. Similarly, CFCs produce the greenhouse effect that is becoming more and more serious as the desire for comfortable living conditions all over the world becomes greater and greater. Finding a type of green technology that can be used in air conditioning and heat pumps is very important with regards to solving the problems caused by traditional compression refrigeration technology.

Another critical problem for refrigeration and heat pumps is energy utilization. Traditional compression refrigerators and heat pumps are commonly driven by electricity. Demands for electricity increase as societies develop. According to data provided by the energy department of the US between 2003 and 2004, the electricity consumed by air conditioners in the summer is 15.4% of the total electricity consumption. In China too, for example, in Shanghai City, in summer electricity consumption by air conditioning reached 45–56% according to data collected from 2010. If we analyze the energy utilized through the electricity generation process we find that energy efficiency for electrical generation is only about 40–50%, and there is a large amount of energy being released into the environment as waste heat at temperatures of around 70–200 °C. Meanwhile solar energy and geothermal heat also exist in large amounts in the environment as a low grade energy. Developing refrigeration and heat pump technologies driven by such low grade heat is a solution for energy conservation.

Sorption refrigeration and heat pump technology which is driven by low grade heat and utilizes the green refrigerants, is coordinated with the sustainable requirements of current energy and environmental developments. Firstly, the sorption technology requires little electricity,

Adsorption Refrigeration Technology: Theory and Application, First Edition. Ruzhu Wang, Liwei Wang and Jingyi Wu.

Companion Website: www.wiley.com/go/wang/refrigeration

secondly, the refrigerants for the sorption refrigeration generally are the substances of water, ammonia, and methanol, and so on, which are green refrigerants with zero ODP (Ozonosphere depletion potential) and zero GWP (Greenhouse warming potential).

As a type of sorption technology, adsorption refrigeration and heat pumps have been paid more and more attention since the 1970s. If compared with other types of sorption technology driven by low grade heat, firstly, adsorption refrigeration has a wide variety of adsorbents, including different physical and chemical adsorbents; which can be used with low grade heat across a large range of temperatures, and generally we find these adsorbents are driven by low grade heat in the range of 50–400 °C. Secondly, adsorption refrigeration doesn't need the solution pump and rectification equipment, and it also doesn't have the problems of refrigerant pollution and solution crystallization that often happens in absorption refrigeration technology. But, generally, adsorption refrigeration is not as efficient as absorption, and it also has the disadvantages of being a large volume system. Because of these advantages and disadvantages, adsorption refrigeration is recognized by academics as an essential complementary technology for absorption refrigeration.

1.1 Adsorption Phenomena

According to the different types of adsorption processes, adsorption is divided into physical adsorption and chemical adsorption [1]. Physical adsorption is driven by the van der Waals force among the molecules, and generally happens on the surface of adsorbents. Physical adsorption is not selective, which means multi-layer adsorption can be formed. The phenomena of physical adsorption can be treated as the condensation process of the refrigerant inside the adsorbents, and for most adsorbents the adsorption heat is similar to the condensation heat of the refrigerant. The molecules for the physical adsorption won't be decomposed in the desorption process.

Chemical adsorption is different to physical adsorption. A chemical reaction will happen between the adsorbent and the adsorbate, and new types of molecules will be formed in the adsorption process. Commonly, the monolayer of the adsorbate will react with the chemical adsorbent, and after this reaction the chemical adsorbents cannot adsorb more layers of molecules. The newly formed molecules will be decomposed in the desorption process. The adsorption/desorption heat produced will be much larger than the physical adsorption heat. The chemical adsorption is selective. For example, H_2 can be adsorbed by W, Pt, and Ni, but cannot be adsorbed by Cu, Ag, and Zn. It is recognized by academics that physical adsorption will happen before chemical adsorption because the effective distance of the van der Waals force is inversely proportional to the power of 7 of distance, and it is much longer than the effective distance for the chemical reaction. Thus, when the adsorbate molecules approach the solid adsorbent the physical adsorption will proceed first, and will transfer into the chemical adsorption when the distance decreases.

The physical adsorption/desorption mainly depends on the heat and mass transfer performances of the adsorbents. For the desorption process, because the pressure is high, correspondingly the mass transfer process will be accelerated by the high pressure, and the heat transfer performance will be the main criterion for the performance. If the heat transfer performance is intensified the main problem for the adsorption systems will be the permeability of the gas inside the adsorbents. Generally, the permeability is higher when the adsorbent granules are smaller. The kinetic reaction rate will also influence the adsorption/desorption rate.

Because the chemical reaction happens in the chemical adsorption process, the chemical adsorption will be influenced by the heat and mass transfer process of the adsorbents, as well as the chemical reaction process and the reaction kinetics of the molecules. Meanwhile, the adsorption hysteresis also exists for the chemical adsorption because the adsorption activated energy is different from the desorption activated energy. The desorption activated energy is always much larger than the adsorption activated energy because it is the sum of the adsorption activated energy and the adsorption heat, and such a phenomenon will lead to a serious hysteresis phenomenon between adsorption and desorption [2].

For adsorption refrigeration most refrigerant molecules are polar molecular gases that can be absorbed under the van der Waals force, such as ammonia, methanol, and hydrocarbons that can be adsorbed by activated carbon, zeolite, and silica gel. For physical adsorption the cycle adsorption quantity is generally from 10 to 20%. The chemical adsorption has greater cycle concentrations than that of physical adsorbents, for example, for $CaCl_2$ the cycle adsorption quantity is always larger than 0.4.

The advantage of chemical adsorption refrigeration is the larger adsorption/desorption quantity, which is essential for the improvement of the specific cooling capacity per kilogram adsorbent (SCP, specific cooling power). But the expansion and agglomeration will happen in the chemical adsorption process, and the expansion space always needs to be kept at two times of the adsorbent volume to ensure high mass transfer performance. In order to improve the heat transfer performance as well as to ensure the mass transfer performance, the solidified compound/composite adsorbents are developed, which uses the porous matrix to keep reasonable permeability of the adsorbent, and then improve the volume filling capacity and volume cooling capacity significantly.

1.2 Fundamental Principle of Adsorption Refrigeration

The fundamental principle of adsorption refrigeration is demonstrated by the solar powered adsorption ice maker in Figure 1.1, and the relative thermodynamic cycle is shown in Figure 1.2.

As shown in Figure 1.1, the solar powered adsorption refrigerator is composed of the adsorber, condenser, evaporator, valve, and refrigerant tank. When the adsorber is cooled at night, the pressure inside the adsorber decreases, and the refrigerant inside the evaporator, which evaporates under the pressure difference between adsorber and evaporator, is adsorbed by

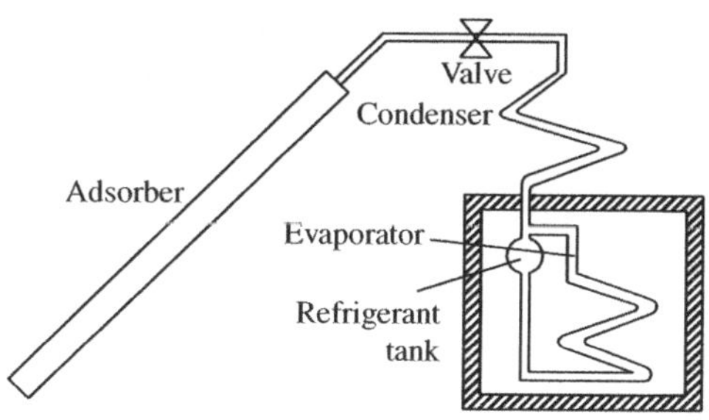

Figure 1.1 The solar powered adsorption refrigeration system

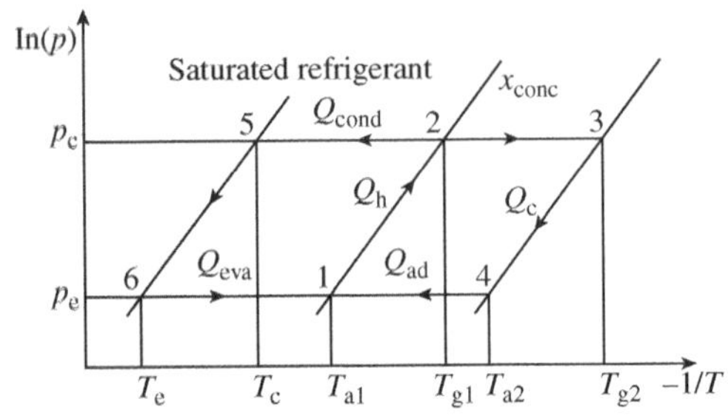

Figure 1.2 The p-T diagram of adsorption refrigeration cycle

the adsorbent inside the adsorber. The evaporation process of the refrigerant generates the refrigeration power. The refrigeration will stop when the adsorbent is saturated. In the daytime, the adsorber is heated by solar energy, and the pressure inside increases. The refrigerant inside the adsorber will be desorbed from the adsorber by the pressure difference between adsorber and condenser, and then will be condensed inside the condenser that was cooled by the environmental air around.

The whole process can be summarized in detail as follows (Figure 1.2):

1. The valve is closed in the morning assuming an environmental temperature T_{a2} of 30 °C. As time passes the adsorber will be heated by solar energy, and the pressure of the adsorber will increase. Finally, the pressure of the refrigerant will be the saturated pressure for the condensing temperature of the refrigerant, which is 30 °C. The temperature of adsorber will be T_{g1} in Figure 1.2.
2. Open the valve and the refrigerant desorbed from the adsorber will be condensed inside the condenser that is cooled by the natural-convection heat transfer method. After that the refrigerant will flow to the evaporator and refrigerant tank and accumulate there. In this phase the final temperature of the adsorber can be as high as T_{g2} (desorbing temperature).
3. The valve is closed in the evening. The temperature of the adsorber begins to decrease because of little or no solar energy outside. The pressure of the adsorber decreases as well, and it will decrease to the saturated pressure for the evaporating temperature, the corresponding temperature of adsorber is T_{a1} (initial adsorption temperature).
4. Open the valve and the refrigerant inside the evaporator will evaporate and be adsorbed by the adsorbent inside the adsorber because of the pressure difference between adsorber and evaporator. The evaporation process of the refrigerant provides the refrigeration power, and the adsorption heat of the adsorber will release to the environment. This phase will proceed till the next morning, and after that a new cycle will begin.

Adsorption refrigeration has two processes, which are the heating-desorbing process and the cooling-adsorbing process. Because of that the simple traditional cycle is a type of intermittent refrigeration cycle, which is a very good feature for the utilization of solar energy because solar energy is also a type of intermittent energy. If the heat source can be provided continually and the continuous refrigeration effect is required, two adsorbers or multi adsorbers need to be designed for an adsorption refrigeration system, for which the heating and cooling processes of multi adsorbers will be complementarily arranged.

1.3 The History of Adsorption Refrigeration Technology

In 1848, Faraday found that the cooling capacity could be generated when AgCl adsorbed NH_3. This is the earliest record of the adsorption refrigeration phenomenon. In the 1920s, G. E. Hulse proposed a refrigeration system in which silica gel-SO_2 was used as the working pair for food storage in a train. It was powered by the combustion of propane and was cooled down by the convection heat transfer of air. The lowest refrigerating temperature could reach 12 °C [3]. R. Plank and J. Kuprianoff also introduced the adsorption refrigeration system with a working pair of activated carbon-methanol [4]. In 1940–1945, the adsorption refrigeration system with working pair $CaCl_2$-NH_3 was used for food storage in the train from London to Liverpool, for which the heat source is the steam at 100 °C. From the 1930s, new technologies, such as the discovery of Freon and the successful development of the totally closed compressor improved the efficiency of the compression refrigeration system significantly. Because of that the adsorption refrigeration technology couldn't compete with the highly efficient CFCs system, it had not been considered by researchers for a long time.

In the 1970s, the energy crisis took hold and it offered a great chance for the development of the adsorption refrigeration technology, mainly because of the fact that the adsorption refrigeration system is driven by a low-grade heat source such as waste heat and solar energy. In the 1990s, environmental pollution became more and more serious, and the shortcomings of the CFCs system had been recognized worldwide as a cause of the ozonosphere depletion and greenhouse warming problems. As a result green refrigeration technology, which is a thermal powered refrigeration technology such as adsorption refrigeration, regained the recognition by the academics. Up until now such a type of technology had been widely researched for heat pump systems, marine refrigeration systems, automobile air conditioning systems [5–7], as well as for the application on aerospace cryogenics because it featured no moving parts, no noise, and had good anti-vibration performance [8, 9].

The research on the adsorption refrigeration originated from Europe. The famous researchers such as F.E. Meunier, M. Pons *et al.* from France [10–12], G. Cacciola *et al.* from Italy [13, 14], R.E. Critoph *et al.* from England [15–17], Shelton *et al.* from America [18–21], and Leonard L. VASILIEV *et al.* from Belarus [22] contributed quite a lot to the development of the technology. In China the research on adsorption refrigeration started during 1980s [23–27]. Shanghai Jiao Tong University (SJTU) started the research in 1991 [28–35] and pursued this work for more than 20 years. The research scopes of SJTU include the adsorption working pairs, adsorption refrigeration cycles, and heat and mass transfer intensification technologies.

From the point of view of its development history, the research on adsorption refrigeration can be summarized according to the research goals, the research contents, and the research methods. In the early years the research started with the performance of the adsorbent-refrigerant working pairs, and most of this research work was performed by chemistry and physics academics instead of refrigeration specialists. The main object was to apply this technology to a real application. The research methods were mostly based on the objects of basic adsorption refrigeration systems, and combined the experimental results with the chemical and physical theories for the analysis of the performance. Such research work improved the basic theory of the adsorption refrigeration, and typical adsorbents and refrigerants were focused mainly on activated carbon, zeolite, silica gel, $CaCl_2$, hydride, and so on, and refrigerants were mainly methanol, ammonia, water, Hydrogen, and so on [36, 37].

The early research work pointed out that the basic adsorption refrigeration cycles needed to be improved in many ways, especially the intermittent refrigeration process. Adsorption/

desorption rate and capacity were related to the properties of the adsorption working pairs and the heat and mass transfer performance in the adsorption bed. Such problems resulted in low COP (coefficient of performance) and low SCP (specific cooling power per kilogram adsorbent). In order to solve these problems, the research concerned many interrelated aspects such as heat transfer, mass transfer, and adsorption properties. Some advanced adsorption cycles, such as continuous heat recovery cycle, thermal wave cycle [28, 38], mass recovery cycle, convective thermal wave cycle [16, 28], and cascading cycle [11], and so on, were proposed and their thermal performances were analyzed at that time. Meanwhile some adsorbents-refrigerants working pairs with better adsorption characteristics, for instance, the composite adsorption working pairs, were proposed in many references [32, 39, 40], for which the adsorption cycles were evaluated as a combination of the adsorption cycle and thermodynamic analysis more than just from the point of view of adsorption capacity.

The references produced up until about 1992, was mostly about the analysis and simulations of different cycles theoretically, especially about how the cycle parameters influenced the performances [41–43]. Those contents were even studied in the last few years. The superiority, feasibility, and enormous potential of some advanced systems were proved [20, 44]. Though the feasibility needed to be proved for some of more advanced cycles such as thermal wave cycle, convective thermal wave cycle, and cascading cycle, the research offered the possibility of continuous refrigeration and provided a bright future for the performance improvement of adsorption refrigeration systems. For the system design, the heat and mass transfer intensification attracted a great deal of attention. As a result, researchers paid more attention to the design of an adsorption bed that could improve heat and mass transfer and achieve better performance of continuous regeneration [13, 45, 46] based on the combination of the theoretical analysis and experimental study. In 1992, the first sorption conference held in Paris brought this technology even more to world's attention. Since then the key research aspects of this technology were uniformly recognized by worldwide researchers [47] because numerous new ideas had been put forward on how to improve the adsorption refrigeration performance.

In the 1990s, the research project of the adsorption refrigeration (JOULE0046F) was listed into the JOULE research plan of the European Union (EU). In that plan the research groups such as Meunier from France (zeolite-water), Critoph from England (activated carbon-ammonia), Cacciola from Italy (zeolite-water), Groll from German (metal hydrides-Hydrogen), Zigler from German, Spinner from France (nickel chloride-ammonia/lithium bromide-water adsorption/absorption) had all studied the adsorption refrigeration technology. The research results had been published in the special issue of *International Journal of Refrigeration* in 1999. The adsorption technology and absorption technology were paralleled in the heat pump plan published by the International Energy Association (IEA). In 1994 the adsorption heat pump was taken as an important issue in the International Absorption Heat Pump Conference (ISHPC) which was held in Louisiana in the United States in 1996, the paper for adsorption refrigeration contributed one-third of all the papers in the ISHPC held in Montreal, Canada. Since 1996 the conference for adsorption heat pumps and absorption heat pumps were combined into sorption heat pump and the conference was renamed ISHPC, which is held every three years. In 1999, adsorption refrigeration was the main topic of the sorption heat pump conference held in Munich, Germany. In the conferences of 1996 and 1999, most of the topics were about the composite adsorbent, polymetallic hydrides for heat recovery cycle, thermal wave cycle, and so on. After that the topics expanded over the following sessions of the conference. For example, ISHPC 2002 was held in SJTU. In this conference, the topics included heat transfer intensification, the multi-stage cycle, thermal wave cycle, heat and mass recovery cycle, triple effect cycle of adsorption/absorption refrigeration, solar adsorption system and locomotive adsorption air conditioner, and so on.

1.4 Current Research on Solid Adsorption Refrigeration

In the last 20 years study on solid adsorption refrigeration and heat pump has been reported from USA, France, Japan, UK, Italy, India, and other countries, and the contents are mainly connected to promoting the development of adsorption refrigeration in the field of adsorption working pairs, heat and mass transfer performance, and adsorption refrigeration cycles, and so on. With the progress of adsorption refrigeration technology, some silica gel-water adsorption chillers have been commercialized successfully in the market. The development of the adsorption refrigeration technology can be summarized more in detail as follows: adsorption working pairs and their mechanism; system structure of adsorption refrigeration; improvement of heat and mass transfer of the adsorption bed, as well as thermal properties of many advanced regenerative cycles.

1.4.1 Adsorption Working Pairs

The adsorption working pair is a key element for the adsorption refrigeration and heat pump system. Thermal properties of working pairs have a great influence on the performance coefficient of the system, the temperature increment velocity of the adsorber, and the initial investment. For efficient refrigeration output, the suitable adsorption working pairs need to be selected according to the heat source temperatures, and the suitable adsorption refrigeration cycles need to be selected according to the actual requirements. The application scope and properties are different for different adsorption refrigeration working pairs. The common adsorption refrigeration working pairs mainly include: activated carbon-methanol, activated carbon fiber-methanol, activated carbon-ammonia, zeolite-water, silica gel-water, metal hydrides-hydrogen, calcium chloride-ammonia, and strontium chloride-ammonia, and so on (physical and chemical adsorption) [48]. Recent studies also show that composite adsorption, which is a type of effective heat and mass transfer intensification technology for a chemical adsorbent, is a prospective technology for refrigeration [32, 39, 40].

For working pairs of physical adsorption, the carbon-methanol working pair has a large adsorption and desorption concentration. Its desorption temperature is around 100 °C, which is not high, and it also has the advantage of low adsorption heat, which is around 1800–2000 kJ/kg. Methanol refrigerant can be applied to make ice because its freezing point is below 0 °C. For activated carbon-methanol working pairs, the highest desorption temperature cannot exceed 120 °C, otherwise methanol will decompose. The advantages of the activated carbon-ammonia system is the low evaporation temperature of the refrigerant which is commonly used for making ice. Characterized by being less sensitive to temperature changes for adsorption capacity, it is generally used for higher heat source temperature. For the working pair of silica gel–water, desorption temperature cannot be too high. If it is higher than 120 °C, silica gel will be destroyed. Thus it is a common adsorbent for the low temperature heat source. The zeolite-water working pair has a wide range of desorption temperature (70–250 °C). Its adsorption heat is about 3200–4200 kJ/kg, and the evaporation latent heat of water is 2400–2600 kJ/kg. Zeolite–water is quite stable and won't be destroyed at a high temperature as happens to silica gel. However, it has the disadvantages of a higher adsorption heat, which will lead to the low COP, as well as an evaporation temperature that needs to be higher than 0 °C, which cannot be utilized for making ice. In addition, the system is a vacuum system, which leads to a high requirement of vacuum sealing; meanwhile the low evaporation pressure also makes the adsorption process slower.

Chemical adsorption working pairs mainly include Hydrides-hydrogen, metal chlorides (salt)-ammonia, metal oxides-water and metal oxides-carbon dioxide, and so on. The metal hydrides-hydrogen system utilizes the adsorption process as well as desorption process between metals or alloys and hydrogen for refrigeration, which is characterized by large adsorption and desorption heat, especially for advanced porous metal hydrides (PMH) or Misch metal (Mm) alloy matrixes including Ni, Fe, La, and Al. Such types of working pairs are generally utilized for the adsorption heat pump because they have high adsorption heat as well as high adsorption concentration. Metal chloride-ammonia working pairs are featured as having a large adsorption capacity. For example, for calcium chloride-ammonia working pair 1 mol of calcium chloride can adsorb 8 mol of ammonia. Simultaneously, the boiling point of ammonia is lower than −34 °C so that can be used for making ice, meanwhile the refrigerator works under the condition of positive pressure, which is a feature of simpler manufacture techniques required for the system. Metal oxides-water and metal oxides-carbon dioxide have the advantages of being able to store high levels of energy in hydration and carbonation processes [49, 50]. Take calcium oxide for example, storage energy in the hydration and carbonation process is 800–900 kJ/kg, which makes it possible to develop efficient heat pump systems by the application of such types of working pairs.

But chemical adsorption has the disadvantages of agglomeration and swelling phenomena, which will lead to problems of low permeability and poor mass transfer performance of adsorbents. In order to overcome this problem, recently the porous heat transfer matrixes were put forward for the improvement of mass transfer as well as the heat transfer (by solidified adsorbents) of chemical adsorbents. Studies on such types of adsorbents mainly focus on the composite adsorbents with the matrixes of expanded natural graphite (ENG), activated carbon fiber, and activated carbon. Research shows that such types of composite adsorbents could improve the volume filling quantity and volumetric cooling capacity [32, 39, 40] of adsorbent.

1.4.2 Heat Transfer Intensification Technology of Adsorption Bed

An important indicator when evaluating the adsorption system is the specific cooling power per kilogram adsorbent (SCP, W/kg), which is defined as [51]:

$$SCP \approx \frac{L\Delta x}{t_c} \tag{1.1}$$

where L is the latent heat of vaporization of the refrigerant, t_c is cycle time, and Δx is cycle adsorption quantity. Equation 1.1 shows that for a given operating condition and a given cycle, the main method used to improve the cooling capacity is to shorten the cycle time. Generally there are two ways to shorten the cycle time; one is to improve the mass transfer performance of an adsorbent in the low pressure system, and another way is to enhance the heat transfer performance of the adsorption bed.

Two main technologies for the heat transfer intensification of the adsorption bed are the performance improvement of adsorbent and adsorber. The former one concentrates on the development of the novel types of adsorbents, and the latter one concentrates on the development of the new type of heat exchangers for the adsorber. If the technologics were summarized more in detail, there are three major ways to improve the overall heat transfer coefficients. Thc first one is to increase the heat transfer area of heat exchanger, the second one is to utilize a compact adsorption bed or coated adsorber, and the last one is to use heat pipe technology.

1.4.2.1 Heat Transfer Intensification by Extending the Heat Transfer Area

The heat transfer area of the adsorber can be extended by the following heat exchangers: finned tube [51], plate heat exchanger, plate-fin heat exchanger. Such technology could shorten the cycle time effectively, such as that a SJTU utilizing plate-fin heat exchanger reduced the cycle time of the system by about 5 minutes. The disadvantage of increasing the heat transfer area is the increment of the heat capacity of the metal materials for adsorbers, thus an advanced cycle is usually required for the recovery of heat among adsorbers. For granular adsorbents, with the application of this technology, the wall heat transfer coefficient generally depends on the granularity of the adsorbent, and a small size adsorbent is believed to be necessary for the improvement of the heat transfer coefficients [51]. For example, Miles and Shelton, using small particle size of adsorbent, shortened the cycle time to 5 minutes [52].

1.4.2.2 Compact Adsorption Bed

This technology is particularly suitable for the occasion when the bulk sorbent is not applicable. Such technology had been used for metal hydrides for a long time [51]. The following study found that combining ENG with the adsorbent can enhance heat transfer performance, which was firstly proposed by Spinner and Le Carbone Lorraine and the thermal conductivity can reach 3000 W/m^2 [40]. The other method is to use aluminum as a heat transfer matrix, and thermal conductivity can reach 12 W/(mK) [40]. Curing the composite adsorbents with the binders is also proposed. By using this technology, SJTU improves the thermal conductivity of activated carbon by 58–100% [53]. The disadvantage of compact adsorbent technology is that the mass transfer performance will be influenced in the adsorption bed, especially for the refrigerant working under the vacuum conditions such as water and methanol, and so on.

1.4.2.3 Coated Heat Exchanger

This technique is particularly suitable for the occasion when COP is not important. The coated adsorbent bed can effectively enhance the thermal conductivity of the adsorber by reducing the contact resistance between heat transfer surface and the absorbent. Dunne utilized zeolite [54] coated on the surface of the metal tube thereby improving the SCP to the level of 1500 W/kg. The disadvantage of this technique is that the metal heat capacity is too high, so usually an efficient heat recovery process is required. Another method of developing coated adsorbers is to insert adsorbents into the ENG plates [55], for which the contact between the heat transfer fluid and the adsorbent is not as close as a coated pipe, but since the diameter of the granular adsorbents is only a few microns, the ratio of the adsorbent heat capacity is greatly improved.

1.4.2.4 Heat Pipe Technology

Meunier put forward a novel idea for the improvement of the heat transfer performance, which used the phase change processes, such as condensation and evaporation processes, for heating and cooling adsorbers to obtain a high heat transfer coefficient [51]. LIMSI has studied this idea, and the heat transfer coefficient is about 10 kW/m^2 [51]. Vasiliev also introduced the concept of pulse heat pipe into the adsorbent bed by using propane as the working medium in the design of the pulse heat pipe, for which the adsorption bed is a tablet-shaped heat pipe made of aluminum, and the width of the heat pipe is only 7 mm [56]. As well as that SJTU applied the heat pipe principle to the marine adsorption ice-making system and chillers driven by low-temperature heat and successfully improved the heat transfer performance [57–59].

1.4.3 Low Grade Heat Utilization

The low grade heat exists abundantly in the environment, and it has great potential for the recovery of such a type of heat for energy conservation. The adsorption refrigeration technology is suitable for the recovery of most low grade heat resources by different adsorption working pairs. For example, the silica gel–water working pair can be utilized as a heat source with a low temperature, whereas the zeolite–water system is applicable for a high temperature heat source. Compared with the liquid absorption refrigeration system, the solid adsorption refrigeration system has the advantage of simple structure and low cost. Moreover, it is believed that adsorption technology is more suitable for the vibratory occasions than absorption technology because the adsorbent is solid. Therefore, the application research of solid adsorption refrigeration has been carried out extensively for low grade heat utilization in recent years.

Suzuki [5] applied the zeolite–water working pair on an adsorption automobile air conditioner and analyzed the performance of the system, and the results showed that the key element for reducing the cycle time and the weight of the system is to improve the heat and mass transfer performance of the adsorption bed effectively. Zhu *et al.* [60] studied the adsorption refrigeration system used for the fish storage on boats. Lavan [61] investigated the probability of the absorption refrigeration system driven by the exhaust gas of trucks. Saha [62] presented a double-stage adsorption refrigeration cycle with four beds driven by a low temperature heat source. Such a double-stage cycle has a higher efficiency than that of single-stage cycle when the heat source temperature is very low (< 54 °C). However, its efficiency decreased dramatically once the heat source temperature is relatively high. Yonezawa Y *et al.* carried out a great deal of research on continuous double-bed adsorption chillers with silica gel–water as the working pair and driven by the waste heat, and obtained a series of patents [63, 64]. Ron M studied the application of metal hydride–hydrogen system in automobile air conditioners. Goetz put forward the concept of resorption on the basis of the utilization of $PbCl_2$ and $MnCl_2$, for which the desorption state of $PbCl_2$ is closely bonded with the $MnCl_2$ adsorption state. The principle for this cycle is to use the desorption process of $PbCl_2$ for refrigeration, and the desorbed ammonia is adsorbed by $MnCl_2$. The pressure and temperature of the adsorption bed need to be controlled in the cooling process, otherwise the desorption state of $PbCl_2$ won't match the adsorption state of $MnCl_2$. This study provides a new idea for adsorption refrigeration [65, 66].

On the utilization of the low grade heat, SJTU [29] designed and manufactured a 5 kW adsorption air conditioner using activated carbon–ammonia as the work pair. SJTU also developed a zeolite–water adsorption air conditioning system with the function of energy storage, which was applied in a locomotive cab. The average cooling power of this system is 5 kW. Meanwhile SJTU developed a physical adsorption ice maker for fishing boats with activated carbon–methanol as the working pair [67]. In addition, combining the heat pipe principle with the adsorption system, SJTU developed the siphon heat pipe type adsorber and a split heat pipe type compound adsorption ice-making system for fishing boats [68].

1.4.4 Solar Energy Utilization

The sorption refrigeration driven by solar energy attracted broad attention because the heat supply and cool demand are very well matched with the season and the heat quantity. Compared with the absorption system, the adsorption system can be driven by the heat sources of lower temperatures, which makes the application of solar energy more feasible on the adsorption system.

The solid adsorption refrigeration technology driven by solar energy has been researched extensively since Tchernev [38] successfully developed the refrigeration system with zeolite–water as the working pair. In France, Pons and Guilleminot studied activated carbon–methanol and zeolite–water adsorption systems driven by solar energy, in which the COP of the activated carbon–methanol ice maker [69] is 0.12–0.14 with a collector area of 6 m^2 (four collectors) and adsorbent mass of 20–24 kg/m^2, and the COP of a zeolite–water refrigerator [70] is about 0.10 with the collector area of 20 m^2 (24 collectors) and the adsorbent mass of 360 kg. K. Sumathy *et al.* investigated an activated carbon–methanol ice maker powered by solar energy, and results showed that the daily ice production is 4–5 kg and the COP is 0.1–0.2 [71] when the area of flat plate collector is 0.92 m^2. Y.K. Tan [23–25] in South China University of Technology and Z.F. Li *et al.* in Guangzhou Institute of Energy Conversion [72] also developed the solid adsorption refrigeration system driven by solar energy, which had a similar performance to the system developed by K. Sumathy.

Different from the refrigeration system with the integrated solar collector–adsorption generator, multi types of solar energy powered adsorption refrigeration systems were developed. Iloeje *et al.* [73, 74] utilized a tubular type of absorber, for which the adsorbent (such as calcium chloride, activated carbon) is filled inside the metal pipes. The concentric tube arranged at the center of the metal pipe served as the mass transfer channel of the refrigerant, and the metal tube is boned on the collector surface. Erhard [75] arranged the condensation part of the horizontal heat pipe inside the adsorbent bed to improve the heat flux density. Headley *et al.* [76] studied the activated carbon–methanol adsorption refrigerating system utilizing the compound parabolic concentrator (CPC) as the heat source. The system could realize refrigeration even if the solar radiation is very feeble, but the efficiency of the refrigeration system is very low. Bansal *et al.* [77] studied the $SrCl_2$-NH_3 adsorption refrigerating system driven by the vacuum tube type collector. Vasiliev [56] developed a continuous adsorption heat pump with heat recovery process driven by solar energy and natural gas, using a parabolic concentrator for collecting the solar energy to heat the circulating water. The system employed solar energy as a main power supply, and the natural gas served as an auxiliary heat source when solar energy is not enough. The system can accomplish continuous refrigeration with the cycle time of 12 minutes Z.Y. Liu [78, 79] put forward the refrigeration system which combined the unit adsorption tube with the collector for the solar energy. For such a design the adsorbent bed can be heated by solar energy directly.

On the topic of solar energy utilization, SJTU [80] developed a compound system of water heater and refrigerator driven by solar energy to improve energy efficiency. Meanwhile, SJTU also developed the silica gel–water adsorption chiller in 2004, which had been applied to the building and grain storage hall with solar energy as the driving power.

1.4.5 Advanced Adsorption Refrigeration Cycle

A prominent problem for adsorption refrigeration is that the COP is low. For a traditional simple cycle under the condition of air conditioning, generally COP is less than 0.4 [51]; if during ice making with a refrigerating temperature lower than 0 °C, commonly COP is less than 0.2 and under some conditions is even lower than 0.1 [53, 67, 81] due to the fact that the adsorption performance usually decreases with the evaporation temperature. The reason for the low COP is mainly caused by the big temperature fluctuation of the adsorption bed under the condition of alternating heating and cooling processes. In order to improve the COP of the adsorption refrigeration system, the concept of heat recovery was proposed. The principle of heat recovery was put forward by Tchernev initially [38], for which the heat transfer fluid was preheated by the adsorption heat, then was heated by the boiler and passed into the adsorber to provide the desorption heat. Nowadays the heat recovery cycles studied by researchers mainly include double-bed heat recovery cycle, cascading cycle, multi-stage cycle, thermal wave cycle, and so on.

1.4.5.1 Double-Bed Heat Recovery Cycle, Cascading Cycle, and Multi-stage Cycle

For such type of cycles in the heat recovery process the heat from a high-temperature adsorption bed is delivered to the low-temperature adsorption bed by the temperature potential. Because the heat recovered is mainly the sensible heat of the adsorbers, the heat cannot be recovered from the low temperature bed to the high temperature bed. Thus the coefficient of heat recovery is limited.

In the research for the continuous refrigeration cycle with double-bed heat recovery process [82–87], the heat recovery coefficient gleaned from experiments is 0.22 [44]. The best result of heat recovery coefficient for the cascading cycle is 0.5 [11, 51], although in the simulation it is as high as 0.63 [88].

Among three types of cycles, i.e., double-bed heat recovery cycle, cascading cycle and multi-stage cycle, the typical cycle is a multi-stage and six-bed adsorption cycle proposed by Saha and Kashiwagi [89]. The three-stage cycle system using silica gel–water as the working pair can decrease the driven temperature effectively. The experimental results showed that it can obtain a chillier water of 12 °C when the heat source temperature is 50 °C and the second law of thermodynamics efficiency is 0.3–0.4. Its driven temperature is lower than that of the LiBr absorption system. Such a technology provided an effective way for the recycling use of low grade heat of 50–60 °C. SJTU also developed a type of two-stage cycle for the freezing conditions, such a type of cycle can be driven by a heat source with the temperature lower than 100 °C, and can generate the cooling power with temperatures as low as −15 °C when the environmental temperature is around 25–35 °C [90].

1.4.5.2 Thermal Wave and Convective Thermal Wave Cycle

Thermal wave cycle was proposed by Shelton [18]. His theory indicated that the heat recovery coefficient of a thermal wave cycle can reach 0.7 and the COP of the heat pump is 1.87.

The principle of the thermal wave cycle is to use the flow of the thermal fluids, which transferred the heat within the adsorbers to form a steep temperature wave. For such technology the heat can be transferred from the low temperature adsorber to the high temperature adsorber [51]. This concept has been applied to chemical heat pumps. Willers *et al.* have studied a multi-hydride–thermal-wave concept [91]. For this cycle, through the combination of low-temperature and high-temperature metal hydride, with the same equilibrium temperature difference for both metal hydrides under the condition of same pressure, a very steep thermal wave can be generated by the accumulating temperature effect in the adsorption bed. Critoph [16] suggested a convective thermal wave cycle. In such a cycle, the refrigerant served as the heat transfer fluid, which could improve the heat recovery efficiency effectively because of the direct contact between the heat transfer fluid and the solid adsorbent.

1.4.5.3 Stages Regeneration Cycle for Dehumidification Refrigeration System

All the refrigeration cycles mentioned above are for the closed adsorption refrigeration cycles. Nowadays there is also a type of stage regenerative dehumidification cycle. The regenerative process of the adsorbent is divided into two stages while the adsorber rotates. The first stage is to heat the adsorbent with the air preheated by the heat of adsorption, and then to heat the adsorbent to the maximum desorption temperature with the air heated by the heat source. For the adsorption phase the dehumidification effect will be achieved by absorbing the water in the environment, and the cooling power can be generated by spraying the water into the dry air in a dehumidification process. The early research can be see in a report from Douglas [92]. By using the rotary beds, the Daikin company in Japan successfully humidified the indoor air by desorbing the water indoor and adsorbing the water outside. Such a mode made the indoor more comfortable during the heat-pump condition of winter.

1.4.5.4 Mass Recovery Cycle

In addition to the heat recovery cycle there is a mass recovery cycle. The mass recovery cycle is that the refrigerant gas in the high pressure generator after desorption is transferred to the low pressure generator of adsorption as a result of the pressure difference between two adsorption beds. It can effectively improve the adsorption/desorption quantity because of the large pressure difference between two beds, thereby enhance the cooling capacity and improving the COP. If compared with the basic cycle, the largest COP increment of the system can reach 100%.

1.4.5.5 Combined Adsorption-Absorption Refrigeration Cycle

Two-stage or three-stage combined adsorption-absorption refrigeration cycle is composed of the adsorption refrigeration system driven by a high temperature heat source and the absorption refrigeration system driven by the exhaust heat from the adsorption system. The total theoretical COP of the combination refrigeration cycle with adsorption chiller of $NiCl_2$-NH_3 (COP = 0.27) or two-bed zeolite–water adsorption chiller (COP = 0.50) as the first stage cycle, and absorption chiller of lithium bromide–water (COP = 0.75) as the second stage cycle is 1.52. The theoretical COP of the combination refrigeration cycle with metal hydride adsorption chiller as the first stage cycle, and silica gel–water or lithium bromide–water absorption chiller as the second stage cycle is expected to reach 1.5. The COP of combination refrigeration cycle is high, but the system is very complicated.

1.4.5.6 Adsorption Refrigeration System Driven by Compressor

Compared with the conventional adsorption refrigeration system for which the adsorption quantity is relative to the temperature difference of the adsorption bed, the difference for this system is that the adsorption is relative to the pressure difference between two adsorption beds, which is formed by the compressor. Thus the adsorption quantity decreased with the pressure of the adsorption bed, and the temperature decreased during the desorption process, for which the cooling power was generated. The main problem for such a system is the high requirements on the pressure ratio of the compressor.

1.4.5.7 Internal Heat Recovery Process and Double Way Cycle

An internal heat recovery process is proposed by P. Neveu and J. Castaing [93]. In this cycle they used two types of salts and recovered the reaction heat of the high temperature salt as the heat source for the desorption process of the low temperature heat source. In such a way, SJTU had established the double way and multi-effect refrigeration cycle, which combined the adsorption and resorption process together for the refrigeration output, as well as combining the internal heat recovery with the sensible heat recovery processes to improve the COP, and the study indicated that the COP could be improved by more than 1 through such a novel cycle [94, 95].

1.4.6 Commercialized Adsorption Chillers

With the rapid development of the adsorption refrigeration technology, adsorption chillers appeared in the market. Nishiyodo Kuchouki Co., Ltd invented the silica gel–water adsorption chiller in 1986, and the schematic diagram was shown in Figure 1.3. The adsorption system used water for heating and cooling. HIJC Company in the United States sold such a type of

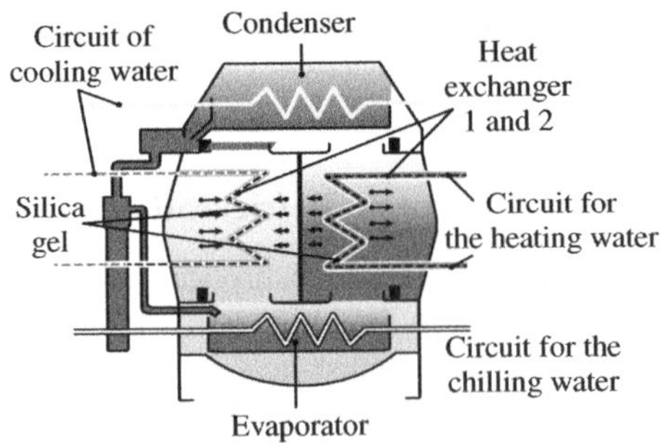

Figure 1.3 Schematic diagram of silica gel-water adsorption system from Nishiyodo Kuchouki Co., Ltd

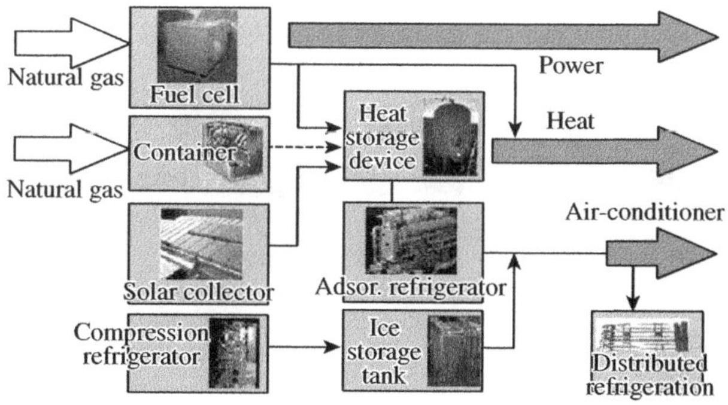

Figure 1.4 CCHP system in Malteser hospital in Germany

adsorption chillers. The chiller produced 3 °C chilling water when the heat source temperature was 50–90 °C.

The Malteser Hospital in Kammenz of Germany were the first to install a CCHP (cogeneration system for cooling, heat, and power) system for which an adsorption chiller was utilized. The system started running from May 2000 and the system diagram is shown in Figure 1.4. The heat collector of the system collected the waste heat from the fuel cell and the low grade heat from solar energy, combined with the adsorption chiller the system supplied the heating and cooling power simultaneously. The cooling power of the adsorption chiller was 105 kW. A complimentary compression chiller is also installed in the system for the regulation of the cooling power.

Macom, a Japanese company, began to produce a silica gel–water adsorption refrigeration chiller since 2003. It can obtain 14 °C chilled water when the driven temperature is 75 °C, and the COP is 0.6.

Tokai Optical Co., Ltd., in Nagoya of Japan, introduced an adsorption CCHP system powered by waste heat in April 2003. A 185 kW diesel engine is used in the system. The waste heat can supply heat, and simultaneously the refrigeration can be generated for dehumidification and cooling. By such a system the annual energy consumption could be reduced by 10%, and CO_2 emissions could be reduced by 12%.

In China, SJTU, South China University of Technology, Guangzhou Institute of Energy Conversion, Chinese Academy of Sciences, Hunan University, and so on, carried out the practical research work on the adsorption refrigeration. Adsorption chillers of series "DY" had been developed by the Hunan University, such as an ice-maker powered by the exhaust heat of the diesel engine on fishing boats and on automobiles. SJTU successfully developed small types of silica gel–water adsorption refrigerators of 10–200 kW using a heat and mass recovery process, which can be driven by the heat source with the temperature of 65 °C. The second generation of the prototype developed by SJTU in 2009 is shown in Figure 1.5. Such a chiller had been successfully utilized for the building and grain storage.

1.4.7 Current Researches on the Adsorption Theory

For physical as well as for chemical adsorption refrigeration, the research direction on the adsorption refrigeration is from the equilibrium adsorption refrigeration with uniform

Figure 1.5 Silica gel–water adsorption refrigerator developed in China

temperature and pressure to the non-equilibrium adsorption refrigeration technology. The dynamic features of the adsorption and desorption is more and more important for the analysis on the adsorption refrigeration theories with the development of the heat and mass transfer intensification technology.

For physical adsorption Critoph proposed a simplified format of the D-A equation [15, 17], for which only the temperature is considered. It is an experiential equation utilized extensively for the equilibrium adsorption performance evaluation, but cannot be utilized for the non-equilibrium adsorption performance analysis. For this problem Sokoda established a model for the adsorption velocity for which the dynamic process of adsorption is considered with the mass transfer process of the gases inside the adsorption systems, and they are:

$$\frac{\mathrm{d}x}{\mathrm{d}t} = K_s a_p (x^* - x) \tag{1.2}$$

$$K_s a_p = \frac{15 D_{so}}{R_p{}^2} \exp(-E_a/RT) \tag{1.3}$$

where x^* is the local adsorption quantity, $K_s a_p$ is the coefficient for the velocity of surface diffusion, D_{so} is the surface diffusion coefficient, E_a is activated energy for the surface diffusion, and R_p is the average diameter of the adsorbent granules. This equation is mainly for the silica gel–water adsorption working pair. The equation can be utilized for other working pairs but needs the amendment of the coefficients in the equations, such as that which E. F. Passos *et al.* [12] had performed on this equation for the activated carbon-methanol working pair.

Compared with physical adsorption the chemical adsorption theories are very complex. There are mainly three categories: local, global, and analytical models. Local models consider mass and heat transfer, and kinetics of small volume that result in partial derivatives equations, which are numerically solved. Global models consider variables and average values of reactor features such as permeability, thermal conductivity, heat capacity, and so on, for simulation. Numerical solutions for the global models give sets of differential equations. Analytical models consider average values of the variables during reaction time and these differential equations are related to the space variable only. Spinner and Rheault [96] researched non-uniform dynamics based on the study of dynamic adsorption rate. Then, based on the achievements of Spinner and Rheault, Mazet *et al.* [97] and Lebrun [98] amended the equation that is suggested by Tykodi [99] and Flanagan [100], it is:

$$\frac{\mathrm{d}x}{\mathrm{d}t} = K_i(1-x)\exp(-A_0/T)\ln\left(\frac{p_c}{p_{eq}(T)}\right) \tag{1.4}$$

where x is adsorption quantity, $\mathrm{d}x/\mathrm{d}t$ is adsorption rate, K_i is dynamic coefficient, subscript i = s for adsorption process, and i = d for desorption process. p_c is the constrained pressure of condenser and evaporator, p_{eq} is equilibrium pressure, and T is adsorption temperature.

Mazet makes a logarithm transformation in Equation 1.4 because the influence of A_0 is not great in the experiments [97]. Based on that Goetz [101] developed a model that considered the mass transfer performance inside the grain, which is

$$\frac{\mathrm{d}N_g}{\mathrm{d}t} = 4\pi r_c{}^2 K_i\left(\frac{p_c - p_{eq}(T)}{p_{eq}(T)}\right)^{Ma} \tag{1.5}$$

where N_g is the molar adsorption quantity, r_c is the diameter of reaction surface, and Ma is the reaction dynamic coefficient.

Another formula [102] for the reaction rate which considered the Darcy equation for reaction surface and grain surface is

$$\frac{\mathrm{d}x}{\mathrm{d}t} = f(x, r_g)\left(\frac{p_c - p_i}{T_c}\right)K_n(m, c) \tag{1.6}$$

where K_n is Knudsen diffusion rate that is related to the diameter of pore and porosity, $f(x,r_g)$ is a function which is related to adsorption quantity x and the radius of grain r_g, and p_i is the pressure inside the pore.

One question that comes out of the chemical adsorption theory is the models for the adsorption and desorption processes. Generally the models for adsorption are also utilized for the desorption process. Furrer once pointed out that there is a quasi-equilibrium region for the solid-gas reaction, and Goetz and Marty [101] had considered this region in his research work [103]. SJTU had studied the chemical and composite adsorption under the condition of non-equilibrium heating and cooling processes, and results showed that a serious hysteresis phenomenon exists for the adsorption and desorption processes [2]. The real refrigeration process is always under the condition of non-equilibrium states, thus such type of hysteresis needs to be considered for the chemical adsorption model.

Another question from the chemical adsorption is the difference between the chemical adsorption models and composite adsorption models. The main adsorbent inside the composite adsorbent is the chemical adsorbent, thus for the simulation of the adsorption process generally the models of chemical adsorbent are utilized for the composite adsorbent. Such a simulation is acceptable for the equilibrium process. For the non-equilibrium process the composite adsorption is complex because it includes the heat and mass transfer processes in chemical adsorbent and porous media. In the reaction phase the volume of chemical adsorbent, as well as the density of porous additive will all be changed, and such a phenomenon will influence the adsorption performance. Thus the heat and mass transfer performances for both chemical and porous materials need to be considered for the non-equilibrium adsorption models.

To summarize the contents above, as a type of energy saving and environmental benign technology the adsorption refrigeration has received more and more attention. Quite a lot of achievements had been made by the researchers with their continuous efforts, and this has established a good foundation for further development. But there is still a long way for the extensive application of the technology. The achievements and problems in the research work will be summarized and analyzed in detail in the following chapters.

References

[1] Zhang, Y.H. (1989) *Adsorption Functions*, Shanghai, Publishing House of Science and Technology, Shanghai, ISBN: 7805134979 (in Chinese).

[2] Wang, L.W., Wang, R.Z., Wu, J.Y. and Wang, K. (2004) Adsorption performances and refrigeration application of adsorption working pair of $CaCl_2$-NH_3. *Science in China, Series E*, **47**(2), 173–185.

[3] Hulse, G.E. (1929) Freight car refrigeration by an adsorption system employing silica gel. *Refrigeration Engineering*, **17**(2), 41–54.

[4] Plank, R. and Kuprianoff, J. (1960) *Die Klein Käeltemaschine*, Springer-Verlag, Berlin.

[5] Suzuki, M. (1993) Application of adsorption cooling systems to automobiles. *Heat Recovery Systems and CHP*, **13**(4), 335–340.

[6] Feng, Y. and Tan, Y.K. (1990) Research on the adsorption refrigeration technology powered by the low temperature heat source. *Energy Conservation*, (2), 40–42, ISSN: 1004–7948 (in Chinese).

[7] You, Q.G. and Wang, R.Z. (1998) The application of solid adsorption refrigeration for the recovery of the waste heat. *New Energy*, **20**(1), 7–12, ISSN: 1002–1418 (in Chinese).

[8] Jones, A.J. and Golben, P.M. (1985) Design, life testing and future designs of cryogenic hydride refrigeration systems. *Cryogenics*, **25**, 212.

[9] Wade, L., Ryba, E., Weston, C. *et al.* (1992) Test performance of a 2 W,137 K sorption refrigerator. *Cryogenics*, **32**(2), 122–126.

[10] Douss, N. and Meunier, F. (1988) Effect of operating temperatures on the coefficient of performance of active carbon-methanol systems. *Heat Recovery Systems and CHP*, **8**(5), 383–392.

[11] Douss, N. and Meunier, F. (1989) Experimental study of cascading adsorption cycles. *Chemical Engineering Science*, **44**(2), 225–235.

[12] Passos, E.F., Escobedo, J.F. and Meunier, F. (1989) Simulation of an intermittent adsorptive solar cooling system. *Solar Energy*, **42**(2), 103–111.

[13] Restuccia, G., Recupero, V., Cacciola, G. and Rothmeyer, M. (1988) Zeolite heat pump for domestic heating. *Energy*, **13**, 333–342.

[14] Cacciola, G. and Restuccia, G. (1994) Progress on adsorption heat pumps. *Heat Recovery Systems and CHP*, **14**(4), 409–420.

[15] Critoph, R.E. (1988) Performance limitations of adsorption cycles for solar cooling. *Solar Energy*, **41**(1), 21–31.

[16] Critoph, R.E. (1994) Forced convection enhancement of adsorption cycles. *Heat Recovery Systems and CHP*, **14**(4), 343–350.

[17] Tamainot-Telto, Z. and Critoph, R.E. (1997) Adsorption refrigerator using monolithic carbon-ammonia pair. *International Journal of Refrigeration*, **20**(2), 146–155.

[18] Shelton, S.V. (1986) Solid adsorbent heat pump system. US Patents 4610148.

[19] Shelton, S.V., Wepfer, W.J. and Miles, D.J. (1988) External fluid heating of a porous bed. *Chemical Engineering Communications*, **71**, 39–52.

[20] Shelton, S.V., Wepfer, W.J. and Miles, D.J. (1989) Square wave analysis of the solid-vapor adsorption heat pump. *Heat Recovery Systems and CHP*, **9**, 233–247.

[21] Shelton, S.V., Wepfer, W.J. and Miles, D.J. (1990) Ramp wave analysis of the solid/vapor heat pump. *Journal of Energy Resources Technology, Transactions of ASME*, **112**, 69–76.

[22] Vasiliev, L.L., Mishkinis, D.A., Antukh, A.A. and Kulakov, A.G. (2004) Resorption heat pump. *Applied Thermal Engineering*, **24**, 1893–1903.

[23] Feng, Y. and Tan, Y.K. (1990) The experimental research on the adsorption refrigeration for the recovery of the waste heat in industries. *Energy Conservation Technology*, (4), 1–3, ISSN: 1002–6339 (in Chinese).

[24] Feng, Y. and Tan, Y.K. (1990) The research for the adsorption refrigeration cycle with activated carbon-methanol as the working pair. *Journal of Refrigeration*, (1), 1–5, ISSN: 0253–4339 (in Chinese).

[25] Feng, Y. and Tan, Y.K. (1991) Research on the heat and mass transfer performance for the adsorption refrigeration systems. *Journal of Chemical Industry and Engineering*, **42**(3), 342–347, ISSN: 0438-1157 (in Chinese).

[26] Yan, A.Z. (1982) Adsorption refrigeration with the working pair of zeolite-water, part 1: the choice of the zeolite. *Journal of Refrigeration*, (4), 24–32, ISSN: 0253-4339 (in Chinese).

[27] Lin, H.Z., Cao, L.P., Zhang, J.G. *et al.* (1994) The experimental research on the dynamic characteristics of zeolite-water adsorption system and its adsorption properties. *Journal of Engineering Thermophysics*, (3), 253–256, ISSN: 0253-231X (in Chinese).

[28] Wang, R.Z. (2001) Performance improvement of adsorption cooling by heat and mass recovery operation. *International Journal of Refrigeration*, **24**(7), 602–611.

[29] Wang, R.Z. (2001) Adsorption refrigeration in Shanghai Jiao Tong University. *Renewable and Sustainable Energy Review*, **5**(1), 1–37.

[30] Zhang, X.J., Dai, Y.J. and Wang, R.Z. (2003) A simulation study of heat and mass transfer in a honeycombed rotary desiccant dehumidifier. *Applied Thermal Engineering*, **23**(8), 989–1003.

[31] Lu, Y.Z., Wang, R.Z., Zhang, M. and Jiangzhou, S. (2003) Adsorption cold storage system with zeolite–water working pair used for locomotive air conditioning. *Energy Conversion and Management*, **44**(10), 1733–1743.

[32] Wang, L.W., Wang, R.Z., Wu, J.Y. and Wang, K. (2004) Compound adsorbent for adsorption ice maker on fishing boats. *International Journal of Refrigeration*, **27**, 401–408.

[33] Zhai, X.Q., Dai, Y.J. and Wang, R.Z. (2005) Comparison of heating and natural ventilation in a solar house induced by two roof solar collectors. *Applied Thermal Engineering*, **25**(5–6), 741–757.

[34] Wang, D.C., Xia, Z.Z., Wu, J.Y. *et al.* (2005) Study of a novel silica gel–water adsorption chiller. Part I. Design and performance prediction. *International Journal of Refrigeration*, **28**(7), 1073–1083.

[35] Wang, D.C., Wu, J.Y., Xia, Z.Z. *et al.* (2005) Study of a novel silica gel–water adsorption chiller. Part II. Experimental study. *International Journal of Refrigeration*, **28**(7), 1084–1091.

[36] Critoph, R.E., Vogel, R. (1986) Possible adsorption pairs for use in solar cooling. *International Journal of Ambient Energy*, 7(4), 183–190.

[37] Feng, Y. and Tan, Y.K. (1989) The principle of adsorption refrigeration and the choice of the adsorption working pairs. *Guangzhou Chemical Industry and Technology*, (1), 48–52, ISSN: 1001-9677 (in Chinese).

[38] Tchernev, D. (1985) Heat pump energized by low-grade heat sources. US Patent PCT/US85/00783.

[39] Aristov, Y.I., Restuccia, G., Caccioba, G. *et al.* (2002) A family of new working materials for solid sorption air conditioning systems. *Applied Thermal Engineering*, **22**, 191–204.

[40] Mauran, S., Prades, P. and Haridon, F.L. (1993) Heat and mass transfer in consolidated reaction beds for thermochemical systems. *Heat Recovery Systems and CHP*, **13**, 315–319.

[41] Luo, L. and Feidt, M. (1992) Thermodynamics of adsorption cycles: a theoretical study. *Heat Transfer Engineering*, **13**(4), 19–31.

[42] Hajji, A., Worek, W.M. and Lavan, Z. (1991) Dynamic analysis of a closed-cycle solar adsorption refrigerator using two adsorbent-adsorbate pairs. *ASME Journal of Solar Energy Engineering*, **113**, 73–79.

[43] Cacciola, G. and Restuccia, G. (1995) Reversible adsorption heat pump: a thermodynamic model. *International Journal of Refrigeration*, **18**(2), 100–106.

[44] Douss, N., Meunier, F. and Sun, L.M. (1988) Predictive model and experimental results for a two-adsorber solid adsorption heat pump. *Industrial and Engineering Chemistry Research*, **27**, 310–316.

[45] Beasley, E.D. and Clark, J.A. (1984) Transient response of a packed bed for thermal energy storage. *International Journal of Heat and Mass Transfer*, **27**(9), 1659–1669.

[46] Guilleminot, J.J. and Meunier, F. (1987) Heat and mass transfer in a non-isothermal fixed solid adsorbent reactor: a uniform pressure-non-uniform temperature case. *International Journal of Heat and Mass Transfer*, **30**(8), 1595–1606.
[47] Wang, R.Z., Wu, J.Y., Teng, Y. *et al.* (1998) The key technologies for the solid adsorption refrigeration. *Acta Energiae Solaris Sinica*, **19**(1), 20–29, ISSN: 0254-0096 (in Chinese).
[48] Srivastava, N.C. and Eames, I.W. (1998) A review of adsorbents and adsorbates in solid-vapour adsorption heat pump systems. *Applied Thermal Engineering*, **18**, 707–714.
[49] Kato, Y., Yamada, M., Kanie, T. and Yoshizawa, Y. (2001) Calcium oxide/carbon dioxide reactivity in a packed bed reactor of a chemical heat pump for high-temperature gas reactors. *Nuclear Engineering and Design*, **210**, 1–8.
[50] Kato, Y., Takahashi, F.U., Watanabe, Y. *et al.* (2001) thermal analysis of a magnesium oxide/water chemical heat pump for cogeneration. *Applied Thermal Engineering*, **21**, 1067–1081.
[51] Meunier, F. (1998) Solid sorption heat powered cycles for cooling and heat pumping applications. *Applied Thermal Engineering*, **18**, 715–729.
[52] Miles, D. and Shelton, S. (1996) Design and testing of a solid-sorption heat-pump system. *Applied Thermal Engineering*, **16**, 389–394.
[53] Wang, L.W., Wu, J.Y., Wang, R.Z. *et al.* (2003) Experimental study of a solidified activated carbon-methanol adsorption ice maker. *Applied Thermal Engineering*, **23**, 1453–1462.
[54] Dunne, S. (1996) Carousel heat exchanger for sorption cooling process. US Patent 5 503 222.
[55] Pons, M. and Dantzer, P. (1994) Heat transfer in hydride packed beds. *International Journal of Research in Physical Chemistry and Chemical Physics*, **183**, 1249–1259.
[56] Vasiliev, L.L. (2005) Heat pipes in modern heat exchangers. *Applied Thermal Engineering*, **25**(1), 1–19.
[57] Wang, L.W., Wang, R.Z., and Wu, J.Y. (2003) The split heat pipe type adsorption ice maker for fishing boats. China Patent 200310108924.5.
[58] Xia, Z.Z., Wang, R.Z., Wu, J.Y., and Wang, D.C. (2004) A new efficient split heat pipe type adsorption refrigerator. China Patent 200410025398.0.
[59] Xia, Z.Z., Wang, R.Z., Wu, J.Y., and Wang, L.W. (2004) The compound alternately heat pipe type generator for the recovery of the low grade heat. China Patent 200410018291.3.
[60] Zhu, R.Q., Han, B.Q., Lin, M.Z. and Yu, Y.Z. (1992) Experimental investigation on an adsorption system for producing chilled water. *International Journal of Refrigeration*, **15**(1), 31–34.
[61] Lavan, Z.A. (1982) Truck exhaust gas operated absorption refrigeration system. *ASHRAE Transactions, Part B*, **2531**, 66–76.
[62] Chua, H.T., Ng, K.C., Malek1, A. *et al.* (1999) Modeling the performance of two-bed, sillica gel-water adsorption chillers. *International Journal of Refrigeration*, **22**(3), 194–204.
[63] Yonezawa, Y., Matsushita, M., Oku, K. *et al.* (1989) Adsorption refrigeration system. US Patent 4881376.
[64] Yonezawa, Y., Ohnishi, T., Okumura, S. *et al.* (1991) Method of operating adsorption refrigerators. US Patent 5024064,1991.
[65] Lepinasse, E., Marion, M. and Goetz, V. (2001) Cooling storage with a resorption process. Application to a box temperature control. *Applied Thermal Engineering*, **21**, 1251–1263.
[66] Goetz, V., Spinner, B., Lepinasse, E. (1997) A solid-gas thermochemical cooling system using $BaCl_2$ and $NiCl_2$. *Energy*, 22(1), 49–58.
[67] Wang, L.W., Wang, R.Z., Wu, J.Y. *et al.* (2004) Adsorption ice makers for fishing boats driven by the exhaust heat from diesel engine: choice of adsorption pair. *Energy Conversion and Management*, **45**, 2043–2057.
[68] Wang, L.W., Wang, R.Z., Wu, J.Y. and Wang, K. (2005) A new type adsorber for adsorption ice maker on fishing boats. *Energy Conversion and Management*, **46**, 2301–2316.
[69] Pons, M. and Guilleminot, J.J. (1986) Design of an experimental solar-powered, solid-adsorption ice maker. *ASME Journal of Solar Energy Engineering*, **108**, 332–337.
[70] Grenier, P., Guilleminot, J.J. and Meunier, F. (1988) Solar powered solid adsorption cold store. *Solar Energy Engineering*, **110**, 192–197.
[71] Sumathy, K. and Li, Z.F. (1999) Experiments with solar-powered adsorption ice maker. *Renewable Energy*, **16**, 704–707.
[72] Li, Z.F., Huang, Z.C. and Liu, G.X. (1991) The principle and experimental research on the solid adsorption refrigeration powered by the solar energy. *Journal of Refrigeration*, **1**, 21–26, ISSN: 0253-4339 (in Chinese).
[73] Enibe, S.O. and Iloeje, O.C. (1997) Design optimization of the flat plate collector for a solid absorption solar refrigerator. *Solar Energy*, **60**(2), 77–87.

[74] Enibe, S.O. and Iloeje, O.C. (1997) Transient analysis and performance prediction of a solid absorption solar refrigerator. *Solar Energy*, **60**(1), 43–59.
[75] Erhard, D., Spindler, K. and Hahne, E. (1998) Test and simulation of a solar powered solid sorption cooling machine. *International Journal of Refrigeration*, **21**(2), 133–141.
[76] Headley, S., Kothdiwal, A.F., McDoom, I.A. *et al.* (1994) Charcoal-methanol adsorption refrigerator powered by a compound parabolic concentrating solar collector. *Solar Energy*, **53**(2), 191–197.
[77] Bansal, N.K., Blumenberg, J., Kavasch, H.J. *et al.* (1997) Performance testing and evaluation of solid absorption solar cooling unit. *Solar Energy*, **61**(2), 127–140.
[78] Liu, Z.Y., Lu, Y.Z. and Zhao, J.X. (1998) Zeolite-active carbon compound adsorbent and its use in adsorption solar cooling tube. *Solar Energy Materials and Solar Cells*, **52**, 45–53.
[79] Liu, Z.Y., Lu, Y.Z. and Wang, Y.T. (2000) Experimental research on a novel solar powered refrigerating pipe. *Acta Energiae Solaris Sinica*, **21**(1), 82–88, ISSN: 0254-0096 (in Chinese).
[80] Wang, R.Z., Li, M., Xu, Y.X. and Wu, J.Y. (2000) An energy efficient hybrid system of solar powered water heater and adsorption ice maker. *Solar Energy*, **68**(2), 189–195.
[81] Wang, S.G., Wang, R.Z., Wu, J.Y. *et al.* (2003) Experimental results and analysis for adsorption ice-making system with consolidated adsorbent. *Adsorption*, **9**(4), 349–358.
[82] Wang, R.Z., Xu, Y.X., Wu, J.Y. *et al.* (1998) The improvement and experiments on a continuous adsorption refrigerator with heat recovery process. *Journal of Engineering Thermophysics*, **19**(3), 275–278, ISSN: 0253-231X (in Chinese).
[83] Poyelle, F., Guilleminot, J.J. and Meunier, F. (1996) Analytical study of a gas-fired adsorptive air-conditioning system. *ASHRAE Transactions*, **102**, 1128–1136.
[84] Cacciola, G., Hajji, A., Maggio, G. *et al.* (1993) Dynamic simulation of a recuperative adsorption heat pump. *Energy*, **18**(11), 1125–1137.
[85] Van Benthem, G.H.W. and Cacciola, G. (1995) Regenerative adsorption heat pumps: optimization of the design. *Heat Recovery Systems and CHP*, **15**(6), 531–544.
[86] Cho, S.H. and Kim, J.N. (1992) Modeling of a silica gel/water adsorption-cooling system. *Energy*, **17**(9), 829–839.
[87] Chua, H.T., Ng, K.C., Malek, A. *et al.* (2001) Multi-bed regenerative adsorption chiller – improving the utilization of waste heat and reducing the chilled water outlet temperature fluctuation. *International Journal of Refrigeration*, **24**, 124–136.
[88] Meunier, F. (1985) Second law analysis of a solid adsorption heat pump operating on cascade cycles. *Heat Recovery Systems and CHP*, **5**, 133–141.
[89] Saha, B.B., Koyama, S., Ng, K.C. *et al.* (2006) Study on a dual-mode, multi-stage, multi-bed regenerative adsorption chiller. *Renewable Energy*, **31**(13), 2076–2090.
[90] Wang, J., Wang, L.W., Luo, W.L. and Wang, R.Z. (2013) Experimental study of a two-stage adsorption freezing machine driven by low temperature heat source. *International Journal of Refrigeration*, **36**(3), 1029–1036.
[91] Willers, E., Wanner, M. and Groll, M. (1999) A multi-hydride thermal wave device for simultaneous heating and cooling. *Journal of Alloys and Compounds*, **293–295**(20), 915–918.
[92] Douglas, K. (1996) Linking indoor air quality (IAQ), ventilation, moisture control, and desiccants to future air conditioning practices. Proceedings of the International Absorption Heat Pump Conference, Montreal, Quebec, pp. 128–137.
[93] Neveu, P. and Castaing, J. (1993) Solid-gas chemical heat pumps: field of application and performance of the internal heat of reaction recovery process. *Heat Recovery Systems and CHP*, **13**(3), 233–251.
[94] Li, T.X., Wang, R.Z., Kiplagat, J.K. and Wang, L.W. (2010) Performance improvement of a combined double-way sorption refrigeration cycle using re-heating process. *AIChE Journal*, **56**, 477–484.
[95] Li, T.X., Wang, R.Z., Kiplagat, J.K. and Ma, L. (2012) Performance analysis of a multi-mode thermochemical sorption refrigeration system for solar-powered cooling. *International Journal of Refrigeration*, **35**, 532–542.
[96] Spinner, B. and Rheault, F. (1985) Kinetics models in solid/gas reactions under imposed pressure and temperature constraints. Proceedings of International Workshop on Heat Transformation and Storage, Ispra, Italy, pp. 9–11.
[97] Mazet, N., Amouroux, M. and Spinner, B. (1991) Analysis and experimental study of the transformation of non-isothermal solid/gas reacting medium. *Chemical Engineering Communications*, **99**, 155–174.
[98] Lebrun, M. (1990) Models of heat and mass transfers in solid/gas reactor used as chemical heat pumps. *Chemical Engineering Science*, **45**, 1743–1753.
[99] Tykodi, R.J. (1979) Thermodynamics of steady state resistance change transitions in steady-state systems. *Bulletin of Chemical Society of Japan.*, **552**(2), 564–568.

[100] Flanagan, T.B. (1978) *Hydrides for Energy Storage*, Oxford Univeristy Press.

[101] Goetz, V. and Marty, A. (1992) A model for reversible solid/gas reactions submitted to temperature and pressure constraints; simulation of the rate of reaction in solid gas reactors used in chemical heat pump. *Chemical Engineering Science*, **47**(17–18), 4445–4454.

[102] Neveu, P. and Castaing-Lasvignottes, J. (1997) Development of a numerical sizing tool for a solid-gas thermochemical transformer – I impact of the microscopic process on the dynamic behaviour of a solid-gas reactor. *Applied Thermal Engineering*, **17**(6), 501–518.

[103] Huang, H.J., Wu, G.B. and Yang, J.*et al.* (2004) Modeling of gas–solid chemisorption in chemical heat pumps. *Separation and Purification Technology*, **34**, 191–200.

2

Adsorption Working Pairs

The adsorption processes include physical adsorption [1–3] and chemical adsorption [4]. Physical adsorption is formed by the van der Waals force between the molecules [2] of the adsorbent and the adsorbate. Physical adsorbents with mesopores can adsorb consecutive layers of adsorbate. Generally physical adsorbents develop the ability to select the adsorbate after they have undergone specific treatments, like when they react under a gas stream or with certain agents. The kind of treatment will depend on the types of sorbents [2].

Chemical adsorption is the reaction between adsorbates and the surface molecules of adsorbents [5]. Electron transfer, atom rearrangement, and fracture or formation of a chemical bond always occurs in the process of chemical adsorption. Only one layer of adsorbate reacts with the surface molecules of chemical adsorbent. The adsorbate and adsorbent molecules after adsorption won't keep their original state, for example, complexation occurs between chlorides and ammonia. Moreover, there are the phenomena of salt swelling and agglomeration, which are critical to heat and mass transfer performance.

Composite adsorbents [6–9] were studied over 20 years ago, and the aim is now to improve the heat and mass transfer performance of the original chemical adsorbents. Such adsorbents are usually obtained through the combination of chemical adsorbents and a porous medium, such as expanded natural graphite, activated carbon, or carbon fiber, and so on.

2.1 Adsorbents

2.1.1 Physical Adsorbents

The common physical adsorbents for adsorption refrigeration are activated carbon, activated carbon fiber, silica gel, and zeolite. In addition, some novel materials have emerged to be used in adsorption heating and cooling applications in recent years.

2.1.1.1 Activated Carbon and Activated Carbon Fiber

The activated carbon is produced by materials such as wood, peat, coal, fossil oil, chark, bone, coconut shell, nut stone, and so on. The microcrystal for the activated carbon is a six element

Adsorption Refrigeration Technology: Theory and Application, First Edition. Ruzhu Wang, Liwei Wang and Jingyi Wu.

Companion Website: www.wiley.com/go/wang/refrigeration

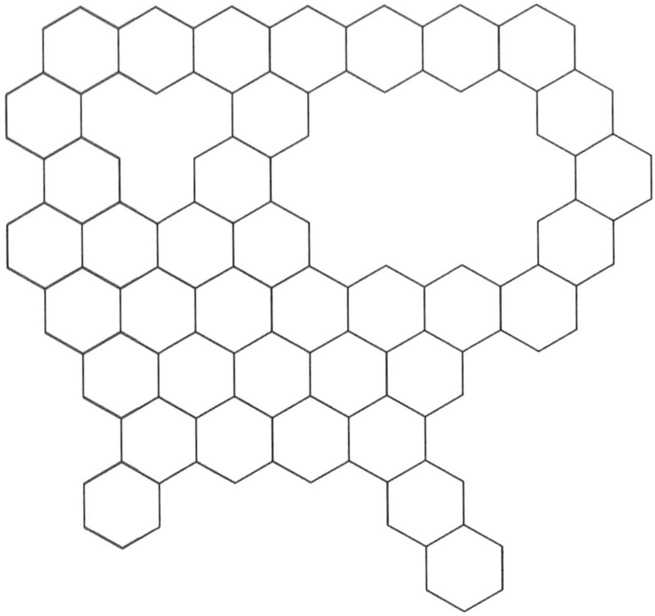

Figure 2.1 Structure of activated carbon

carboatomic ring [10], and generally the size of the microcrystal is 2.3×0.9 nm, as shown in Figure 2.1 [2]. The surface area of activated carbon is commonly between 500 and 1500 m^2/g. The activated carbon will be different if the original carbonaceous material or the production technique is different, which will influence the adsorption performance. For example, the activated carbon produced by the petroleum residue or charred coal has small micro pore, large surface area, and high density; whereas the activated carbon produced from the brown coal has large micro pore, small surface area, and low density. The adsorption performance of activated carbon is influenced by the functional groups that are connected to the carboatomic ring. For example, the arene group increases the adsorption performance, whereas the sulfonic group will decrease it. The acidic functional group will increase adsorption selectivity.

The net structure of activated carbon pores is composed of irregular channels, which have larger pore area at the surface of the grain, and narrow pore area within the grain. The difference between activated carbon and other types of adsorbent is the surface feature. The whole surface of activated carbon is covered by an oxide matrix and by some inorganic materials, and therefore, it is non-polar or has a weak polarity, which also leads to a lower desorption heat than other adsorbents because of the more simple decomposition process.

Activated carbon fiber is generally used for producing fabric, such as cloth, tissue, and so on. Compared with granular activated carbon, carbon fiber has a better mass transfer performance. The specific surface area of activated carbon fibers is larger than that of activated carbon, the pores of activated carbon fiber are more uniform than those of activated carbon, and the heat transfer performance of activated carbon fibers is also larger than that of activated carbon. If comparing the adsorption refrigeration performance of carbon fiber with that of activated carbon, generally the *COP* can be improved by 10–20%, and the cycle adsorption quantity can be improved by two to three times. The disadvantages of activated carbon fibers are the anisotropic thermal conductivity, and the higher contact thermal resistance between the fiber and the adsorber wall, when compared with granular activated carbon. A lower density is also a drawback of carbon fiber because it will decrease the filling quantity of the adsorbent inside the adsorber, which will lead to a large volume in the adsorption system.

2.1.1.2 Silica Gel

Silica gel is a type of amorphous synthetic silica. It is a rigid, continuous net of colloidal silica (Figure 2.2), and is composed of very small grains of hydrated SiO_4. The hydroxyl in the structure is an important component for adsorption because it is polar and it can form hydrogen bonds with polar oxides, such as water and alcohol. The adsorption ability of silica gel increases when the polarity increases. One hydroxyl can adsorb one molecule of water [10].

Each kind of silica gel has only one type of pore, which usually is confined in narrow channels. The pore diameters of common silica gel are 2 nm, 3 nm (A type), and 0.7 nm (B type), and the specific surface area is about 100–1000 m^2/g. Silica gel is widely used for desiccation because of its high adsorption ability. Type A silica gel could be used for all desiccation conditions, but type B silica gel can only be used when the relative humidity (RH) is higher than 50%.

2.1.1.3 Zeolite

Zeolite is a type of aluminosilicate crystal composed of alkali or alkali soil. The chemical formula of zeolite is

$$M_{y/n}[(AlO_2)_y(SiO_2)_m]zH_2O \tag{2.1}$$

where y and m are all integer and m/y is equal to or larger than 1. n is the chemical valence of positive ion of M and z is the number of water molecules inside a crystal cell unit.

A crystal cell unit of zeolite is shown in Figure 2.3. The positive ion must have its electric charge balanced with the electric charge of aluminum atom. The net electric charge of each aluminum atom is −1. Water can be removed by heating. The porosity of the aluminosilicate

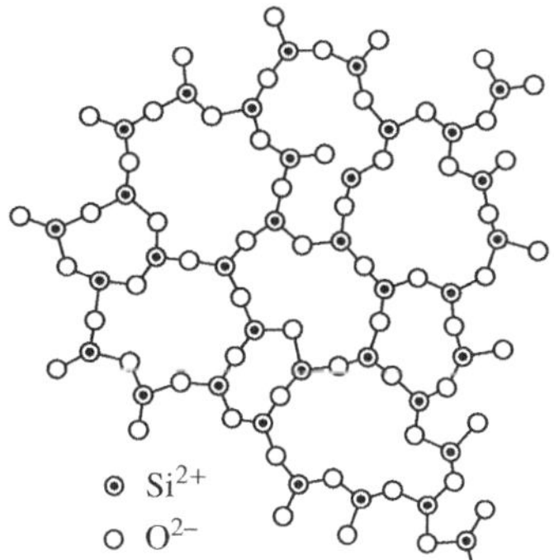

Figure 2.2 Array of SiO_4 in silica gel

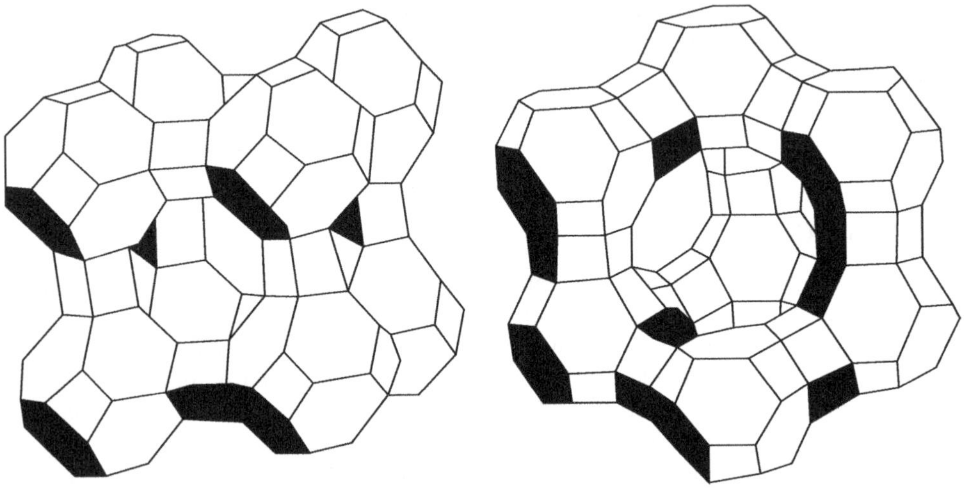

Figure 2.3 Crystal cell unit of zeolite. (a) Crystal cell unit of type A zeolite and (b) crystal cell unit of type X, Y zeolite or faujasite

framework is between 0.2 and 0.5. The aluminosilicate framework has a cage format, and it is usually connected by six casement sections, which can adsorb a large amount of extra molecules [10].

There are about 40 types of natural zeolites, and the main types for adsorption refrigeration are chabazite, sodium chabazite, cowlesite, and faujasite. About 150 types of zeolites can be artificially synthesized, and they are named using one letter or a group of letters, such as type A, type X, type Y, type, ZSM Zeolites, and so on [2].

Artificially synthesized zeolites are more expensive than natural zeolites, but they have higher bulk specific weight and better heat transfer performance. The adsorption ability of zeolites is dependent on the proportion between Si and Al, and the adsorption ability is higher when the proportion is smaller.

The pore size of zeolites determines the selectivity of the adsorption process, and the cage structure of the micropore means that the adsorption process could proceed in a small range, thus the zeolite is also known as a zeolite molecular sieve. Artificial synthesized zeolite molecular sieves have micropores with uniform size, and different sizes can be obtained by different manufacturing methods. 4A, 5A, 10X, and 13X zeolite molecular sieves are the main types used for adsorption refrigeration. The adsorption and desorption heat of zeolite pairs is high, and the desorption temperature of these pairs is also high, which is about 250–300 °C. Most zeolite molecular sieves can be destroyed at temperatures higher than 600–700 °C, however mercerized zeolites can withstand a temperature of 800 °C. The zeolites are usually employed in adsorption air conditioner systems that have a heat source of between 200 and 300 °C.

2.1.1.4 Novel Porous Materials

The aforementioned physical adsorbents have been studied and commercialized for a long time. With the rapid development of material science since the 1980s, several new classes of porous adsorbents have been discovered and proposed for adsorption cooling applications, which mainly include aluminophosphates (AlPOs), silico-aluminophosphates (SAPOs), and metal organic frameworks (MOFs).

The syntheses of AlPO and SAPOs molecular sieves were first reported in 1982 [11, 12], representing the first family of framework oxides to be synthesized without silica. These materials are zeolite-like and often named "zeotype materials," as they exhibit similar frameworks and pore systems. These molecular sieves have a very narrow range of chemical composition (i.e., rather invariant ratio of P/Al compared with the wide range of Si/Al ratio in zeolites), but exhibit a rich diversity of framework structures. The chemical composition of AlPO [13] is:

$$xRA \cdot Al_2O_3 \cdot 1.0 \pm 0.2P_2O_5 \cdot yH_2O \tag{2.2}$$

where *RA* is an amine or quaternary ammonium ion. The average of the ionic radii of Al^{3+} (0.39 Å) and P^{5+} (0.17 Å) is 0.28 Å, which is similar to the ionic radius of Si^{4+} (0.26 Å). This similarity apparently is responsible for the narrow range of chemical composition (i.e., $P/Al \approx 1$).

The AlPOs have a moderate intracrystalline pore volume, from 0.05 to 0.35 cm^3/g. Many of the AlPOs exhibit excellent thermal stability as they undergo calcination at 400–600 °C during synthesis [14]. Due to more complex synthesis, they are more expensive than aluminosilicate zeolites or silica gels.

Substituting different metals into the frameworks gives the possibility of efficient adjustment of the structure and adsorption properties of aluminosilicates. This tuning is formed by silicon, aluminum, phosphorous, and oxygen atoms in tetrahedral coordination, with uniform pore channels in molecular dimension. The most famous family of substituted AlPOs is presented by SAPOs materials (Si inserted instead of P).

Generally speaking, the affinity to water for the AlPOs and SAPOs is less than that for the zeolites but more than that for silica gels. They are neutral hydrophilic adsorbents because of containing no extra-framework cations in their framework. This hydrophilic property is considered a consequence of the difference in electronegativity between aluminum and phosphorus. It is said that a subtle balance between hydrophilic and hydrophobic surface properties leads to the S-shaped water sorption isotherms. Possible explanations for such isotherms mainly include a transition from the crystalline phase during hydration and a capillary condensation in 12-membered ring channels. The S-shaped isotherms and relatively low desorption temperature make these materials very promising for adsorption heat transformation (AHT) applications.

Beside the classes of AlPOs and SAPOs, another novel class of microporous materials, namely the MOFs or porous coordination polymers (PCPs) has emerged. In contrast to zeolites, MOFs are not purely inorganic, but inorganic-organic hybrid materials based on metal ions or metal ion clusters as nodes, which are linked by organic, at least bidentic ligands. One of the first three-dimensional porous MOFs, namely 3D-$\{[Cu_3(btc)_2(H_2O)_3]\cdot \sim 10H_2O\}$s (btc − benzene-1, 3, 5-tricarboxylate), also called HKUST-1 or just Cu-BTC, was evaluated for use in heat transformation applications [15]. HKUST-1 consists of a basic building unit containing two central Cu2+ ions that are coordinated by four trimesate molecules through their carboxylate groups to form the paddlewheel-like structure of copper acetate $Cu_2(CH_3COO)_4(H_2O)_2$ (see Figure 2.4). MOFs possess unique features such as huge surface area, large pore volume, and an unprecedented geometric, chemical, and physicochemical variability, which are due to their tunable composition.

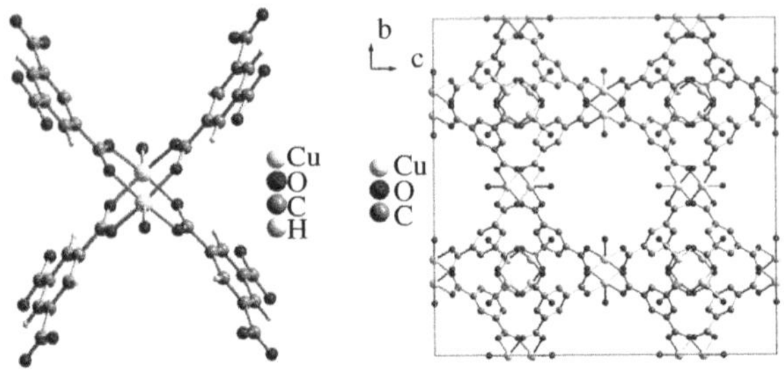

Figure 2.4 $[Cu_2(btc)_4]$ building unit and packing diagram with the cubic unit cell of one of the first three-dimensinal MOFs [15]

2.1.2 Chemical Adsorbents

Chemical adsorbents mainly include metal chlorides, metal hydrides and metal oxides [16], and salt hydrates.

2.1.2.1 Metal Chlorides

The metal chlorides for adsorption refrigeration are mainly calcium chloride, strontium chloride, magnesium chloride, barium chloride, and so on. The adsorption reaction between metal chlorides and refrigerants is a complexation reaction, and the complex compound is also called coordinated compound.

The coordinated compound is different if the element is located in a different position of the periodic system of chemical elements. According to the theory of coordinate bond, the center atom provides a free hybrid orbit for a lone electron pair to form a coordinate bond between the center atom and a ligand. For second periodical elements, such as Li(I), Be(II), B(III), the valence electron layer has four free orbits, which are 2s, $2p_x$, $2p_y$, and $2p_z$, and they could form three types of hybrid orbits, which are line type, plane triangle type or regular tetrahedron. The adsorbents for adsorption refrigeration, such as K(I), Rb(I), Cs(I), Ca(II), Sr(II), are mainly the elements in the fourth, fifth, and sixth periods, and the reaction between adsorbents and adsorbates are more complex because they are transition metals. For these metal chlorides, which have a regular dodecahedron structure, the sp^3d^4 hybrid orbit can occur [17].

Ammonia is the usual adsorbate for metal chlorides. During the adsorption process, salt swelling and agglomeration will happen, which will influence the heat and mass transfer performance.

2.1.2.2 Metal Hydrides

Hydrogen can react with almost all elements, and can form four types of hydrides. The first type is composed of salt type hydrides, such as LiH and CaH_2, and which can be formed by the reaction between hydrogen and the elements of IA and IIA subgroup. Metal hydrides are the second type of hydrides due to their low electronegativity and high chemical activity. The whole reaction process is: a hydrogen atom enters the crystal lattice of a parent metal when it reacts with transition metals, and forms metal hydrides. The other types of hydrides are the covalent high-polymerized hydrides, and the non-metal molecular hydrides.

Salt hydrides and metal hydrides can be utilized for adsorption refrigeration. The density of salt hydrides are larger than the density of pure metals, but the density of metal hydrides is smaller than the density of pure metals because the volume and the mass of the former type don't increase proportionally in the adsorption process, whereas the latter one expands on a large scale [18].

2.1.2.3 Metal Oxides

The metal oxides are usually employed as catalysts for oxidation and deoxidation reactions. When the metal oxides are used as adsorbent in adsorption heat pumps, oxygen is the refrigerant. On the surface of metal oxides, the elements that influence the adsorption performance are the coordination number of the metal ion, the unsaturated degree of coordination, the direction of the chemical bond on the surface of the chemical material, the symmetrical characteristic of the transition metal ligand field, and the arrangement of the active centers, and so on [19].

The swelling and agglomeration also occur during the adsorption process for metal oxides.

2.1.2.4 Salt Hydrates

Salt hydrates are the most familiar inorganic compounds in inorganic chemistry. They began to attract the attention of some researchers over a decade ago as promising candidates for the use in long-term thermal storage. Salt hydrates refer to the kind of inorganic salts which contain water molecules combined in a definite ratio as an integral part of the crystal. These water molecules are either bound to a metal center or crystallized with the metal complex. Such hydrates are also said to contain water of crystallization or water of hydration. A colorful example of salt hydrates is cobalt chloride, which turns from blue to red upon hydration, and can therefore be used as a water indicator.

Salt hydrates are often formed in a crystallization process for saturated salt solutions. Different hydrates will be produced at different temperatures. For example, when in equilibrium with a saturated $CaCl_2$ solution, the crystallization product of the $CaCl_2$-H_2O system is formed by hydrates with 6, 4, 2, 1, and 1/3 mol of water per mol salt as the salt concentration in the solution increases, and by anhydrous $CaCl_2$ at the highest concentrations. In addition, there exist three different crystalline modifications of tetrahydrate denoted as α, β, and γ. Hydration reactions between different hydrates are considered as a promising solution for sorption thermal storage systems.

2.1.3 Composite Adsorbents

Composite adsorbents are developed and studied with mainly two goals:

1. Improve the heat and mass transfer performance of chemical adsorbents, especially for those which have swelling and agglomeration phenomena in the adsorption process, which will seriously influence the heat and mass transfer performance. The additives inside the composite adsorbents, such as expanded graphite and activated carbon, generally have a

porous structure and a high thermal conductivity, which can improve the heat and mass transfer performance successfully.

2. Improve the adsorption quantity of physical adsorbents. If compared with the chemical adsorbents the adsorption quantity of physical adsorbents is much lower, but it has the advantages of abundant micropores and better mass transfer performance. Based on the features of chemical and physical adsorbents mentioned above, the composite adsorbent is developed by mixing the chemical and physical adsorbents together, and it will improve the cycle adsorption quantity of physical adsorbents effectively.

The composite adsorbents made by porous media and chemical sorbents are commonly a combination of metal chlorides and porous materials or physical adsorbents, such as expanded graphite, activated carbon, activated carbon fiber, silica gel, and zeolite.

The methods for producing composite adsorbents are mainly as follows:

1. **Simple mixture.** In such a process, the chemical adsorbent and the additive are mixed in a defined mass or volume ratio [4].
2. **Impregnation.** This method is mainly used for the additives of activated carbon fiber, graphite fiber, or expanded graphite. The procedures are: firstly, the chemical adsorbent is dissolved in the water or other solvent, secondly, the additive is put in the solution, and lastly, the adsorbent is dried to remove the solvent. One advantage of such a type of adsorbent is the large porosity, which benefits the mass transfer performance. If carbon fiber is used as the additive such a type of adsorbent can have high thermal conductivity in the longitudinal direction, but it will have poor thermal conductivity in the radial direction and also will have high thermal resistance between the fiber and reactor wall [20, 21].
3. **Mixture or impregnation and consolidation.** Consolidated adsorbent can be produced by compressing the composite powder prepared by mixture or impregnation, as explained above, or by firstly compressing the additive or the physical sorbent, and then, impregnating it with the salt solution, with posterior drying to remove the solvent. The advantage of such a type of adsorbent is the high thermal conductivity in a perpendicular direction to the compression, and the disadvantage is related to the complex developing process of the adsorbent. The mass transfer performance is seriously influenced by the ratio between the components and bulk density; therefore these parameters must be carefully chosen [4].

2.2 Refrigerants

2.2.1 Most Common Refrigerants

Adsorption technology can be used not only for air conditioning and refrigeration but also to upgrade heat with thermal transformers, and the types of refrigerant should be selected according to the application.

The requirements for a suitable refrigerant are generally as follows: (i) high latent heat of vaporization per volume unit or mass unit, (ii) thermal stability, (iii) environmental harmless, (iv) non-flammable, (v) innoxious, and (vi) saturation pressure between 1 and 5 atm under the condition of the working temperatures (a perfect value would be close to 1 atm). Unfortunately, there are no refrigerants that have all the characteristics above, and the common refrigerants for

Table 2.1 Some physical properties of common refrigerants for adsorption systems

Refrigerants	Chemical formula	Normal boiling point (°C)	Molecular weight	Latent heat of vaporization L (kJ/kg)	Density ρ (kg/m^3)	$\rho \times L$ (MJ/m^3)
Ammonia	NH_3	−34	17	1368	681	932
Water	H_2O	100	18	2258	958	2163
Methanol	CH_3OH	65	32	1102	791	872
Ethanol	C_2H_5OH	79	46	842	789	665

adsorption refrigeration system are ammonia, water, and methanol. Some physical properties of refrigerants for adsorption systems are shown in Table 2.1.

Refrigerants with a boiling point below −10 °C at 1 atm are positive pressure refrigerants, whereas the others are vacuum refrigerants. Ammonia is an example of refrigerant with positive pressure, and it can be used for the adsorbents of chlorides, activated carbon, and activated carbon fiber. The saturation pressure of ethanol and methanol are similar, but the latent heat of the former one is about 30% lower than that of the latter one. Methanol is normally used in association with activated carbon or activated carbon fiber. Water could be considered a perfect refrigerant, except for its extreme low saturation pressure and for the impossibility of freezing conditions below 0 °C. Water is normally employed in a pair with silica gel or zeolite.

2.2.2 *Other Refrigerants*

Hydrogen and oxygen are examples of other refrigerants for adsorption refrigeration and heat pump systems.

Hydrides are the adsorbents for hydrogen, which is produced from the decomposition of water, and it is inflammable, explosive, and requires extreme precaution when handled [19].

The types of oxygen that can be adsorbed by oxides are O_2, O_2^-, O^-, and O^{2-}. The reaction between oxides and oxygen has large enthalpy, thus, it is usually employed in chemical heat pumps [19]. However, this pair is also suitable for cryogenic systems with temperatures below 120 K.

Other refrigerants, such as R134a, R22, R407c, and ethanol can be utilized for the adsorbents of activated carbon or activated carbon fiber. However, compared with methanol, their cooling power per unit mass is smaller due to their small adsorption quantity or due to their low latent heat of vaporization. Besides the limitations described above, HCFCs hydrochlorofluro-carbonand HFCs also have the drawback of relatively high GWP (global warming potential) values [22, 23].

2.3 Adsorption Working Pairs

2.3.1 *Physical Adsorption*

The adsorption forces involved in physical adsorbents are intermolecular forces (van der Waals force), which mainly include dispersion force, Debye force, and orientation force. They do not involve a significant change in the electronic orbital patterns of the species involved.

The most studied physical adsorption working pairs are activated carbon (or activated carbon fiber)–methanol, activated carbon (or activated carbon fiber)–ammonia, silica gel–water, and zeolite–water.

2.3.1.1 Activated Carbon or Activated Carbon Fiber and Methanol or Ammonia

The adsorption processes of activated carbon–methanol and activated carbon–ammonia are similar, and the process can be looked as a filling and condensation process of adsorbate inside adsorbent pores. The adsorption mainly occurs in micro pores, which have specific volume of about 0.15–0.50 cm^3/g. The surface area of micropores is about 95% of the whole surface area of activated carbon. The function of middle pores and large pores is mainly to transport the adsorbate molecules to micropores.

Activated carbon–methanol is one of the most common working pairs due to the large adsorption quantity and lower adsorption heat, which is about 1800–2000 kJ/kg. As the main heat consumption in the desorption phase is due to the adsorption heat, low values of adsorption heat are beneficial to the coefficient of performance (*COP*). The activated carbon–methanol is also a suitable working pair for using solar energy as a heat source due to the low desorption temperature, which is about 100 °C. Temperatures higher than 120 °C should be avoided because the decomposition of methanol into other compounds occurs above this temperature. Activated carbon–methanol has the disadvantage that it operates under sub-atmospheric pressure. The necessity of vacuum inside a machine increases the manufacturing complexity, and reduces the reliability of the system, as even a small air infiltration can seriously compromise the machine's performance.

Activated carbon–ammonia is another common working pair. Compared with activated carbon–methanol, both pairs have similar adsorption heat, but the former pair has the advantage of higher working pressure, which is about 16 bar at a condensation temperature of 40 °C. Due to the higher operation pressure of the activated carbon–ammonia pair, the mass transfer performance is better, and the cycle time can be reduced. Another advantage of the activated carbon–ammonia pair when compared with the activated carbon–methanol pair is the possibility of using heat sources at 200 °C or above. The disadvantages are the toxicity and pungent odor of ammonia, the incompatibility between ammonia and copper, and the cycle adsorption quantity that is smaller when compared to the value obtained with activated carbon–methanol under the same working conditions.

2.3.1.2 Silica Gel and Water

In the adsorption process between water and silica gel, the water molecule is connected with the silica alcohol group = Si-OH … OH_2 while the surface coverage degree is low. As the surface coverage degree increases, the hydrogen bond becomes the main connecting force. The adsorption heat for this pair is about 2500 kJ/kg and the desorption temperature can be very low, but above 50 °C [10].

There is about 4–6% mass water connected with a single hydroxyl group on the surface of silica atom, which cannot be removed, otherwise the silica gel would lose the adsorption capability. Thus the desorption temperature cannot be higher than 120 °C, and it is generally lower than 90 °C.

Researchers in Japan developed a three-stage adsorption system with silica gel/water working pair, and such a system can be powered by heat sources with a temperature of 40–50 °C. The lowest driven temperature for the silica gel–water working pair taken from experiments is about 55 °C [24, 25]. Such a low desorption temperature is very suitable for solar energy utilization [26].

One disadvantage of silica gel–water working pair is the low adsorption quantity, which is about 0.2 kg/kg. Another disadvantage is the impossibility of producing evaporation temperatures below 0 °C.

2.3.1.3 Zeolite-Water

The structure and adsorption mechanisms of different zeolites are different. For example, type A and type X and Y zeolites have the structure of truncated octahedron, and such unit crystals are known as the cage structure of sodalite zeolites. There are 24 water molecules which could be adsorbed in the center cages or the pores of unit crystal and in the cages or pores of eight sodalite zeolite crystals. The skeleton structure of type X and Y zeolites is similar to that of natural zeolites. The volume of pores for type X and Y zeolites are larger than the volume of other types of zeolites, and their void ratio can be as high as 50% when there is no water adsorbed. One crystal unit can have 235 molecules of water after adsorption, and most of the molecules would accumulate in the center pore [10].

The zeolite–water pair can be utilized in the dehumidification cooling system and adsorption refrigeration system. The adsorption heat for the zeolite–water pair is higher than that of the silica gel–water pair, and it is about 3300–4200 kJ/kg. The zeolite–water is stable at high temperatures; hence, this pair can be used to recover heat above 200 °C. The adsorption isotherm is quite insensitive to the condensation pressure, thus, the system can operate with similar performance in a large range of condensation temperature. Due to the large adsorption heat and high desorption temperature, the performance of the zeolite–water pair is worse than that of the activated carbon–methanol pair at middle and low temperature heat sources (lower than 150 °C), but the former pair can have higher *COP* and *SCP* (Specific Cooling Power) if the temperature of the heat source is higher than 200 °C.

The disadvantages of this pair are similar to the disadvantages of silica gel–water: the impossibility of producing evaporation temperatures below 0 °C and bad mass transfer performance due to the low working pressure. Due to the high value of adsorption heat and high desorption temperature, for a heat source with the same power, the cycle time for the pair of zeolite–water is longer than that of other pairs.

2.3.2 Chemical Adsorption Working Pairs

For the adsorption between chemical adsorbents and refrigerants, the force of chemical adsorption working pairs mainly includes the function of complexation, coordination, hydrogenation, and oxidization.

Chemical adsorption working pairs mainly include metal chlorides–ammonia, metal hydrides–hydrogen, and metal oxides–oxygen.

2.3.2.1 Metal Chlorides and Ammonia

The force between metal chlorides and ammonia is complexation force [16, 17, 27, 28]. For example, the reaction between calcium chloride and ammonia can be written as:

$$CaCl_2 \cdot n_1NH_3 + n_2\Delta H_r \leftrightarrow CaCl_2 \cdot (n_1 - n_2)NH_3 + n_2NH_3 \tag{2.3}$$

where ΔH_r is the reaction enthalpy, (J/mol), the numbers of n_1 and n_2 could be 2, 4, and 8.

The first advantage of metal chlorides–ammonia is the large adsorption quantity, which is higher than 1 kg/kg for most chlorides. The second advantage is the working pressure that is higher than the atmosphere pressure, which is important for ensuring the reasonable mass transfer performance. The metal chlorides working pairs also have the feature of a large variety that is suitable for a large range of driven heat temperatures. The adsorption heat is related to the types of the chlorides, and different chlorides will have different values of adsorption heat.

The disadvantage of metal chlorides–ammonia working pair is the swelling and agglomeration phenomena during adsorption, which will influence the heat and mass transfer.

2.3.2.2 Metal Hydrides–Hydrogen

The refrigeration process of metal hydrides–hydrogen is generally different from the common refrigeration process, which depends on the desorption heat of the working pairs, that is, it is resorption process other than the adsorption process. Advanced porous metal hydrides (PMHs), or the misch metal (Mm) matrix alloys, including the alloys with Ni, Fe, La, and Al, have very high reaction heat and adsorption quantity.

The reaction process between metal hydrides and hydrogen is:

$$M(s) + n/2H_2(g) \rightarrow MH_n(s) \tag{2.4}$$

Generally the hydrogen doesn't react with the hydrides under the environmental conditions, and the hydrogen will be adsorbed by the metal hydride if the temperature inside the reactor rises. The desorption process will need the heat input, which provides the refrigeration power output.

The advantage of the metal hydride–hydrogen working pairs is the wide range of driven temperatures that can be chosen for the reaction process, and it is –100–500 °C. It also has fast reaction rate, large reaction heat, and big density, all of these features are helpful for decreasing the volume of the reaction system. The disadvantages are that the hydrogen is inflammable and explosive, as well as metal hydrides being expensive, thus such working pairs are rarely utilized for common refrigeration occasions. But they are the optimal choice for very high temperature heat pumps or cryogenic conditions that cannot be fulfilled by common adsorption working pairs [29, 30], or just utilized for the hydrogen storage process [18].

2.3.2.3 Metal Oxides and Oxygen

There are two types of oxygen, the molecular oxygen and the atomic oxygen, that can be adsorbed by a metal. The oxygen atom enters the metal lattice to form metal oxides. The type of oxygen, molecular or atomic, adsorbed by metal depends on the external conditions and on the types of metal. Generally, after oxygen molecules are adsorbed, in the heating and desorption process, some oxygen molecules are desorbed, some changed into the stable oxygen atom inside metal, and this transition process needs activated energy [19].

Such types of working pairs are not common for the refrigeration process. Some can be utilized for the heat pump or simply for energy storage process.

2.3.2.4 Salt Hydrates and Water

The interests in using hydration reactions for heat storage application mainly focus on the hygroscopic salts such as magnesium chloride ($MgCl_2$) [31–33], sodium sulfide (Na_2S) [34, 35], strontium bromide ($SrBr_2$) [36], and magnesium sulfate ($MgSO_4$) [37, 38]. Usually, the products of hydration reactions of these salts are assumed to be higher hydrates with more crystal water molecules. However, in some cases, the RH pressure is so high that the product of the hydration is a saturated salt solution, rather than a salt hydrate. This process is called deliquescence, which is an important solid-water interaction phenomenon. Deliquescence is defined as a first order phase transformation of the solid to a saturated solution when the RH reaches a certain threshold value, namely, the deliquescence relative humidity (DRH). DRH is also the equilibrium RH above the saturated solution. The two cases can be represented by the following equations [39].

$$\text{Salt(s)} + H_2O(g) \leftrightarrow \text{Salt hydrate(s)} \quad \text{for RH} < \text{DRH} \tag{2.5}$$

$$\text{Salt(s)} + H_2O(g) \leftrightarrow \text{Salt solution(l)} \quad \text{for RH} > \text{DRH} \tag{2.6}$$

The value of DRH depends on the properties of the salt and the temperatures. Some problems will be encountered if the deliquescence phenomenon happens. The forming of liquid film on the surface of salt crystal will not only prevent the hydration reaction, but also cause corrosion problems due to the dripping of solution to other metal components.

The DRH of LiCl and LiBr are only 11.3 and 6.2% at 30 °C, implying that it's very easy for their solids to form solutions in most situations. Thus LiCl and LiBr should not be considered for solid/gas hydration reactions. Though the DRH of $MgCl_2$ (32.4% at 30 °C) is not as low as those of LiBr and LiCl, it is also regarded as a deliquescent salt, meaning that treating $MgCl_2$ in sorption storage systems needs special care. Compared with $MgCl_2$, $MgSO_4$ is a hydrothermally stable salt with a high DRH (90% at 30 °C).

2.3.3 The Heat and Mass Transfer Intensification Technology and Composite Adsorbents

The heat and mass transfer intensification technology mainly focuses on decreasing the thermal resistance of adsorbers [40]. The total heat transfer coefficient α of adsorber is

$$\frac{1}{\alpha A_f} = \frac{1}{\alpha_f A_f} + \frac{1}{\alpha_w A_{eff}} + \frac{e_{eff}}{\lambda_{eff} A_{eff}} \tag{2.7}$$

where A_f and A_{eff} are the variables of the heat transfer area of the heat exchanger at the fluid side and solid adsorbent side, respectively. α_f and α_w are the heat transfer coefficients of the heat exchangers by the fluid side and solid adsorbent side, separately. e_{eff} and λ_{eff} are the effective thickness and thermal conductivity of the adsorbent. The total heat transfer performance is influenced by the conditions as follows:

1. The thermal conductivity of granular adsorbents, which generally is low. For example, the thermal conductivity of granular zoelite is about 0.1 W/(mK) [41], the thermal conductivity of granular chlorides and activated carbon is about 0.3–0.5 W/(mK) [8, 42], and the thermal conductivity of granular metal hydrides is about 1 W/(mK) [40, 43].
2. The low heat transfer coefficient between the adsorbent and the wall of heat exchanger. This value generally is very low because it mainly depends on the thermal conductivity of the adsorbents, and the thermal resistance between the adsorbent and the wall will further decrease the heat transfer performance [40].
3. If the laminar flow happens for the fluid side the heat transfer coefficient between the fluid and the metal wall will also influence the total heat transfer coefficient.

For improving the total heat transfer coefficient of α, one method is to develop the solidified adsorbents with the heat transfer intensification matrix such as the metal powder, metal foam, and expanded graphite, which could improve the thermal conductivity λ_{eff} and the heat transfer coefficient by solid side, that is, α_w, thus the last item in the Equation 2.7 will decrease, and the total heat transfer coefficient will be improved.

For the physical adsorbents, the common heat transfer intensification technology is to develop the solidified adsorber or coated adsorber, and the method is to solidify the composite materials of adsorbent, agglomerant, and water, and then dry the composite adsorbent in the oven [44]. The coated adsorber generally means bonding the adsorbent on the outside of the heat transfer pipe or heat transfer fins by agglomerant [45].

Composite adsorbents are proposed for the intensification process of chemical adsorbents, such as metal chlorides. The adsorption principle involves the chemical adsorption and the mass transfer process inside the micropores of porous matrix. For example, if the carbon fiber is utilized as the porous matrix of calcium chloride, the calcium chloride will serve as the adsorbent for adsorption process, and the carbon fiber will serve as the porous material for mass transfer process.

Composite adsorption working pairs have the advantages of both chemical adsorbents and porous matrixes, and can improve the heat and mass transfer performance effectively, as well as improving the stability of chemical adsorption performance. The problem of composite adsorbents is the complex developing process, which required the reasonable ratio between the chemical adsorbent and the porous matrix, reasonable filling quantity of the adsorbents inside the adsorbers, as well as the reasonable compact density if the composite adsorbents are solidified in the developing process.

2.4 Equilibrium Adsorption Models

The equilibrium adsorption quantity is the amount of refrigerant adsorbed by the sorbent when the reaction time is an infinite value, and it is an important parameter for adsorption working pairs. The equilibrium adsorption models for physical adsorption and chemical adsorption are different. For the equilibrium state in physical adsorption there are two independent properties, that is, generally pressure and temperature need to be considered. While for the equilibrium state in chemical adsorption the independent property is only one, that is, temperature or pressure will be enough.

2.4.1 *Equilibrium Models for Physical Adsorption*

The equilibrium model for physical adsorption is a function of adsorption temperature and pressure, and it is written in a generic form as:

$$x_{\infty} = f(p, T) \tag{2.8}$$

Models with different fundamental processes are as follows:

$$x_{\infty} = f(p)_T \quad \text{Isothermal} \tag{2.9}$$

$$x_{\infty} = f(T)_p \quad \text{Isobaric} \tag{2.10}$$

$$p = f(T)_x \quad \text{Isosteric} \tag{2.11}$$

Adsorption isotherm models are utilized only for fitting the experimental data with small adsorbent mass, because the isothermal conditions cannot be achieved if the adsorbent mass is large, because the heat production of the adsorption reaction will be large and it will be difficult to enable the isothermal conditions. Adsorption isobar models are more suitable for designing adsorption refrigeration systems because the quantity by which the refrigerant mass desorbed and adsorbed during a cycle can be easily calculated. Adsorption isosteric models are convenient to calculate the adsorption heat and usually utilized for the selection of working pairs.

Inside an adsorption refrigeration system the adsorption and desorption processes are all isobaric process, as shown in Figure 2.5. The solid adsorbent desorbs the refrigerant when it was heated by the heat source, and then the refrigerant condenses into the liquid and flows to the condenser. Inversely the solid adsorbent will adsorb the refrigerant vapor from the evaporator when it is cooled by the heat sink under the pressure difference between the adsorber and the evaporator. In the adsorption process the latent heat of evaporation will provide the refrigeration effect.

For example, for the activated carbon-methanol system, under conditions of evaporating pressure and adsorption temperature the adsorption process will proceed, for which the adsorption concentration is x_{conc} as shown in Figure 2.6. Under conditions of condensing pressure

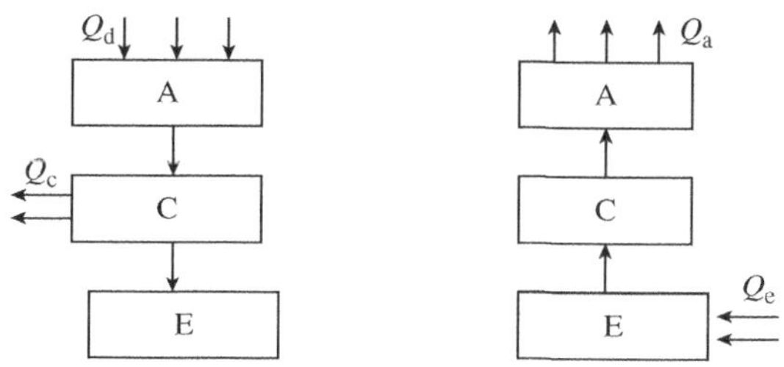

Figure 2.5 Adsorption and desorption processes

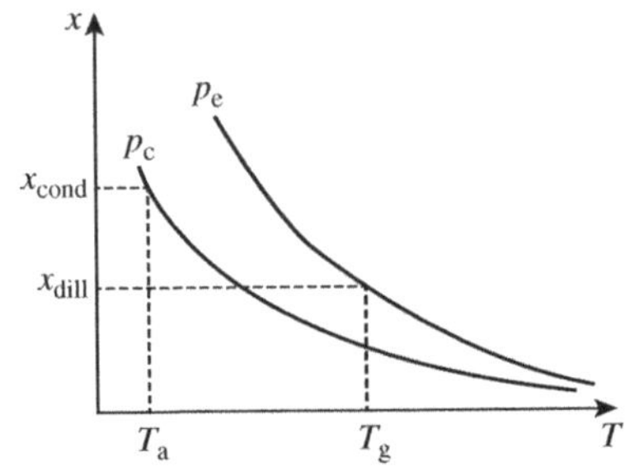

Figure 2.6 Adsorption and desorption isobars

and desorption temperature the adsorbent desorbs the refrigerant, and the adsorption quantity is x_{dil}. The cycle adsorption quantity will be $\Delta x = x_{conc} - x_{dil}$, which is an important parameter for evaluating the adsorption-desorption performance.

Because two processes, that is, desorption–condensation and adsorption–evaporation are involved in one cycle, and only the adsorption process can output the refrigeration effect, thus the simple adsorption cycle is the intermittent cycle. The continuous refrigeration process can only be fulfilled by more than two adsorbers in one adsorption refrigeration system.

For the adsorption refrigeration cycle based on the equilibrium states the *COP* of the refrigeration cycle and the refrigerating power are dependent on the adsorption properties of adsorption working pairs, as well as the adsorption temperature, desorption temperature, evaporating temperature, and condensing temperature.

2.4.2 *Equilibrium Models for Chemical Adsorption*

The phase equilibrium equations for chemical adsorption are different from that of physical adsorption. As mentioned above, the physical adsorption has two independent properties, whereas the chemical adsorption only has one independent property.

The chlorides are the common chemical adsorbents, for which the complexation process is adsorption process, and the decomposition process is the desorption process. The chemical reaction equation is:

$$M_yCl_z \cdot n_1NH_3 + n_2\Delta H_r \leftrightarrow M_yCl_z \cdot (n_1 - n_2)NH_3 + n_2NH_3 \tag{2.12}$$

where ΔH_r is chemical reaction enthalpy (J/mol); n_1 and n_2 are values of 2, 4, 8 (or 6), respectively. M_yCl_z is the metal chloride.

The equilibrium adsorption model for chemical adsorption is

$$p = f(T)_{adsorbent} \quad x_\infty = x_n \tag{2.13}$$

where n is the mole number of refrigerant adsorbed by the adsorbent.

The change of adsorption quantity in chemical adsorption is stepwise, as shown in Figure 2.7 [46]. Each reaction with defined stoichiometric coefficients is described by one equilibrium condition, and the maximum amount of refrigerant adsorbed cannot be higher than the stoichiomeric value. This implies that, once the system reaches the equilibrium at a certain condition of temperature, the increase of the pressure won't increase the amount of refrigerant

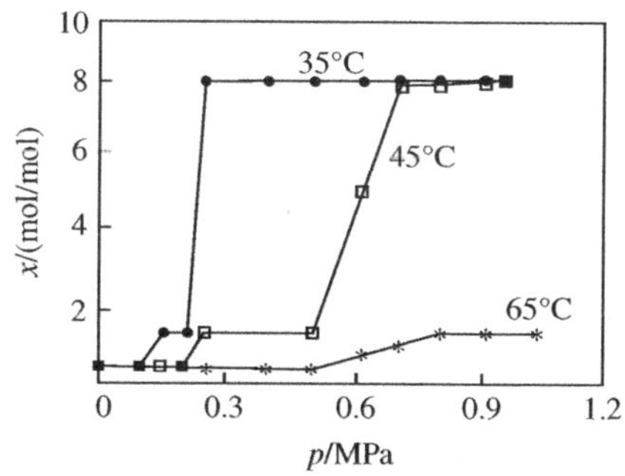

Figure 2.7 Adsorption curves of strontium chloride–ammonia [46]

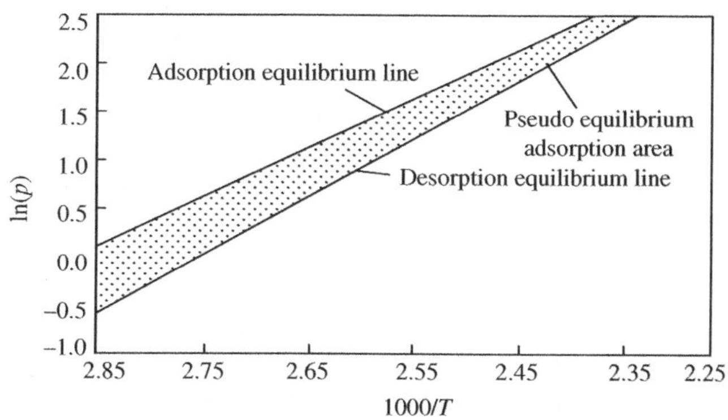

Figure 2.8 Clapeyron diagram of $MnCl_2$-NH_3 [47]

adsorbed, unless the equilibrium condition of other reaction with other stoichiometric coefficients is reached.

Another difference between the chemical adsorption and physical adsorption is the pseudo equilibrium adsorption area between the lines for equilibrium adsorption and desorption, just as shown in Figure 2.8 for manganese chloride. In the pseudo equilibrium adsorption area the adsorption and desorption rates are all 0 [47].

2.5 Methods to Measure Adsorption Performances

The adsorption quantity needs to be measured under equilibrium and non-equilibrium conditions to assess the maximum amount that can be adsorbed–desorbed under a certain condition, and to assess how long it takes to reach this value. Such studies are essential for the estimation of refrigeration performances. The measurement of the equilibrium adsorption quantity can be done by volumetric, gravimetric, and chromatography methods. The adsorption heat is usually measured by calorimetry. Volumetric and gravimetric methods [48] are traditionally used to measure adsorption rate, although calorimetry can be also employed. Calorimetry is often utilized to get thermodynamic and kinetic parameters for different adsorption working pairs.

The adsorption quantity can be assessed by gas chromatography due to the variation of heat conductivity of the gas with the density, which changes during the adsorption or desorption. However, such a method is rarely used because of its complexity.

In one of the volumetric methods [49], the variation of mass is calculated by measuring the variation of pressure in a vessel with known volume. The measuring principle of this method is shown in Figure 2.9 [49]. A thermostatic bath controls the temperature of a saturated fluid located in the vapor generator, and the vapor produced flows to a storage chamber. A coil water pipe controls the temperature of this chamber. The adsorption and desorption temperature of the adsorbent is controlled by another thermostatic bath. The adsorption quantity is calculated using the values of p_1 and p_2 during adsorption and desorption, respectively, and the value of dead volume of the adsorption equipment. In order to get precise results, the volume of the vapor chamber, the volume of the vapor inside the adsorber, and the mass of adsorbent should be small. The adsorbent mass measured in Figure 2.9 was only 3 g.

Another type of volumetric method can be used to assess the adsorbed or desorbed mass by measuring the variation of the volume of the refrigerant inside a vessel. The changes in the volume can be identified by measuring the level changes by visual observation or by the utilization of more precise equipments, like magnetostrictive sensor [50].

A test rig using a level sensor is shown in Figure 2.10 [50]. In this method, the liquid level in the condenser/evaporator is measured by the level sensor, and this data, together with the value of the useful area of the condenser/evaporator and the density of the refrigerant, are used to calculate the adsorption/desorption quantity. The precision in the measurement increases with the mass of adsorbent, because the amount desorbed/adsorbed increases, and thus, the variation of the liquid level inside condenser/evaporator also increases.

For the gravimetric method, the adsorption performance is calculated by the mass change of the adsorbent during adsorption or desorption. One of the commonly used instruments is the suspension thermogravimetric balance from Rubotherm Co (see Figure 2.10). The magnetic suspension balance allows the changes in force and mass which act on samples under controlled environments, to be measured with high accuracy. Instead of hanging directly at the

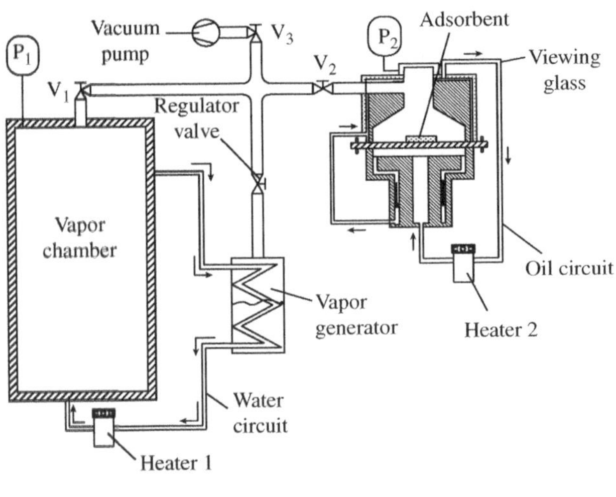

Figure 2.9 Test unit for volumetric method [49]

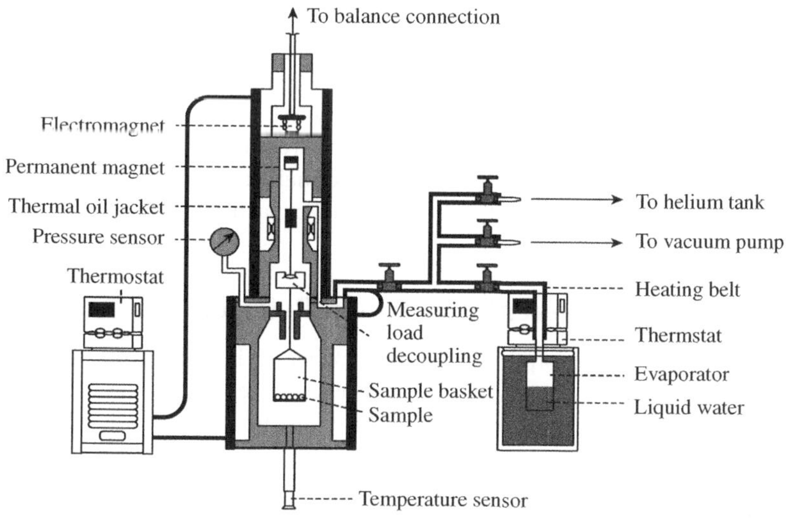

Figure 2.10 Test unit for gravimetric method: Rubotherm thermogravimetric balance

balance the sample to be investigated is linked to a so-called suspension magnet which consists of a permanent magnet, a sensor core and a device for decoupling the measuring load (sample). Using this magnetic suspension coupling the measuring force is transmitted contactlessly from the measuring chamber to the microbalance, which is located outside the chamber under ambient atmospheric conditions. Thus, this instrument is able to measure adsorption properties of corrosive fluids like ammonia. The gravimetric method has been adopted in many previous studies to investigate adsorption isotherms and isobars for various working pairs [50–53].

The volumetric method that uses the variation of the liquid level to assess the adsorption performance (Figure 2.11) is better than the volumetric method that employs variation of pressure and the gravimetric method described above because of the operation of the equipments involved is easier, the precision is higher, and the test rig is simpler.

Two types of calorimeters can be used on the test of adsorption heat: one type is isothermal and the other is adiabatic. The adsorption heat can be assessed in isothermal calorimetry by melting ice or other solid chemical material, such as phenol (melting point of 40.6 °C) or

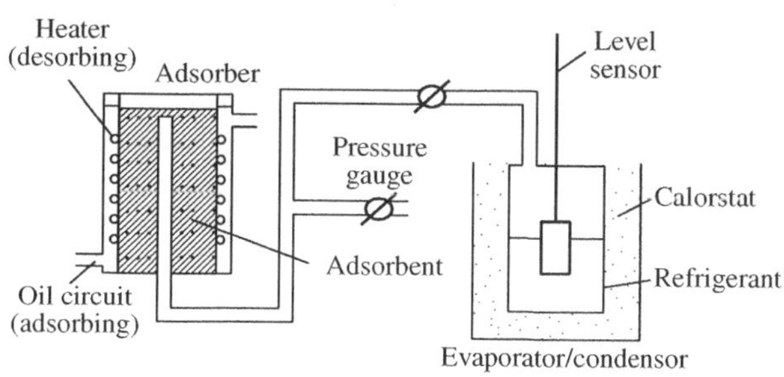

Figure 2.11 Test unit for liquid level measuring method [20]

glacial acetic acid (melting point of 16.7 °C), and measuring the mass of the liquid. To ensure high precision in this method, it is important that the amount of liquid produced is much larger than the possible amount of liquid drops that can stay connected to the solid material or to the adsorber wall. This method is not a convenient way to measure the adsorption equilibrium conditions because each condition would require a different melting material.

In adiabatic calorimetry, the adsorber needs to be thermally insulated, and the adsorption heat is calculated using the temperature lift and the thermal capacity of the heat transfer media.

2.6 Comparison of Different Adsorption Refrigeration Pairs

The main heat sources for adsorption machines are waste heat and solar energy. Physical adsorption working pairs are usually preferred when solar energy is the heat source. Silica gel–water is a suitable working pair for solar energy due to the low desorption temperature,

Table 2.2 Performance comparison of different adsorption working pairs

Evaporation temperature (°C)	Adsorption working pair	*COP*	*SCP* (W/kg)	Characteristics	Data source
8	Activated carbon-NH_3	–	1000	Convective thermal wave cycle	Calculation[a] [54]
1	Activated carbon fiber/$CaCl_2$-NH_3	0.6	330	Composite adsorbent, heat pipe heating	Experiment[b] [6]
3	Activated carbon-NH_3	0.67	557	Convective thermal wave cycle	Calculation [55]
−10	$SrCl_2$-NH_3	0.32	230	Single effect system	Experiment [56]
−25	($MnCl_2$ + $NiCl_2$)-NH_3	0.4	70 × 2[c]	Double effect system	Calculation [56]
−10	Metal hydride-hydrogen	0.43	25 × 2	Thermal wave cycle	Experiment [29]
3	Graphite/silica gel-water	–	35 × 2	Composite adsorbent	Experment [57]
10	Silica gel-water	0.4	85	Split heat pipe type evaporator	Experiment [24, 25]
5	Zeolite-water	0.9	125 × 2	Intermittent convective thermal wave cycle	Calculation [58]
−15	$CaCl_2$/activated carbon –ammonia	0.41	731	Composite adsorbent, heat pipe type heating and cooling	Experiment [59]
−25	$CaCl_2$/activated carbon –ammonia	0.36	627.7	Composite adsorbent, heat pipe type heating and cooling	Experiment [59]

[a]Calculation: data calculated from the adsorption performances of working pairs or from the simulation of an adsorption refrigeration cycle.

[b]Experiment: data obtained experimentally.

[c]The *SCP* in the original literature was calculated based on the total cycle time, with same adsorption and desorption time. To allow a fair comparison, the original value was multiplied by two.

but it can only be applied for the air conditioning due to the impossibility of producing sub-zero temperature.

Activated carbon–methanol pair can be used for freezing applications, and it can be driven by heat sources with temperatures lower than 120 °C. Silica gel-water and activated carbon-methanol are also suitable working pairs for low temperature waste heat. Suitable pairs for high temperature waste heat are zeolite–water, activated carbon–ammonia, metal chlorides–ammonia, and composite adsorbents–ammonia. Adsorption refrigeration systems with silica gel–water, activated carbon–methanol and the zeolite–water operate under vacuum conditions, and leak-proof machines are essential for maintaining the performance of the system. Systems that utilize ammonia as a refrigerant have positive pressure, and the manufacturing and maintenance are much easier if compared with vacuum systems.

There are two main parameters to evaluate the performance for adsorption refrigeration: *COP* and *SCP*. *COP* can be improved by an advanced adsorption refrigeration cycle, such as heat recovery cycle and mass recovery cycle, and *SCP* can be improved by advanced adsorber technology and by using adsorbent with high heat and mass transfer performance.

Table 2.2 shows that some of the promising experimental performances were obtained with composite adsorbents–ammonia working pairs. The *SCP* obtained with the pair activated carbon–$CaCl_2$–ammonia is as high as 731 W/kg and the *COP* is about 0.4–0.5, when the evaporating temperature was −15 °C. However, some calculations indicate that *SCP* higher than 1000 W/kg can be obtained with convective thermal wave cycle using activated carbon–ammonia working pair.

References

[1] Ponec, V., Knor, Z. and Cerny, S. (1974) *Adsorption on Solids*, Butterworth Group, Oxford.
[2] Zhang, Y.H. (1989) *Adsorption Action*, Shanghai Press of Science and Technology, Shanghai, ISBN: 7805134979, 9787805134970 (in Chinese).
[3] Srivastava, N.C. and Eames, I.W. (1998) A review of adsorbents and adsorbates in solid-vapour adsorption heat pump systems. *Applied Thermal Engineering*, **18**, 707–714.
[4] Wang, L.W. (2005) Performances, mechanisms, and application of a new type compound adsorbent for efficient heat pipe type refrigeration driven by waste heat. Chinese PhD Thesis. Shanghai Jiao Tong University, Shanghai, China.
[5] Gasser, R.P.H. (1987) *An Introduction to Chemisorption and Catalysis by Metals*, Clarendon Press, Oxford.
[6] Vasiliev, L.L., Mishkinis, D.A., and Vasiliev, L.L. Jr., (1996) Multi-effect complex compound/ammonia sorption machines. Proceedings of International Sorption and Heat Pump Conference, Montreal, Canada, pp. 3–8.
[7] Wang, L.W., Wang, R.Z., Wu, J.Y. and Wang, K. (2004) Compound adsorbent for adsorption ice maker on fishing boats. *International Journal of Refrigeration*, **27**, 401–408.
[8] Mauran, S., Prades, P. and Haridon, F.L. (1993) Heat and mass transfer in consolidated reaction beds for thermochemical systems. *Heat Recovery Systems and CHP*, **13**, 315–319.
[9] Aristov, Y.I., Restuccia, G., Caccioba, G. *et al.* (2002) A family of new working materials for solid sorption air conditioning systems. *Applied Thermal Engineering*, **22**, 191–204.
[10] Yang, R.T. (1991) *Gas separation by Adsorption Technology*, Chemical Engineering Press, Beijing, ISBN: 7 5025 0924 0 (in Chinese).
[11] Wilson, S.T., Lok, B.M., Messina, C.A. *et al.* (1982) Aluminophosphate molecular sieves: a new class of microporous crystalline inorganic solids. *Journal of the American Chemical Society*, **104**(4), 1146–1147.
[12] Lok, B.M., Messina, C.A., Patton, R.L. *et al.* (1984) Silicoaluminophosphate molecular sieves: another new class of microporous crystalline inorganic solids. *Journal of the American Chemical Society*, **106**(20), 6092–6093.
[13] Yang, R.T. (2003) *Adsorbents: Fundamentals and Applications*, Wiley-Interscience, Hoboken, NJ.
[14] Aristov, Y.I. (2013) Challenging offers of material science for adsorption heat transformation: a review. *Applied Thermal Engineering*, **50**(2), 1610–1618.

[15] Henninger, S.K., Jeremias, F., Kummer, H. and Janiak, C. (2012) MOFs for use in adsorption heat pump processes. *European Journal of Inorganic Chemistry*, **2012**(16), 2625–2634.

[16] Peng, S.P., Wang, B. and Luo, Z.J. (1984) *Structure of Atoms and Molecules, Complex Compounds, Colloid Chemistry*, People's Press of Sichuan Province, Chengdu, ISBN: 7118.814 (in Chinese).

[17] Ci, Y.X. and Zhou, T.Z. (1999) *The Multiple Complex Compounds in the Analytical Chemistry*, Science Press, Beijing, ISBN: 7-03-007128-X/O·1069 (in Chinese).

[18] Yasuaki, O. (1990) *The Property and Application of Metal Hydride*, Chemical Engineering Press, Beijing, ISBN: 9787502507510 (in Chinese).

[19] Tetsurō, S. (1991) *Metal Oxides and its Catalysis*, Chemical Engineering Press, Beijing, ISBN: 7-312-00210-2 (in Chinese).

[20] Dellero, T., Sarmeo, D. and Touzain, P. (1999) A chemical heat pump using carbon fibers as additive. Part I: enhancement of thermal conduction. *Applied Thermal Engineering*, **19**, 991–1000.

[21] Dellero, T. and Touzain, P. (1999) A chemical heat pump using carbon fibers as additive. Part II: study of constraint parameters. *Applied Thermal Engineering*, **19**, 1001–1011.

[22] Tan, Z.C. (1999) Fundamental research on the adsorption refrigeration systems driven by the waste heat of the engine on cars. Chinese PhD thesis. Shanghai Jiao Tong University, Shanghai, China.

[23] Riffat, S.B., Williams, M.D. and Stuart, C. (1997) Adsorption heat pump using HFC refrigerants. *International Journal of Energy Research*, **21**, 481–494.

[24] Wang, D.C., Wu, J.Y., Xia, Z.Z. *et al.* (2005) Study of a novel silica gel–water adsorption chiller. Part II. Experimental study. *International Journal of Refrigeration*, **28**(7), 1084–1091.

[25] Wang, D.C., Xia, Z.Z., Wu, J.Y. *et al.* (2005) Study of a novel silica gel–water adsorption chiller. Part I. Design and performance prediction. *International Journal of Refrigeration*, **28**(7), 1073–1083.

[26] Saha, B.B., Boelman, E.B. and Kashiwaki, T. (1995) Computational analysis of an advanced adsorption-refrigeration cycle. *Energy*, **20**(10), 983–994.

[27] Peng, C.H. and Zhang, X.Y. (1981) *Principle of Complexometric Titration*, Publishing Press of Beijing University, Beijing, ISBN: 7-301-13209-9-1 (in Chinese).

[28] Dai, A.B. (1987) *Coordination Chemistry*, Volume 12 of Series of Inorganic Chemistry, Science Press, Beijing, ISBN: 13031.4 (in Chinese).

[29] Willers, E. and Groll, M. (1999) Evaluation of metal hydride machines for heat pumping and cooling applications. *International Journal of Refrigeration*, **22**(1), 47–58.

[30] Kemal, A. (1991) Metal hydride heat pumps. *International Journal of Energy Research*, **15**(7), 549–560.

[31] Zondag, H., Kikkert, B., Smeding, S. *et al.* (2013) Prototype thermochemical heat storage with open reactor system. *Applied Energy*, **109**, 360–365.

[32] Zondag, H.A., Essen, V.M.V., Bleijendaal, L.P.J. *et al.* (2010) Application of $MgCl_2$•$6H_2O$ for thermochemical seasonal solar heat storage. Proceedings of 5th International Renewable Energy Storage Conference (IRES2010), Berlin, Germany.

[33] Zondag, H.A., Kikkert, B.W.J., Smeding, S.F., and Bakker, M. (2011) Thermochemical seasonal solar heat storage with $MgCl_2$•$6H_2O$: first upscaling of the reactor. Proceedings of International Conference for Sustainable Energy Storage, Belfast, Ulster.

[34] Boer, R.D., Haije, W.G., Veldhuis, J.B.J., and Smeding, S.F. (2004) Solid-sorption cooling with integrated thermal storage: the SWEAT prototype. Proceedings of the 3rd International Heat Powered Cycles Conference (HPC 2004), Larnaca, Cyprus.

[35] Lammak, K., Wongsuwan, W., and Kiatsiriroj, T. (2004) Investigation of modular chemical energy storage performance. Proceedings of the Joint International Conference on Energy and Environment, Hua Hin, Thailand.

[36] Lahmidi, H., Mauran, S. and Goetz, V. (2006) Definition, test and simulation of a thermochemical storage process adapted to solar thermal systems. *Solar Energy*, **80**(7), 883–893.

[37] Iype, E., Nedea, S.V., Rindt, C.C. *et al.* (2012) DFT Study on characterization of hydrogen bonds in the hydrates of $MgSO_4$. *The Journal of Physical Chemistry C*, **116**(35), 18584–18590.

[38] Essen, V.M., Zondag, H.A., Cot Gores, J. *et al.* (2009) Characterization of $MgSO_4$ hydrate for thermochemical seasonal heat storage. *Journal of Solar Energy Engineering*, **131**(4), 0410141–0410147.

[39] Posern, K. and Kaps, C. (2010) Calorimetric studies of thermochemical heat storage materials based on mixtures of $MgSO_4$ and $MgCl_2$. *Thermochimica Acta*, **502**(1–2), 73–76.

[40] Meunier, F. (1998) Solid sorption heat powered cycles for cooling and heat pumping applications. *Applied Thermal Engineering*, **18**, 715–729.

[41] Sahnoune, H. and Grenier, P.H. (1989) Mesure de la conductivitéthermique d'une zeolithe. *The Chemical Engineering Journal*, **40**, 45–54.

[42] Gurgel, J.M. and Grenier, P. (1990) Mesure de la conductivité thermique du charbon actif AC-35 en presence de gaz. *The Chemical Engineering Journal*, **44**, 43–50.
[43] Pons, M. and Dantzer, P. (1994) Heat transfer in hydride packed beds. *Zeitschrift für Physikalische Chemie*, **183**, 1249–1259.
[44] Wang, S.G., Wang, R.Z., Wu, J.Y. *et al.* (2003) Experimental results and analysis for adsorption ice-making system with consolidated adsorbent. *Adsorption*, **9**(4), 349–358.
[45] Dunne, S. (1996) Carousel heat exchanger for sorption cooling process. US Patent 5 503 222.
[46] Chen, L., Dong, P.F., Fu, W.U. and Sun, L.M. (2003) The experimental research on the refrigeration performance of compound adsorbent of strontium chloride. *Fluid Machinery*, **31**, 201–203ISSN: 1005-0329 (in Chinese).
[47] Goetz, V. and Marty, A. (1992) A model for reversible solid/gas reactions submitted to temperature and pressure constraints; simulation of the rate of reaction in solid gas reactors used in chemical heat pump. *Chemical Engineering Science*, **47**(17–18), 4445–4454.
[48] Kitagawa, H., Suzuki, K. (1983) *Foundation and Design for Adsorption*. Chemical Engineering Press, Beijing, ISBN: 15063.3488 (in Chinese).
[49] Dawoud, B. and Aristov, Y.I. (2003) Experimental study on the kinetics of water vapor sorption on selective water sorbents, silica gel and alumina under typical operating conditions of sorption heat pumps. *International Journal of Heat and Mass Transfer*, **46**, 273–281.
[50] Nunez, T., Henning, H.M., and Mittelbach, W. (1999) Adsorption cycle modeling: Characterization and comparison of materials. Proceedings of International Sorption Heat Pump Conference, Munich, Germany, pp. 209–218.
[51] Zhong, Y., Critoph, R.E., Thorpe, R.N. *et al.* (2007) Isothermal sorption characteristics of the $BaCl_2$-NH_3 pair in a vermiculite host matrix. *Applied Thermal Engineering*, **27**(14–15), 2455–2462.
[52] Henninger, S.K., Schicktanz, M., Hügenell, P.P.C. *et al.* (2012) Evaluation of methanol adsorption on activated carbons for thermally driven chillers part I: thermophysical characterization. *International Journal of Refrigeration*, **35**(3), 543–553.
[53] Wang, L.W., Metcalf, S.J., Critoph, R.E. *et al.* (2012) Development of thermal conductive consolidated activated carbon for adsorption refrigeration. *Carbon*, **50**(3), 977–986.
[54] Critoph, R.E., Tamainot-Telto, Z., and Davies, G.N.L. (1997) Design of an adsorption generator utilising a novel carbon-aluminium laminate. Proceedings of International Heat Pump Conference, Nottingham, UK, pp. 349–358.
[55] Critoph, R.E. (1999) Forced convection adsorption cycle with packed bed heat regeneration. *International Journal of Refrigeration*, **22**(1), 38–46.
[56] Pons, M., Meunier, F., Cacciola, G. *et al.* (1999) Thermodynamic based comparison of sorption systems for cooling and heat pumping. *International Journal of Refrigeration*, **22**, 5–17.
[57] Eun, T.H., Song, H.K., Han, J.H. *et al.* (2000) Enhancement of heat and mass transfer in silica-expanded graphite composite blocks for adsorption heat pumps, Part II: Cooling system using the composite blocks. *International Journal of Refrigeration*, **23**, 74–81.
[58] Lai, H.M. (2000) An enhanced adsorption cycle operated by periodic reversal forced convection. *Applied Thermal Engineering*, **20**, 595–617.
[59] Wang, L.W., Wang, R.Z., Lu, Z.S. *et al.* (2006) Split heat pipe type compound adsorption ice making unit for fishing boats. *International Journal of Refrigeration*, **29**, 456–468.

3

Mechanism and Thermodynamic Properties of Physical Adsorption

The common physical adsorbents, such as activated carbon, silica gel, and zeolite, and so on, are porous media, thus generally the adsorption behavior is the capillary condensation process inside the adsorbents.

It is mentioned in Chapter 2 that pore diameter and pore configuration of different porous media are different. According to the size of different pores, the adsorbents can be divided into macropore medium, mesopore medium, and micropore medium. Macropore doesn't have the ability for adsorption. If mesopore and micropore are regarded as the cylindrical capillary pore with different diameters, the adsorbate vapor should be condensed in the micro pores firstly because the saturation pressure of the concave liquid surface in a cylindrical pore with a small diameter is relatively small. With increasing pressure, the adsorbate vapor should be condensed in larger pores. This is the capillary condensation inside the porous medium [1]. This phenomenon can be shown by the adsorption hysteresis phenomenon.

The adsorption hysteresis phenomena of activated carbon–ammonia were studied by measuring the adsorption performance with the evaluation of the level change (shown in Figure 2.9). When the temperatures of evaporator/condenser are 25, 30, and 35 °C (corresponding saturation pressures are 1.013, 1.167, and 1.361×10^6 Pa), respectively, the adsorption hysteresis phenomena are shown in Figure 3.1.

Figure 3.1 shows that for isobaric adsorption/desorption processes, the temperature/pressure of evaporation/condensation influences the adsorption hysteresis phenomena significantly. If the temperature of evaporation/condensation is 25 °C, the desorption temperature is larger than the adsorption temperature when the adsorption quantity for adsorption and desorption process is the same, and the area of adsorption hysteresis circle is relatively large. If the condensation/evaporation temperature is increased to 30 °C (the corresponding isobaric adsorption/desorption pressure is 1.617×10^6 Pa), the area of adsorption hysteresis circle is decreased while the adsorption quantity for both adsorption and desorption remains the same. When the condensation/evaporation temperature is improved to 35 °C there is almost no desorption hysteresis phenomenon. It is hypothesized that this phenomenon can be explained by contact angle hysteresis model of Zigmondy. As activated carbon microcrystal is a six-carbon ring, the structure of the pore can be seen as unparalleled construction, and the adsorbate vapor

Adsorption Refrigeration Technology: Theory and Application, First Edition. Ruzhu Wang, Liwei Wang and Jingyi Wu.

Companion Website: www.wiley.com/go/wang/refrigeration

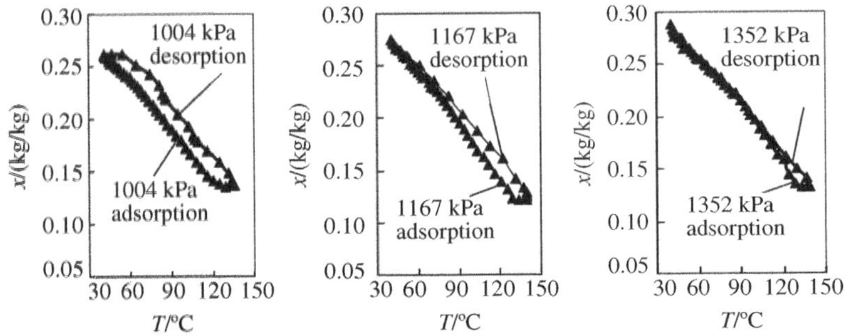

Figure 3.1 The adsorption hysteresis phenomena of activated carbon-ammonia [1]. (a) Evaporation/condensation temperature is 25 °C; (b) evaporation/condensation temperature is 30 °C; and (c) evaporation/condensation temperature is 35 °C

would be condensed in the micropores in the adsorption process. Suppose that the condensation angle is θ_a for the adsorption process when the gas goes into the pores and condenses, and in desorption the contact angle is θ_d when the liquid in pores evaporates and is desorbed, θ_a and θ_d are related to the evaporation/condensation pressure (the saturation pressure of the vapor outside of the pores). θ_a and θ_d change according to the evaporation/condensation temperature. In this process, the changes might be different. For this case the different adsorption hysteresis phenomena occur.

For physical adsorption, the adsorption hysteresis between adsorption and desorption is not serious. So the desorption performance can be simulated by adsorption performance.

3.1 Adsorption Equations

3.1.1 Polanyi Adsorption Potential Theory and Adsorption Equation

For a physical adsorption working pair, the adsorption equations can be divided into three kinds based on the literatures:

1. The adsorption equations for adsorption rate. For this aspect the Langmuir equations are generally for the monolayer adsorption process and the adsorption energy is commonly constant. The Langmiur equations are common in an earlier research stage, but they cannot be utilized extensively because there are too many limitations for the application of the equations.
2. The thermodynamic equations for adsorption processes. They are based on the Polanyi adsorption potential theory and the adsorption theory in micropores proposed by M.M. Dubininl. It is regarded as the best way to present the equilibrium adsorption process between the adsorbate vapor and activated carbon.
3. The adsorption equations set up for the condensation process of the adsorbate vapor inside the micropores. It generally ignores the impact of the energy distribution on the surface

of adsorbents. The pores of the adsorbent are considered as capillary pores, such as the Kelvin equation.

Besides those three kinds of equations, there is a type of equation derived from the experience, which is:

$$x = f(T, p) = f_1(T)f_2(p) \tag{3.1}$$

where T is the adsorption temperature, p is the adsorption pressure.

By doing experiments referring to the changes of the adsorption capacity relating to temperature and pressure in Equation 3.1, the empirical formulae can be solved. By fitting those formulae the better equations are obtained. However, it requires a lot of experimental data, and generally it cannot fit the characteristics of adsorbent working pairs well.

Adsorption potential theory is proposed by Polanyi in 1914 [2]. It is based on the adsorption potential field on the surface of the adsorbent and describes the theoretical model of multilayer adsorption. This theory has been studied from the viewpoint of thermodynamics, also the change of surface Gibbs functions caused by adsorption is considered. However, it has not described the physical adsorption mechanism in more detail. Because of the achievement by M.M. Dubinin [3] on the development of adsorption theory, this theory is also called Dubinin-Polanyi theory [4].

Adsorption potential is determined by the components and the pore structure of adsorbents. The surface of the adsorbent can be seen as being composed of points with different potential. The adsorption center is the point which has the largest potential. Those points with the same potential compose the equi-potential surface, while those points whose distances are the same from the equi-potential surface compose a new equi-potential surface. When the distance increases to r_{max}, the potential is decreased to zero.

According to the Polanyi adsorption potential theory, the adsorbed gas is compressed so that there is an adsorptive force between the adsorbent surface and the space around. There is a gradient change when the gas density changes from the adsorbent surface to the adsorption potential surface of zero. Polanyi thought that the potential ε can be expressed by the isothermal compression work of the adsorbed gas, which is:

$$\varepsilon = RT \ln \left(\frac{p_s}{p} \right) \tag{3.2}$$

where p_s is the saturation pressure corresponding to the adsorption temperature. p is the adsorption equilibrium pressure, which corresponds to the saturation temperature T_{sat} of the refrigerant in the evaporator/condenser.

Dubinin *et al.* found that one kind of adsorbent had similar adsorption characteristic curves even when the adsorbates were different. In the same adsorption phase:

$$\varepsilon_l = \frac{\varepsilon}{\beta} \tag{3.3}$$

where β is the affinity coefficient. It shows the molar volume ratio of the refrigerant to the reference adsorbate (normally benzene) when under the same temperature T. β is only related to the adsorbate.

According to the Dubinin-Radushkevich theory, the adsorption potential of activated carbon that is made up of the micropores with the efficient radius in the range of $18–20\times10^{-10}$ m (it is the first type of activated carbon according to the Dubinin classification) obeys the Gauss Distribution. As a result, the volume of the refrigerant V_c adsorbed, the total pore volume of the adsorbate V_0 and the adsorption potential ε have a relationship as follows:

$$V_c = V_0 \exp\left[-B\left(\frac{\varepsilon}{\beta}\right)^2\right] \tag{3.4}$$

where B is the parameter of the pore structure of the adsorbent. It decreases with the increasing porosity of the adsorbent.

Substitute Equation 3.2 to Equation 3.4, the equation is:

$$V_c = V_0 \exp\left[-B\left(\frac{RT}{\beta}\ln\frac{p_s}{p}\right)^2\right] \tag{3.5}$$

Equation 3.5 shows that if p_s equals to p, then V_c equals to V_0. Similarly different p corresponds to different V_c. V_0 changes with p. According to the Dubinin-Radushkevich theory, V_0 should be unrelated to p if V_0 stands for the pore diameter. As a result, V_0 is defined as the extreme volume filled by the refrigerant under pressure p, that is, V_0 is determined by the working pair, and it is related to p.

When the adsorption quantity is constant and the temperature change range is small, the relationship between saturation pressure and temperature satisfies the Clausius-Clapeyron equation:

$$\ln p = A - \frac{C}{T_{sat}} \tag{3.6}$$

Substitute p in Equation 3.5 by p in Equation 3.6:

$$V_c(T, T_{sat}) = V_0(T_{sat}) \exp\left\{-B\left[\frac{RC}{\beta}\left(\frac{T}{T_{sat}} - 1\right)\right]^2\right\} \tag{3.7}$$

The adsorption potential function (Equation 3.4) is based on Gauss distribution. To expand its application scope, the index of 2 can be changed into n, and the equilibrium adsorption rate is set as x.

$$x = \frac{M_{ref}}{M_a} = \frac{\rho_f V_c}{M_a} \tag{3.8}$$

where ρ_f is the density of the liquid membrane, M_{ref} is the mass of the refrigerant adsorbed and M_a is the mass of the adsorbent. Equation 3.9 can be obtained [5, 6]:

$$x(T, T_{sat}) = x_0(T_{sat}) \exp\left[-K\left(\frac{T}{T_{sat}} - 1\right)^n\right] \tag{3.9}$$

where n reflects the distribution of the pore diameter of the adsorbent. In other words, it means how close the micropore diameter is to the molecular diameter ($18–20\times10^{-10}$ m). When $n=2$, it means that micropores whose pore diameters are in the range of $18–20\times10^{-10}$ m are dominant in the adsorption process. When $n<2$, it means that the micropores whose pore diameters are larger than 20×10^{-10} m are dominant. When $n>2$, it means that the micropores whose pore diameters are smaller than 18×10^{-10} m are the leading micropores for adsorption. K is determined by a working pair, and it is unrelated to temperature. The experimental results show that K is normally in the range of 10–50. The format of equation 3.9 is determined by the properties of working pair, as well as is related to adsorption pressure p (the corresponding saturation temperature is T_{sat}).

In some references x_0 in Equation 3.9 is regarded as a constant value [5, 7]. In fact it is related to adsorption pressure. For instance, in the desorption-condensation process, T_{sat} is condensation temperature T_c corresponding to condensation pressure, while in the cooling-adsorption process T_{sat} is evaporation temperature T_e corresponding to evaporation pressure.

The isothermal (isobaric) adsorption equation is utilized extensively for the design and simulation of adsorption system, whereas D-A and D-R equations are isobaric equations widely used for the analysis of the micropore adsorption process. D-A and D-R equations are simple and suitable for a wide range of temperature and pressure if the adsorbent has a porous surface or its pore diameters are uniform.

For the D-R equation of Equation 3.4, it can also be presented in the form of an adsorption rate as follows [8]:

$$x = x_0 \exp\left(-k\left(\frac{\varepsilon}{\beta}\right)^2\right) \tag{3.10}$$

where x_0 is the maximum adsorption rate, k is a coefficient determined by the adsorbent structure and is unrelated to the refrigerant. β is the affinity coefficient, and it is determined by adsorbent and refrigerant.

Equation 3.10 is suitable for the activated carbon with the pore diameter in the range of $18-20\times10^{-10}$ m. (It is called as the first type micropore by Dubinin.) In the experiments, it is found that three types of activated carbon have deviations from Equation 3.10. To explain this phenomenon, Dubinin *et al.* proposed Equation 3.11 to describe two different groups of pores of activated carbon.

$$x = x_{01} \exp\left(-k_1\left(\frac{\varepsilon}{\beta}\right)^2\right) + x_{02} \exp\left(-k_2\left(\frac{\varepsilon}{\beta}\right)^2\right) \tag{3.11}$$

where four constants x_{01}, x_{02}, k_1, k_2 present the maximum adsorption rates and structure parameters of two groups of pores.

After that Dubinin and Astakhov proposed the Gauss distribution equation which is suitable for any kind of adsorbents with a uniform pore diameter [8]

$$x = x_0 \exp\left(-\left(\frac{\varepsilon}{E}\right)^n\right) \tag{3.12}$$

There are three constants in the equation. Dubinin thought that n in Equation 3.12 is the integer of 2–6. E is the specific adsorption power, and it is determined by the energy characteristics of the adsorption system. The latter experiments pointed out that n can be decimals, which enlarged the equation's application scope. The form of this equation is simple; however, it is only suitable for the adsorbents with a smooth surface and the adsorbent with a weak polarization. Also, some shortcomings were pointed out by Dubinin and other scholars:

1. The adsorption quantity cannot fit the Henry law (i.e., adsorption quantity is proportional to adsorption pressure) under low pressure.
2. The hypothesis that the specific curves are unrelated to temperature is unsuitable for many systems. The errors are especially large when the adsorbent is polar material.
3. The explanation of the adsorbent with a non-uniform pore is unsatisfactory.

3.1.2 *The Improved Adsorption Equation*

3.1.2.1 The Polarity of the Adsorbent

The common adsorbates are methanol, water, ammonia, and ethanol. Their dipole moments are 1.7, 1.8, 1.6, and 1.7 D, respectively, so they are all highly polar gases and have permanent dipole moments. Their interaction forces include dispersion force, induction force, and relatively strong electrostatic force. Besides, there is also hydrogen bonding which can be merged into electrostatic force (like water). For such gases, when the fugacity of the gas is calculated, the state equation relating to polar gases should be used.

The most reliable state equation for polar gases is the Matin-Hou Equation. Its constants are simple and widely used. The relative error is smaller than 1% when the gases are non-hydrocarbon. It is more precise for water, ammonia, and alcohols. The general form of the equation is:

$$p=\sum_{i=1}^{5}\frac{f_i(T)}{(V_c-b)^i}=\frac{RT}{V_c-b}+\frac{A_2+B_2T+C_2\exp(-K_MT/T_c)}{(V_c-b)^2}+\frac{A_3+B_3T+C_3\exp(-K_MT/T_c)}{(V_c-b)^3}+\frac{A_4}{(V_c-b)^4}+\frac{A_5+B_5T+C_5\exp(-K_MT/T_c)}{(V_c-b)^5} \quad (3.13)$$

where $K_M = 5.475$. There are 11 constants A_2, A_3, A_4, A_5, B_2, B_3, B_5, b, C_2, C_3, and C_5.

The constants of the M-H equation of methanol, water, and ammonia are seldom introduced, and they can be found in [9], and the results are shown in Table 3.1.

3.1.2.2 Amendments on the Ideal Gas Model

The common refrigerants are methanol, water, and ammonia. Those gases are different from the idea gases. As a result, Equation 3.2 should be revised by fugacity for the

Table 3.1 The calculating constants in M-H equation for methanol, water, and ammonia

Constants in M-H equation	Methanol	Water	Ammonia
$A_2\times10^7$	−1.181884	−0.7347924	−0.51407794
$A_3\times10^9$	0.9823941	0.3367519	0.25238453
$A_4\times10^{10}$	−4.911565	−0.6221217	−0.71432305
$A_5\times10^{11}$	−10.96979	−0.9647009	−0.64793943
$B_2\times10^4$	0.9605366	0.4867358	0.47427270
$B_3\times10^8$	−5.55264	−2.162056	−1.1981187
$B_5\times10^8$	41.58169	2.432953	1.1981187
$b\times10^2$	0.1265582	0.06601304	0.10464551
$C_2\times10^8$	−4.55933	−2.37388	−1.1226807
$C_3\times10^{10}$	5.609259	1.365615	0.79905689
$C_5\times10^{13}$	−8.947510	−0.4708418	−0.40828723
K_M	5.475	5.475	5.0

corresponding pressure instead of the pressure. The adsorption potential per mole of real gas ε_r is:

$$\varepsilon_r = RT\ln\left(\frac{f_0}{f}\right) \tag{3.14}$$

where f_0 and f are the fugacity under p_s and p, respectively. That is the amendment on the adsorption potential for the real gas.

The fugacity can be calculated by Equation 3.15:

$$\ln\frac{f}{p} = \frac{d-d^*}{RT} - (1-Z) \tag{3.15}$$

where Z is the gas compression factor. For M-H equation, we can obtain:

$$d^* - d = RT\ln\frac{V-b}{V} - \sum_{i=2}^{5}\frac{f_i(T)}{(i-1)(V-b)^{i-1}} + RT\ln\frac{V}{V^*} \tag{3.16}$$

where "*" presents the ideal state in Equation 3.16.

3.1.2.3 The Consideration of the Pore Diameter Distribution on the Adsorbent Surface

The D-R equation is achieved by using the fugacity.

$$x = x_0\exp\left(-k\left(\frac{\varepsilon_r}{\beta}\right)^2\right) = x_0\exp\left(-B\left(\frac{R^2T^2}{\beta^2}\ln^2\left(\frac{f_0}{f}\right)\right)\right) \tag{3.17}$$

Considering most of the adsorbents have non-uniform surfaces the adsorption can be regarded as occurring in many kinds of the micropore groups; $f(B)$ is defined as probability function of the micropore distribution.

$$x = \sum_j x_{0j}f(B)\exp\left(-k_j\left(\frac{\varepsilon_j}{\beta}\right)^2\right) = \sum_j x_{0j}f(B)\exp\left(-B_j\left(\frac{R^2T^2}{\beta^2}\ln^2\left(\frac{f_0}{f}\right)\right)\right) \tag{3.18}$$

where $B_j(k_j)$ and x_{0j} are corresponding to a certain kind of micropore, and different x_{0j} corresponds to different B_j (k_j), so B presents the pore structure.

From the former introduction on adsorbents, it can be concluded that on the surface of different adsorbents there are different pore diameter distributions because of the processing technique and raw material. The probability density $f(B)$ of the pore structure of different kinds of adsorbents is normalized when it is processed, and $f(B)$ satisfies the equation below:

$$\int_0^\infty f(B)\mathrm{d}B = 1 \tag{3.19}$$

For the micropore groups which have the pore diameter characteristics B_j, the adsorption equation can be described as:

$$x_j = x(T, f, B_j) \tag{3.20}$$

Thus, for any surfaces, the whole adsorption rate can be presented by the distribution function like this:

$$x = \int_0^{\infty} x(T,f,B)f(B)\mathrm{d}B \tag{3.21}$$

Two kinds of adsorbents with different pore diameter distributions are as follows:

1. For the adsorbents which have a narrow range of pore diameter distribution, for instance, activated carbon fiber, the pore diameter distribution function is as follows:

$$f(B) = \mathrm{const}(B_1 < B < B_2), \text{and } \Delta B = B_2 - B_2 \to 0 \tag{3.22}$$

that is, B follows even distribution between B_1 and B_2, which can be regarded as uni-modal distribution (shown in Figure 3.2).

$$f(B) = \frac{1}{B_2 - B_1} \tag{3.23}$$

Equation 3.23 is substituted in Equation 3.21, then:

$$x = \int_{B_1}^{B_2} \frac{1}{B_2 - B_1} x(T,f,B)\mathrm{d}B \tag{3.24}$$

$\Delta B \to 0$ while pore diameter distribution is in a narrow range. As a result, the characteristic function B can be regarded as a constant in this narrow range. At this time the adsorption quantity is equal to that calculated by a pore characteristic of B. If the maximum adsorption rate of adsorbent is x_{i0}, the corresponding characteristic adsorption power is E_i, and then the equation can be concluded by Equations 3.12 and 3.14:

$$x = x_i = x_{i0} \cdot \exp\left[-\left(\frac{RT\ln(f_0/f)}{E_i}\right)^{n_i}\right] \tag{3.25}$$

2. For the adsorbents which have a relatively wide range of surface pore diameter distribution, such as activated carbon and zeolite, most of them have non-uniform surfaces. The adsorption on those surfaces can be regarded as the contribution of the common adsorption of many kinds of micropore groups. The pore diameter distribution of normal adsorbents is randomly distributed, which can be described by the normal distribution function with a half width Δ (shown in Figure 3.3):

$$f(B) = \frac{1}{\sqrt{2\pi}\Delta}\exp\left[-\frac{(B - B_0)^2}{2\Delta^2}\right] \tag{3.26}$$

where B_0 is the constant.

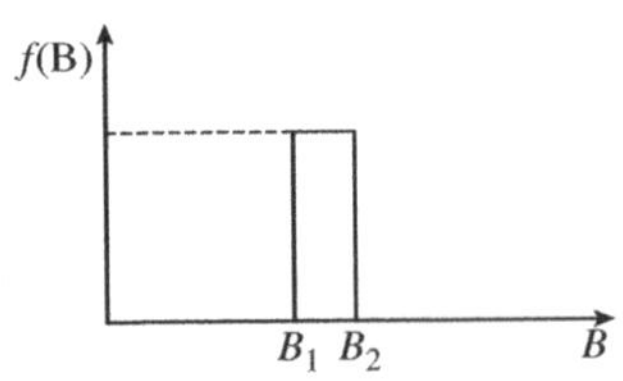

Figure 3.2 The uni-modal distribution of the uniform pore diameter (similar to the distribution function of Dirac)

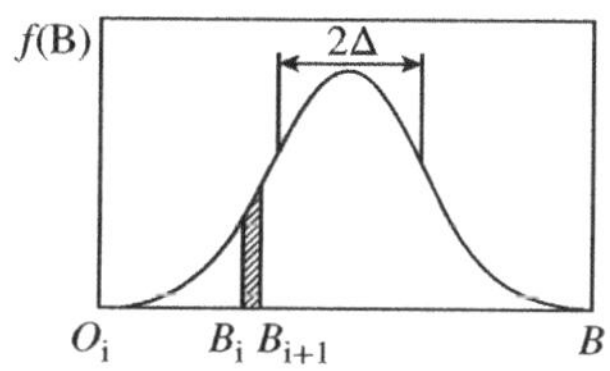

Figure 3.3 Gauss distribution of diameters of non-uniform pores

x corresponding to B can be presented by the D-R equation:

$$x(T, p, B) = x_{B0} \exp\left(-k\left(\frac{\varepsilon}{\beta}\right)^2\right) \tag{3.27}$$

Then

$$x = \int_0^\infty f(B) x_{B0} \exp(-By) \mathrm{d}B \tag{3.28}$$

The general maximum adsorption rate is:

$$x_0 = \int_0^\infty x_{B0} f(B) \mathrm{d}B \tag{3.29}$$

where

$$y = (RT/\beta)^2 \ln^2(f_0/f) \tag{3.30}$$

It can be concluded that Equation 3.28 is the Laplace integral based on the D-R equation.

Equation 3.26 is substituted in Equation 3.28, then

$$x = x_0 \exp(-B_0 y) \cdot \exp(y^2 \Delta^2/2) \cdot [1 - \mathrm{erf}(z)]/2 \tag{3.31}$$

where $z = (y - B_0/\Delta^2)\Delta/\sqrt{2}$, erf is the Error Function. There are three constants in Equation 3.31, i.e., x_0, B_0, and Δ, the values are determined by the experiments on the adsorption working pairs.

For the adsorbents of activated carbon, activated carbon fiber, and zeolite in Table 3.2, by measuring their adsorption rate with methanol or water, the relevant parameter can be reached by Gaussian distribution or uniform pore diameter distribution. It proves the feasibility of Equation 3.31 for adsorbents with non-uniform surfaces and Equation 3.12 for adsorbents with uniform pore diameter while the adsorbates are polar gases. The adsorption parameters of several adsorbents achieved from Equations 3.31 and 3.12 are listed in Tables 3.3 and 3.4.

Research [8, 9] shows that the adsorption model presented by Equations 3.9, 3.10, and 3.12 is suitable for adsorbents with a uniform pore diameter such as activated carbon fiber. However, the model shown in Equation 3.11 that divides the adsorbents into two kinds according to pore diameter is not ideal for different working pairs. Equation 3.31 that supposes the adsorbent pore diameter is a Gaussian distribution is correct for activated carbon-methanol, activated carbon-ammonia, and zeolite molecular sieve-water, and so on, and one tested adsorption isobar can be perfectly applied to other isobars and a variety of isothermal lines.

Table 3.2 Parameters of several adsorbents [8, 9]

Adsorbent	Specific surface area (m^3/g)	Density (kg/l)	Mesh numbers	Material	Origin
Coconut-shell activated carbon (YKAC)	1200	0.62	8–20	Coconut-shell	Shanghai
Coal-based activated carbon in Shanxi (SXAC)	1100	0.65	7–15	Coal	Shanxi
Activated carbon fiber (SYACF)	>1200	0.246	–	Mucilage glue	Shenyang
Activated carbon fiber (NTACF)	>1200	0.243	–	–	Nantong
Activated carbon fiber (JIAACF)	>1200	–	–	Fabric	SJTU
Zeolite (PSO3-HP)	–	0.72	8–12	Zeolite	Shanghai

Table 3.3 The adsorption parameters of three kinds of working pairs derived from Equation 3.31 [8, 9]

Working pair	x_0 (kg/kg)	B_0 ($\times 10^{-6}\,K^{-2}$)	Δ ($\times 10^{-6}\,K^{-2}$)
YKAC-methanol	0.294	1.033	0.289
SXAC-methanol	0.265	1.273	0.251
Zeolite-water	0.203	1.152	0.310

Table 3.4 The adsorption parameters of three kinds of working pairs derived from Equation 3.12 [8, 9]

Working pair	x_0	E (kJ/mol)	n
JIAACF-methanol	0.342	6.703	1.346
SYACF-methanol	0.606	3.904	0.904
NTACF-methanol	0.602	7.674	1.272

3.1.3 Simplified D-A Equation and Its Application

The D-A equation can be written in another form using Equation 3.5:

$$V_a = V_0 \exp\left[-D\left(\frac{T}{\beta}\ln\frac{p_s}{p}\right)^2\right] \tag{3.32}$$

where V_a and V_0 were defined as adsorption volume and maximum pore volume (m^3); T was adsorbent temperature (K); p was the pressure of refrigerant; and p_s was the saturation pressure of the refrigerant corresponding to T. The affinity coefficient β of several refrigerants was defined by comparing the substances to benzene, and was shown in Table 3.5.

Generally, the affinity coefficient β and D can be combined into D', while the index is set as n instead of 2:

$$\frac{\rho_L V_a}{\rho_L V_0} = \exp\left[-D'\left(T\ln\frac{p_s}{p}\right)^n\right] \tag{3.33}$$

Table 3.5 The affinity coefficient β [10] of several refrigerants

Refrigerant	Chemical formula	β
Methanol	CH_3OH	0.40
Ammonia	NH_3	0.28
Sulfur dioxide	SO_2	0.471
Nitrogen oxide	NO_2	0.656

Table 3.6 The parameters of activated carbon–methanol and zeolite–water working pairs

Adsorbent	V_0 (l/kg)	D/D' ($\times 10^{-7}$)	n	Working pair	Equation
AC-35	0.425	5.02	2.15	Activated carbon-methanol	Equation 3.33
207C (cocoanut active charcoal)	0.289	0.608	2.0		Equation 3.32
207E (cocoanut active charcoal)	0.334	1.54	2.0		Equation 3.32
BPL (particle)	0.414	163	1.45		Equation 3.32
Shanghai18# (AC)	0.300	1878	1.33		Equation 3.33
Jiangxi809 (AC)	0.416	2260	1.30		Equation 3.33
Molecular sieve	0.269	1.80	2.0	Zeolite-water	Equation 3.33

where ρ_L is the liquid density of the adsorbate according to the adsorption temperature. x is the adsorption rate (kg/kg). With Equations 3.32 and 3.33, the parameters of common working pairs like activated carbon-methanol [2] and molecular sieve-water [11] were researched by Critoph and Meunier as shown in Table 3.6.

Critoph *et al.* utilized the simplified D-A equation [7, 10] for the simulation, and the equation is:

$$x = x_0 \exp\left[-K\left(\frac{T}{T_s} - 1\right)^n\right] \tag{3.34}$$

In Equation 3.34 [10]:

$$x_0 = x(p_s) = x(T_s) \tag{3.35}$$

where p_s was the saturation pressure corresponding to T_s.

If the adsorption isobars for condensing pressure and evaporation pressure are tested, then the maximum adsorption rate x_1 for evaporating pressure can be determined by the adsorption temperature while the maximum adsorption rate x_2 for the condensing pressure can be determined by the desorption temperature. Then the cycle adsorption mass M_{ref} is:

$$M_{ref} = M_a \times (x_1 - x_2) \tag{3.36}$$

where M_a is the mass of adsorbent (kg/kg). If the sensible heat of the circulating refrigerant, which decreased from condensing temperature to evaporation temperature, is neglected, then:

$$Q_{ref} = M_{ref} \times L_e \tag{3.37}$$

The parameters of D-A Equation 3.34 for several adsorption working pairs are listed in Table 3.7.

Table 3.7 Parameters of D-A equations of several common working pairs

Working pair		T_s (K)	x_0	K	n	Notes
Activated carbon/ activated carbon fiber-methanol	18#AC	298.1	0.238	13.30	1.33	R.Z. Wang *et al.* [12, 13], R.Z. Wang *et al.* [14]
	YKAC	288.3	0.284	10.21	1.39	
	Jiangxi809	298.0	0.333	12.436	1.30	
	Eshland AC	295.5	0.266	11.57	1.41	
	ACF0	290.9	0.400	17.19	1.66	
	ACF1	297.2	0.682	10.84	1.21	
	ACF2	295.5	0.662	10.94	1.31	
	ACF3	287.2	0.516	15.13	1.49	
Activated carbon/ activated carbon fiber-ammonia	208C particle	30–120 °C, 0.1–2 MPa	0.252	8.572	1.832	R.E. Critoph [10], Z. Tamainot-Telto [7]
	BPL particle		0.277	5.674	1.281	
	SC 2 particle		0.283	6.936	1.312	
	AX21powder		0.605	6.095	1.607	
	AX31 (particle by AX21)		0.465	12.416	1.900	
	ACF CC200		0.304	4.611	1.468	
	ACF CC250		0.315	5.569	1.602	
	Shaped activated carbon plate		0.138	4.600	2.0	
	Compact AS12	30–250 °C, 0.1–3 MPa	0.318	10.226	1.99	
	SS13		0.243	8.834	1.756	
	LM127		0.3629	3.6571	0.94	
	LM128		0.3333	3.6962	0.99	
Molecular sieve-water	NaY	–	0.314	5.89	2	A.Z. Yan *et al.* [15]
	5A		0.244	3.57	2	
	13X		0.331	2.99	2	
	13X particle (20 wt% binder)		0.302	4.40	2	

3.1.4 p-T-x Diagram for Gas-Solid Two Phases Equilibrium

The adsorption rate is always related to equilibrium pressure and temperature. The adsorption refrigeration processes can be shown on the *p-T-x* diagram. Figure 3.4a and 3.5a show the adsorption isosteres of zeolite-water. Generally an adsorption refrigeration cycle involves two isosteres and two isobars.

With the *p-T-x* diagram, when the evaporation pressure and condensation pressure are known parameters, the ideal thermodynamic cycle can be drawn for different adsorption and desorption temperatures. The change of adsorption rate can be determined by the diagram, and the cycle cooling quantity can be determined.

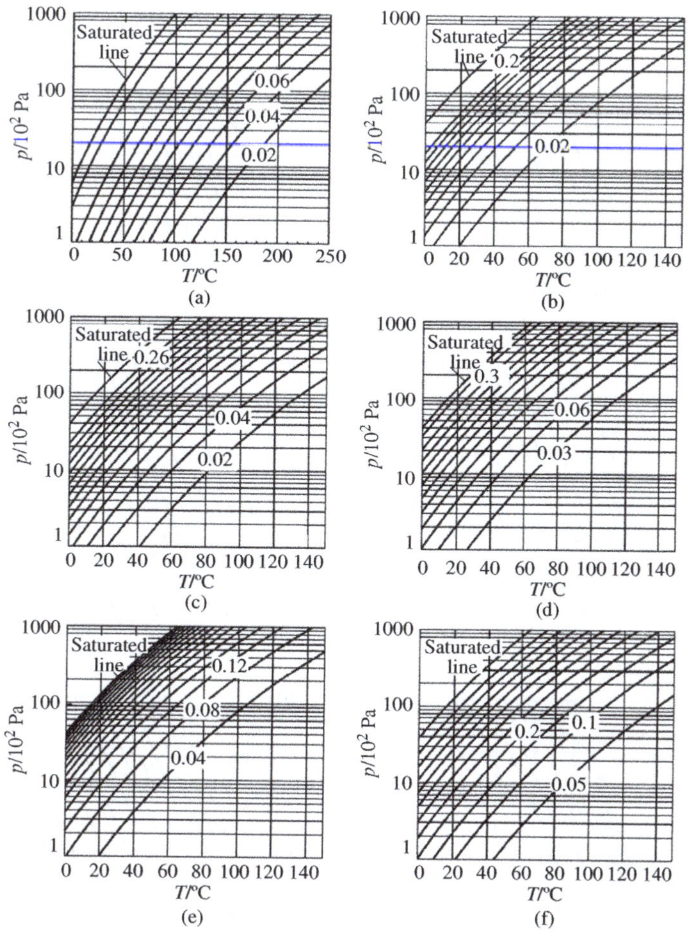

Figure 3.4 *p-T-x* diagram. (a) Zeolite-water; (b) SXAC-methanol; (c) YKAC-methanol; (d) JIAACF-methanol; (e) SYACF-methanol; and (f) NTACF-methanol

Because adsorption pressure p corresponds with saturation temperature T_{sat}, p-T-x diagram can be changed into T_{sat}-T-x diagram easily, which is convenient for optimizing working pairs and evaluating the cycle (T_{sat} is evaporation temperature or condensing temperature).

By solving Equation 3.6, the trend of pressure variation according to temperature can be determined. Figure 3.4a–f are p-T-x diagrams for zeolite-water, SXAC-methanol, YKAC-methanol, JIAACF-methanol, SYACF-methanol, and NTACF-methanol, respectively. The adsorption characteristics of a variety of activated carbon (SXAC, SYAC, YK-AC, JIAACF, SYACF, and NTACF) on methanol can be found in reference [14].

If the vapor pressure is changed into the corresponding saturated temperature, the T_{sat}-T-x diagram (as Figure 3.5a–f shown) can be obtained. The vapor pressure equation in the calculation is the Antonine equation that was recommended by reference [16].

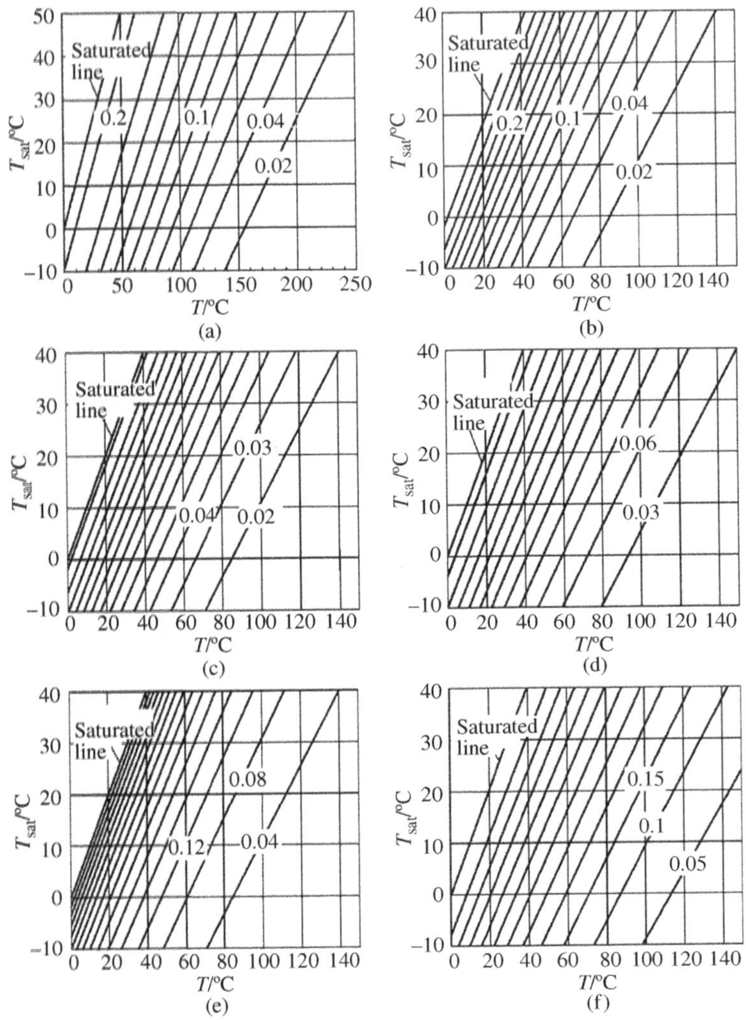

Figure 3.5 T_{sat}-T-x. (a) Zeolite-water; (b) SXAC-methanol; (c) YKAC-methanol; (d) JIAACF-methanol; (e) SYACF-methanol; and (f) NTACF-methanol

3.2 Adsorption and Desorption Heat

As mentioned before, physical adsorption is a condensation process of vapor inside the micro pores, and there is heat transfer in this process. By the second law of thermodynamics, when adsorption occurs for a certain temperature and certain pressure, the adsorbed molecules changed from dispersed state into condensation state. The degree of freedom of adsorbed molecules is decreased. So the entropy which stands for the disordering degree of the system is smaller. In this process the vapor condenses on the surface of the solid adsorbent. The free enthalpy on the surface of the solid also decreases. The free enthalpy is:

$$G = H - TS \tag{3.38}$$

If the temperature and the pressure are constant, then

$$\Delta H = \Delta G + T\Delta S < 0 \tag{3.39}$$

The adsorption process releases heat, that is, adsorption heat.

Adsorption heat is made of two parts, one part is the condensation heat of the vapor, which occurs because of the van der Waals' force between the adsorbed molecules; the other part is called surface energy. When the adsorbed molecules and the molecules on the surface of the adsorbent attract each other, the freedom degree is decreased. The heat would be released in this process. The first part can be found in the steam table while the second part is hard to measure. Fortunately the adsorption heat can be calculated by the adsorption equation.

3.2.1 Thermodynamic Derivation of the Adsorption Heat

Assuming vapor 2 is adsorbed by adsorbent 1 and the adsorption reaches the equilibrium state, then the surface chemical potential μ_f of the adsorbate equals the chemical potential μ_g of the adsorbed gas.

$$\mu = f(T, p, \Gamma) \tag{3.40}$$

where Γ is the adsorption quantity per unit area of solid surface. It is proportional to the degree of coverage.

When the adsorption quantity remains constant and the temperature changes, the chemical potential will change with the temperature. When it reaches the new equilibrium state,

$$\mathrm{d}\mu_\mathrm{f} = \mathrm{d}\mu_g \tag{3.41}$$

According to the basic formulas of thermodynamics, the chemical potential on the surface of adsorbent 2 is:

$$\mathrm{d}\mu_{\mathrm{f}2} = \frac{\partial \mu_{\mathrm{f}2}}{\partial T}\mathrm{d}T + \frac{\partial \mu_{\mathrm{f}2}}{\partial p}\mathrm{d}p + \frac{\partial \mu_{\mathrm{f}2}}{\partial \Gamma}\mathrm{d}\Gamma = -\overline{S_2}\mathrm{d}T + \overline{V_2}\mathrm{d}p + \frac{\partial \mu_{\mathrm{f}2}}{\partial \Gamma}\mathrm{d}\Gamma \tag{3.42}$$

where the partial molar entropy is:

$$\overline{S_2} = \left(\frac{\partial S}{\partial n_2^s}\right)_{T,p,n_j} \tag{3.43}$$

The partial molar volume is:

$$\overline{V_2} = \left(\frac{\partial V}{\partial n_2^s}\right)_{T,p,n_j} \tag{3.44}$$

For the pure gas:

$$\mathrm{d}\mu_g = -\overline{S}_g\mathrm{d}T + \overline{V}_g\mathrm{d}p \tag{3.45}$$

where molar entropy is:

$$\overline{S}_g = S_g/N_g \tag{3.46}$$

Molar volume is:

$$\overline{V}_g = V_g/N_g \tag{3.47}$$

Then:

$$-\overline{S_2}\mathrm{d}T + \overline{V_2}\mathrm{d}p + \frac{\partial \mu_{f2}}{\partial \Gamma}\mathrm{d}\Gamma = -\overline{S}_g\mathrm{d}T + \overline{V}_g\mathrm{d}p \tag{3.48}$$

When the adsorption quantity is constant (the degree of coverage θ is constant), the equation above can be changed into:

$$\left(\frac{\partial P}{\partial T}\right)_x = \frac{\overline{S}_g - \overline{S_2}}{\overline{V}_g - \overline{V_2}} \tag{3.49}$$

If $\overline{V}_2$ is so small in the adsorption process that it can be ignored compared with $\overline{V}_g$, and when the gas is ideal:

$$\overline{V}_g = RT/p \tag{3.50}$$

According to the adsorption equilibrium conditions, that is:

$$\Delta G = \Delta H - T\Delta S = 0 \tag{3.51}$$

It can be concluded that:

$$\Delta S = \overline{S}_g - \overline{S_2} = \frac{\overline{H_g} - \overline{H_2}}{T} \tag{3.52}$$

Then the isosteric heat is:

$$q_{st} = \overline{H_g} - \overline{H_2} \tag{3.53}$$

And:

$$\left(\frac{\partial\,(\ln p)}{\partial T}\right)_M = \frac{q_{st}}{RT^2} \tag{3.54}$$

The equation above is called the differential enthalpy of adsorption under constant adsorption quantity. For solving this equation the adsorption rate equation is required.

3.2.2 *Simplified Formula of Adsorption and Desorption Heat*

The following equation can be derived from Equation 3.54:

$$\left[\frac{\mathrm{d}}{\mathrm{d}T}(\ln p)\right]_{x=const.} = \frac{h}{RT^2} \tag{3.55}$$

where p is the pressure in the adsorption bed, R is the constant of the refrigerant gas. T is the adsorbent temperature. Obviously, h is the function of T and p, and it is the heat released or absorbed by the unit mass of refrigerant in the adsorption or desorption process.

If Equation 3.9 is taken as the adsorption rate equation, then

$$\frac{1}{T_{sat}} = \frac{1}{T}\left[\left(\frac{\ln x_{\max}(T_{sat}) - \ln x(T, T_{sat})}{K}\right)^{\frac{1}{n}} + 1\right] \tag{3.56}$$

From Clausius-Clapeyron's Equation 3.6, we can obtain:

$$\ln p = A - \frac{C}{T}\left[\left(\frac{\ln x_{\max}\left(T_{sat}\right) - \ln x(T, T_{sat})}{K}\right)^{\frac{1}{n}} + 1\right] \tag{3.57}$$

$$\left[\frac{\mathrm{d}\left(\ln p\right)}{\mathrm{d}T}\right]_{x=const.} = \frac{C}{T^2}\left[\left(\frac{\ln x_{\max} - \ln x}{K}\right)^{\frac{1}{n}} + 1\right] = \frac{C}{T_{sat}\cdot T} \tag{3.58}$$

Then the adsorption heat corresponding to the evaporation pressure p_e (the saturation temperature T_{sat} equals to T_e) is:

$$h_a(T, T_e) = RC\cdot\frac{T}{T_e} \tag{3.59}$$

The desorption heat corresponding to the condensing pressure p_c (the saturation temperature T_{sat} equals to T_c) is:

$$h_d(T, T_c) = RC\cdot\frac{T}{T_c} \tag{3.60}$$

The adsorption and desorption heat of a cycle is:

$$H_a = \int_{x_2}^{x_1} m_a h_a \mathrm{d}x = \int_{T_{a1}}^{T_{a2}} m_a h_a \frac{\partial x(T, T_e)}{\partial T}\mathrm{d}T \tag{3.61}$$

$$H_d = \int_{x_1}^{x_2} h_d m_a \mathrm{d}x = \int_{T_{g1}}^{T_{g2}} h_d m_a \frac{\partial x(T, T_c)}{\partial T}\mathrm{d}T \tag{3.62}$$

According to the Clausius-Clapeyron Equation 3.6, the constant C can be obtained with $\ln p \sim 1/T$ diagram which is drawn by the adsorption quantity. The $\ln p \sim 1/T$ diagrams of different working pairs are shown in Figure 3.6a–f.

3.3 Equilibrium Adsorption and Adsorption Rate

3.3.1 The Equilibrium Adsorption and Non-equilibrium Adsorption Process

It is essential to improve the adsorption properties and increase the cooling quantity for the adsorption refrigeration cycle by different methods, such as decreasing the adsorption temperature, improving the evaporating pressure, increasing the desorption temperature, and decreasing the condensing pressure, and so on. Generally the temperature and pressure inside the adsorbents change in one cycle, and the performances of adsorption and desorption processes are theoretically related to temperature and pressure.

In one cycle the adsorption process releases heat while the desorption process absorbs heat. The partial pressure of the adsorbate on the adsorbent surface will influence the diffusion process of the adsorbate in the adsorbent. The adsorption performance is also influenced by the heat and mass transfer performance in the adsorber.

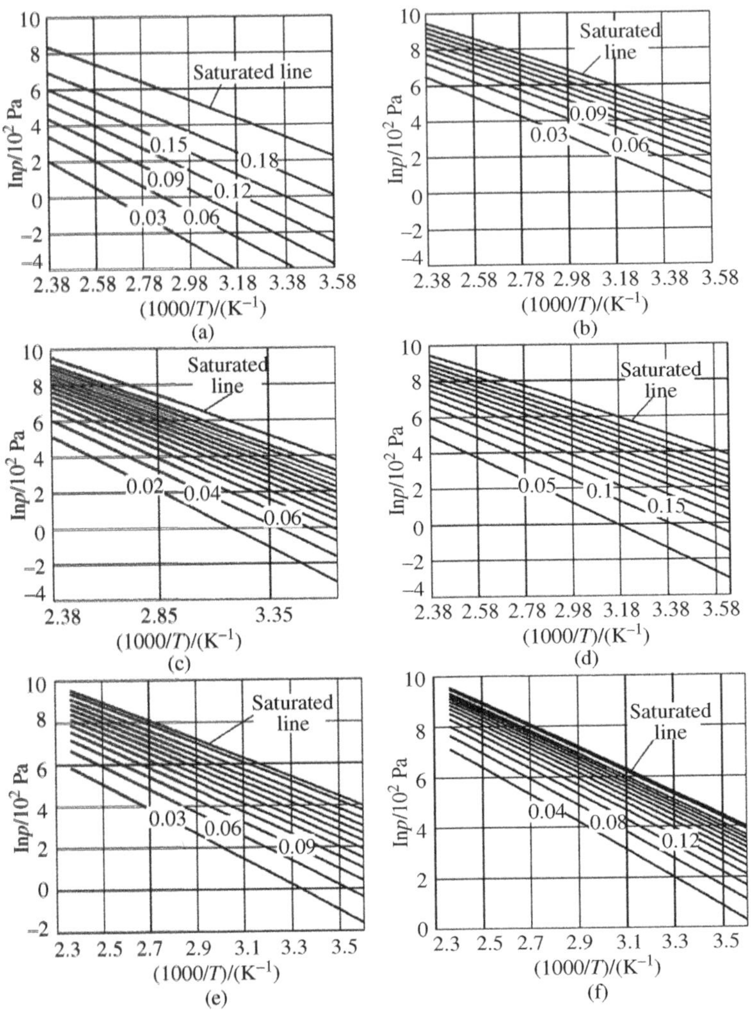

Figure 3.6 ln $p \sim 1/T$ diagram. (a) Zeolite-water; (b) SXAC-methanol; (c) YK-methanol; (d) NTACF-methanol; (e) JIAACF-methanol; and (f) SYACE-methanol

Theoretically, the adsorption capacity is a state function, but it is always influenced by the process because of the non-equilibrium adsorption which commonly happens in the real adsorption refrigeration process.

For an equilibrium state, the quantity of adsorbate adsorbed is equal to that desorbed on the adsorbent surface. If there is a significant difference between the quantities of adsorbate for two processes it is called non-equilibrium state. The adsorption (desorption) is influenced by the mass transfer in the micropores, which is the main reason non-equilibrium is discussed here. The characteristic is that when the temperature field of adsorbent or the concentration filed (pressure) of adsorbate changes, the process will reach the equilibrium state after a non-equilibrium process. If the heat and mass transfer performance is reasonable, the adsorbate will be diffused and condensed quickly in adsorbent, that is, the adsorber will reach the equilibrium state quickly.

3.3.2 Diffusion Process of Adsorbate Inside Adsorbent

For the physical adsorbent (for instance, activated carbon and molecular sieve), 95% of the surface area where the adsorption process happening is the micropore surface. However, the volume of the micropore is not large. For instance, the micropore volume of activated carbon is in the range of 0.15–0.5 cm^3/g [14, 17]. In the adsorption process the adsorbate goes through the mesopores and macropores firstly and then arrives at the interior of the adsorbent, and then it is adsorbed by the micropore. In the desorption process, the desorbed gas firstly escapes from the surface of the micropore, and then it goes through the mesopores and macropores, and finally leaves the adsorbent. In all those steps, the adsorption or desorption rate is far faster than the diffusion rate in all of the pores, and the main resistance comes from the diffusion in the micropores.

The diffusion processes of adsorbate in porous media are divided into four forms: molecular diffusion, Knudsen diffusion, surface diffusion, and limited diffusion [18–20].

Molecular diffusion happens when the capillary pore diameter is far smaller than the average free distance of the adsorbate molecule, molecular collision occurs, and the collision resistance between the molecules and the surface of the pores can be neglected. The effective diffusion coefficient is not only related to the general molecular coefficient, it is also related to the porosity and the complexity of the capillary pore.

Knudsen diffusion happens when the gas pressure is very low or the capillary pore diameter is very small. When the average free distance of the adsorbate molecule is larger than the capillary pore diameter, the collision between the molecule and the surface occurs more easily than the collision between molecules. According to the molecular kinematics, the Knudsen diffusion coefficient is related to the average capillary pore diameter and the speed of the movement of molecules.

Surface diffusion happens when the adsorbate is adsorbed by the solids and there is a concentration difference of adsorbate on the adsorbent surface. The surface diffusion depends on the adsorption layer. It is a mass transfer in two dimensions. Generally the adsorption layer is supposed to be very thin, so the decrease of the pore area in the gas diffusion is limited. It is generally acknowledged that surface diffusion and the gas diffusion do not occur at the same time.

Limited diffusion occurs if the micropore diameter of the adsorbent is close to that of the adsorbate. For such a process the diffusion is nearly a filling process of adsorbate inside the adsorbent micropores. The molecules in this process can only move along the direction of the pore, and the diffusion resistance is large. It is not easy for the adsorbate to be desorbed for the adsorption heat will be very high because the free energy of the adsorbate molecules decreases seriously.

If the adsorbent particles are regarded as a ball with radius r, when the adsorbate diffuses in the pores, according to the law of conservation of mass, the equation is:

$$\varepsilon_a \frac{\partial c}{\partial t} + \rho_s \frac{\partial x}{\partial t} = D_i \left(\frac{\partial^2 c}{\partial r^2} + \frac{2}{r} \frac{\partial c}{\partial r} \right) \quad (3.63)$$

where c is the concentration of adsorbate, ε_a is the porosity of adsorbent, ρ_s is the apparent density of adsorbent, and D_i is the effective diffusion coefficient. Because:

$$\frac{\partial x}{\partial t} = \frac{\partial x}{\partial c} \frac{\partial c}{\partial t} \quad (3.64)$$

The Equation 3.63 can be transformed into:

$$\frac{\partial c}{\partial t} = \frac{D_i}{\varepsilon_a + \rho_s\left(\frac{\partial x}{\partial c}\right)}\left(\frac{\partial^2 c}{\partial r^2} + \frac{2}{r}\frac{\partial c}{\partial r}\right) \tag{3.65}$$

The diffusion of the adsorbate in the adsorbent can be divided into two parts: the diffusion in the micropore (micropore diffusion), the double diffusion on the surface of the micropore (surface diffusion) [1]. If these two processes proceeded separately, then:

$$\varepsilon_a\frac{\partial c}{\partial \tau} = D_e\left(\frac{\partial^2 c}{\partial r^2} + \frac{2}{r}\frac{\partial c}{\partial r}\right) \tag{3.66}$$

$$\rho_s\frac{\partial x}{\partial t} = D_s\rho_s\left(\frac{\partial^2 x}{\partial r^2} + \frac{2}{r}\frac{\partial x}{\partial r}\right) \tag{3.67}$$

where D_e is the diffusion coefficient in the micropore and D_s is the surface diffusion coefficient.

According to the Law of Henry, the adsorption capacity is proportional to the pressure (or concentration) [17]:

$$x = \beta_1 c$$

where β_1 is a constant, $\frac{\partial x}{\partial c}$ is also a constant and can be expressed as β_1. The Equation 3.67 can be transformed into:

$$\rho_s\frac{\partial x}{\partial c}\frac{\partial c}{\partial t} = D_s\rho_s\frac{\partial x}{\partial c}\left(\frac{\partial^2 c}{\partial r^2} + \frac{2}{r}\frac{\partial c}{\partial r}\right) \tag{3.68}$$

Add Equation 3.66 with Equation 3.68, then:

$$\frac{\partial c}{\partial t} = \frac{D_e + \rho_s D_s\frac{\partial x}{\partial c}}{\varepsilon_a + \rho_s\frac{\partial x}{\partial c}}\left(\frac{\partial^2 c}{\partial r^2} + \frac{2}{r}\frac{\partial c}{\partial r}\right) \tag{3.69}$$

Equation 3.69 is compared with Equation 3.65 and the result is:

$$D_i = D_e + \rho_s D_s\frac{\partial x}{\partial c} \tag{3.70}$$

Generally, $\rho_s\frac{\partial x}{\partial c} >> \varepsilon_a$, and Equation 3.70 can be expressed as:

$$\frac{\partial c}{\partial t} = \frac{D_i}{\rho_s\left(\frac{\partial x}{\partial c}\right)}\left(\frac{\partial^2 c}{\partial r^2} + \frac{2}{r}\frac{\partial c}{\partial r}\right) \tag{3.71}$$

3.3.3 The Adsorption Rate and the Mass Transfer Coefficient Inside the Adsorbent

For a type of adsorbent, if γ is the filling density, a_v(m^2/m^3) is the surface area per unit volume of the adsorbent, a_p(m^2/kg) is the surface area per unit mass of the adsorbent, the equation of adsorption rate is [1, 17–23]:

$$\frac{dx}{dt} = \frac{k_F a_v}{\gamma}(c - c_i) = k_F a_p(c - c_i) = \frac{k_s a_v}{\gamma}(x_i - x) = k_s a_p(x_i - x) \tag{3.72}$$

where c is the concentration of the liquid adsorbate, c_i is the concentration of the adsorbate on the surface of the adsorbent. k_F is the mass transfer coefficient of the adsorbate, x_i is the equilibrium adsorption capacity corresponding to the concentration c_i, k_s represents the format of the film of the solid form, and is called as the mass transfer coefficient inside the solid phase film.

According to the research result of E. Glueckauf, *et al.* [21, 22]:

$$k_s a_p = \frac{AD_i}{\beta R^2} = \frac{AD_i'}{R^2} \tag{3.73}$$

when the adsorption time $t \to \infty$, A is 15 (generally A is 15 in the references).

The measurement and calculation of c_i and x_i are difficult. If the concentration of the liquid is c, the equilibrium adsorption capacity is expressed as x^*, and the equilibrium concentration corresponding to the adsorption capacity x is expressed as c^*, by Law of Henry, Equation 3.72 can be transformed as:

$$\begin{aligned}\frac{dx}{d\tau} &= k_F a_p (c - c_i) = \frac{k_F a_p}{\beta_1}(x^* - x_i) = k_s a_p (x_i - x)\\ &= \frac{x^* - x}{\dfrac{\beta_1}{k_F a_p} + \dfrac{1}{k_s a_p}} = \frac{c - c^*}{\dfrac{1}{k_F a_v} + \dfrac{1}{\beta_1 k_s a_v}} = K_s a_p (x^* - x)\end{aligned} \tag{3.74}$$

where K_s is the overall mass transfer coefficient. Obviously, $K_s a_p$ is related to the geometric characteristic of adsorbent surface and the acting force between the adsorbent and the adsorbate. Sometimes $K_s a_p$ is called as the surface diffusion rate coefficient.

3.3.4 Typical Model of Adsorption Rate

According to the reaction kinetics, A. Sokoda, M. Suzuki *et al.* [24] thought that for the silica gel-water adsorption system, the adsorption rate is influenced by the surface diffusion. The adsorption rate for the silica gel-water system is:

$$\frac{dx}{dt} = K_s a_p (x^* - x) \tag{3.75}$$

$$K_s a_p = \frac{15 D_{so}}{R_p^2} \exp(-E_a/RT) \tag{3.76}$$

where x^* is the equilibrium adsorption capacity. $K_s a_p$ is the surface diffusion rate coefficient. The surface diffusion coefficient D_{so} is 2.54×10^{-4} m^2/s, the surface diffusion activation energy E_a is 4.2×10^4 J/mol, the average diameter of the adsorbent particles R_p is 7.1×10^{-4} m.

E.F. Passos, J.F. Escobedo *et al.* [25] presented the reference value of the parameters of the relation for activated carbon-methanol refrigeration system according to the experiment data of solar ice maker, which are $15D_{so}/R_p^2 = 7.35 \times 10^{-3}$ 1/s, $E_a/R = 978$ K.

In an adsorption refrigeration system, both adsorption rate and non-equilibrium adsorption have a strong effect on the adsorption characteristics of the adsorber, especially for the systems that are operated under low pressure and continuous cycles with multi-beds and short cycle

Table 3.8 Comparison of the performance of the cycle with two different models [26]

Cycle time (min)	COP for equilibrium adsorption	The length of the adsorber for 1 kW cooling power (m)	COP for non-equilibrium adsorption	The length of the adsorber for 1 kW cooling power (m)
20	0.209	14.4	0.026	129.6
60	0.302	21.5	0.109	78.2
120	0.312	38.3	0.188	77.9
240	0.312	81.7	0.256	99.3
360	0.312	122.5	0.278	131.0
480	0.312	163.3	0.285	167.3

time. In reference [26], the adsorber is supposed to be cylinder and the adsorbent was filling as a form of ring with the thickness of 5 mm. The working pair is activated carbon–methanol. Table 3.8 shows the comparison of the characteristics. The data of the non-equilibrium adsorption is from the reference [25]. Table 3.8 shows that the influence of the non-equilibrium adsorption cannot be neglected for obtaining the reasonable refrigeration performance.

References

[1] Zhang, Y.H. (1989) *Adsorption Action*, Shanghai Press of Science and Technology, Shanghai, ISBN: 7805134979, 9787805134970 (in Chinese).

[2] Polanyi, M. (1914) Über die Adsorption vom Standpunkt des dritten Wärmesatzes. *Verhandlungen der Deutsche Physikalische Gesellschaft*, **16**, 1012–1016.

[3] Dubinin, M.M. (1975) in *Progress in Surface and Membrane Science* (ed D.A. Cadenhead), Academic Press, New York.

[4] Rand, B. (1976) On the empirical nature of the Dubinin-Radushkevich equation of adsorption. *Journal of Colloid and Interface Science*, **56**(2), 337–345.

[5] Teng, Y., Wang, R.Z. and Wu, J.Y. (1997) Study of the fundamentals of adsorption systems. *Applied Thermal Engineering*, **17**(4), 327–338.

[6] Teng, Y., Wang, R.Z. and Wu, J.Y. (1997) The analysis on the thesis of adsorption refrigeration/heat pump. *Acta Energiae Solaris Sinica*, **18**(1), 22–30 ISSN: 0254–0096 (in Chinese).

[7] Tamainot-Telto, Z. and Critoph, R.E. (1997) Adsorption refrigerator using monolithic carbon-ammonia pair. *International Journal of Refrigeration*, **20**(2), 146–155.

[8] Wang, R.Z. and Wang, Q.B. (1999) The adsorption mechanism of the adsorption working pair and the improvement of the adsorption rate equation. *Acta Energiae Solaris Sinica*, **20**(3), 259–269 ISSN: 0254-0096 (in Chinese).

[9] Wang, R.Z. and Wang, Q.B. (1999) Adsorption mechanism and improvement of adsorption equation for adsorption refrigeration pairs. *International Journal of Energy Research*, **23**(10), 887–898.

[10] Critoph, R.E. (1988) Performance limitations of adsorption cycles for solar cooling. *Solar Energy*, **14**(1), 21–31.

[11] Meunier, F. and Douss, N. (1990) Performance of adsorption heat pumps:active carbon-methanol and zeolite-water pairs. *ASHRAE Transactions*, **96**, 267–274.

[12] Wang, R.Z., Dai, W., Zhou, H.X. and Jia, J.P. (1995) The research of the adsorption refrigeration characteristics of activated carbon –methanol. *Acta Energiae Solaris Sinica*, **16**(2), 155–161 ISSN: 0254-0096 (in Chinese).

[13] Wang, R.Z., Jia, J.P., Teng, Y. *et al.* (1997) A promising adsorption refrigeration working pair: activated carbon fiber-methanol. *Acta Energiae Solaris Sinica*, **18**(2), 222–227 ISSN: 0254-0096 in Chinese).

[14] Wang, R.Z., Jia, J.P., Zhu, Y.H. *et al.* (1997) Study on a new solid adsorption refrigeration pair: active carbon fiber-methanol. *Journal of Solar Energy Engineering, Transactions of the ASME*, **119**(3), 214–218.

[15] Yan, A.Z., Bao, S.L., Yan, Y.C. and Wang, R. (1982) The adsorption refrigeration with zeolite and molecular sieve: the chosen of their system. *Journal of Refrigeration*, (4), 24–31 ISSN: 0253-4339 (in Chinese).

[16] Song, W.R., Xiao, R.J. and Fang, D.Y. (1991) *The Methanol Engineering*, Chemical Industry Press, Beijing, ISBN: 7502509232, 9787502509231 (in Chinese).
[17] Kitagawa, H. and Suzuki, K. (1983) *Foundation and Design for Adsorption*, Chemical Engineering Press, Beijing, ISBN: 15063.3488 (in Chinese).
[18] Ruthven, D.M. (1984) *Principles of Adsorption and Adsorption Processes*, John Wiley & Sons, Inc., New York.
[19] Ponec, V., Knor, Z. and Cerny, S. (1974) *Adsorption on Solids*, Butterworth and Company Limited.
[20] Zhen, D.X. and Liu, F.R. (1998) *The Separation of Multi-Component Gases*, Xi'an Jiaotong University Press, Xi'an, ISBN: 7-5605-0129-X (in Chinese).
[21] Glueckauf, E. (1955) Theory of chromatography. Part 10. – Formulæ for diffusion into spheres and their application to chromatography. *Transactions of the Faraday Society*, **51**, 1540–1551.
[22] Glueckauf, E. and Coates, J.I. (1947) Theory of chromatography. Pt.4. The influence of incomplete equilibrium on the front boundary of chromatograms and the effectiveness of separation. *Journal of the Chemical Society*, **149**, 1315–1321.
[23] Jaroniec, M. and Madey, R. (1997) *The Physical Adsorption on Non-Uniform Solids*, Chemical Industry Press, Beijing, ISBN: 7502519858, 9787502519858 (in Chinese).
[24] Sokoda, A. and Suzuki, M. (1984) Fundamental study on solar powered adsorption cooling system. *Journal of Chemical Engineering of Japan*, **17**(1), 52–57.
[25] Passos, E.F. and Escobedo, J.F. (1989) Simulation of an intermittent adsorptive solar cooling system. *Solar Energy*, **42**(2), 103–111.
[26] Wang, W. and Wang, R.Z. (2001) The investigation of heat transfer character in adsorbent bed with consideration of non-equilibrium adsorption. *Journal of Engineering Thermophysics*, **22**(2), 215–218 ISSN: 0253-231X (in Chinese).

4

Mechanism and Thermodynamic Properties of Chemical Adsorption

Chemical adsorption is based on the interaction force of the molecules, such as complexation, coordination, hydrogenation, oxidization, and so on. The chemical adsorption working pairs mainly include metal chloride–ammonia [1, 2], metal hydride–hydrogen [3, 4], and metal oxide–oxygen [5]. The refrigeration performances are different when the adsorption working pairs are different. For example, the lowest refrigeration temperature of metal chlorides is −40 to −10 °C, whereas the lowest refrigeration temperature of metal oxides is −30 to 0 °C [6]. The metal oxide–oxygen working pair is mostly used for heat pumps. As far as the proper selection of working pairs, Lebrun and Neveu [7] proposed the following criteria: the investment (including the price of the adsorbents and the generator); performances (the refrigeration temperature, the cooling quantity per kilogram absorbent, and the total mass of the adsorbent); comprehensive consideration of the investment and the performance (the refrigeration quantity per unit mass of the total system and per unit heat transfer area; the change of the temperature per unit area of the heat exchangers). Considering the aforementioned criteria, of all the working pairs for chemical adsorption refrigeration, metal chloride–ammonia is the optimal working pair. As a kind of environmental benign refrigerant, the ammonia has been designated as the substitution of CFCs and HCFCs by the International Institute of Refrigeration [8]. The adsorption refrigeration capacity of metal chloride–ammonia working pair is far higher than that of the metal hydride–hydrogen. Moreover, the price of the metal chloride is also far lower than metal hydride. Most commonly used metal chloride adsorbents are calcium chloride, strontium chloride, magnesium chloride, barium chloride, and so on, in which 1 mol calcium chloride can complex 8 mol ammonia, with the adsorption quantity as high as 1.225 kg/kg due to the minimum molecular mass. Meanwhile, relative to other metal chloride adsorbents, calcium chloride has the lowest price. Therefore, calcium chloride is a promising adsorbent for refrigeration.

4.1 The Complexation Mechanism of Metal Chloride–Ammonia

The complex mechanism of metal chloride–ammonia can be demonstrated by the hysteretic phenomena. With the same saturation pressure of refrigerants, the isobaric adsorption and desorption characteristics of the calcium chloride–ammonia working pair can be demonstrated

Adsorption Refrigeration Technology: Theory and Application, First Edition. Ruzhu Wang, Liwei Wang and Jingyi Wu.

Companion Website: www.wiley.com/go/wang/refrigeration

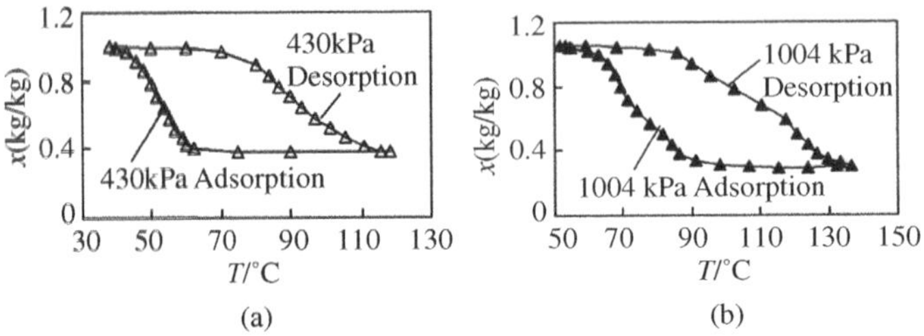

Figure 4.1 Isobaric adsorption hysteresis phenomenon of $CaCl_2$-NH_3 [9]. (a) Evaporating/condensing pressure of 430 kPa and (b) evaporating/condensing pressure of 1004 kPa

in Figure 4.1 [9]. Figure 4.1 shows that the calcium chloride–ammonia working pair has the serious adsorption hysteretic phenomenon. In contrast to Figure 3.1, Figure 4.1 indicates that the adsorption hysteresis phenomenon of the $CaCl_2$–NH_3 working pair exceeds that of activated carbon–NH_3, which can't be explained by the capillary condensation phenomena of physical adsorption.

On the one hand, the adsorption hysteresis phenomenon of the $CaCl_2$–NH_3 working pair is closely associated with the stability constant of complexations in terms of chemical adsorption mechanism. Presumed $CaCl_2$–NH_3 complexation can form two molecular, four molecular, and eight molecular complexes corresponding to the stability constant of k_1, k_2, and k_3, respectively. On the contrary, the unstable constant of octa-ammoniate calcium chloride is $1/k_3$. As for the ammoniate complex with coordinating number 4 and 2, the unstable constant are $1/k_2$ and $1/k_1$, respectively. The order for the ammoniate synthesis process (i.e., adsorption process) is 2, 4, 8, and the order for the ammoniate decomposition process is 8, 4, 2 [10]. For example, at the fixed temperature of 60 °C, Figure 4.1 shows that the adsorption process transit from tetra–ammoniate to octa–ammoniate, the stability constant is k_3, and the total stability constant is $k_3 \times k_2 \times k_1$ [10, 11]. The desorption process transit from octa–ammoniate to tetra–ammoniate, the unstable constant is $1/k_3$. The unstable constant is not equal to the stable constant, therefore, the decomposition rate is not equal to the synthesis rate, even showing a large difference, which brings about further difference from the adsorption isobar to the desorption isobar.

On the other hand, this phenomenon can also be explained by using the activated energy [12]. Assuming the required activated energy for the chemical adsorption is E_a, and the activated energy needed for the desorption process is E_d, both of them have significant difference [13]. The required activated energy for the desorption process of calcium chloride equals to the sum of the activated energy and the reaction heat of the adsorption process. The activated energy needed for the adsorption process of calcium chloride is very small. Under this circumstance desorption activated energy is approximately equal to adsorption heat. The difference of both activated energies could lead to the distinct difference between the adsorption and desorption processes [9].

4.2 The Clapeyron Equation of Metal Chloride-Ammonia

4.2.1 The General Clapeyron Equations

The universal reaction formula for the complex reaction between metal chlorides and NH_3 is as follows [14]:

$$M_aX_b(NH_3)_n + (m-n)NH_3 \Longleftrightarrow M_aX_b(NH_3)_m \tag{4.1}$$

where M represents metal elements, X represents Cl, and *a, b, n, m* are the reaction equilibrium constants.

The standard reaction free enthalpy change ΔG^0 can be described as:

$$\Delta G^0 = (m - n)(\Delta H^0 - T\Delta S^0) \tag{4.2}$$

where ΔH^0 and ΔS^0 are the changes between the standard enthalpy and entropy for complexing per mole ammonia. When the reaction reaches the equilibrium state, the reaction free enthalpy is:

$$\Delta G = \Delta G^0 + RT \ln K = 0 \tag{4.3}$$

where K is the equilibrium constant of the reaction.

Combining Equations 4.2 and 4.3, the following equation is obtained:

$$\ln K = (m - n)\left(-\frac{\Delta H^0}{RT} + \frac{\Delta S^0}{R}\right) \tag{4.4}$$

According to the relations between the reaction equilibrium constants and the concentration of the reactants and resultants, together with the gas pressure relationship, assume that the activity constant of solid ammoniate is 1, and regarding ammonia as an ideal gas, K is given below:

$$K = \left(\frac{a\left(\mathrm{M}_a\mathrm{X}_b(\mathrm{NH}_3)_m\right)}{a(\mathrm{M}_a\mathrm{X}_b(\mathrm{NH}_3)_n)p_{\mathrm{NH3}}^{-(m-n)}}\right) = p_{\mathrm{NH3}}^{-(m-n)} \tag{4.5}$$

Combining Equations 4.4 and 4.5, the Clausius-Clapeyron equation is obtained as follows:

$$\ln p_{NH3} = \frac{\Delta H^0}{RT} - \frac{\Delta S^0}{R} \tag{4.6}$$

The Equation 4.5 depicts the case for the standard state ($p = 101325$ Pa and $T = 298.15$ K).

Under the conditions of non-standard states, the enthalpy and the entropy is the function of the temperature, and thereby considering the heat capacity of the adsorbent, the equation can be written as:

$$\ln p_{NH3} = \frac{\Delta H^0}{RT} + \frac{1}{RT}\int_0^T \Delta C_a \mathrm{d}T - \frac{\Delta S^0}{R} - \frac{1}{R}\int_0^T \frac{\Delta C_a}{T}\mathrm{d}T \tag{4.7}$$

Based on this equation, Biltz and Huttig put forward an experiential formula [15].

$$\ln p_{NH3} = \frac{\Delta H^0}{RT} + 1.75 \ln T + aT + 3.3 \ln 10 \tag{4.8}$$

In this formula, according to the different value of m and n, the parameter "a" is between 0.0017 and 0.0024.

Touzain summarized 350 kinds of complex reactions about metal chloride and ammonia among the last 300 years literatures, and investigated the Equation 4.6. By calculating the mean deviation of the results in all the literatures, he got the average value of the deviation of −130 kJ/mol. Considering the error resulting from this value the calculation formula of the enthalpy is:

$$\Delta H^0_{cal} = \Delta H^0_{\mathrm{exp}} - T_{\mathrm{exp}}(\Delta S^0_{\mathrm{exp}} + 130) \tag{4.9}$$

When the pressure of ammonia is 1 bar, the corresponding reaction equilibrium temperature is:

$$T(^{\circ}\mathrm{C}) = \left(\Delta H^{0}_{cal}/\Delta S^{0}_{cal}\right) - 273 \tag{4.10}$$

When the reaction described by chemical Equation 4.1 occurs and releases the heat ΔH_{cal} per hour, the corresponding energy (W) per kilogram ammonia complex is:

$$W_{cal} = -\left(\frac{\Delta H^{0}_{cal}(m-n)}{3.6M_{(MaXb(NH3)m)}}\right) \tag{4.11}$$

Based on the fundamental cycle and neglecting sensible heat of the adsorption bed, *COP* (coefficient of performance) depends on the reaction heat and latent heat of the adsorbents. The calculated formula is:

$$COP = \frac{\Delta H_{eva}}{\Delta H^{0}_{cal}} \tag{4.12}$$

where ΔH_{eva} is the enthalpy under the evaporation temperature of 0 °C. Its value is 21.4 kJ/mol.

The relationship between the energy density W_{cal} and refrigeration energy density is:

$$W^{cold}_{cal} = W_{cal} \times COP \tag{4.13}$$

According to Equation 4.13, refrigeration performance is proportional to the energy density. The adsorbents with reasonable adsorption refrigeration performance are listed in Table 4.1 [14] together with the energy level per kilogram sorbent. Table 4.1 also lists the reaction steps and the temperature gradient, which provides the beneficial reference for the selection of available adsorbent. Take the $BeCl_2$ as an example, Table 4.1 shows that the reaction is divided into two steps for adsorption refrigeration, one is 2–4 coordination ions, and another is 4–6 coordination ions. The temperature gradient reaches 310 °C between two steps, which is very critical for the application of the absorbent.

4.2.2 *The Principle and Clapeyron Diagram of Metal Chloride-Ammonia Adsorption Refrigeration*

The basic principle of chemical adsorption refrigeration is illustrated by Figure 4.2a. The Clapeyron diagram is shown in Figure 4.2b. The adsorption refrigeration cycle includes four processes: the decomposition reaction (i.e., desorption), condensation, evaporation, and synthesis reaction [16]. In the decomposition process, the heat Q_1 at the temperature of T_H is supplied for the desorption (point 1). The desorbed gaseous refrigerant condenses into liquid under the temperature T_m in the condenser, releasing the heat Q_2 (point 2). In the adsorption process, liquid refrigerant evaporates under temperature T_L, absorbing heat to produce a cooling effect (point 3). The gas flows into the reactor to synthesize with the adsorbent, releasing the heat Q_4 at T_m (point 4).

Chemical reaction process is a single variable process, which means that if the reaction equilibrium temperature is known, the pressure is also determined. During the condensation and evaporation process, the temperature is dependent on the pressure because the vapor is under the saturated state [17]. The vapor-fluid equilibrium equation is:

$$n_1\mathrm{NH_3(gas)} \leftrightarrow n_1\mathrm{NH_3(liq)} + n_1\Delta L_{\mathrm{am}} \tag{4.14}$$

Table 4.1 The ammoniate chlorides with reasonable refrigeration performance [14]

Complex n-m	Energy Q_{cal} (Wh)	Step	Temperature gradient (°C)	Complex n-m	Energy Q_{cal} (Wh)	Step	Temperature gradient (°C)
$MgCl_2$/0-6	528	3	240	$NiCl_2$/2-6	272	1	0
$NiCl_2$/0-6	467	3	210	$CaCl_2$/2-8	270	2	10
$CoCl_2$/0-6	444	3	240	$SrCl_2$/1-8	266	1	0
$FeCl_2$/0-6	431	3	240	$CoCl_2$/2-6	258	1	0
$CaCl_2$/0-8	409	4	200	NH_4Cl/2-6	251	1	0
$MnCl_2$/0-6	408	3	270	$AlCl_3$/2-6	250	1	0
$BeCl_2$/0-6	324	2	310	LiCl/2-5	249	3	70
NaCl/0-5	317	1	0	$FeCl_2$/2-6	244	1	0
$SrCl_2$/0-8	308	2	30	$CuCl_2$/2-6	235	3	30
$MgCl_2$/2-6	306	1	0	LiCl/0-1	235	1	0
$CdCl_2$/0-6	305	3	230	$BaCl_2$/0-8	234	1	0
NH_4Cl/0-3	279	1	0	$HgCl_2$/0.67-9.5	231	3	280
CuCl/0-3	278	3	180	$MnCl_2$/2-6	227	1	0
$BeCl_2$/2-4	275	1	0				

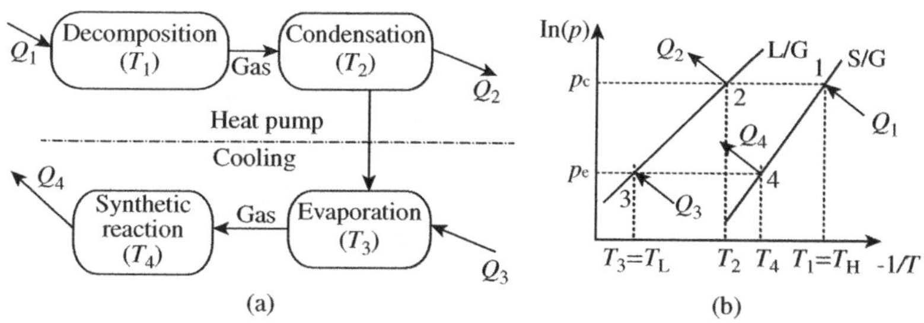

Figure 4.2 The chemical adsorption refrigeration process and the Clapeyron diagram. (a) The chemical adsorption refrigeration process and (b) the Clapeyron diagram

where ΔL_{am} is the vaporization latent heat of the ammonia.

The Clapeyron equation of chemical adsorption is generally appropriate for the single step of chemical synthesis and decomposition reaction. The reaction equilibrium line of the Clapeyron diagram (Figure 4.2) is defined by the different reaction steps [18]. Meanwhile, there exists a pseudo adsorption equilibrium phenomenon in the Clapeyron diagram of the chemical adsorption process [18, 19]. Take the $CaCl_2$ and CH_3NH_2 as the example, with coordination numbers from 6 to 4 and from 4 to 2, the corresponding reaction equations are:

Step 1: $CaCl_2 \cdot 6CH_3NH_2 \Longleftrightarrow CaCl_2 \cdot 4CH_3NH_2 + 2CH_3NH_2$
Step 2: $CaCl_2 \cdot 4CH_3NH_2 \Longleftrightarrow CaCl_2 \cdot 2CH_3NH_2 + 2CH_3NH_2$.

The corresponding Clapeyron diagram of two different reaction steps is shown in Figure 4.3. The pseudo adsorption equilibrium phenomenon of the second reaction is described by the zone between B and C [18], where the reaction rate is zero [19].

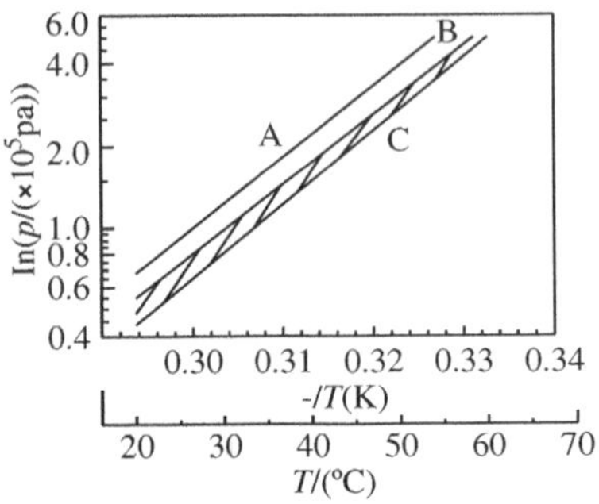

Figure 4.3 The Clapeyron diagram of different reaction steps and the pseudo adsorption equilibrium phenomenon

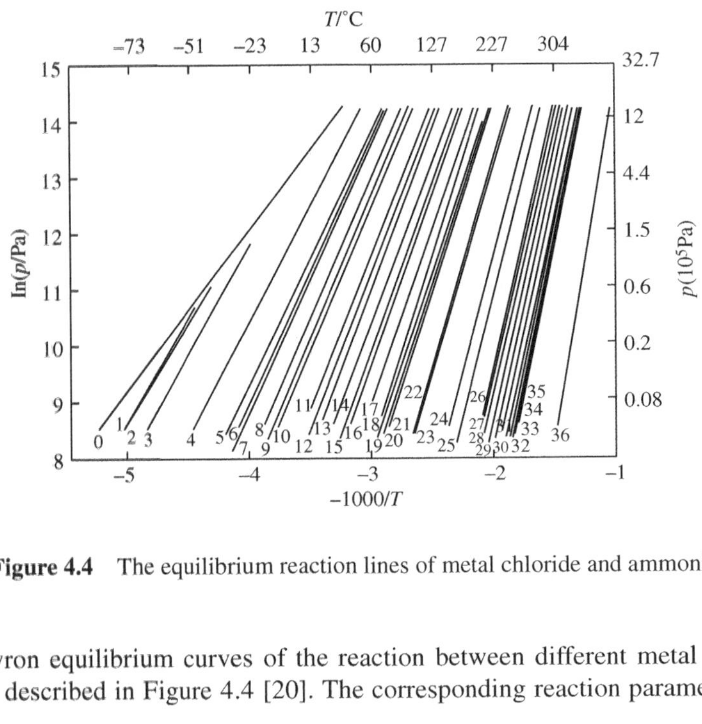

Figure 4.4 The equilibrium reaction lines of metal chloride and ammonia

The Clapeyron equilibrium curves of the reaction between different metal chlorides and ammonia are described in Figure 4.4 [20]. The corresponding reaction parameters are given in Table 4.2.

4.3 Chemical Adsorption Precursor State of Metal Chloride–Ammonia

Because the van der Waals force is inversely proportional to the seventh power of distance, which is much longer than the effective reaction distance between chemical molecules, thus for the molecules in the chemical adsorption process, physical adsorption [13], which is called precursor state of chemical adsorption, generally occurs firstly. The chemical adsorption curves of $CaCl_2$–NH_3 are shown in Figure 4.5 [13]. The potential energy decreases to the lowest value

Table 4.2 The reaction parameters of metal chloride and ammonia (corresponding to the curves in Figure 4.4)

Number	Substance	ΔH (J/mol)	ΔS (J/(mol K))	C_p (J/(mol K))
0	NH_3	23 366	150.52	80.27 (liquid)
1	Zn10-6	29 588	219.23	71.27
2	Cu10-6	31 387	227.72	71.81
3	Sn9-4	31 806	224.86	70.60
4	Pb8-3.25	34 317	223.76	70.05
5	Ba8-0	37 665	227.25	75.10
6	Sn4-2.5	38 920	229.82	70.60
7	Pb3.25-2	39 339	230.27	70.05
8	Ca8-4	41 013	230.30	72.52
9	Sr8-1	41 431	228.80	75.53
10	Ca4-2	42 268	229.92	72.52
11	Zn6-4	44 779	230.24	71.27
12	Pb2-1.5	46 035	230.89	70.05
13	Pb1.5-1	47 290	232.50	70.05
14	Mn6-2	47 416	228.07	72.86
15	Zn4-2	49 467	230.24	71.27
16	Cu5-3.3	50 241	230.75	71.81
17	Fe6-2	51 266	227.99	76.57
18	Cu3.3-2	56 497	237.22	71.81
19	Co6-2	53 986	228.10	78.41
20	Pb1-0	55 660	231.04	70.05
21	Mg6-2	55 660	230.63	71.31
22	Ni6-2	59 217	227.75	71.60
23	Ca2-1	63 193	237.34	72.52
24	Ca1-0	69 052	234.14	72.52
25	Mn2-1	71 019	232.35	72.86
26	Mg2-1	74 911	230.30	71.31
27	Fe2-1	76 167	231.91	76.57
28	Co2-1	78 134	232.17	78.41
29	Ni2-1	79 515	232.17	71.60
30	Zn2-1	80 352	229.72	71.27
31	Mn1-0	84 202	233.18	72.86
32	Fe1-0	86 880	233.01	76.57
33	Mg1-0	87 048	230.88	71.31
34	Co1-0	88 303	232.80	78.41
35	Ni1-0	89 810	233.01	71.60
36	Zn1-0	104 625	227.79	71.27

when the precursor state of chemical adsorption is stable, and then it will increase when the distance between Ca^{2+} and NH_3 is shortened because of the repulsive force between NH_3 molecules. The reaction between Ca^{2+} and NH_3 will transit into a chemical reaction when the potential curve of physical adsorption intersects with the chemical reaction curve (point X); at that time, the required activated energy for chemical adsorption is E_a.

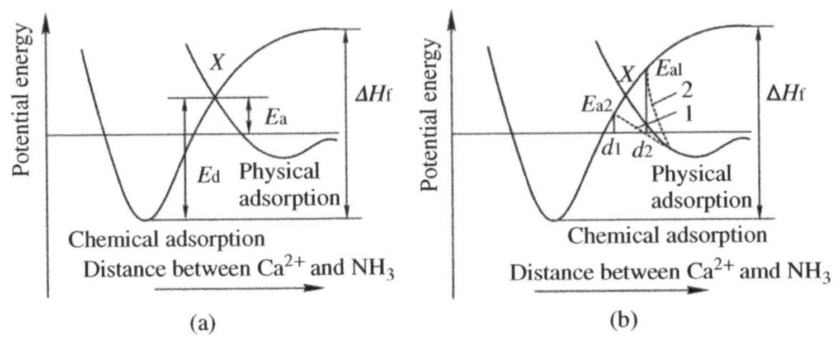

Figure 4.5 The chemical adsorption principle. (a) Potential curve of chemical adsorption and (b) precursor states of chemical adsorption under different conditions

Figure 4.5a shows that the chemical adsorption precursor state is a key factor for the adsorption refrigeration process. The shield factor, which is the ratio between repulsive force and attractive force, increases when the coordination number of NH_3 around Ca^{2+} reaches a certain number in the formation process of ammoniates. Thus it is difficult for NH_3 to approach the effective reaction range around Ca^{2+} under the influence of shield factor if the distance between Ca^{2+} and NH_3 is great. For example, if the distance between Ca^{2+} and NH_3 is d_2 in Figure 4.5b, the chemical adsorption precursor state will be curve 2, and then the activated energy will be more than the activated energy of the precursor state in Figure 4.5a. The transition from precursor state to chemical adsorption will be difficult under such a condition. If the distance between Ca^{2+} and NH_3 is shorter, for example, if the chemical adsorption precursor state is curve 1 in Figure 4.5b, then the condition will be different. This phenomenon can be shown by the adsorption performance and activated energy of $CaCl_2$ for different expansion space of $CaCl_2$ in adsorption/desorption process.

4.3.1 Chemical Adsorbent with Different Expansion Space

Chemical adsorbents generally have a large cycle adsorption concentration. Adsorption performance utilizes the testing method of level measurement in Figure 3.9. The volume of the adsorption bed is generally large. Under such conditions, heat and mass transfer of the adsorbent bed should be considered. One typical design of adsorption bed is shown in Figure 4.6 and is used in the study on metal chloride adsorbent for different expansion space.

The adsorber (Figure 4.6) consists of two half-part columns that are fully welded between the fin and adsorption bed to ensure the good heat transfer performance. Two half part columns are connected with flange seals. Flange utilizes groove and convex to ensure sealing well. Adsorption bed and experimental device are connected by the flange. A folding is designed at the edge of the fin to prevent the adsorbent falling down from the fin. Mass transfer is realized by a channel between fins. For $CaCl_2$, since the thickness of chemical adsorbent on the fin is different, expansion space as well as the gap between Ca^{2+} and Ca^{2+} is different, which will influence the chemical adsorption precursor state.

Corresponding to the design of Figure 4.6, when volume ratio between expansion space and adsorbent is 5 : 1, the state of ammoniate $CaCl_2$ in the adsorption bed after adsorption is shown in Figure 4.7. Because there are still at least 2 mol ammonia coordinates in 1 mol adsorbent after desorption, the adsorbent is swelling and full between fins even after desorption.

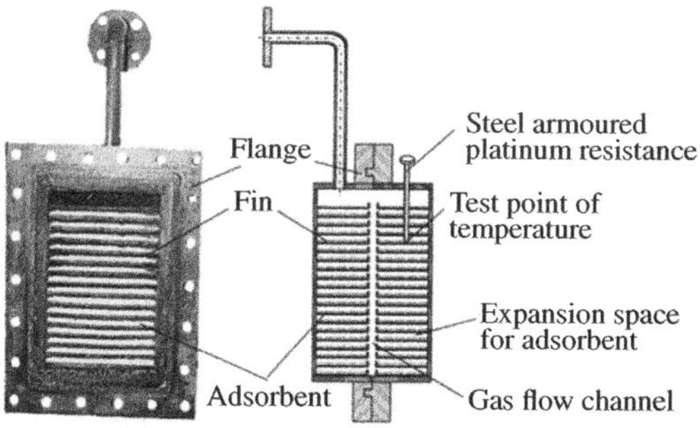

Figure 4.6 Bulk adsorbent bed [9]

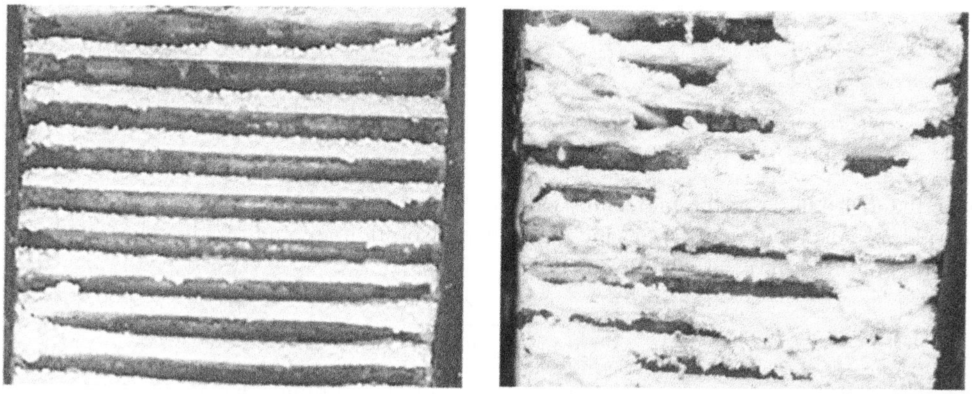

(a) (b)

Figure 4.7 States of ammoniate $CaCl_2$ after adsorption and after desorption [21]. (a) After desorption and (b) after adsorption

The ammonia coordinated in the adsorbent increases in the adsorption process, the adsorbent swells seriously and enters into the mass transfer channel of adsorber after adsorption. The distribution of adsorbent can be assumed as full between fins in the process of adsorption and desorption because the mass transfer space is very small, which is about 5.8% of effective space for adsorbent (space between fins).

The ratio between expansion space and volume of adsorbent is defined as r_{as}. The chosen r_{as} are 5 : 1 (sample 1), 3 : 1 (sample 2), 2 : 1 (sample 3), and 1.4 : 1 (sample 4). For $CaCl_2$, assuming that the adsorbent evenly distributes in the adsorber, and assuming that the distribution of Ca^{2+} of sample 1 in the adsorption process is shown in Figure 4.8a, the distances between Ca^{2+} for sample 2, sample 3, and sample 4 are respectively 1.5, 2, and 2.5 times of that of sample 1 according to different r_{as}, distribution of Ca^{2+} for sample 2, sample 3, and sample 4 are shown in Figure 4.8b–d. Figure 4.8 shows that distribution of Ca^{2+} is different for the adsorbents with different values of r_{as}. The distribution of Ca^{2+} is loose for sample 1, under this condition the shield factor of NH_3 is large because the concentration of NH_3 is large, thus it will be difficult for NH_3 to enter the effective reaction range of Ca^{2+}

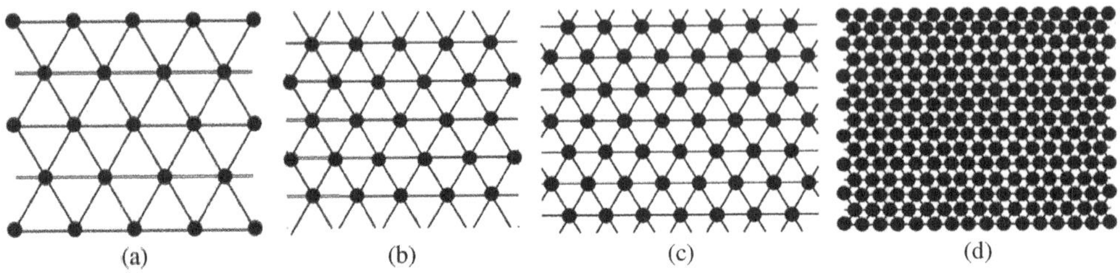

Figure 4.8 Distribution of Ca^{2+} for adsorbent in the process of adsorption. (a) Sample 1 (r_{as} is 5 : 1); (b) sample 2 (r_{as} is 3 : 1); (c) sample 3 (r_{as} is 2 : 1); and (d) sample 4 (r_{as} is 1.4 : 1)

and the required activated energy will be large. The distance between Ca^{2+} is too short for sample 4, and the permeation of NH_3 will be influenced under the condition of low evaporation pressure.

4.3.2 Attenuation Performance of the Adsorbent and Its Chemical Adsorption Precursor State

Since the distribution of the molecules will be different when the expansion space of metal chloride the adsorbent is different, the molecular adsorption amount of adsorbents under the condition of different r_{as} is as follows:

$$N_g = \frac{x \times m_C}{m_N} \tag{4.15}$$

where m_c is the molecular mass of metal chloride, m_N is the molecular mass of NH_3, which is 17.

When r_{as} is different, the swelling and agglomeration phenomena of the adsorbent after adsorption and desorption will be different. When r_{as} is large due to large expansion space the swelling phenomenon is serious. When r_{as} is small serious expansion will lead to agglomeration of the adsorbent. Pictures of $CaCl_2$ with different r_{as} after the adsorption and desorption are shown in Figure 4.9.

Figure 4.9 Adsorbents after adsorption and desorption [21]. (a) r_{as} is 5 : 1; (b) r_{as} is 2 : 1; and (c) r_{as} is 1.4 : 1

The reaction formulas for the complex reaction between $CaCl_2-NH_3$ are as follows:

$$CaCl_2 \cdot 8NH_3 + \Delta H_1 \leftrightarrow CaCl_2 \cdot 4NH_3 + 4NH_3 \quad \text{at the temperature of } T_{e1} \tag{4.16}$$

$$CaCl_2 \cdot 4NH_3 + \Delta H_2 \leftrightarrow CaCl_2 \cdot 2NH_3 + 2NH_3 \quad \text{at the temperature of } T_{e2} \tag{4.17}$$

$$CaCl_2 \cdot 2NH_3 + \Delta H_3 \leftrightarrow CaCl_2 + 2NH_3 \quad \text{at the temperature of } T_{e3} \tag{4.18}$$

where ΔH_1, ΔH_2, and ΔH_3 are enthalpies of transformation for reactions (J/mol), and T_{e1}, T_{e2}, and T_{e3} are equivalent temperatures for reactions.

For 2 mol ammonia complex, the complex formed is linear mode through sp orbital. For $CaCl_2 \cdot 4NH_3$, the complex formed is regular tetrahedron mode through sp^3 hybrid orbital. Compared with $CaCl_2 \cdot 4NH_3$, for $CaCl_2 \cdot 6NH_3$ and $CaCl_2 \cdot 8NH_3$, the ammonia is occupied in d orbital. The complex is formed as a regular octahedron, dodecahedron structure. When the adsorbents are in the complexion process, the complex structure constantly adjusts from linear mode to dodecahedron mode [2]. For a large expansion space, the adjustment of adsorbent pore must have an influence on the concentration change of ammonia around Ca^{2+}, thus will strengthen or weaken the repulsive force of anion and instability of adsorption performance. But when the expansion space of the adsorbent is limited, the adsorbent tends to connect with each other in adsorption and desorption process. This adjustment has little influence on the structure of the adsorbent, and the adsorption performance will be stable after the second adsorption. This phenomenon can be gained from the performance attenuation curves of calcium chloride–ammonia working pair.

Figure 4.10 shows that the performance of sample 1 is similar to that of sample 2. They all have performance attenuation, and they all have similar adsorption quantities after the performance attenuation. The performance attenuation does not exist for the curves for sample 3 and sample 4, and adsorption quantities become stable after the second time for adsorption. The stable cycle adsorption quantity of sample 3 is about 0.71 kg/kg. The cycle adsorption quantity of sample 1 degenerates from 0.75 to 0.57 kg/kg; the largest attenuation value is 31.6% compared with the adsorption quantity after attenuation. For sample 4, a small expansion space leads to the serious agglomeration phenomena and limitation of forming $CaCl_2 \cdot 8NH_3$. Therefore its adsorption performance is lower than that of sample 3.

In the adsorption process, the adsorption rate is faster than desorption rate, and net adsorption rate K_v is

$$K_V = \frac{d\theta}{dt} = K_a(1-\theta)p - K_d\theta \tag{4.19}$$

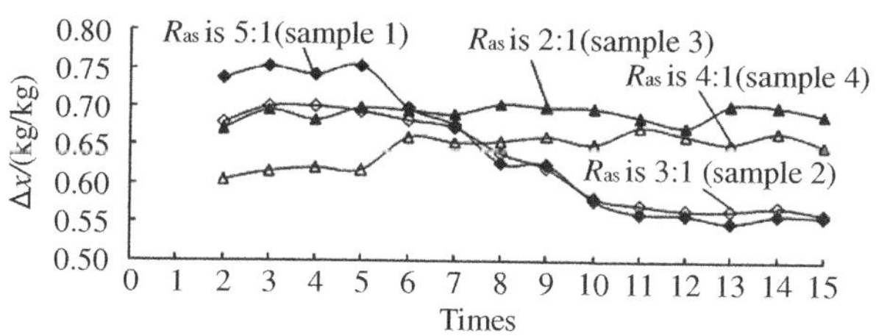

Figure 4.10 Adsorption performance attenuation curves [21]

where θ is the surface coverage, which is the ratio between adsorption quantity x and the largest adsorption quantity x_{max} that is 1.225 kg/kg, which is corresponding to mole adsorption quantity of 8 mol/mol. According to the formula of Arrhenius activated energy [3, 15] the constant of reaction rate K_a in adsorption and the constant of reaction rate K_d in desorption are respectively:

$$K_a = A_{fe} \exp \frac{-E_a}{RT} \tag{4.20}$$

$$K_d = A_{fe} \exp \frac{-E_d}{RT} \tag{4.21}$$

where A_{fe} is an anterior factor and R is a universal gas constant. E_a is adsorption activated energy, E_d is desorption activated energy, which is the sum of adsorption activated energy E_a and adsorption heat ΔH_r [14]. T is adsorption temperature. Use Equations 4.20 and 4.21 to substitute K_a and K_d in Equation 4.19, and make logarithmic transformation, then

$$\ln K_V = \ln A_{fe} - \frac{E_a}{RT} + \ln\left[(1-\theta)p - \theta \exp\left(\frac{\Delta H_r}{RT}\right)\right] \tag{4.22}$$

where anterior factor A_{fe} is generally a constant and ΔH_r is adsorption heat. E_a is generally inversely proportional to the net adsorption rate.

Two typical curves, on average $\ln K_v$ for the attenuation curves of sample 1 and sample 3, are analyzed under the condition of that anterior factor cannot be determined. Results are shown in Figure 4.11.

$\ln K_v$ of sample 3 does not change very much in Figure 4.11 in repeated experiments; that is, activate energy does not change very much according to the relation between E_a and $\ln K_v$. The reason is analyzed and the result is the distance between Ca^{2+} and NH_3. The distance between Ca^{2+} and NH_3 is limited and doesn't change very much under the condition of agglomeration in adsorption, thus the required activated energy will be stable according to the chemical adsorption principle in Figure 4.5. $\ln K_v$ of sample 1 in Figure 4.11 decreases in the experiments of anterior 10 times, and it becomes stable from the 11th experiment; that is, the activated energy increases in the anterior 10 times experiments and becomes stable from the 11th experiment. This result is also coincident with the chemical adsorption principle (Figure 4.5). The distance between Ca^{2+} and NH_3 continually increases in the experiments of anterior 10 times for sample 1 because r_{as} is large and there is enough expansion space, thus the activated energy that is required for the transition from precursor state to chemical adsorption increases. For the 11th experiment, sample 1 is full between fins because of swelling, and then the distance between Ca^{2+} and NH_3 does not change very much because the further swelling of adsorbent is limited by the space between fins, thus activated energy also becomes stable after the 11th experiment.

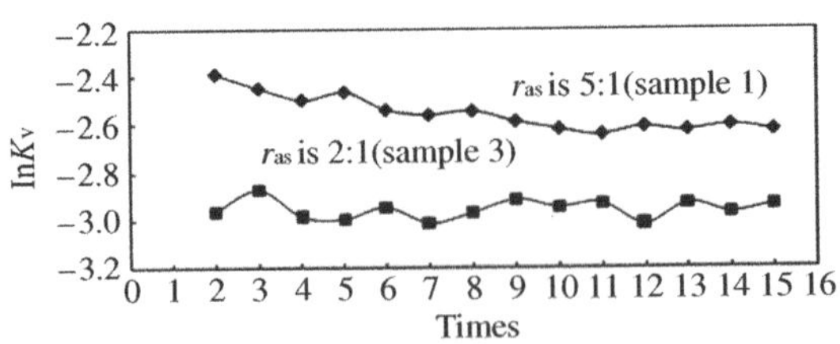

Figure 4.11 The $\ln K_v$ for different experiments [21]

4.3.3 Isobaric Adsorption Performance and Activated Energy

Activated energy is a constant when the temperature is a constant, and the anterior factor is also a constant [12, 13]. Differentiate ln K_v to T in Equation 4.22, and the result of activated energy is shown in Equation 4.23.

$$E_a = \left[\frac{1}{K_v}\frac{dK_v}{dT} - \frac{\Delta H_r}{RT^2}\exp\left(\frac{\Delta H_r}{RT}\right)\frac{1}{(1-\theta)p - \theta\exp\left(\frac{\Delta H_r}{RT}\right)} \right] T^2 R \qquad (4.23)$$

In the process of adsorption refrigeration, activated energy that this complex requires is negative, and the greater the absolute value is, the better the cooling effect is. In order to make the results of the activated energy more intuitive, it is generally possible to use the absolute value of activated energy. For the metal chloride-ammonia working pairs, when evaporation pressure is 430 kPa, the activated energy of adsorbent with different r_{as} is calculated and shown in Figure 4.12a. In the adsorption process, the activated energy required decreases when the temperature increases, which is mainly related to the stability and instability constant for the adsorption/desorption process. The adsorption reaction is exothermic reaction and stability constant will increase when the temperature decreases [10]. Activated energy required for such an adsorption process decreases with decreasing temperature. In Figure 4.12a, the average activated energy required for sample 3 and sample 4 is lower than that of sample 1 and sample 2, which is in accord with Figure 4.5. Ca^{2+} distribution of sample 3 and sample 4 are concentrated, and therefore even considering the shielding factors, the longest distance between NH_3 gas molecule and Ca^{2+} is within the normal range for chemical adsorption, for which activated energy should be less than that of sample 1 and sample 2. In Figure 4.12a, activated energy required for sample 4 is different from the chemical adsorption theory. For chemical adsorption theory, when r_{as} decreases, the distance between NH_3 molecule and Ca^{2+} is influenced by the concentrated distribution of Ca^{2+}, which should be within the reaction

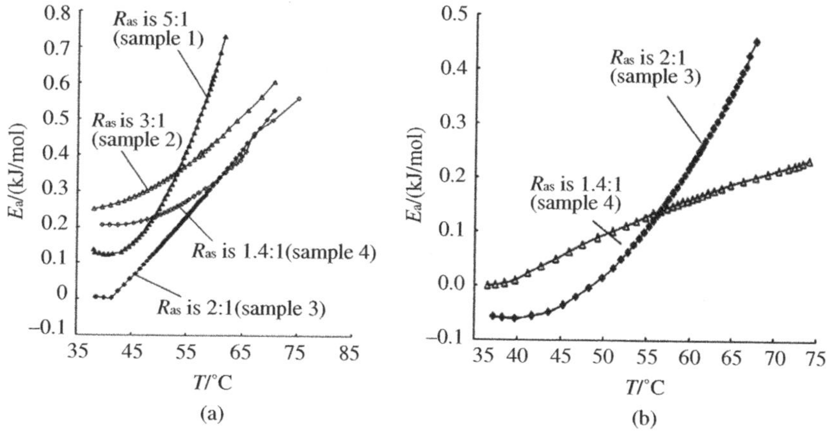

Figure 4.12 Activated energy of different samples [21]. (a) Evaporation pressure is 430 kPa and (b) evaporation pressure is 595 kPa

range of Ca^{2+} and the activated energy is small. But in Figure 4.12a, the activated energy of sample 4 is higher than that of sample 3. This is because mass transfer problems happen in the process of adsorption, which leads to difficult permeability inside the adsorbent, lowering the adsorption rate. Increasing evaporation pressure of refrigerant to 595 kPa, thereby pressure difference between the refrigerant and adsorbent is increased which is shown in Figure 4.12b. At this time, activated energy of sample 4 and sample 3 are similar, which shows that chemical adsorption is influenced both by the chemical adsorption precursor state and mass transfer performance. If the distance between the complex molecules isn't small enough to affect mass transfer, the chemical adsorption precursor state determines the chemical adsorption. If the distance between the complex molecules is too small to mass transfer as well as the saturate pressure of refrigerant not being high enough to solve the problem of mass transfer due to the small molecule distance, both chemical adsorption precursor state and mass transfer performance are important for chemical adsorption process. According to the principle of minimum activated energy, as shown in Figure 4.12, when the evaporation pressure is 430 kPa, the best molecular distance is for the sample with r_{as} of 2 : 1, and when evaporation pressure is 595 kPa, the best molecular distance is for the sample with r_{as} of 1.4 : 1.

4.4 Reaction Kinetic Models of Metal Chlorides–Ammonia

Stitou and Crozat classified the chemical adsorption kinetic models into three categories, that is, local, global, and analytical models [6, 22]. The local model derives the partial derivative equations concerning the heat and mass transfer, as well as dynamics of the adsorbents with small volumes. Local equations can be numerically solved. The global model takes into account the parameters such as permeability, thermal conductivity, thermal capacity, and the average parameters and variables of the reactor. The result of the numerical solution of these equations is a series of partial differential equations. The analytical model takes the average values of variables into account, which are only related to the average values of variables during the reaction time [6].

The partial and global model can be solved by using the model for the grain, for which the reaction interface was changed in the adsorption process (a chemical reaction and heat and mass transfer are coupled). This helps to define the parameters, such as the hydraulic radius of the grain, the porosity of the grains, the adsorption/desorption kinetic coefficients, and so on.

The basic model of chemical adsorption rate is [23]:

$$v = \frac{dx}{dt} = k(p, T)f(x) \tag{4.24}$$

where x is the adsorption quantity, $k(p,T)$ is the specific adsorption rate that is related to the difference between the temperature and pressure with the equilibrium parameters. $f(x)$ is the change of the reaction, concerning the changes related with the reaction process and the physical structure of the reactants. For the models based on the phenomena, $f(x)$ represents the changes of the reaction interface. For the models based on the essence and global reaction, the models need to describe the dynamic model through the comparison of the dynamic characteristics between similar substances. At present, the models based on the phenomena are widely used.

4.4.1 The Model Based on Phenomena and Proposed by Tykodi

The model proposed by Tykodi [24] is:

$$f(x) = (1 - x)^n \tag{4.25}$$

$$k(p, T) = C_0 \exp\left(\frac{-E_p}{T}\right) \ln\left(\frac{p}{p_{eq}}\right) \tag{4.26}$$

where E_p is pseudo-activated energy (J/mol), p is reaction pressure, p_{eq} is reaction equilibrium pressure, and n is reaction order.

Flanagan, Rudman, and Goodell researched on this model, and they thought that different factors will influence the reaction process, as well as the specific adsorption rate [23].

1. When the mass transfer rate of the vapor layer on the metal surface is the main limiting factor of adsorption properties,

$$k(p, T) = C_1 \exp\left(\frac{-E_{p1}}{T}\right)(p - p') \tag{4.27}$$

   where p' is the pressure on the metal chloride's surface.
2. When the limiting factor is the process of chemical reaction,

$$k(p, T) = C_2 \exp\left(\frac{-E_{p2}}{T}\right) \ln\left(\frac{p_h}{p_{eq}}\right) \tag{4.28}$$

   where p_h is the pressure of reaction interface, p_{eq} is equilibrium pressure at the reaction temperature of T.
3. When the limiting factor is the mass transfer performance,

$$k(p, T) = C_3 \exp\left(\frac{-E_{p3}}{T}\right) \ln\left(\frac{p'}{p_e}\right) \tag{4.29}$$

4.4.2 The Global Reaction Model Proposed by Mazet

Mazet used m_x instead of the value of n in Equation 4.25 to express the order of reaction and reaction quantity x. The equation is:

$$f(x) - (1 - x)^{m_x} \tag{4.30}$$

where x is the product of the first step reaction.

y is defined as the product of the second step reaction. The equation is:

$$f(y) = [(1 - y)x]^{m_y} \tag{4.31}$$

According to Arhenius' law, $k\ (p,T)$ can be expressed as follows:

$$k = s\exp(-E_p/RT) \tag{4.32}$$

where s is a constant. According to Arhenius' law, taking into account the differences between the reaction pressure and equilibrium pressure, as well as the difference between the reaction temperature and equilibrium temperature, $k(p,T)$ can be expressed as:

$$k(p,T) = s\exp\left(\frac{-E_p}{RT}\right)f'(p,T) \tag{4.33}$$

According to the simulation results of adsorption quantity by the global model, Mazet defined:

$$f'(p,T) = \frac{p_c - p_{eq}(T)}{p_c} \tag{4.34}$$

where p_{eq}(T) is calculated according to the Clapeyron equation. As standard conditions p_0 and T_0 are known, the formula of the equilibrium pressure is:

$$p_{eqx,y}(T) = p_0\exp\left\{-\frac{\Delta H_{x,y}}{R}\left(\frac{1}{T} - \frac{1}{T_{0x,y}}\right)\right\} \tag{4.35}$$

where $p_0 = 1$ bar, $T_{0x} = 299$ K, $T_{0y} = 305.4$ K.

In order to distinguish different reaction processes of two steps, the constant is replaced by Ar_x and Ar_y. For a cylindrical adsorbent bed, Mazet proposed the global kinetic model is:

$$\frac{dx(t,r)}{dt} = [1 - x(t,r)]^{m_x} \cdot A_{rx} \cdot \frac{p_c - p_{eqx}[T(t,r)]}{p_c} \tag{4.36}$$

$$\frac{dy(t,r)}{dt} = \{[1 - y(t,r)]x(t,r)\}^{m_y} \cdot A_{ry} \cdot \frac{p_c - p_{eqy}[T(t,r)]}{p_c} \tag{4.37}$$

where r is the radius of adsorbent, t is time.

4.4.3 The Model Based on the Phenomena and Proposed by Goetz

Taking into account a model based on the phenomena, the grain theory could determine the properties of the reactants. Goetz and Marty [19] had studied the dynamics of the mixed adsorbent of $MnCl_2$ and inert matrix which were based on the grain theory.

The research included two steps. The first step is to determine the reaction rate of the metal chloride, which is based on constraints of the temperature and pressure. In this research thermal gravimetric analysis and micro-calorimetric method were used by measuring several milligrams of adsorbent for eliminating the influence of heat and mass transfer of the adsorbent grains. The second step couples the reaction process with the heat and mass transfer models to determine the global reaction rate of the adsorption bed.

Goetz compared the curves for the permeability properties and chemical kinetics (Figure 4.13), and the results showed that the experimental curve is between the curve of permeability and the curve of chemical kinetics.

Taking into account the vapor diffusion process and chemical kinetics, the model proposed by Goetz is shown in Table 4.3.

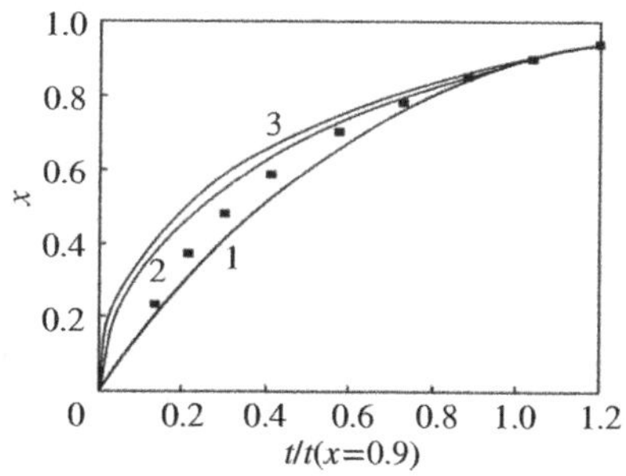

Figure 4.13 Characteristic curves of the prevailing regime and example of an experimental curve. (■) Experimental curve in adsorption, with $p_c = 7.5$ bar and $T_c = 363$ K; (1) chemical regime; (2) diffusion regime in product layer, with grain of constant size; and (3) diffusional regime in product layer, with grain of changing size with $Z_c = \frac{V_{m((Mn(NH_3)6Cl_2)}}{V_{m((Mn(NH_3)2Cl_2)}} = 2.1$ [19]

Table 4.3 The formula of the reaction crystal [19]

	Adsorption process		Desorption process	
Mass transfer	$\frac{dN_g}{dt} = \pm \frac{r_g r_c}{r_g - r_c} \frac{4\pi}{RT_c} K_s (p_c - p_i)$			(4.38)
Chemical kinetics	$\frac{dN_g}{dt} = 4\pi r_c^2 K_a \left(\frac{p_i - p_{ea}(T_c)}{p_{ea}(T_c)} \right)^{Ma}$	(4.39)	$\frac{dN_g}{dt} = 4\pi r_c^2 K_d \left(\frac{p_{ed}(T_c) - p_i}{p_{ed}(T_c)} \right)^{Md}$	(4.40)
Variation in grain size	$r_g^3 = r_c^3 + (r_g^3{}_{[Mn(NH_3)_2Cl_2]} - r_c^3) \times \frac{V_{m[Mn(NH_3)_6Cl_2]}}{V_{m[Mn(NH_3)_2Cl_2]}}$	(4.41)	$r_g^3 = r_c^3 + (r_g^3{}_{[Mn(NH_3)_2Cl_2]} - r_c^3) \times \frac{V_{m[Mn(NH_3)_6Cl_2]}}{V_{m[Mn(NH_3)_2Cl_2]}}$	(4.42)
Advancement	$x = 1 - \left\{ \frac{r_c}{r_{g[Mn(NH_3)_2Cl_2]}} \right\}^3$	(4.43)	$x = 1 - \left\{ \frac{r_c}{r_{g[Mn(NH_3)_6Cl_2]}} \right\}^3$	(4.44)

In Table 4.3, N_g is the molar adsorption quantity (mol/mol), r_g is the grain radius (m), r_c is the radius of the reaction surface (m), T_c is the constraint temperature (K), and K_s is the permeability (m^2/s), mainly related to the flow rate. In the case of viscous fluid, K_s is calculated according to Darcy's law; p_c is constraint pressure (Pa), p_i is the pressure of the vapor reactant interface (Pa), K_a and K_d are dynamic coefficients of adsorption and desorption (1/(m^2s)), p_{ea} and p_{eq} are the equilibrium pressures of adsorption and desorption (Pa), Ma and Md are power coefficients of adsorption and desorption, V_m is the molar volume (m^3/mol), and x is the adsorption rate.

For the complex of $MnCl_2$ and NH_3, coefficients obtained by Goetz are as follows: $r_{g[Mn(NH_3)_6Cl_2]} = 4.9 \times 10^{-4}$m, $K_a = 0.17$ (vapor's mol/(m^2s)), $Ma = 0.77$, $K_d = 0.38$(vapor's mol/(m^2s)), $Md = 2.2$, $V_{m[Mn(NH_3)_6]Cl_2} = 159 \times 10^{-6}\ m^3/mol$, and $V_{m[Mn(NH_3)_2]Cl_2} = 79 \times 10^{-6}$ m^3/mol.

Based on grain theory, if mass transfer is the major limiting factor for adsorption, simulation was established based on the Equation 4.38. If mass transfer isn't a major limiting factor, simulation was established based on the chemical adsorption kinetics model (Equation 4.39) and the heat transfer model (Equation 4.40).

Lu *et al.* conducted a further study on the model of grain theory proposed by Goetz. The mass transfer equation is similar to Equation 4.38, but the chemical kinetics equation is different from the model created by Goetz. The equation is [25]:

$$\frac{dN_g}{dt} = 4\pi r_c^{\ 2} K_r \exp\left(-\frac{M_r}{RT_c}\right)\left(\frac{p_i - p_{eq}(T_c)}{p_{eq}(T_c)}\right) \tag{4.45}$$

where K_r and M_r are reaction kinetic constants.

Lu *et al.* proposed that the adsorption is influenced by both heat transfer and reaction kinetics when the reactants are at the level of grain, and the change of the grain is shown in Equations 4.41 and 4.42. If the reaction is at the level of pellet, for example, calculating the chemisorption of the graphite matrix in a reactor, it needs to combine both heat and mass transfer equations. Lu *et al.* simplified the continuous reaction interface to the sharp reaction interface. In this simplified process, removing the impact of the grain level on the chemical reaction process, it can create a model, in which the chemical reaction process has no effect on it, only considering the factors of heat and mass transfer [25].

Shanghai Jiao Tong University studied the model established by Goetz [26] by considering the impact of the spacing between the molecules of chemisorption. The experimental data of the adsorption and desorption process is shown in Figures 4.14 and 4.15. Both figures show that the space between the molecules has a great impact on adsorption but has little effect on desorption. Even at the grain level, if the space between molecules is too small, it will be a problem for mass transfer in the process of adsorption [26].

Taking into account the impact of the distance among molecules on the chemisorption process, for Equation 4.39, the model is rectified by adding an item related to the distance between

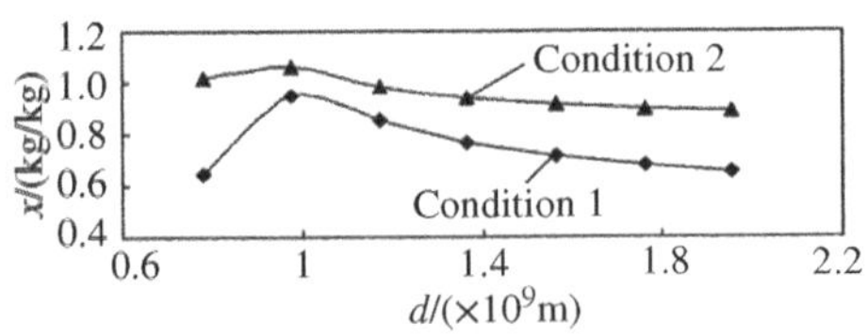

Figure 4.14 Chemisorption process under the conditions with different molecular distance [26]

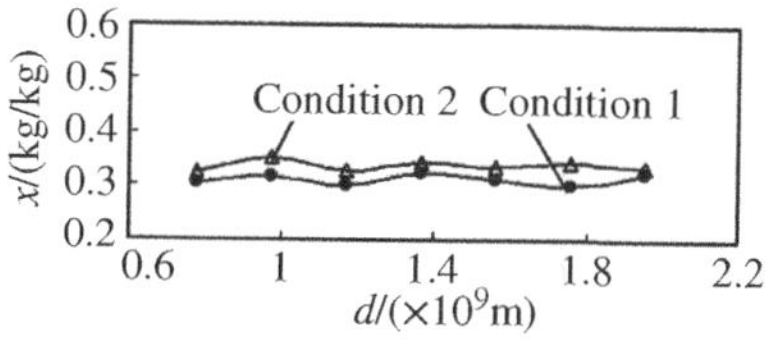

Figure 4.15 Desorption process under the conditions with different molecular distance [26]

the molecules. The revised model is:

$$\frac{dN_g}{dT} \times \frac{dT}{dt} = \frac{\ln[K_{ms} \times (d - r_g)]}{K_{md} \times d^3} 4\pi r_c{}^2 K_a \left(\frac{p_c - p_{eq}(T)}{p_{eq}(T)} \right)^{Ma} \tag{4.46}$$

where K_{ms} is the coefficient of the mass transfer, K_{md} is the coefficient for the influence of chemical kinetics on the reaction. Equation 4.46 applies only for the adsorption process, and for the desorption process the experimental results showed there is little influence caused by the molecular distance.

For the chemisorption kinetic model the most difficult part is the selection of coefficients. Equation 4.45 is fitted by the experimental results of calcium chloride-ammonia adsorption performance. According to activation energy results of Equations 4.20, 4.22, and 4.23, $K_a = 0.334$, $K_{ms} = 3.96 \times 10^9$, $K_{md} = 7.6 \times 10^{11}$, $Ma = 0.512$, $V_{m[Ca(NH_3)_6]Cl_2} = 209.6 \times 10^{-6}$ m^3/mol, and $V_{m[Ca(NH_3)_2]Cl_2} = 130.98 \times 10^{-6}$ m^3/mol. Divide Equation 4.21 by Equation 4.20, the result is:

$$K_d = \frac{\exp\left(\frac{\Delta H_r}{RT}\right)}{K_a} \tag{4.47}$$

4.4.4 *Other Simplified Chemisorption Models*

In addition to the classical chemisorption kinetic models mentioned above, there are some other simplified kinetic models. One of the typical models for the chemisorption process based on the Langmuir's monolayer adsorption theory which is proposed by Spinner is [27]:

$$x = K_1(1 - x)^{n_1} \frac{p_c - p_{eqa}}{p_c} \quad \text{Adsorption process} \tag{4.48}$$

$$x = K_2 x^{n_2} \frac{p_{eqd} - p_c}{p_c} \quad \text{Desorption process} \tag{4.49}$$

where x is the adsorption rate which is the ratio of adsorption quantity and the maximum adsorption capacity.

Another simplified model applied in the simulation by Iloeje *et al.* is [28]:

$$\frac{dx}{dt} = K_x(T_a - T_{eq})(x_{max} - x) \tag{4.50}$$

where T_a is the temperature of adsorbent (metal chloride salts), T_{eq} is the equilibrium reaction temperature, and K_x is the reaction coefficient (°C/s).

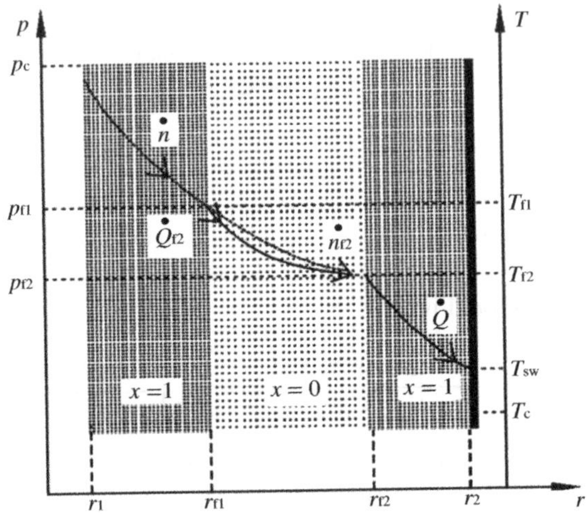

Figure 4.16 Temperature and pressure gradient changes of adsorption synthesis process [29]

Equation 4.50 is derived by the relation between the reaction rate and the reaction time, which is relatively simple, but K_x is very hard to determine. For calcium chloride–ammonia, the equation for selecting the coefficient is:

$$K_x = \frac{(0.4p_c + 1.7)}{6 \times 10^5} \tag{4.51}$$

Mauran optimized the application of the composite adsorbent, which was developed by metal chlorides and graphite, in the adsorption heat pump [29]. Mauran studied the adsorbers with cylindrical structure, surrounded by a heat transfer channel, and centered a channel for the mass transfer. The mass transfer process is from the center channel to the heat transfer channel, and the heat transfer process is from the heat transfer channel to the mass transfer channel. In half of the radial cross section, the change of temperature and pressure gradient in the adsorption process is shown in Figure 4.16 [29]. The total molar $\dot{n}$ flows through the first layer ($x = 1$) to the mass transfer surface ($r = r_{f1}$), and only part of the molar $\dot{n}_{f2}$ flows to another heat transfer surface ($r = r_{f2}$). The heat transfer process is the opposite. The total heat $\dot{Q}$ flows through the first layer ($x = 1$) to the heat transfer surface ($r = r_{f2}$), and only part of the heat $\dot{Q}_{f1}$ passes. The restrictions of equilibrium conditions (temperature T_f and pressure p_f) are Clapeyron equations as follows:

$$\ln(p_{f1}) = -\frac{\Delta H}{RT_{f1}} + \frac{\Delta S}{R} \quad \text{(mass transfer surface)} \tag{4.52}$$

$$\ln(p_{f2}) = -\frac{\Delta H}{RT_{f2}} + \frac{\Delta S}{R} \quad \text{(heat transfer surface)} \tag{4.53}$$

The overall reaction rate of x is related to r_{f1}, r_{f2}, r_1, and r_2. Mauran thought that for the reaction of $\langle S_1 \rangle + nG(G) \Leftrightarrow \langle S_2 \rangle$, metal chloride salts S_1 exists only for $x = 0$, which is between r_{f1} and r_{f2}, but the salt S_2 completing the reaction isn't related to r_{f1} and r_{f2}, only part of salt S_1 may react. The reaction rate is:

$$x = 1 - \frac{(r_{f2}^2 - r_{f1}^2)}{(r_2^2 - r_1^2)} \tag{4.54}$$

Simulation process was established by Equation 4.54, the initial conditions are $r_{f1} = r_1$ and $r_{f2} = r_2$.

4.5 Refrigeration Principle and Van't Hoff Diagram for Metal Hydrides–Hydrogen

In a similar way to physical adsorption and metal chlorides, the adsorption process of metals and alloys adsorb or desorb hydrogen depends mainly on temperature and pressure. The adsorption and hydrogenation process are exothermic, and the desorption process is endothermic. Especially for advanced porous metal hydride (PMH), or misch metal (Mm) matrix alloys, including Ni, Fe, La, Al, and H, they have high adsorption heat and adsorption rate when the hydrogen is as the adsorbate.

Using metal hydrides–hydrogen for refrigeration, its disadvantage is the smaller cooling capacity per unit mass. The advantage is the adsorbent bed works as the condenser or evaporator alternatively, and the density of the metal hydride is big ($\rho = 6.5-8$ kg/l), so its volume cooling capacity is big, which can be used for the places without the requirement on the weight of the system, but has strict requirements of the space.

4.5.1 Adsorption Refrigeration Characteristics and Van't Hoff Diagram

Adsorption refrigeration cycles of the metal hydrides–hydrogen working pair include basic single-stage cycle, two-stage cycle, and multi-metal hydride thermal wave cycle [30]. The cycles' common characteristics are as follows:

1. The heat released into the environment by the refrigeration cycle is related to the adsorption process of low-temperature metal hydride (refrigerant). Compared with the solid adsorption process of the working fluid with the evaporating and condensing processes, the evaporation process is the desorption process of the low-temperature adsorption bed, and the condensation process is the adsorption process of the low-temperature adsorbent. This is an important reason why the system is heavy for metal hydride. Compared with the ordinary adsorption refrigeration system with condenser and evaporator, the weight of the metal hydrides–hydrogen system is doubled. High-temperature metal hydride is the regeneration component, that is, the desorption process of the high temperature salt is the regeneration process of the system.
2. Adsorption hysteresis existed for most of metal hydrides–hydrogen working pairs and there is also a pressure bevel (in Figure 4.17). The horizontal length of the pressure inclined plane is defined as $\Delta x/x$, which is the amount of adsorption or desorption of hydrogen between two reactors. Due to this characteristic of metal hydrides–hydrogen, it needs higher heating and cooling temperature than refrigerant working pairs which don't have adsorption hysteresis. So, it generally requires special heat treatment in the adsorption or

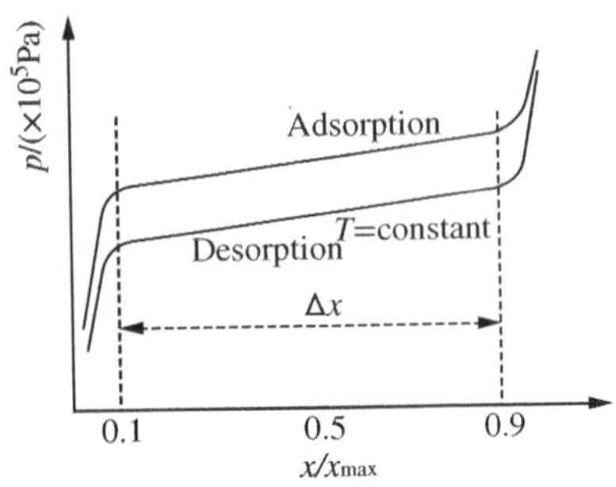

Figure 4.17 Adsorption hysteresis and pressure for metal hydride–hydrogen working pair [30]

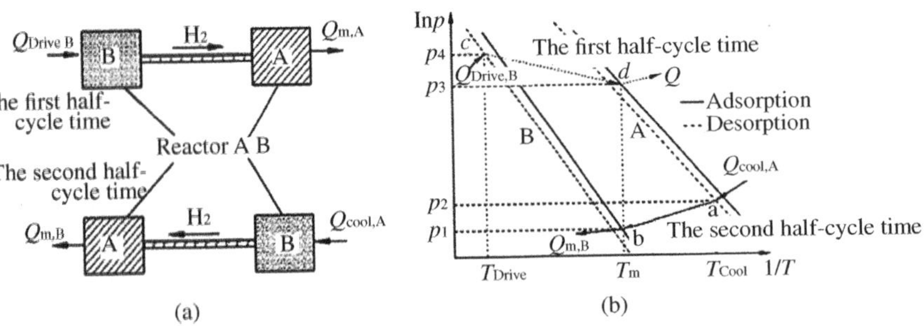

Figure 4.18 Adsorption schematic and Van't Hoff diagram of basic single-stage cycle [30]. (a) Schematic of the cycle and (b) Van't Hoff diagram

desorption process. This is also a major factor of lower thermal power efficiency for metal hydrides–hydrogen.

3. The refrigerating performances of these cycles are related to the characteristics of the metal hydride. This can be obtained through refrigeration schematic of basic cycle (Figure 4.18a). At the cooling phase, the *COP* is calculated as:

$$COP_C = \frac{Q_{Cool}}{Q_{Drive}} \quad \text{Cooling} \tag{4.55}$$

$$COP_H = \frac{Q_{Heat}}{Q_{Drive}} = \frac{Q_{m,A} + Q_{m,B}}{Q_{Drive}} \quad \text{Heat Pump} \tag{4.56}$$

The basic single-stage cycle generally consists of two adsorption beds, and its working principle and van't Hoff diagram are shown in Figure 4.18. Working pressure is calculated as [31]:

$$R \ln P = \frac{\Delta H}{T} - \Delta S \tag{4.57}$$

The metal hydride B desorbs under the condition of heat of Q_{Drive} and temperature of T_{Drive} during the first half-cycle. Under the driven conditions of differential pressure, the hydrogen desorbs into the generator which filled with metal hydride A. The heat may be released into the environment or the users for effective use based on different T_m. The second half-cycle is under the low-pressure conditions. Due to desorption of the metal hydride A, the generator produced cold Q_{Cool} under the conditions of temperature T_{Cool}, and the hydrogen is adsorbed by adsorption bed B in which the heat of adsorption can serve as a useful heat or waste heat released into the environment. As a heat pump, each half-cycle released heat, but the cooling capacity is produced just in a half-cycle of the entire cycle. To get a continuous cooling capacity, it needs two working pairs with four generators.

4.5.2 The Novel Adsorption Refrigeration Theory of Metal Hydrides–Hydrogen

For metal hydrides–hydrogen adsorption systems, some scholars [31, 32] proposed the concept of a cross-type Van't Hoff line. This concept is shown in Figure 4.19.

In the cross-type Van't Hoff diagram, two metal hydride lines intersect. The temperature of *d* point is higher than that of *c* point. The heat released by metal hydride A at the point *d* can be completely absorbed by the metal hydride B at the point *c*. The theoretical *COP* under these conditions is:

$$COP_C = \frac{Q_{Cool}}{Q_{Drive} - Q_{m1,A}} \quad (4.58)$$

Obviously this value is higher than performance parameter values in Equation 4.55.

The performances of different working pairs are calculated throughthe cross-type Van't Hoff line. The results are shown in Table 4.4.

Compared with traditional metal hydrides–hydrogen refrigeration, the main disadvantage of this new theory for the application is that adsorbent regeneration requires higher temperature and higher pressure, which will lead to an increase in the inclination of the pressure bevel in Figure 4.17. An increase in the inclination of the pressure bevel will reduce the horizontal width of the inclined plane and the adsorption rate, which leads to the decline of the cooling efficiency.

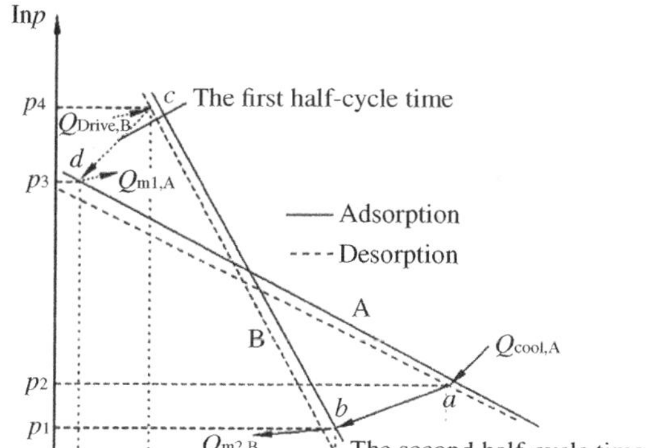

Figure 4.19 Cross-type Van't Hoff diagram [31]

Table 4.4 The performance of metal hydride with crossing Van't Hoff lines [31]

Working pairs refrigerant/regeneration refrigerant	ΔH (kJ/mol H_2)	ΔS (J/mol H_2)	*COP*	T_e (°C)	T_h (°C)
$LaNi_5/V_{0.855}Ti_{0.095}Fe_{0.05}$	31.0 43.2	109 140.6	2.5	5	137
$Fe_{0.9}Mn_{0.1}Ti/V_{0.846}Ti_{0.104}Fe_{0.05}$	26.6 42.9	99.2 148.5	1.6	14	71
$MmNi_{3.98}Fe_{1.04}/V_{0.846}Ti_{0.104}Fe_{0.05}$	27.3 42.9	105.6 148.5	1.8	2	107
$MmNi_{3.98}Fe_{0.85}/VCr_{0.05}$	25.1 37.4	104.6 139.3	2.0	10	101

In addition, increased temperature will lead to increased losses of the metal heat capacity in the system, which also will influence the *COP*. Taking the working pair of $LaNi_5/V_{0.865}Ti_{0.095}Fe_{0.05}$ in Table 4.4 as an example, the *COP* of traditional system and improved system are 0.71 and 2.5, respectively, when it doesn't take the metal heat capacity into account. The *COP* of traditional system and improved system are 0.64 and 0.83, respectively, when it takes the metal heat capacity into account.

References

[1] Dai, A.B. (1987) *Coordination Chemistry*, Volume 12 of Series of Inorganic Chemistry, Science Press, Beijing, ISBN: 13031.4 (in Chinese).

[2] Ci, Y.X. and Zhou, T.Z. (1999) *The Multiple Complex Compounds in the Analytical Chemistry*, Science Press, Beijing, ISBN: 7-03-007128-X/O·1069 (in Chinese).

[3] Gasser, R.P.H. (1987) *An Introduction to Chemisorption and Catalysis by Metals*, Clarendon Press, Oxford.

[4] Yasuaki, O. (1990) *The Property and Application of Metal Hydride*, Chemical Engineering Press, Beijing, ISBN: 9787502507510 (in Chinese).

[5] Tetsurō, S. (1991) *Metal Oxides and its Catalysis*, Chemical Engineering Press, Beijing, ISBN: 7-312-00210-2 (in Chinese).

[6] Wongsuman, W., Kumar, S., Neveu, P. and Meunier, F. (2001) A review of chemical heat heat pump technology and applications. *Applied Thermal Engineering*, **21**, 1489–1519.

[7] Lebrun, M. and Neveu, P. (1991) Conception, simulation, dimensioning and testing of an experimental chemical heat pump. *ASHRAE Transactions*, **98**, 420–429.

[8] Spinner, B. (1993) Ammonia-based thermochemical transformers. *Heat Recovery Systems and CHP*, **13**(4), 301–307.

[9] Wang, L.W., Wang, R.Z., Wu, J.Y. and Wang, K. (2004) Adsorption performances and refrigeration application of adsorption working pair of $CaCl_2$-NH_3. *Science in China, Series E*, **47**(2), 173–185.

[10] Peng, S.P., Wang, B. and Luo, Z.J. (1984) *Structure of Atoms and Molecules, Complex Compounds, Colloid Chemistry*, People's Press of Sichuan Province, Chengdu, ISBN: 7118.814 (in Chinese).

[11] Peng, C.H. and Zhang, X.Y. (1981) *Principle of Complexometric Titration*, Publishing Press of Beijing University, Beijing, ISBN: 7-301-13209-9-1 (in Chinese).

[12] Xia, S.W. (1993) *Activated Energy and its Calculation*, High Education Press of Beijing, Beijing, ISBN: 7040036312 (in Chinese).

[13] Zhang, Y.H. (1989) *Adsorption Action*, Shanghai Press of Science and Technology, Shanghai, ISBN: 7805134979, 9787805134970 (in Chinese).

[14] Touzain, P. (1999) Thermodynamic values of ammonia-salts reactions for chemical sorption heat pumps. Proceedings of International Sorption Heat Pump Conference, Munich, Germany, pp. 24–26.
[15] Biltz, W. and Huttig, G.F. (1920) Uber die auswertung von dissoziationsmessungen bei ammoniakaten nach dem theorem von nernst mit hilfe von nomogrammen. *Zeitschrift Fur Anorganische Und Allgemeine Chemie*, **109**, 111–125.
[16] Mbaye, M., Aidoun, Z., Valkov, V. *et al.* (1998) Analysis of chemical heat pumps (CHPS): basic concepts and numerical model description. *Applied Thermal Engineering*, **18**, 131–146.
[17] Llobet, J. and Goetz, V. (2000) Rotary system for the continuous production of cold by solid-gas sorption: modeling and analysis of energy performance. *International Journal of Refrigeration*, **23**, 609–625.
[18] Mazet, N., Amouroux, M. and Spinner, B. (1991) Analysis and experimental study of the transformation of a non-isothermal solid/gas reaction medium. *Chemical Engineering Communications*, **99**, 155–174.
[19] Goetz, V. and Marty, A. (1992) A model for reversible solid-gas reactions submitted to temperature and pressure constraints: simulation of the rate of reaction in solid-gas reactor used as chemical heat pump. *Chemical Engineering Science*, **47**(17–18), 4445–4454.
[20] Neveu, P. and Castaing, J. (1993) Solid-gas chemical heat pumps:field of application and performance of the internal heat of reaction recovery process. *Heat Recovery Systems and CHP*, **13**(3), 233–251.
[21] Wang, L.W., Wang, R.Z., Wu, J.Y. and Wang, K. (2005) Research on the chemical adsorption precursor state of $CaCl_2$-NH_3 for adsorption refrigeration. *Science in China, Series E*, **48**(1), 70–82.
[22] Stitou, D. and Crozat, G. (1997) Dimensioning nomograms for the design for fixed-bed solid-gas thermochemical reactors with various geometrical configurations. *Chemical Engineering and Processing*, **36**, 45–48.
[23] Lebrun, M. and Spinner, B. (1990) Models of heat and mass transfers in solid-gas reactors used as chemical heat pumps. *Chemical Engineering Science*, **45**(7), 1743–1753.
[24] Tykodi, R.J. (1979) Thermodynamics of steady states: resistance change transitions in steady-state systems. *Bulletin of the Chemical Society of Japan*, **522**, 564–567.
[25] Lu, H.B., Mazet, N., Coudevylle, O. and Mauran, S. (1997) Comparison of a general model with a simplified approach for the transformation of solid-gas media used in chemical heat transformers. *Chemical Engineering Science*, **52**(2), 311–327.
[26] Wang, L.W. (2005) Performances, mechanisms, and application of a new type compound adsorbent for efficient heat pipe type refrigeration driven by waste heat. PhD Thesis. Shanghai Jiao Tong University, Shanghai, China (in Chinese).
[27] Choi, H.K., Neveu, P., and Spinner, B. (1996) System modeling and parameter effects on design and performance of ammonia based stelf thermochemical transformer. Proceedings of the International Absorption Heat Pump Conference, Quebec, Montreal, pp. 505–512.
[28] Iloeje, O.C., Ndili, A.N. and Enibe, S.O. (1995) Computer simulation of a $CaCl_2$ solid-adsorption solar refrigerator. *Energy*, **20**(11), 1141–1145.
[29] Mauran, S., Coudevylle, O., and Lu, H.B. (1996) Optimization of porous reactive media for solid sorption heat pumps. Proceedings of the International Absorption Heat Pump Conference, Quebec, Montreal, pp. 401–408.
[30] Willers, E. and Groll, M. (1999) Evaluation of metal hydride machines for heat pumping and cooling applications. *International Journal of Refrigeration*, **22**, 47–58.
[31] Libowitz, G.G., Feldman, K.T. Jr., and Stein, C. (1997) Thermodynamic properties of metal hydrides for a novel heat pump configuration. *Journal of Alloys and Compounds*, **253–254**, 673–676.
[32] Stein, C. and Redding, E.M. (1996) Hydride heat pump. US Patent 5 497 630.

5

Adsorption Mechanism and Thermodynamic Characteristics of Composite Adsorbents

Composite adsorbents can be obtained by mixing chemical absorbents and porous media together. Composite adsorption has a series of advantages. Composite adsorption can improve the heat and mass transfer performance of chemical adsorbents by the advantages of the high thermal conductivity and porous structure of the porous media. Meanwhile, the composite adsorbents keep the advantages of chemical sorbents, that is, they possess large adsorption and desorption quantities.

The porous media matrixes that can be employed in the composite adsorption include activated carbon, active carbon fiber, silica gel, and graphite.

5.1 The Characteristics of Porous Media

In the adsorption refrigeration, when the silica gel is used as a type of matrix to prepare composite sorbents, the goal is to use the chemical adsorbents to improve the cycle adsorption quantity of silica gel. While the active carbon is used as a matrix, its purpose is to use the rich micro pore structure of activated carbon to enhance the mass transfer performance of chemical adsorbents. As a typical physical adsorbent the characteristics of silica gel and active carbon has been introduced in Chapter 2. This chapter will mainly introduce the characteristics of activated carbon fiber (ACF), graphite, and graphite/carbon fiber.

The development of carbon fiber and graphite fiber can be dated back to the 1960s. The carbon fiber and graphite fiber can improve the strength of the material, so they have been rapidly developed and widely used. The most prominent research work, which applies the graphite as additive to the adsorption refrigeration, should be attributed to M. Groll and S. Mauran. The aim is to use the high thermal conductivity of graphite to improve the heat and mass transfer

Adsorption Refrigeration Technology: Theory and Application, First Edition. Ruzhu Wang, Liwei Wang and Jingyi Wu.

Companion Website: www.wiley.com/go/wang/refrigeration

performance of metal chlorides [1–5]. Subsequently, T. Dellero [6, 7], L.L. Vasiliev [8, 9] *et al.* employ ACF not only to improve successfully the chemical adsorption performance of metal chloride, but also suppress the expansion and agglomerate phenomenon of chemical adsorbent.

5.1.1 Activated Carbon Fiber

ACF, which is made of the fibrous precursor body through the carbonization and activation, is also called fibrous activated carbon. Compared to the activated carbon, it has the better adsorption performance. Meanwhile, it is an environmentally benign engineering material [10]. More than 50% of the carbon atoms are situated inside and outside the surface, forming unique adsorptive structure, and therefore called superficiality solid. It has good thermal conductivity. Moreover, it has large specific surface area and narrow pore size distribution, which makes it possess the higher adsorption/desorption speed and larger adsorption capacity.

1. **Structure characteristics**
   ACF is a kind of typical microporous activated carbon (MPAC). It is considered to be the combination of "ultra micron particle, irregular surface structure, and very narrow space" with a diameter of 10–30 μm. The pore is located in the fiber surface. The ultramicro particle is combined in various ways, forming the rich nano space. The nanometer space scales have the same order of magnitude as the size of the ultramicro particle, and result in a large specific surface area. It contains many irregular structures, such as heterocyclic structure or the microstructure of the surface functional groups. The interaction between the micro pore and molecular in pore side will create a strong molecular field due to the great surface energy, and thereby providing molecules under the sorption state with the high pressure system of the physical and chemical change. As a result, the diffusion path to the vacancy of the adsorbates is shorter than that of activated carbon, the driving force is larger and pore size distribution is very concentrated. It is the main reason that ACF has larger specific surface area, a faster desorption rate and a higher adsorption efficiency than activated carbon.
2. **Functional methods**
   Using pore interstitial structure control and surface chemical modification can achieve the efficient adsorption transformation of the specific materials. ACF is usually suitable for the adsorption of gas and liquid molecules with low relative molecular mass (relative molecular mass $M_w < 300$). When adsorbent pore size is twice as much as the critical size of the adsorbate molecular, the adsorbate is easily adsorbed. Aperture adjustment can make the micro pores of ACF match molecular size of adsorbent. There are usually three methods: (i) Using the activation process or changing the activate degree to the nanometer level to adjust aperture; (ii) Adding metal compounds either in the raw fiber or ACF and activation can adjust the pore size. Besides, the carbonization and activation of some other materials can change the pore size. The raw fibers need to have relative big pore size; and (iii) Using the pyrolyzation of hydrocarbons and deposition in cell wall, post-processing under high temperature condition can decrease pore size. Surface chemical modification can vary the acid and alkaline of ACF surface by the introduction or removal some surface functional groups. After the high temperature or the hydrogenation, the surface oxygen groups can be removed (reduction). Through the gas phase oxidation and liquid-phase oxidation, the

acid surface can be obtained. The influence of the physical and chemical structure should be considered during the modification process.

In the adsorption refrigeration applications, composite adsorbents of ACF and chemical adsorbents have a series of advantages. On the one hand, using its good thermal conductivity can enhance the heat transfer performance of chemical sorbents; on the other hand, using its rich micro porous structure can suppress the expansion and agglomerate phenomenon of chemical adsorbent to improve the mass transfer performance.

5.1.2 The Characteristics of Graphite

The performance of the graphite is similar to the pure carbon. Generally, during the transition process from hydrocarbons to the pure carbon under high temperature and high pressure conditions, carbon of graphite structures will be generated [11].

There are two types of crystal structure of carbon: diamond and graphite structure. Besides, amorphous carbon is another crystal structure. Through the modern X-ray diffraction crystal structure analysis, amorphous carbon also has graphite crystal structure. But the tiny crystals appear in the irregular gathered state. Graphite is cumulate hexagonal grid levels and form a crystal, which belongs to the hexagonal system structure. The others belong to the rhombohedral system. For the hexagonal system structure of graphite, crystal cell has four carbon atoms. The density of ideal graphite crystal is 2266 kg/m^3. The ideal graphite crystals usually don't exist. The monocrystal graphite can be found from the natural flake graphite of a very high crystallization degree, but only a few millimeters. The artificial graphite is not the ideal monocrystal graphite in the modern carbon industry, but the polycrystal graphite with irregular arrangement of atoms. Because of the different raw material sources, the crystal gathered state of the polycrystal graphite is irregular.

For the monocrystal graphite with the regular crystal structure, its performance has obvious anisotropy. The performance of this kind of graphite will show isotropic characteristics because of the irregular arrangement of tiny single grains. Obviously, the artificial graphite not only shows anisotropic, but also is a little isotropous.

According to the crystallization degree, graphite crystal is divided into carbonaceous or graphitic structure. The carbon with low crystallization degree, which has incomplete crystal structure, is called carbon structure. So, carbonaceous is also called amorphous carbon or non-crystalline carbon. In fact, the crystal structure of carbon and graphite has no clear boundary. After further heat treatment, many carbonaceous carbons can be converted into graphite to make the crystal structure perfect.

For applications the graphite has the following characteristics:

1. **Lubricity.** Because of the weak interaction force between the graphite layers, when the friction between graphite and metal takes place, the metal surface will easily form graphite film to reduce friction. For polished steel surface, the dynamical friction coefficient of high-strength graphite under room temperature and atmospheric pressure is about 0.35. Therefore, graphite is often referred to as a type of lubricant to manufacture graphite bearing and forge colloidal graphite, and so on.
2. **Small thermal expansion performance.** During 20–200 °C, thermal expansion coefficient is $(1-2)\times 10^{-6}$/°C along the suppressed direction of graphite products, whereas it is perpendicular to the suppressed direction for $(2-3)\times 10^{-6}$/°C. Thermal expansion

coefficient of expanded graphite is larger. The values are 5 and 100×10^{-6}/°C along the parallel and perpendicular direction, respectively. Graphite products have high thermal shock resistance. For example, the graphite electrode in a steelmaking furnace will undertake urgent cold and urgent hot, and so on.

3. **Good thermal and electrical conductivity.** The conductivity of crystal face direction is larger than that of vertical direction of crystal face. But the thermal conductivity and resistance of graphite are also affected by temperature. The resistance coefficient is negative from 700 to 900 K, whereas it is positive when the temperature is more than 900 K. The thermal conductivity will reach the maximum in a certain temperature. Its good thermal and electric conductivity show anisotropic, and therefore graphite is used as refractory materials, heat insulation materials, and graphite electrode.
4. **Wide applicability.** (The graphite can be applied for a wide application temperature range.) The melting point of graphite is 38.50 °C, and the boiling point is 4250 °C, so it can be used for the temperature range of –200 to 450 °C in the air. It can be used for the temperature range of –200 to 3000 °C in vacuum or reduction atmosphere. The strength of the graphite will rise with an increase of temperature. Graphite is a brittle material at room temperature. However, graphite begins to creep in temperatures above 1700 °C, but the creep deformation is very small, therefore it is often used under very high temperature conditions for modern industries.
5. **Stable chemical performance and non-toxic.** Graphite is non-toxic, and begins to produce oxidation for temperatures of 400 °C. Graphite can react with steam and CO_2, when the temperature is more than 700 and 900 °C. Graphite can react with the hydrogen only for a temperature of over 1000 °C. In addition to chloroazotic acid, chromium acid, strong sulfuric acid, and nitric acid, graphite can't react with acid, alkali, and organic solvent. Under high temperature, graphite can react with many metallics and non-metallics or their oxides. The resistance to radiation and thermal neutron radiation section of graphite is small, which makes it become the sole slow change materials available in a nuclear reactor.
6. **Good thermal shock resistance.** The graphite can undertake severe temperature change without damage under high temperature. What's more, when the temperature forms the mutations, the volume of graphite changes very little and won't produce crack.
7. **Other properties.** Graphite can be coated and encapsulated. The graphite is very light and easily processed into the formation. Besides, it is an important resource of carbon, and can offer high purity carbon for various kinds of materials in many different applications.

At present, the graphite is mainly used as amaterial for refractory, conductive, wear-resisting, sealing, corrosion resistant, anti-radiation objects, or occasions. In adsorption refrigeration, the expanded graphite is used to improve the heat transfer performance of chemical adsorbents.

5.1.3 Expanded Natural Graphite (ENG)

The expanded natural graphite (ENG) has good heat and mass transfer performance, as well as anisotropic thermal conductivity and permeability. In the experiments, the ENG is prepared by heating untreated natural graphite in an oven at a temperature of 700 °C for 12–15 minutes [12]. The graphite is manufactured in the Shanghai YiFan Graphite Company in China, with parameters of 50–80 mesh and percentage purity larger than 99%.

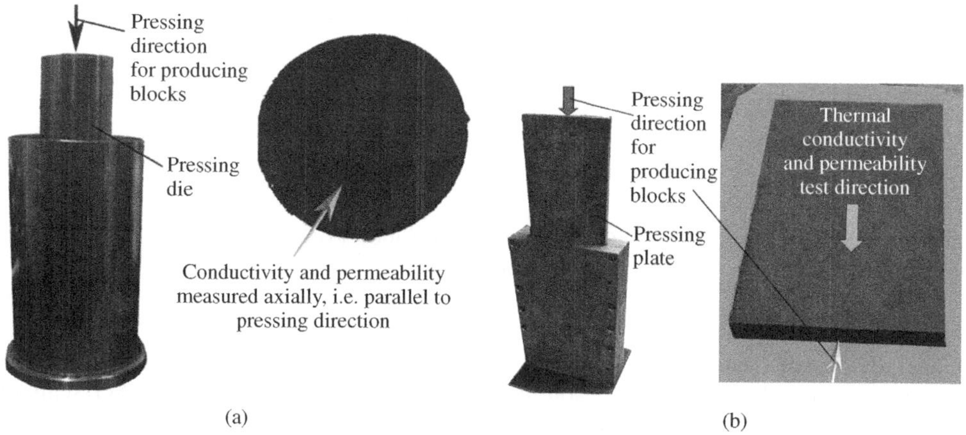

Figure 5.1 (a) The rig and sample for the DCENG and (b) the rig and sample for the PCENG

The disc compacted expanded natural graphite blocks (DCENG) and the plate compacted expanded natural graphite blocks (PCENG) are produced (Figure 5.1) in order to investigate the anisotropic thermal conductivity and permeability.

5.1.3.1 Thermal Conductivity of DCENG

DCENG of different densities are produced, and the average thermal conductivities at different values of heat flux are calculated. The relation between thermal conductivities and densities are shown in Figure 5.2. Figure 5.2 shows an interesting phenomenon, which is that there is a range where the thermal conductivity decreases while the density increases. In Figure 5.2 while the density of the compacted disc is lower than 300 kg/m^3, the thermal conductivity increases while the density of the disc increases. The values of thermal conductivity vary very slightly while the values of the density of the discs range from 343 to 576 kg/m^3, and the optimal thermal conductivity, which is 1.70 W/(mK), is also obtained in this range. The thermal conductivity decreases when the density of the adsorbent is higher than 658 kg/m^3, and the lowest thermal conductivity of 1.21 W/(mK) is obtained when the density of the adsorbent is 698 kg/m^3. After that the thermal conductivity increases again while the density increases, and the value is 1.318 W/(mK) for 730 kg/m^3.

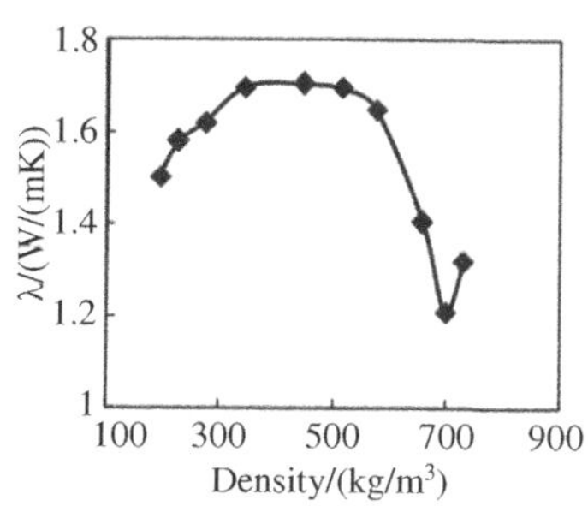

Figure 5.2 The average thermal conductivity of DCENG vs. density

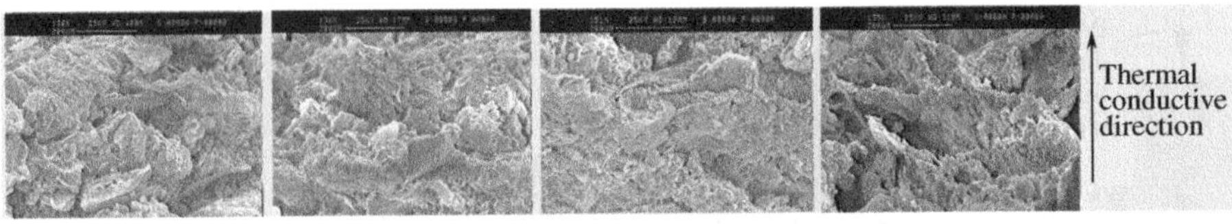

Figure 5.3 SEM pictures of DCENG with different densities. (a) 343 kg/m^3, 136×; (b) 446 kg/m^3, 136×; (c) 658 kg/m^3, 151×; and (d) 730 kg/m^3, 135×

Generally for any type of material, the thermal conductivity always increases when the density increases. In order to find the reason for this abnormal phenomena of DCENG, scanning electron microscope (SEM) pictures were taken of different samples, as shown in Figure 5.3.

From Figure 5.3 we can see that the layers of ENG are formed under pressure, and they are perpendicular to the pressing direction as the ENG is compressed. For the DCENG the thermal conductive directions measured are parallel to the pressing direction, that is, the thermal conductive direction will be perpendicular to most micro layers. Figure 5.3a,b shows that when the density is lower, for example, when the density of the compacted DCENG is 343 kg/m^3, the micro layers inside the sample are distributed in a disorderly way because the compressive force needed to make the DCENG with lower density is less. Such a structure is helpful for the thermal conductive process because some layers that are parallel to the thermal conductive direction exist in the sample. When the density of the sample increase, for example, when the density is 446 kg/m^3 (Figure 5.3b), although the layers are more uniformly distributed along a horizontal direction and some perpendicular layers are destroyed by the larger pressing force, there are also some perpendicular layers remaining, so that the thermal conductivity of the sample shown in Figure 5.2 is also useful. But when the density of the adsorbent is larger than 658 kg/m^3, from Figure 5.3c,d we can see that the layers are all distributed uniformly along the horizontal direction, and the perpendicular layers that are essential for the thermal conductive process in the perpendicular direction are disrupted seriously by the larger compressive force, thus the thermal conductivity in the perpendicular direction decreases although the density increases. After most of the perpendicular layers are destroyed, the thermal conductivity will increase again while the density increases; this is mainly because the heat resistance between horizontal layers decreases for the same reason as with the higher density produced by the higher compressive force.

Figure 5.3d shows that the micro layer is distributed very uniformly under the effect of higher pressure. For such a structure the thermal conductivity will not be good in the direction perpendicular to the layers, but the optimal heat transfer performance will be obtained in the direction parallel to the layers. In order to research such conditions, the PCENG is studied.

5.1.3.2 Thermal Conductivity of PCENG

Samples of PCENG with different densities were produced. They were cut into circular shapes, and then the thermal conductivities were measured. The experimental results are shown in Figure 5.4. It can be seen that the thermal conductivity keeps increasing with increasing density.

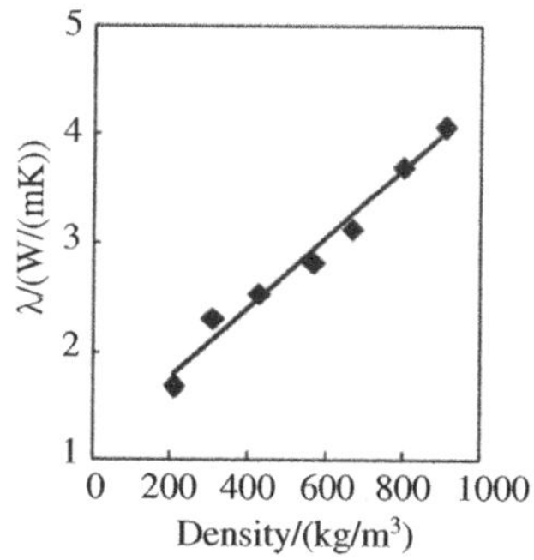

Figure 5.4 The average thermal conductivity of PCENG vs. densities

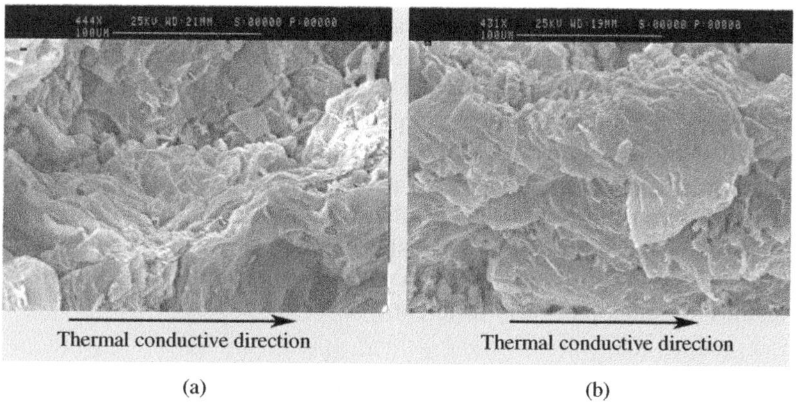

Figure 5.5 SEM pictures of PCENG with different densities, (a) 557 kg/m^3, 444× and (b) 700 kg/m^3, 431×

The trend of the thermal conductivities in Figure 5.4 is analyzed, and it is also related to the distribution of the micro layers inside the samples, which is caused by different pressures applied on the sample. The SEM pictures of PCENG are shown in Figure 5.5. For PCENG the direction of thermal conductivity tested by the test unit is perpendicular to the direction of compression. Under such a condition, because the layers of the expanded graphite formed under the effect of pressure are also perpendicular to the pressing direction, the thermal conductive direction is parallel with the distribution of layers, just as Figure 5.5a,b shows. The layers are distributed more uniformly when the density that is achieved by larger compressive forces is higher, thus the thermal conductivity is also higher when the density of the sample increases.

Comparing the results of Figures 5.2 and 5.4, the PCENG has a much larger value of thermal conductivity than that of the DCENG when the densities of these two types of blocks are similar, and the difference is larger while the density is higher. For example, when the density is between 210 and 220 kg/m^3, the thermal conductivities of PCENG and DCENG are 1.67 and 1.58 W/(mK), respectively, the value of PCENG is only improved by 6%. When the density is about 660–670 kg/m^3, the thermal conductivities of PCENG and DCENG are 3.13

and 1.40 W/(mK), respectively, the value of PCENG improved by a factor of about 2. These phenomena are also related to the distribution of micro layers, which is disorderly for smaller density and uniform for larger density for the reason of different values of pressures, and there is higher thermal transfer resistance in the direction perpendicular to the distribution of layers at a large density that is caused by increased pressure.

5.1.3.3 Anisotropic Permeability

The micro layers inside the compacted expanded graphite not only influence the thermal conductivity, but also influence the permeability.

Four types of blocks were chosen to compare the anisotropic permeability, which are two blocks of PCENG and DCENG with a similar density of 430–450 kg/m^3, and another two blocks with a similar density of 650–700 kg/m^3.

The values of permeability are calculated by the experimental results, and they are shown in Table 5.1. In Table 5.1 we can see that for the same type of blocks, the permeability decreases while the density of the block increases. For a different type of blocks with similar densities, the PCENG has higher permeability than that of DCENG. For example, when the samples with similar density of about 440 kg/m^3, the permeability of PCENG is three times higher than that of DCENG. It is mainly related to the micro layers inside the sample that are caused by different pressures, as shown in Figures 5.3 and 5.5. For DCENG because the mass transfer direction of the gas is perpendicular to the distribution of most micro layers, the uniform layers will have a larger resistance for the mass transfer process, and then the permeability will be lower. For PCENG the mass transfer direction of the gas is parallel to the distribution of the micro layers, thus the situation is very different.

5.1.4 Expanded Natural Graphite Treated by the Sulfuric Acid (ENG-TSA)

The ENG-TSA (treated by the sulfuric acid) is manufactured by Mersen (previously Carbone Lorraine) in France [13]. The sample is made from natural graphite that is soaked in sulfuric acid, which becomes intercalated in the layered structure of the graphite. Finally the sample was exfoliated by heating in a flame, forming expanded graphite with a much lower density than normal ENG while the intercalated acid was removed.

The consolidated ENG-TSA has the anisotropic thermal conductivity and the permeability. In order to measure the anisotropic properties, two types of samples with the compression direction either in the plane of a disk or along its axis are produced, which is similar with the PCENG and DCENG shown in Figure 5.1. They are called disk samples and plate samples,

Table 5.1 Permeability of different samples

Type of block	Density (kg/m^3)	Permeability (m^2)	Characteristics of gas transfer direction
DCENG	446	2.076×10^{-12}	Parallel to the direction of compression and perpendicular to the micro layers
	698	1.873×10^{-12}	
PCENG	437	8.788×10^{-12}	Perpendicular to the direction of compression and parallel to the micro layers
	653	4.495×10^{-12}	

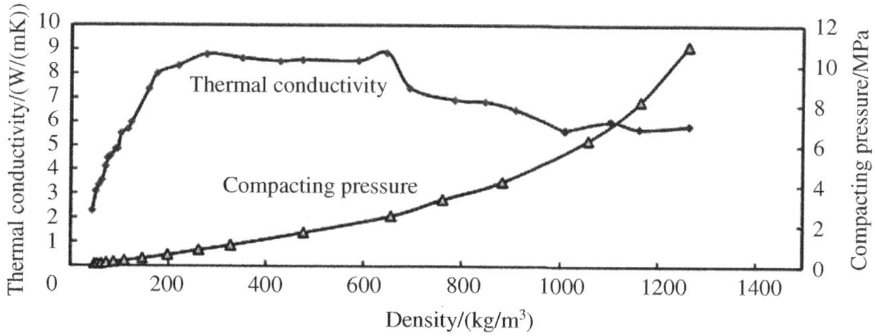

Figure 5.6 Thermal conductivity and compacting pressure vs. density of consolidated disks of ENG-TSA

respectively. The two types of sample are both heated in an oven at 150 °C for 4 hours to make sure there is no retained water before they were compressed.

5.1.4.1 The Anisotropic Thermal Conductivity of the Consolidated ENG-TSA

The thermal conductivity of 31 consolidated samples of disk is tested, and the results are shown in Figure 5.6. The thermal conductivity increases rapidly as density increases, then stabilizes, and finally decreases. The thermal conductivity stabilizes at a value of about 5–6 W/(mK) when the density is between 1000 and 1200 kg/m^3. The thermal conductivity of the consolidated disks of ENG-TSA is much higher than that of consolidated disks of ENG [12], and the highest value for ENG-TSA is 8.9 W/(mK), which is about five times more than the highest value for ENG [12]. The corresponding compacting pressure needed for producing consolidated disks of ENG-TSA is shown in Figure 5.6 as well. We can see there is a relationship between thermal conductivity and compacting pressure or density. The thermal conductivity of the material decreases when the density is higher than 600 kg/m^3. The relationship between compacting pressure and density is roughly linear when the density is lower than 600 kg/m^3, and increases rapidly and non-linearly when the density is higher than 600 kg/m^3. It means that the consolidated ENG-TSA is much more difficult to compress when the density is higher than 600 kg/m^3. Figure 5.6 also shows that the density beyond which compaction pressure increases rapidly is the same density beyond which conductivity decreases.

The consolidated plates of ENG-TSA had much higher thermal conductivity than disks of consolidated ENG-TSA due to the orientation of the micro layers. The thermal conductivity of the consolidated plates is shown in Figure 5.7. Thermal conductivity increases rapidly as the density increases. The highest thermal conductivity is 337 W/(mK) when the density of the sample is 831 kg/m^3. It is almost 100 times higher compared to the values for the consolidated plates of ENG [12], and almost 50 times higher than the samples of disk with similar density. For the samples of plate, the compacting pressure increases almost linearly with the density while density is lower than 600 kg/m^3. The trend is similar to that of consolidated samples of disk with density lower than 600 kg/m^3. It increases rapidly when the density is higher than 600 kg/m^3. Figure 5.7 also shows that the density beyond which the compacting pressure is required increases rapidly is somewhat higher than the density beyond which conductivity increases rapidly. This is in complete contrast to the trends shown for disk samples (conductivity parallel to direction of compression) above.

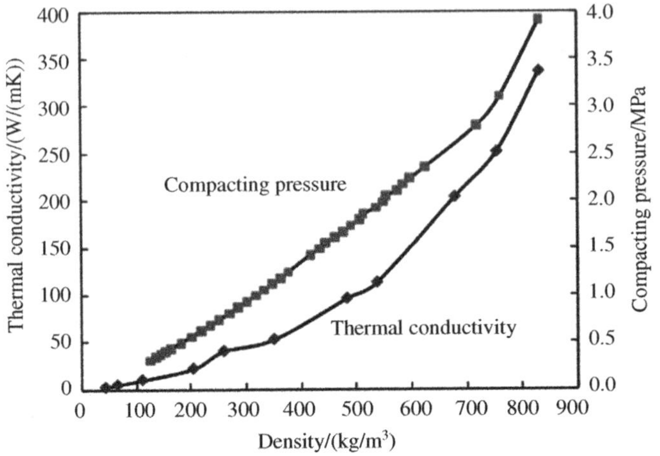

Figure 5.7 Thermal conductivity and compacting pressure vs. density for plates of consolidated ENG-TSA

Table 5.2 Permeability of different samples

Type of block	Density (kg/m^3)	Permeability (m^2)	Direction of gas flow
Consolidated samples of disk	75	5.30×10^{-14}	Parallel to the compressive direction
	164	1.12×10^{-15}	
	212	3.89×10^{-16}	
	309	9.15×10^{-17}	
	475	1.49×10^{-17}	
Consolidated samples of plate	111	1.12×10^{-11}	Perpendicular to the compressive direction
	211	1.17×10^{-14}	
	303	2.01×10^{-15}	
	487	3.96×10^{-16}	
	539	1.64×10^{-16}	

5.1.4.2 The Anisotropic Permeability of the Consolidated ENG-TSA

Five consolidated samples of disk and five consolidated samples of plate were chosen for the permeability measurements, and the results are shown in Table 5.2.

In Table 5.2 the permeability decreases rapidly with increased density for both samples of disk and plate. The permeability of samples of plate (perpendicular to the direction of compression) is at least 30 times higher than those of the disk (parallel to compression). When the density is lower than 200 kg/m^3, the permeability of the samples of plate is more than 200 times higher than that of the disk.

5.1.4.3 Scanning Electronic Microscope (SEM) Pictures of Consolidated Samples

Measurement shows that the consolidated samples of plate (perpendicular to compression) not only have higher thermal conductivity, but also have higher permeability than those of the

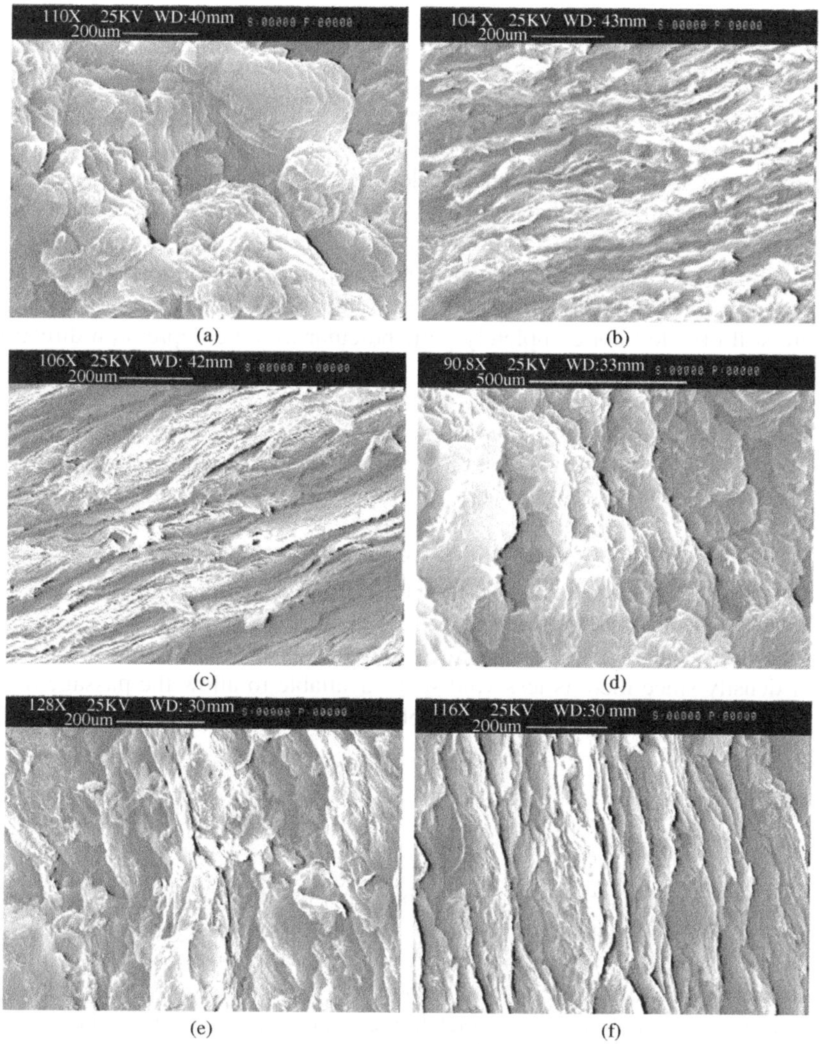

Figure 5.8 SEM pictures of consolidated samples with different values of density. (a) Sample of disk, 65 kg/m^3, 110×; (b) Sample of disk, 379 kg/m^3, 104×; (c) Sample of disk, 1084 kg/m^3, 106×; (d) Sample of plate, 65 kg/m^3, 90.8×; (e) Sample of plate, 363 kg/m^3, 128×; and (f) Sample of plate, 702 kg/m^3, 116×

disks (parallel to compression). These results are mainly due to the micro layers formed by the compressive force. SEM pictures of consolidated samples of disk and consolidated samples of plate are shown in Figure 5.8. All magnifications in Figure 5.8 fall within 90–130×.

Figure 5.8 shows that when the density of the samples is lower, for example, Figure 5.8a,d, the micro structure of the samples has a worm-like structure, which is similar to that of consolidated ENG. But when the density is higher, for example, when the density is higher than 300 kg/m^3, the worm structure of the samples is replaced by layers, and the layers of graphite appear to be distributed much more uniformly than ENG with similar density [12].

It is suggested that this uniformity accounts for consolidated ENG-TSA showing a much higher thermal conductivity compared to consolidated ENG.

The SEM pictures indicate possible reasons for the trend of thermal conductivity in Figures 5.6 and 5.7. For the disk samples (Figures 5.6 and 5.8a–c), the thermal conductivity is influenced by two factors. One is the contact resistance between layers and the other is conduction through the layers, both parallel and perpendicular to the graphite flakes. It is suggested that as the disk density increases (up to 250 kg/m^3) there is an initial rise in conductivity because the contact resistance is reduced. Within the layers there is not complete alignment of flakes perpendicular to the compression direction and so a significant contribution to the total conductivity is made by conduction in the plane of the flakes, many of which are still oriented not completely perpendicular to the compression direction. With further compression (density 250–650 kg/m^3), the thermal conductivity keeps at a stabilized value of about 8–9 W/(mK) because the positive influence of reduced resistance between flakes balances with the negative influence of layers aligning perpendicular to the heat transfer direction. With almost complete alignment (density above 650 kg/m^3), the thermal resistance stabilizes at a lower level with all heat transfer in the graphite perpendicular to the plane of the flakes.

For the plate samples (Figure 5.8d–f) the thermal conductivity always increases with increasing compacting pressure because the heat transfer is all in the plane of the graphite flakes and the better the alignment the better the heat transfer (Figure 5.7).

For the permeability in Table 5.2, the samples of both disk and plate have lower permeability with higher density since there is less void space available to allow the passage of gas. The samples with highest density would be unsuitable for use as matrices because of the poor mass transfer performance.

5.1.5 Graphite Fiber

Graphite fiber is a kind of carbon fiber. The carbon fiber has high strength, high modulus fiber, and high temperature resistance with a carbon content of more than 90%. It's only 6–8 μm in diameter. It is composed of many micro crystals with a thickness of 4–10 nm and length of 10–25 nm. Each micro crystal consists of about 12–30 layers. Similar to graphite in structure, its axial strength and modulus are much higher than graphite, whereas the radial strength and modulus are lower than graphite. Carbon fiber includes three categories: viscose, acrylic, and asphalt. Carbon fiber possesses excellent heat resistance and high temperature resistance (resistant to temperatures of 2000 °C), and the thermal expansion coefficient is almost 0. Its sublimation temperature is as high as 3650 °C. But graphite fiber produces obvious oxidation when the air temperature is higher than 400 °C. Carbon fiber is also self-lubricating, has a small friction coefficient, excellent abrasion resistance, and impact resistance. The corrosion resistance of carbon fiber is higher than glass fiber. By adding carbon fiber into the metal the composite material can greatly improve the wear resistance, impact resistance, and resistant to fatigue, whilst keeping it very light in weight.

When carbon fiber has a carbon content of more than 99% it is called graphite fiber. Graphite fiber is mainly composed of carbon atoms through the heat treatment under the high temperature. For graphite fiber, three-dimensional, highly ordered crystal structure can be observed by X-ray diffraction [14]. Graphite fiber is of high modulus and strength, and is a good electrical and thermal conductor. In the graphitization process, because the crystal structures of carbon may not be as complete as graphite crystal, namely the structure may include some amorphous carbon area, the graphite fiber is commonly known as carbon/graphite composite fiber.

Carbon/graphite fiber has a series of advantages, such as low density, high intensity, high modulus, high temperature resistance, chemical erosion resistance, low resistance, high thermal conductivity, low thermal expansion, resistance to chemical radiation. In addition, the carbon/graphite fiber is also soft and plastic, and the strength and modulus is superior to other reinforced fibers. At present, graphite fiber is mainly used in the aerospace industry, aviation industry, transportation, sports equipment and building construction industry, and so on.

Relative to the ACF, the production process of the graphite fiber is difficult, and the price is more expensive, so there are few applications of it in the adsorption refrigeration.

5.2 The Preparation and Performance of the Composite Adsorbent

According to the preparation method, the preparation of the composite/compound adsorbent includes simple mixture, impregnation, and consolidation. These methods have been introduced in the Chapter 2. This chapter will mainly discuss specific examples of composite/compound adsorbent using different porous media as the additive.

5.2.1 Composite Absorbents Using the Graphite as the Matrix

The S. Mauran research group [1, 3–5] used graphite matrix in the heat pump of metal chloride adsorbent–ammonia so as to improve the permeability, heat conductive property, and stability of the metal chloride adsorbent.

Before the preparation of the compound absorbent using the graphite as an additive, the graphite needs to be preprocessed, that is, to expand the graphite. The processes are: firstly place the graphite powder into the oven drying for 10 hours in a temperature of 60 °C, then expanding the dry graphite powder in a temperature of 400–700 °C. The standard method in the patent is used to calculate the density of the expanded graphite in order to ensure its accuracy. The development processes of composite adsorbent are described as follows:

1. Cool the expanded graphite to room temperature in the dry oven.
2. Pour the expanded graphite into the cylinder with a diameter of 4 mm through a funnel; the free-falling height of the graphite powder retains 40 cm.
3. Measure the height of the graphite inside the cylinder, calculate the volume and weight, and then calculate the density of the graphite.

For the production of the graphite block, the S. Mauran group has combined impregnation with pressing-block. The procedures are divided into the following aspects:

1. Compress the graphite block with φ 20 × 10 mm, and make the salt solution with 20% $CaCl_2$ and 80% water.
2. Place the graphite block into the salt solution.
3. Evacuate the container filled with salt solution into a vacuum, which ensures the graphite block is immersed as much as possible into the salt solution.

4. Dry the graphite block which has been immersed in the salt solution for 5 hours in the oven at a temperature of 100 °C.
5. Keep for 3 h under vacuum conditions of 200 °C and 0.02 bar.

Following the aforementioned procedures, the newly formed graphite block is called IMPEX. The manufacturing process of the IMPEX is shown in Figure 5.9.

The SEM scanning electron microscopy of IMPEX is shown in Figure 5.10. The needle-like structure of calcium crystal is shown in this figure. Through Figure 5.10, it can be seen that the distribution of $CaCl_2$ in the graphite block is very uniform due to the combination of vacuum means and impregnation.

When it reacts with ammonia, the characteristic value and dimension stability of IMPEX is seen in Table 5.3. The parameters in Table 5.3 are described as follows:

The volume density of the graphite:

$$\rho_b = \frac{M_g}{V_b} \tag{5.1}$$

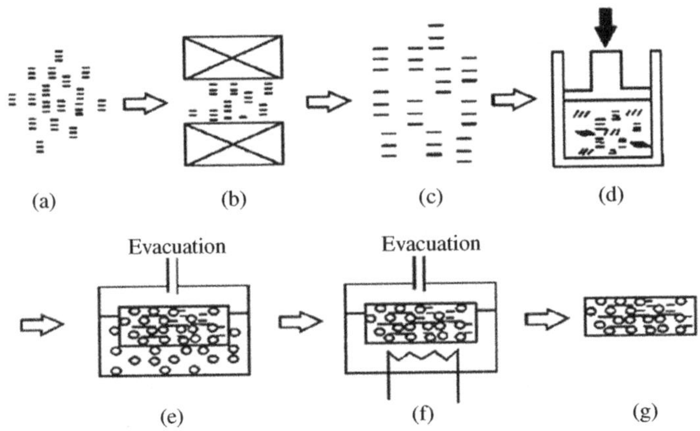

Figure 5.9 The manufacture process of IMPEX. (a) The graphite powder; (b) heat treatment of graphite powder for expansion; (c) expanded graphite powder; (d) pressing process; (e) immersed process of graphite into $CaCl_2$ salt solution; (f) evacuate to vacuum and dry process; and (g) formed IMPEX

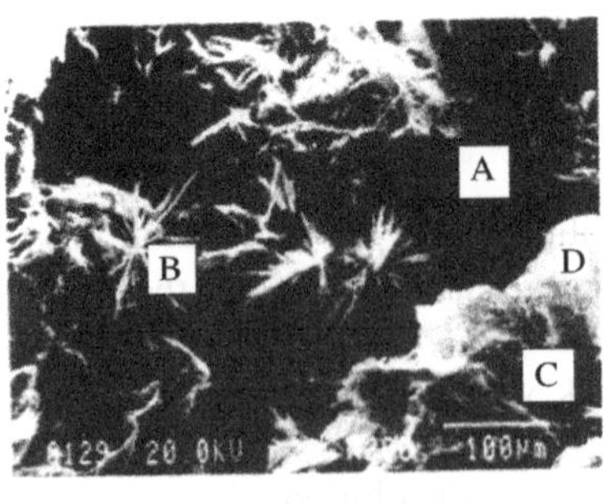

Figure 5.10 The salt distribution in IMPEX block [1]. (A) Space inside the graphite; (B) Ca crystals; (C) the graphite; and (D) the distribution of Cl

Table 5.3 The characteristic values of the IMPEX reacted with NH_3 gas [1]

ρ_b (kg/m^3)	w (%)	ε_{IMPEX}	Radial expansion value ($\Delta d/d_0$)	Axial expansion value ($\Delta h/h_0$)	Volume expansion value ($\Delta V/V_0$)
69	23	0.63	0.08	0.27	0.49
74	67	0.63	0	0.057	0.057
111	48	0.47	0.013	0.066	0.093
116	55	0.48	0	0.035	0.035
121	67	0.51	0.001	0.023	0.025
153	50	0.45	0.014	0.092	0.12
156	71	0.45	0	0	0
189	56	0.41	0	0.038	0.038

The proportion of graphite:

$$w = \frac{M_g}{M_C + M_g} \tag{5.2}$$

Porosity:

$$\varepsilon_{IMPEX} = \frac{V_p}{V_C + V_p} \tag{5.3}$$

where M_g is the mass of the graphite, M_C is the mass of $CaCl_2$, V_p is the internal porosity volume of the IMPEX, V_C is the volume of $CaCl_2$ solid.

From Table 5.3, it can be seen that the volume expansion rate of IMPEX is 0 when the volume density of the graphite is 156 kg/m^3. The size of the compound adsorbent will not change even if the expansion and agglomerate phenomenon of the chemical adsorbent occurs during the adsorption and desorption process.

Kai Wang in Shanghai Jiao Tong University [15] and R.G. Oliveira [16] investigate the composite absorbent of expanded graphite and $CaCl_2$. The preparation of compound adsorbent is different with the Mauraun group. The main difference lies in the manufacture of the absorbent block. The Mauraun group manufactures the graphite block first, and then immersing the graphite into the salt solution. Kai Wang and R.G. Olivera immersed the expanded graphite in the salt solution first, and then compress the composite adsorbent into the block. The methods are shown as follows:

1. Heat the 80 mesh graphite for 2 minutes using a high temperature of 800 °C to expand the graphite.
2. Immerse the expanded graphite into the salt solution of 14%, and then heat the sample for 22 hours under a temperature of 110 °C to remove the free water and ensure $CaCl_2 \bullet nH_2O$ are embedded into the graphite simultaneously.
3. Keep the mixture of the graphite and water under a temperature of 270 °C for 8 hours to ensure the stabilization of the product $CaCl_2 \bullet nH_2O$.
4. It takes 10 seconds to press the composite absorbent into the block under a pressure of 10 MPa in the mold. The solidification of the composite absorbent can improve the thermal conductivity of the absorbent.

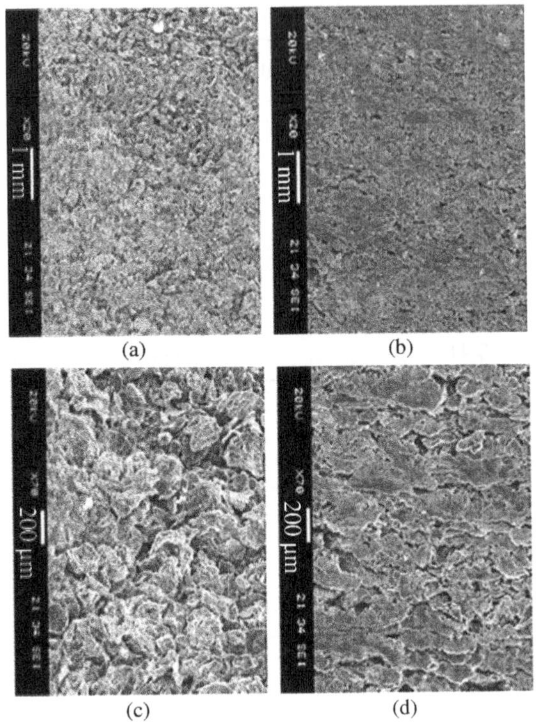

Figure 5.11 The SEM figure of solidified adsorbent. (a,c) Parallel to the compression direction and (b,d) vertical to the compression direction

From the scanning electronic image Figure 5.11 of the adsorbent block, it can be seen that the graphite block appears like a flake-like parallel layer in the radius direction, which is vertical to the compression direction. Moreover, this direction is also the direction of the heat and mass transfer. However, the special organization form can't be observed along the axial direction of the expanded graphite.

Before putting the solidified adsorbent into the adsorber, the density of the adsorbent is measured and the value is 290 kg/m^3. The absorbent being filled into the adsorber is shown in Figure 5.12. The cooling performance is estimated using the performance measurement results of the adsorbent. When the evaporation temperature is 10–20 °C and the cooling temperature is 20–30 °C, the SCP and the volume cooling capacity of the absorbent are higher

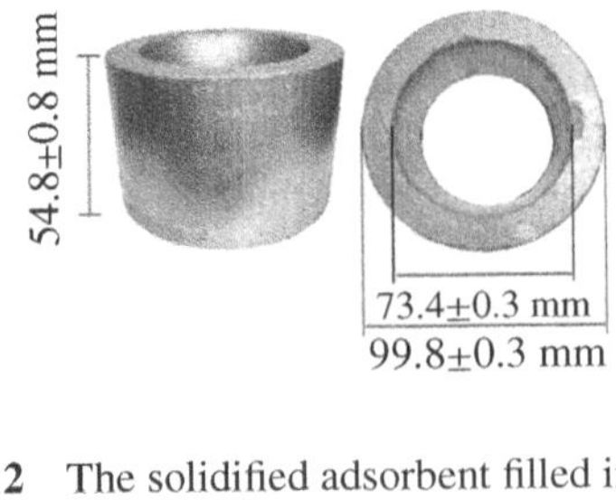

Figure 5.12 The solidified adsorbent filled into adsorber

than 1000 W/kg $CaCl_2$ and 290 kg/m^3, respectively. Testing the heat transfer performance of the adsorbent in the form of complex $CaCl_2 \cdot 2NH_3$, the heat transfer coefficient reaches 787 W/(kg °C). The calculated COP (coefficient of performance) is 0.35.

5.2.2 Composite Adsorbent with ENG-TSA as Matrix

5.2.2.1 Development of Adsorbents

Considering the anisotropic thermal conductivity and permeability of consolidated expanded graphite matrix [13], the plate samples, for which the thermal conductive and mass transferring direction is perpendicular to the compressing direction, were developed for the experiments [17].

The whole process for the preparation of the compound absorbent included firstly mixing AC, water, and ENG-TSA, and then compressing the composite adsorbent using a pressing machine. Two types of composite adsorbents were developed, one type was the composite adsorbents with a large grain size of AC (30–40 mesh), and another type was the composite adsorbents with a smaller size of AC (80–100 mesh), both types of AC were from Chemviron.

The density of flake ENG-TSA is only about 6 kg/m^3, and is very different from bulk granular AC, which is more than 300 kg/m^3.This led to a difficult mixing and consolidating process, and experiments showed that it was difficult to mix the adsorbent evenly and or to compress it effectively with large AC grain size. Experiments showed that the cracks easily occurred in the sample with smaller density or larger density, or larger grain of AC, especially with a larger proportion of AC. For example, the samples with an AC of 80–100 mesh and 67% proportion are shown in Figure 5.13a–c. Results show that no cracks happened for the sample with a density of 338 kg/m^3 (Figure 5.13b). Whereas the cracks occurred for the sample with a density of 249 kg/m^3 (Figure 5.13a), and the cracks also existed for a density of about 448 kg/m^3 (Figure 5.13c).

In the experiments 27 composite samples with different proportions of AC, different grain size of AC, and different density were developed. The parameters for development are shown in Table 5.4.

The bulk density of granular AC of 80–100 mesh is 369 kg/m^3, and the bulk density of granular AC of 30–40 mesh is 306 kg/m^3. The bulk density of AC for composite adsorbents

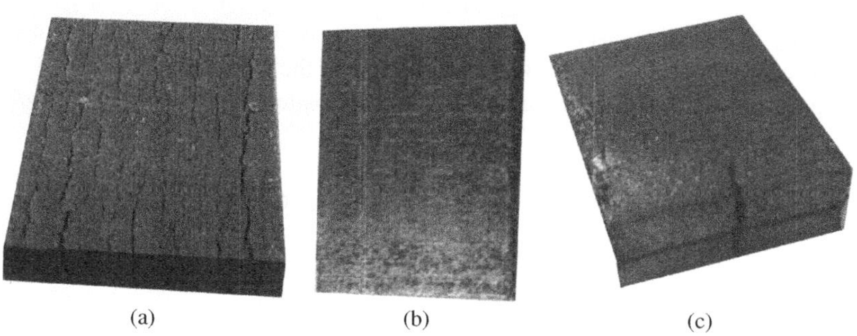

Figure 5.13 Consolidated composite adsorbents. (a) Density of 249 kg/m^3; (b) density of 388 kg/m^3; and (c) density of 448 kg/m^3

Table 5.4 Parameters of the samples developed for the research

Serial No.	Ratio of AC (%)	Grain size of AC (mesh)	Density of sample 1 (kg/m^3)	Density of sample 2 (kg/m^3)	Density of sample 3 (kg/m^3)	Density of sample 4 (kg/m^3)	Density of sample 5 (kg/m^3)
1	33	80–100	278	374	430	–	–
2	40	80–100	215	260	322	419	–
3	50	80–100	264	289	302	412	500
4	60	80–100	231	315	401	467	–
5	67	80–100	250	308	388	448	–
6	71	80–100	190	255	335	372	–
7	50	30–40	244	303	356	–	–

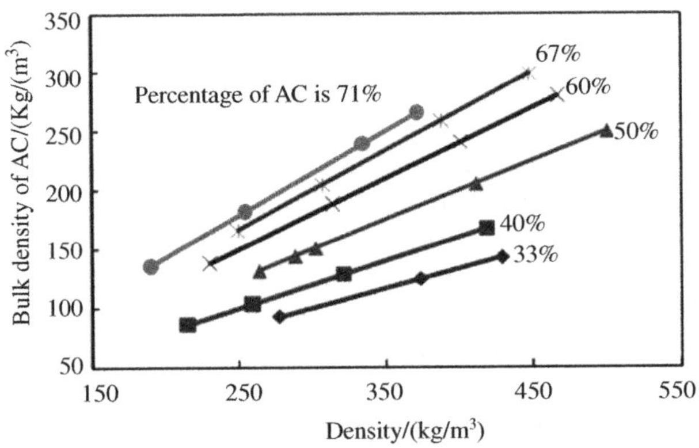

Figure 5.14 Relation between bulk density of AC, percentage of AC, and density of composite adsorbents

was calculated by dividing the total volume of composite adsorbents with the mass of AC in composite adsorbents. The relation between bulk density of AC, density of composite adsorbent, and ratio of AC is shown in Figure 5.14. The maximum bulk density of AC was 298 kg/m^3 while the percentage of AC was 67% and the density of composite adsorbent was 448 kg/m^3. The bulk density of AC decreased when the percentage of AC and the density of composite adsorbent decreased. The smallest bulk density of AC in the experiments was 93 kg/m^3 when the percentage of AC was 34% and the density of composite adsorbent was 278 kg/m^3.

5.2.2.2 Thermal Conductivity of Composite Adsorbents

The thermal conductivity of samples was tested, and the relation between thermal conductivity and the bulk density of AC is shown in Figure 5.15, in which the parameters for the samples with different serial numbers are shown in Table 5.4.

Figure 5.15 shows that for the same bulk density of AC, the thermal conductivity always increases while the ratio of AC decreases. For the same ratio, generally the thermal

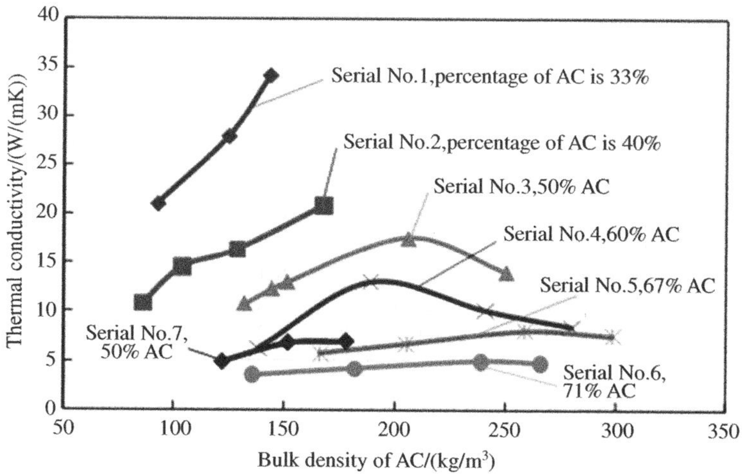

Figure 5.15 Thermal conductivity vs. bulk density of AC for consolidated composite adsorbents

conductivity increases with the increasing bulk density of AC. For some samples, such as the samples for Serial No.3 and Serial No.4, the thermal conductivity decreases while the bulk density of AC is too high mainly because the cracks occurred in such samples. The highest thermal conductivity of the consolidated composite adsorbent is as high as 34.15 W/(mK), which is improved by about 150 times if compared with the data of granular AC, which is 0.23 W/(mK).

Figure 5.15 also shows that the adsorbents which have a small size of AC have much better performance than that with a large size of AC. For example, while the ratio of AC is 50%, and the bulk density of AC is around 150 kg/m^3, the composite adsorbent with AC of 30–40 mesh has the thermal conductivity of 6.84 W/(mK), and the composite adsorbent with AC of 80–100 mesh has the thermal conductivity of 13.02 W/(mK). The data of the sample with AC of 80–100 mesh has been improved by over 90% if compared with the data of the sample with AC of 30–40 mesh.

5.2.2.3 Permeability of the Adsorbents

The permeability of the samples was tested by using a specially designed test unit shown in [13]. Since the samples tested were porous media with very low gas velocities, the Ergun model was applied in the research [13].

The results of permeability are shown in Table 5.5. Table 5.5 shows that the permeability decreases while the ratio of AC in the sample decreases. The value is generally equal to or higher than 10^{-11} while the ratio of AC is larger than 40%, and it decreases seriously while the ratio of AC is 33%.

Table 5.5 also shows that somehow the permeability increases while the density of the sample increases, and it is different from the results had been obtained from the compact ENG [18]. In order to get a general understanding for such a phenomenon, the relation between the permeability and the bulk density of AC was analyzed, and results are shown in Figure 5.16.

Table 5.5 Permeability of consolidated composite adsorbents

Ratio of AC (%)	Sample 1		Sample 2		Sample 3		Sample 4	
	ρ (kg/m^3)	K (m^2)	ρ (kg/m^3)	K (m^2)	ρ (kg/m^3)	K (m^2)	ρ (kg/m^3)	K (m^2)
33	278	1.24×10^{-14}	374	3.03×10^{-14}	430	3.6×10^{-14}	–	–
40	215	1.89×10^{-11}	260	1.50×10^{-11}	322	1.33×10^{-11}	–	–
50	264	9.89×10^{-11}	302	9.29×10^{-11}	412	3.30×10^{-11}	–	–
60	231	1.05×10^{-10}	315	9.03×10^{-11}	401	4.44×10^{-10}	–	–
67	250	1.43×10^{-10}	308	1.21×10^{-10}	388	6.57×10^{-11}	448	7.81×10^{-10}
71	255	1.68×10^{-10}	335	1.1×10^{-10}	372	1.44×10^{-10}	–	–

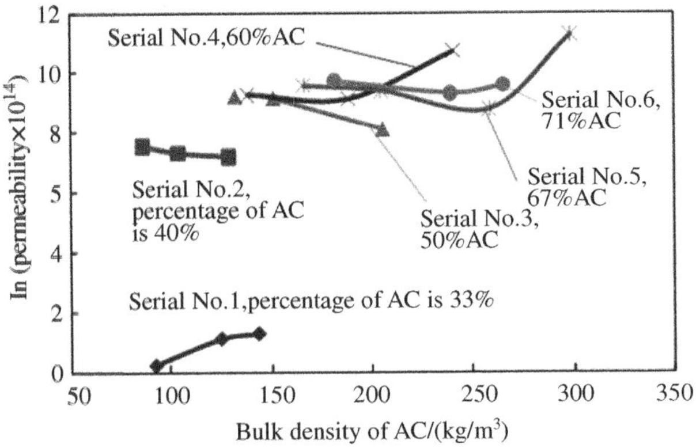

Figure 5.16 Permeability of composite adsorbents

Figure 5.16 shows that while the permeability is very low, that is, for the sample with 33% AC, the permeability increases slightly when the bulk density of AC increases. It is mainly because the continuous structure of ENG-TSA had been destroyed by the AC, and such a phenomenon will be helpful for the improvement of the mass transfer process. While the percentage of AC is equal to or larger than 40%, the permeability of composite adsorbent is improved significantly. For such a situation the permeability firstly decreases while the bulk density of AC is smaller because the tight connection between AC and ENG-TSA had prevented the gas transfer process. With the increase in bulk density of AC, the permeability will increase because the AC cannot be compressed very much and it will resist the compressing process of ENG-TSA, and the resisting force will destroy the continuous structure of ENG-TSA. Such a result was analyzed by the SEM pictures of the adsorbents.

5.2.2.4 SEM Pictures of Consolidated Composite Adsorbents

The SEM pictures were observed, and results are shown in Figure 5.17. For consolidated composite adsorbents, the grains of AC are embedded in the ENG-TSA. The structure of ENG-TSA is worm-like when the density of the samples is small (Figure 5.17a,b), whereas it distributes as organized layers when the density is larger (Figure 5.17c,d).

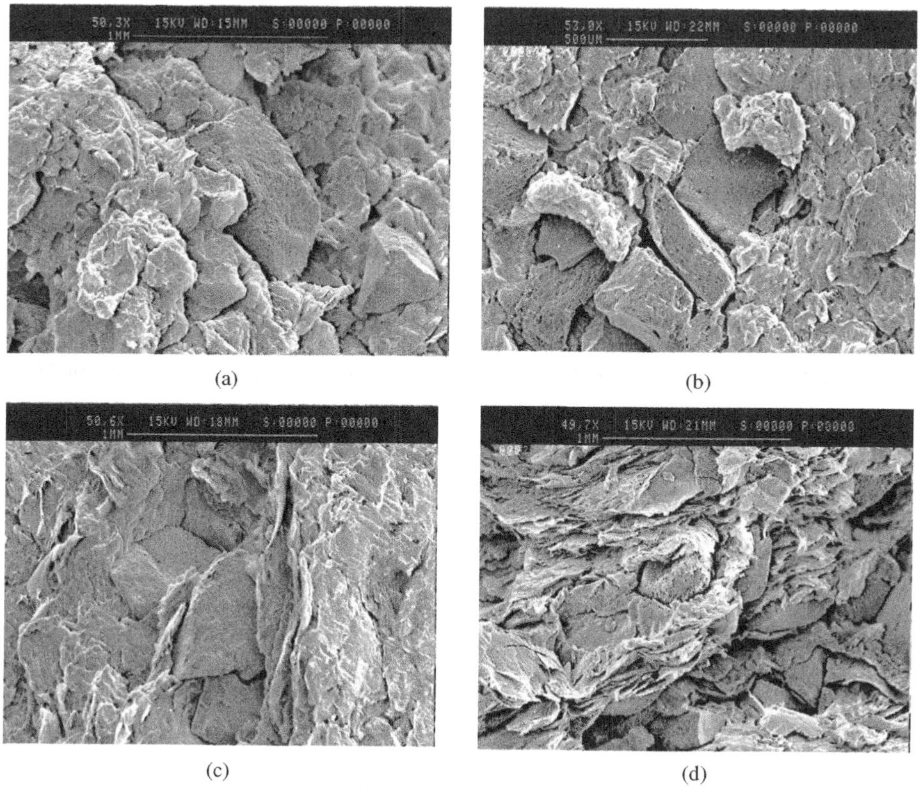

Figure 5.17 SEM pictures of consolidated composite adsorbents. (a) AC percentage of 33%, 278 kg/m^3, 50.3×; (b) AC percentage of 67%, 250 kg/m^3, 53.8×; (c) AC percentage of 33%, 430 kg/m^3, 50.6×; and (d) AC percentage of 67%, 448 kg/m^3, 49.7×

Inside the consolidated samples which have larger density values, the thermal conductivity increases while the density increases because the structure of ENG-TSA is much more organized in a larger density. But if the density is too high the larger bulk density of AC will influence the thermal conductivity because more grains of AC in a fixed volume will cause a larger thermal resistance for the heat transfer process. Inversely for some samples larger bulk density of AC will be helpful for the mass transfer process, especially for the samples with smaller proportion of AC. It is mainly because of the fact that more AC grains in a fixed volume disconnect more connections among ENG-TSA layers, and makes the micro mass transfer channels larger inside the samples.

5.2.2.5 Experiments on Equilibrium Adsorption Performance

For the adsorption working pair with AC–ammonia as a working pair, the D-A equation [19, 20] is applicable, and it is as follows:

$$x = x_0 \exp\left[-K\left(\frac{T}{T_s} - 1\right)^n\right] \tag{5.4}$$

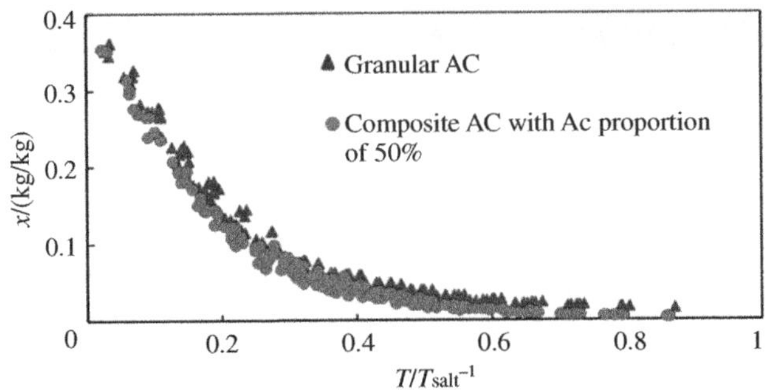

Figure 5.18 The adsorption performance of granular AC and composite adsorbent of AC

where x is the adsorption quantity (kg/kg), T is the temperature of adsorbent (K), T_s is the saturated temperature of refrigerant (K), x_0 is the maximum adsorption quantity, K and n are coefficients.

The equilibrium adsorption performance was tested, and the relation between adsorption quantity x and $(T/T_{sat} - 1)$ is shown in Figure 5.18. Results show that the performance of composite adsorbent is not influenced by the addition of ENG-TSA, and it is similar with granular AC.

The performance of AC and composite adsorbent of AC was shown by exponential equations, which are as follows:

$$x = 0.4655 \times \exp\left[-4.282 \times \left(\frac{T}{T_{sat}} - 1\right)^{0.81}\right], R^2 = 0.9698 \text{ Granular AC} \tag{5.5}$$

$$x = 0.4703 \times \exp\left[-5.551 \times \left(\frac{T}{T_{sat}} - 1\right)^{0.89}\right], R^2 = 0.9872 \text{ Composite AC} \tag{5.6}$$

The coefficient of determination (R^2) in Equations (5.5) and (5.6) are 0.9698 and 0.9872, respectively, for granular AC and composite AC, and it means the data obtained from composite AC has slightly higher precision. It is mainly because of the higher heat transfer performance for the composite adsorbent. The thermal conductivity of granular AC is much lower than the composite adsorbent of AC, and then the temperature difference between granular AC and the heat source will be slightly larger than that of composite adsorbent of AC. Such a phenomenon somehow will cause a slightly higher error between the experimental data and the real data. It should be noted that the error caused by heat transfer is very small and can be neglected. Just as Figure 5.18 shows, the relative difference between the data of granular AC and composite AC is less than 9%.

5.2.3 Composite Adsorbents with Activated Carbon as Matrix

Liwei Wang *et al.* apply activated carbon as a matrix to metal chloride–ammonia working pairs, aimed at improving the mass transfer of chemical adsorbent by adding activated carbon. Heat transfer will be further improved if the composite is consolidated [21].

Compared with an adsorbent with graphite and ACF as a matrix, an adsorbent with activated carbon as a matrix has the advantages of the hard activated carbon particles as well as the simple preparation process of composite adsorbents. Due to hard activated carbon particles, in the application the chemical adsorbent in activated carbon will not detach from the porous media and accumulates at the bottom of the adsorption bed, which generally occurs for the matrix of ENG and ACF. Such a process means that the chemical adsorbent will be evenly distributed in the activated carbon particles.

For different application occasions, composite adsorbents of activated carbon and $CaCl_2$ can be chosen as bulk mixture as well as consolidated composite. Bulk mixed adsorbent is generally used for the adsorption bed with limited filling space such as the plate adsorption bed because consolidated mixed adsorbent is hard to fill. Consolidated composite adsorbent can be used for the adsorption bed with large filling space such as the tube adsorption bed.

For the preparation of bulk adsorbent, there is little difference between the composites made by impregnation and the simple mixing process. Thus, generally, the simple mixing procedure is used for the preparation of bulk adsorbent. The process is to mix the activated carbon and $CaCl_2$ together firstly, and then dry it for the later application. To prepare the consolidated adsorbent, $CaCl_2$ is dissolved into the water firstly, and then cement and activated carbon are added to the solution, lastly the composite will be consolidated and dried. The simply mixed composite adsorbent and consolidated adsorbent prepared in a mold are shown in Figure 5.19 [22].

Adsorption performance of $CaCl_2$ is tested, showing that the performance of adsorbent is closely related with the reserved expanding space [23, 24] of the adsorbent. Due to the phenomenon of agglomeration and swelling in the adsorption process, serious agglomeration and swelling will cause the problem of mass transfer and will influence the adsorption and refrigeration performance when adsorbent expanding space is limited.

For the environmental temperature of 30 °C and air conditioning temperature of 9 °C, $CaCl_2$ is tested and the optimal filling volume is only 41.6% in the adsorption bed. If filling quantity is more than 41.6%, the adsorption performance of the adsorbent will degrade due to the problem of mass transfer. For the same condition, the results of simply mixed adsorbent as well as the consolidated adsorbent with different volume ratio are shown in Tables 5.6 and 5.7. For

(a) (b)

Figure 5.19 Composite adsorbent of activated carbon and $CaCl_2$ [22]. (a) Simply mixed composite adsorbent and (b) consolidated composite adsorbent

Table 5.6 Performance of simply mixed composite adsorbent for air conditioner (compared with pure $CaCl_2$ adsorbent) [22]

	$CaCl_2$ ratio (%)	Activated carbon ratio (%)	Increment of filling volume (%)	Increment of volume cooling capacity (%)
Condition 1	70	30	29	29
Condition 2	79	21	32	32
Condition 3	82	18	35	26

Table 5.7 Performance of consolidated composite adsorbent for air conditioner (compared with pure $CaCl_2$ adsorbent) [22]

	$CaCl_2$ content (%)	Activated carbon ratio (%)	Water ratio (%)	Cement ratio (%)	Increment of filling volume (%)	Increment of volume cooling capacity (%)
Condition 1	55.3	24	13.8	6.9	29	29
Condition 2	60.2	17.2	15.1	7.5	35	35
Condition 3	57.8	20.5	14.5	7.2	32	32

condition 3 in Table 5.6, due to the high volume percentage of $CaCl_2$, the increment of volume cooling capacity of the adsorber is lower than that of volume filling quantity, which shows agglomeration and swelling influence the mass transfer of adsorbent as well as refrigeration performance of the adsorbent.

For a environmental temperature of 28 °C and a freezing temperature lower than −10 °C, experiments show that, if the filling volume of $CaCl_2$ is larger than the optimal data, i.e. 33.3%, then the mass transfer performance will be critical. The performance of simply mixed adsorbent as well as consolidated adsorbent with a different ratio is shown in Tables 5.8 and 5.9. Condition 2 of Table 5.8 is only suitable for the condition when the refrigeration temperature is −10 °C. If the refrigeration temperature is lower than −10 °C, problems of mass transfer will happen due to the low evaporation pressure in the adsorption process.

Table 5.8 Performance of simply mixed composite adsorbent for ice making condition (compared with pure $CaCl_2$) [22]

	$CaCl_2$ ratio (%)	Activated carbon ratio (%)	Increment of filling volume (%)	Increment of volume cooling capacity (%)	Refrigeration temperature (°C)
Condition 1	60	40	29	29	−15
Condition 2	70	30	40	40	−15
Condition 3	80	20	51	38	−15

Table 5.9 Performance of consolidated composite adsorbent for ice making condition (compared with pure $CaCl_2$) [22]

	$CaCl_2$ ratio (%)	Activated carbon ratio (%)	Water ratio (%)	Cement ratio (%)	Increment of filling volume (%)	Increment of volume cooling capacity (%)	Refrigeration temperature (°C)
Condition 1	49	32.7	12.2	6.1	29	29	−15
Condition 2	55.4	23.8	13.9	6.9	40	40	−10
Condition 3	52.2	28.3	13	6.5	34	34	−15

5.2.4 Composite Adsorbent with Activated Carbon Fiber as Matrix

If the ACF is chosen as a matrix of composite adsorbent the sample with high thermal conductivity is preferred, this will improve the heat transfer of chemical adsorbent effectively. Generally, the higher graphitization degree of the ACF will lead to better heat transfer performance. However, since the price of ACF is quite high, the optimal ACF should always have a comparatively low price as well as high graphitization degree.

Currently, two types of successful composite adsorbents with ACF as the additive are the composites of ACF and $MnCl_2$ (impregnated carbon fiber (ICF)) [6, 7], as well as ACF and $CaCl_2$ (graphite fibers intercalation compound (GFIC)) [8, 9, 25].

5.2.4.1 ICF and GFIC Composite Adsorbents

1. ICFs (with $MnCl_2$) [6, 7]
   The procedures for preparing the composite adsorbents of ACF and $MnCl_2$ by the impregnation method are as follows: firstly, dilute the $MnCl_2$ with alcohol, and then impregnate the activated carbon into the solution of $MnCl_2$, after that, heat the composite adsorbent to dry the sample, which is simple because alcohol evaporates more easily than water. For such a type of adsorbent (ICF) the combination between $MnCl_2$ and ACFs is more fixed after the adsorbent is prepared. The preparation time of ICF is short, for example, only 3–4 hours. But because ICF and $MnCl_2$ only remain at the micro-molecular level, after several cycles of heating desorption and cooling adsorption, the combination between salts in the ICF and ACF will break, and then salt will fall from the ACF and will accumulate at the bottom of the reactor, which will influence the adsorption and desorption performance.
2. GFICs [6, 7]
   $MnCl_2$ formed an intercalated structure in the ACF, synthesized by the following steps: Firstly, ACF is graphitized at high temperatures (thermal conductivity after graphitization is higher than 600 W/(mK), and then under the conditions of chlorine gas temperature of 500 °C $MnCl_2$ intercalated in carbon fibers. To gain a high intercalated rate, in the process of preparing composite adsorbent, generally a small amount of other types of salt will be added such as $FeCl_3$ and $CuCl_2$. Preparation time of the GFIC is long; generally it takes about one week. GFIC in $MnCl_2$ particles of ACF keeps at the atomic level, even if some of the salt is inclined to leave the ACF sandwich, the combination of all the salt and ACFs is very solid.

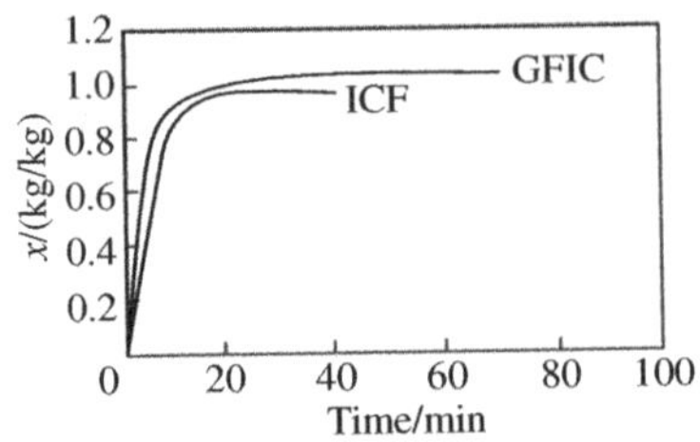

Figure 5.20 Adsorption performance of ICF and GFIC [6, 7]

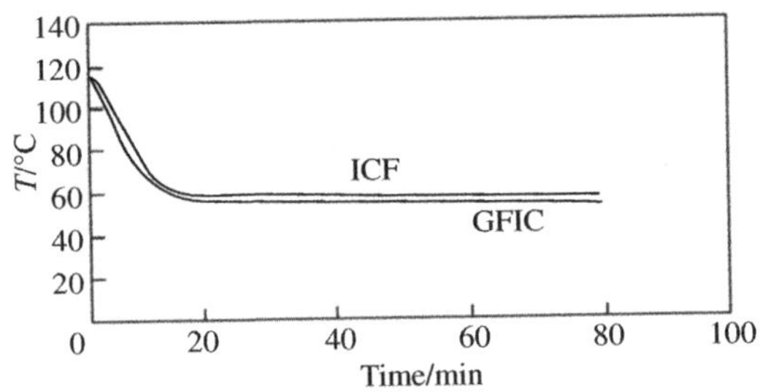

Figure 5.21 Temperature evolution of ICF and GFIC for the refrigeration process [6, 7]

ACF makes up about 25% in ICF and GFIC, respectively. In IFC, if the salt didn't leak from the ACF and if the ammonia is used as a refrigerant, the adsorption performance and cooling process of ICF and GFIC are shown in Figures 5.20 and 5.21. Performance of both working pairs is similar.

The results of comparing the performance of ICF and GFIC are shown in Table 5.10.

Table 5.10 Advantages and disadvantages between GFIC and ICF [6]

Composites	Advantages	Disadvantages
GFIC	Good dynamic performance and adsorption performance	Preparing time is long
	Salt particles evenly distribute in the activated carbon fiber	Intercalation salt is hard to control
		Requires high degree of graphitization of the activated carbon fiber
IFC	Good dynamic performance and adsorption performance	Adsorption performance slightly lower than GFIC
	Easier preparing process	Salt particles is easy to fall down from the activated carbon fiber
	Easy to control the proportion of salts in the activated carbon	
	Doesn't require high degree of graphitization of the activated carbon fiber	

Table 5.11 Composite adsorbent in the resorption system with ammonia as refrigerant [8, 25]

Adsorption bed	Chemical adsorbent	Composite mass (kg)	Proportion of inorganic salts (%)	Proportion of activated carbon fiber (%)	Performance
High temperature adsorption bed	$NiCl_2$	135	51.8	48.2	Power output 1.5 kW, temperature increment 100 °C, and *COP* of 1.43–1.62 (two adsorption beds)
Low temperature adsorption bed	$BaCl_2$	120	50	50	
High temperature adsorption bed	$NiCl_2$	430	41.9	58.1	Temperature increment 50 °C, heat pump can operate at 0 °C (two adsorption bed)
Low temperature adsorption bed	$BaCl_2$	610	44.3	55.7	
High temperature adsorption bed	$MnCl_2$	–	–	–	Highest temperature of heat pump is 120 °C, combined heat pump and refrigeration *COP* is 1.2–1.4 (four adsorption bed)
Low temperature adsorption bed	$BaCl_2$	610	44.3	55.7	

5.2.4.2 Other Composite Adsorbents

Currently, for ACF and metal chloride composite adsorbents, the usual method is the ordinary water impregnated preparation method. Vasiliev *et al.* use Bosofit ACF and metal chloride to produce composite adsorbent, for which a chemical adsorbent can be evenly distributed in 2–3 μm thin layers of ACF. This adsorbent is mainly used for composite adsorbent in the resorption system with ammonia as a refrigerant [8, 25]. The preparation and performance are shown in Table 5.11.

5.2.5 Composite Adsorbents with Silica Gel as Matrix

5.2.5.1 Composite Adsorbents of Silica Gel and $CaCl_2$

Composite adsorbents with silica gel as a matrix are generally prepared by the impregnation method [26–29], for which silica gel is soaked in salt solutions (such as $CaCl_2$) with a fixed solubility, and then the silica gel is dried to obtain a composite adsorbent with a strong adsorption ability. Actually, composite adsorbent with silica gel as a matrix is used to impregnate the metal chloride chemical adsorbent into the microspores of the silica gel to get a strong adsorption ability. Adsorption and desorption reactions may happen between composite adsorbent and water. In actual use, a special processing technique is needed to prevent a salt solution from flowing out of the pores of silica gel after adsorbing water. The characteristics of the composite adsorbent can be improved by the following procedures [30]:

1. **Change the porous structure of the matrix.** In silica gel–metal chloride composite adsorbent, each atom is generally closely surrounded by atoms with a characteristic diameter. If the diameter of salt atoms in the solution is far less than the atom diameter of silica gel, the micro pores of silica gel won't affect the distribution of the salt, and the adsorption performance of composite adsorbent is similar to the inorganic salts. Otherwise its adsorption performance will be different from that of inorganic salt.
2. **Change the chemical characteristic of intercalation salts.** For different inorganic salts, the complexion reaction between the salts and water will be different. Performance of composite adsorbents will show different trends.
3. **Change the quantity of salts inside micro pores.** Generally, agglomeration and the swelling phenomenon of adsorbent will become serious when the quantity of salts in composite adsorbent increases. This will enhance the heat transfer performance as well influencing mass transfer performance.

Take the preparing processes of silica gel and $CaCl_2$ composite adsorbent in Shanghai Jiao Tong University as an example [31]:

1. Preparation for $CaCl_2$ aqueous solution
   Dissolve the $CaCl_2$ into the water with the designated concentration of salt. If the required concentration of solution is not very high, $CaCl_2$ almost instantly dissolves. For a calcium chloride aqueous solution with high concentration it commonly takes about 30 minutes to prepare.
2. Sample preparation
   Soak the silica gel as the matrix in $CaCl_2$ aqueous solution, and then leave the sample open to the environment for 12 hours. Such a process makes $CaCl_2$ impregnate into the micropores. Then filter the silica particles which have taken up $CaCl_2$ with the sieve, the rest of the low concentration of $CaCl_2$ solution won't be used. Heat and dry the silica gel particles at 80 °C in a constant temperature and humidity oven. In the drying process, weigh the samples at regular intervals, and stop the drying process when the weight of the sample is no longer decreasing or the degree of reduction can be ignored.

Aristov produced a composite adsorbent for testing thermal conductivity, for which the preparation method is as follows [30]:

1. Mix silica gel powder and $CaCl_2$ solutions to a 40% concentration.
2. Solidify the powder into a mold which is the shape of square brick sized $7 \times 3 \times 1.5\ cm^3$.
3. Bring the brick into contact with the water vapor until a predetermined water adsorption value is obtained and then test its thermal conductivity.

According to the test results, the highest adsorption capacity of silica gel–$CaCl_2$ composite adsorbent with water can reach 0.7–1.5 kg/kg, which is greater than that of zeolite to water as well as silica gel to water.

Figure 5.22 SEM (scanning electron microscope) picture for composite adsorbent [31]

But there is always a negative effect for the preparation process of silica gel–metal chloride composite adsorbent, that is, the destruction of the solid skeleton of the silica gel particles by the impregnation of $CaCl_2$. This is mainly because adsorption heat leads to the rapid rise of temperature of silica gel. Local temperature can even reach more than 120 °C which will cause the destruction of the structure of silica gel. Figure 5.22 shows the inner structure of composite adsorbent after adsorption and desorption through an electron microscope. Results show that the higher the concentration of $CaCl_2$ is the greater the degree of fragmentation. Because of this, pure silica gel will show a better adsorption performance than the composite adsorbents when relative humidity is low. If an adsorption performance of the composite adsorbent is required to be higher than that of silica gel, on the one hand, the concentration of a $CaCl_2$ aqueous solution should be higher which can improve the adsorption performance of silica gel by the additive of salts; on the other hand, environmental relative humidity should be lower than 70% for composite adsorbent, otherwise the liquefaction of $CaCl_2$ will influence the performance of composite adsorbent.

Figure 5.23 shows the adsorption performance of composite adsorbent with different ratios of salts under the environmental condition, in which S is the sample of adsorbent, the number behind S is the concentration of calcium chloride solution for preparing composite adsorbent, for example, S0 is the silica gel matrix, S40 is composite adsorbent with 40% $CaCl_2$ solution concentration. From Figure 5.23, when salts concentration is more than 40%, increment of equilibrium adsorption quantity is not great. Considering that the higher the concentration of $CaCl_2$ is, the easier the liquefaction phenomenon becomes, the optimal composite adsorbent is sample S40.

5.2.5.2 The Composite Adsorbent of Silica Gel and LiCl

Mass concentration of the impregnating solutions and the pore structure of the matrix are two key parameters to be considered which will have a strong impact on the characteristics of the

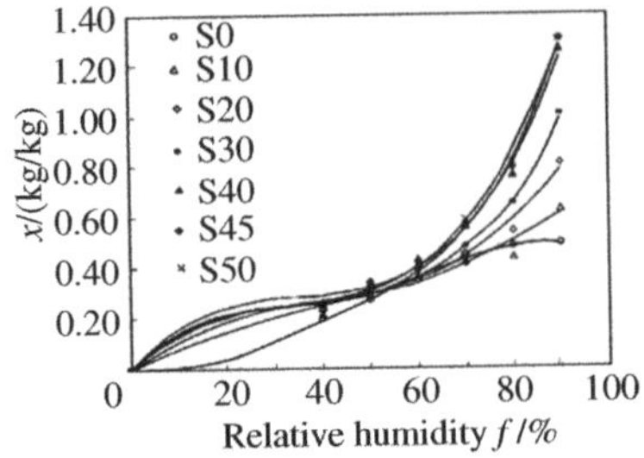

Figure 5.23 Adsorption isothermal curves of different samples [25, 31]

Table 5.12 LiCl mass concentration in the composite sorbents prepared by different silica gel pore sizes and impregnating LiCl mass concentrations

Impregnating LiCl mass concentration in solution		10%	20%	30%	40%
LiCl mass concentration in composite sorbent	Silica gel type A (pore size 2–3 nm)	6.5%	19.8%	24.3%	25.6%
	Silica gel type C (pore size 8–12 nm)	11.4%	24.0%	35.1%	43.6%

prepared composite sorbents. As shown in Table 5.12, two types of mesoporous silica gels (type A and C) and four concentrations of LiCl solutions (10, 20, 30, and 40 wt%) have been used to prepare eight samples. For both silica gel type A and type C, the LiCl mass concentrations in the composites increase with the LiCl concentrations in the impregnating solution, in accordance with expectation. Due to larger pore size and pore volume, silica gel type C is able to carry more LiCl salt crystal in its internal space, especially at high concentrations of 30 and 40 wt% when it seems that the pores of silica gel type A are filled with LiCl crystal particles. Since more salt contents in silica gels would cause higher water sorption potentials, the composite sorbents developed by silica gel type C were chosen as the research samples. The four composites with four concentrations were named SLi10, SLi20, SLi30, and SLi40 for short; the last two numbers represented the impregnating LiCl mass concentration. The pure silica gel type C was called SG and its properties are also studied for comparison with the composites.

Water sorption isobars at 0.88, 1.66, 2.88, and 4.45 kPa on SG and the four silica gel-LiCl composites are shown in Figure 5.24. In Figure 5.24a, isobars for SG are smooth and divariant. Water uptake is a function of both temperature and pressure. Unfortunately, this type of mesoporous silica gel exhibits a poor water sorption property under experimental conditions – the maximum water uptake is 0.088 g/g at 30 °C and 2.88 kPa. The fact suggests that this type of silica gel is inadequate for use of water sorption alone. Isobars for the composites in Figure 5.24b–e show a similar tendency: a sudden change of slope could be found at high temperatures, revealing some sort of transformation happening in this range. After the transition region, the water uptake will drop to nearly zero. Nonetheless, no plateau indicating the formation of the salt hydrate which has been discovered for composites impregnated with other salts like $CaCl_2$ [32, 33] are observed in our research. It is useful to learn that the water uptake increases as the salt content in the composite goes up. For SLi10, the maximum water

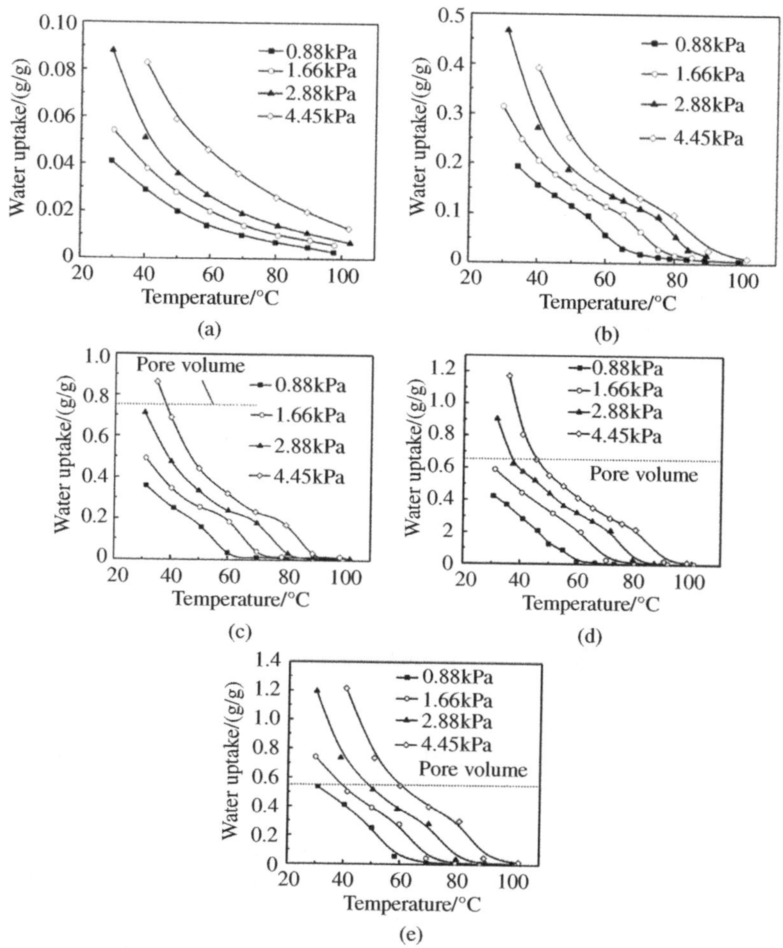

Figure 5.24 Water sorption isobars on the sorbents: (a) SG; (b) SLi10; (c) SLi20; (d) SLi30; and (e) SLi40

uptake is 0.467 g/g, lower than its pore volume of 0.91, so there is no need to be concerned about the carryover issue. The maximum water uptake for SLi40 exceeds 1.2 g/g under a condition of 40 °C and 4.45 kPa, but it should be noted that a carryover problem is encountered in this situation since the pore volume of SLi40 is only 0.55 cm^3/g. SLi20 and SLi30 will also meet the same problem as SLi40 does. When referring to a specific working condition such as a closed sorption system a selection of these sorbents could be carried out to find the most suitable sorbent. If the maximum water uptake takes place at the lowest temperature of 30 °C and the highest pressure of 1.88 kPa regarding an evaporation temperature of 15 °C, it can be concluded that SLi30 is the best option as it has the largest water uptake and is free of the worrying issue of carryover. Another advantage of the composite sorbents is that complete water desorption can be reached at relatively low temperatures, in a range from 60 to 100 °C, which means that they can be regenerated by conventional low-temperature heat sources.

5.3 Adsorption Kinetics of Composite Adsorbents

Under the conditions of different porous medium matrixes the dynamic characteristics of composite adsorbents will be different. For the composite adsorbents with the matrix of expanded graphite, owing to graphite having no adsorption effect on refrigerants, the dynamic equations adopt the chemical adsorption dynamics equations [4, 5, 34, 35] combined with the mass transfer process of expanded graphite matrix, which were introduced in Chapter 4. For the composite adsorbents with physical adsorbents as a matrix, such as activated carbon and silica gel, the current research indicates that, because of the combining functions of chemical and physical adsorption processes, as well as the mass transfer process inside the porous media, the adsorption kinetics characteristics is different from the chemical adsorption kinetics. But it is difficult to separate the physical adsorption from the chemical adsorption for the composite adsorption processes, thus it is generally not easy for the establishment of the kinetic equations with such a type of composite adsorption process involving the chemical and physical adsorption processes.

5.3.1 Dynamics Characteristics of Composite Adsorbents with the Matrix of Silica Gel

For the composite adsorbent with the silica gel as matrix (Aristov defined it as selective water sorbent SWS), the chemical adsorbent is impregnated in the micropores of the porous medium which had an adsorption reaction, showed the characteristics between a porous silica gel and pure hygroscopic salt [26, 30].

Figure 5.25 shows the adsorption properties of the composite adsorbent with the silica gel having an average pore radius of 7.5 nm and $CaCl_2$ proportion of 33.7% in SWS. Figure 5.25 shows that the SWS adsorption performance has been greatly improved if compared with silica gel. The maximum adsorption quantity of SWS and silica gel are 0.75 and 0.1 kg/kg, respectively. In addition, it also shows that for low adsorption capacities, SWS highlights the chemical adsorption phenomenon which is different from physical adsorption. When the adsorption quantity is 0.11 kg/kg, there is an adsorption platform, which indicates that the adsorption capacity during this process is only associated with pressure but has nothing to do with the adsorption temperature. That is, it is controlled by a single parameter. While the adsorption capacity is higher than 0.11 kg/kg, the adsorption process is controlled by the dual variables of constraint pressure and adsorption temperature.

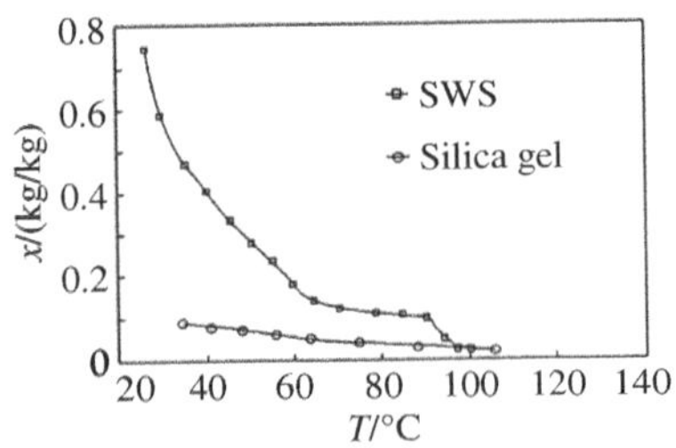

Figure 5.25 Comparison of adsorption performance between SWS and silica gel under the condition of 25 mbar vapor pressure [26]

The adsorption platform in Figure 5.25 states that $CaCl_2$ embedded in the porous silica gel has no effect on the adsorption characteristics, but is different from the pure $CaCl_2$. The adsorption characteristics of solid $CaCl_2$ crystal hydrate changes because of salts impregnated inside the micro pores of silica gel. For the SWS, the double hydrate and three hydrates can be formed at a very low vapor pressure. Compared with the formation process of $CaCl_2$ hydrate, the vapor pressure in the formation process of hydrate for SWS is one order lower.

Using other chemical adsorbents such as LiBr, LiCl, and so on, it will show the common characteristics similar with $CaCl_2$ for the performance of composite adsorbent, namely:

1. The monohydrate forms during the phase of small adsorption capacity.
2. When the adsorption was higher, compared with the phase of small adsorption quantity, the characteristics of monohydrate will change significantly, which was mainly caused by the impregnating process of the inorganic salts in the porous matrix.
3. The impregnation process of the salt in the porous medium won't influence the adsorption performance of salt very much.

In general, the adsorption and desorption dynamics of composite adsorbents are better than the pure silica gel. Taking the S40 composite adsorbent studied by Daou in SJTU as an example its kinetic characteristics are compared with pure silica gel, and the results are shown in Figure 5.26. It can be seen that the adsorption and desorption rate of a composite adsorbent is much faster than that of the pure silica gel.

5.3.2 Dynamics Characteristics of Composite Adsorbents with the Matrix of Activated Carbon Fiber

For the composite adsorbent of ACF and metal chloride, it also showed a combination of physical and chemical adsorption characteristics during the adsorption process of ammonia.

In the application of resorption system, adding the ACF in the metal chloride, the composite adsorbent could react with the refrigerant at a very low constraint pressure for that the ACF adsorption of refrigerant is controlled by a single variable and the porous medium can get a rapid reaction process. So it effectively reduced the differential pressure of the adsorption

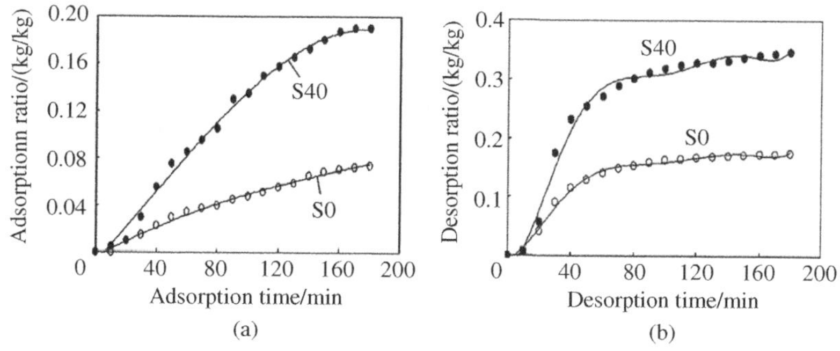

Figure 5.26 Performance comparison of composite adsorbents and silica gel [31]. (a) Adsorption process under the conditions of 10 °C freezing water and 30 °C cooling water and (b) desorption process under the conditions of 80 °C heat source

refrigeration process and desorption process, which could improve the *COP* of the system. In the experiments, ACF could react with ammonia rapidly at the beginning of the heating/cooling process (5 minutes), and terminate the adsorption process before it reaches the end, so the reactor's pressure changes very fast, that is, the pressure changes before the salt reacts with ammonia [8, 25].

For the composite adsorbent of ACF and chemical adsorbent, to study the dynamic characteristics the most complex part is the migration process of the refrigerant inside the adsorbent. Due to the interaction between the salt and ACFs, during the adsorption and desorption process, the refrigerant transfers from the ACF to salt, which reacts with the salt, then part of the refrigerant transfers from the salt to ACF, which is a physical adsorption reaction. This refrigerant migration makes the kinetic study very difficult.

5.3.3 Dynamics Characteristics of Composite Adsorbents with the Matrix of Activated Carbon

For the composite adsorbents with activated carbon as matrix the performance is also the combination of physical adsorption and chemisorption.

Figure 5.27 is the isobaric adsorption performance of chemical adsorbent ($CaCl_2$) and composite adsorbent (composite of $CaCl_2$ and activated carbon) under the conditions of the same mass of chemical adsorbent and the same filling volume. The physical adsorbent inside the composite sample 1 is the 14–28 mesh-activated carbon produced by Hainan coconut shell, and the volume ratio of the calcium chloride and activated carbon is 2 : 1 (mass ratio 4 : 1). This fully guarantees the adsorption time for different temperatures to exclude the heat transfer and mass transfer influence on isobaric adsorption performance. If the dynamic characteristic of composite adsorbent is supposed as a simple combination of chemical adsorption and physical adsorption, the adsorption quantity should be the sum of chemical adsorption quantity and physical adsorption quantity, and the chemical adsorption curve of $CaCl_2$ inside the composite adsorbent should be the same or similar at least to the chemical adsorption curve. However, Figure 5.27 indicates that the adsorption curve of pure $CaCl_2$ at 55 °C was close to the abscissa of the vertical line, that is, it is a chemical adsorption process controlled by a single variable, but for the composite adsorption, due to the addition of physical adsorbent, it appeared obvious

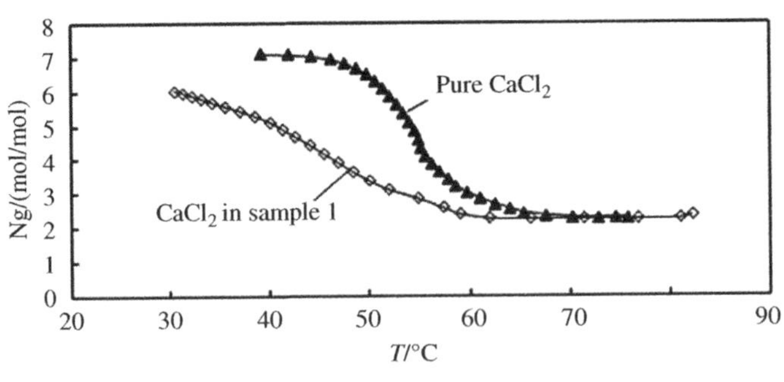

Figure 5.27 Isobaric adsorption performance of sample 1 and $CaCl_2$ [22]

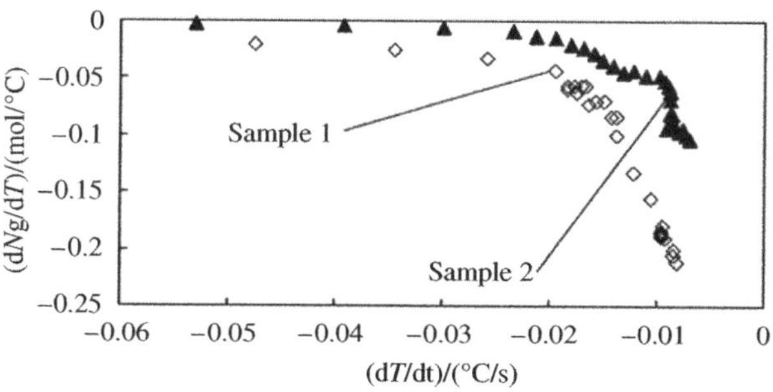

Figure 5.28 Performance comparison of sample 1 and sample 2

characteristics controlled by two variables, that is, its adsorption performance is influenced both by the adsorption temperature, and the constraint pressure.

The deviation of chemisorption process of $CaCl_2$ inside a composite adsorbent is mainly caused by the physical adsorbent in the composite adsorbent. In the adsorption process of the composite adsorbent, the combination of the capillary condensation process of the refrigerant in the physical adsorbent and the chemical adsorption process made the chemical adsorption precursor states deviate from the chemisorption theory. In order to have a clear understanding of that, the composite sample 2 is produced by 20–40 mesh-activated carbon and $CaCl_2$. The mass ratio of additives and the filling volume inside the bed is the same as that of sample 1. Figure 5.28 showed the two samples' kinetic results under the condition of the evaporating pressure of 430 kPa (corresponding to the saturated evaporating temperature of 0 °C).

Figure 5.28 shows that different composite adsorbents have different performances. The performance of the composite sample 1 was better than the one of the composite sample 2. This indicates that the composite adsorption kinetics should combine with chemical adsorption kinetics and mass transfer kinetics of the porous medium. But it is difficult to separate the adsorption kinetics of chemical adsorption process from the mass transfer process and the adsorption performance of different types of granular activated carbon.

References

[1] Han, J.H., Cho, K.W., Lee, K.H. *et al.* (1996) Characterization of graphite-salt blocks in chemical heat pumps. Proceedings of International Ab-sorption Heat Pump Conference, Montreal, Canada, pp. 67–73.

[2] Groll, M. (1992) Reaction beds for dry sorption machines. Proceedings of Symposium of Solid Sorption Refrigeration, Paris, France, pp. 208–214.

[3] Mauran, S., Lebrun, M., Prades, P. *et al.* (1994) Active composite and its use as reaction medium. US Patent 5.283.219.

[4] Mauran, S., Coudevylle, O., and Lu, H.B. (1996) Optimization of porous reactive media for solid sorption heat pumps. Proceedings of International Ab-sorption Heat Pump Conference, Montreal, Canada, pp. 3–8.

[5] Mauran, S., Prades, P. and Haridon, F.L. (1993) Heat and mass transfer in consolidated reaction beds for thermochemical systems. *Heat Recovery Systems and CHP*, **13**, 315–319.

[6] Dellero, T., Sarmeo, D. and Touzain, P. (1999) A chemical heat pump using carbon fibers as additive. Part I: enhancement of thermal conduction. *Applied Thermal Engineering*, **19**, 991–1000.

[7] Dellero, T. and Touzain, P. (1999) A chemical heat pump using carbon fibers as additive. Part II: study of constraint parameters. *Applied Thermal Engineering*, **19**, 1001–1011.

[8] Vasiliev, L.L., Mishkinis, D.A., Antukh, A.A. and Kulakov, A.G. (2004) Resorption heat pump. *Applied Thermal Engineering*, **24**, 1893–1903.

[9] Vasiliev, L.L., Mishkinis, D.A., and Vasiliev Jr.,, L.L. (1996) Multi-effect complex compound/ammonia sorption machines. Proceedings of International Ab-sorption Heat Pump Conference, Montreal, Canada, pp. 3–8.

[10] Wang, M.Z. and He, F. (1984) *Manufacture, Property, and Application of Carbon Fiber*, Science Press, Beijing, ISBN: 15030.585 (in Chinese).

[11] Xie, Y.Z. (1988) *Technique of Carbon Graphite Materials*, The Publishing Press of Hunan University, Changsha, ISBN: 9787314002647 (in Chinese).

[12] Wang, L.W,. Tamainot-Telto, Z., Metcalf, S.J., Critoph, R.E., Wang, R.Z. (2010) Anisotropic thermal conductivity and permeability of compacted expanded natural graphite. *Applied Thermal Engineering*, **30**(13), 1805–1811.

[13] Wang, L.W., Metcalf, S.J., Thorpe, R. *et al.* (2011) Thermal conductivity and permeability of consolidated expanded natural graphite treated with sulphuric acid. *Carbon*, **49**(14), 4812–4819.

[14] Delmonte, J. (1987) *The Technology for the Composite Materials of Carbon Fiber and Graphite Fiber*, Science Press, Beijing, ISBN: 15081.858 (in Chinese).

[15] Wang, K., Wu, J.Y., Wang, R.Z. and Wang, L.W. (2006) Composite adsorbent of $CaCl_2$ and expanded graphite for adsorption ice maker on fishing boats. *International Journal of Refrigeration*, **29**, 199–210.

[16] Oliveira, R.G., Wang, R.Z. and Wang, C. (2007) Evaluation of the cooling performance of a consolidated expanded graphite-calcium chloride reactive bed for chemisorption icemaker. *International Journal of Refrigeration*, **30**(1), 103–112.

[17] Wang, L.W., Metcalf, S.J., Thorpe, R. *et al.* (2012) Development of thermal conductive consolidated activated carbon for adsorption refrigeration. *Carbon*, **50**, 977–986.

[18] Wang, L.W., Tamainot-Telto, Z., Thorpe, R. *et al.* (2011) Study of thermal conductivity, permeability, and adsorption performance of consolidated composite activated carbon adsorbent for refrigeration. *Renewable Energy*, **36**, 2062–2066.

[19] Tamainot-Telto, Z. and Critoph, R.E. (1997) Adsorption refrigerator using monolithic carbon-ammonia pair. *International Journal of Refrigeration*, **20**(2), 146–155.

[20] Critoph, R.E. (1988) Performance limitations of adsorption cycles for solar cooling. *Solar Energy*, **14**(1), 21–31.

[21] Wang, L.W., Wang, R.Z., Wu, J.Y. and Wang, K. (2004) Compound adsorbent for adsorption ice maker on fishing boats. *International Journal of Refrigeration*, **27**(4), 401–408.

[22] Wang, L.W. (2005) Performances, mechanisms, and application of a new type compound adsorbent for efficient heat pipe type refrigeration driven by waste heat. PhD thesis. Shanghai Jiao Tong University, Shanghai, China (In Chinese).

[23] Wang, L.W., Wang, R.Z., Wu, J.Y. and Wang, K. (2004) Adsorption performances and refrigeration application of adsorption working pair of $CaCl_2$–NH_3. *Science in China, Series E*, **47**(2), 173–185.

[24] Wang, L.W., Wang, R.Z., Wu, J.Y. and Wang, K. (2005) Research on the chemical adsorption precursor state of $CaCl_2$–NH_3 for adsorption refrigeration. *Science in China, Series E*, **48**(1), 70–82.

[25] Vasiliev, L.L., Mishkinis, D.A., Antuh, A. *et al.* (1999) Multisalt-carbon chemical cooler for space applications. Proceedings of International Absorption Heat Pump Conference, Munich, Germany, pp. 579–583.

[26] Aristov, Y.I., Restuccia, G., Caccioba, G. *et al.* (2002) A family of new working materials for solid sorption air conditioning systems. *Applied Thermal Engineering*, **22**, 191–204.

[27] Tokarev, M., Gordeeva, L., Romannikov, V. *et al.* (2002) New composite sorbent $CaCl_2$ in mesopores for sorption cooling/heating. *International Journal of Thermal Science*, **41**, 470–474.

[28] Levitskij, E.A., Aristov, Y.I., Tokarev, M.M. *et al.* (1996) Chemical heat accumulators: a new approach to accumulating low potential heat. *Solar Energy and Solar Cells*, **44**, 219–235.

[29] Restuccia, G., Freni, A., Vasta, S. and Aristov, Y.I. (2004) Selective water sorbent for solid sorption chiller: experimental results and modeling. *International Journal of Refrigeration*, **27**, 284–293.

[30] Aristov, Y.I., Tokarev, M.M., Parmon, V.N., *et al.* (1999) New working materials for sorption cooling/heating driven by low temperature heat: properties. Proceedings of International Sorption Heat Pump Conference, Munich, Germany, pp. 24–26.

[31] Daou, K. (2006) The development, experiment, and simulation of a new type of efficient composite adsorbent driven by the low grade heat source. PhD Thesis. Shanghai Jiao Tong University, Shanghai, China (in Chinese).

[32] Aristov, Y.I. (2007) New family of solid sorbents for adsorptive cooling: material scientist approach. *Journal of Engineering Thermophysics*, **16**, 63–72.

[33] Cortés, F.B., Chejne, F., Carrasco-Marín, F. *et al.* (2012) Water sorption on silica- and zeolite-supported hygroscopic salts for cooling system applications. *Energy Conversion and Management*, **53**, 219–223.

[34] Lebrun, M. and Spinner, B. (1990) Models of heat and mass transfers in solid-gas reactors used as chemical heat pumps. *Chemical Engineering Science*, **45**(7), 1743–1753.

[35] Lu, H.B., Mazet, N., Coudevylle, O. and Mauran, S. (1997) Comparison of a general model with a simplified approach for the transformation of solid-gas media used in chemical heat transformers. *Chemical Engineering Science*, **52**(2), 311–327.

6

Adsorption Refrigeration Cycles

The adsorption refrigeration cycle system can roughly be summarized in Figure 6.1 through the development history of adsorption refrigeration as well as the study status of scholars international.

By the working principle the adsorption refrigeration cycle can be divided into intermittent cycles and continuous cycles. For the intermittent adsorption refrigeration cycles there was always one bed utilized in the system. For the continuous refrigeration cycles there are generally two or more beds running alternately, which can provide continuous refrigeration output. For an adsorption refrigeration cycle if only the heating and desorbing process as well as the cooling and adsorbing process are involved, it is generally defined as the basic adsorption refrigeration cycle. The advanced adsorption refrigeration cycles are commonly referred to as the cycles with heat and mass recovery processes, such as two-bed heat recovery process, multi-bed heat recovery process, mass recovery process, thermal wave, and convective thermal wave cycles. According to the adsorption system characteristics and temperature source selections, the advanced refrigeration cycles of multi-stage and cascading refrigeration system can also be constructed.

6.1 Basic Adsorption Refrigeration Cycles

6.1.1 The Basic Intermittent Adsorption Refrigeration Cycle and Its Clapeyron Diagram

A basic intermittent adsorption refrigeration cycle [1] consists of an adsorbent bed, a condenser, a reservoir, and an evaporator. The system's principle is shown in Figure 6.2. The adsorption bed is also known as a reactor in the chemisorption refrigeration cycle.

For different adsorbents the Clapeyron diagrams of adsorption refrigerating cycles are different. For example, activated carbon, silica gel, and zeolite are physical adsorbents, and the metal chlorides are chemical adsorbents. The Clapeyron diagrams of physical and chemical adsorbents are different.

The Clapeyron diagram of an ordinary physical adsorption working pair as is shown in Figure 6.3a. The Clapeyron diagram of a metal chloride–ammonia adsorption working pair is shown in Figure 6.3b (taking calcium chloride–ammonia, for example). Comparing two

Adsorption Refrigeration Technology: Theory and Application, First Edition. Ruzhu Wang, Liwei Wang and Jingyi Wu.

Companion Website: www.wiley.com/go/wang/refrigeration

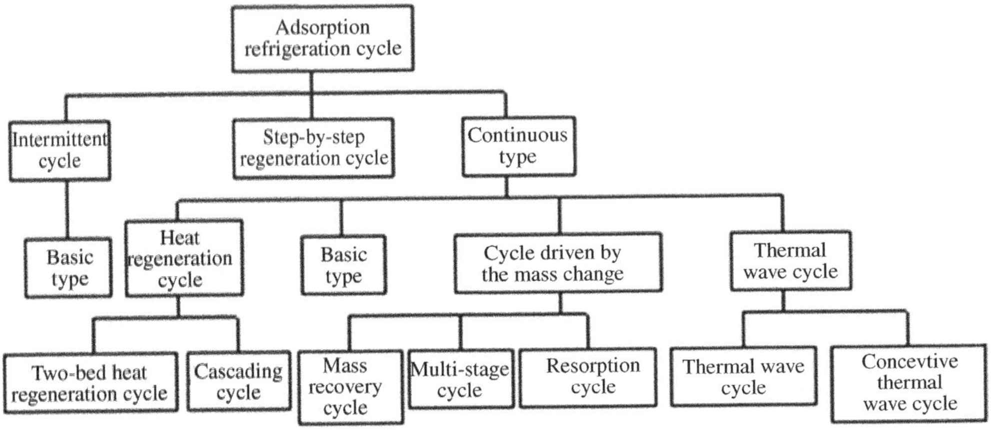

Figure 6.1 Classification of adsorption refrigeration cycle

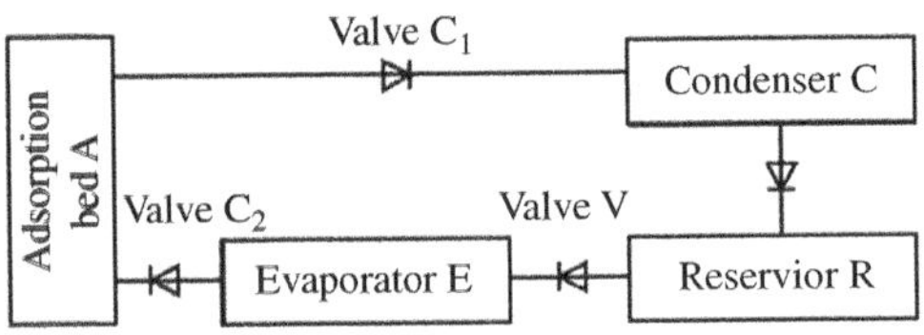

Figure 6.2 Schematic diagram of basic adsorption cycle

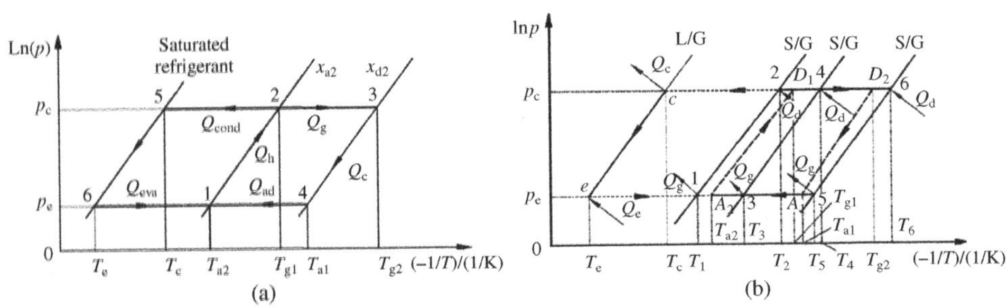

Figure 6.3 Basic adsorption refrigeration cycle diagrams. (a) The Clapeyron diagram of physical adsorption and (b) the Clapeyron diagram of chemical adsorption for $CaCl_2$-NH_3

figures the difference is that x in the physical adsorbent is determined by two independent variables, while for metal chlorides the adsorption process is determined by a single independent variable.

The working processes of physical adsorbents can be shown in Figure 6.3a in detail:

1. **1-2**: In the adsorbent bed, the adsorber and the adsorbent, which occurs after adsorption and is saturated, is heated, and the temperature rises from T_{a2} to T_{g1}, and the pressure of

the refrigerant in the adsorption bed rises from p_e to p_c during the process. It needs to be emphasized here that p_e and p_c are determined by the evaporating temperature and the condensing temperature. For the heating process, the valve C_1, which is shown in Figure 6.2, is closed at the beginning, so generally we assume that desorption doesn't occur until the pressure reaches p_c, thus for this process the volume doesn't change, that is, it is an isometric heating process. Because of this the mass of the refrigerant gas in the mass transfer channels of the bed and in the micropores of adsorbent is very small relative to the mass of the refrigerant; in the heating process the sensible heat consumed by the refrigerant gas in the adsorbent bed is generally ignored.

2. **2-3**: The adsorbent inside the bed is continually heated until its temperature reaches the maximum desorption temperature T_{g2}. At the same time, the adsorbed refrigerant is desorbed. Because the pressure is mainly controlled by the condensing pressure in this phase, the process can be looked as an isobaric process with the pressure of p_c. This process assumes that the refrigerant gas is condensed into the condenser immediately when it is desorbed.
3. **3-4**: It is similar with the process 1-2. When the adsorbent inside the bed is desorbed completely, the bed will be cooled and the temperature dropped from T_{g2} to T_{a1} as well as the pressure of the refrigerant being reduced from p_c to p_e. In this process because the valve linked the evaporator and the bed is closed, the volume can be looked upon as a constant, that is, the process can be treated as an isometric process.
4. **4-1**: When the adsorbent is cooled to the adsorption temperature T_{a2}, the valve between the evaporator and the bed will be open. The adsorbent will adsorb the refrigerant inside the evaporator, and pressure will be controlled by the evaporating pressure. Thus the process can be analyzed by an isobaric process with a pressure of p_e. This process will be completed when the adsorbent is restored to the state 1.

The refrigerant gas desorbed from the adsorbent bed is condensed in the condenser by an isobaric process with a pressure of p_c. Condensing pressure p_c is determined by the condensation temperature T_c. The initial temperature of the refrigerant gas from the bed is assumed to be the same as the temperature of the gas entering the condenser. T_{a2} and T_{g2} are the final temperatures of the adsorption and desorption processes, respectively. T_e and T_c are the evaporation temperature and the condensation temperature, separately (corresponding saturated pressure are p_e and p_c). T_{a1} and T_{g1} are initial temperatures of adsorption and desorption processes, separately.

As shown in Figure 6.3b, when the adsorbents are different, the equilibrium reaction curves will be different. However, for the same chemical adsorbent, if the equilibrium temperature of the reaction is determined, the equilibrium pressure of the reaction is also determined. Taking the 3-4 curve ($CaCl_2{\cdot}2NH_3 \leftrightarrow CaCl_2{\cdot}4NH_3$) for example, the reaction processes are as follows:

1. **3-4**: It is the heating process of the reactor (adsorbent bed). In this process, the valve C_1, which is shown in Figure 6.2, is shut off, and the pressure of the adsorption bed rises while it is heated by the external heat source. When the temperature reaches T_4, the corresponding pressure reaches condensing pressure p_c, which is the condensing pressure.
2. **Point 4**: It is the desorption point of the reactor. Open valve C_1 in Figure 6.2, then the adsorbent bed begins to desorb at temperature T_4 and pressure p_c. In this process both the temperature and pressure are constant.

3. **4-3**: It is the cooling process of the reactor. In this process, valve C_2 in Figure 6.2 is closed, and the temperature and the pressure of the reactor decrease as a result of the heat exchange with an external cooling source. When the temperature of the reactor drops to T_3, the corresponding pressure reaches the evaporation pressure of p_e.
4. **Point 3**: It is the adsorption point of the reactor. In this process valve C_2 in Figure 6.2 is open, and the refrigerant in the evaporator evaporates and produces a cooling effect under the conditions of temperature T_3, the corresponding pressure is p_e. In this process, both temperature and pressure are constant.

Chemisorption reaction of the metal chloride is the complexion reaction between the metal chloride and the refrigerant. The stability constant is raised as the temperature decreases. So for any reaction equilibrium line in Figure 6.3b, the decomposition reaction generally occurs to the right of the reaction equilibrium line. Meanwhile the synthesis reaction generally occurs at the reaction equilibrium line or to the left of the reaction equilibrium line.

Figure 6.3b shows the adsorption refrigeration cycle of A_2-D_1-D_2-A_1-A_2, the specific working processes are:

1. $\boldsymbol{A_2}$**-**$\boldsymbol{D_1}$: The heating process of the adsorbent bed. The adsorbent bed is heated by the external heat source, and its temperature and pressure increases.
2. $\boldsymbol{D_1}$**-4**: When the adsorbent bed's pressure reaches the condensation pressure, the valve between the bed and the condenser is open, and the refrigerant vapor condenses in the condenser. D_1-4 is at the right side of the reaction line 1-2, which is the decomposition process from $CaCl_2{\cdot}8NH_3$ to $CaCl_2{\cdot}4NH_3$.
3. **4-**$\boldsymbol{D_2}$: Point 4 is at the curve of 3-4, and 4-D_2 is at the right side of the reaction line 3-4, which is the decomposition process from $CaCl_2{\cdot}8NH_3$ to $CaCl_2{\cdot}4NH_3$.
4. $\boldsymbol{D_2}$**-**$\boldsymbol{A_1}$: The cooling process of the adsorbent bed. The adsorbent bed is cooled by the external cold source, and its temperature and pressure decreases.
5. $\boldsymbol{A_1}$**-3**: When the adsorbent bed's pressure is decreased to the evaporation pressure, the valve between the evaporator and the adsorbent bed is open, and the bed adsorbs the refrigerant from the evaporator. A_1-3 is to the right of the reaction line 5-6, which is the synthesis process from $CaCl_2$ to $CaCl_2{\cdot}2NH_3$.
6. **3-**$\boldsymbol{A_2}$: Point 3 is at the curve of 3-4, 3-A_2 is to the left of the reaction line 3-4, which is the synthesis process from $CaCl_2{\cdot}2NH_3$ to $CaCl_2{\cdot}4NH_3$.

For the basic cycle process of A_2-D_1-D_2-A_1-A_2, since the reaction lines 1-2 and 5-6 are outside of the cycle of A_2-D_1-D_2-A_1-A_2, the synthesis process of $CaCl_2{\cdot}4NH_3$ to $CaCl_2{\cdot}8NH_3$ and the decomposition process of $CaCl_2{\cdot}2NH_3$ to $CaCl_2$ could not be completed. So the actual reaction process only involves $CaCl_2{\cdot}2NH_3 \leftrightarrow CaCl_2{\cdot}4NH_3$, which means only 2 mol of ammonia are involved in the reaction.

In an adsorption system, an evaporator, a throttle, and the liquid reservoir can be viewed as a subsystem. In this subsystem, the liquid refrigerant's temperature drops from T_c to T_e (evaporation temperature), then evaporates from the evaporator under the constant pressure of p_e, which is the evaporating pressure.

6.1.2 *Continuous Adsorption Refrigeration Cycle*

The basic type of continuous adsorption refrigeration cycle is generally composed of two or more adsorbent beds, which can ensure that at any time at least one adsorbent bed is at the cooling and adsorption stage, for which the cooling power can be output continually.

The two-bed operating system is generally used in the cycle when the optimum desorption time is the same as the optimum adsorption time. This can take a moderate average time as a half-cycle period both for adsorption and desorption processes. The multi-bed system can be used in the cycle for which the optimum adsorption time varies significantly from the optimum desorption time. For example, if the optimal adsorption time is twice of the optimal desorption time, the three-bed system can be used. For such a system when two beds are under the condition of cooling and adsorption another will be under the condition of the heating and desorption process.

The diagram of two-bed refrigeration cycle is shown in Figure 6.4. Taking the physical adsorption cycle for example, in Figure 6.4 when the adsorbent bed 1 is heated and desorbed the process is 1-2-3 in Figure 6.3a. During the process of 2-3, valve C_3 is open, the refrigerant gas desorbing from the adsorbent bed will be condensed in the condenser C. At the same time, adsorbent bed 2 is at the cooling and adsorption state, the change process is 3-4-1 in Figure 6.3a. During the process of 4-1, valve C_4 is open for adsorbent bed 2 and the refrigerant evaporated in the evaporator as a result of the adsorption in adsorbent bed 2. Such a process produces cooling power.

Assuming that the adsorbent bed is saturated, the operation processes of two-bed and multi-bed cycles are shown in Tables 6.1 and 6.2, respectively. t is desorption time for an adsorbent bed, and it is also a half-cycle time for the two-bed system.

If compared with the multi-bed system the two-bed system is relatively simple. In the two-bed system at the time 0 the adsorbent bed 1 is switched to be heated (SH) for desorption. Then at time t, when the adsorbent bed 1 is switched to be cooled (SC) for adsorption, the adsorbent bed 2 starts to be heated for desorption. A cycle ends when the cycle time is $2t$. Under normal operating conditions, $2t$ to $4t$ will be the next cycle for the two-bed system.

For the multi-bed system, it will need n adsorbent beds if the optimum adsorption time is $n-1$ times of the optimum desorption time. Then when one adsorbent bed is under the condition of heating and desorption, the other $n-1$ adsorbent beds will be under the condition of cooling and adsorption refrigeration. However, it will also need n adsorbent beds if the

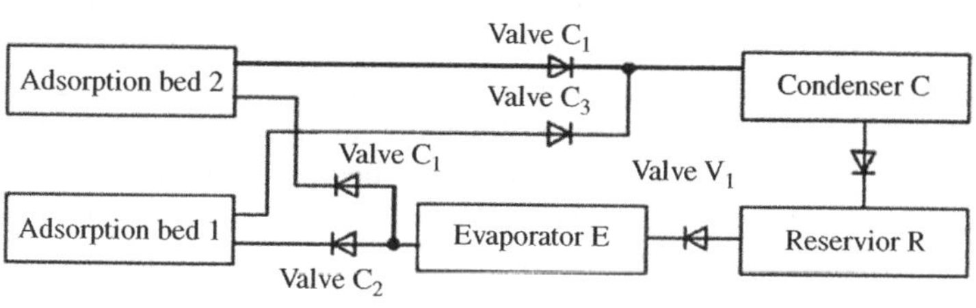

Figure 6.4 Two-bed continuous adsorption refrigeration cycle

Table 6.1 The operation process of two-bed system

Time (minutes)	Adsorbent bed 1	Adsorbent bed 2
0	SH[a]	–
t	SC[b]	SH
2t	SH	SC
3t	SC	SH
4t	SH	SC

[a]SH, switched to be heated.
[b]SC, switched to be cooled.

Table 6.2 The operation processes of multi-bed system

Time (minutes)	Adsorbent bed 1	Adsorbent bed 2	Adsorbent bed 3	...	Adsorbent bed (n – 1)	Adsorbent bed n
0	SH[a]					
t	SC[b]	SH				
2 t	SC	SC	SH	...		
...	...	...	...	...		...
(n – 1)t	CC[c]	CC	CC	...	SC	SH
nt	SH	CC	CC	...	CC	SC
(n + 1)t	SC	SH	CC	...	CC	CC
(n + 2)t	CC	SC	SH	...	CC	CC
...	...	...	...	...		...
2nt	SH	CC	CC	...	CC	SC

[a]SH, switched to be heated.
[b]SC, switched to be cooled.
[c]CC, continue to be cooled.

optimum desorption time is $n - 1$ times of the adsorption time. Then when one bed is under the condition of cooling and adsorption, the other $n - 1$ adsorbent beds will be under the condition of heating and desorption. Assume that for a system the optimum adsorption time is $n - 1$ times of the optimum desorption time, the operation processes of a multi-bed system are shown in Table 6.2. When the system works, the adsorbent bed 1 begins to be heated and the desorption proceeds. After the time of t, bed 1 is switched to the condition of cooling and adsorption, and bed 2 will be switched to the condition of heating and desorption. At the time of $2t$, bed 1 will continue the process of cooling and adsorption, bed 2 will be switched to be cooled for adsorption process, and bed 3 will start to be heated for desorption process. When the adsorbent bed n is heated and desorbs completely, that is, at the beginning of time nt, one cycle completes. The new cycles will be proceeded as shown in Table 6.2 as from time nt to $2nt$.

In addition to the above cycles, another basic type of continuous refrigeration cycle is the multi-bed cycle for which the optimum adsorption time and the optimum desorption time have no multiple relation. Taking a two-bed adsorption refrigeration system, for example, when the

Table 6.3 The operation processes of two-bed system with irregular optimal adsorption and desorption time

Time (minutes)	Adsorbent bed 1	Adsorbent bed 2
0	SH[a]	–
20	SC[b]	–
30	CC[c]	SH
50	SH	SC
70	SC	CC
80	CC	SH
100	SH	SC

[a]SH, switched to be heated.
[b]SC, switched to be cooled.
[c]CC, continue to be cooled.

optimal cooling and adsorbing time of the adsorbent bed is 30 minutes and the optimal heating and desorbing time is 20 minutes, two-bed cycle's operation processes are shown in Table 6.3, and the cycle time is 50 minutes. Apparently the cycle will be simpler if the heating time is equal to the cooling time, such as when we take the equal half cycle times of 25, or 20 and 30 minutes, respectively.

For a basic continuous cycle without considering the heat recovery processes, the operation processes of a multi-bed system are mainly determined by the optimum adsorption and desorption time. The factors which will affect the adsorption and desorption time mainly include the adsorption and desorption characteristics of the adsorbent, the heat transfer performance of the adsorbent bed, and the mass transfer of the adsorbent bed, and so on.

The most widely used cycle for the two-bed system is the system with equal adsorption and desorption time. In the 1920s, based on the concept of continuous cycle, the two-bed system driven by the open gas flame for heating and desorption process was developed. The heat transfer process is generally the natural convection process for cooling the adsorbent bed and condenser. Commonly such a system has the merits of simple structure and reliable performance. Under ideal conditions, the refrigeration performances can be calculated by the desorption heat of the working pairs and the latent heat of vaporization of the refrigerant, and the metal heat capacity of the adsorbent bed can be neglected. Under such a condition the coefficient of performance (COP) is a theoretical value and always will be much higher than the actual applications. In the real application the COP is generally less than 0.4 due to the heat capacity and other irreversible factors [2].

6.1.3 Thermodynamic Calculation and Analysis of a Basic Cycle

From the perspective of heat transfer performance, as shown in Figure 6.3, a basic cycle involves seven kinds of heat. Assuming that the working fluids for heating and cooling processes are the same in the adsorbent bed (i.e., the same pipes are used both for heating and cooling processes), and taking no account of the accumulated fluid heat capacity in the adsorbent bed during the heating and cooling processes, the various heats are calculated as follows:

1. Q_h – sensible heat in the adsorbent bed during the isometric heating process (the point 1-2 in Figure 6.3a and the point of A_2-D_1 in Figure 6.3b).

$$Q_h = \int_{T_{a2}}^{T_{g1}} C_a(T)M_a\mathrm{d}T + \int_{T_{a2}}^{T_{g1}} C_{Lc}(T)M_a x_{a2}\mathrm{d}T + \int_{T_{a2}}^{T_{g1}} C_m(T)M_{madb}\mathrm{d}T \qquad (6.1)$$

where $C_a(T)$ is the specific heat capacity of the adsorbent; $C_{\mathrm{Lc}}(T)$ is the specific heat capacity of the liquid refrigerants; $C_m(T)$ is the specific heat of metal in the adsorbent bed; M_a, $M_a x_{a2}$, and M_{madb}, are the mass of the adsorbent, liquid refrigerants, metal of the adsorbent bed, respectively. x_{a2} is the adsorption quantity of the adsorbent bed at the end of the adsorption process at temperature T_{a2}. The first part of the formula is the sensible heat of the adsorbent, the second part is the sensible heat of refrigerants, and the third part is the sensible heat of the metal of the adsorbent bed.

2. Q_d – desorption heat in the desorption process (point 2-3 in Figure 6.3a and point D_1-D_2 in Figure 6.3b)

$$Q_d = \int_{T_{g1}}^{T_{g2}} C_a(T)M_a\mathrm{d}T + \int_{T_{g1}}^{T_{g2}} C_{Lc}(T)M_a x\mathrm{d}T + \int_{T_{g1}}^{T_{g2}} C_m(T)M_{madb}\mathrm{d}T - \int_{T_{g1}}^{T_{g2}} M_a h_d \frac{\mathrm{d}x}{\mathrm{d}T}\mathrm{d}T \qquad (6.2)$$

where dx is negative in the desorption process, h_d is desorption heat, x is the adsorption quantity of the adsorbent. The first part of the formula is the sensible heat of the adsorbent, the second part is the sensible heat of the refrigerant in the adsorbent bed, the third part is the sensible heat of the metal heat capacity, and the last part is the desorption heat.

3. Q_c – the sensible heat during the cooling process of adsorbent bed (point 3-4 in Figure 6.3a and point D_2-A_1 in Figure 6.3b)

$$Q_c = \int_{T_{a1}}^{T_{g2}} C_a(T)M_a\mathrm{d}T + \int_{T_{a1}}^{T_{g2}} C_{Lc}(T)M_a x_{d2}\mathrm{d}T + \int_{T_{a1}}^{T_{g2}} C_m(T)M_{madb}\mathrm{d}T \qquad (6.3)$$

where x_{d2} is the adsorption quantity at the end of the desorption at temperature T_{g2} adsorbent.

4. Q_{ad} – the heat exhausted to the surroundings in the cooling process for adsorption by the cold source (the process 4-1 in Figure 6.3a and the process of A_1-A_2 in Figure 6.3b).

$$Q_{ad} = \int_{T_{a2}}^{T_{a1}} C_a(T)M_a\mathrm{d}T + \int_{T_{a2}}^{T_{a1}} C_{Lc}(T)M_a x\mathrm{d}T + \int_{T_{a2}}^{T_{a1}} C_m(T)M_{madb}\mathrm{d}T + \int_{T_{a2}}^{T_{a1}} M_a h_a\mathrm{d}x$$
$$- \int_{T_{a1}}^{T_{a2}} C_{pc}(T)M_a(T - T_e)\frac{\mathrm{d}x}{\mathrm{d}T}\mathrm{d}T \qquad (6.4)$$

where $C_{pc}(T)$ is the specific heat at a constant pressure of the gaseous working fluid and the h_a is the adsorption heat. The last part is the adsorbed sensible heat of the gaseous working fluid from the evaporation temperature to T_{a2}.

5. Q_{eref} – the cooling power from the evaporation heat of the refrigerant

$$Q_{\text{eref}} = M_a \Delta x\, L \tag{6.5}$$

where L is the latent heat of vaporization of the refrigerant.

6. Q_{cond} – the heat released in the condensation process (assuming the condenser is at a constant temperature, and ignoring the sensible heat of the metal of the condenser)

$$Q_{cond} = M_a L \Delta x + \int_{T_{g1}}^{T_{g2}} C_{pc}(T) M_a (T - T_c) \frac{\mathrm{d}x}{\mathrm{d}T} \mathrm{d}T \tag{6.6}$$

where the first part is latent heat of vaporization and the second part is the sensible heat of the vapor working fluid released during the condensation process. T_g is the temperature of the adsorbent bed in the desorption process.

7. Q_{evas} – the sensible heat released to the surroundings by the liquid refrigerant dropped from T_c to the evaporation temperature T_e.

$$Q_{evas} = \int_{Tc}^{Te} C_{Lc}(T) M_a \Delta x \mathrm{d}T \tag{6.7}$$

The above formulas are all the theoretical equations. Actually, due to the complexity of the physical properties of the working fluid and various types of losses, it's relatively difficult to calculate the heat accurately. Under this condition we can use the above formulas to analyze the cycle and guide the design of the system theoretically.

COP can be used to evaluate the cycle, and the formula is as follows:

$$COP = \frac{Q_{eref} - Q_{evas}}{Qh + Q_d} \tag{6.8}$$

In the calculation, T_{g1} is associated with T_c, and T_e is associated with T_{a2}. Solving D-A equation on both sides of x_{a2} the following formula can be obtained:

$$T_{g1} = \frac{T_c \times T_{a2}}{T_e} \tag{6.9}$$

Similarly, solving D-A equation on both sides of x_{d2} the following formula it can be calculated:

$$T_{a1} = \frac{T_e \times T_{g2}}{T_c} \tag{6.10}$$

For the physical adsorption, the desorption heat can be obtained by the Clausius-Clapeyron equation:

$$H_r = \frac{R \times A \times T}{T_c} \tag{6.11}$$

where T is the adsorbent bed temperature, T_c is the condensing temperature, p_c is the corresponding saturation temperature, R is the universal gas constant, A is the Clausius-Clapeyron equation coefficient, for the refrigerant of methanol, $A = 4432$.

For the system of the working pair of metal chloride–ammonia, the reaction heat (adsorption/desorption heat) has the same magnitude, and the typical value $\Delta H_r = 50 \pm 15$ kJ/mol. Generally, the value of the condensation heat of vaporization is half of the reaction heat.

6.2 Heat Recovery Concept Introduced in the Adsorption Refrigeration Cycle

The adsorption refrigeration has a prominent problem of low COP, which is generally less than 0.4 in most cases [3] of the basic cycles. The main reason of this problem is due to the adsorbent bed which is under the condition of heating and cooling alternatively, that is, the adsorbent bed will be switched between the high temperature and low temperature frequently. During the adsorption process, the adsorbent bed needs to release the sensible heat and adsorption heat, whereas in the desorption process the adsorbent bed needs to absorb the sensible heat and the desorption heat from the external heat sources. Such a process will make the sensible heat loss large and therefore will influence the energy coefficient. For a single-bed refrigeration system, the above operation is required and sensible heat loss cannot be avoided. But for a multi-bed system, we can consider the heat recovery process, which is generally operated at the switch time for the adsorption and desorption processes and could recover the sensible heat somehow to improve the efficiency.

The heat recovery concept is firstly introduced in the adsorption refrigeration system for improving COP based on the principle of recoverer. The internal space of the heat recoverer is similar to the adsorbent bed, and in alternating heating and cooling processes the thermal energy can be stored in the heat recoverer and then released [4]. The heat recoverer used in the solid-gas adsorption system is the adsorbent bed itself, so the adsorption system with the heat recoverer is also called a heat regeneration cycle. The typical cycle for the heat regeneration process is proposed by Tchernev, which is shown in Figure 6.5 [5]. At the beginning of the heating process of the adsorbent bed the heat transfer fluid is preheated by the adsorption heat of another adsorbent bed that is at the beginning of cooling, then heated by the boiler, and at last flows into the adsorbent bed for the desorption process.

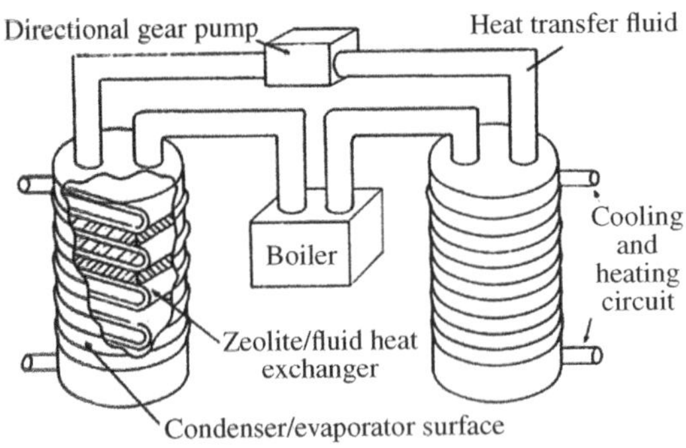

Figure 6.5 Heat regeneration cycle proposed by Tchernev [5]

Because part of the sensible heat of the adsorber can be recovered successfully in the heat recovery process, the heat of q_{reg} provided by the external heat source for heating the adsorbent bed reduces in the heat regeneration cycle, and the recovery coefficient r_{hc} (also known as the heat recovery rate), which is defined as the heat ratio of heat recovered from the adsorbent bed in a heat recovery process and the heat required by the adsorbent bed without heat recovery process:

$$r_{hc} = \frac{q_{reg}*}{q_{reg}} \tag{6.12}$$

where q_{reg} is the required heat of the adsorbent bed without heat recovery process (the heat consumed in the desorption process of a basic adsorption refrigeration cycle), $q_{reg}*$ is the heat recovered from a heat recovery process of the adsorbent bed. Then for a cycle with the heat recovery process the required heat in the desorption process will become $(1-r_{hc})\, q_{reg}$.

Obviously, for an ideal cycle, without heat recovery process $r_{hc}=0$, and with heat recovery process $r_{hc}>0$.

The cooling energy coefficient COP_{int} in the basic intermittent cycle is defined as:

$$COP_{\text{int}} \approx \frac{L\Delta x}{M_a\Delta x.\Delta H_r + \sum(M_a + M_a.x)C_p\Delta T} \tag{6.13}$$

The COP with the heat recovery process is:

$$COP = \frac{COP_{\text{int}}}{1-r} \tag{6.14}$$

Due to the heat recovery efficiency r is bigger than 0, the refrigeration cycle's COP with heat recovery process is significantly larger than that of the cycle without the heat recovery process.

6.3 The Heat Recovery Process of Limited Adsorbent Bed Temperature

In the heat recovery process heat will be transferred from a hot adsorbent bed to a cold adsorbent bed. In this process the temperature of the cold adsorbent bed could not be heated to higher than the hot adsorbent bed, and the coefficient of heat recovery is limited by thermodynamic constraints related to the operating conditions. Generally the heat recovery cycles include two-bed continuous heat regeneration cycle and cascading cycle.

6.3.1 Two-Bed Heat Regeneration Cycle

In order to fulfill the heating and cooling processes the valves and pumps need to be provided in the adsorption refrigeration system. When the adsorbent bed is switched from the cooling and adsorption mode to the heating and desorption mode two beds need to be connected for the heat recovery processes. The heat recovery process generally is fulfilled by the heat transfer fluid flows from the high temperature adsorbent bed to the low temperature adsorbent bed. The working process and Clapeyron diagram of a heat regeneration cycle [6, 7] are shown in Figure 6.6.

In Figure 6.6a the operation direction of two beds are 180° inversed. The heat recovery proceeds between the hot bed after desorption and the cold bed after adsorption at the switch time. For the first phase, the sensible heat and desorption heat required by the cold bed 1,

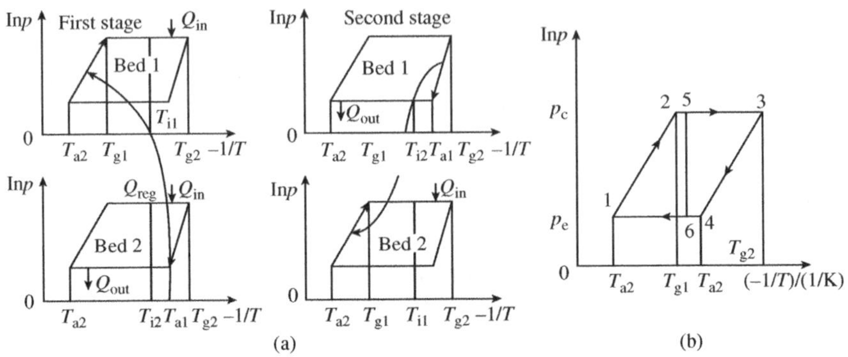

Figure 6.6 Two-bed heat regeneration adsorption refrigeration cycle. (a) Working process of the heat regeneration cycle and (b) Clapeyron diagram of the heat regeneration cycle

which has just switched from the cooling process to heating process, will be provided by the hot bed 2 just after desorption. After the heat recovery in the second phase heat Q_{in} will be provided by the high-temperature heat source. In the heat recovery process hot bed 2 releases the heat to cold bed 1, such a process could reduce the heat exhausted to the surroundings by hot bed 2, that is, part of the heat released by bed 2 is absorbed by the bed 1, which provides the heat needed by bed 1 for the heating and desorption process. The other part of heat Q_{out} is released to the environment. At the second stage, the situation is exactly the opposite, it saves the energy required by the heating process of cold bed 2 and reduces the waste heat emission to the environment by hot bed 1.

The Clapeyron diagram of the heat regeneration cycle is shown in Figure 6.6b. Assuming that the working state of the system is switched, bed 1 in Figure 6.6a is point 3 in Figure 6.6b, and bed 2 is at the state of point 1 in Figure 6.6b. During the heat recovery process, the temperature of adsorbent bed 1 would be dropped to point 6 along the 3-4-6, and the temperature of adsorbent bed 2 would be raised to point 5 along 1-2-5. In the continuous heat regeneration cycle, the formulas of heat change $Q_1(T)$ of bed 1 and $Q_2(T)$ of bed 2 are as follows:

$$Q_1(T) = Q_h|_{T_{a2}}^{T_{g1}} + Q_g|_{T_{g1}}^{T_5} \tag{6.15}$$

$$Q_2(T) = Q_c|_{T_{a1}}^{T_{g2}} + Q_{ad}|_{T_6}^{T_{a1}} \tag{6.16}$$

where Q_h and Q_g are calculated by Equations 6.1 and 6.2, respectively; T_5 was the temperature of point 5 in Figure 6.6b. Q_c and Q_{ad} are calculated by Equations 6.3 and 6.4, respectively; T_6 is the temperature of point 6 in Figure 6.6b. T_6 is equal to T_5.

$$Q_1(T) = Q_2(T) \tag{6.17}$$

The regenerative temperature T_{reg} [8, 9] of the ideal condition could be obtained from Equation 6.17 (T_5 and T_6). The ideal regenerative heat was:

$$Q_{reg} = Q_1(T_{reg}) = Q_2(T_{reg}) \tag{6.18}$$

It should be noted that the regenerative temperature T_{reg} could be at a different point between T_{a2} and T_{g1} in Figure 6.6, and it accordingly could be characterized by a variety of situations such as the adsorption heat recovery, the sensible heat recovery of the adsorbent bed, and so on. For most cases T_{reg} is at the typical states as shown as the typical points in Figure 6.6. In the actual situation, taking into account the temperature difference of heat transfer and the time for heat recovery, generally T_5 is smaller than T_6, and the difference between them commonly is 5–10 °C.

6.3.2 The Examples for the Thermodynamic Calculation of Two-Bed Heat Regenerative Adsorption Refrigeration Cycle

To demonstrate the characteristics of two-bed heat regenerative adsorption refrigeration cycle, the working pair of the activated carbon–methanol is chosen for the calculation of the *COP* for the refrigeration and air conditioning conditions of an adsorption refrigeration cycle. The physical properties used in the simulation are shown in Table 6.4.

Because the specific heat of carbon fiber cannot be found in the literature, the specific heat of the YAK is chosen as the data of the carbon fiber.

Different designs of the adsorbent bed, such as the different types of the adsorbent bed of the shell and tube type and plate-fin type, will have different values of metal heat capacity, and consequently will have different refrigeration performances. In order to exclude the influences caused by the design of the adsorbent beds on calculation results, only the physical properties of the adsorbents are used for the performance comparisons of different adsorption refrigeration cycles. For example, the two-bed heat regenerative adsorption refrigeration cycle and the basic cycle are compared in this chapter, and in the comparison we only consider the heat capacity of the adsorbent and adsorbate, regardless of the design of the adsorbent bed. The conditions for the numerical calculation are as follows:

1. The conditions for the refrigeration are: evaporation temperature $T_e = -10\,°C$, condensing temperature $T_c = 30\,°C$, and adsorption temperature $T_{a2} = 30\,°C$.
2. The conditions for air conditioning are: evaporation temperature $T_e = 5\,°C$, condensing temperature $T_c = 30\,°C$, and adsorption temperature $T_{a2} = 30\,°C$.

Table 6.4 The equations for the properties used in the simulation

Properties	Equation
Specific heat of liquid methanol (kJ/(kg·K))	$C_{Lc} = 0.78019 + 0.005862T$
Specific heat of gas methanol (kJ/(kg·K))	$C_{pc} = 0.66 + 0.221 \times 10^{-2}T + 0.807 \times 10^{-6}T^2 - 0.89 \times 10^{-9}T^3$
Specific heat of the YK activated carbon (kJ/(kg·K))	$C_a = 0.805 + 0.00211T$
Latent heat of vaporization of methanol (heat of condensation) (kJ/kg)	$L = 1252.43 - 1.59593T - 0.00881551T^2$[a]

[a]Note: In the equation of the vaporization heat the unit of the temperature is °C, and for other equations the unit for the temperature is K.

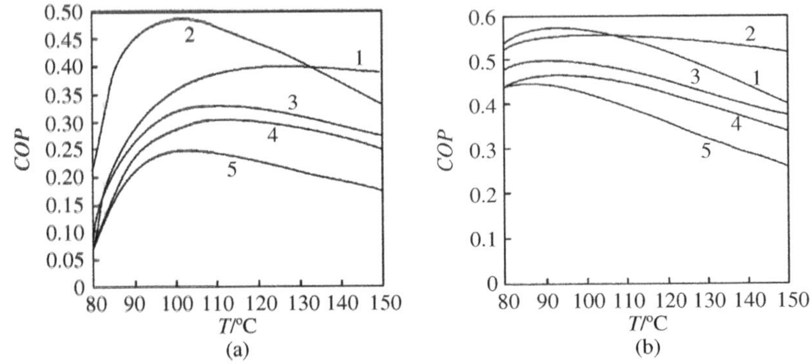

Figure 6.7 Relationship between the COP of basic cycle and the desorption temperature. (a) Refrigeration cycle and (b) air conditioning cycle

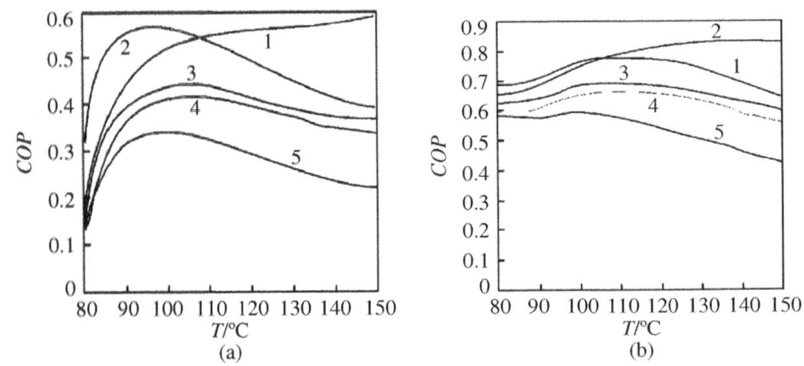

Figure 6.8 Relationship between the *COP* of heat regenerative cycle and the desorption temperature. (a) Refrigeration cycle and (b) air conditioning cycle

3. The relationship between the highest desorption temperature and the *COP* for both the basic cycle and two-bed regenerative cycle are shown in Figures 6.7 and 6.8. Curve 1 is for the working pair of NTACF–methanol; Curve 2 is for the working pair of SYACF–methanol; Curve 3 is for the working pair of JIAACF–methanol; Curve 4 is for the working pair of YKAC–methanol; Curve 5 is for the working pair of SXAC–methanol. The models of various adsorbents have been listed in Table 3.2.

Comparing Figure 6.7 with Figure 6.8, the following conclusions can be obtained:

1. For the working pairs for the cycles with regenerative processes, the *COP* is obviously increased relative to the basic cycle.
2. *COP* will increase significantly when the evaporation temperature increases.
3. There is a great difference among the working pairs. Basically, the *COP* of the adsorbent of activated carbon fiber is higher than that of the activated carbon. For the activated carbon fiber under a moderate heating temperature, the regularity of *COP* is: *COP* for the

NTACF > the *COP* for SYACF > the *COP* for JIAACF. While the temperature is high, SYACF has a better performance, but it does not help the performance of the adsorption system because we don't always use too high a temperature for the methanol refrigerant considering that the methanol will decompose under that condition.

The influence of evaporation temperature T_e and the maximum desorption temperature T_{g2} on the cycle characteristics represents the influence of the cycle conditions on the cycle characteristics. In general, the adsorption temperature T_{a2} and the condensing temperature T_c were limited by the ambient temperature. For the design of the system, it is reasonable to make it sufficiently close to the ambient temperature. The combination of the evaporation temperature T_e and the maximum desorption temperature T_{g2} represents the combination of the application condition and the driving heat source. The suitable driving heat source under certain application conditions or some suitable applications under the known driving heat source can be obtained from the following simulation results. Figure 6.9a indicates the influence of evaporation temperature and the maximum desorption temperature on the COP of a basic cycle for the working pair of the activated carbon–methanol. Figure 6.9b showed the comparison of *COP* between the regenerative cycle and the basic cycle for the working pair is the activated carbon YKAC–methanol. It could clearly be seen in Figure 6.9b that the COP of the regenerative cycle is about 20% higher if compared with that of the basic cycle.

If the impact of the metal heat capacity of the adsorbent bed and the heat fluid were considered for the evaluation on cycle characteristics in the above simulation, the results would be different. The influence of the metal heat capacity and fluid on the performance and design optimization of the adsorption refrigeration system will be analyzed in detail in Chapter 8.

6.3.3 *Cascading Cycle*

Generally two or more types of working pairs will be needed for the cascading cycle. For different working pairs the equilibrium adsorption/desorption temperature will be different,

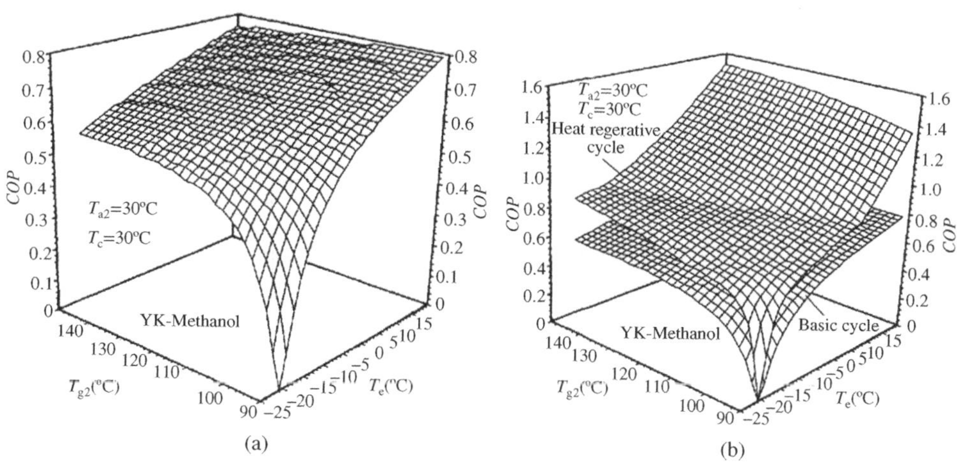

Figure 6.9 The influence of evaporation temperature T_e and the maximum desorption temperature T_{g2} on the COP of the basic cycle. (a) Basic cycle and (b) comparison of heat regenerative cycle and basic cycle

and the heat recovery process will be operated between different working pairs. The main characteristic of the cycle is that the heat will be transferred from the working pair with a higher equilibrium adsorption/desorption temperature to the working pair with the lower equilibrium adsorption/desorption temperature. Taking the working pairs of activated carbon–methanol and zeolite–water as an example, the detailed process of a cascading cycle will be demonstrated.

The properties for the working pairs of zeolite–water and activated carbon–methanol have already been described in detail in the second chapter. The working pair of zeolite–water can be used for the recovery of waste heat under the condition of equal to or higher than 200 °C. But the highest desorption temperature of activated carbon–methanol can't be higher than 120 °C, otherwise methanol will decompose. Using the working pairs of the activated carbon–methanol and zeolite–water for the construction of cascading cycles, two-stage cascading double effect cycle and two-stage cascading triple effect refrigeration cycle can be achieved by these two working pairs.

6.3.3.1 Two-Stage Cascading Double Effect Adsorption Refrigeration Cycle (Adsorption Heat Utilization)

Two-stage cascading double effect adsorption refrigeration cycle (adsorption heat utilization) is shown in Figure 6.10, which uses the working pair of zeolite–water for the recovery of the high temperature/middle temperature heat source (250 to 100 °C), and activated carbon–methanol working pair worked for the recovery of the middle temperature/low temperature heat source (100 to 35 °C). For the low-temperature adsorbent bed the driving heat was provided in full by the sensible heat and adsorption heat by cooling the high-temperature adsorbent bed. As shown in Figure 6.10b, when the zeolite adsorbent bed was heated by an external heat source, the activated carbon adsorbent bed is in the cooling state by the external cold source, and the adsorption of the activated carbon produces refrigeration output. When the

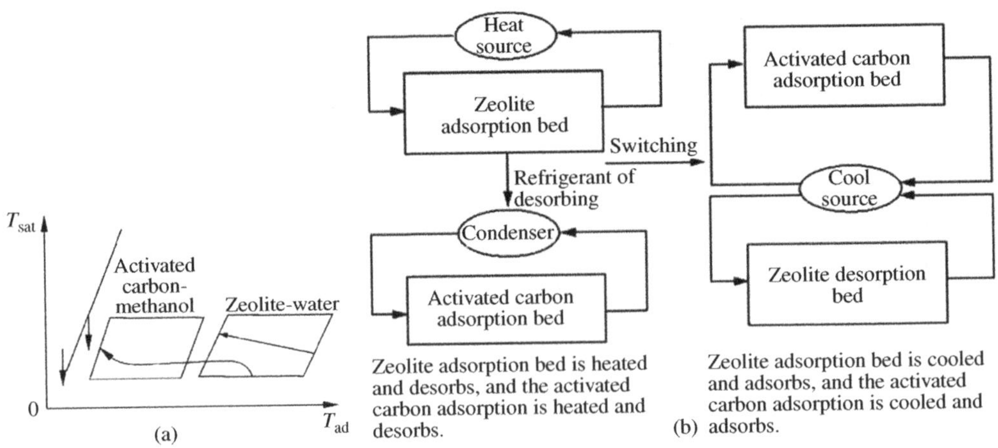

Figure 6.10 Two-stage cascading double effect adsorption refrigeration cycle (adsorption heat utilization). (a) Diagram of the temperature for the cycle and (b) the heating and cooling processes of adsorbent bed

desorption inside the zeolite adsorbent bed is completed, the zeolite adsorbent bed is cooled for the adsorption process, and in this process the zeolite bed releases the heat to the bed of activated carbon, which provides the heat for the desorption process of the activated carbon bed. In the cooling process of the zeolite bed we need to control the outlet temperature of the bed at 100 °C or so, and then let the outlet water of the zeolite bed flow to the bed of activated carbon adsorbent. In this cycle the double effect heat utilization includes the heat from the outside high-temperature heat source (the first effect heat utilization) and the heat from the cooling process for high-temperature adsorbent bed (sensible heat and the adsorption heat) that is used to drive the low-temperature adsorbent bed (the second effect heat utilization).

In Figure 6.10, the two-stage cascading cycle could be coupled with heat recovery and mass recovery processes. In the cycle two types of adsorption working pairs, that is, high temperature and low-temperature adsorption working pairs are adopted, and at switch time, the heat recovery and mass recovery processes proceed between two high-temperature adsorbent beds and two low-temperature adsorbent beds, respectively. After the heat recovery the heating and cooling processes will be operated as shown in Figure 6.10b.

In Figure 6.10, assume that the evaporating temperature of the system was T_e, the condensation temperature was T_c, and the heat source temperature was T_h, then the *COP* was:

$$COP = COP_Z + COP_{AC} \tag{6.19}$$

where Z was the zeolite; AC was the activated carbon.

6.3.3.2 Two-Stage Cascading Double Effect Adsorption Refrigeration Cycle (Condensation Heat Utilization)

The principle diagram and flow chart of a two-stage cascading double effect adsorption refrigeration cycle (condensing heat utilization) are shown in Figure 6.11. Figure 6.11a shows that

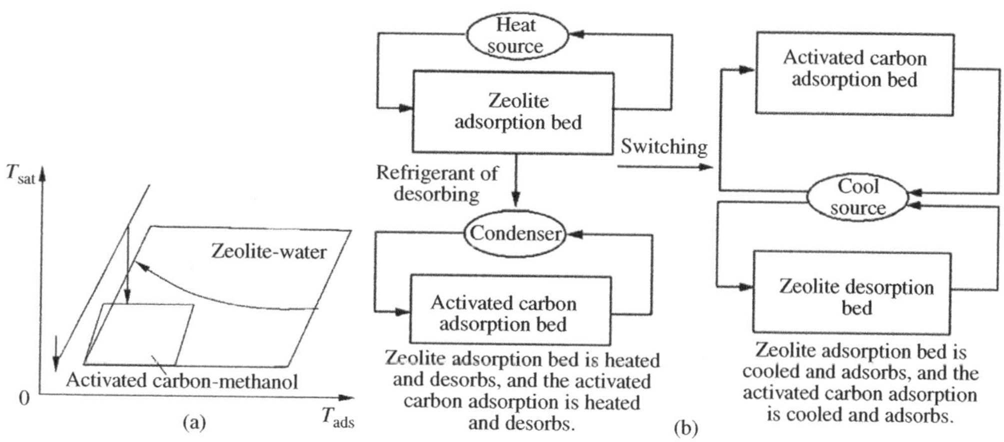

Figure 6.11 Two-stage cascading double effect adsorption refrigeration cycle (condensation heat utilization). (a) The diagram of the temperature for the cycle and (b) heating and cooling processes of adsorbent bed

in the cycle the working pair of zeolite–water works at high temperature/low temperature (250 to 35 °C), and the activated carbon–methanol working pair works at middle temperature/low temperature (100 to 35 °C). The heat source of the low-temperature adsorbent bed is provided by the condensation heat of liquid desorbed from the high-temperature adsorbent bed. As shown in Figure 6.11b, at the beginning the zeolite adsorbent bed is heated by an external heat source. In this process the refrigerant desorbed from the high temperature bed will also have the high temperature and will be condensed in the condenser. By controlling the balance of condensation heat and the external cold source for the condenser, the temperature for the outlet of the condenser can be controlled at a temperature of 100 °C. This heat can be used for the heating and desorption process of the activated carbon adsorbent bed. When the heating and desorption processes for the zeolite adsorbent bed and activated carbon adsorbent bed are completed, the external cold source will be used to cool the zeolite and activated carbon adsorbent bed to complete the adsorption refrigeration process of the system. The operation of the whole system needs the heat input to be provided by an external heat source (the first effect) and the internal vapor condensation heat (the second effect).

The two-stage cascading cycle could also be used for two zeolite adsorbent beds and two activated carbon adsorbent beds in order to achieve the heat and mass recovery processes between adsorbent beds.

In Figure 6.11, assume that the evaporating temperature of the system is T_e, the condensation temperature is T_c, and the heat source temperature is T_h, then the COP is:

$$COP = COP_Z + COP_Z \times COP_{AC} \tag{6.20}$$

6.3.3.3 Two-Stage Cascading Triple Effect Adsorption Refrigeration Cycle

The schematic and working processes of a two-stage cascading triple effect adsorption refrigeration cycle are shown in Figure 6.12. Figure 6.12a shows that, in this cycle, the zeolite–water working pair works at high temperature/middle temperature area (250 to 100 °C), and the working pair of activated carbon–methanol works at middle temperature/low temperature area (100 to 35 °C). The triple effect adsorption refrigeration cycle generally needs four or more

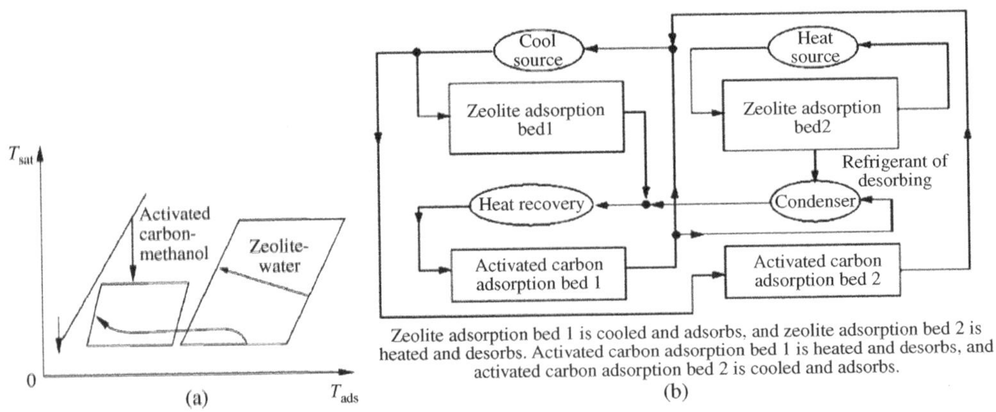

Figure 6.12 Two-stage cascading triple effect adsorption refrigeration cycle. (a) The diagram of the temperature for the cycle and (b) heating and cooling processes of adsorbent beds

adsorbent beds. The working processes of a four-bed system are shown in Figure 6.12b. While the zeolite adsorbent bed 1 is at the cooling and adsorption state, the zeolite adsorbent bed 2 is at the heating and desorption state. There is a heat recoverer recovering the sensible heat of zeolite adsorbent bed 1, the adsorption heat of zeolite adsorbent bed 1, and the condensation heat of zeolite adsorbent bed 2 in the condenser, by which the recoverer will heat the activated carbon bed 1. At this time, the activated carbon bed 2 is cooled by the cold source, and the adsorption of the zeolite bed 1 and activated carbon bed 2 produces a cooling capacity. When the above working states are completed as shown in Figure 6.12b, the cycle will be switched to the state where the zeolite adsorbent bed 1 is heated for the desorption process, the zeolite adsorbent bed 2 is cooled for the adsorption process, activated carbon adsorbent bed 1 is cooled for the adsorption process, and activated carbon adsorbent bed 2 is heated for the desorption process. Then the heat source for activated carbon adsorbent bed 2 is provided by the condensation heat of zeolite adsorbent bed 1, the sensible heat of zeolite adsorbent bed 2, and the adsorption heat of zeolite adsorbent bed 2. For the triple effect thermodynamic cycle the operation of the whole system needs an external heat source (the first effect), the external sensible and adsorption heat from zeolite beds (the second effect), and internal vapor condensation heat (the third effect).

Assuming that the two-stage cycle has the same evaporation temperature, the *COP* of the cycle can be obtained under the conditions of the working temperatures, such as evaporation temperature, the condensation temperature, environmental temperature, and the external heat source temperature, that is:

$$COP = COP_Z + (1 + COP_Z)COP_{AC} \tag{6.21}$$

There is also a state for the incomplete utilization of the heat in a two-stage cascading triple effect adsorption refrigeration cycle. For such a state the schematic of temperature for the cycle is shown in Figure 6.13. Compared with Figure 6.12, the utilization process of the heat for the two-stage cascading triple effect adsorption refrigeration cycle in Figure 6.13 is the same as that in Figure 6.12, and the working processes are shown in Figure 6.12a. The heat source of the low-temperature adsorbent bed is provided by the condensation heat of the refrigerant desorbed from the high-temperature adsorbent bed, the sensible heat for cooling the bed, and the adsorption heat. The difference is mainly for the area of the working temperature for the high-temperature adsorbent bed. The high-temperature adsorbent bed of the two-stage cascading triple effect adsorption refrigeration cycle works at the high temperature/low temperature area (250 to 35 °C). In this case, the utilization of the sensible heat and the adsorption heat for cooling the high-temperature adsorbent bed is limited by the low-temperature area of the adsorbent bed, so the heat utilization of the low temperature bed can't match the heat released by the high temperature bed perfectly.

In the two-stage cascading triple effect adsorption refrigeration cycle, the heat and mass recovery processes can also be used for high-temperature adsorbent beds as well as low-temperature adsorbent beds to improve the system's adsorption refrigeration performance.

6.3.4 *The System Design of a Cascading Cycle, Working Process Analysis, and the Derivation for the COP of Triple Effect Cycles*

The example for the design of a cascading system is shown in Figure 6.14a. In the system the refrigerant for the adsorbent bed A, B, C, D is the same. In the figure, 1 is the condenser and 2 is the evaporator. A and B are the high-temperature adsorbent beds (used for zeolite–water).

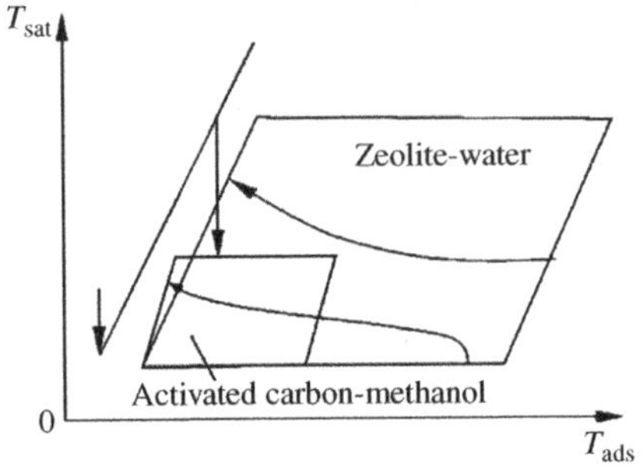

Figure 6.13 Two-stage cascading triple effect adsorption refrigeration cycle (incomplete heat utilization)

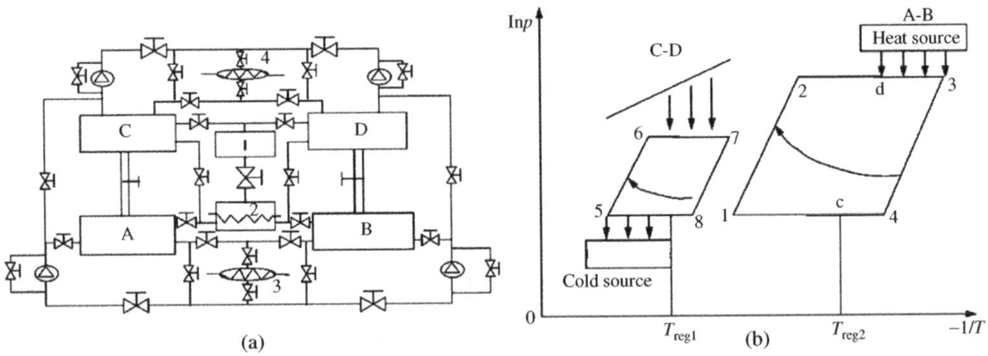

Figure 6.14 Four-bed double effect/triple effect cascading adsorption refrigeration system. (a) System design and (b) Clapeyron diagram

The external heat source is required for the desorption processes of A and B. C and D are the low-temperature adsorbent beds (used for the working pair of silica–water), which is heated by the high-temperature vapor desorbed from the adsorbers of A and B that directly flows into the adsorbers of C and D, as well as the sensible heat and the adsorption heat of the adsorbent beds A and B. The adsorption pressure of A, B, C, D is the same. The heater is 3 and the cooler is 4. The working processes are as follows:

1. Regeneration processes between beds of A and B as well as beds of C and D. After the adsorption processes for adsorbers A and C completes, and the desorption processes for the adsorbers of B and D finishes, the regeneration process will proceed between the beds by the control of the valves in the system. For the regeneration process between beds of A and B, the temperature after regeneration is T_{reg1}, and the temperature after the regeneration process between beds of C and D is T_{reg2}, which are shown in Figure 6.14b.
2. The heat transfer process between beds B and C and the heating process to bed C by the vapor desorbed from the bed A. As shown in Figure 6.14b, the process of c-1 is the adsorption process of bed B, which is an exothermic process. The exothermic heat includes the

sensible heat and the adsorption heat of the adsorbent, which is used for the heating process of bed C. At the same time, bed A desorbs, and the desorbed high-temperature vapor flows into bed C. After releasing a part of the sensible heat, the vapor will be mixed with the desorbed vapor from C and all the vapor will flow into the condenser, then the adsorption process of bed B is completed. The required desorption heat for *d*-3 process of bed A is supplied by the external heat source. Meanwhile the adsorption process of bed D is completed, and the heat is released to the environment.

The working processes of the system are shown in Figure 6.14a. The desorption and adsorption processes of two groups of adsorbers should be synchronous. In order to let the vapor in the bed of A flow into the bed of C, the desorption pressure of bed A should be higher than the desorption pressure of bed C.

For the operation of the cycle, if the sensible heat and the adsorption heat of the high-temperature adsorbent bed are used the cascading system is the double effect cycle. For such a cycle the high temperature adsorbent bed will be heated by the external heat source (the first effect), and the low temperature adsorbent bed will be heated by the sensible heat and the adsorption heat of the high temperature bed. If the desorbed vapor can be utilized for the heating process of the low temperature adsorbent bed, the third effect energy can be utilized, but for such an occasion the desorption temperature/pressure of the high temperature adsorbent bed needs to be improved.

For the theoretical heat recovery process, the temperature after the heat recovery will be the intermediate temperature $T_1 = T_7$, as shown in Figure 6.14b. In the figure the first stage cycle is 5-6-7-8 in the diagram (low-temperature stage), and the second stage cycle is 1-2-3-4 (high-temperature stage).

Taking the triple effect cascading cycle for example, the first stage cycle (low-temperature stage), and the second stage cycle (high-temperature stage) can be seen as two continuous regeneration cycles, but the first stage cycle is driven by the sensible heat and adsorption heat of the second stage cycle after the heat recovery process, as well as the condensing heat of the refrigerant. Assuming that the refrigeration coefficient is COP_1 for the first stage cycle and is COP_2 for the second stage cycle, then

$$COP_1 = \frac{Q_{ref1}}{Q_{hg1} - Q_{reg1}} \tag{6.22}$$

$$COP_2 = \frac{Q_{ref2}}{Q_{hg2} - Q_{reg2}} \tag{6.23}$$

where Q_{ref} is the cooling capacity and Q_{hg} the heat from the heat source.

Due to only the high temperature adsorbent bed using a heat input that is from an external heat source, and assuming that in the cycle the total cooling capacity is Q_{ref}, the refrigeration quantity for the high-temperature and low-temperature stages are Q_{ref2} and Q_{ref1}, respectively. For the double effect cascading cycle, the *COP* is:

$$COP = \frac{Q_{ref}}{Q_{hg2} - Q_{reg2}} = \frac{Q_{ref2} + Q_{ref1}}{Q_{hg2} - Q_{reg2}} = COP_2 + COP_1 \cdot \frac{Q_{hg1} - Q_{reg1}}{Q_{hg2} - Q_{reg2}} \tag{6.24}$$

For the cycle of the high-temperature adsorbent beds, according to the energy conservation:

$$Q_{hg2} + Q_{ref2} = Q_{ad2} + Q_2 \tag{6.25}$$

where Q_2 is the exothermic heat by the refrigerant vapor at the high-temperature stage, and the Q_{ad} is the heat released to the environment by the adsorption process.

Ideally, the heat from the vapor of the high-temperature adsorbent beds can be used completely for driving the low-temperature adsorbent beds. According to the working processes of the double effect cascading cycle, we can get:

$$Q_{hg1} - Q_{reg1} = (Q_{ad2} - Q_{reg2}) + Q_2 \tag{6.26}$$

Substitute Equation 6.25 into Equation 6.26, we can get:

$$Q_{hg1} - Q_{reg1} = Q_{hg2} - Q_{reg2} + Q_{ref2} \tag{6.27}$$

Substitute Equation 6.27 into Equation 6.24, we get:

$$\begin{aligned} COP &= COP_2 + COP_1 \cdot \frac{Q_{hg1} - Q_{reg1}}{Q_{hg2} - Q_{reg2}} = COP_2 + COP_1 \cdot \frac{Q_{hg2} - Q_{reg2} + Q_{ref2}}{Q_{hg2} - Q_{reg2}} \\ &= COP_2 + COP_1 \cdot (1 + COP_2) = COP_1 + COP_2 + COP_1 \cdot COP_2 \end{aligned} \tag{6.28}$$

For example, in a cycle the zeolite–water is used as the working pair for the first-stage, and the silica gel–water is used as the working pair for the second stage, under conditions where the evaporation temperature is 5 °C, condensing temperature is 35 °C, the degree of subcooling is 5 °C, and the condensation pressure of the second stage is the saturated pressure of water at 50 °C. Then in the ideal case, the double effect cascading cycle can be simulated by the computer. Take the intermediate temperature of 100 °C and the maximum heating temperature of 200 °C, COP_1 calculated by the computer is 0.65, COP_2 calculated by the computer is 0.41, and the total refrigeration *COP* is 1.2. In this case, the condensation pressure in the second-stage cycle is higher than that in the first stage, so there is a certain loss. In the ideal case, two stages should have the same pressure and the total refrigeration *COP* should be greater than 1.2.

In an ideal case of the triple effect adsorption refrigeration cycle, if the *COP* of high-temperature and low-temperature stages is 0.6, the total *COP* of the two-stage cascading system is up to 1.56 [10, 11].

6.4 Thermal Wave Cycles

It was Shelton who first proposed using thermal wave in the adsorption system [12, 13]. Compared with the heat recovery process that is limited by the temperature of the beds, the merit of a thermal wave cycle [14–19] is that the heat can be transferred from the hot adsorbent bed to cold adsorbent bed because the temperature difference is great.

6.4.1 The Principle of the Basic Thermal Wave Cycle

The comparison of a basic continuous cycle and the thermal wave cycle is shown in Figure 6.15. The basic continuous solid adsorption refrigeration system only has two adsorbent beds, and the adsorption and desorption processes proceed alternatively. That is, when one bed is heated by the external heat source for desorption, another bed is cooled by the

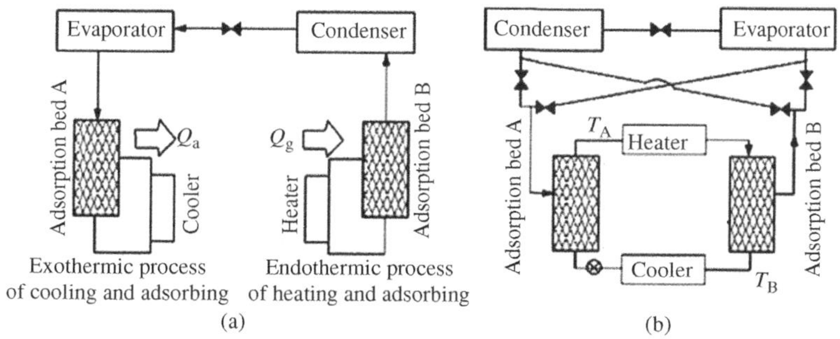

Figure 6.15 Schematic of heat flow. (a) Basic cycle and (b) thermal wave cycle

external cooling source for adsorption. The processes are shown in Figure 6.15a. The object for the design of a thermal wave cycle is to try to use the exothermic heat Q_a completely and reduce the heat Q_g required by the desorption process from the external heat source. Such a process could improve the performance of a system significantly. Figure 6.15b is a typical system for the thermal wave cycle. The basic principle of such a cycle is: using a single heating and cooling fluid circuit to connect two adsorbent beds, the cooler, and the heater. The circuit of the fluid could transfer the released heat from the adsorption bed to the desorption bed, and to recover the adsorption heat for the improvement of the energy efficiency of the system. The efficiency is analyzed by the second law, and the 80% adsorption heat can be recovered for the desorption process in a thermal wave cycle, that is, the heat recoveries rate $r = 0.8$.

The thermal wave cycle requires that the temperature at the outlet (T_B) of the desorption bed is low, while the temperature at the outlet of the adsorbing bed is relatively high. Otherwise, a big amount of the heat will be released at the cooler, and for such a process it is difficult to recover the heat back to the heater effectively. Especially for the condition where $T_A < T_B$, it is impossible to recover heat. So, for a thermal wave cycle, a large temperature difference in the two adsorbent beds is essential.

For a thermal wave cycle, the thermal wave means that the temperature of the fluid drops or rises rapidly in the adsorbent bed, forming a large temperature difference, as a steep waveform (shown in Figure 6.16). Such a process could transfer the heat between two beds.

Combined with four basic cycles, an entire thermal wave cycle is shown in Figure 6.17. When bed A is heated for desorption and bed B is cooled for desorption, a thermal wave cycle includes the following two procedures:

1. The pressure increasing process of bed A and the pressure decreasing process of bed B under the condition of constant volume.

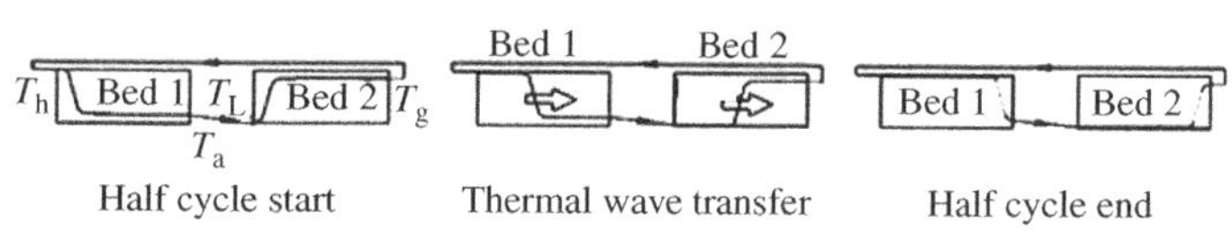

Figure 6.16 Schematic of heat flow of semi-cycle process

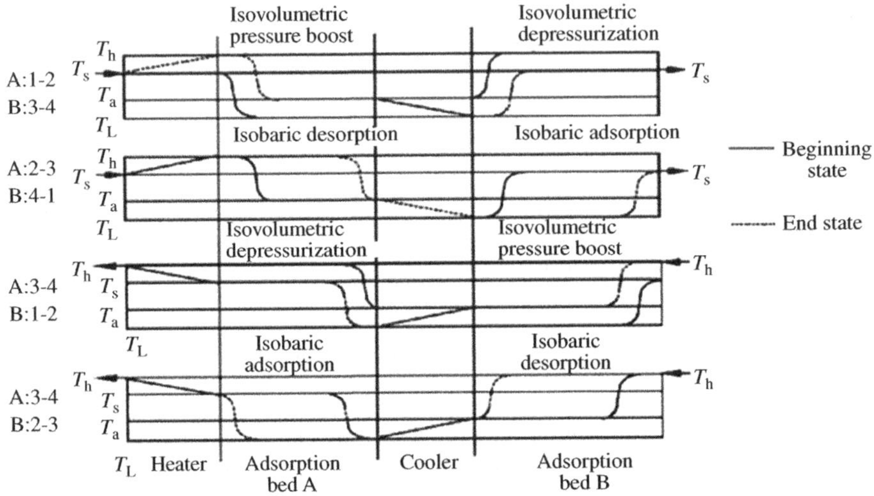

Figure 6.17 The diagram for the energy transfer of a thermal wave cycle

The initial state of bed A is saturation adsorption state (T_L,p_e), which is ready for desorption; For the initial state of bed B the desorption process (T_h,p_c) is complete, and will be switched for the cooling and adsorption process. The hot fluid is heated by the heater to a temperature of T_h, and then flows into bed A to exchange the heat with bed B. By this heat transfer process a steep thermal wave will be formed at the left side of bed A, and except for the left side other parts are heated up to T_a. At the outlet a cooler will cool the fluid to a temperature of T_L. After that the fluid will flow into the adsorbent bed B, similarly by the heat transfer process at the left end of bed B a thermal wave with a large temperature difference from the other parts of the bed will be formed, and the bed is cooled to the temperature of T_s. Then the fluid will flow back to the heater again for the next cycle. In this process when the temperature of bed A increases, a part of the adsorbent will desorb, and the pressure in the bed will rise to the condensing pressure p_c; similarly for bed B the temperature decreases, so that a part of the adsorbent adsorbs the refrigerant, and the pressure of the bed decreases to p_e.

2. The isobaric desorption process of Bed A, and the isobaric adsorption process of bed B.

   When the pressure of bed A and bed B reaches p_c and p_e, respectively, open the valve between bed A and the condenser, as well as opening the valve between bed B and the evaporator. Because the thermal wave will go forward and each part of bed A will be heated to the temperature of T_h, which is ready for the isobaric desorption; similarly each part of bed B will be cooled to T_L, which is ready for isobaric adsorption. When the thermal waves inside bed A and B move to the right end of the two adsorbers, the desorption in bed A is completed, and the adsorption in bed B is also complete.

When the above two processes are completed, bed A reaches the starting state of bed B, and bed B reaches the starting state of bed A. Then the flowing direction of the fluid will be switched to an opposite direction, accordingly the working processes of two beds will also be changed.

Bed A begins to be cooled for adsorption and bed B begins to be heated for desorption until bed A and B go to the initial state again.

6.4.2 Calculation of the Thermal Wave Cycle

On analyzing the theoretical performance of the thermal wave cycle, Shelton used oblique wave and square wave methods to obtain reasonable results [12, 13]. However, the simulation didn't go into the conditions for the establishment of thermal wave. After that Pons did some experiments and the prototype research on the thermal wave cycle [17, 18]. In addition, a large number of literatures studied the *COP* of the cycle, and a few people studied other parameters such as the specific cooling power of unit mass of adsorbent (*SCP*).

A key factor which needs to be considered for the calculation of the thermal wave cycle is the formation of the thermal wave and its effective movement. In order to achieve this the heat transfer performance in the adsorbent bed is essential. The thermal wave is just a visual description of the fluid temperature field, which completely depends on the heat transfer characteristics of adsorbent beds. Thus in analyzing the thermal wave we need to analyze the heat transfer performance that can affect the thermal wave. By that we can get the establishment conditions, system performance, and the energy density under these conditions. As well as that the feasible thermal wave for the solid adsorption refrigeration cycles can be obtained.

6.4.2.1 The Establishment of the Heat Transfer Models of the Adsorbent Bed

The plate-fin type heat exchanger is selected as the adsorbent bed for the calculation of the heat transfer performance. The adsorbent bed is shown in Figure 6.18. The bed is composed of a series of plate-fin heat exchanger units, and the internal heat transfer process is the convective heat transfer process. The one-dimensional heat transfer model of the adsorbent bed is shown in Figure 6.19.

The heat transfer process mainly relates to the heat transfer fluid, the metal walls of the bed, and the adsorbents. Various heat transfer methods are involved and need to be considered, and they are as follows:

1. The heat transfer process by the flowing process of the thermal fluid.
2. Convective heat transfer process between the heating (cooling) fluid and the adsorbent bed.
3. The thermal conductive process of the fluid.
4. The thermal conductive process of the adsorbent bed along the direction of fluid.

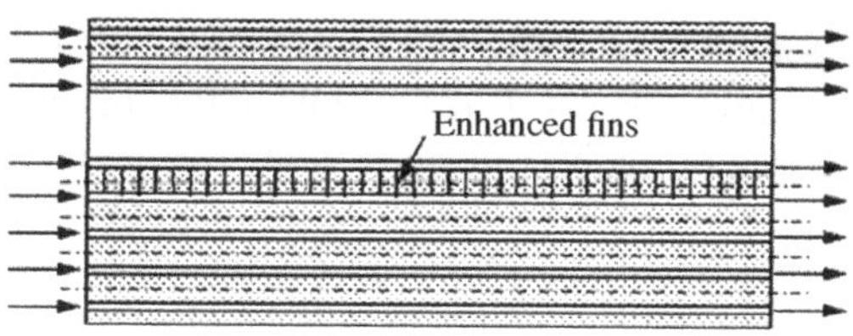

Figure 6.18 Plate-fin type adsorbent bed schematic

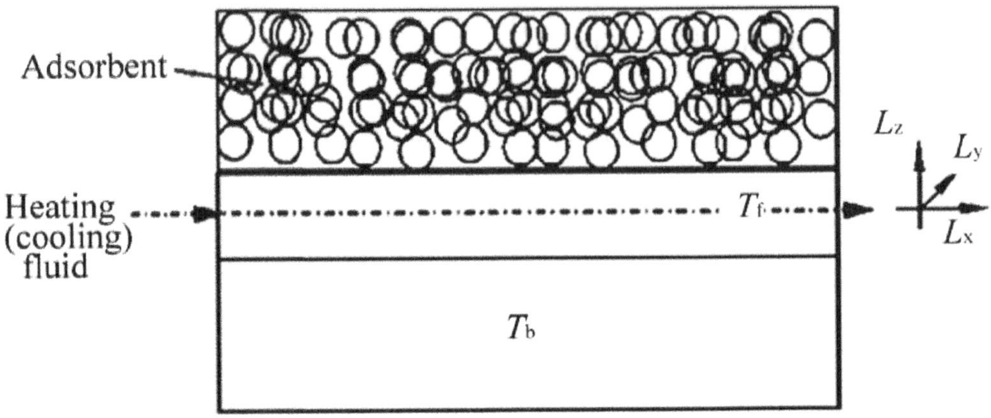

Figure 6.19 One-dimensional heat transfer model of adsorbent bed

The different heat transfer processes will influence each other. The heat transfer process also needs to be considered with the mass transfer process of the refrigerant gas. Taking the heat transfer fluid, metal walls, and the adsorbent as the researching objects the heat transfer equations can be established.

Heating (Cooling Fluid)

Three factors will influence the internal energy exchange of the heat transfer fluid, and they are the heat transfer by the flowing process of the fluid, the convective heat transfer process between the fluid and the adsorbent bed, and the thermal conductivity of the fluid.

In the calculation process, assuming that the width of the fluid channel inside the plate-fin heat exchanger (L_y direction) is infinitely large, the convective heat transfer between the fluid flow and adsorbent bed is equivalent to an external heat source of the convective heat transfer system. Taking into account the thermal conductive process of the top and bottom plates, and assuming that the external heat source is averagely distributed by the direction of L_z, then the heat in a micro unit of $dL_x \times dL_y \times dL_z$ is:

$$\begin{aligned} dQ_{hs} &= \frac{2\alpha_f \times L_{ad} \times L_b \times (T_f - T_w)}{L_{\mathrm{ad}} \times L_b \times L_m} \times dL_x \times dL_y \times dL_z \times dt \\ &= \frac{2\alpha_f \times (T_f - T_w)}{L_m} \times dL_x \times dL_y \times dL_z \times dt \end{aligned} \tag{6.29}$$

where Q_{hs} is the heat quantity for convective heat transfer process, α_f is the heat transfer coefficient between the fluid and metal walls, T_f is the temperature of fluid, T_w is the temperature of metal wall, L_m is the height of the heat medium along the direction of L_z, t is time, L_{ad} is the length of the adsorbent bed, L_b is the width of the adsorbent bed along the direction of L_y.

The equation of dQ_{hs} is substituted into the energy conservation equation of convective heat transfer process, then the heat transfer equation will be obtained from the energy balance equation:

$$\frac{\partial T_f}{\partial t} + u_f \frac{\partial T_f}{\partial L_x} = \xi_f \frac{\partial^2 T_f}{\partial {L_x}^2} - \frac{2\alpha_f}{\rho_f C_{pf} L_m}(T_f - T_w) \tag{6.30}$$

where u_f is the velocity of the heat transfer fluid, ξ_f is the thermal diffusivity of the fluid, ρ_f is the density of the fluid, and C_{pf} is the thermal capacity of the fluid.

Metal Walls

The internal energy change of the metal wall is caused by the heat transfer process among both sides of the metal walls, heat transfer fluid, and the adsorbent. The heat transfer equation of the metal walls can be obtained from the energy conservation equation.

$$\frac{\partial T_w}{\partial t} = \xi_w \left(\frac{\partial^2 T_w}{\partial {L_x}^2} \right) + \frac{\alpha_f}{\rho_w C_{pw} L_{bw}} (T_f - T_w) - \frac{\alpha_b}{\rho_w C_{pw} L_{bw}} (T_w - T_b) \tag{6.31}$$

where ξ_w is the thermal diffusivity of the metal walls, ρ_w is the density of the metal walls, C_{pw} is the thermal capacity of the metal walls, L_{bw} is the thickness of the wall, α_b is the heat transfer coefficient of the wall and the adsorbent bed, T_b is the temperature of the space inside the adsorbent bed.

Heat and Mass Transfer in the Adsorbent Bed

The mass conservation equation can be listed if we take a micro unit of the adsorbent in the bed as the research object. There are two main factors influencing the change of adsorbate density in the micro unit, and they are the flow of the gas inside the bed and the adsorption/desorption mechanics of adsorbent [13].

On the calculation of the mass transfer performance for the flowing process of the gas inside the adsorbent, the influence of the velocity for convective mass transfer process on the mass change of adsorbate is neglected. We only consider the impact of the mass conduction between the adsorbate and adsorbent in the mass transfer process, and assuming that the mass transfer space and the flow direction of the heat medium are perpendicular, that is, the mass transfer direction is L_y. Then the mass conservation equation is:

$$\frac{D_{ms} \partial^2 \rho_{ad}}{\partial L_y^2} + [\varepsilon_b - (1 - \varepsilon_b)\varepsilon_a] \frac{\partial \rho_{ad}}{\partial t} = (1 - \varepsilon_b)(1 - \varepsilon_a) \frac{\partial x_V}{\partial \tau} \tag{6.32}$$

where D_{ms} is the mass diffusion coefficient, ρ_{ad} is the density of the adsorbent, L_y is the thickness of the adsorbent for the direction perpendicular to the channel of mass transfer, ε_a and ε_b are the porosity of the adsorbent particles and the adsorbent bed, x_V is the volume adsorption amount.

The adsorbate gas density ρ_{refg} is calculated by the formula:

$$\rho_{refg} = \frac{p}{RT} \tag{6.33}$$

There are five main factors influencing the internal energy change of adsorbent, one is endothermic desorption (or exothermic adsorption), and the second is the endothermic (or exothermic) convective heat transfer process of the fluid. The third is the sensible heat for heating the adsorbent; the fourth is the flowing process of adsorbate gas and heat transfer process; the fifth is the thermal conductivity of the bed along the flowing direction of the fluid. Ignoring the influence of the convective heat transfer process of the adsorbate on the temperature of the bed, the energy balance equation is:

$$\lambda_a\frac{\partial^2 T_b}{\partial L_x^2}+\frac{2\alpha_b}{L_a}(T_w-T_b)=-(1-\varepsilon_b)(1-\varepsilon_a)\frac{\partial x_V}{\partial t}|\Delta H_r|+\frac{x_V C_{ref}\partial T_b}{\partial t}+\frac{\rho_{ad}C_a\partial T_b}{\partial t} \quad (6.34)$$

where the first part is thermal conductive performance of the adsorbent, λ_a is the thermal conductivity of the adsorbent; the second part is the convective heat transfer performance, and there are two thermal conduct metal walls connecting with the adsorbent, so we multiply by two at the beginning of the second part. L_a is the adsorbent thickness along the direction of L_y. The third part is the adsorption and desorption heat, and the fourth part is the temperature change of the adsorbed adsorbate, and C_{ref} is the specific heat capacity of the refrigerant. The last part is the temperature change of the adsorbent, where ρ_a is the density of the adsorbent, and C_a is the specific heat capacity of the adsorbent.

Obviously, it is a complex model. The heat transfer process of the fluid, the thermal conductivity of the wall, the thermal conductivity of the adsorbent, and the mass transfer performance in the beds interact with each other and restrict each other. It is a typical coupling problem involving a set of differential equations to be solved. Generally it is very difficult to solve these equations thoroughly. Taking into account the most important parts of the equations: the temperature change of the heat transfer fluid and the adsorbent bed, especially the tendancy for the fluid temperature to fluctuate and the heat transfer performance of the convection process between the fluid and the adsorbent bed, a simplified model can be established. The model will consider the convective heat transfer process between the fluid and the adsorbent bed, as well as the heat conductive process inside the beds. Taking the adsorbent bed, adsorbent and adsorbate as one unit, the equation is established by the description of the heat transfer process in the bed by the lumped parameters:

$$\frac{\partial T_f}{\partial t}+u_f\frac{\partial T_f}{\partial L_x}=\xi_f\frac{\partial^2 T_f}{\partial L_x^2}-\frac{2\alpha_t}{\rho_f C_{pf} L_m}(T_f-T_b) \quad (6.35)$$

$$\frac{\partial T_b}{\partial t}=a_b\frac{\partial^2 T_b}{\partial L_x^2}+\frac{2\alpha_z}{\rho_{bt} C_{pw} L_a}(T_f-T_b) \quad (6.36)$$

where ξ_b, ρ_{bt}, and C_{pw} are the thermal diffusion coefficient, density, and heat capacity, respectively, for one unit concerning the adsorbent bed, the adsorbent, and adsorbate. α_t is the total heat transfer coefficient from the fluid to the adsorbent bed.

Making the variable in Equation 6.35 dimensionless, we need to define:

$$\Phi=\frac{T_f-T_0}{T_{in}-T_0};\Theta=\frac{T_b-T_0}{T_{in}-T_0};L_x{}^*=\frac{L_x}{L_{xt}};\tau^*=\frac{t}{t_R};t_R=\frac{L_{xt}}{u_f} \quad (6.37)$$

where L_{xt} is the total length along the direction of L_x.

Then we get:

$$\frac{\partial\Phi}{\partial\tau^*}+\frac{\partial\Phi}{\partial L_x{}^*}=\frac{\xi_f}{u_f L_{xt}}\frac{\partial^2\Phi}{\partial L_x{}^{*2}}-\frac{2\alpha_t L_{xt}}{u_f\rho_f C_{pf} L_m}(\Phi-\Theta) \quad (6.38)$$

$$\frac{\partial\Theta}{\partial\tau^*}=\frac{\xi_b}{u_f L_{xt}}\times\frac{\partial^2\Theta}{\partial L_x{}^{*2}}+\frac{2\alpha_t L_{xt}}{u_f\rho_{bt} C_{pw} L_a}(\Phi-\Theta) \quad (6.39)$$

The initial conditions:

$$\Phi(L_x{}^*, \tau^* = 0) = 0; \Theta(L_x{}^*, \tau^* = 0) = 0 \tag{6.40}$$

The boundary conditions (heating process):

$$\Phi|_{L_x{}^*=0} = \Phi_{in}(t), \frac{\partial \Phi}{\partial L_x{}^*}|_{L_x{}^*=1} = 0 \tag{6.41}$$

$$\frac{\partial \Theta}{\partial L_x{}^*}|_{L_x{}^*=0} = \frac{\partial \Theta}{\partial L_x{}^*}|_{L_x*=1} = 0 \tag{6.42}$$

The solution of the above differential equations refers to the calculation and analysis of the thermal wave phenomena and laws from Yang [20].

6.4.2.2 The Analysis on the Characteristics of the Thermal Wave

According to the analysis of Shelton, dimensionless thermal wavelength is defined as [13]:

$$F_{WL} = L_x{}^*(\Phi = 0.1, t) - L_x{}^*(\Phi = 0.9, t) \tag{6.43}$$

It is the distance of the dimensionless temperature of the fluid that is reduced from 0.9 to 0.1.

The ideal thermal wave cycle requires that the thermal wave is small and does not increase with time, that is, the thermal wave moves forward inside the bed slowly. However, the thermal wave illustrates the fluid temperature change in the adsorbent bed, which is mainly affected by the influence of the heat transfer conditions in the adsorbent bed. The factors influencing the conditions include the velocity of the fluid, the equivalent heat transfer coefficient of the fluid and the adsorbent bed, and the ratio of the volume heat capacity between the fluid and the adsorbent bed. The influence of the above factors on the thermal wave is tested, and the experimental conditions are shown in Table 6.5. In Table 6.5, λ_f is the thermal conductivity coefficient of the fluid, λ_{ad} is the thermal conductivity coefficient of the adsorbent bed, and Bi is the Biot number.

1. The influence of the flow velocity of the fluid on the thermal wave.

   The flow velocity of fluid is an important factor affecting thermal wave properties. When the parameters are constant, the wave will become very flat when the flow velocity

Table 6.5 The data of the experiments done by Shelton *et al.* [12]

Parameter	Value
Peclet number of the fluid $Pe = \frac{\rho_f C_{pf} u_f L_{ad}}{\lambda_f}$	17 800
The ratio of the thermal conductivity between the adsorbent bed and fluid $KA = \frac{\lambda_{ad} L_a}{\lambda_f L_m}$	63
$NTU_f = Bi \times K_A$	19 500 × 63
Ratio of thermal diffusivity between the adsorbent bed and fluid $DR = \frac{\xi_b}{\xi_f}$	10
Velocity of the fluid u_f	3.44 m/h
Length of the adsorbent bed L_{ad}	1.42 m

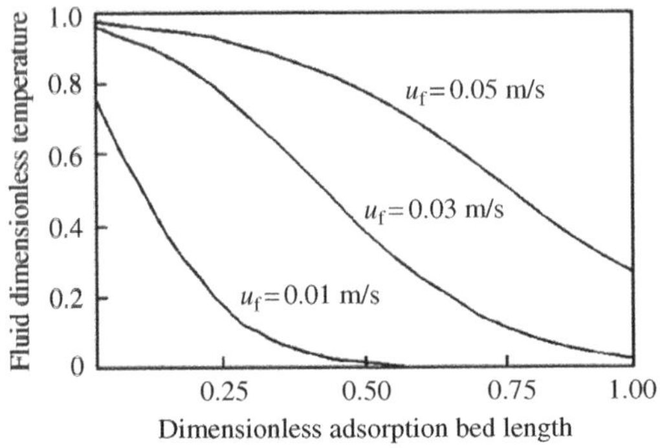

Figure 6.20 The influence of the flow velocity influence on the thermal waveform (40th second)

increases, while the formation of a steep thermal wave is rather obvious when the flow velocity is very small. The flow velocity determines the heat transfer process along the direction of the fluid. When the flow velocity increases, the heat transfer is concentrated along the direction of fluid, and the temperature of the fluid will rise quickly. Such a process obviously isn't good for the formation of the thermal wave with a large temperature difference.

Figure 6.20 listed the formation of the thermal wave under the conditions of the different flow velocity of u_f, and they are 0.05, 0.03, 0.01 m/s, respectively. The time for the flow in Figure 6.20 is 40th second. Obviously, when u_f is 0.01 m/s, the thermal wave length is very short and the outlet temperature is relatively low, which is an ideal waveform. When the flow velocity increases, the waveform changes to be flat and the fluid outlet temperature increases significantly, this is very different to the ideal thermal wave. It can also be seen in the figure: although when u_f is 0.01 m/s the waveform is good, but after 40th seconds, the distance that the fluid flows is only one-third of the whole adsorbent bed. The influence of the flow velocity on the outlet temperature is shown in Figure 6.21. When the flow velocity (u_f = 0.01 m/s) is low, the outlet temperature rises gently, and when the flow velocity

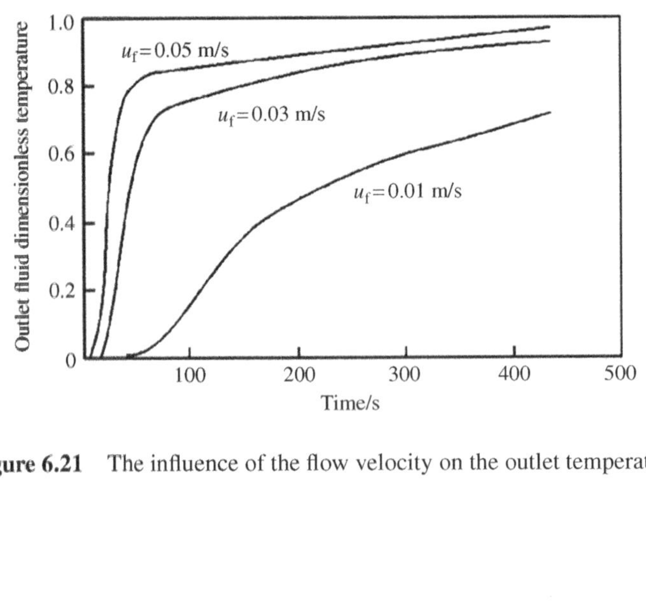

Figure 6.21 The influence of the flow velocity on the outlet temperature

increases ($u_f = 0.03$, 0.05 m/s), the outlet fluid temperature rises rapidly. Such a condition clearly does not meet the requirements of the thermal wave cycle.

2. The influence of the equivalent heat transfer coefficient on the thermal wave.

The flowing process can transfer the heat along the flowing direction of the fluid, whereas the convective heat transfer process will transfer the heat between the fluid and adsorbent bed, that is, the heat transfer direction is perpendicular to the flow direction of the fluid. If we slow down the heat transfer process by the flow direction of the fluid, we can reduce the temperature change of the fluid, which is good for the formation of the thermal wave.

The heat transfer coefficient on the side of the fluid is related to the velocity of the fluid, and won't change very much if the flow velocity is controlled in a certain range. The thermal contact resistance between the adsorbent and the metal wall is affected by several factors, such as the tightness of the contact between the adsorbent particles and the walls; the thermal resistance of the metal walls, the material of the metal walls, and the thickness of the metal walls. Due to the structure of the adsorbent being much looser compared with metal, generally the thermal resistance of the adsorbent is the most important part, which is mainly affected by the adsorbent thickness, that is, the width of the channel filling with the adsorbent.

The trend of the outlet temperature under different conditions is shown in Figure 6.22. It can be seen from the diagram that the increment velocity of the outlet temperature will be reduced when the flow velocity and the increasing equivalent heat transfer coefficient increase. With the same equivalent heat transfer coefficient, the two temperature curves will show a large difference when the velocity is different. Under the condition of same flow velocity, the temperature change curves will have little difference when the equivalent heat transfer coefficients are different. So, the influence of the flow velocity on the temperature change is much greater than that of the equivalent heat transfer coefficient.

The essential condition for the formation of the thermal wave is the low flow velocity. But the drawback for the low flow velocity is a worse heat transfer performance. For example, when the flow velocity decreases to only a few centimeters per second, the fluid flow will be the laminar flow, and the heat transfer performance at the fluid side will be significantly

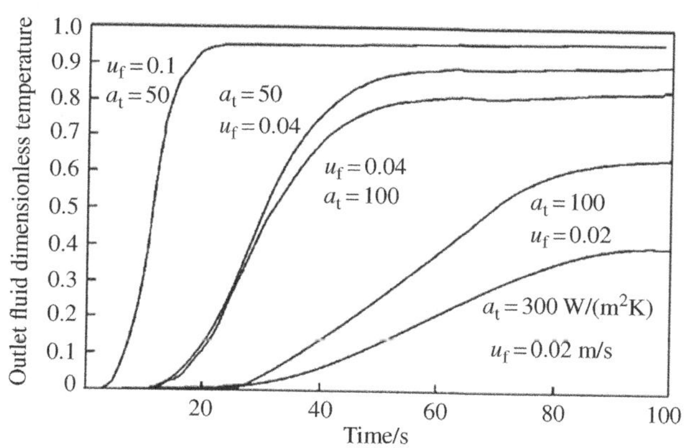

Figure 6.22 The influence of flow velocity and equivalent heat transfer coefficients on the outlet temperature of the fluid

weakened, and will consequently reduce the equivalent heat transfer coefficient. For the influences of the thermal resistances on the equivalent heat transfer coefficient the contact thermal resistance between the granular adsorbent and the metal wall and the thermal conductive resistance of adsorbent are the main factors. Compared with the influences of the thermal resistance the impact of the fluid side that is a result of the flow velocity is small. Thus reducing the fluid velocity is the main method for achieving an ideal thermal wave.

3. The impact of the structure of the adsorbent bed on the equivalent heat transfer coefficient and the influence of the ratio of volume heat capacity.

   Equivalent heat transfer coefficient and the ratio of volume heat capacity have a relationship with the adsorbent bed structure. Widening the adsorbent flow channel or reducing the width of the flow channel of the fluid side will improve the ratio of volume heat capacity, but the former method will increase the thermal resistance of the adsorbent and consequently will decrease the equivalent heat transfer coefficient. The latter method will increase the flow velocity of the fluid. Neither of these two methods will help to improve the shape of the thermal wave.

   On the other hand, widening the adsorbent flow channel or reducing the width of the flow channel of the fluid side will both increase the density of the heat exchanger unit, that is, will increase the metal ratio of the adsorbent bed, and consequently will increase the irreversible losses in the heating and cooling process. Of course, the thermal conductivity ratio K_A between the adsorbent bed and the fluid, and the thermal diffusivity ratio D_R between the adsorbent bed and the fluid can have a compromise value, that is, the optimal structure size of the adsorbent bed, but it is difficult to achieve significant improvement for forming the ideal thermal wave with small wavelength.

6.4.2.3 Performance Calculation of the Thermal Wave Cycle

As shown in Figure 6.23, a thermal wave cycle connects the heater, adsorbent bed A, cooler, and the adsorbent bed B directly. The temperature gradient of the fluid will be formed by the heat transfer processes between the four parts. The ideal temperature gradient is shown in Figure 6.23. For the ideal cycle the outlet temperature of bed A is low, the outlet temperature of bed B is high, and by such a relation the fluid that is released from bed A can absorb the heat in bed B effectively before it flow back to the heater, which will reduce the heat provided by the external heat source.

The thermal wave cycle transfers energy between two beds by the forward and reversed flow through the fluid circuit. The fluid absorbs heat Q_g in the heater, releases heat Q_c to

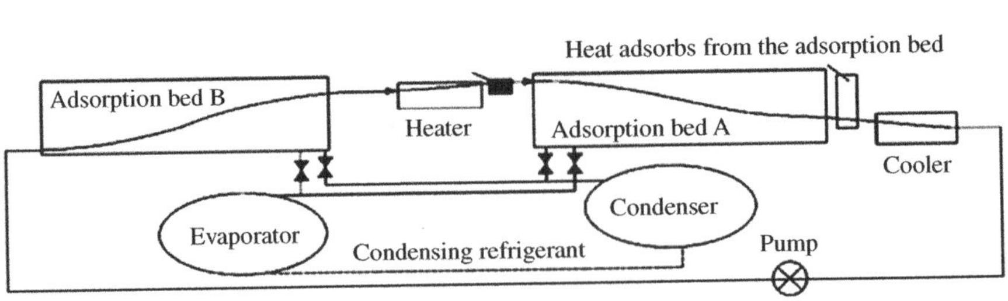

Figure 6.23 The temperature distribution and heat transfer process in ideal thermal wave cycle

the environment in the cooler, and the refrigerant vapor desorbed from high-temperature and high-pressure bed A condenses and releases condensation heat Q_{cond}. Bed B adsorbs the vapor that is provided by the evaporation inside the evaporator, which outputs the cooling power of Q_{ref}. The key point of the thermal wave cycle is that the fluid transfers heat between bed A and bed B, recycles the sensible heat, and the adsorption heat of bed B as much as possible to improve the heat recovery rate.

For a heating and cooling process Q_d and Q_c can be obtained by the fluid temperature difference between the inlet and the outlet of bed A and B:

$$Q_d = \sum \rho_f C_{pf} u_f A_f \Delta t(\Phi_{Ain} - \Phi_{Bout}) \times (T_{in} - T_0) \tag{6.44}$$

$$Q_c = \sum \rho_f C_{pf} u_f A_f \Delta t(\Phi_{Aout} - \Phi_{Bin}) \times (T_{in} - T_0) \tag{6.45}$$

$$Q_{ref} = M_a \Delta x L_e \tag{6.46}$$

$$Q_{cond} = M_a \Delta x L_c \tag{6.47}$$

where A_f is the cross-sectional area of fluid, the subscript "in" is the heater inlet, B_{out} is the outlet of bed B, A_{out} is the outlet of bed A, B_{in} is the inlet of bed B. M_a is the adsorbent mass, Δx is the cycle adsorption quantity, L_e is the evaporation latent heat of the refrigerant in the evaporator, L_c is the condensation latent heat of the refrigerant in the condenser.

If the heating process for the bed tends to be an ideal process, the temperature will rise, and the outlet temperature of the bed is high. In this case a major amount of heat will be lost in the cooler, apparently it is bad for improving the system performance. Thus the cycle needs to be stopped when the thermodynamic perfectness reaches a certain value, and generally the thermodynamic perfectness is taken as 80–85%. Here the thermodynamic perfectness refers to the dimensionless average temperature of the whole bed.

The COP of the adsorption refrigeration system (COP_{ref}) and the heat pump system (COP_{hp}) can be calculated by:

$$COP_{hp} = \frac{Q_c + Q_{cond}}{Q_d} \tag{6.48}$$

$$COP_{ref} = \frac{Q_{ref}}{Q_d} \tag{6.49}$$

Energy density corresponding to the cycle time of the system is:

$$\text{Heat pump:} SHP = (Q_c + Q_{cond})/(t_c \cdot M_a) \tag{6.50}$$

$$\text{Refrigeration:} SCP = Q_{ref}/(t_c \cdot M_a) \tag{6.51}$$

where t_c is the cycle time.

The influence of the ratio of equivalent heat transfer coefficient and the flow velocity on the system performance is shown in Figure 6.24a. It can be seen from the diagram that the performance of the system improves when the α_t/u_f increases. When the α_t/u_f is bigger, the thermal wave will tend to be more ideal, and the system performance will be more improved. It is difficult for the small wavelength thermal wave to form when the α_t/u_f decreases. If the thermal wavelength increases rapidly, the outlet temperature will rise quickly. According to the principle of the thermal wave cycle a lot of the heat will be exhausted in the cooler, and the

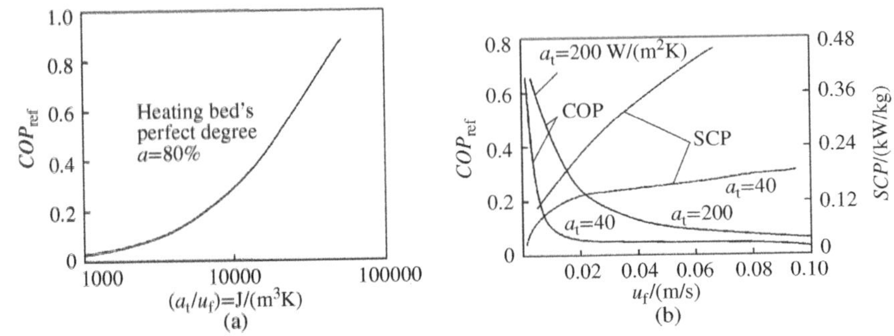

Figure 6.24 The influence of the velocity and heat transfer coefficient on the performance and energy density of the system. (a) COP_{ref} vs. α_t/u_f and (b) the relationship among the velocity, heat transfer coefficient, and the performance

large temperature difference between the high temperature fluid and cooling water will cause a large amount of irreversible loss in the cooler, which will reduce the performance of the system.

Also the flow velocity cannot be decreased unlimitedly. The flow velocity is lower, the amount of heat transported in the cycle will be less, and the heating (or cooling) power will be significantly decreased. Accordingly, the equivalent heat transfer coefficient will decrease, and the cooling and the heating process will be prolonged, consequently the cycle time will be increased and the energy density of the system will decrease. As shown in Figure 6.24b, the flow velocity decrement improves the thermal performance of the system, but the energy density of the system decreases significantly at the same time [14].

Two main requirements for the ideal thermal wave cycle are the formation of ideal thermal wave and its effective movement. The formation of the thermal wave depends on the heat and mass transfer performance in the bed, and it also can be looked as a visual description of the temperature field, which is different from the usual physical wave. It is very difficult to form the ideal thermal wave. Even though reducing the flow velocity and increasing the heat transfer performance can improve the waveform, the outlet temperature of the flow will still rise rapidly, and will have significant heat loss. Besides, limited by the improvement of the equivalent heat transfer coefficient, if we reduce the flow velocity we can get a reasonable thermal wave, but the energy density of the system will be decreased. Because it is difficult to get both ideal thermal waveform and the reasonable energy density of the system, it is difficult to apply the thermal wave cycle in the solid adsorption refrigeration system.

6.4.3 Convective Thermal Wave Cycle

Based on the study of the basic thermal wave cycle [12–14, 21–24], R.E. Critoph proposed a novel method, which is convective thermal wave cycle [25]. Such a cycle used the forced convection between the refrigerant gas and adsorbent, that is, the adsorbent is directly heated or cooled by the high pressure refrigerant vapor to obtain a higher heat density.

The schematic of a convective thermal wave cycle is shown in Figure 6.25. In Figure 6.25a, the refrigerant gas outside the heat exchanger is heated to a certain temperature and then flows into the adsorbent bed. In the adsorbent bed, the convective heat transfer process is

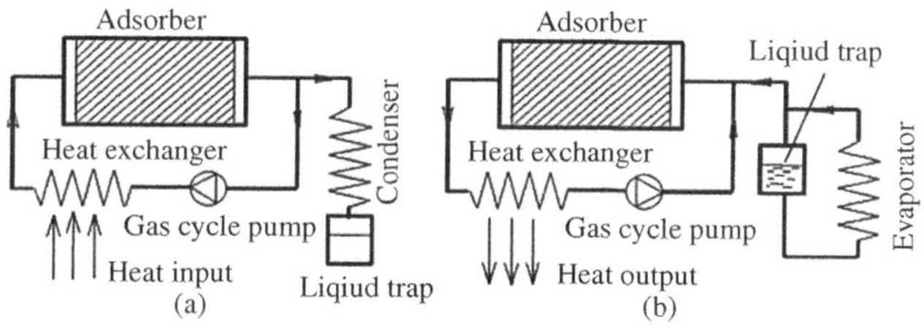

Figure 6.25 The schematic of the convective thermal wave cycle. (a) Desorption process and (b) adsorption process

conducted between the refrigerant gas and the solid adsorbent (heat transfer coefficient can be 102–103 W/(m^2·K)), and the refrigerant gas releases heat that provides the desorption heat for adsorbent. The desorbed gas flows out of the adsorbent bed together with the gas for the heat transfer process. The temperature for the gas flowing out of the adsorbent bed drops. One part of the gas is sent to the heat exchanger by the gas pump and is heated there, then the gas will flow to the adsorbent bed to provide the desorption heat for the adsorbent again. The other part of the gas goes into the condenser and condenses into liquid which flows into the fluid collection device. After completing the heating and desorption process, the system switches to the adsorption process, which is shown in Figure 6.25b. The gas pump transports the gas in the opposite direction, conveys the low temperature gas to the adsorbent bed, and absorbs. The adsorption heat is released by the adsorbent in the adsorption process of the refrigerant gas. The mass decrement of the gas at the outlet is compensated by the evaporated gas from the evaporator. The gas flowing out from the adsorbent bed is cooled by the heat exchanger and is sent to the adsorbent bed by the cycle pump. Either in the heating process or cooling process the steep temperature gradient appears along the direction of gas flow in the bed. With time , the temperature gradient (temperature wave) is moved along the direction of the gas flow. This phenomenon is similar to the thermal wave proposed by S.V. Shelton *et al.*, but the formation mechanism is slightly different, thus the cycle is known as convective thermal wave cycle. The temperature waveform and the trends are related to gas flow velocity, gas flow properties, the convective heat transfer coefficient, and the adsorbent bed heat capacity, and so on.

When the two beds operate together, the heat released from the heat exchanger in the adsorption process is used to heat the gas which flows out of the adsorbent bed in the desorption process for heat recovery. In order to obtain the higher heat recovery efficiency, it is essential to control the operation of the two beds to match the desorption process and adsorption process.

6.4.4 Mathematical Model of Convective Thermal Wave Cycle

6.4.4.1 Mathematical Model's Establishment

Taking the adsorbent bed of activated carbon fiber as an example, the mathematical model of convective thermal wave cycle is shown in Figure 6.26. As shown in the figure the sheet-like (or cloth-like) activated carbon fibers are filled in the adsorbent bed along the axial direction of the bed, which leaves a significant cross-section gas flow passage between the layers of the activated carbon fiber. The convective heat transfer process occurs between ammonia gas flow through the channel and the activated carbon fiber in the channel on both sides.

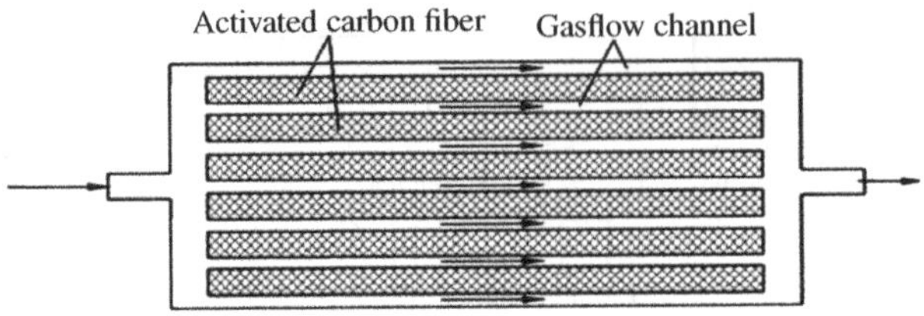

Figure 6.26 The structure of activated carbon fiber bed for convective thermal wave cycle

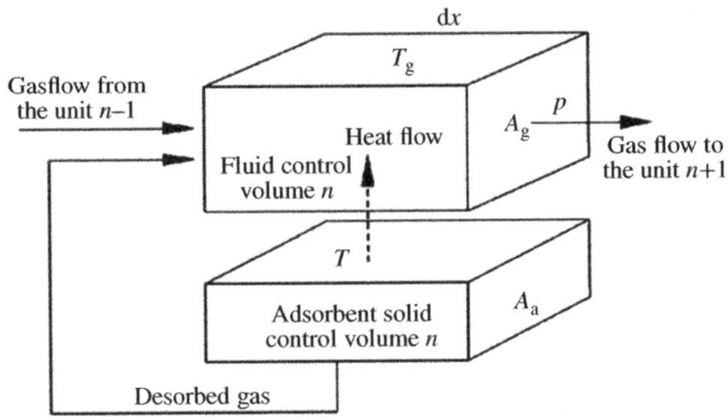

Figure 6.27 The relationship between the adsorbent and the heating gas flow

The entire adsorbent bed shown in Figure 6.26 is divided along the direction of the gas flow into several units, and each unit includes a gas flow control volume and an adsorbent control volume, and then the heat and mass transfer relationship is shown in Figure 6.27.

The flow of the gas obeys the mass conservation law, that is, the increment of the gas mass in the gas flow control volume n is equal to the gas flow mass which flows into the $n-1$ unit subtracts the gas flow mass flow to the $n+1$ unit, and plus the mass of the gas desorbed from the adsorbent solid control volume. The equation is:

$$\frac{\partial \dot{m}}{\partial L_x}\mathrm{d}L_x + \frac{\partial x}{\partial t}\mathrm{d}M_a + A_g \mathrm{d}L_x \frac{\partial \rho_g}{\partial \tau} = 0 \tag{6.52}$$

where $\dot{m}$ is the gas flow rate from a unit to the next unit (kg/s), $\mathrm{d}L_x$ is the length of the unit along the direction of the gas flow (m), x is the adsorption capacity of adsorbent (kg/kg), t is time (s), $\mathrm{d}M_a$ is the adsorbent mass (activated carbon fiber) in the unit (kg), A_g is the gas flow cross-sectional area in the unit (m^2), and ρ_g is the density of gas flow (kg/m^3).

Using the D-A equation to fit the adsorption capacity of activated carbon fiber for the refrigerant of ammonia:

$$x = x_0 \exp\left[-K\left(\frac{T}{T_s} - 1\right)^n\right] \tag{6.53}$$

where x_0 is the maximum adsorption capacity of adsorbent to the refrigerant (kg/kg), K and n are the constants relating to the adsorption properties of the working pair, T is the adsorbent (activated carbon fiber) temperature (K), T_s is the saturated adsorption temperature corresponding to the adsorption pressure of the refrigerant (K).

To simplify the modeling process, assuming that the temperature and the pressure of the control unit of the gas and the control volume of the adsorbent are uniform, also assumes there isn't heat and mass transfer in the control volume. The change of the adsorption and desorption rates depends mainly on the convective heat transfer performance between the control volume of the adsorbent and the control volume of the gas. Meanwhile postulate:

1. Ammonia pressure is uniform along the direction of the gas flow in the adsorbent bed.
2. Assuming the density is constant for each unit in a very short time.

Then the increment of internal energy in the control volume of the gas comes from three parts:

1. The heat from the activated carbon fiber by the heat convective process.
2. The enthalpy of the gas that flows into the control volume subtracts the enthalpy of the gas that flows out of the control volume.
3. The enthalpy of the ammonia desorbed from the activated carbon fiber.

$$\frac{\mathrm{d}Q_d}{\mathrm{d}t} = C_{vg}A_g\mathrm{d}L_x\frac{\partial(\rho_g T_{gas})}{\partial t} + \frac{\partial(\dot{m}h_f)}{\partial L_x}\mathrm{d}L_x + \frac{\partial x}{\partial t}\mathrm{d}M_a h_{gT} \quad \frac{\partial x}{\partial \tau} < 0 \tag{6.54}$$

$$\frac{\mathrm{d}Q_d}{\mathrm{d}t} = C_{vg}A_g\mathrm{d}L_x\frac{\partial(\rho_g T_{gas})}{\partial t} + \frac{\partial(\dot{m}h_f)}{\partial L_x}\mathrm{d}L_x + \frac{\partial x}{\partial t}\mathrm{d}M_a h_f \quad \frac{\partial x}{\partial \tau} > 0 \tag{6.55}$$

where $\mathrm{d}Q_d/\mathrm{d}t$ is the heat transferred from the adsorbent to the control volume of the fluid (W), C_{vg} is the specific heat of the refrigerant gas (ammonia) (J/(kgK)), T_{gas} is the temperature of the refrigerant gas (K), h_f is the specific enthalpy of the refrigerant liquid (J/kg), h_{gT} is the specific enthalpy of the refrigerant gas at the temperature of T (J/kg). The heat transferred from the adsorbent through a convective heat transfer process to the gas flow is calculated by the following formula:

$$\frac{\mathrm{d}Q_d}{\mathrm{d}t} = -\alpha_{ac}L_B\mathrm{d}L_x(T_{gas} - T) \tag{6.56}$$

where α_{ac} is the convective heat transfer coefficient between the activated carbon fiber and ammonia flow (W/(m^2 K)), L_{B} is the unit lateral equivalent width (m).

The heat adsorbed from the unit adsorbent is mainly from two parts:

1. The heat transferred from gas flow to the adsorbent through convective heat transfer process.
2. The heat conducted from the adjacent unit of the adsorbent.

As shown in the following equation:

$$\frac{\mathrm{d}Q_c}{\mathrm{d}t} = \alpha_{ac}L_B\mathrm{d}L_x(T_{gas} - T) + \frac{\partial^2 T}{\partial {L_x}^2}\lambda_{ad}A_a\mathrm{d}L_x \tag{6.57}$$

where $\mathrm{d}Q_c/\mathrm{d}t$ is the heat adsorbed by the adsorbent (W), λ_{ad} is the thermal conductivity coefficient of the adsorbent, A_a is the adsorbent cross-sectional area in the unit (m^2).

A part of the heat that the adsorbent adsorbs is used to heat the adsorbent and the adsorbed gas on the adsorbent surface, another part of the heat is used to provide the desorption heat for the desorption process, the formula is:

$$\frac{\partial q_c}{\partial t} = \frac{\partial T}{\partial t}(C_a + x \cdot C_{vf}) - \frac{\partial x}{\partial t} L_{sat} \frac{T}{T_S} \tag{6.58}$$

where q_c is the heat adsorbed by the adsorbent (activated carbon fiber) (J/kg), C_a is the isobaric specific heat of the adsorbent (activated carbon fiber) (J/(kg·K)), C_{vf} is the specific heat at a constant volume of the liquid refrigerant (ammonia) at the adsorption state. L_{sat} is the evaporation latent heat of the refrigerant at the temperature of T_s (J/kg).

6.4.4.2 Examples

For the calculation of the adsorbent bed in a convective thermal wave cycle, it assumes that each adsorbent bed is filled with 1 kg activated carbon fiber. The results are shown in Figure 6.28. The adsorbent bed is heated by the ammonia gas flow with a pressure of 13.7 bar and the heating temperature of 165 °C, and the gas flow through the activated carbon fibers at the inlet with a speed of 1 m/s. The convective heat transfer coefficient between the ammonia and activated carbon fiber is about 1050 W/(m^2.K) [25]. After the activated carbon fibers are heated up, the ammonia gas is desorbed from the adsorbent and goes into the condenser, and then condenses into liquid. The condensation temperature is 35 °C. The curves in the diagram are the adsorbent bed's temperature and the adsorption capacity for the time of 1, 5, 10, 15, 20, … 75 seconds. The results showed that the ammonia desorbs completely in a time of 75 seconds.

In Figure 6.29, the adsorbent bed is cooled by the ammonia gas flow with the pressure of 2.7 bar and the temperature of 40 °C, and the gas flow flows in a direction opposite to that of the heating process with the inlet speed of 1 m/s, and the convective heat transfer coefficient between the ammonia and activated carbon fiber is about 433 W/m^2 K [25]. After the activated carbon fiber cooled and adsorbed ammonia, the mass decrement of the ammonia gas at the outlet is supplied by the ammonia evaporated from the evaporator at the evaporation temperature of −8 °C. The curves in the diagram are the bed temperature and the simulation results of adsorption capacity for the time of 1, 10, 20, 30, 40, 50, … 320 seconds. We can see

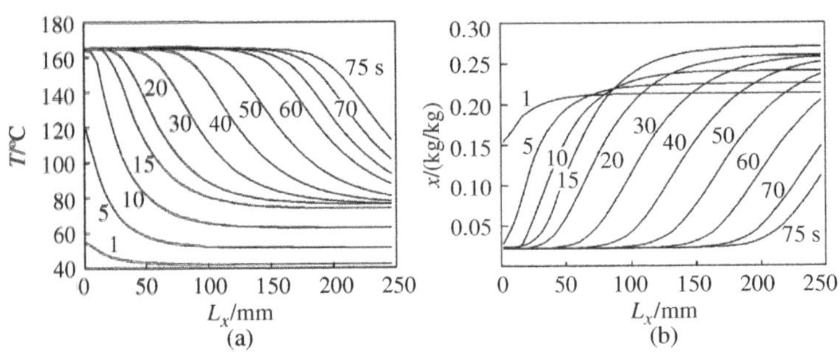

Figure 6.28 The temperature of the adsorbent bed and the adsorption quantity of the convective thermal wave cycle in the heating process. (a) The change of the temperature and (b) the change of the adsorption quantity

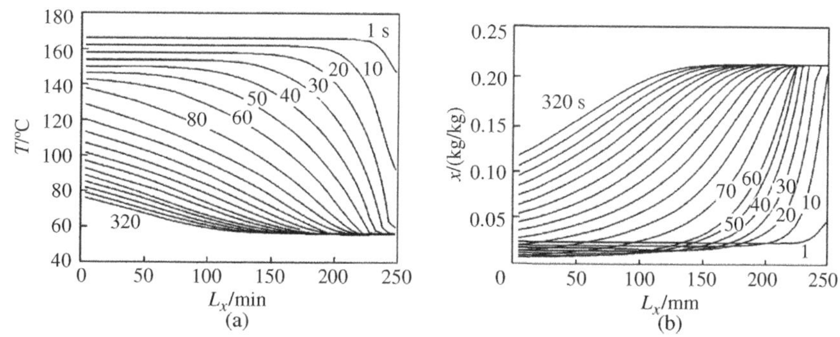

Figure 6.29 The temperature of the adsorbent bed and the adsorption quantity of the convective thermal wave cycle in the cooling process. (a) The change of the temperature and (b) the change of the adsorption quantity

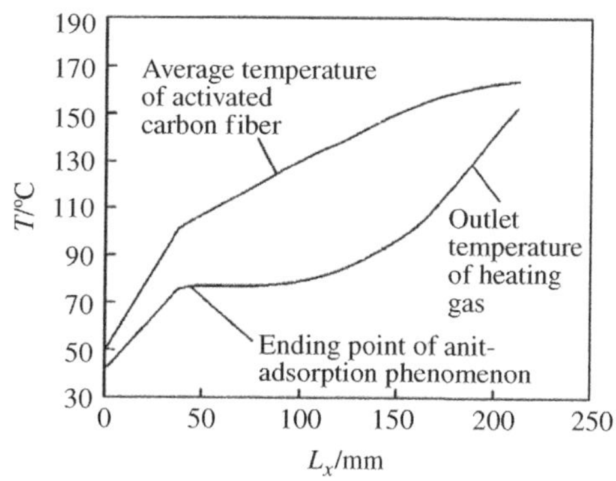

Figure 6.30 The average temperature of the bed and the outlet temperature of gas flow in the heating process

from Figures 6.28 and 6.29 that the time for the heating process doesn't match with that for the cooling process, and this phenomenon can be solved by controlling the energy flow of the system, such as controlling the gas flow in the heating process.

The convective thermal wave cycle will have an anti-adsorption phenomenon, which is shown in Figure 6.30. Before the start of the heating process, the pressure of the activated carbon fiber is low and also saturated. For this condition when we switch the system to the heating process, the adsorbent bed pressure increases. The front end of the adsorbent bed will be quickly heated to the equilibrium desorption temperature by the gas flow at a high temperature. But the temperature of the rear end of the bed changes less. In this case, a high adsorption pressure will accelerate the adsorption process of the adsorbent, and the released adsorption heat makes the average temperature of the bed increase rapidly (as shown in Figure 6.30). In general, the anti-adsorption phenomenon of activated carbon fiber will last for 20–30 seconds.

When heat transferred by the gas flow to the adsorbent bed in the heating process is:

$$Q_h = \int_0^{t_c/2} \dot{m}_{in}(h_{in} - h_{out})\mathrm{d}t \tag{6.59}$$

where t_c is the cycle time, h is the specific enthalpy of the gas flow.

The heat released to the gas flow by the adsorbent bed in the cooling process is:

$$Q_c = \int_0^{t_c/2} \dot{m}_{out}(h_{out} - h_{in})\mathrm{d}t \tag{6.60}$$

The heat released to the condenser in a refrigeration process is:

$$Q_{cond} = \int_0^{t_c/2} (\dot{m}_{out} - \dot{m}_{in}) \cdot [h_{out} - h_L(T_{cond})]\mathrm{d}t \tag{6.61}$$

where h_L is the specific enthalpy of the ammonia liquid at the condensation temperature.

The refrigeration power generated by the evaporator in a refrigeration process is:

$$Q_{ref} = \int_0^{t_c/2} (\dot{m}_{in} - \dot{m}_{out}) \cdot L(T_{evp})\mathrm{d}t \tag{6.62}$$

The refrigeration coefficient is:

$$COP_{ref} = \frac{Q_{ref}}{Q_h - Q_{reg}} \tag{6.63}$$

Heat pump efficiency:

$$COP_H = \frac{Q_c - Q_{reg} + Q_{cond}}{Q_h - Q_{reg}} \tag{6.64}$$

Energy density:

$$SCP = \frac{Q_{ref}}{\tau \cdot M_a} \tag{6.65}$$

$$SHP = \frac{Q_{cond} + Q_c - Q_{reg}}{\tau \cdot M_a} \tag{6.66}$$

where $\dot{m}_{\mathrm{in}}$, $\dot{m}_{\mathrm{out}}$ are the flow rate of the inlet and outlet gas for heating the adsorbent bed in Equations 6.59 and 6.61. $\dot{m}_{\mathrm{in}}$, $\dot{m}_{\mathrm{out}}$ are the flow rate of inlet and outlet gas for cooling the adsorbent bed in Equations 6.60 and 6.62; Q_{reg} is the recovered heat; τ is the heating and cooling time; M_a is the adsorbent mass.

Figures 6.28 and 6.29 showed that, when the speed of the ammonia passed through the activated carbon fibers is 1 m/s at the entrance of the adsorbent bed, the time required for the heating and desorption process is far less than the time required for cooling the adsorbent bed.

Table 6.6 Calculation conditions and performance parameters of convective thermal wave cycle

Parameter	Value	Parameter	Value
Gas flow temperature for heating	165 °C	Gas flow temperature for cooling	40 °C
Evaporation temperature	−8 °C	Condensation temperature	35 °C
Joined heat from heating the adsorbent bed	494.97 kJ	Released heat from cooling the adsorbent bed	470.93 kJ
Released heat from condenser	271.73 kJ	Adsorbed heat from evaporator	251.60 kJ
Heat of the heat recovery	192.904 kJ	Heat recovery rate	40%
Time for cooling and heating	330 s	Heat pump coefficient COP_{hp}	1.78
Refrigeration coefficient COP_{ref}	0.7872	Energy density of heat pump system SHP	1616 W/kg
Energy density of refrigeration system SCP	760 W/kg		

It is because the heating gas flow has high pressure and large density, in unit time the flow rate of the gas through the adsorbent bed and the heat taken in are large. If the difference for the mass flow between heating and cooling processes is great it will be bad for the heat recovery process in the heat recoverer. Thus the flow rate of the gas fluid for heating the adsorbent bed needs to be reduced (in Table 6.6, gas flow velocity for the heating process is 0.25 m/s), and the heating time needs to be prolonged. The cooling gas flow outlet temperature is below 60 °C when the time for cooling the adsorbent bed is 330 seconds, then switches the system at this time. Based on the above formulas, the system's performance parameters are calculated and shown in Table 6.6. In the diagram the heat recovery rate is up to 0.40 between two beds, the refrigeration coefficient of the refrigeration system is 0.78, the *COP* of the heat pump system is up to 1.78, and the energy density of the system is high. If the design of the adsorbent bed and the operation of the system is optimized the system will get a higher *COP* and energy density.

The cycle time of a convective thermal wave cycle can be shortened by several methods [25]:

1. The mass of gas flowed through the system and the heat taken away by the gas flow can be increased if the gas flow rate increases. Such a method has a drawback that will influence the formation of the thermal wave if the heat transfer coefficient is constant, and *COP* will also decrease. But the energy density of the system may be increased.
2. If we increase the cross-section area of the gas flow we will get the same results as mentioned above, and the cycle time will be decreased, which is not good for forming the steep thermal wave, and the *COP* may be decreased. The system energy density will be increased.
3. Improving the pressure of the gas flow for cooling/heating processes we can shorten the cycle time, but it depends on the evaporation temperature and the condensation temperature.

Actually if we consider the components of the system, the biggest difficulty of the thermal wave cycle lies in the gas cycle pump that is used for driving the refrigerant vapor. The thermal wave cycle has a very high requirement on the cycle pump. At first, the pump needs to be able to withstand a high pressure that is up to 15–20 kg/cm^2, it needs to be a dry gas pump (compressor), otherwise the pump lubricating oil will get in the adsorbent bed and the adsorption properties of the adsorbent will be influenced.

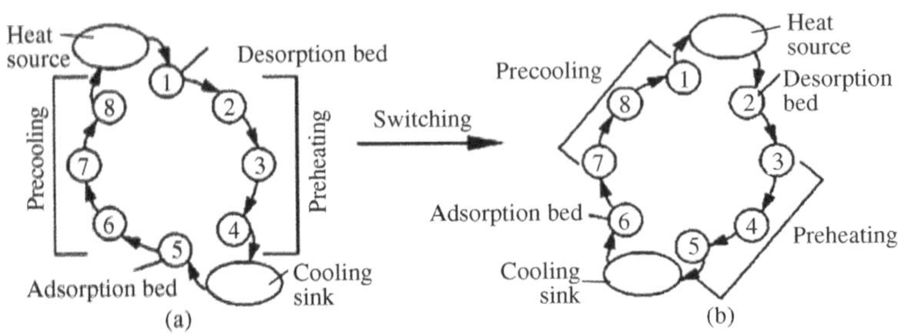

Figure 6.31 (a,b) Heat recovery cycle of multi-bed adsorption refrigeration system

6.4.5 Thermal Wave Heat Recovery Cycle for Multi-Bed Systems

For multi-bed systems, if the temperature of the heat source after the heat recovery among beds reaches the temperature of the cold source or close to the cold source temperature, the heat recovery process can be used throughout the whole working process of the system. Figure 6.31 shows the eight-bed heat recovery refrigeration cycle. When adsorbent bed 1 is in the process of heating and desorbing, the heating fluid will flow into adsorbent bed 2, adsorbent bed 3, and adsorbent bed 4, consecutively. When fluid flows out from adsorbent bed 4, the consumption of the desorption heat and the sensible heat of the beds makes the temperature of the fluid close to the cold source, then fluid is cooled by the cold source before it flows into adsorbent bed 5. When the fluid flows through adsorbent bed 5, the fluid will be heated by the sensible and adsorption heat of the bed. Similarly the temperature of the fluid will rise significantly when it flows through adsorbent beds 6, 7, 8. Lastly, the fluid which has a temperature similar to that of the heat source, and then flows into adsorbent bed 1 to start a new cycle.

Compared with the simple two-bed regenerative cycle, the major difference of the multi-bed heat recovery system is that the heat can be transferred from the low-temperature adsorbent bed to the high-temperature adsorbent bed. Taking Figure 6.31a, for example, the fluid flowing out from adsorbent bed 8 adsorbs the adsorption heat, even if its temperature is lower than the temperature of adsorbent bed 1, through heating by an external heat source, and the heat still can be transferred to adsorbent bed 1. According to the characteristics of the multi-bed heat recovery cycle, it is classified as the thermal wave recovery cycle.

6.4.6 The Properties of Multi-Bed Thermal Wave Recovery Cycle

For a multi-bed thermal wave cycle, the heat recovered in the heat recovery process is related to the number of the adsorbent beds. The more the adsorbent beds is, the more the recovered heat is.

Figure 6.32 shows that $\mathrm{d}Q/\mathrm{d}T$ changes with the temperature of the adsorbent bed. $\mathrm{d}Q/\mathrm{d}T$ can be regarded as equivalent specific heat of the adsorbent bed at the temperature T. At different stages, it reflects the change of temperature by the influences of comprehensive specific heat of adsorbent, adsorption rate, specific heat of refrigerant, the differential adsorption heat, and differential desorption heat. The upper curves in the diagram are the changes in the heating stage, and the lower curves are the changes for the exothermic stage. The areas surrounded by the upper and lower curves and the T axis are the heat absorbed and released, respectively.

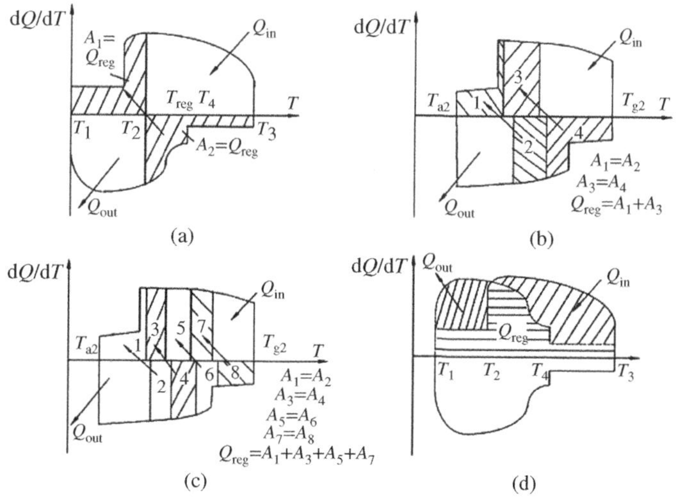

Figure 6.32 The diagram for the calculation of the recovered heat and temperature of multi-bed heat recovery systems. (a) Basic heat recovery cycle of two-bed system; (b) thermal wave heat recovery cycle of four-bed system; (c) thermal wave heat recovery cycle of eight-bed system; and (d) thermal wave heat recovery cycle of infinite-bed system

The working process of the adsorbent beds can be explained by Figure 6.32a. Figure 6.32a is the diagram for the ordinary heat recovery process between two beds. The heat is transferred completely from adsorbent bed 2 after desorption to adsorbent bed 1 after adsorption. The adsorbent bed 2 is cooled, and the temperature is decreased from the highest desorption temperature T_3 to the regeneration temperature of T_{reg} through T_4, and the heat is directly used by the heating process of adsorbent bed 1, which makes the temperature of adsorbent bed increase from the adsorption temperature T_1 to the regeneration temperature T_{reg} through T_2. In Figure 6.32 the change rate of the sensible heat for heating or cooling processes can be regarded as constant, while in the desorption or adsorption process the change rate will increase sharply. With the recovery heat of Q_{reg}, the heating load of the cycle is reduced to Q_{in}, and the cooling load of the cycle decreases to Q_{out}. A four-bed thermal wave recovery cycle is shown in Figure 6.32b. For the calculation of the area, shadow area of 1 is equal to that of 2, and shadow area of 3 is equal to that of 4, so the recovered heat is the sum areas of shadow 1 and 3 or shadow 2 and 4. Similarly, the heat recovery process of a eight-bed thermal wave recovery cycle is shown in Figure 6.32c.

When the number of beds increases to an infinite number, by multi-bed heat recovery method the heat recovered will increase to a maximum value. In Figure 6.32d, the value of dQ/dT for the exothermic process is flipped to the upper side of the T-axis, and the area surrounded by the exothermic and endothermic processes is the maximum heat (as shown by the transverse shadow lines on the figure).

Under the conditions of the thermal wave heat recovery cycle of the multi-bed system, using the working pair of the activated carbon–methanol, the relation between *COP* and the number of beds is shown in Figure 6.33 under the conditions of the evaporation temperature of −10 °C, the condensing temperature of 30 °C, and the adsorption temperature of 30 °C. The theoretical

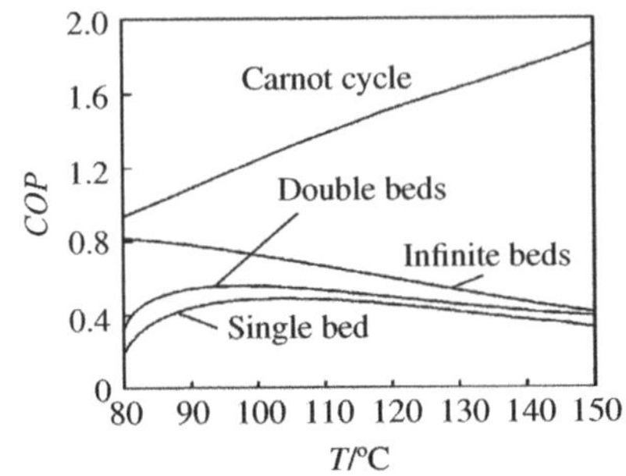

Figure 6.33 *COP* vs. maximum desorption temperature for different cycles [27]

efficiency of the Carnot cycle is also calculated under the same conditions in the diagram by the formulas in document [26]:

$$COP_{carnot} = \frac{1 - \frac{T_c}{T_{g2}}}{\frac{T_{a2}}{T_e} - 1} \qquad (6.67)$$

where T is shown in Figure 6.3a.

It can be seen from Figure 6.33 that the heat recovered is relatively sufficient when the number of beds increases, and consequently *COP* improves. When the temperature of T_{g2} is 100 °C, *COP* calculated by the Carnot cycle for a single bed system, two-bed system, and infinite-bed system are 39.22%, 45.46%, and 59.07%, respectively.

Meunier calculated the heat recovery coefficient for various multi-bed systems when the activated carbon–methanol is taken as the working pair for the air conditioning and refrigeration conditions. The results showed that, when the number of the beds increased to the infinite number the *COP* could be 1.852 [28].

6.5 The Optimized Cycle Driven by the Mass Change

Unlike the heat recovery device to achieve a heat regeneration cycle, the optimized cycle driven by the mass change uses different flow types of the adsorbate (refrigerant) to optimize the performance of an adsorption refrigeration cycle. Such types of cycles include mass recovery cycle, multi-stage cycle, and resorption cycle.

6.5.1 Mass Recovery Cycle

The mass recovery cycle is carried out between two or among more adsorbent beds in the adsorption system at the switch time. The cooled adsorbent bed before switch time connected with the evaporator, and its pressure is close to the evaporation pressure at the switch time, while it was much lower than the condensing pressure. Meanwhile the hot adsorbent bed before the switch time connects with the condenser, and its pressure is close to the condensing pressure that is much higher than the evaporation pressure. Under this condition, connecting

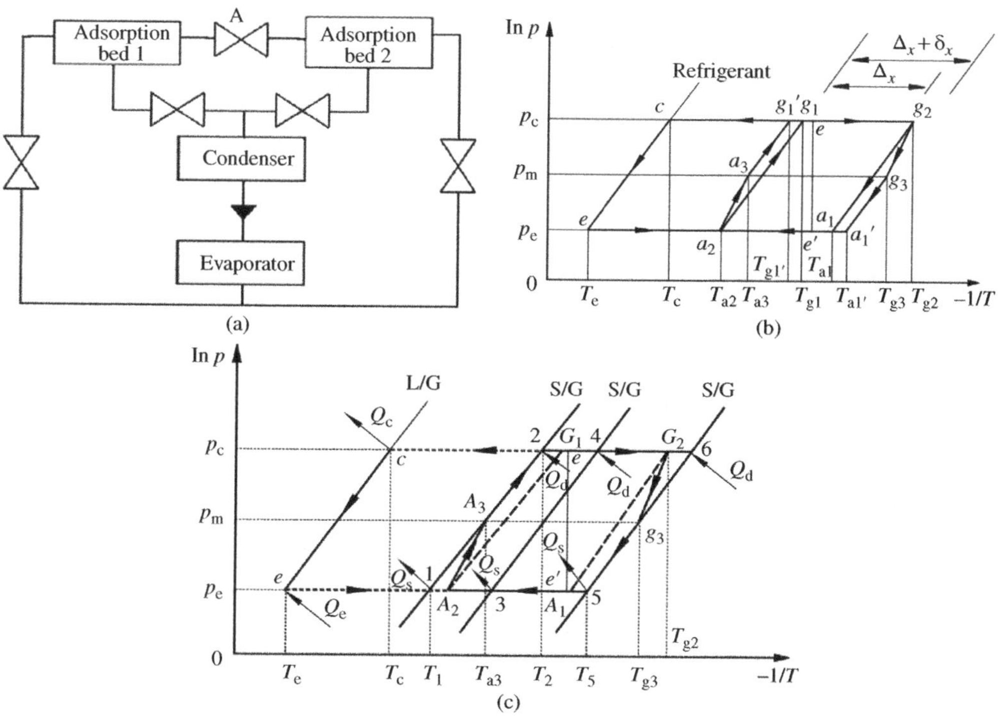

Figure 6.34 Clapeyron diagram of mass recovery cycle for physical and chemical adsorption systems. (a) Principle of the mass recovery cycle; (b) Clapeyron diagram of physical adsorption; and (c) Clapeyron diagram of chemical adsorption of $CaCl_2$-NH_3

the hot bed with the cold bed at the switch time can greatly increase the desorption rate of the hot bed, which will be helpful for the improvement of the adsorption quantity of the hot bed in the next half cycle for the cooling and adsorption process, and thereby it will improve the cooling capacity. The Clapeyron diagram of physical and chemical adsorption mass recovery cycle was shown in Figure 6.34.

The principle of the mass recovery cycle is shown in Figure 6.34a. When the desorption process of adsorbent bed 1 completes, the adsorption process of adsorbent bed 2 also finishes. Then close the valves connecting the adsorption beds, evaporator, and condenser, and open valve A for mass recovery between the high-pressure adsorbent bed (hot bed) and low-pressure adsorbent bed (cold bed). The mass recovery will proceed between two beds.

The Clapeyron diagram of physical adsorption is shown in Figure 6.34b. When the desorption process of adsorbent bed 1 (generator) completes, it is under the conditions of temperature T_{g2} and the condensing pressure p_c. When the adsorption of adsorbent bed 2 finishes, it is under the conditions of the adsorption temperature T_{a2} and the evaporation pressure p_e. For the mass recovery process we connect two adsorbent beds, the balance pressure of adsorbent bed 1 and 2 is p_m. It means that the temperature and pressure of adsorbent bed 1 decreases to the point g_3, while the temperature and pressure of adsorbent bed 2 increases to the point a_3. If coupling the mass recovery process and the heat recovery process together, that is to proceed the heat recovery process after the mass recovery process, the heat recovery process of adsorbent bed 1 is g_3-$a_1{}'$-e' in Figure 6.34b. Similarly the heat recovery process of adsorbent bed 2 was a_3-$g_1{}'$-e in Figure 6.34b. Figure 6.34b also showed that by the mass recovery or heat and mass recovery cycle the adsorption quantity of the adsorbent could be improved from Δx to $\Delta x + \delta x$ (a_2-g_1-e-g_2-a_1-e'-a_2), that is, the cooling capacity is improved. If the adsorption heat is the same as the desorption heat under the same pressure, the cycle *COP* will be increased with the increase in the cooling capacity.

Figure 6.34c is the Clapeyron diagram of the chemisorption mass recovery cycle. As discussed in the previous chapters about the chemical adsorption properties, the chemical reaction which may occur in the basic cycle is only $CaCl_2{\cdot}2NH_3 \leftrightarrow CaCl_2{\cdot}4NH_3$ when the reaction line 1-2 and 5-6 are all located outside the basic cycle of A_2-G_1-G_2-A_1-A_2 as shown in Figure 6.34c. For such a case the cycle adsorption amount is only 2 mol ammonia/mol calcium chloride. However, for the mass recovery cycle, the temperature and the pressure of high-temperature and high-pressure adsorbent bed decrease, while the temperature and pressure of low-temperature and low-pressure adsorbent bed increase. Assuming that after the mass recovery the state of the adsorbent bed 2 is located at the point G_3, then $CaCl_2 \leftrightarrow CaCl_2{\cdot}2NH_3$ that corresponds to the equilibrium reaction line of 5-6 will proceed. Assuming that the adsorbent bed 1 is after the mass recovery process, the state of the bed is located at point A_3, then $CaCl_2{\cdot}4NH_3 \leftrightarrow CaCl_2{\cdot}8NH_3$ that corresponds to the equilibrium reaction line of 1-2 will proceed. Due to the mass recovery cycle A_2-A_3-2-G_2-G_3-5-A_2 covering 1-2, 3-4, 5-6 curves; the maximum cycle adsorption quantity can be as high as 8 mol/mol. If compared with the basic cycle A_2-G_1-G_2-A_1-A_2 the maximum cycle adsorption quantity is improved by 6 mol/mol. Chemisorption mass recovery cycle also can be coupled with heat recovery cycle. Similarly the mass recovery process will be operated firstly, and then the heat recovery process will be operated. The principle of the heat and mass recovery cycle is the same as that of the physical adsorption cycle.

Taking the physical adsorption refrigeration cycle as the example to establish the models for the mass recovery cycle, and assuming that the beds thermal insulate with the outside in the mass recovery process, the mass recovery between two beds will correspond to a_2-a_3 and g_2-g_3 in Figure 6.34b. Due to the pressure difference, part of the refrigerant gas of the high-pressure adsorbent bed is transferred to the low pressure adsorbent bed, and the pressure decrement of the generator is equal to the pressure increment of the low-pressure adsorber.

$$\delta x_{a2-a3} = \delta x_{g2-g3} \tag{6.68}$$

Due to the beds in the mass recovery process being adiabatic, the temperature of the high-temperature and high-pressure adsorbent bed drops due to the desorption heat, while the temperature of the low-temperature and low-pressure adsorbent bed rises due to the adsorption heat. Then:

$$(C_{\rm a} + xC_{\rm pL})(T_{\rm a3} - T_{\rm a2}) = \Delta h \delta x_{\rm a2-a3} \tag{6.69}$$

$$(C_{\rm a} + xC_{\rm pL})(T_{\rm g3} - T_{\rm g2}) = \Delta h \delta x_{\rm g2-g3} \tag{6.70}$$

where C_a is the specific heat of the adsorbent; $C_{\rm pL}$ is the specific heat of the refrigerant liquid, and Δh is the adsorption/desorption heat.

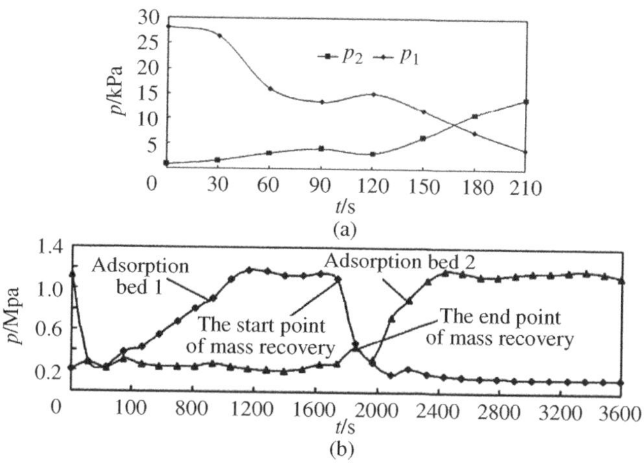

Figure 6.35 The pressure change of the adsorbent beds in the mass recovery process. (a) Activated carbon–methanol adsorption refrigeration system and (b) composite adsorbent of $CaCl_2$ and the activated carbon–NH_3 adsorption refrigeration system

The pressure of the two beds is the same after the mass recovery process, that is:

$$p_{g3} = p_{a3} \tag{6.71}$$

The final pressure of the cycle must satisfy the mass conservation equation of Equation 6.68.

The adsorption refrigeration systems include the low pressure systems (refrigerants such as methanol and water) and high pressure systems (refrigerants such as ammonia). Due to the pressure difference between the high-pressure adsorbent beds and low-pressure adsorbent beds being different, the influence of the mass recovery process on performance is different. When the cooling water temperature is 25 °C and the evaporation temperature is about −10 °C, the pressure changes of activated carbon–methanol adsorption refrigeration system and composite adsorbent-NH_3 system are shown in Figure 6.35 [29].

Figure 6.35 showed that for the working pair of the activated carbon–methanol, the pressure difference between two adsorbent beds in the mass recovery process is only about 26 kPa, while the pressure difference of two adsorbent beds of composite adsorbent-ammonia working pair is about 0.9 MPa. A larger pressure difference of composite adsorbent also accelerates the mass recovery process. The mass recovery process of composite adsorbent only lasts for 47 seconds, while the mass recovery process of physical adsorption lasts for about 170 seconds.

For chemisorption it has already been mentioned that the highest cycle adsorption quantity can be increased from 2 to 8 mol/mol in the mass recovery process. Its cooling capacity can be increased about three times compared with the basic cycle. For physical adsorption, Figure 6.36 showed the *COP* of air-conditioning conditions of a typical basic cycle (two-bed and no heat recovery) and the heat and mass recovery cycle [30], and the data in the diagram are simulation data under the conditions of neglecting the metal heat capacity.

It can also be seen from Figure 6.36 that the mass recovery process has significant impact on the *COP* of the adsorption refrigeration cycle, especially under the conditions of the lower

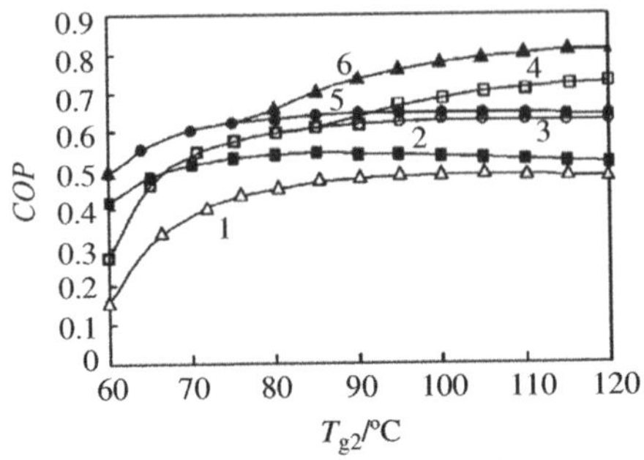

1 - Basic cycle; 2 - Mass recovery cycle; 3 - Sensible heat recovery cycle;
4 - Sensible heat and adsorption heat recovery cycle;
5 - Mass recovery and sensible heat recovery cycle;
6 - Mass recovery, sensible heat recovery, and adsorption heat recovery cycle

Figure 6.36 *COP* of different kinds of adsorption refrigeration cycle ($T_e = 5\,°C$, $T_c = T_a = 30\,°C$)

desorption temperature. The *COP* increment is as high as about 100% when the desorption temperature is 60 °C, which shows that the mass recovery after the heat recovery will significantly improve the system *COP*. In the actual adsorption refrigeration system, there are two types of heat recovery processes, the first type is the sensible heat recovery process, and the second type is the heat recovery process of both sensible heat and subsequent adsorption heat. Figure 6.36 showed the *COP* of different cycles in the ideal system (thermal capacity of metal materials and heat transfer fluid is neglected) [30]. Apparently, the adsorption refrigeration system with the mass recovery and the following heat recovery (sensible heat and adsorption heat) has the best performance. When the temperature of the heat source is 80 °C, the *COP* of the activated carbon–methanol adsorption air-conditioning system is up to 0.6. The *COP* can be as high as about 0.8 when the heat source temperature is 120 °C.

The experimental Clapeyron diagrams for physical [29, 31] and chemical adsorption [32, 33] with the mass recovery process are shown in Figure 6.37. Figure 6.37 shows that the area surrounded by the curves for the mass recovery cycle is bigger than that of the basic cycle. Such a phenomenon shows that the mass recovery improved the thermodynamic performance of the

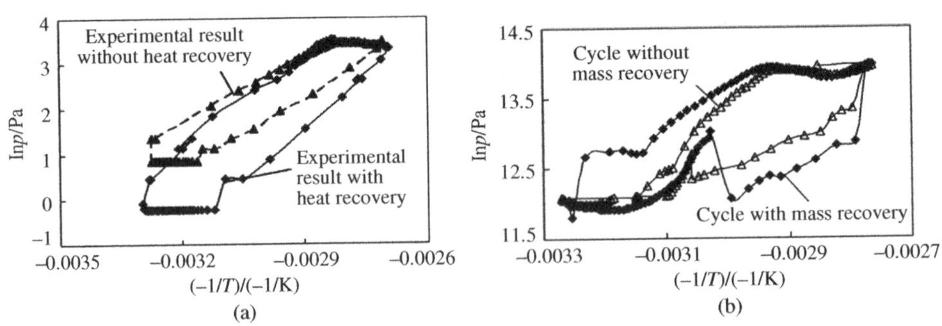

Figure 6.37 Experimental Clapeyron diagram of mass recovery cycle [29]. (a) The working pair of the activated carbon–methanol and (b) the working pair of composited adsorbent–ammonia

adsorption refrigeration significantly. Under such conditions, the cycle quantity of refrigerant would also be significantly increased [34].

6.5.2 *Multi-Stage Cycle*

The multi-stage cycle needs two or more adsorbent beds, and its principle is to couple the desorption process of an adsorbent bed with the adsorption process of another adsorbent bed, so as to effectively reduce the requirements on the temperature of the heat source by an adsorption refrigeration system.

The principle of multi-stage refrigeration cycle is shown in Figure 6.38, in which six adsorbent beds are used to achieve the three-stage adsorption refrigeration process [35, 36]. In the figure the refrigeration adsorbent beds are the adsorbent bed 3 and the adsorbent bed 6, and the adsorbent bed 1, 2, 4, and 5 are used as condensers. In the diagram the adsorbent bed 1 is heated for the desorption process, which desorbs the refrigerant to the condenser. The adsorbent bed 2 serves as the condenser for the adsorbent bed 3, and in this process bed 3 is heated and desorbed the refrigerant to bed 2, which is cooled by the external cooling source and adsorbs the refrigerant desorbed from bed 3. The adsorbent bed 6 connects with the evaporator, and it is cooled by the external cooling source and adsorbs the refrigerant that evaporated from the evaporator. Such a process outputs the refrigeration power. Bed 4 connects with bed 5, and bed 4 serves as the condenser for bed 5, that is, bed 5 is heated and desorbs, while bed 4 is cooled and adsorbs. To use bed 2 as a condenser for bed 3 in the diagram, such a process could decrease the constrict pressure for bed 3 because the pressure of bed 2 is much lower than the condensing pressure, then under the condition of the same minimum adsorption amount of the cycle the temperature of the heat sources needed by bed 3 will be effectively reduced. Similarly, for the desorption process of bed 2 bed 1 will serve as the condenser, thus the heat source temperature required by bed 2 in the desorption process will also be reduced. For bed 1 because it adsorbs the refrigerant vapor from bed 2, which has much higher pressure than

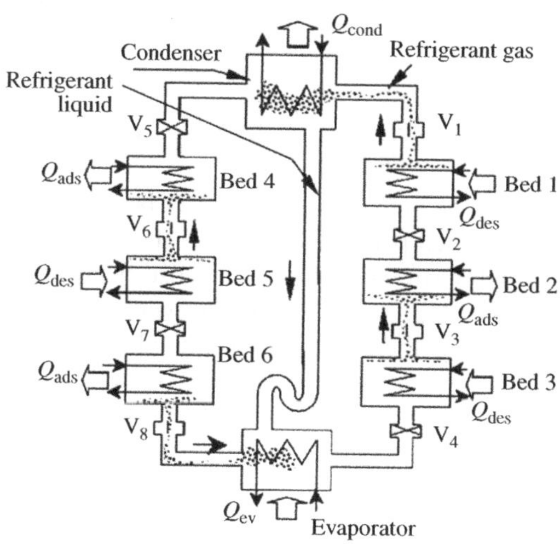

Figure 6.38 The diagram of three-stage cycle [35, 36]

the evaporator, thus the maximum adsorption amount of bed 1 will be effectively improved. Thus, under the conditions of the same cycle adsorption quantity the required heat source temperature of bed 1 will again be effectively reduced. When the working process as shown in Figure 6.38 finishes, bed 3 will be cooled and adsorb the refrigerant from the evaporator, such a process generates the cooling power. Bed 4 will be heated and desorbs the refrigerant to the condenser. The adsorbent bed 5 will serve as the condenser for the adsorbent bed 6, and the adsorbent bed 6 is heated for the desorption process. The adsorbent bed 1 will serve as the condenser for the adsorbent bed 2, and the adsorbent bed 2 is heated and completes the desorption process. Generally, all kinds of cycles need the heat source temperature to be at least 80 °C, but in Figure 6.38 the three-stage cycle [35, 36] can make use of the waste heat of about 50 °C.

For the model of a multi-stage cycle, the energy balance equations of the adsorption and desorption process are the same as the basic cycle. When heating and cooling processes proceed in two adsorbent beds, respectively, it is necessary to control the pressure for controlling the switch time [36, 37]. For example, when the cooling and heating processes proceed in adsorbent beds of 3 and 2, separately, the pressure for controlling the system for adsorption and desorption balance is:

$$p_{3des} = p_{2ads} \tag{6.72}$$

Taking the theoretical model established by Saha, for example, if we ignore the balance equation for the gaseous refrigerant, the conservation equation for mass and heat of a multi-stage cycle is:

$$\frac{dM_{water}}{dt} + M_a \times \left(\frac{dx_{des}}{dt} + \frac{dx_{ads}}{dt} \right) = 0 \tag{6.73}$$

where M_a is the adsorbent mass in the adsorbent bed (kg); M_{water} is the mass (kg) of the refrigerant liquid; x_{des} is the desorption amount (kg/kg); x_{ads} is the adsorption amount (kg/kg), t is time (s).

The equations for the adsorption and desorption heat are:

$$T_{waterout} = T_{adb} + (T_{waterin} - T_{adb}) \times \exp[-\alpha_{adb} \times A_{adb}/(m_{water} \times C_{pwater})] \tag{6.74}$$

$$\frac{d}{dt}[M_a \times (C_a + C_{Lc} \times x) + (C_{pm} \times M_{madb})] \times T_{adb}$$
$$= Q_{st} \times M_a \times \frac{dx}{dt} + m_{water} \times C_{pwater} \times (T_{waterin} - T_{waterout}) \tag{6.75}$$

where $T_{waterout}$ is the outlet temperature of the adsorbent bed (°C); T_{adb} is the adsorbent bed temperature; $T_{waterin}$ is the inlet temperature of the adsorbent bed (°C); α_{adb} is the adsorbent bed's heat transfer coefficient (W/(m^2 °C)); A_{adb} is the adsorbent bed's heat transfer area (m^2); m_{water} is the water flow (kg/s); C_{pwater} is the specific heat (kJ/(kg °C)); M_{madb} is the metal weight of the adsorbent bed; Q_{st} is the isobaric adsorption heat.

The balance equation between the evaporator and the condenser is:

$$T_{waterout} = T_{eva,cond} + (T_{waterin} - T_{eva,cond}) \times \exp[-\alpha_{eva,cond} \times A_{eva,cond}/(m_{water} \times C_{pwater})] \tag{6.76}$$

where the subscript "eva" represents the evaporator, "cond" is condenser. When the evaporator is calculated the subscript "eva" is used, whereas for calculating the condenser the subscript "cond" is taken.

$$\frac{d}{dt}[(C_{Lc} \times M_{ew} + C_{pm} \times M_{meva,cond}) \times T_{eva,cond}] = -L_{ref} \times M_a \times \frac{dq_{ads,des}}{dt}$$

$$+m_{water} \times C_{pwater} \times (T_{waterin} - T_{waterout}) + C_{ref} \times T_{cond} \times M_a \times \frac{dq_{des}}{dt} \quad (6.77)$$

where M_{ew} is the mass of the refrigerant liquid inside the evaporator; $M_{\text{meva,cond}}$ is the metal mass of the evaporator/condenser; L_{ref} is the latent heat of vaporization of the refrigerant. When the condenser is analyzed the subscripts of "cond" and "des" (q_{des} is the desorption heat) are used. Because for the condensing process there is no liquid accumulated in the condenser, and the M_{ew} is taken as 0 for the calculation of the condenser. For the calculation of the evaporator the subscripts "eva" and "ads" are taken, and in this process the desorption heat is taken as 0 by the fact that the desorption doesn't happen in the evaporating process.

According to the formulas above, a three-stage cycle is simulated and the results are verified by the experimental results for the working pair of silica gel–water. The experimental results and the simulation results for the adsorbent beds, condenser, and evaporator are shown in Figure 6.39.

The conditions for the simulation and for the experiments in Figure 6.39 are all shown in Table 6.7. The experiments are completed by a 1 kW refrigerating machine when the inlet temperatures of hot water, cooling water, and chilled water are all constant, and the flow rate also doesn't change. For the calculation of the condenser and the evaporator the relative error between simulation and experiments is less than 5%, for the adsorbent bed the relative error is 7%, and for the desorption bed the error is 15%. The error for the desorption bed is biggest mainly due to the heat dissipating to the environment in this process, the heat loss is very big because the adsorbent bed is switched from the cold situation to the hot situation and the sensible heat will be lost in the process.

When the hot water inlet temperature is controlled at 40 and 50 °C respectively, and the chilled water inlet temperature is controlled at 12 °C, the cooling capacity and *COP* of the multi-stage cycle is shown in Figure 6.40. When the cooling water temperature increases, the cooling capacity is reduced from 2.25 to 0.09 kW, and *COP* is reduced from 0.22 to about 0.04. Because the multi-stage cycle greatly improves the adaptive capacity of the adsorbent

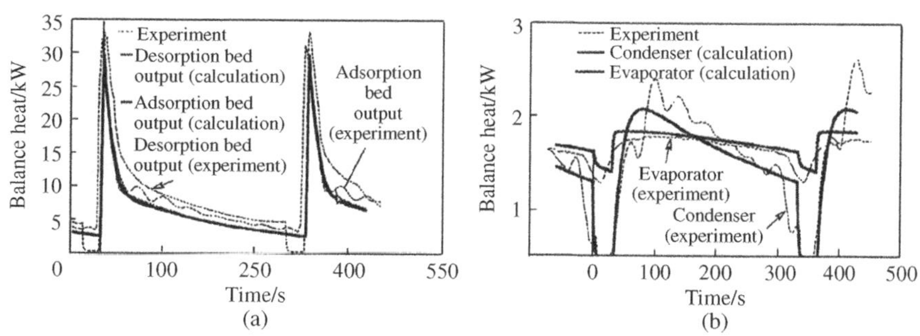

Figure 6.39 Simulation and experimental results of the multi-stage cycle. (a) Adsorbent bed and (b) evaporator and condenser

Table 6.7 Standard condition

Hot water inlet		Cooling water inlet		Chilled water inlet			
T (°C)	Flow rate (kg/s)	*T* (°C)	Flow rate (adsorbent bed + condenser, kg/s)	*T* (°C)	Flow rate (kg/s)	Adsorption-desorption cycle	The time for the switch
50	0.58	30	0.91(0.58 + 0.33)	12	0.06	300 s	30 s

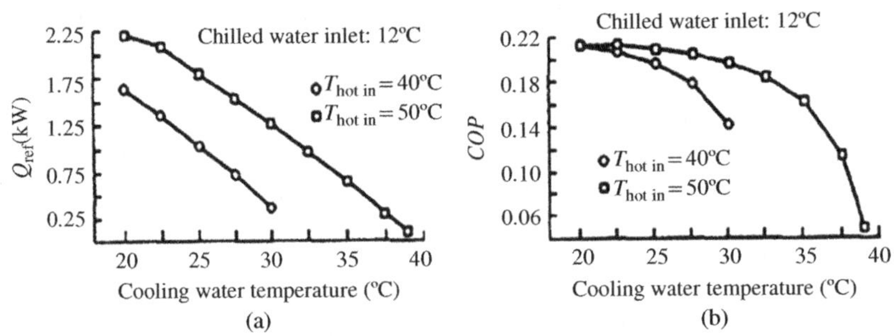

Figure 6.40 The performance of the cycle vs. the temperature of the cooling water [37]. (a) Cooling capacity vs. the temperature of the cooling water and (b) *COP* vs. the temperature of the cooling water

bed for the heat source, the cooling capacity and *COP* of the adsorption system were about 1.25 kW and 0.2 under the conditions of the cooling water temperature of 30 °C and the hot water temperature of 50 °C. Figure 6.41 also shows that a multi-stage adsorption cycle can be driven by a low temperature heat source. The cooling capacity and *COP* of the adsorption system were about 1.6 kW and 0.2 under the conditions of the cooling water temperature of 30 °C and the heat source of 55 °C. According to the cooling capacity the optimal cycle time is determined by Figure 6.42, and the results showed that when the adsorption-desorption cycle time is 300 seconds the cooling capacity of the system reaches the maximum value, and corresponding COP is close to 0.2.

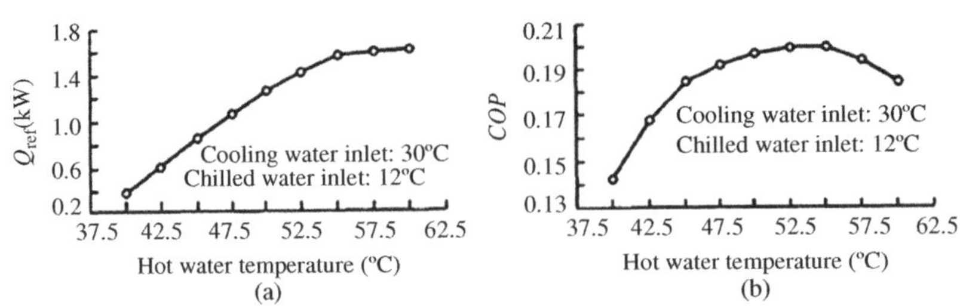

Figure 6.41 The performance of the cycle vs. the temperature of the hot water [37]. (a) The cooling capacity vs. the temperature of the hot water and (b) *COP* vs. the temperature of the hot water

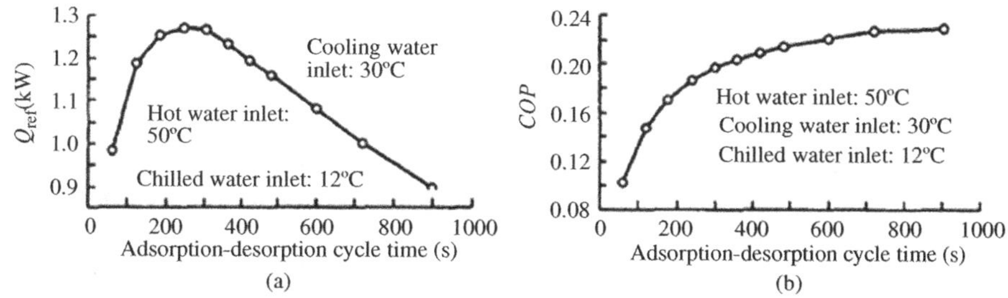

Figure 6.42 The performance of the cycle vs. cycle time [37]. (a) The cooling capacity vs. cycle time and (b) *COP* vs. cycle time

6.5.3 Resorption Cycle

6.5.3.1 Single-Effect Resorption Process

Resorption refrigeration cycle [38] requires two or more adsorbent beds, and its characteristic is that the adsorbent bed also can be used as the condenser and evaporator, but the system needs two or more adsorption working pairs (commonly the working pair of metal chloride-ammonia is used). The resorption refrigeration cycle can be used for both refrigeration and heat pump and can also be used for upgrading the temperature of the heat source.

The single-effect resorption refrigeration cycle generally uses two adsorption working pairs [39]. The Clapeyron diagram and schematic of single-effect resorption refrigeration cycle for both refrigeration and heat pump is shown in Figure 6.43. In Figure 6.43 two adsorption working pairs are used. The adsorbent bed 1 is the low-temperature working pair, corresponding to L_2-L_1 line in the Clapeyron diagram, and the adsorbent bed 2 is the high-temperature working pair, corresponding to the H_1-H_2 line in the Clapeyron diagram. In the diagram Q_{Hd} is the desorption heat for the high-temperature adsorbent bed that is provided by an external heat source. Q_{Hs} is the complexation reaction heat of the high-temperature adsorbent bed and the refrigerant. Q_s is the complexation reaction heat of the low-temperature adsorbent bed and the refrigerant. Q_d is the desorption heat of the low-temperature adsorbent bed.

The working processes of the cycle are as follows:

1. The adsorbent bed 2 desorbs at a high temperature, and the adsorbent bed 1 adsorbs at a high temperature (heat pump). The working process is that the adsorbent bed 2 is heated and

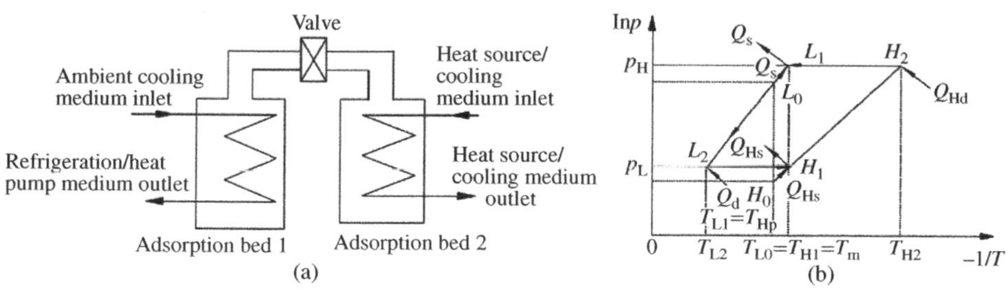

Figure 6.43 The diagram of the resorption refrigeration/heat pump cycle. (a) Resorption refrigeration/heat pump system and (b) Clapeyron diagram of the cycle

desorbs at the heat source temperature of T_{H2}, and the desorbed refrigerant flows through the valve to the adsorbent bed 1, and the adsorbent bed 1 adsorbs at the environmental temperature of T_m (T_{L0}). The adsorption heat improves the temperature of bed 1 to T_{L1} (T_{HP}), and the heat output is used for the heat pump.
2. The adsorbent bed 2 adsorbs at a low temperature (heat pump), and the adsorbent bed 1 desorbs and produces refrigeration. The working process is that the adsorbent bed 2 is cooled and adsorbs at the ambient temperature of T_{H2}, and the adsorption heat makes its temperature rise to T_{H1} (T_{HP}). The adsorption heat is used for the heat pump. The adsorbent bed 1 is cooled by the heat transfer fluid. Under the function of adsorption by bed 2 bed 1 will desorb, and the desorption heat makes it decrease from a temperature of T_{L1} to T_{L2}. The desorption heat of the process is used to provide the refrigeration output of the system.

According to the Clapeyron diagram in Figure 6.43b, the formulas of system *COP* (coefficient of cooling performance) and *COA* (heat pump COP) are:

$$COP = \frac{Q_d}{Q_{Hd}} \quad COA = \frac{Q_s + Q_{Hs}}{Q_{Hd}} \tag{6.78}$$

The single-effect resorption system also can be used for upgrading the heat source temperature. The system and the system Clapeyron diagram are shown in Figure 6.44. For upgrading the temperature three heat sources will be required, that is, a high-temperature heat source, a middle-temperature heat source, and an external cooling source. In Figure 6.44a the system working processes are as follows:

1. The adsorbent bed 2 is heated at the high-temperature heat source and the initial temperature is T_{H0}. The adsorbent bed 1 desorbs and the state point of adsorbent bed 2 migrates from H_0 to H_1, correspondingly the temperature is increased from T_{H0} to T_{H2} by the influence of the pressure of bed 1 and the adsorption heat of bed 2.
2. The adsorbent bed 1 adsorbs at ambient temperature T_m, and the adsorbent bed 2 desorbs at middle-temperature heat source. The adsorption effect of adsorbent bed improves the temperature from T_{L0} to T_{L1}, and provides the heat output for the heat pump under the condition of this temperature.

The resorption cycle for upgrading the temperature is mainly used for the occasions when the required temperature is higher than the highest temperature of the heat source. Corresponding to Figure 6.44b, without considering the heat loss as well as the fluid heat capacity, the

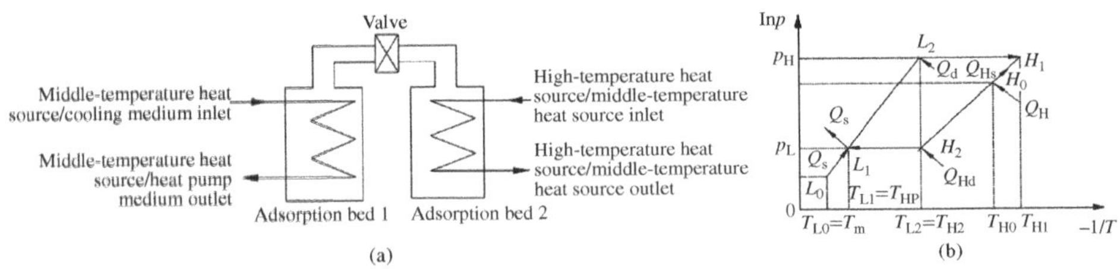

Figure 6.44 The resorption cycle for upgrading the heat source temperature. (a) The system and (b) Clapeyron diagram of the system

temperature increment is:

$$\Delta T = \frac{Q_{Hs}}{M_{Ha}C_{Ha}+M_{madb}C_m + M_{Ha}xC_{Lc}} \tag{6.79}$$

where M_{Ha} is the adsorbent mass of the high-temperature adsorbent bed; C_{Ha} is the adsorbent heat capacity in the high-temperature adsorbent bed; M_{madb} and C_m are the metal mass and heat capacity of the adsorbent bed, respectively; x is the adsorption amount; C_{Lc} is the heat capacity of the liquid refrigerant.

The *COA* of the heat pump for the cycle is:

$$COA = \frac{Q_{Hs}}{Q_d + Q_{Hd} + Q_H} \tag{6.80}$$

6.5.3.2 Double Effect Resorption Process

To construct the double effect resorption system we generally need to use three different salts in four adsorbent beds. Then at the switch time the heat recovery process can be used to reduce the heat demand of the system from the outside [39]. The work processes are shown in Figure 6.45.

1. As shown in Figure 6.45a, the third salt (S3) in the adsorbent bed 3 desorbs when it is heated at the temperature of T_h and the heat of Q_{Hd}. S1 (the first salt) in the adsorbent bed 1 adsorbs when it is cooled at temperature of T_m. S2 (the second salt) that is in the adsorbent bed 2 also adsorbs when it is cooled at the temperature of T_m, and the S1 that is in the adsorbent bed 4 desorbs at the temperature of T_{L}. The desorption heat Q_d of bed 4 provides the cooling capacity.

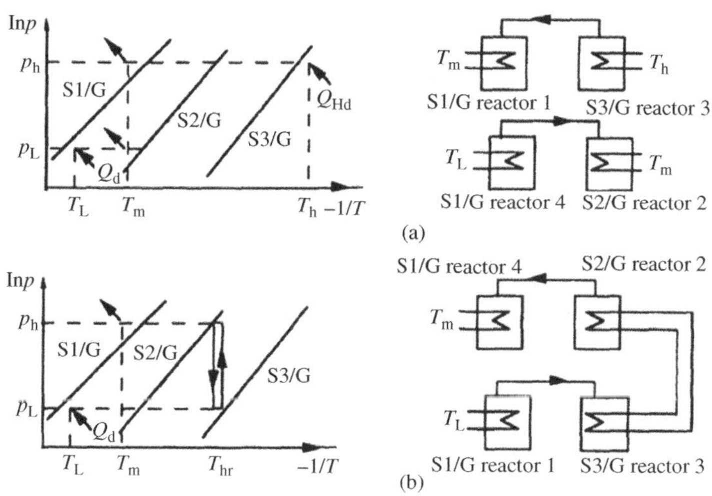

Figure 6.45 The diagram of the double effect resorption cycle [39]. (a) First half cycle and (b) second half cycle

2. As shown in Figure 6.45b, in the second half cycle, S3 in the adsorbent bed 3 and S2 in the adsorbent bed 2 adopt the heat recovery process. The adsorption heat of S3 at the temperature of T_{hr} provides the desorption heat of S2 at the temperature of T_{hr}. The adsorbent bed 4 is cooled and adsorbs at the temperature of T_m. The adsorbent bed 1 desorbs at the temperature of T_L, and the desorption heat Q_d provides cooling capacity.

6.5.3.3 The Selection of Adsorption Working Pairs and the Example of Performance Analysis

The selection of the resorption working pairs is related to the required refrigeration temperature, the cooling water temperature, and the heating temperature, which need to be selected based on the Clapeyron diagram of different salts. One example of the Clapeyron diagram of different salts–ammonia working pairs is shown in Figure 6.46.

Assuming that the preset refrigerant temperature is 273–278 K, and the ambient temperature is 313 K, Figure 6.46 shows that only $BaCl_2$ and $PbCl_2$ can be decomposed under such a low temperature. For the adsorption of $PbCl_2$ at the temperature of 313 K, due to the reaction the equilibrium temperature is close to the saturation temperature of ammonia, the refrigerant ammonia will be easily condensed and accumulated in the reactors and the corresponding pipe, which will increase the risk of the system. Under this condition, the ideal low-temperature adsorption working pair is $BaCl_2$–NH_3. The reaction equation is:

$$BaCl_2 + 8NH_3 \Longleftrightarrow BaCl_2(8NH_3) + \Delta H_r, \Delta H_r = 37000\ \mathrm{J/mol} \tag{6.81}$$

After the low-temperature working pair was selected, the high-temperature working pairs selection is unlimited by ambient temperature and cooling temperature. For the resorption process because the consumption of the gas in the reactor for synthesis process is equal to that in the adsorbent bed for decomposition process, the low-temperature reactor and the high-temperature reactor will be balanced mainly by the working pressure. The energy released by the adsorbent bed is mainly related to the gradient of the equilibrium pressure and the equilibrium temperature, and they are:

$$\Delta p_{eq} = p_c - p_{eq}(T_c) \text{ and } \Delta T_{eq} = T_c - T_{eq}(p_c) \tag{6.82}$$

If the $NiCl_2$ is chosen as the high temperature adsorbent it will be in a relatively high temperature zone in the Clapeyron diagram, the reaction equation is:

$$NiCl_2(2NH_3) + 4NH_3 \Longleftrightarrow NiCl_2(6NH_3) + \Delta H_r, \Delta H_r = 62000\ \mathrm{J/mol} \tag{6.83}$$

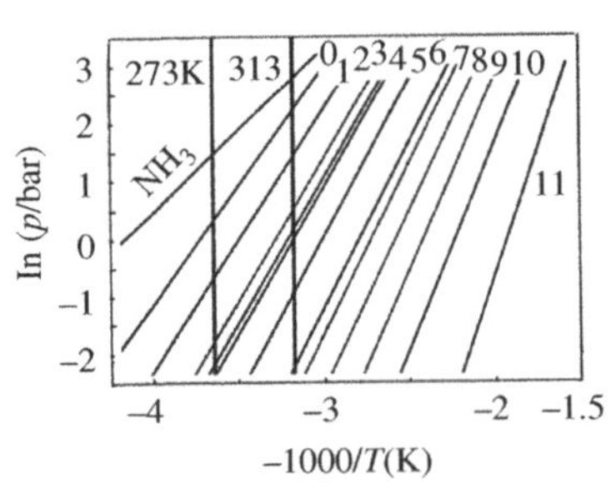

Figure 6.46 The reaction equilibrium of metal chloride–ammonia [40]

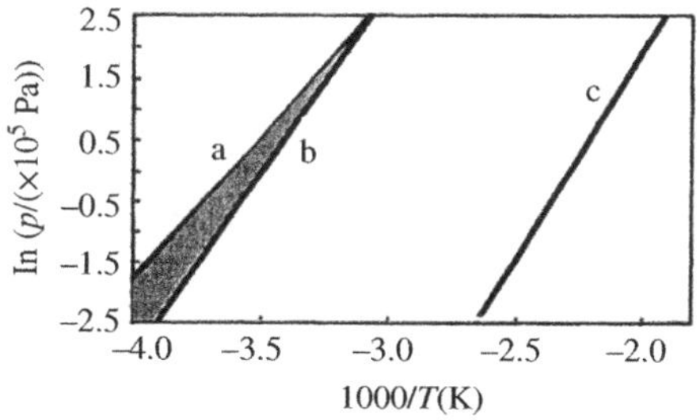

Figure 6.47 Reaction equilibrium lines of $BaCl_2/NH_3$ and $NiCl_2/NH_3$ in Clapeyron diagram [40]

For $BaCl_2$ and $NiCl_2$, consideration of the pseudo adsorption equilibrium zone the equilibrium adsorption lines are shown in Figure 6.47, and the equilibrium equations are [40]:

$$\text{Synthesis of } BaCl_2{:}\, p_{\text{eqs}}(T) = 4.68\exp[-36903/R(1/T - 1/308)] \tag{6.84}$$

$$\text{Decomposition of } BaCl_2{:}\, p_{\text{eqd}}(T) = 4\exp[-49496/R(1/T - 1/308)] \tag{6.85}$$

For the synthesis and decomposition of $NiCl_2$, the equation is:

$$p_{\text{eqs}}(T) = p_{\text{eqd}}(T) = 0.55\exp[-56160/R(1/T - 1/308)] \tag{6.86}$$

For the resorption process, the heat conservation equations are similar to the ordinary adsorption refrigeration process. The difference is that the resorption process works through the gas state of a two solid-gas reactor, and the working pressure has a relationship with the amount of gas adsorbed and desorbed at the same time. Therefore for researching the dynamic interaction between two adsorbent beds the pressure calculation is clearly necessary. The pressure is generally calculated by the enthalpy of the reaction gas and mass conservation. Assuming that the gas is ideal gas and the temperature of the adsorbent bed is equal to the temperature of the adsorbent, we can get:

$$dN_{gas}/dt = dN_{des}/dt - dN_{syn}/dt \tag{6.87}$$

$$dT_{gas}/dt = dN_{des}/dt[(T_{des} - T_{gas})/N_{gas}] \tag{6.88}$$

$$dp_{gas}/dt = (RT_{gas}/V_{gas})(dN_{gas}/dt) + (RN_{gas}/V_{gas})(dT_{gas}/dt) \tag{6.89}$$

where N_{gas} is the mole number of the gas in the space for the gas; N_{des} is the mole number of the desorbed gas; N_{syn} is the mole number of the gas taking part in the synthesis reaction; t is the time; T_{gas} is the temperature of the gas; T_{des} is the desorption temperature; p_{gas} is the pressure of the gas; R is the gas constant; V_{gas} is the gas volume.

Under low pressure, for the balance of the pressure it requires that the rate of decomposition of $BaCl_2$ ($8NH_3$) is equal to the rate of synthesis of $NiCl_2$ ($6NH_3$), that is:

$$m_{\text{amBaCl}_2} = -m_{\text{amNiCl}_2} \tag{6.90}$$

where m_{am} is the mass flow rate of the ammonia.

By coupling the equations above for the simulation of $BaCl_2$ and $NiCl_2$ resorption system, the simulation and experimental results are shown in Figure 6.48. Figure 6.48 shows that the experimental results fit with the simulation results very well.

The performance of the resorption system with the adsorbents of $BaCl_2$ and $NiCl_2$ obtained from Gotez is shown in Figure 6.49. In diagram T_c ($BaCl_2$) is the desorption temperature

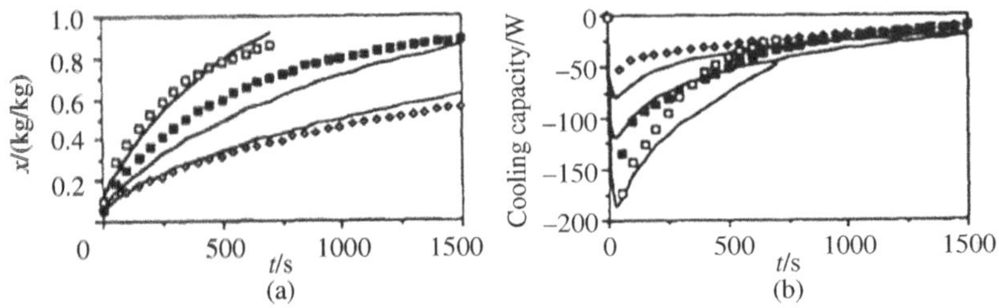

Figure 6.48 Comparison of simulation and experimental results [40]. (a) The adsorption quantity of $BaCl_2$ vs. time and (b) the cooling capacity vs. time. – :simulation value; ◇ $T_c(BaCl_2) = 273\,K$; ■ $T_c(BaCl_2) = 288\,K$; □ $T_c(BaCl_2) = 313\,K$

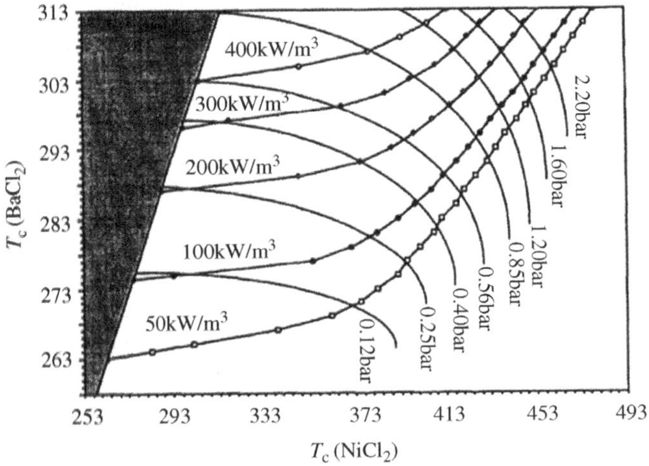

Figure 6.49 The relationship among $T_c(BaCl_2)$, $T_c(NiCl_2)$, volume cooling capacity of $BaCl_2$, and the equilibrium pressure [40]

of barium chloride, and T_c ($NiCl_2$) is the adsorption temperature of $NiCl_2$. Under the different adsorption temperatures of $BaCl_2$ and the different adsorption temperatures of $NiCl_2$ the performance of the adsorption refrigeration system for different states can be obtained. For example, when the desorption temperature (refrigeration temperature) of $BaCl_2$ is 273 K, the adsorption temperature of $NiCl_2$ is 298 K, and the equilibrium pressure of the system is about 0.12 Bar, the corresponding performance of the system is $100\,kW/m^3$.

6.6 Multi-Effect and Double-Way Thermochemical Sorption Refrigeration Cycle

6.6.1 Solid-Gas Thermochemical Sorption Refrigeration Cycle with Internal Heat Recovery Process

6.6.1.1 Double-Effect Refrigeration Cycle

For the above-mentioned solid sorption refrigeration cycles with heat management strategy, there is a common problem that it is difficult to keep the effective cascaded match of working temperature between different beds during the heat recovery phase when different sorption beds are filled with the same sorbents. The reclaimable thermal energy is usually the sensible heat from regeneration temperature to adsorption temperature of the sorption bed, and only small partial adsorption heat can be reutilized in the heat recovery process due to the low temperature difference of heat exchange and the limitation of heat transfer inside the bed. Thus, the regenerative efficiency is usually low for this kind of heat recovery sorption refrigeration cycle with the same sorbents.

To further improve the energy reutilization efficiency, Neveu and Castaing [41] proposed a new internal heat recovery technology for the performance improvement of a solid-gas thermochemical sorption refrigeration cycle. In this case, two reactors were filled with two different reactive salts, and the heat production of one reactor during the pre-cooling and adsorption phase was recovered and used to supply the heat consumption of the other reactor during the pre-heating and desorption phase. Moreover, the later heat consumption could be completely supplied by the former heat production, and thus the additional heat input from an external driving heat source was not required. During the heat recovery process, the reclaimable thermal energy not only includes the sensible heat of the reactor but also the reaction heat (adsorption heat) of reactive salt. Thus, this kind of cycle is referred to as double-effect solid-gas thermochemical sorption refrigeration cycle. In fact, the heat recovery strategy of the advanced double-effect thermochemical sorption cycle is similar to that of the cascaded physiosorption refrigeration cycle as in the aforementioned description.

The operation principle and the Clapeyron diagram of the proposed double-effect thermochemical sorption refrigeration cycle with internal heat recovery process is shown in Figure 6.50. The system consists of two solid-gas reactors, a condenser and an evaporator. Two different reactive salts are filled with solid-gas reactor 1 (R1) and reactor 2 (R2). S/G1 is

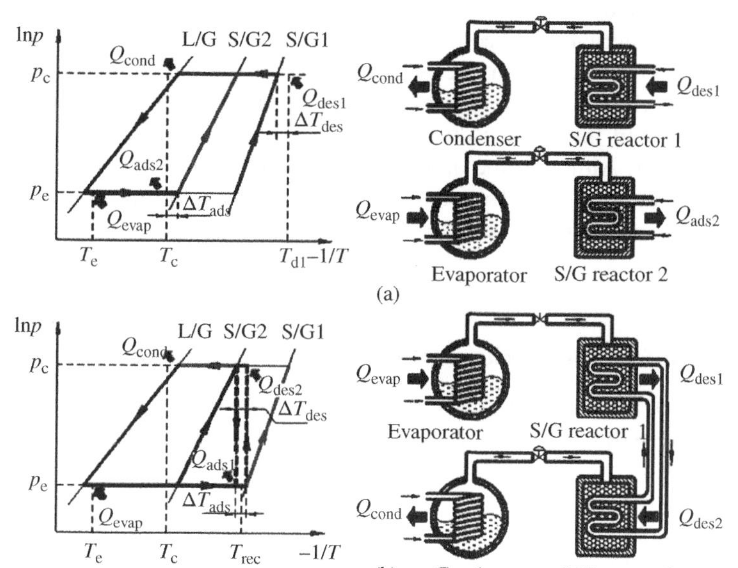

Figure 6.50 Operation principle and Clapeyron diagram of the double-effect thermochemical sorption refrigeration cycle with internal heat recovery processes [41]. (a) Adsorption and desorption process and (b) internal heat recovery

the equilibrium line of the high-temperature salt (HTS) inside R1 and S/G2 is the equilibrium line of the middle-temperature sorbent (MTS) inside R2. The HTS has a higher equilibrium temperature than the MTS at the same constraint pressure.

During the first working phase, the HTS reactor 1 is heated by an external heat source at a high temperature T_{d1} to perform decomposition in the desorption process of HTS. The refrigerant gas desorbs from the HTS reactor, flows to the condenser and becomes liquid refrigerant by rejecting condensation heat to heat sink at a temperature of T_c. At the same time, the MTS reactor 2 is cooled by a heat sink fluid to perform synthesis chemical reaction in the adsorption process of MTS. The refrigerant evaporates and then flows from the evaporator to the MTS reactor, and the vaporization heat of the refrigerant produces the first useful cooling-effect for the end user. During this working phase, the synthesis heat released by the MTS is removed by the heat sink, while the decomposition heat consumed by the HTS is supplied by an external heat source.

During the second working phase, the working modes of the reactors are interchanged, in which the MTS reactor 2 performs synthesis chemical reaction and the HTS reactor 1 undergoes a decomposition chemical reaction. An internal heat recovery process is introduced between the two reactors to improve energy utilization efficiency. The reaction heat released by the HTS during the synthesis phase is recovered and used to regenerate the MTS reactor. Moreover, the synthesis heat of the HTS is higher than the decomposition heat of the MTS, and thus the additional heat consumption from the external heat source is not required in the regeneration process of the MTS. During this working phase, the MTS is heated to desorb refrigerant gas to the condenser, and the HTS is cooled to adsorb the refrigerant from the evaporator. The vaporization heat of the refrigerant produces the second useful cooling effect for the end user.

It can be concluded that two useful cooling effects produced by the evaporation heat of the refrigerant could be obtained at the expense of only one high-temperature heat input for the HTS reactor during one double-effect thermochemical sorption cycle. Thus, the double-effect sorption cycle with internal heat recovery process could improve the system performance significantly when compared with the conventional heat recovery sorption cycle. Later, Sorin *et al.* [42] developed a combined heat recovery method where both the reaction heat of one reactive salt and the condensation heat of refrigerant were reclaimed to regenerate the other reactive salt. Although the amount of reclaimable thermal energy increased, this kind of double-effect sorption refrigeration cycle is affected by high operating pressures, which causes a large proportion of reactor thermal capacity and reduces the reliability of the sorption machine.

6.6.1.2 Double-Effect Thermochemical Resorption Refrigeration Cycle

Based on the basic resorption refrigeration cycle, a double-effect solid-gas chemi-resorption refrigeration system is developed by Spinner [39] and Goetz *et al.* [40] using a method similar to the internal heat recovery described above. To accomplish such a task, four reactors and three different reactive salts are used in the proposed sorption system. The reaction heat of the reactor filled with a high-temperature sorbent is utilized in the regeneration process of the reactor filled with a MTS. The operation principle and Clapeyron diagram of the double-effect thermochemical resorption refrigeration cycle with internal heat recovery process is shown in Figure 6.51.

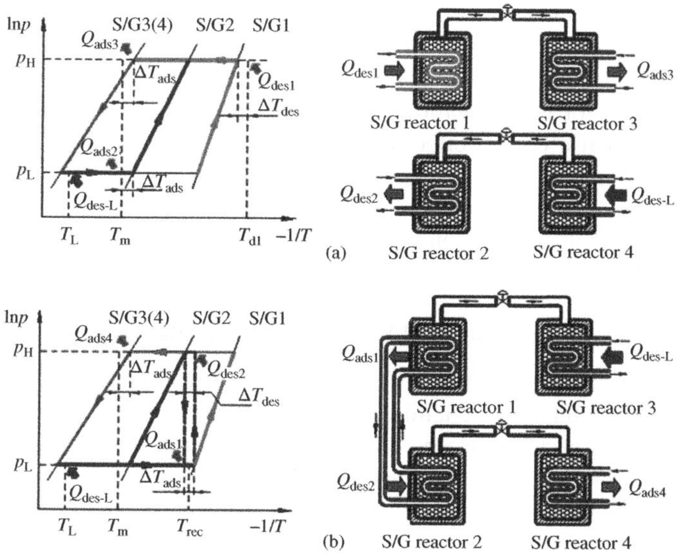

Figure 6.51 Operation principle and Clapeyron diagram of the double-effect thermochemical resorption refrigeration cycle with internal heat recovery process [40]. (a) Adsorption and desorption processes and (b) internal heat recovery

Different from the above-mentioned double-effect thermochemical sorption cycle, it can be seen that the condenser and evaporator are replaced by two low-temperature reactors in the double-effect thermochemical resorption cycle. To produce the cold continuously, three different reactive salts are filled with four solid-gas reactors. Reactor 1 (R1) is filled with a high-temperature sorbent (HTS) and reactor 2 (R2) is filled with a MTS. Reactor 3 (R3) and reactor 4 (R4) are filled with low-temperature salts (LTS). S/G1 is the equilibrium line of the high-temperature sorbent (HTS) inside R1, S/G2 is the equilibrium line of the MTS inside R2, and S/G3 is the equilibrium line of the low-temperature sorbent (LTS) inside R3 and R4. The HTS has the highest equilibrium temperature and the LTS has the lowest equilibrium temperature at the same constraint pressure among these different reactive salts.

During the first phase, the HTS reactor 1 is heated by an external heat source to desorb the refrigerant to the LTS reactor 3, and the synthesis reaction heat releases by the LTS reactor 3 is removed by heat sink. At the same time, a resorption refrigeration process occurs between the MTS reactor 2 and the LTS reactor 4, in which the refrigerant gas is desorbed from the LTS reactor 4 and flows to the MTS reactor. The decomposition reaction heat consumed by the LTS at a low temperature is used to produce the first cooling-effect, and the synthesis reaction heat released by the MTS reactor is removed by heat sink.

During the second phase, the working modes of the different reactors are interchanged. A resorption refrigeration process occurs between the HTS reactor 1 and the LTS reactor 3, in which the refrigerant gas desorbed by the latter reactor flows to the former reactor. The second cooling effect is produced by the decomposition reaction heat consumed by the LTS at low temperature. Simultaneously, the reaction heat released by the HTS reactor 1 is recovered and used to regenerate the MTS reactor 2 during the internal heat recovery process. The decomposition heat consumed by the MTS could be supplied completely by the synthesis heat released by the HTS. The refrigerant desorbed from the MTS reactor flows into the LTS reactor 4, and synthesis reaction heat released by the LTS reactor 4 is removed by heat sink.

Thus, two cold productions could also be obtained at the expense of only one high-temperature heat input for HTS reactor during one double-effect thermochemical resorption cycle. However, in such a double-effect resorption cycle, the cooling effects are produced by the reaction heat consumed by the LTS at a low temperature during the decomposition phase. The double-effect resorption cycle with internal heat recovery process could improve the system performance in comparison with the conventional heat recovery sorption cycle.

6.6.1.3 Multi-Effect Thermochemical Sorption Refrigeration Cycle

Based on the above-mentioned double-effect sorption and resorption refrigeration cycle, an advanced multi-effect solid thermochemical sorption refrigeration cycle was proposed [43] to further improve system performance by employing two internal heat recovery processes. In such a multi-effect cycle, three cold productions could be obtained at the expense of only one high-temperature heat input during one sorption cycle. The operation principle and Clapeyron diagram of the multi-effect thermochemical sorption refrigeration cycle with two internal heat recovery processes is shown in Figure 6.52.

It mainly consists of three solid-gas reactors, a condenser, and an evaporator. In order to implement the two internal heat recovery processes among the reactors, three kinds of reactive salts are utilized in the multi-effect sorption cycle. One reactor is filled with a high-temperature sorbent (HTS), the second reactor is filled with a MTS, and the third reactor is filled with a low-temperature sorbent (LTS). The operating mode of the multi-effect sorption cycle can be divided into two phases:

1. During the first phase, the decomposition reaction of the HTS in reactor 1 occurs by supplying a heat input from an external heat source at a high temperature, and the desorbed refrigerant gas flows into the condenser and becomes a liquid refrigerant by rejecting the condensation heat. At the same time, an internal heat recovery process is performed between the MTS and the LTS, where the reaction heat released during the synthesis reaction of the MTS is recovered and used to regenerate the LTS in reactor 3. The LTS is heated to desorb the refrigerant gas to the condenser, and the MTS is cooled to adsorb the refrigerant from the evaporator. The evaporation heat of the adsorbed refrigerant produces the first cooling effect for the end user.
2. During the second phase, the working modes of three reactors are interchanged. Both the HTS and the LTS reactors perform synthesis chemical reaction while the MTS reactor undergoes a decomposition chemical reaction. The LTS in reactor 3 is cooled by heat sink to adsorb the refrigerant from the evaporator and the vaporization heat of the refrigerant produces the second cooling effect for the end user. At the same time, the second internal heat recovery is performed between the HTS and MTS, in which the reaction heat released during the synthesis reaction of the HTS is transferred and utilized in the regeneration process of the MTS. The MTS is heated to desorb the refrigerant gas to the condenser, and another cooling effect is also obtained for the end user during the adsorption process of the HTS in reactor 1. During the two internal heat recovery processes, the LTS and the MTS reactors are heated by using the recovered thermal energy, and no additional external heat inputs are required during the desorption processes of the LTS and the MTS.

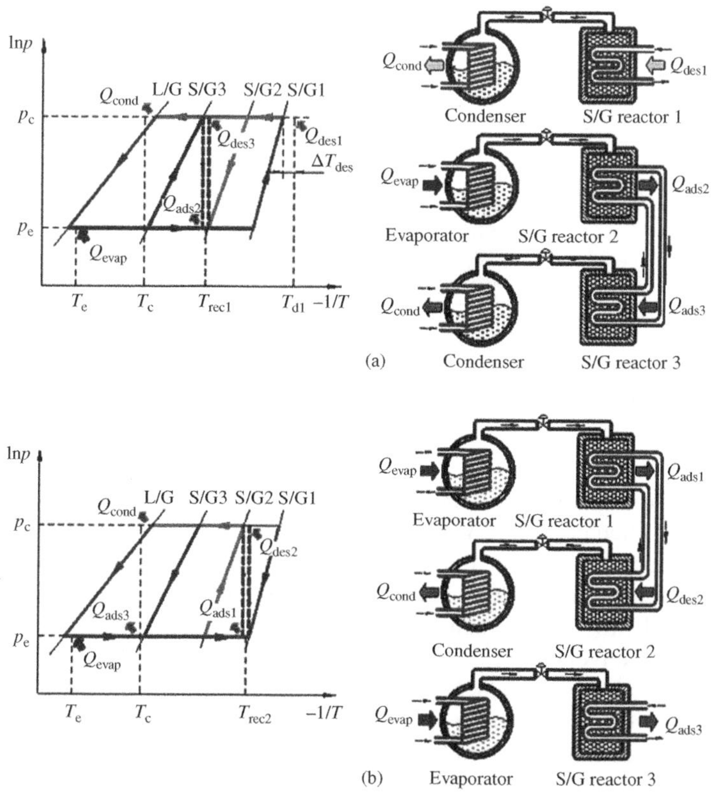

Figure 6.52 Operation principle and Clapeyron diagram of the multi-effect thermochemical sorption refrigeration cycle with two internal heat recovery processes [43]. (a) First half cycle and (b) second half cycle

In comparison with the double-effect sorption cycle, the proposed multi-effect sorption cycle has a more rigorous match between different sorption working pairs due to the implementation of the two internal heat recovery processes. The proposed internal heat recovery processes are only possible if the cascaded match of working temperatures and the amount of energy are established during the chemical reaction processes:

1. The total amount of heat released by the HTS/MTS during the synthesis phase should be higher than the total heat input required during the regeneration phase of the MTS/LTS.
2. The adsorption temperature of the HTS/MTS should be higher than the regeneration temperature of the MTS/LTS. The theoretical results showed that the amount of synthesis reaction heat released by the $NiCl_2/MnCl_2$ was higher than the decomposition reaction heat consumed by the $MnCl_2/SrCl_2$ during the two internal heat recovery processes. The result suggests that the proposed two internal heat recovery processes between different salts are feasible.

The performances of the single-effect, double-effect, and multi-effect thermochemical sorption refrigeration cycle were evaluated. The ideal COP_i of the single-effect cycle, the double-effect

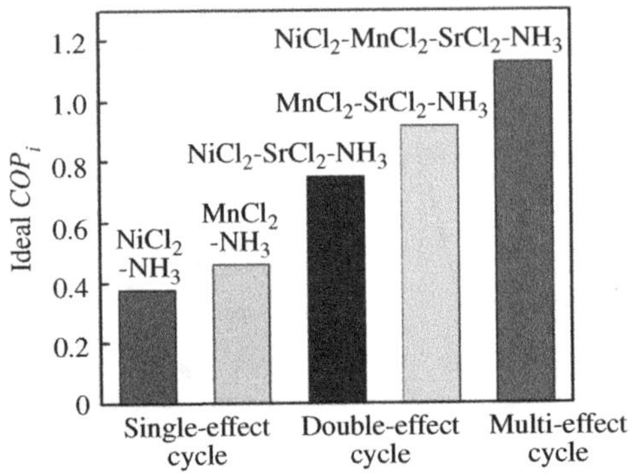

Figure 6.53 Ideal COP_i of the single-effect, double-effect, and multi-effect thermochemical sorption refrigeration cycle [43]

cycle, and the multi-effect cycle is shown in Figure 6.53. The double-effect sorption cycle has a higher COP_i when compared with the single-effect sorption cycle, and the multi-effect sorption refrigeration cycle has the highest COP_i among the three different kinds of sorption cycles. The improvement in the ideal COP_i obtained with the multi-effect sorption cycle varies between 23–50 and 146–200% when compared with the double-effect and the single-effect sorption cycles, respectively [43].

Figure 6.54 shows the theoretical *COP* variation with different mass ratio (*R*) when the sensible heat of the reactive salts, the refrigerant, and the metallic part of the reactors are considered. The ratio (R) is defined as the mass ratio between the metallic part of the reactor and the reactive salt. The *COP* decreases significantly with increasing the mass ratio R. In the range of mass ratio R from 0 to 15, the *COP* of the multi-effect cycle varies between 0.97 and 0.75. The system performance is significantly improved by the application of the multi-effect sorption cycle as compared with the conventional single and double-effect sorption cycles.

In the multi-effect thermochemical sorption refrigeration cycle, only the HTS in the reactor requires a high-temperature heat supply from an external heat source due to the two internal heat recovery strategies. In comparison with a conventional sorption refrigeration cycle, the

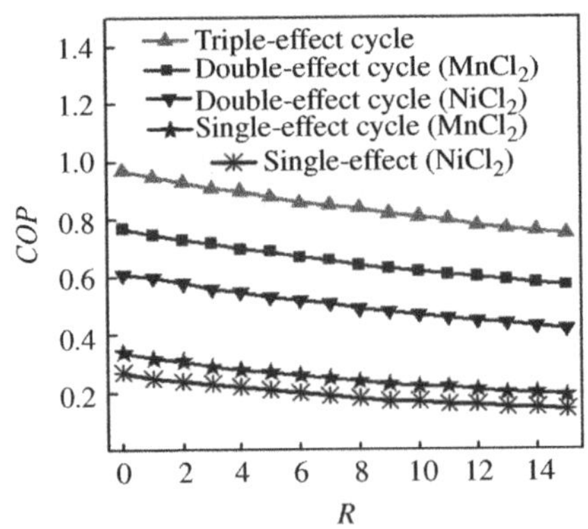

Figure 6.54 Performance of the single-effect, double-effect, and multi-effect thermochemical sorption refrigeration cycle [43]

multi-effect sorption cycle has the distinct advantage of a larger cooling capacity. This is because the multi-effect system could have three cold productions at the expense of only one heat input at a high temperature.

6.6.2 *Combined Double-Way Thermochemical Sorption Refrigeration Cycle Based on the Adsorption and Resorption Processes*

6.6.2.1 Working Principle of the Combined Double-Way Sorption Cycle

The aforementioned double/multi-effect thermochemical sorption refrigeration cycles can significantly improve the system performance due to the implementation of internal heat recovery. However, these advanced sorption cycles usually have a high thermal capacity of metal and they are complicated because the additional heat exchanger coils are required for the internal heat recovery process.

An innovative combined double-way thermochemical sorption refrigeration cycle is proposed by Li *et al.* [44, 45] to improve the sorption performance, in which both adsorption refrigeration and resorption refrigeration processes are combined to improve the cooling capacity. The operation principle and Clapeyron diagram of the combined double-way thermochemical sorption refrigeration cycle based on the adsorption and resorption process is shown in Figure 6.55.

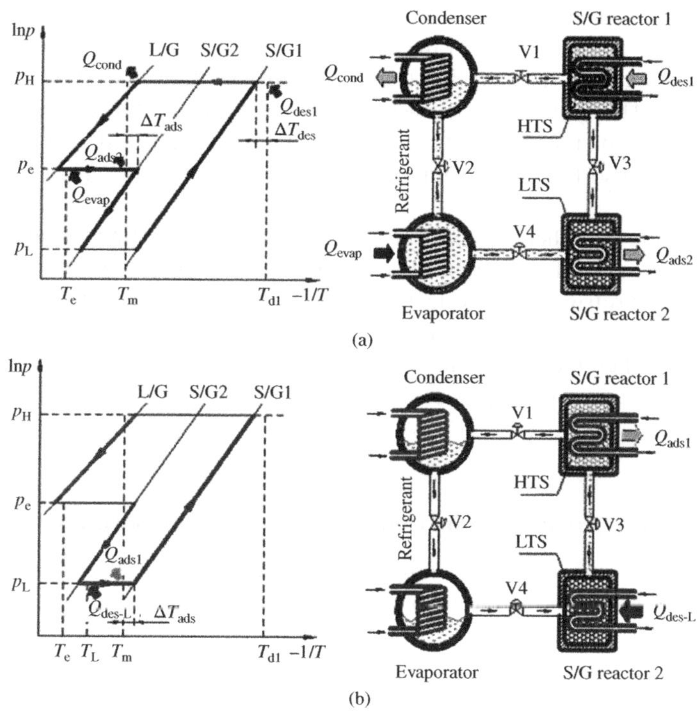

Figure 6.55 Operation principle and Clapeyron diagram of the combined double-way thermochemical sorption refrigeration cycle based on adsorption and resorption process [44]. (a) First half cycle and (b) second half cycle

It mainly consists of a reactor operating at a high temperature (R1), a reactor operating at a low temperature (R2), a condenser, and an evaporator. The former reactor is filled with a reactive salt with a higher equilibrium temperature, and another salt filled inside the latter reactor under the same operating pressure. Thus, the two reactive salts were referred to as the high-temperature salt (HTS) and the low-temperature salt (LTS), respectively. The working mode of the combined double-way sorption cycle can be divided into two phases.

1. During the first working phase, the HTS in reactor 1 performs a decomposition chemical reaction while the LTS in reactor 2 undergoes a synthesis chemical reaction. The HTS inside reactor 1 is heated by an external heat source at a high temperature to desorb the refrigerant gas to the condenser at a high-pressure p_{H}. The refrigerant gas becomes a liquid refrigerant by rejecting condensation heat to the environment and then flows into the evaporator. At the same time, an adsorption refrigeration process occurs between reactor 2 and the evaporator, in which the LTS inside reactor 2 is cooled by a heat sink to adsorb the refrigerant from the evaporator at a middle-pressure p_e. In the course of the adsorption refrigeration process, the reaction heat released by the LTS was removed by heat sink, and the vaporization heat of the refrigerant produced the first cooling effect by extracting heat from a chilled medium.
2. In the second working phase, the initial working modes of the two reactors are interchanged, in which reactor 1 performs a synthesis chemical reaction while reactor 2 undergoes a decomposition chemical reaction. The HTS inside reactor 1 is cooled by heat sink and then connected to reactor 2. At the same cooling temperature, the working pressure of the HTS was much lower than that of the LTS. Thus, the refrigerant gas would be transferred from reactor 2 to reactor 1 due to the high driving pressure drop as two reactors were connected. A resorption refrigeration process occurs between the HTS and the LTS at a low-pressure p_{L}. The synthesis reaction heat released by the HTS is removed by heat sink, and the decomposition reaction heat consumed by the LTS is utilized to produce the second cooling effect by absorbing heat from the chilled medium at a low temperature.

The combined double-way thermochemical sorption cycle has a distinct advantage of a larger cooling capacity per unit of heat input when compared with the conventional single-effect sorption or resorption refrigeration cycle, because two cooling effects could be obtained at the expense of only one heat input at a high temperature. The first production occurs during the adsorption process when the refrigerant vaporizes in the evaporator, and the second production occurs during the resorption process when the low-temperature sorbent absorbs heat at a low temperature.

6.6.2.2 Performance Analysis of the Combined Double-Way Sorption Cycle

In the conventional thermochemical sorption or resorption cycle, the operation process is carried out at two levels of pressure: high pressure during the regeneration phase of the reactive salt and low pressure during the cold production phase. For the combined double-way thermochemical sorption cycle, the operation process is constrained by three levels of working pressure as shown in Figure 6.56. The high-pressure p_{H} during the regeneration phase of the HTS, the middle-pressure p_e during the cold production by evaporation process, and the low-pressure p_{L} during the second cold production by the desorption heat of LTS. Moreover, the operating pressure during the resorption process is much lower than that during the

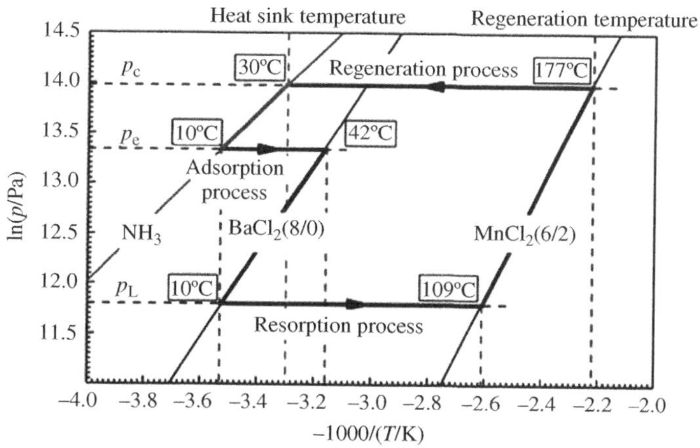

Figure 6.56 Clapeyron diagram of the combined double-way sorption thermodynamic cycle [45]

adsorption process when kept at the same refrigeration temperature. This suggested that the mass transfer performance inside the reactive materials during the resorption cooling is more important than that during the adsorption cooling due to the low operating pressure. The combined double-way cycle would create a vacuum if a low refrigeration temperature is required during the resorption process [46].

Figure 6.57 shows the performance of the combined double-way thermochemical refrigeration cycle based on the adsorption and resorption process. A group of working pair $MnCl_2$-$BaCl_2$-NH_3 is selected in order to carry out adsorption refrigeration and resorption refrigeration, in which $BaCl_2$ and $MnCl_2$ are used as the reactive salts and ammonia is the refrigerant. As shown in Figure 6.57a, the combined double-way sorption refrigeration cycle based on adsorption and resorption processes has the highest ideal COP_i of 1.24 among

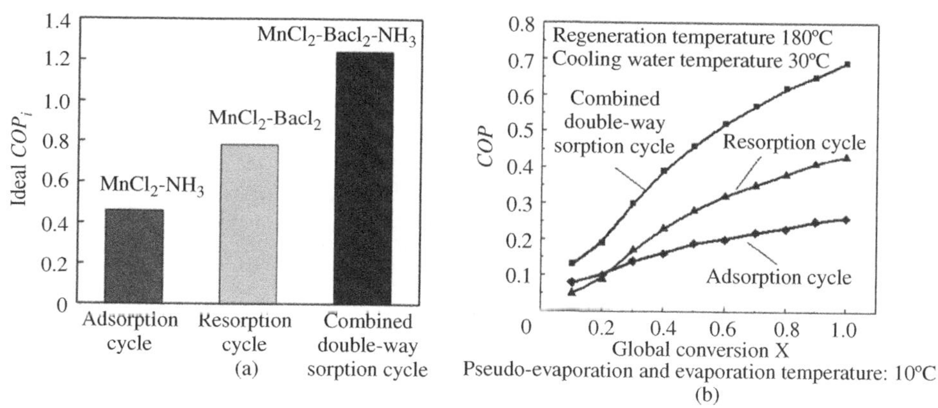

Figure 6.57 Performance of the combined double-way thermochemical refrigeration cycle based on the adsorption and resorption process [44]. (a) Ideal COP_i and (b) COP with different global conversion

the conventional solid adsorption and resorption refrigeration cycles. At the heat source temperature of 180 °C, cooling water temperature of 30 °C, evaporation temperature of 10 °C, and pseudo-evaporation temperature (regeneration temperature of LTS) of 10 °C. The theoretical *COP* of adsorption cycle ($MnCl_2/NH_3$), resorption cycle ($MnCl_2/BaCl_2$), and combined double-way sorption cycle ($MnCl_2/BaCl_2/NH_3$) are calculated based on different global conversions, and the result is shown in Figure 6.57b.

The theoretical *COP* of three kinds of cycles increases by increasing the global conversion. The increase rate of the combined double-way cycle is higher than those of the other two cycles. Moreover, the combined double-way cycle has the highest *COP* among the three kinds of cycles, ranging between 0.13 and 0.69 when global conversion *X* varies from 0.1 to 1. However, the theoretical *COP* is lower than the ideal COP_i of 1.24 mentioned previously. This is due to a large proportion of thermal capacity of the metallic part of the reactor in the simple experimental test unit. Usually, a high global conversion requires a considerable long cycle time owing to the fact that the reaction rate decreases with the amount of conversion, and the cooling power of the system could be largely reduced. Assuming a more realistic instance where the reactions are 85% completed, then the calculated *COP* of the combined double-way sorption cycle would be as high as 0.64. This value represents an improvement of 167 and 60% in the *COP* when compared with the adsorption cycle and the resorption cycle, respectively [44].

Later, a re-heating process is proposed that is aimed at improving the system performance of a combined double-way thermochemical sorption thermodynamic cycle [47]. Analysis of the experimental data showed that the reaction rate during the resorption refrigeration phase is lower than that during the adsorption refrigeration phase due to the different driving equilibrium differences. The temperature evolution of the LTS reactor with $BaCl_2$ during one combined double-way sorption cycle of the re-heating process is shown in Figure 6.58. The first useful cold is produced by the evaporation heat of the refrigerant inside the evaporator during the adsorption phase of the LTS, and the synthesis reaction heat released is removed by heat sink fluid. During the resorption phase, the LTS reactor works as a pseudo-evaporator

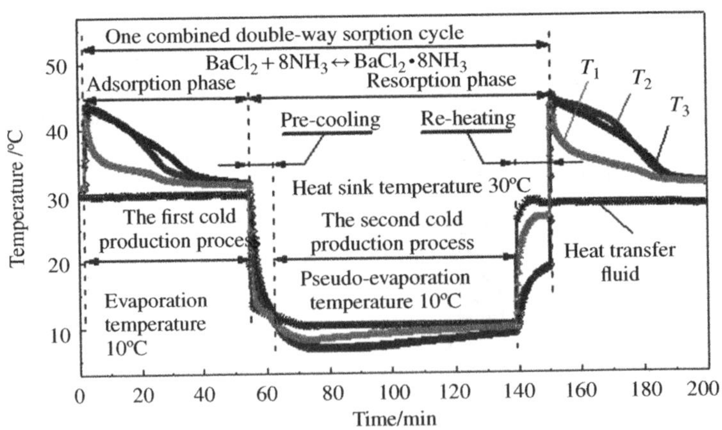

Figure 6.58 Temperature evolution of the LTS reactor with $BaCl_2$ during one combined double-way sorption cycle with re-heating process [47]

as decomposition reaction occurs, and the reaction heat consumed by the sorbent during the decomposition reaction from $BaCl_2 \cdot 8NH_3$ to $BaCl_2$ produces a useful cooling effect. At the end of the resorption refrigeration phase, the proposed re-heating process is introduced in order to increase the cycled mass of the refrigerant. Consequently, a prompt increase in the reactor temperature is observed during the re-heating process. The high driving equilibrium temperature difference ΔT enhances the reaction rate and thus improves the global conversion of the sorbents. Therefore, the system performance could be improved by the proposed re-heating process.

It was proved that the proposed re-heating process is an effective technique for improving the performance of the combined double-way cycle. The improvement in the *COP* ranges between 12% and 48% according to the different constraint conditions, when compared with the combined double-way cycle without re-heating [47]. Moreover, low pseudo-evaporation temperature and high heat sink temperature can further improve the system performance.

6.6.3 *Combined Double-Effect and Double-Way Thermochemical Sorption Refrigeration Cycle*

6.6.3.1 Working Principle of the Double-Effect and Double-Way Sorption Cycle

Based on the previous works, an advanced double-effect and double-way thermochemical sorption refrigeration cycle is further developed [48] to improve the working performance of a solid sorption refrigeration system by combining the internal heat recovery technology and combined double-way thermochemical sorption refrigeration. The operation principle and Clapeyron diagram of the double-effect and double-way thermochemical sorption refrigeration cycle with internal heat recovery process is shown in Figure 6.59.

For double-effect and double-way sorption cycle, it consists of four solid-gas reactors, a condenser, and an evaporator. Three different reactive salts are used as the sorbents in the double-effect and double-way sorption cycle. One reactor is filled with a HTS, and another is filled with a MTS, while another two reactors are filled with LTS. The cycle operation has five thermodynamic processes, including pre-heating and desorption, pre-cooling and adsorption, resorption refrigeration, and internal heat recovery. The cycled processes can be divided into two sub-cycles according to the refrigerant flow.

During the first working phase as shown in Figure 6.59a, the HTS in reactor 1 is heated to desorb the refrigerant gas to the condenser by opening valve 1, and the desorbed refrigerant condensed in the condenser and then flowed into the evaporator. The decomposition chemical reaction heat consumed by reactor 1 is supplied by an external heat source at a high temperature in the regeneration process of the HTS. Simultaneously, the LTS in reactor 4 is cooled by a heat sink to adsorb the refrigerant from the evaporator by opening valve 6. The synthesis chemical reaction heat released by reactor 4 is removed by external heat sink, and the vaporization heat of the refrigerant produces the first cooling effect by extracting heat from a chilled medium during the adsorption process. At the same time, the MTS in reactor 2 is cooled and connected to the LTS reactor 3 by opening valve 5, and the refrigerant gas is transferred from the LTS reactor 3 to the MTS reactor 2 due to the driving pressure drop. Thus, a resorption process between the MTS reactor 2 and the LTS reactor 3 occurs. The synthesis chemical reaction heat released by the MTS reactor 2 is removed by external heat sink, and the second cooling-effect is produced

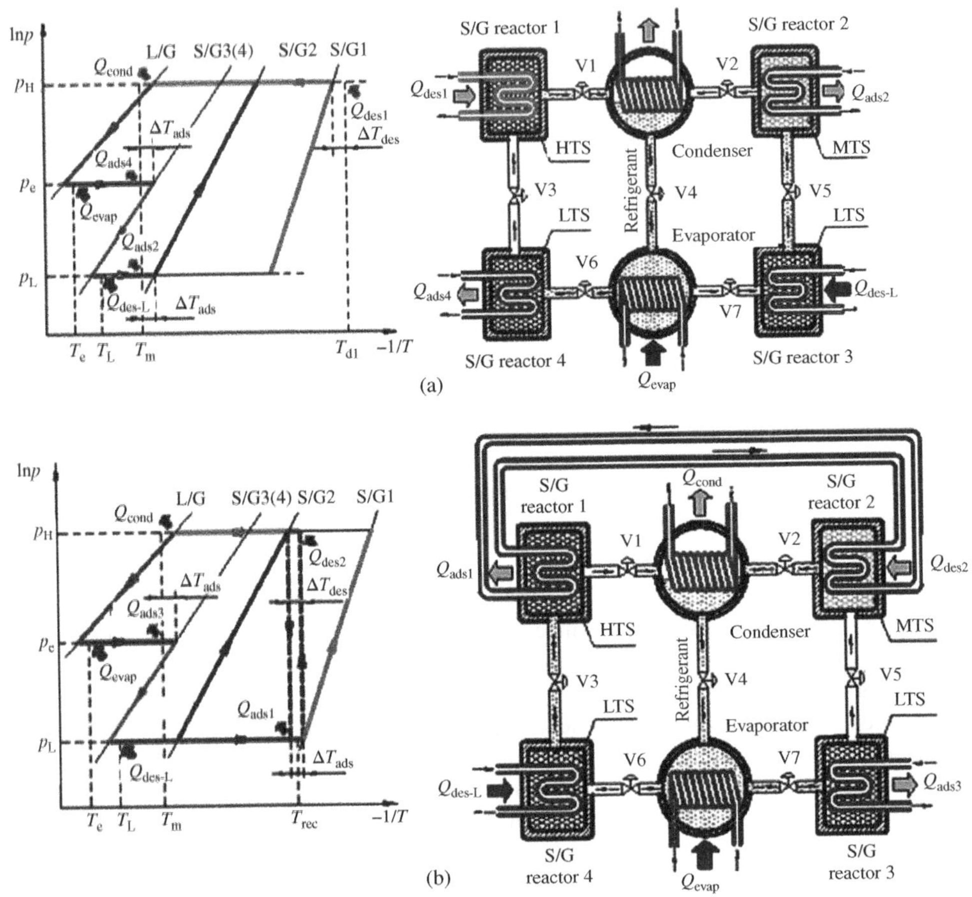

Figure 6.59 Operation principle and Clapeyron diagram of the double-effect and double-way thermochemical sorption refrigeration cycle with internal heat recovery process [48]. (a) First half cycle and (b) second half cycle

by the reaction heat consumed by the LTS at a low temperature during the decomposition process of the LTS reactor 3.

During the second working phase as shown in Figure 6.59b, the working modes of the reactors are interchanged. The reactors 1 and 3 are in adsorption modes while reactors 2 and 4 are under desorption modes. The HTS reactor 1 is connected to the LTS reactor 4 by opening valve 3, in which the reactor 4 works as a pseudo-evaporator to produce the third cooling effect during the resorption refrigeration process between the HTS and the LTS. The refrigerant gas flows from the LTS reactor 4 to the HTS reactor 1 and reacts with the high-temperature salt, and a large amount of synthesis reaction heat is released by the HTS in the adsorption process. In this case, an internal heat recovery is performed between the HTS reactor 1 and the MTS reactor 2, in which the reaction heat released by the HTS is reutilized for the regeneration of the MTS. The refrigerant desorbed from the MTS reactor 2 flows to the condenser by opening valve 2. In this process, no additional external heat source is required during the regeneration process of the MTS. At the same time, the LTS reactor 3 is cooled by a heat sink to adsorb the refrigerant from the evaporator. The evaporation heat of the refrigerant produces the fourth cooling effect for the end user.

The proposed advanced double-effect and double-way thermochemical sorption cycle has the distinct advantage of a larger cooling capacity per unit of heat input when compared with other kinds of solid sorption refrigeration cycles, due to the four cooling effects obtained at the expense of only one high-temperature heat input for the HTS. During every working phase, the cold production consists of the adsorption refrigeration based on the evaporation process and the desorption refrigeration based on the resorption process.

6.6.3.2 Performance Analysis of the Double-Effect and Double-Way Sorption Cycle

A group of working pair NH_3, $BaCl_2$, $CaCl_2$, and $NiCl_2$ is selected to assess the performance of the proposed double-effect and double-way sorption refrigeration cycle, in which $NiCl_2$ is used as HTS, $CaCl_2$ is utilized as MTS, and $BaCl_2$ is employed as LTS. The Clapeyron diagram of the double-effect and double-way thermochemical sorption cycle with internal heat recovery is shown in Figure 6.60.

The decomposition heat of $NiCl_2$ was supplied by an external heat source during the regeneration process, and the decomposition heat consumed by the $CaCl_2$ is supplied by the synthesis reaction heat released by $NiCl_2$ by introducing the internal heat recovery strategy between the HTS and the MTS. The useful cold is produced by the evaporation heat of the refrigerant during the adsorption phase of $BaCl_2$ and the chemical reaction heat consumed by $BaCl_2$ at a low temperature during the decomposition phase of the LTS. Although the working refrigeration temperature (10 °C) during the resorption refrigeration process is higher than that (0 °C) during the adsorption refrigeration process, the operating pressure during the former process is much lower than that during the latter process. Such a result indicates that the mass transfer

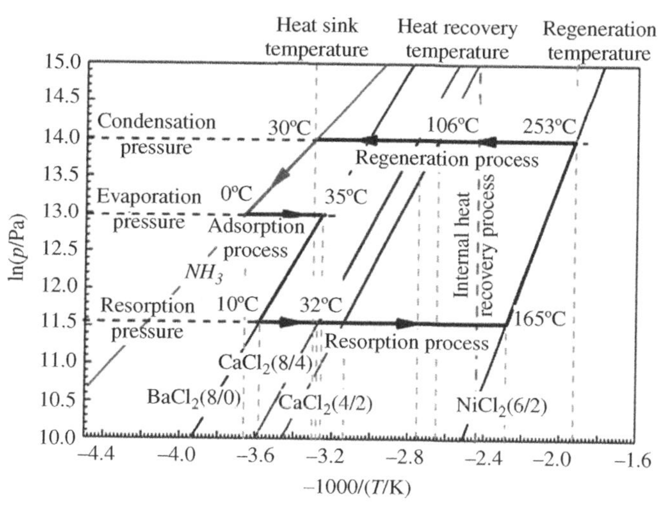

Figure 6.60 Clapeyron diagram of the double-effect and double-way thermochemical sorption refrigeration cycle with internal heat recovery process [48]

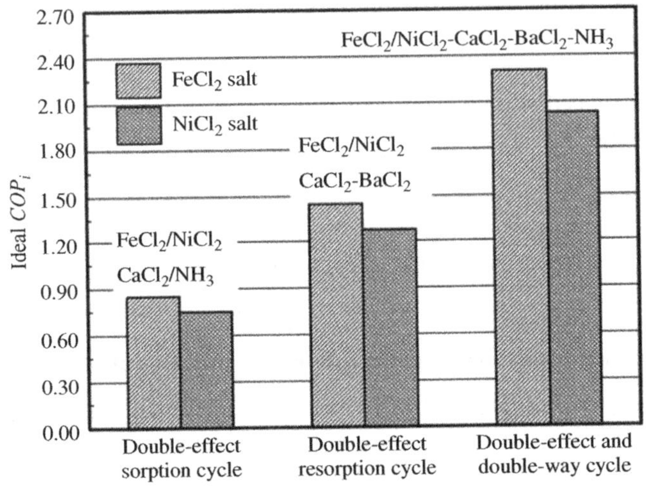

Figure 6.61 Ideal COP_i of the double-effect and the double-way sorption cycle [48]

performance inside the reactive materials is a major constraining factor during the resorption refrigeration process, and it would have a strong influence on the system performance of the double-effect and double-way thermochemical sorption refrigeration cycle.

The ideal COP_i of the double-effect and the double-way thermochemical sorption cycle with internal heat recovery process is shown in Figure 6.61. The double-effect and the double-way sorption refrigeration cycle based on the adsorption and resorption has the highest COP_i among the three kinds of sorption cycles. In comparison with the double-effect sorption cycle ($FeCl_2/NiCl_2$-$CaCl_2$-NH_3) based on the evaporation process and the double-effect resorption cycle ($FeCl_2/NiCl_2$-$CaCl_2$-$BaCl_2$) based on the resorption process, the proposed double-effect and the double-way thermochemical sorption refrigeration cycle ($FeCl_2/NiCl_2$-$CaCl_2$-$BaCl_2$-NH_3) could improve the ideal COP_i by more than 169 and 59%, respectively [48].

At a heat sink temperature of 30 °C, evaporation temperature of 0 °C, and the pseudo-evaporation temperature (regeneration temperature of LTS) of 10 °C, the *COP* variation with different mass ratio (*R*) between the metallic part of the reactor and the reactive salt is shown in Figure 6.62, in which the sensible heat of the reactive salts, the refrigerant, and the metallic part of the reactors is considered in the calculation *COP* of the double-effect and the double-way thermochemical sorption cycle. The *COP* decreases significantly as mass ratio *R* increases. In the range of mass ratio *R* from 0 to 15, the *COP* varied between 1.08 and 1.80 for the working pair of $FeCl_2$-$CaCl_2$-$BaCl_2$-NH_3. For most of the solid-gas sorption systems with optimized design, the mass ratio (*R*) between the metallic part of the reactor and the reactive salt is about 5. The corresponding *COP* obtained with the double-effect and double-way sorption cycle ranges from 1.50 to 1.26 [48]. Thus, the system performance could be significantly improved by the presented double-effect and double-way sorption cycle when compared with the conventional single-effect and double-effect sorption cycles.

The feasibility and working performance of the proposed double-effect and double-way thermochemical sorption refrigeration cycle is experimentally investigated [49] using a group of

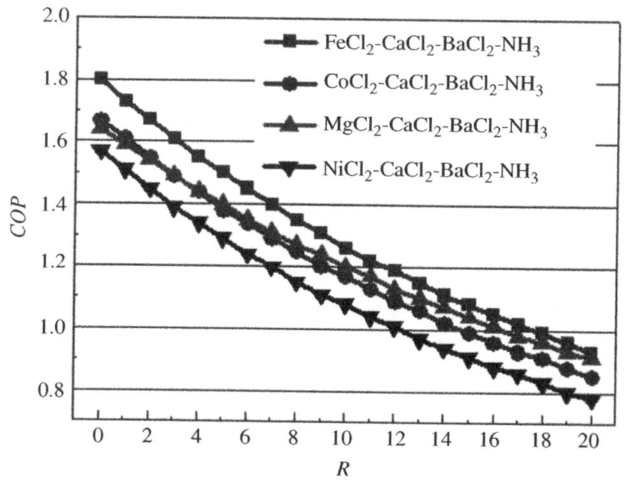

Figure 6.62 *COP* variation with *R* in the double-effect and double-way sorption cycle [48]

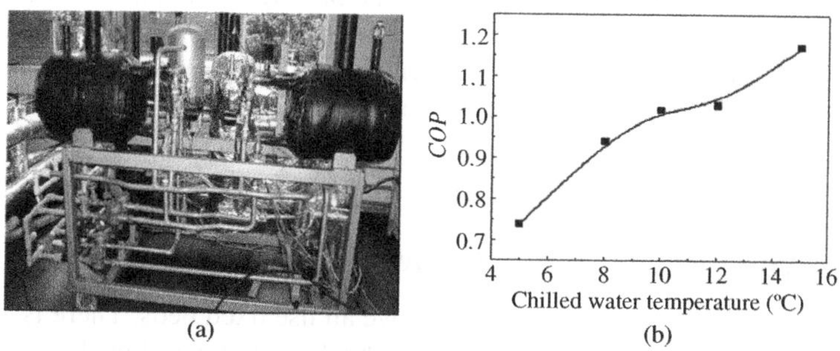

Figure 6.63 Experimental test unit and the COP of the proposed double-effect and double-way thermochemical sorption refrigeration system [49]. (a) Experimental test unit and (b) *COP* vs. chilled water temperature

sorption working pair of $NiCl_2$-$MnCl_2$-$BaCl_2$-NH_3. The experimental test unit of the proposed double-effect and double-way thermochemical sorption refrigeration system is shown in Figure 6.63a. In the test unit, three different metal chlorides, $NiCl_2$, $MnCl_2$, and $BaCl_2$, are used as HTS, MTS, and LTS, respectively. The experimental results show that the proposed double-effect and double-way thermochemical sorption refrigeration cycle is feasible in the field of air-conditioning and refrigeration, and it can produce four useful cooling effects during one cycle by using only one heat input at high temperature. Figure 6.63b shows that the experimental *COP* was higher than 1.0 at the chilled water temperature of 10–15 °C, heat source temperature of 260 °C, and heat sink temperature of 30 °C [49].

Later, the performance comparisons for different kinds of thermochemical sorption thermodynamic cycles were analyzed [50]. These advanced thermodynamic sorption cycles included the single-effect sorption cycle ($NiCl_2$-NH_3), the single-effect resorption cycle ($NiCl_2$-$BaCl_2$), the double-effect sorption cycle ($NiCl_2$-$MnCl_2$-NH_3), the double-effect resorption cycle

($NiCl_2$-$MnCl_2$-$BaCl_2$), the combined double-way sorption cycle ($NiCl_2$-$BaCl_2$-NH_3), and the double-effect and double-way sorption cycle ($NiCl_2$-$MnCl_2$-$BaCl_2$-NH_3). Simulation results showed that the combined double-way sorption cycle based on the adsorption and resorption process had higher *COP* when compared with the single-effect sorption cycle and the single-effect resorption cycle without internal heat recovery. The *COP* obtained with the combined double-way sorption cycle, single-effect resorption cycle, and single-effect sorption cycle were 0.47, 0.32, and 0.16 respectively. In comparison with the single-effect sorption cycle and single-effect resorption cycle, the combined double-way sorption cycle could improve the *COP* by more than 194 and 47%, respectively.

When the internal heat recovery process is employed between the different reactive salts, the *COP* and the energy utilization efficiency are improved significantly due to the implementation of heat recovery process. Generally, the double-effect thermodynamic cycles have higher *COP* than their corresponding single-effect thermodynamic cycles. For example, the *COP* improvement is about 125% in the double-effect resorption cycle when compared with the single-effect resorption cycle, and it is about 138% for the double-effect sorption cycle. Moreover, the double-effect and double-way sorption cycle with internal heat recovery process has the highest COP among these thermochemical cycles, and the *COP* was as high as 1.10. This figure represented seven times the value of *COP* obtained with the conventional single-effect sorption cycle and three times the value of *COP* obtained with the double-effect sorption cycle [50]. In addition, the double-effect and double-way sorption cycle usually had a long cycle time due to the low driving equilibrium drop. This means that the heat and mass transfer inside reactors is the key factor to the double-effect and double-way thermochemical sorption cycle.

6.7 Step-by-Step Regeneration Cycle

The described adsorption refrigeration cycles above all use fixed beds. There is an adsorption dehumidification cycle using a rotating bed (the wheel). This type of cycle has already been successfully used for dehumidification, refrigeration, and air conditioning. This cycle also can be used for producing refrigeration [51, 52] using the method of dehumidification.

The principle of the dehumidification cycle was shown in Figure 6.64. The humidification process is as follows: the right half of the cylinder adsorbs water vapor from the wet air to

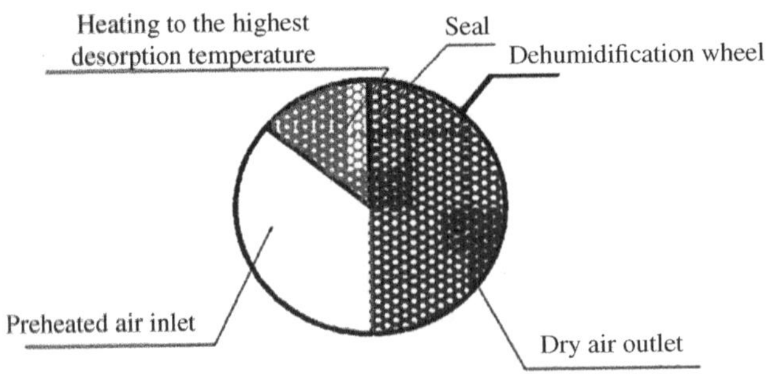

Figure 6.64 Step-by-step regeneration cycle

produce dry air, and then the adsorbent rotates to the left half of the cylinder for desorption process by the rotation of the wheel. The basic idea of the step-by-step desorption is that only a part of the heat medium of the air is heated to a maximum temperature, which could save the energy for heating process effectively. The desorption process is divided into two steps. For the first step the adsorbent adsorbs the desorption heat from the preheated air. (It doesn't consume additional thermal energy for the preheating process and the thermal energy is from the adsorption heat released during the adsorption process.) For the second step the desorption heat is provided by the air which is heated to a maximum desorption temperature by the external heat source. If we spray the water into the dehumidified dry air the humidification process with the constant enthalpy will generate the cooling power.

The dehumidification air conditioner by the rotary wheel is an important way of adsorption refrigeration. Compared with the above mentioned cycles with fixed beds, the rotary wheel cycle is an open adsorption refrigeration cycle and generally could output the refrigeration through the water evaporation process and the adsorption function of the wheel.

6.7.1 Desiccant Cooling Refrigeration

The dehumidification refrigeration is the combination of dehumidification process and the evaporation refrigeration process. Desiccant materials have a strong adsorption and dehumidification ability. Generally commercial desiccants can adsorb the moisture of 10–1100% of its own weight [53]. When the surface vapor pressure of the desiccant is equal to the partial pressure of the wet air the adsorption process stops. The hot air with a temperature of 50–260 °C flowing through the desiccant surface can take away the moisture adsorbed by the desiccant, and it is known as the regeneration process. Cooling the regenerative desiccant by the external cooling source could restore its ability for absorption dehumidification. If we install an evaporation refrigeration component on a dehumidification device an open-type coupling system of adsorption refrigeration and dehumidification will be constructed.

The energy consumed by the dehumidification refrigeration system is mainly thermal energy and mechanical energy. The thermal energy is used to regenerate the desiccant, and the mechanical energy is used to drive the fan and the wheel. The performance of the dehumidification system depends on its structure and the characteristics of the desiccant. The structure of the dehumidification system affects the pressure drop of the airflow, the size, and the cost, and also has an impact on the performance and cooling capacity of the system. Desiccant characteristics include desiccant specific heat, the shape of the adsorption isotherms, adsorption heat, physical and chemical stability as well as the coefficient of heat and mass transfer. Generally a rational selection of the desiccant and structure of the system can effectively reduce the cost of the system and improve the performance.

The dehumidification refrigeration system that uses air as the working fluid and water as a refrigerant operates in an open environment, so its structure is much simplified if compared with the closed system, and it also doesn't pollute the environment. Thus it is one of the ideal alternatives for a traditional compressor refrigeration system. Since the 1960s scientists have begun to research the dehumidification refrigeration system, and the technology has developed rapidly since that time [54].

According to the types of desiccant the dehumidification refrigeration system is divided into two major categories, and they are the solid adsorption direct cooling system (SDCS) and the liquid absorption direct cooling system (LDCS). Commonly solid adsorbents used for the dehumidification refrigeration include silica gel, alumina, zeolite, and so on, and the liquid absorbents mainly include lithium chloride, calcium chloride solution, and triethylene glycol.

Compared with the performance of solid adsorbent, the velocity for the liquid absorption for dehumidification and refrigeration process is relatively slow. Moisture absorbed by the liquid desiccant is on the surface. For the liquid absorption the liquid on the surface is very easily mixed up with the liquid inside. Usually, for liquid absorption the gas contact is made with the liquid by a falling film, and the exposed area is fairly large, which is why the liquid desiccant process is identified as the absorption process. The solid dehumidification adsorbent also incorporates a process for water to be diffused from the surface of the adsorbent to the inside of the adsorbent, but the diffusion effect is relatively weak. Both adsorption and absorption dehumidification are accompanied by a thermal effect. As the thermal effect is greater than the latent heat of vaporization of water, the two dehumidifying processes can be considered to be the reaction process between the moisture and the desiccant, while the regeneration process can be considered as the decomposition process of the hydrate. This section mainly focuses on the solid adsorption dehumidification refrigeration cycle.

According to the source of the working airflow the dehumidification refrigeration cycle is divided into ventilation type and recycling type. For the design of the system according to the working status of the adsorbent bed the system can be divided into the fixed bed and rotary bed types. As the fixed bed system works intermittently (regeneration and adsorption processes need to be switched), the rotating bed system in recent years has received more and more attention. The dry cooling system has the following distinctive features:

1. The air and water that are used as the working fluid is harmless to the environment.
2. The energy transfers directly and the indoor air temperature is dependent on the dryness of the air.
3. Since the first step is dehumidification in a dehumidification refrigeration process, the treatment on the latent heat load is particularly effective.
4. The energy-saving effect is remarkable. The power consumption is greatly reduced if compared with the conventional refrigeration systems. The system can be driven by a low-grade heat source (65–85 °C, such as solar energy, waste heat, natural gas, etc.) [55].
5. The desiccant can effectively adsorb the pollutants in the air to improve indoor air quality.
6. The system can be operated in an atmosphere environment. It has the merits of low noise, easy operation, and easy maintenance.
7. The refrigeration system can be used as a heat pump in winter, which can save the energy further.

6.7.2 The Ideal Solid Adsorbents for Adsorption Dry Cooling Process

The ideal adsorbent material suitable for the dehumidification refrigeration system should have the following characteristics:

1. Physical and chemical prosperities are stable. The adsorbent material does not dissolve into the liquid solution in the adsorption process, and the cycle doesn't have hysteresis phenomenon.

2. Adsorption rate is high, such as the hygroscopic capacity of adsorbent of per unit weight is large. The high adsorption rate can reduce the amount of adsorbent, thereby will reduce the size of the equipment.
3. The adsorption capacity is high when the partial pressure of vapor is low. Absorbent capacity does not decrease when the vapor pressure of the water is low, which can improve the dryness of the processed air and reduce the power consumption of the fan.
4. The adsorption heat is small, which ensures the improvement of *COP*.
5. The adsorption isotherm is ideal and the renewable energy consumption can be reduced, consequently the *COP* can be improved.

As an other type of materials, the desiccant generates adsorption heat when it adsorbs moisture. For most materials the adsorption heat is greater than the vaporization heat of water (condensing heat), which means that actual dehumidification and regeneration process are not an isenthalpic process. For most desiccants the adsorption heat is only 5–10% bigger than the latent vaporization heat of water (condensing heat). Generally, the maximum adsorption heat is 1.25–1.5 times that of the latent vaporization heat of water, and the minimum adsorption heat is the latent heat of vaporization.

For the adsorbents usually used in the actual applications, such as silica gel and zeolites, the cooling capacity and the adsorption heat per unit weight of adsorbent are almost the same [56]. The adsorption rate of silica gel is sensitive to temperature, and the adsorption rate decreases rapidly with increasing temperature. The adsorption rate of silica gel also depends on the partial pressure of water vapor when the temperature is low. Zeolites are not sensitive to temperature change and the partial pressure of the water vapor. The adsorbent of silica gel is easily regenerated under the condition of low heat source temperature, while the zeolite performs well in the adsorption dehumidification process. In addition, to meet the general requirements for the choice of the adsorbents we should consider the requirements of dryness, the cost, as well as the overall performance of the dehumidification refrigeration process. The adsorption isotherm lines of silica gel, alumina, and molecular sieves are shown in Figure 6.65.

Collier *et al.* [56] studied the properties of adsorbent that are required for optimal performance and found that the characteristics of adsorbent influences the cycle cooling capacity and

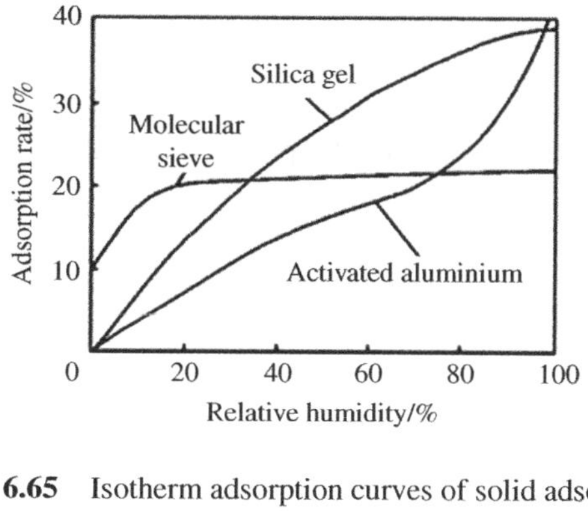

Figure 6.65 Isotherm adsorption curves of solid adsorbents

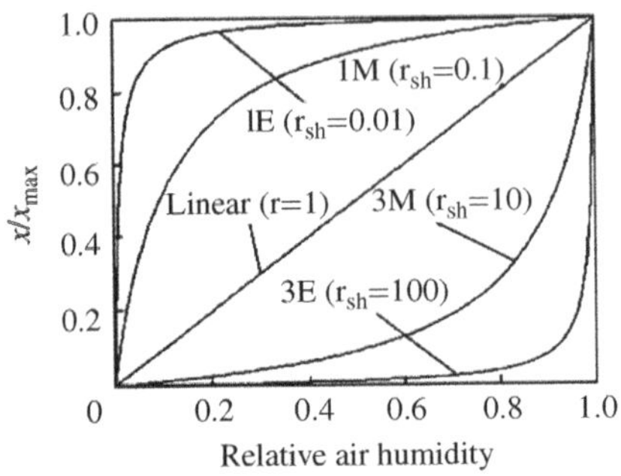

Figure 6.66 Isotherm adsorption curves of several ideal adsorbents

COP significantly. The ideal characteristics of adsorbents can be expressed by the following equation:

$$\frac{x}{x_{\max}} = \frac{R_H}{r_{sh} + (1 - r_{sh})R_H} \tag{6.91}$$

where x is the adsorption rate; $x_{\max}$ is the maximum adsorption rate; R_{H} is the relative humidity; r_{sh} is the shape factor of isothermal adsorption process of ideal adsorbent material. Figure 6.66 listed five kinds of adsorption isotherm lines of 1E, 1M, linear, 3M, and 3E, which correspond to $r_{sh} = 0.01$, 0.1, 1.0, 10, and 100, respectively.

Figure 6.66 shows that the 1M type adsorbent is the perfect adsorbent. From the isothermal adsorption curves in Figure 6.65 the curve for the adsorbent of 1M is between that of silica gel and molecular sieve. For wet air with relative low humidity the adsorption rate of molecular sieve was very high, and the adsorption rate increases slowly with the increasing humidity. The relation between the adsorption rate of silica gel and the relative humidity of air is close to linear. The type 1M adsorbent maintains the merit of the zeolite of a higher adsorption rate at relative low humidity, and also has the advantages of the silica gel, for which the adsorption rate increases with the increasing humidity. Collier *et al.* [56] indicated that the COP of the dehumidification refrigeration system with the adsorbent of type 1M can improve the thermal COP from 0.85–1.05 to 1.3–1.4. Further analysis of the system showed that the cost for the operation of the cycle with such a type of adsorbent can be decreased by 20% if compared with the conventional materials, meanwhile the size of the system and auxiliary power requirements are greatly reduced. Nowadays the research direction for the desiccant materials is to find a type of material that has a better performance than that of silica gel. The ideal adsorbents should have a performance close to the ideal type 1M material.

6.7.3 The Development of Solid Adsorption Dehumidification Refrigeration

In 1955 Pennington [57] obtained the first patent on a dehumidification refrigeration system. The technology he used is to impregnate the hygroscopic adsorbent for the dehumidification refrigeration. In the 1960s, Dunkle [54] established a dehumidification refrigeration cycle completely based on the solid wheel. Since then, the Institute of Gas Technology (IGT) in the USA developed a wheel adsorption unit that impregnated the desiccant in the molecular sieve. AiResearch developed an adsorption wheel with silica gel particles as adsorbent. Exxon and

America Solar King also successfully researched the dehumidification refrigeration system, respectively [58].

The desiccant refrigeration system was originally proposed by Pennington and was utilized for the ventilation mode, and the air stream was completely from the outside. The outside air flew into the solid wheel, where the moisture was adsorbed by the adsorbent, and the humidity reduced. Because the adsorption heat released in the adsorption process, the temperature of the air and the adsorbent increased. The air at the outlet of the system entered a heat exchanger driven by the fan, and the sensible heat was taken away there, so the temperature decreased. When the air went through the direct evaporation cooler, the water evaporated, and absorbed the heat there, therefore the humidity of the air increased and the temperature decreased to the ideal condition before it flowed into indoors. After that the indoor air first went directly into a direct evaporation cooler, then as the cooling medium it flowed into the heat exchanger driven by the fan. After recovering the sensible heat in the heat exchanger, the temperature of the airflow increased, and then was heated by a low-grade heat source to a regeneration temperature. Then the hot air went through the adsorbent wheel to regenerate the adsorbent, and after that the hot and moist air were discharged outside. The thermodynamic coefficient of this cycle is about 0.8–1.0. The Tecogen company developed a prototype of dehumidification refrigeration system with cooling capacity of 10.5–17.8 kW, and the *COP* reached 1.0 [53]. Schultz *et al.* [59] used the direct solar radiation to regenerate the desiccant, and the results showed that the *COP* of the system was lower than the system regenerated by the hot air.

The early improved mode based on the ventilation cycle developed by Pennington is shown in Figure 6.67. The difference between the two schemes is that the gas flow of the latter one is from the indoors, and the airflow circulated among the components in the system; meanwhile the regenerated gas flow is from the outside environment, and it released to the atmosphere and then the desiccant is regenerated. The *COP* (*ARI*) of this system is generally lower than 0.8 [53]. The Dunkle cycle, which is shown in Figure 6.68, combined the advantage of the recycle mode in providing greater cooling capacity and the merit of a ventilation mode for providing the cooling airflow with a lower temperature for the heat exchanger. The method for the combination is to add one more solid wheel heat exchanger. The drawback of the combination is the lack of the fresh air if compared with the recycle mode.

Maclaine cross proposed a simplified advanced solid desiccant cycle, namely, SENS cycle in 1974 [60]. Under the ideal condition the COP of the cycle can be higher than 2.0. The diagram of the cycle is shown in Figure 6.69. The air from the outside is firstly dehumidified by a solid

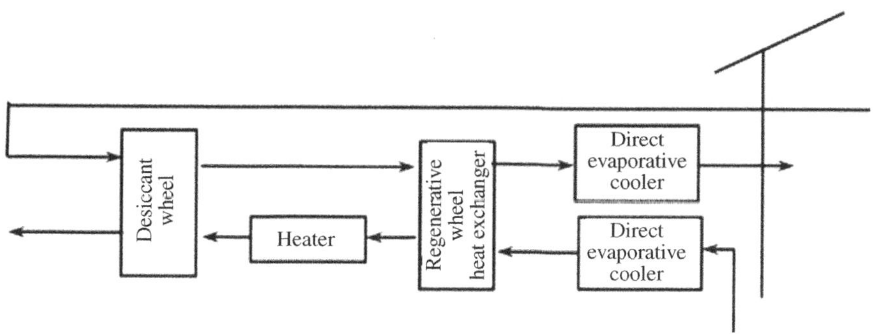

Figure 6.67 Recycle-type dehumidification refrigeration system

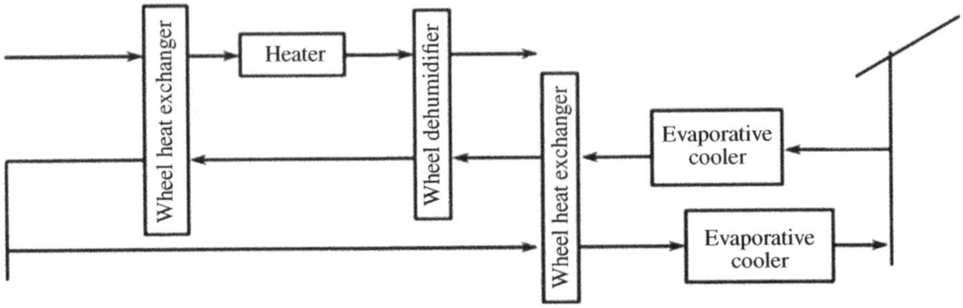

Figure 6.68 Dunkle cycle for dehumidification refrigeration system

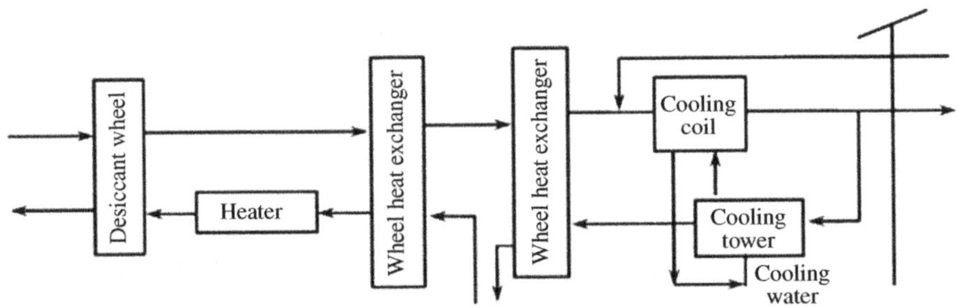

Figure 6.69 SENS dehumidification cooling system

adsorbent wheel, and then is cooled by the wheel type exchanger. After that the air will be mixed with the air from the refrigeration cycle, and then will flow into a fin-type gas-liquid heat exchanger (a cooling coil). The cooling medium of the heat exchanger is the cooling water from a small cooling tower. A part of the cooling air from the cooling coil is sent into the room for air conditioning, and the other part of it is sent to the cooling tower, where it is cooled by the cooling water from the heat exchanger and then released to the atmosphere. Meanwhile the air for the regeneration is from the outside and is heated firstly by the heat exchanger. After that the air will be sent into the solar (or waste heat) heat exchanger, and it is heated to the regeneration temperature there. Lastly the air is sent to the desiccant wheel for the working process of regeneration. The hot humid air after regeneration is released to the outdoor. The Solar Energy Applications Laboratory (SEAL) in Colorado State University tested the cycle under the conditions of the ambient temperature of 26 °C and relative humidity of 26%, and its *COP* is 2.45 [61].

The researchers [62] in Texas A&M proposed the combined cycle for the direct and indirect evaporative cooling system. If compared with the ventilation mode, the difference is that the system substituted a direct evaporative cooler with combined direct and indirect evaporative cooler in the ventilation system (Figure 6.70). According to the reports, the cycle *COP* (ARI) is up to 1.6.

Collier and Cohen [63] proposed the cycle with the multi-stage regeneration (Figure 6.71a,b) processes. In the first stage, a part of unheated airflow from the heat exchanger is sent directly into the dehumidifier to regenerate the desiccant there. In the second stage, the other part of

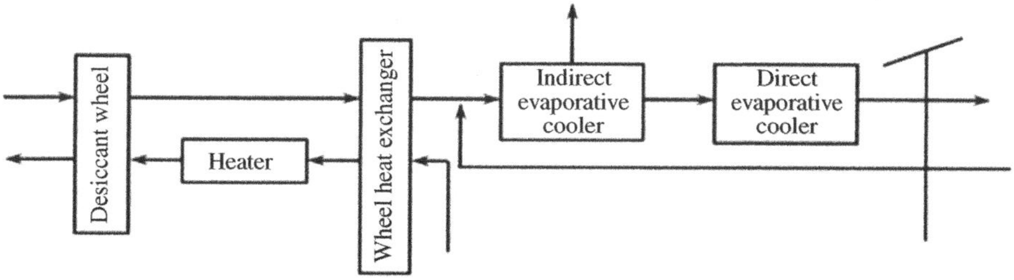

Figure 6.70 The combined cycle for the direct and indirect evaporative cooling system

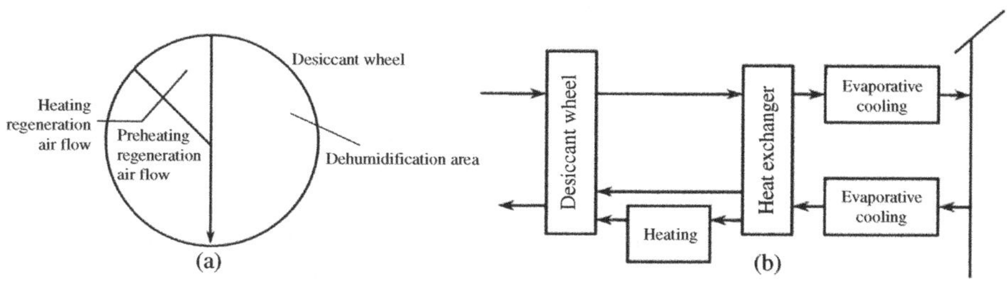

Figure 6.71 The diagram of the dehumidification refrigeration system with the multi-stage regeneration process. (a) Principle and (b) the working process

the airflow is heated by the heat exchanger to the regeneration temperature for heating and regenerating the desiccant. For the system with silica gel, as the desiccant research shows, the scheme is better than the scheme that added the inert gas in the desiccant, and it also could significantly improve the *COP* and the cooling capacity of the system. Their research also indicates that the combination of a multi-stage regeneration process with the desiccant with low thermal capacity is the optimal way to improve the performance of the dehumidification wheel.

E. van den Bulck *et al.* [64] analyzed the wheel type dehumidifier by the second law of thermodynamics, and discussed how to improve the dehumidification performance by improving the reversibility of the dehumidification process. Z. Lavan *et al.* [55] studied the reversibility of the ventilating dehumidification refrigeration system comprehensively by the second law of thermodynamics, including the dehumidifier, heat exchanger, and evaporative coolers. Results showed that the theoretical *COP* of reversible dehumidification refrigeration system was up to 4.66 (ARI), and it tended to be an infinite value when the moisture content of the inlet air from the outside is small enough.

6.7.4 The Evaporative Cooling Process of the Dehumidification Refrigeration System

The dehumidification refrigeration cycle mainly includes the evaporative cooling process and the dehumidification process by the solid adsorption process. The evaporative cooling is where the dehumidified gas flows on the surface of the water; consequently the evaporation of the water absorbs the latent heat and produces a cooling effect. Evaporative cooling can be used to maintain a certain relative humidity such as in a room, greenhouse, animal barn, as well as the engine room, textile workshop, and so on.

Evaporative cooling is the first type of refrigeration method used by people around the world. There are usually two ways, one way is the direct evaporative cooling, and the other way is the indirect evaporative cooling. Evaporative cooling has the advantages of low energy consumption, low cost, simple structure, easy operation, and easy maintenance, and so on. This technology uses water as the refrigerant, which is a type of green refrigerant and has no damaging effects on the ozone layer. In addition, the direct evaporative cooling can effectively improve the air quality by getting rid of suspended matter such as smoke, pollen, and dust in the air, as well as the soluble harmful gas of sulfur dioxide, and so on.

Under typical dry and hot conditions, the direct evaporative cooler can be used as an air conditioner which humidifies and cools the air down to a comfortable range, and such a process could regulate the temperature and the humidity of the dry environment. Thus it is known as the "desert air conditioner." In some cities in the southwestern states of the United States and the countries and regions of the Arabian Peninsula, such a type of device is effectively used as an air conditioner and refrigerator the whole year round. There are two main drawbacks of the direct evaporative cooling method, one is that the water directly evaporates into the air which makes the humidity of the air too high, and the other is that the cooling capacity is limited by the wet bulb temperature of inlet air (the outlet temperature is higher than the wet bulb temperature).

The way to overcome the first issue is to use an indirect evaporative cooling method. The indirect evaporative cooling process combined the direct evaporative cooling with the heat exchanger device. Such a method could use the low-temperature and wet airflow provided by a direct evaporative cooler to cool the working airflow through the heat exchanger device, and consequently avoid a direct humidifying process on the airflow. A typical indirect evaporative cooling apparatus is shown in Figure 6.72. A heat exchanger device uses the plate gas–gas heat exchanger with cross-flow. Because in the indirect evaporative cooling process the temperature difference between the cold gas flow and working airflow is small, Scofield *et al.* [65] proposed a method that is to use the heat pipe heat exchanger replacing the cross-flow plate heat exchanger, and achieved good results. The method to overcome the second issue is to use a two-stage evaporative cooling process (also known as combined evaporative cooling process). At the first stage of such a process it is best to use the indirect evaporative cooling method to reduce the wet bulb temperature of the working airflow under conditions where the

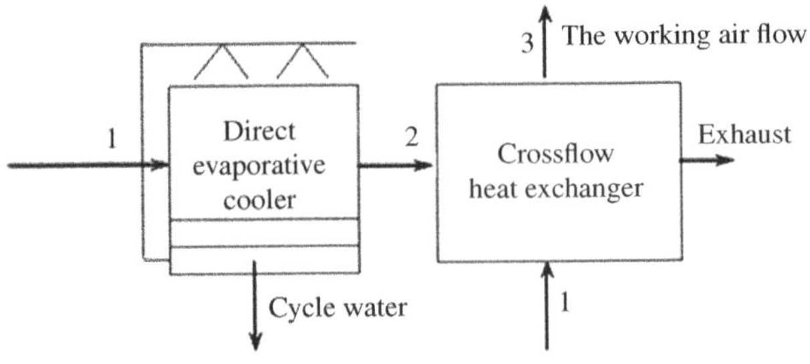

Figure 6.72 Indirect evaporative cooling apparatus

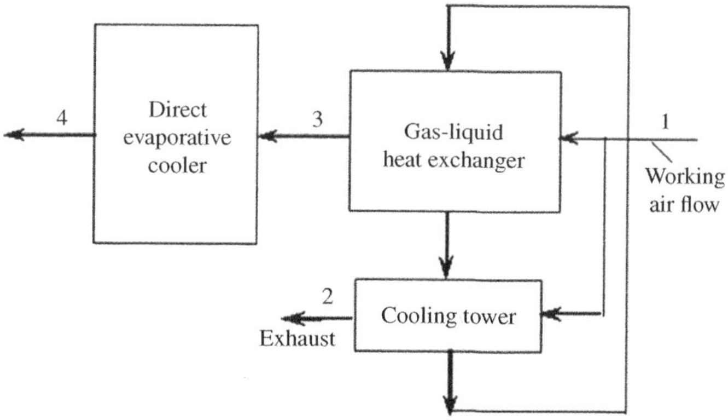

Figure 6.73 Combined evaporative cooling system with cooling tower

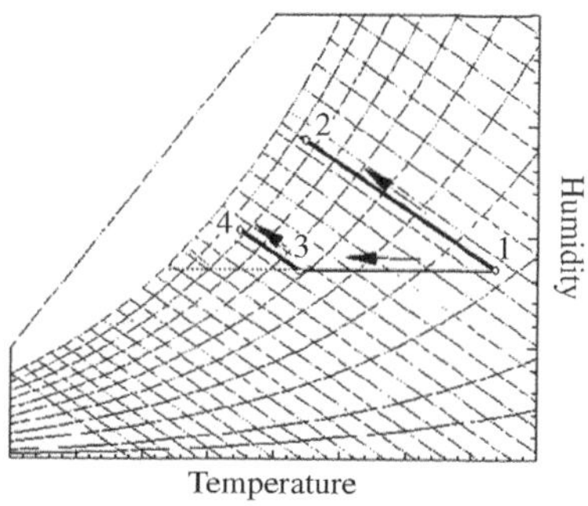

Figure 6.74 The thermodynamic diagram of indirect and combined evaporative cooling processes

air humidity doesn't increase. The second stage uses the direct evaporative cooling method to further reduce the temperature of the work airflow. For the ideal situation such a process could reduce the air temperature to its dew point temperature. Figure 6.73 showed a combined evaporative cooling device. The thermodynamic processes of two apparatus were shown in Figure 6.74.

Point 3 corresponds to the state of the outlet air of an indirect evaporative cooling device, and point 4 corresponds to the outlet state of the combined evaporative cooling device in Figure 6.74. 1-2 process is the adiabatic humidification process reflecting the change state of the wet air in the direct evaporation cooling device. The actual process is that the temperature of the air and the circuit water are both reduced. From the energy balance point moist air undergoes a process for which the enthalpy increases. For a general evaporative cooling system the enthalpy increment of the humid air is not obvious, and can be considered to be an isenthalpic process. For a single-stage direct evaporative cooling device it can be seen from the humidity diagram that the outlet temperature of airflow is the wet bulb temperature of inlet airflow under an ideal situation. A combined evaporative cooling system with cooling tower was shown in Figure 6.73. The cooling water from the cooling tower is precooled by the working airflow, and such a process reduces the temperature of the working airflow to the wet-bulb temperature. After that the airflow will go through the direct evaporative cooler, and such a process can achieve a lower outlet temperature. According to this principle the precooling devices of three or more stages may be set up and used. The cold source is the cold water from the cooling tower, which is cooled by the precooling airflow with the evaporative cooling process. If we ignore the heat loss, under the ideal situation the multi-stage regenerative evaporative cooling device can make the air temperature drop to its dew point temperature. But there is one big drawback of this type of system which is its large size, furthermore, because the cooling airflow is from the cooling tower of each stage, the inlet gas flow rate is big and consequently the fan load is great.

The performance of a direct evaporative cooler is commonly evaluated by evaporative cooling efficiency ε_{ev},

$$\varepsilon_{ev} = \frac{T_{out} - T_{in}}{T_{out} - T_{web}} \tag{6.92}$$

where T_{in}, T_{out}, and T_{web} are the inlet air temperature, the outlet air temperature, and the inlet air wet bulb temperature, respectively.

The relation between the outlet temperature of the industrial direct evaporative cooler, the temperature of inlet air, and the relative humidity of the inlet air is shown in Figure 6.75, for which all other parameters are kept constant. The evaporative cooler has a honeycomb structure, which is constructed by the channels of 0.8 m high, 0.1 m thick, and 0.48 m wide. The equivalent spacing between the channels is 6 mm. When the cooling water flow is 2000 kg/h and the air volume is 10 000 m^3/h, the cooling water temperature is 4 °C lower than the temperature of airflow. The results showed that the inlet parameters of airflow have a great impact on the performance of a direct evaporative cooler. Under relatively dry ambient conditions, the evaporative cooling method can replace the traditional cooling method or largely alleviate the refrigeration load of the conventional cooling equipment.

6.7.5 *Drying Dehumidification Process of Dehumidification Refrigeration Cycle*

The dehumidification process of adsorption dehumidification refrigeration cycle is that the solid adsorbent adsorbs the moisture of the wet air, so the humidity of the air reduces.

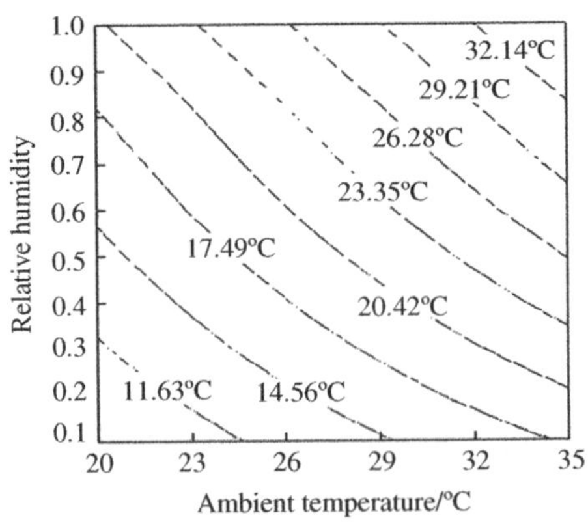

Figure 6.75 The influence of the temperature and relative humidity of inlet air on T_{out}

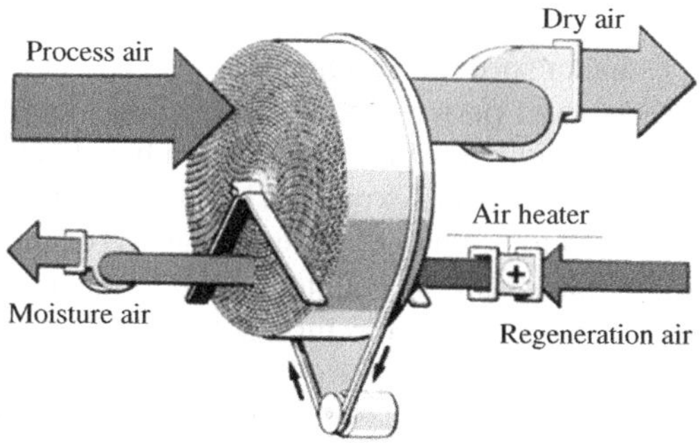

Figure 6.76 Wheel dehumidifier

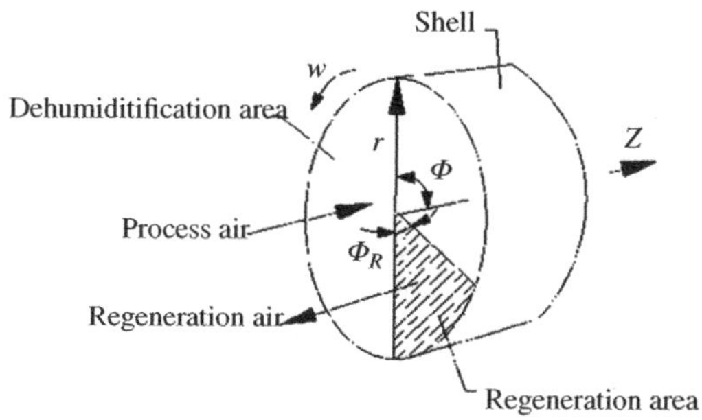

Figure 6.77 The coordinate schematic diagram of wheel dehumidifier

The solid adsorbent dehumidifier mainly includes the types with fixed bed and the wheel. Generally the intermittent and continuous cycle are involved in the systems with a fixed bed. Compared with the fixed bed type system, the desiccant wheel has the advantages of easy operation, easy maintenance, and continuous dehumidification process.

Figure 6.76 was a commercial wheel dehumidifier, and the corresponding diagram is shown in Figure 6.77. One side of the wheel is the dehumidification area, and the other side of the wheel is the regeneration zone. The processed air after dehumidification is transformed into the dry air, and then it is sent to the room. The air for the regeneration is firstly heated to the regeneration temperature by an air heater, and then is sent to the regeneration zone to regenerate the desiccant there. Usually the rotational speed of the wheel is about 10 r/h, and the regeneration zone occupies one-third to one-half of the entire cross-sectional area of the wheel.

6.7.5.1 The Numerical Solution of Desiccant Wheel

A variety of mathematical methods to evaluate the performance of the dry wheel had been developed, and the one normally used is the model established on the basis of the mass and energy conservation, which can describe the complex heat and mass transfer phenomena in the adsorption process. These models generally need to be solved by numerical methods. Due to the rapid development of computation technology, the numerical methods developed rapidly in recent years. For example, the DESSIM program edited by RS Barlow of American Solar

Energy Research Institute (SERI) is used by academics. By this program the desiccant wheel is divided into a number of smaller parts, and each part processed a small portion of the wet air, after that the heat and mass transfer performances are simulated by the heat and mass transfer equations. The program used Nusselt and Sherwood numbers to express the heat and mass transfer effect, and the relations of them are expressed by a Lewis number. In a program Collier [63] introduced an iterative loop to accelerate the convergence of the mass transfer process, and further expanded the program from the countercurrent to the Time and Tide calculation. In addition, there are also models reported from Jurinak [66], W.M. Worek [67, 68], and so on. Domestic Northwestern Polytechnical University [69] established a relative complete mathematical model for a desiccant wheel and developed the RDCS program. The program can be used for the analysis of the performance of the fixed bed, rotating bed, steady-state, and transient performances.

Kang and Maclaine-cross [69] summarized the potential functions-performance analysis methods to calculate the performance of the dehumidification refrigeration system and its components. It summarized the equilibrium adsorption properties of the desiccant into a potential function diagram for F_1-F_2. F_1 which looks like the isenthalpic curve in the humidity diagram and F_2 looks likes the relative humidity curve. According to the types of inlet conditions and desiccants, the temperature of wet air and humidity of dehumidifier outlet can be determined, consequently the assessment on the performance of the desiccant dehumidification and dehumidification refrigeration system can be completed. Maclaine-cross developed the potential function diagram for how the lithium chloride and calcium chloride were used as desiccants. Jurinak developed the potential function diagram in which the silica gel was used as adsorbent [70].

The wheel dehumidifier can be divided into many micro adsorbent channels with a honeycomb structure. Its performance can be obtained by solving the mathematical model of heat and mass transfer of the dehumidifier wheel, and the mathematical models were made by using following assumptions:

1. The speed is very low, so we can ignore the influence of centrifugal force on heat and mass transfer.
2. No leakage in the dehumidification district and the regeneration zone.
3. The wheel shell is adiabatic.
4. The pressure loss of the airflow in the axial direction is ignored.
5. Adsorbent is uniformly distributed within the entire wheel.

The wet air moisture mass conservation equation is:

$$\frac{\partial Y}{\partial t} + \omega\frac{\partial Y}{\partial \phi} + \frac{m_i}{\rho_i f_S}\frac{\partial Y}{\partial Z} = \frac{k_y f_V}{\rho_i f_S}(Y_W - Y) \tag{6.93}$$

where Y is the moisture content of the air (kg water/kg dry air), Y_W is the moisture content of the air on the surface of the adsorbent. m_i is the airflow through the unit cross-sectional area of wheel (kg/m^2/s), subscript 1 denotes the dehumidified air, and subscript 2 denotes the regeneration air. ρ_i is the air density (kg/m^3); f_V is the surface area of unit volume of adsorbent (m^2/m^3); f_S is the ratio between the area of airflow area and area of the cross-section area of wheel (m^2/m^2); k_y is the convection mass transfer coefficient (kg/m^2/s); ω is the speed of the wheel (rad/s); t is the time; ϕ, Z are the coordinates shown in Figure 6.77.

The energy conservation equation of moist air is:

$$\frac{\partial T_{air}}{\partial t} + \omega \frac{\partial T_{air}}{\partial \phi} + \frac{m_i}{\rho_i f_S} \frac{\partial T_{air}}{\partial Z} = \frac{\alpha f_V}{\rho_i f_S (C_{pair} + Y C_{pwater})} (T_{sa} - T_{air}) \tag{6.94}$$

where T_{air} is the temperature of air (°C); T_{sa} is the temperature of adsorbent surface (°C); C_{pair} is the specific heat of air (J/(kg °C)); C_{pwater} is the specific heat of water vapor (J/(kg °C)); α is the convective heat transfer coefficient (W/(m °C)).

The mass conservation equation of the moisture content inside the adsorbent is:

$$\frac{\partial x}{\partial t} + \omega \frac{\partial x}{\partial \phi} - D_i (1 - f_S) \left[\frac{2 \ln (r_2/r_1)}{{r_2}^2 - {r_1}^2} \frac{\partial^2 x}{\partial \phi^2} + \frac{\partial^2 x}{\partial Z^2} \right] = \frac{k_y f_V}{M_{av}} (Y - Y_W) \tag{6.95}$$

where x is the adsorption rate (kg water/kg adsorbent). Di is the effective diffusion coefficient of adsorbent (m^2/s). r_1 and r_2 are the hub diameter and outer diameter of the wheel (m), respectively. M_{av} is the adsorbent mass in unit volume (kg/m^3).

The energy conservation equation of adsorbent is:

$$\frac{\partial T_a}{\partial t} + \omega \frac{\partial T_a}{\partial \phi} - \frac{\lambda_{eff}(1 - f_S)}{[M_a (C_{pa} + W C_{pwater}) + M_m C_{pm}]} \left[\frac{2 \ln (r_2/r_1)}{{r_2}^2 - {r_1}^2} \frac{\partial^2 T_a}{\partial \phi^2} + \frac{\partial^2 T_a}{\partial Z^2} \right]$$

$$= \frac{1}{[M_a (C_{pa} + W C_{pwater}) + M_m C_{pm}]} [\alpha f_V (T_{air} - T_a) + k_y f_V (Y - Y_W) H_a] \tag{6.96}$$

where T_a is the temperature of adsorbent (°C); λ_{eff} is the effective thermal conductivity (W/(m °C)); M_m is the mass of support body in the unit volume (kg/m^3); C_{pwater} is the specific heat of water (J/kg/°C); C_{pa} is the specific heat at constant pressure of adsorbent (J/kg/°C), C_{pm} is the specific heat of the support body (J/kg/°C); H_a is the adsorption heat (kJ/kg).

In addition, the equations reflected the relationship between the adsorption equilibrium of the adsorbent surface and the wet air moisture, and the relationship between the convective heat transfer coefficient and mass transfer coefficient also needs to be given. These relationships change when the adsorbents are different [71].

1. The convective heat transfer coefficient α and the convective mass transfer coefficient k_y:

$$k_y = 0.704 m_i \mathrm{Re}^{-0.51} \quad \mathrm{kg/(m^2 \cdot s)} \tag{6.97}$$

$$\alpha = 0.683 m_i \mathrm{Re}^{-0.51} C_{pair} \quad \mathrm{W/(m^2 \cdot K)} \tag{6.98}$$

2. Adsorption heat:

For RD silica gel (normal density),
$$\left.\begin{aligned} H_a &= -12400W + 3500, & W &\le 0.05 \\ H_a &= -1400W + 2950, & W &> 0.05 \end{aligned}\right\} \mathrm{kJ/kg\ water} \tag{6.99}$$

ID silica gel (middle density),
$$\left.\begin{aligned} H_a &= -300W + 2950, & W &\le 0.15 \\ H_a &= 2050 & W &> 0.15 \end{aligned}\right\} \mathrm{kJ/kg\ water} \tag{6.100}$$

3. The diffusion coefficient:

   For RD silica gel, the surface diffusion is dominant, and the general diffusion and Knudsen diffusion can be ignored [71]. The surface diffusion coefficient formula is:

$$D_S = D_0 \exp[-0.974 \times 10^{-3} \times (H_a/(T + 273.15))] \quad \mathrm{m^2/s} \tag{6.101}$$

   The literature [72] gave that $D0$ is 0.8×10^{-6} m²/s.

For ID silica gel, the surface diffusion and Knudsen diffusion played an important role and the general diffusion can be ignored [71].

Knudsen diffusion coefficient is:

$$D_k = 22.86 d_{ave}(T + 273.15)^{1/2} \ \mathrm{m^2/s} \tag{6.102}$$

where d_{ave} is the average pore diameter (m). The formula of the surface diffusion coefficient is the same as that of RD silica gel.

The boundary conditions are:

$$\text{Regeneration zone, } 2\pi - \phi_R \le \phi < 2\pi, Y_{in} = Y_2, \quad T_{in} = T_2 \tag{6.103}$$

$$\text{Dehumidification zone, } 0 \le \phi < 2\pi - \phi_R, Y_{in} = Y_1, \ T_{in} = T_1 \tag{6.104}$$

The periodic boundary conditions are:

$$\begin{aligned} &Y(0,Z,t) = Y(2\pi,Z,t), \quad T(0,Z,t) = T(2\pi,Z,t), \\ &x(0,Z,t) = x(2\pi,Z,t), \quad T_a(0,Z,\tau) = T_a(2\pi,Z,\tau) \end{aligned} \tag{6.105}$$

Considering the transient problems as well as the initial conditions, the equations are:

$$\text{For the desiccant, } x(\phi,Z,\ 0) = x_0, \ T_a(\phi,Z,\ 0) = T_0 \tag{6.106}$$

$$\text{For processed air, } 0 \le \phi < 2\pi - \phi_R, \quad Y(\phi,Z,\ 0) = Y_1, \ T(\phi,Z,\ 0) = T_1 \tag{6.107}$$

$$\text{For regeneration air, } 2\pi - \phi_R \le \phi < 2\pi, \quad Y(\phi,Z,\ 0) = Y_2, \ T(\phi,Z,\ 0) = T_2 \tag{6.108}$$

where T_1 and T_2 are the inlet air temperature at the dehumidification zone and the regeneration zone, respectively. T_0 and Y_0 are the temperature and humidity at the initial moment, separately. Y_1 and Y_2 are the inlet air humidity of the dehumidifying zone and the regeneration zone.

Equations 6.93–6.96 can be transformed into the discrete equations using the finite difference method, and consequently we can achieve the numerical solution of these equations.

6.7.5.2 Waveform Analysis

If we recorded the state of the wet air at the outlet of a desiccant wheel along a circumferential direction, we can get a waveform curve on a humidity diagram. According to the characteristics of the heat and mass transfer of the adsorbent and the wet air, the states of the wet air at the outlet that changed in one cycle could predict the change of the adsorbent in the cycle, which is the waveform analysis. Collier and Cohen [63] introduced the waveform analysis

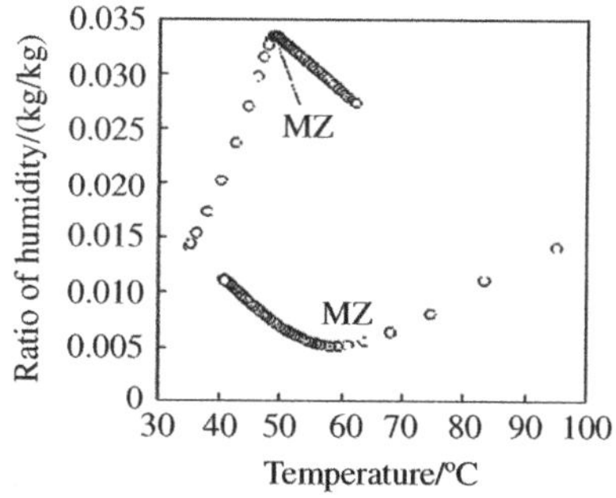

Figure 6.78 The wave diagram of wheel dehumidifier

method to improve the performance of dehumidification refrigeration system, which could explain the principle for the improvement on the system performance by segmented regeneration technology. Such a technology received extensive attention. Lavan *et al.* also analyzed the influence of adsorbent material properties on the performance of the system by using the waveform analysis.

The changing process of the state of the air at the outlet of the wheel can be expressed by two basic wave fronts on the air hygrogram. Corresponding to two wave fronts, the states of the air at the outlet of the dehumidifier is shown in Figure 6.78. Each discrete point in the diagram denotes the state of the outlet air for the dehumidification and regeneration processes relating to time change. The time interval of various points is taken to be equal. For the dehumidification process, the thermal wave is represented by a series of points located between the state of regeneration air and the state at the lowest absolute humidity point. The air generally is kept in the control area of the thermal wave for a period of time, and then at the outlet the air will shift from the point with minimum humidity to the control area of the concentration wave. This region covers the range of the point in minimum humidity area and states of the dehumidified air inlet.

The minimum humidity point on hygrogram is known as MZ point or central state, and it is the turning point of the two waves. For the regeneration process, there are two wave fronts and turning points that are the same as the dehumidification process. The faster thermal wave is located between the states of inlet air and the point for the highest outlet humidity. The outlet air will stay in the control region of the thermal wave for a short time before it is transferred to the control region of the concentration wave. The region covered the area between MZ point and the states of inlet regeneration air, and the state of the air is maintained constant in this area for a long time. The performance of the system is determined by the final position on the hygrogram of two outlet air states (dehumidification and regeneration) related to the processes of heat and mass transfer in the adsorbent bed. The thermodynamic optimal state should correspond to two MZ points which are determined by the dehumidification and regeneration processes.

One of the wave fronts is a thermal wave. The characteristic of thermal wave is the short duration, and it is also significantly influenced by the total heat capacity of the dehumidifier. During the dehumidification process of the thermal wave, the adsorbent is released from the regeneration zone, and its temperature is high. Consequently the surface equilibrium vapor pressure is high and the adsorbent capacity is weak. The temperature change is bigger than the moisture change. For the regeneration process of the thermal wave, the adsorbent comes from the dehumidification zone and its temperature is not high. Accordingly the surface equilibrium vapor pressure is small and the desorption ability is weak, so the temperature change is bigger. For the concentration wave, it lasts a long time, and it is decided mainly by the thermal effects associated with the adsorption process. It is called a concentration wave because the change of the moisture content of the adsorbent mainly concentrated in this region, and the temperature change that is mainly affected by the thermal effect is small in this region.

For the dehumidification process, the place for the states of the average outlet air on the hygrogram is critical for the performance of the system. The lower the absolute humidity is, the bigger the dehumidification amount of the system is. The higher the temperature is, the higher the preheating temperature of the regenerated air is. As a result of that the required heat for regenerative desiccant can be reduced, and the thermal *COP* can be improved. Because the wet air corresponding to the thermal wave has high temperature and high humidity, the velocity of the thermal wave decreases, and it makes the average temperature and humidity of the outlet air increase. As a result of that the increasing slope will make the outlet average temperature increase. The decrement of the change range, speed, and slope of the concentration wave will increase the average humidity of outlet and decrease the temperature. To optimize the performance we should reduce the range of the concentration wave and make it as flat as possible, meanwhile we should maximize the speed and slope of the thermal wave so that the average outlet air will be near to the MZ point. There is a similar situation for the regeneration process. At a certain regeneration temperature, in order to let the dehumidifier outlet air satisfy the requirements of low humidity and high temperature, we can change the shape and position of the thermal waves and the concentration wave by adjusting the adsorbent characteristic parameters as well as some of the operating parameters. Such a method will make the outlet state as close as possible to the MZ point, and consequently will improve *COP* and achieve optimal performance.

The standard for evaluating the performance of the desiccant wheel by waveform analysis method is: the thermal wave is faster, the performance is better. The slope of the thermal wave is larger, the performance is better. For getting a better performance the concentration wave should be in the area near the point of the MZ. The velocity and the range for the wave change are smaller, the performance for the states near the MZ point area is better. For the dehumidification process, the MZ points are lower and are closer to the right region, which has a high outlet temperature, the performance will be better. For the regeneration process, MZ points are higher and are closer to the left region, for which the outlet temperature is low, the performance will be better.

6.8 Adsorption Thermal Storage Cycles

6.8.1 Mechanism and Basic Cycle

The operation procedures of an energy storage system can be generally divided into two stages: the charging stage in which heat is stored and the discharging stage in which heat is released. With respect to adsorption thermal storage, desorption and adsorption processes represent charging stage and discharging stage respectively.

The mechanism of adsorption thermal storage process can be represented by the following equation:

$$\mathrm{A} \bullet (m+n)\mathrm{B} + \mathrm{Heat} \underset{\text{discharging}}{\overset{\text{charging}}{\rightleftharpoons}} \mathrm{A} \bullet m\mathrm{B} + n\mathrm{B} \tag{6.109}$$

where A is the sorbent and B is the adsorbate.

A/B is called an adsorption working pair or adsorption couple. For a physical adsorption process, A•$(m+n)$B denotes the enrichment of B on the surface of A as $(m+n)$ mole B is adsorbed. For a chemical adsorption process, A•$(m+n)$B and A•mB signify a compound of 1 mol A with $(m+n)$ mole B and m mole B respectively. During the charging process, when heat is added to A•$(m+n)$B, the binding force between A and B is broken and a part of B is released from A. Energy is stored in terms of chemical potential which the mass fraction of B decreases. During the exothermal discharging process, A•mB contacts with B to form A•$(m+n)$B again and the chemical potential is transferred into thermal energy. In short, the adsorption technology owes its storage function to a difference in the amount of the adsorbate attracted onto/into the adsorbent, accompanied by vast amounts of heat.

According to the system configurations, adsorption storage systems can be divided into open and closed systems. Closed adsorption systems, which are isolated from the atmospheric environment, have long been studied for refrigeration and heat pump applications. Closed systems are attractive choices in small-scale applications where compact and highly efficient devices are needed. In closed systems, not the adsorbate itself but the entropy is released/absorbed to the environment via a heat exchanger. The operation principle of a closed adsorption thermal storage system is presented in Figure 6.79. The system is mainly composed of two vessels: a reactor where reactive sorbent is located and a condenser/evaporator where liquid water is collected. The vessels are connected by a conduct as a passage for vapor. The charging process consists of a desorption reaction in the reactor and a gas-liquid phase change reaction in the condenser. When a high temperature heat collected by a solar collector is added to the reactor, the adsorbate which clings to the adsorbent, starts to escape from the binding force between the adsorbate and the adsorbent. Through the duct, the vapor turns into its liquid state in a condenser at a low temperature level. The heat of condensation is taken away and

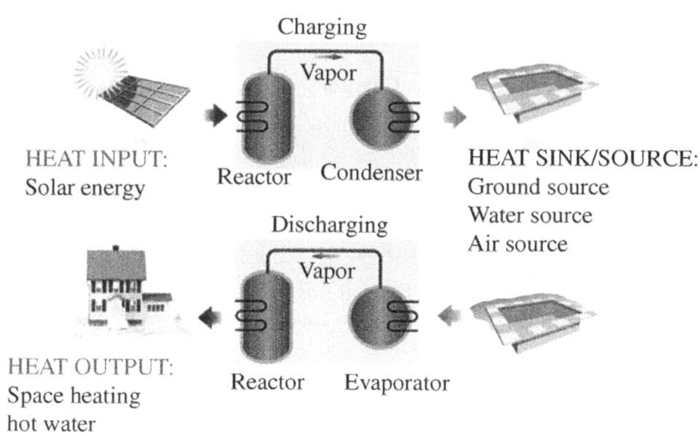

Figure 6.79 Operation principle of closed adsorption thermal storage system

released to the heat sink. After the charging process is finished, the reactor and the condenser are separated from each other. If heating or cooling demands are needed, the reactor and the condenser/evaporator are connected again. The discharging process works in a reverse direction: it includes an adsorption reaction in the reactor and a liquid-gas phase change reaction in the evaporator. Depending on the practical requirement, a cooling effect can be produced by the evaporator or a heating effect can be created by the reactor. This feature makes the adsorption thermal storage processes able to offer a "cold storage" function in summer and a "heat storage" function in winter.

Discharging of closed systems requires an additional heat source to provide the heat of evaporation, making the choice of heat source a critical issue. Air is normally proposed as a heat source for it is always available, easy to design a heat exchanger and free of location restrictions [73]. However, when the ambient temperature is too low to drive the evaporation, it will be quite difficult for the discharging process to proceed. Thus, air source is suitable for areas with a warm climate in winter. Compared with air, temperature of the ground is higher and more stable. Ground source heat exchangers are adopted in some sorption storage projects [4] to extract heat from the earth. The main drawback of ground source is that construction of heat exchangers is time-consuming and costly. If the storage system is just near waters or swimming pools, water source is a decent option. Adsorption storage systems using water as the sorbate are unable to operate under 0 °C, so ammonia or methanol should then be considered for cold areas.

Open systems, as the name implies, are connected with the ambient environment to allow the release and adsorption of the sorbate. Thus, only water can be used in those systems. Figure 6.80 depicts the operation principle of an open adsorption thermal storage system. In the charging process, a dry air stream, after heating by a heat source like solar energy, becomes a dry hot stream and enters a reactor filled with sorbent. Water adsorbed by the adsorbent is extracted by the hot air and exits the bed. The air then becomes wetter and cooler. During discharging, a humid, cool air stream goes into the previously desorbed reactor. Part of the water vapor in the air is attracted by the adsorbent. The released heat of adsorption makes the air become hotter and the hot air could be used for heating. The lower cost of investment, coupled with better heat, and mass transfer conditions (compared with closed systems), provides compelling reasons for practical projects employing open adsorption systems to storage thermal

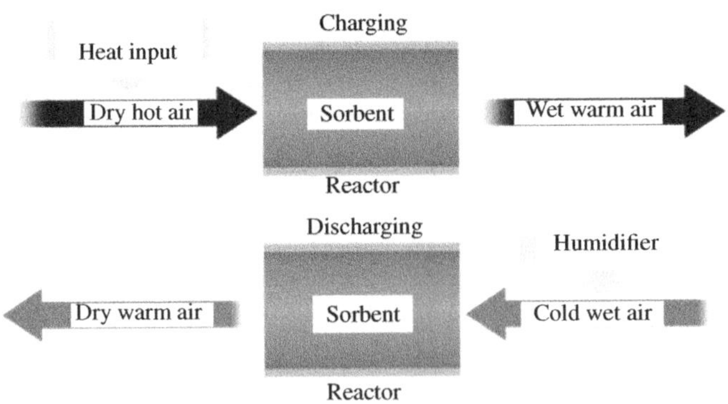

Figure 6.80 Operation principle of open adsorption thermal storage system

energy [75, 76]. However, before employing the open systems, an analysis should be carried out to define whether the ambient moisture is sufficiently high for a good discharging rate. Otherwise an additional humidifier is required to make the air wet enough. Furthermore care has to be taken to reduce or to limit the pressure drop when blowing humid air through the reaction system to keep the electricity demand for the blower on a low level [77].

From a design point of view, in contrast with closed adsorption, open adsorption has many benefits – free of many troubles like condensers, evaporators, water storage reservoirs, maintenance of system pressure, and complex process control strategy. Therefore, open systems are considered in many recent research related to adsorption thermal storage [78–81].

6.8.2 Thermodynamic Analysis

For adsorption thermal storage systems, the energy for charging Q_{char} includes three different parts:

$$Q_{char} = Q_{sens} + \underbrace{Q_{cond} + Q_{bind}}_{Q_{des}} \tag{6.110}$$

The sensible heat Q_{sens} is a prerequisite energy to heat up the reactor to a required desorption temperature. This heat is subject to the temperature difference, heat ratio between thermal mass of adsorbent and auxiliary components, and heat losses. The sensible heat can be partially retrieved for short-term thermal storage, but not for long-term storage due to thermal losses. The heat of condensation Q_{cond} is defined as the heat of liquid-gas phase change at a specific temperature (normally condensation temperature at the condenser), which is assumed constant. The heat of binding Q_{bind} is used to denote the difference between the heat required for desorption Q_{des} and the condensation heat Q_{cond}. The binding heat Q_{bind} is contributed by the adsorption forces between the adsorbent and the adsorbate, and when referred to the unit mass of adsorbate it is often called the differential heat of adsorption. The definitions of Q_{cond} and Q_{bind} are illustrated in Figure 6.81.

For physical adsorption, the differential heat of adsorption is affected by many factors, such as temperature, pressure, and adsorbate concentration (adsorbate uptake on/in the adsorbent in g/g). But it is generally accepted that adsorbate concentration plays the most significant

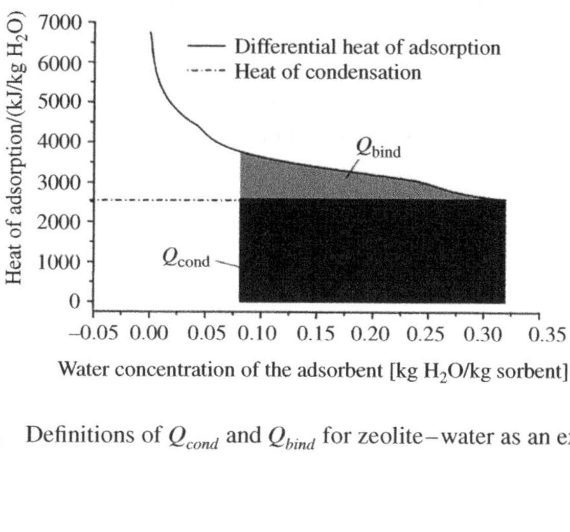

Figure 6.81 Definitions of Q_{cond} and Q_{bind} for zeolite–water as an example [81]

part and the differential heat of adsorption can be seen as a function of the adsorbate concentration over some ranges of temperature. As a rule, the differential heat will increase as the concentration decreases, in accordance with the fact that more heat is required for desorption at a low adsorbate concentration. For chemical adsorption, the value of reaction heat is always presumed invariable, calculated from the formation enthalpies of materials involved in the reactions, the reaction equilibrium equations, or measured from calorimetric methods.

As shown in Figure 6.81, integrating the curve of the differential heat of adsorption from the lowest concentration at the desorption process to the highest concentration at the end of the adsorption process gives the integral heat of adsorption/desorption (Q_{des}). Q_{char} is the sum of Q_{des} and Q_{sens}. The ratio of heat of condensation to the total charging heat (Q_{cond}/Q_{char}) is an important parameter to show the contribution of the heat of condensation and then to evaluate the thermal storage potential of adsorption materials. Higher values of Q_{cond}/Q_{char} mean lower contributions of the binding heat, coming along with lower desorption temperatures, as is the case for zeolite–water [83]. For cold storage, the value of Q_{cond}/Q_{char} almost equals the cooling COP_c. Therefore, greater COP_c could be expected with higher Q_{cond}/Q_{char}.

The characteristics of storage materials – especially the energy density – are prerequisite to obtaining compact and efficient adsorption thermal storage applications. Energy density is defined as the amount of energy stored in a given system or region of space per unit volume or per unit mass. Energy density of adsorption materials in terms of mass and volume are both widely applied in previous works. In the preliminary search of advanced adsorption materials, energy density by mass (ρ_{Q-m}), is adopted to evaluate the investment on storage materials, defined as follows:

$$\rho_{Q-m} = \frac{heat\ storage\ capacity\ output}{mass\ of\ storage\ material} \tag{6.111}$$

The volume of the storage material could also be considered to achieve a compact thermal storage system. Since most adsorption systems include several essential components, energy density by volume (ρ_{Q-V}), is a more practical parameter to calculate the heat storage potential of prototypes.

$$\rho_{Q-V} = \frac{heat\ storage\ capacity\ output}{volume\ of\ storage\ material\ or\ prototype} \tag{6.112}$$

The higher the value of the ρ_{Q-V} is, the smaller the volume of the prototype required for the energy storage. It is essential for the occasions where space for the energy storage system is limited, such as the transportation process of the energy storage vessels from one site to another site. For example, in the chemical engineering factories the waste heat is abundant. If the waste heat is stored in the phase change materials, and then transported to the places which require the heat, the transportation process will be easier when the volume of the heat storage unit is smaller.

References

[1] Hulse, G.E. (1924) Freight car refrigeration by an adsorption system employing silica gel. *Refrigerating Engineer*, **17**, 2.

[2] Meunier, F. (1985) Second law analysis of solid adsorption heat pump operation on reversible cascade cycles: application to the zeolite/water pair. *Heat Recovery Systems and CHP*, **5**(2), 131–141.

[3] Meunier, F., Kaushik, S.C., Neveu, P. and Poyelle, F. (1996) A comparative thermodynamic study of sorption systems: second law analysis. *International Journal of Refrigeration*, **19**(6), 414–421.

[4] Walker, G. and Bingham, E.R. (1994) *Low-capacity Cryogenic Refrigeration*, Oxford Science Publication, Oxford.

[5] Tchernev, D.I. and Emerson, D.T. (1988) High-efficiency regenerative zeolite heat pump. *ASHRAE Transactions*, **14**, 2024–2032.

[6] Douss, N., Meuiner, F. and Sun, L.M. (1988) Predictive model and experimental results for a two-adsorber solid adsorption heat pump. *Industrial and Engineering Chemistry Research*, **27**(2), 310–316.

[7] Tchernev, D.I. and Emerson, D. (1988) Closed cycle regenerative heat pump. 2nd International Workshop on Research Activities on Advanced Heat Pumps, Graz, Austria.

[8] Teng, Y. and Wang, R.Z. (1995) Research on convective heat regenerative adsorption refrigeration cycle of activated carbon-methanol. *Cryogenics*, (3), 48–52, ISSN: 1000-6516 (in Chinese).

[9] Douss, N. (1993) Experimental study of adsorption heat pumps cycles. *International Chemical Engineering*, **33**(2), 207–214.

[10] Wang, R.Z. (1999) Study on a four-beds cascade adsorption refrigeration cycle capable of COP over 1.1. 20th International Congress of Refrigeration, Sydney, Australia.

[11] Tan, Z.C. and Wang, R.Z. (1998) Research on the double effect adsorption refrigeration cycle. *Acta Energiae Solaris Sinica*, **19**(2), 156–160, ISSN: 0254-0096 (in Chinese).

[12] Shelton, S.V., Weeper, W.J. and Miles, D.J. (1998) External fluid heating of a porous bed. *Chemical Engineering Communications*, **71**, 39–52.

[13] Shelton, S.V. (1990) Ramp wave analysis of the solid/vapor heat pump. *ASME Journal of Energy Resource Technology*, **112**(3), 69–78.

[14] Yang, L.M., Wang, R.Z. and Wu, J.Y. (1997) Thermal wave's research and analysis on solid adsorption refrigeration. *Journal of Engineering Thermophysics*, **18**(6), 673–676, ISSN: 0253-231X (in Chinese).

[15] Critoph, R.E. (1996) Evolution of alternative refrigerant-adsorbent pairs for refrigeration cycles. *Applied Thermal Engineering*, **16**, 891–900.

[16] Sun, L.M., Feng, Y. and Pons, M. (1997) Numerical investigation of adsorption heat pump system with thermal wave heat regeneration under uniform-pressure conditions. *International Journal of Heat and Mass Transfer*, **40**(2), 281–293.

[17] Pons, M. and Szarzynski, S. (1999) An adsorption cooling system with heat-regeneration: experiment and numerical study. Proceedings of the International Sorption Heat Pump Conference, Munich, Germany, pp. 625–630.

[18] Pons, M. and Poyelle, F. (1999) Adsorptive machines with advanced cycles for heat pumping or cooling applications. *International Journal of Refrigeration*, **22**(1), 27–37.

[19] Willers, E. and Groll, M. (1999) Evaluation of metal hydride machines for heat pumping and cooling applications. *International Journal of Refrigeration*, **22**(1), 47–58.

[20] Yang, L.M. (1997) Research on heat regenerative solid adsorption refrigeration system. Master Thesis. Shanghai Jiao Tong University, Shanghai, China (in Chinese).

[21] Zheng, W. and Worek, W.M. (1995) Effect of operating conditions on the performance of the two-bed closed cycle solid sorption heat pump system. *Journal of Solar Energy Engineering*, **117**, 171–181.

[22] Harkonen, M. and Aittomaki, A. (1991) The principle of internal regeneration as applied to the adsorption heat pump process. *Heat Recovery Systems and CHP*, **11**(4), 239–248.

[23] Wang, R.Z., Jia, J.P., Zhu, Y.H. *et al.* (1997) Study on a new solid adsorption refrigeration pair: active carbon fiber -methanol. *Journal of Solar Engineering, the Transaction of ASME*, **119**(3), 214–218.

[24] Cheng, J. and Wang, R.Z. (1998) Analysis on the solid-sorption cycle with forced convective heat and mass transfer inside absorbent bed. *Acta Energiae Solaris Sinica*, **19**(4), 437–443, ISSN: 0254-0096 (in Chinese).

[25] Critoph, R.E. (1994) Forced convection enhancement of adsorption cycles. *Heat Recovery Systems and CHP*, **14**(4), 343–350.

[26] Miles, D.J., Sanborn, D.M., Nowakokwski, G.A., and Shelton, S.W. (1992) Gas fired sorption heat pump development. Proceedings of Symposium on Solid Sorption Refrigeration, Paris, France, pp. 91–96.

[27] Wang, R.Z., Wu, J.Y., Dai, Y.J. *et al.* (2001) *Adsorption Refrigeration*, China Machine Press, Beijing, ISBN: 7-111-09532-4 (in Chinese).

[28] Critoph, R.E. (1990) Refrigeration in developing countries – the renewable options, in *1st World Renewable Energy Conference*, Pergamon, Reading, pp. 63–104.

[29] Wang, L.W. (2005) Performances, mechanisms and application of a new type compound adsorbent for efficient heat pipe type refrigeration driven by waste heat. PhD thesis. Shanghai Jiao Tong University, Shanghai, China (in Chinese).

[30] Wang, R.Z. (2000) Performance improvement of adsorption heat pump by heat and mass recovery operations. Proceedings of 4th Minsk International Seminar on Heat Pipes, Heat Pumps and Refrigerators, Minsk, Berlaus, pp. 84–91.

[31] Wang, L.W., Wu, J.Y., Wang, R.Z. *et al.* (2003) Experimental study of a solidified activated carbon-methanol adsorption ice maker. *Applied Thermal Engineering*, **23**, 1453–1462.

[32] Wang, L.W., Wang, R.Z., Lu, Z.S. *et al.* (2006) Split heat pipe type compound adsorption ice making unit for fishing boats. *International Journal of Refrigeration*, **29**(3), 456–468.

[33] Wang, L.W., Wang, R.Z., Lu, Z.S. *et al.* (2006) Comparison of the adsorption performance of compound adsorbent in a refrigeration cycle with and without mass recovery. *Chemical Engineering Science*, **61**(11), 3761–3770.

[34] Wu, J.Y. (2000) Experimental research and the cycle properties of a continuous heat regenerative adsorption heat pump. PhD thesis. Shanghai Jiao Tong University, Shanghai, China (in Chinese).

[35] Saha, B.B. and Kashiwagi, T. (1997) Experimental investigation of an advanced adsorption refrigeration cycle. *ASHRAE Transactions*, **103**, 51–58.

[36] Saha, B.B., Akisawa, A. and Kashiwagi, T. (1997) Silica gel water advanced adsorption refrigeration cycle. *Energy*, **22**(4), 437–447.

[37] Saha, B.B., Boelman, E.C. and Kashiwagi, T. (1995) Computational analysis of an advanced adsorption-refrigeration cycle. *Energy*, **20**(10), 983–994.

[38] Vasiliev, L.L., Mishkinis, D.A., Antukh, A.A. and Kulakov, A.G. (2004) Resorption heat pump. *Applied Thermal Engineering*, **24**, 1893–1903.

[39] Spinner, B. (1993) Ammonia-based thermochemical transformers. *Heat Recovery Systems and CHP*, **13**(4), 301–307.

[40] Goetz, V., Spinner, B. and Lepinasse, E. (1997) A solid-gas thermochemical cooling system using $BaCl_2$ and $NiCl_2$. *Energy*, **22**(1), 49–58.

[41] Neveu, P. and Castaing, J. (1993) Solid-gas chemical heat pumps: field of application and performance of the internal heat recovery. *Heat Recovery Systems and CHP*, **13**, 233–251.

[42] Sorin, M., Spinner, B. and Stitou, D. (2002) Thermodynamic techniques for the conceptual design of thermochemical refrigerators using two salt materials. *Chemical Engineering Science*, **57**, 4243–4251.

[43] Li, T.X., Wang, R.Z., Kiplagat, J.K. *et al.* (2009) A conceptual design and performance analysis of a triple-effect solid–gas thermochemical sorption refrigeration system with internal heat recovery. *Chemical Engineering Science*, **64**, 3376–3384.

[44] Li, T.X., Wang, R.Z., Oliveira, R.G. *et al.* (2009) A combined double-way chemisorption refrigeration cycle based on adsorption and resorption processes. *International Journal of Refrigeration*, **32**, 47–57.

[45] Li, T.X., Wang, R.Z., Kiplagat, J.K. *et al.* (2009) Thermodynamic study of a combined double-way solid-gas thermochemical sorption refrigeration cycle. *International Journal of Refrigeration*, **32**, 1570–1578.

[46] Xu, L., Wang, R.Z., Li, T.X. and Wang, L.W. (2011) Experimental study on a combined double-way chemisorption refrigeration system. *International Journal of Refrigeration*, **34**, 914–921.

[47] Li, T.X., Wang, R.Z., Kiplagat, J.K. and Wang, L.W. (2010) Performance improvement of a combined double-way sorption refrigeration cycle using re-heating process. *AIChE Journal*, **56**, 477–484.

[48] Li, T.X., Wang, R.Z., Oliveira, R.G. and Wang, L.W. (2007) Performance analysis of an innovative multi-mode, multi-salt and multi-effect chemisorption refrigeration system. *AIChE Journal*, **53**, 3222–3230.

[49] Xu, L., Wang, R.Z., Li, T.X. and Wang, L.W. (2011) Experimental study on the performance of double-effect and double-way thermochemical sorption refrigeration cycle. *Applied Thermal Engineering*, **31**, 3658–3663.

[50] Oliveira, R.G., Wang, R.Z. and Li, T.X. (2008) Transient analysis of a chemisorption air conditioning system operating under different kinds of cycle. *Industrial and Engineering Chemistry Research*, **47**, 1102–1110.

[51] Douglas, K. (1996) Linking indoor air quality (IAQ), ventilation, moisture control, and desiccants to future air conditioning practices. Proceedings of International Conference of Absorption refrigeration, Montreal, Canada, pp. 128–137.

[52] Collier, R.K. and Cohen, B.M. (1991) An analytical examination of methods for improving the performance of desiccant cooling systems. *Journal of Solar Energy Engineering*, **113**, 157–163.

[53] Waugaman, D.G., Kini, A. and Kettleborough, C.F. (1993) A review of desiccant cooling systems. *Journal of Energy Resources Technology*, **115**, 1–8.

[54] Dunkle, R.V. (1965) A method of solar air conditioning. *Mechanical and Chemical Engineering Transactions*, **1**, 73–78.

[55] Lavan, Z., Monnier, J.B. and Worek, W.M. (1982) Second law analysis of desiccant cooling systems. *Journal of Solar Energy Engineering*, **104**, 229–236.

[56] Collier, R.K., Barlow, R.S. and Arnold, F.H. (1982) An overview of open-cycle desiccant cooling systems and materials. *Journal of Solar Energy Engineering*, **104**, 28–34.

[57] Pennington, N.A. (1955) Humidity changer for air conditioning. US Patent 2700537.

[58] Yu, J.D., Luo, G. and Zhang, H.F. (1995) Dry wheel dehumidifier's new model and RDCS program. *Acta Energiae Solaris Sinica*, **16**(4), 368–378, ISSN: 0254-0096 (in Chinese).

[59] Schultz, K., Barlow, B., Pesaran, A. and Kreith, K. (1987) An analysis of a direct radiation solar dehumidification system. *Journal of Solar Energy Engineering*, **109**, 15–21.

[60] Maclaine-cross, I.L. (1974) *A theory of combined heat and mass transfer in regenerations* (PhD thesis). Monash University, Australia.

[61] Appleyard, J. and Lof, G.O.G.(1988) Preliminary performance of simplified advanced solar desiccant cooling system employing air-water exchanger coil and partial air recirculation. Proceedings of the 10th Annual ASME Solar Energy Conference, Denver, CO, pp. 145–151.

[62] Waugaman, D.G. and Kettleborough, C.F. (1987) Combining direct and indirect evaporative cooling with a rotating desiccant wheel in residential applications. Proceedings of the ASME International Thermal and Solar Energy Conference, Honolulu, Hawaii.

[63] Collier, R.K. and Cohen, B.M. (1991) An analytical examination of methods for improving the performance of desiccant cooling systems. *Journal of Solar Energy Engineering*, **113**, 151–163.

[64] Bulck, E.V.D., Klein, S.A. and Mitchell, J.W. (1988) Second law analysis of solid desiccant rotary dehumidifier. *Journal of Solar Energy Engineering*, **110**, 2–9.

[65] Al-Juwayhel, F.I., Al-Haddad, A.A., Shaban, H.I. and El-Dessouky, H.T.A. (1997) Experimental investigation of the performance of two-stage evaporative coolers. *Heat Transfer Engineering*, **18**(2), 22–32.

[66] Jurinak, J.J. and Mitchell, J.W. (1984) Effect of matrix properties on the performance of a counterflow rotary dehumidifier. *Journal of Heat Transfer*, **106**, 638–645.

[67] Charoensupaya, D. and Worek, W.M. (1988) Parametric study of an open-cycle adiabatic, solid, desiccant cooling system. *Energy*, **13**(9), 739–749.

[68] Zheng, W. and Worek, W.M. (1993) Numerical simulation of combined heat and mass transfer process in a rotary dehumidifier. *Numerical Heat Transfer Part A*, **23**, 211–232.

[69] Kang, T.S. and Maclaine-cross, I.L. (1989) High performance solid desiccant open cooling cycles. *Journal of Solar Energy Engineering*, **111**, 176–183.

[70] Bulk, E.V.D., Mitchell, J.W. and Klein, S.A. (1985) Design theory for rotary heat and mass exchangers-I. Wave analysis of rotary heat and mass exchangers with infinite transfer coefficients. *International Journal of Heat and Mass Transfer*, **28**, 1575–1586.

[71] Pesaran, A.A. and Mills, A.F. (1987) Moisture transport in silica gel packed beds-I: theoretical study. *International Journal of Heat and Mass Transfer*, **30**, 1037–1049.

[72] Pesaran, A.A. and Mills, A.F. (1987) Moisture transport in silica gel packed beds-II: experimental study. *International Journal of Heat and Mass Transfer*, **30**, 1051–1060.

[73] Lahmidi, H., Mauran, S. and Goetz, V. (2006) Definition, test and simulation of a thermochemical storage process adapted to solar thermal systems. *Solar Energy*, **80**, 883–893.

[74] Stitou, D., Mazet, N. and Mauran, S. (2011) Experimental investigation of a solid/gas thermochemical storage process for solar air-conditioning. *Energy*, **41**, 261.

[75] Hauer, A. (2007) Adsorption systems for TES-Design and demonstration projects, in *Thermal Energy Storage for Sustainable Energy Consumption*, Springer, pp. 409–427.

[76] Hauer, A. and Avemann, E. (2007) Open absorption systems for air conditioning and thermal energy storage, in *Thermal Energy Storage for Sustainable Energy Consumption*, Springer, pp. 429–444.

[77] Kerskes, H., Mette, B., Bertsch, F. *et al.* (2011) Development of a thermo-chemical energy storage for solar thermal applications. Proceedings of ISES Solar World Congress, Kassel, Germany.

[78] Hauer, A. (2007) Evaluation of adsorbent materials for heat pump and thermal energy storage applications in open systems. *Adsorption*, **13**, 399–405.

[79] Kerskes, H., Mette, B., Bertsch, F. *et al.* (2012) Chemical energy storage using reversible solid/gas-reactions (CWS)–results of the research project. *Energy Procedia*, **30**, 294–304.

[80] Wu H.J., Dai, Y.F., and Zhang, X.Q. (2009) Simulation and evaluation of an open-type thermal storage system using composite sorbents for building efficiency. Proceedings of International Conference on Energy and Environment Technology, Guilin, China, pp. 313–316.

[81] Tanguy, G., Marias, F., Rouge, S. *et al.* (2012) Parametric studies of thermochemical processes for seasonal storage. *Energy Procedia*, **30**, 388–394.
[82] Hauer, A. (2007) Sorption theory for thermal energy storage, in *In Thermal Energy Storage for Sustainable Energy Consumption*, Springer, pp. 393–408.
[83] Bales, C. (2005) Thermal Properties of Materials for Thermo-Chemical Storage of Solar Heat. IEA-SHC Task 32 Report B2 of Subtask B "Chemical and Sorption Storage". Solar Energy Research Center SERC, Högskolan Dalarna, Borlänge.

7

Technology of Adsorption Bed and Adsorption Refrigeration System

The key component of the adsorption refrigeration system is the adsorption bed, and it will directly influence the performance of the whole system. Taking the two-bed continuous adsorption refrigeration system [1] as an example, it mainly consists of two units. The first unit includes two adsorption beds, heater, and cooler. The adsorption bed is equivalent to the compressor of the traditional refrigeration system. The adsorption bed, which is under the condition of the desorption process, desorbs the high temperature and high pressure refrigerant vapor to the condenser. Whereas the adsorption bed in the case of the adsorption state adsorbs the low temperature and low pressure refrigerant vapor from the evaporator. The refrigerant will continuously vaporize as it generates the refrigeration effect. The second part includes condenser, flow control valve, and evaporator, which is similar to the conventional refrigeration system. The desorbed refrigerant vapor flows to the condenser and is condensed there. Thus, the refrigerant becomes the low temperature and low pressure liquid via the flow control valve. The liquid refrigerant enters the evaporator and is evaporated. The vaporized refrigerant is adsorbed by the bed that is under the adsorption condition once again.

Compared to the vapor compression refrigeration system, the adsorption bed plays a role similar to the compressor, but it is driven by thermal energy. However, one adsorption refrigeration system always has several adsorption beds. The heat and mass transfer performance of the adsorption bed reflects the performance of the thermal driven compressor.

7.1 The Technology of Adsorption Bed

There are two key parameters for evaluating the performance of the adsorption system, one is the *COP* (Coefficient of Performance) and the other is the *SCP* (Specific Cooling Power). The *COP* of the system can be greatly improved by means of using the heat and mass recovery process. The *SCP* of the system is closely related with the design of the adsorption bed.

Adsorption Refrigeration Technology: Theory and Application, First Edition. Ruzhu Wang, Liwei Wang and Jingyi Wu.

Companion Website: www.wiley.com/go/wang/refrigeration

The definition of SCP is:

$$SCP \approx \frac{L\Delta x}{t_c} \tag{7.1}$$

where L is the latent heat of vaporization, t_c is the cycle time.

Under a given operating condition and for a certain cycle, the method of increasing the refrigerating capacity is mainly to shorten the cycle time. There are two kinds of technology used in an adsorption bed to reduce the cycle time. One kind is to improve the mass transfer of the adsorbent under low pressure conditions; another is to enhance the heat transfer performance of the adsorption bed.

As for the chemical adsorbent (such as metal chloride), the agglomeration phenomenon will occur during the adsorption process when the chlorides adsorb the steam or ammonia. The agglomeration phenomenon might result in serious mass transfer deterioration under the low pressure circumstance. As far as the chlorides used for the chemical adsorbent are concerned the important issue is to enhance the mass transfer under low pressure conditions. The researchers in Shanghai Jiao Tong University proposed that employing the compound adsorbent can enhance the mass transfer [2, 3]. In addition, as for the adsorption refrigeration system using water or methanol as an adsorbent under negative pressure, the mass transfer channel is required inside the adsorption bed to shorten the mass transfer time.

Aimed to decrease the thermal resistance of the adsorption bed, the enhancement of heat transfer receives great attention [4]. The overall heat transfer coefficient α of the adsorption bed is:

$$\frac{1}{\alpha A_f} = \frac{1}{\alpha_f A_f} + \frac{1}{\alpha_w A_{eff}} + \frac{e_{eff}}{\lambda_{eff} A_{eff}} \tag{7.2}$$

where A_f and A_{eff} are the area for the heat exchanger at the side of the fluid and solid adsorbent, respectively. α_f and α_w are the heat transfer coefficients for the side of fluid and solid adsorbent. e_{eff} is the effective thickness of adsorption bed, and λ_{eff} is the thermal conductivity. The overall heat transfer is mainly restricted by the following factors:

1. The thermal conductivity of the granular adsorbent is low. The thermal conductivity of zeolite is approximate to 0.1 W/(m K) [5]. Generally speaking, the thermal conductivity of the metal chlorides and activated carbon is in the range of 0.3–0.5 W/(m K) [6, 7]. The thermal conductivity of metal hydrides is about 1 W/(m·K) [4, 8].
2. The heat transfer coefficient between the adsorbent and the heat exchanger is low. The heat transfer coefficient is relatively low in the absence of the convection. Moreover, when the refrigerant with the low evaporation pressure, such as water or methanol, is employed, it will be limited by the Knudsen condition, that is, the mean free path of the gas is greater than the distance among the particles under rarefied gas circumstances. The heat transfer coefficient will be relatively low on this occasion.
3. The heat transfer coefficient of the fluid is very low, especially when the fluid is under the condition of the laminar flow in the heat exchanger.

There are three main technologies to enhance the overall heat transfer coefficient α. The first is to increase the heat transfer area. The second technique is to use the consolidated adsorption bed or coated surface to improve λ_{eff} and α_w. Both of them make the last item of Equation 7.2 reduce. The third technology is to improve α_f by using heat pipe technology.

7.1.1 The Heat Transfer Intensification Technology of Adsorption Bed Using the Extended Heat Exchange Area

The following two effects can be obtained by increasing the heat exchange area:

1. $\alpha_w A_{eff}$ can be adjusted to a very high value even if the α_w is relatively small because A_{eff} is still very large.
2. Using the extended surface area can reduce the thickness of the adsorbent, and therefore can get a high heat transfer coefficient.

To date numerous methods for increasing the heat exchange area have been used. These methods include the finned tube [9], capillary equipment, plate heat exchanger, plate-fin heat exchanger. The common drawback of increasing the heat exchanger area is to increase simultaneously the heat capacity of the adsorption bed, which reduces the *COP* of the system, so increasing the heat exchange area as well as adopting the advanced cycle can improve the energy utilization efficiency effectively.

Employing the extended heat exchange area will improve the heat transfer coefficient of the adsorbent on the following two occasions. One is when the wall heat transfer coefficient is not very low, and the expansion phenomenon of the adsorbent doesn't occur in the adsorption process, which will change the wall heat transfer coefficient. The second is when the operation pressure is high enough in order to avoid the Knudsen effect. In this case, the wall heat transfer coefficient depends on the size of adsorbent particles. The small granular adsorbent can improve the heat transfer coefficient. As Miles and Shelton's investigation [10] showed the cycle time can be shortened to 5 minutes and the optimal *COP* is obtained when the adsorbent with the small size has been employed. In the case of the low pressure refrigerant (such as water or methanol), if the adsorbent particle size is too small, the Knudsen effect will influence the wall heat transfer coefficient. In this case, even increasing the heat exchange area will not significantly increase the comprehensive heat transfer coefficient α of the adsorption bed.

So far for a vacuum system, such as using water as a refrigerant, the most effective ways of extending the heat exchange area is by employing the plate-fin heat exchanger [11, 12] as an adsorption bed. The plate-finned adsorption bed is composed of a series of plate-fin heat exchange units (Figure 7.1). Both the fluid side and adsorbent side of the adsorption bed have thin fins to extend the heat exchange area. The adsorbent is divided into many small units by the fins of the adsorbent side. The adsorbent is surrounded by the plate wall and the fin, which intensifies effectively the heat transfer performance of the adsorbent side. The pores on the fins by the fluid side enhance the turbulence and thus destroy the flow boundary layer and thermal boundary layer, which enhances the heat transfer. In addition, the pressure difference between the adsorbent side and fluid side reduce the contact thermal resistance. Taking the mass transfer into consideration, the plate-fin adsorption bed isn't fully filled with the adsorbent, but has reserved some channels through the interval metal piece. These channels are used for the diffusion of adsorbates and therefore enhance the mass transfer performance along the depth direction. The diffusion proceeds into both of the sides through the pores of the fins. Hence, the mass transfer performance of the entire adsorption bed can be enhanced.

The heat transfer performance of the plate-fin heat exchanger is very good and the cycle time of the system is very short. Using silica gel–water as a working pair, the research in Shanghai Jiao Tong University adopts the plate-fin heat exchanger together with the heat recovery cycle thus improving the heat transfer performance effectively. The cycle time is shortened to only

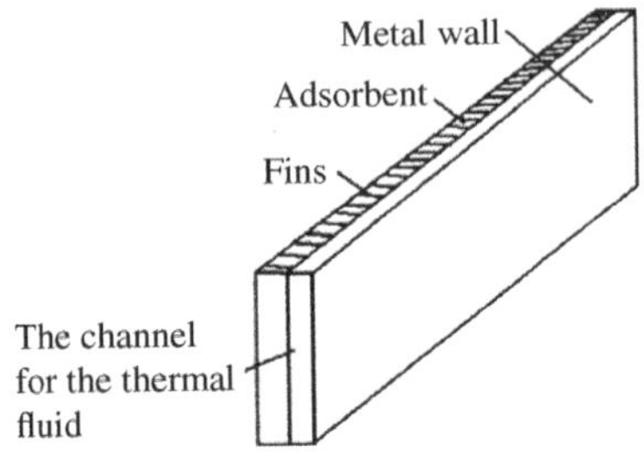

Figure 7.1 The heat exchanging unit of the plate-fin type heat exchanger [11, 12]

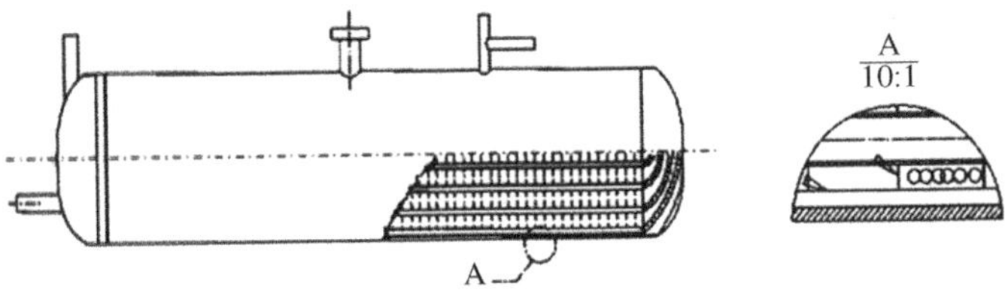

Figure 7.2 The special design of shell and tube type adsorption bed with the fins [13]

about 5 minutes. The plate-fin heat exchanger has a large heat capacity for the metal and the heat transfer fluid. This is the reason why the *COP* of the system is relatively low.

Because the plate-fin type heat exchanger leaks easily when the pressure inside the adsorption bed is positive, to avoid this leakage we utilize the shell and tube heat exchanger as the adsorption bed to increase the heat exchange area for the high pressure system, such as using ammonia as the refrigerant. For such a condition the heat exchange performance can be enhanced by increasing the numbers of the fins inside the adsorption bed. When the shell and tube heat exchanger is used for the adsorption bed, the fluid in the tube side is the cooling media and heating media of the adsorbent. The adsorbent is filled by the shell side. The outer shell of the heat exchanger is made of a seamless steel tube to prevent leakage under high pressure. The unique design of shell and tube type adsorption bed is shown in Figure 7.2. Generally speaking, as far as the shell and tube heat exchanger are concerned, the thickness of the outer shell and tube plate is as much as dozens of millimeters in order to undertake enough pressure. The fin thickness inside the adsorption bed is only 0.3 mm. So increasing the number of fins can greatly enhance the heat exchange area, whereas the entire metal heat capacity increases little if compared with the heat capacity of the shell and tube heat exchanger. This can effectively reduce the cycle time.

7.1.2 The Technology for the Heat Transfer Intensification in the Adsorption Bed

The performance of the adsorption bed can be improved by heat transfer intensification technologies, among which the consolidated adsorbent is widely used due to its high conductivity. Such a type of adsorbent is particularly suitable for the occasions where the bulk adsorbent cannot be used. The research is initially applied to the metal hydride over a very long period, and the adsorbents were mainly sintered during the preparation process of the adsorbent. At present, the ways of heat transfer intensification can be mainly divided into the following categories:

1. The adsorbents with metal powder or foam metal as matrix. By such a method the thermal conductivity of zeolite can reach 12 W/(m·K) [4].
2. Because of the high thermal conductivity, the graphite can serve as the matrix for the heat transfer intensification. Spinner and Le Carbone Lorraine were the first to put forward this method and used it for the ammonia salt. After that the technology had been applied to other adsorbents [14]. A very good heat transfer performance can be achieved by this technology, and the effective thermal conductivity depends on the density of the mixture. The obtained wall heat transfer coefficient of such adsorbents reached 500–3000 W/m^2 [7]. The consolidation process of compound adsorbent using the graphite as the matrix has already been introduced in Chapter 5.
3. Bonding directly the adsorbent with the binder to consolidate the adsorbent can strengthen the heat transfer performance. Shanghai Jiao Tong University bonded the activated carbon together with the binder, and then consolidated the adsorbent by using the mold. Compared to the bulk adsorbent, thermal conductivity of such adsorbent can be increased by 100% [15]. The mass transfer performance of the adsorption bed will be influenced when the consolidated adsorbent is adopted. For such an occasion the mass transfer channel is required in order to improve the performance.
4. Another emerging method for increasing the heat transfer coefficient of the adsorption bed is the coating of the heat exchanger surface with the adsorbents. This method stands for the recent trend of the development of adsorption bed design and it has been successfully applied in some commercial adsorption machines.

Several coating methods have been reported in the literature. Figure 7.3 gives a classification of coating methods for zeolite-based systems according to the thermal contact of the adsorbent with the thermal conductive metal substrate. In terms of coated adsorbers, heat and mass transfer through the adsorbent layer depends on the layer thickness, density, or accessibility of pores, and hence on the coating method. Owing to the nature of the coating method, ex situ and in situ coating techniques can be differentiated.

In ex situ methods adsorbent production and heat exchanger coating are realized in several independent steps. The coating can be achieved either by gluing adsorbent pellets on a substrate surface or by dip or slurry coating the surface in a suspension containing prefabricated zeolite powder and an organic or inorganic binder (e.g., polyvinyl alcohol or Al_2O_3). For example, a silica gel coating by gluing silica gel particles on heat exchanger lamellas with the help of a resin was developed by Sortech AG [17]. A compact heat exchanger design can be realized by this method thus improving the specific power. Dawoud *et al.* [18] investigated a zeolite-coated sample. A polymer was used as an adhesive for the ex situ coating. The authors reported an improved wall heat transfer coefficient for the zeolite-coated sample compared with conventional fixed beds.

One advantage for an ex situ coated heat exchanger is a bonded thermal contact between the adsorbent and the thermal conductive metal surfaces. Another important merit is associated with its flexibility, meaning that nearly every metal can be coated with an adsorbent by

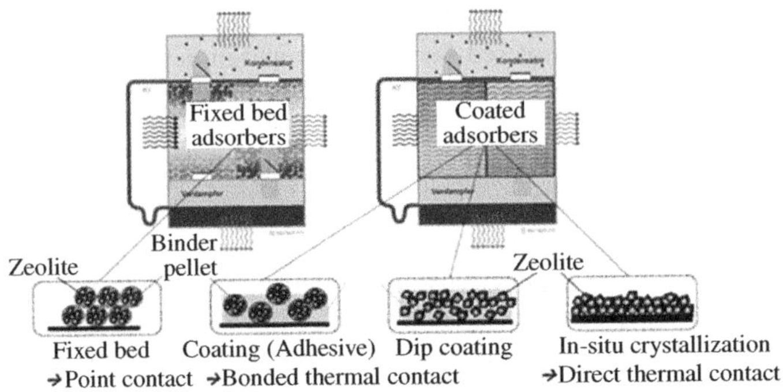

Figure 7.3 Classification of adsorption heat exchanger coating methods according to the thermal contact [16]

ex situ methods. There still exist several drawbacks for ex situ methods. High amounts of binder result in pore plugging of the adsorbent and hence loss of accessibility and adsorption capacity. Furthermore, in vacuum systems, the non-condensable gas released by organic binders will lead to problems about pressure increase which will also affects the adsorption performance.

In situ methods mean that the synthesis and adsorbent coating are realized in one step. In situ coating without or with reaction of the support can be classified. In the first approach zeolite layers are grown on inert supports (e.g., ceramics or stainless steel) from gels or solutions containing all desired reactants for the zeolite formation. In the second approach at least one reactant for the growth of the zeolite layer (generally silicon or aluminum) is taken from the support either by extraction from an inert support matrix or by partial support dissolution while all other reactants are supplied from a reactive solution. For heat transformation mainly in situ coating on inert supports has been adopted. An important feature using zeolite in situ coating is the synthesis of relatively thick zeolite layers to ensure sufficient adsorption capacity. A range of 25–150 mm has been predicted as the optimum layer thickness for in situ coated systems depending on operating conditions and heat exchanger configuration.

A key feature of a in situ coating technique is the direct thermal contact of adsorbent and metal surface, providing an optimum heat transfer from the heat conductive metal to the adsorbent. An extraordinarily high SCP is possible with this approach. Although no binders are used, the mechanical stability can be very high depending on the synthesis method. Dunne studied the coated tube using a single crystal of zeolite that has developed from the metal surface. The SCP of the system can reach 1500 W/kg [19]. Bauer *et al.* [16] prepared an in situ coated aluminum substrate using AlPO-type zeolite. A thickness of more than 100 μm was realized by one step partial support transformation (PST) method. A first test of a coated heat exchanger showed that a SCP of 350 W/m^3 heat exchanger (or 560 W/kg heat exchanger) could be obtained under a condition of 85 °C (desorption temperature)/30 °C (adsorption temperature)/15 °C (evaporation temperature).

A common shortcoming of coated heat exchangers is the fact that it is sensitive to dead mass and volume ratio which may reduce the COP value. To overcome this limitation and to improve dynamic properties, optimization of the heat exchanger design – for example, copper/aluminum foam structures or sintered fibers – has been suggested.

7.1.3 The Heat Pipe Technology

The heat transfer intensification technologies mentioned above can get very good overall effect for the adsorbent side. When the entire heat transfer effect of the adsorbent bed is good the heat transfer effect of heat exchange fluid will become the main impact factor for the heat transfer enhancement.

The convection flowing processes for the heat transfer of the adsorption bed include the laminar flow and turbulent flow. For certain cycles such as the thermal wave cycle a large temperature difference between inlet and outlet of the adsorption bed is required, and consequently the fluid flow adopted in the cycle is always the laminar flow. For such an occasion, using the capillary or reducing the space between the plates can improve the heat transfer coefficient of the fluid. For example, when the diameter of the tube is 1 and 0.3 mm, the α_f is 300 and 1000 W/m^2, respectively. Using the capillary and activated carbon powder with the diameter of about 50 μm, Menuier had obtained the α_f with the value as high as 1000 W/m^2 by using the ammonia as the adsorbate under conditions of high pressure. The application of capillary pipes reduced the thickness of the adsorbent (that is less than 10 mm). This is very beneficial to the thermal wave cycle especially when activated carbon–ammonia is used as the working pair.

When the laminar flow is not necessary, the turbulent flow can produce a very good heat exchange performance, and such a type of flow can be used for the systems that used the tubes with a large diameter. But the turbulent flow will lead to very high energy consumption. Using the two-phase heat exchange process can solve this problem. In fact, it is just the principle of the heat pipe. Generally the heat transfer fluid of the heat pipe is water and the heat source is the steam generator. The generated steam is used to heat the adsorption bed, and is thereby condensed in the heat exchanger. The liquid water flows into the heat exchanger of the adsorption bed, and the adsorption bed is cooled due to the evaporation of water. In this way the very high heat transfer coefficient can be obtained. LIMSI in France has studied this scheme, and the highest heat transfer coefficient is about 10 kW/m^2 [9]. Vasiliev has introduced the concept of the pulse heat pipe in adsorption bed. The designed pulse heat pipe adsorption bed is shown in Figure 7.4.

The propane is used as the working medium in the heat pipe in Figure 7.4. The adsorption bed is made of the flat-plate aluminum tubes with a width of 7 mm, and the width of the heat pipe is only 7 mm [20]. Shanghai Jiao Tong University also applied the heat pipe technology on the adsorption bed in the ice making system on fishing boats, as well as on the adsorption chillers driven by a low temperature heat source. A series of patents of alternating heat pipe and separation heat pipe adsorption systems [21–23] had also been obtained by Shanghai Jiao Tong University. The detailed design of the heat pipe type adsorption bed is introduced in Chapter 8.

7.1.4 Other Types of Adsorption Bed with Special Design

In addition to the above typical technologies for adsorption beds, the beds also can be designed for particular occasions. As an example two typical schemes are used for the special design of adsorption beds.

The first scheme is the spiral plate heat exchanger used as the adsorption bed, which is shown in Figure 7.5 [24, 25]. The heat transfer medium flows inside the splints among the spiral plates, and the adsorbent is filled in the spiral spaces among spiral plates. When the spreading area

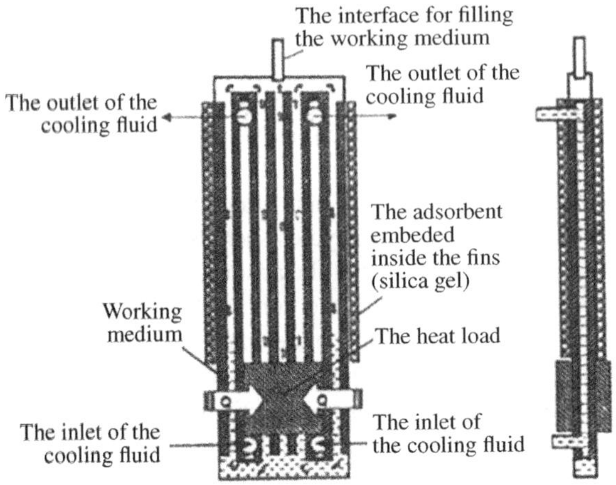

Figure 7.4 The pulse heat pipe type adsorption bed with miniature fin of multi-channel [20]

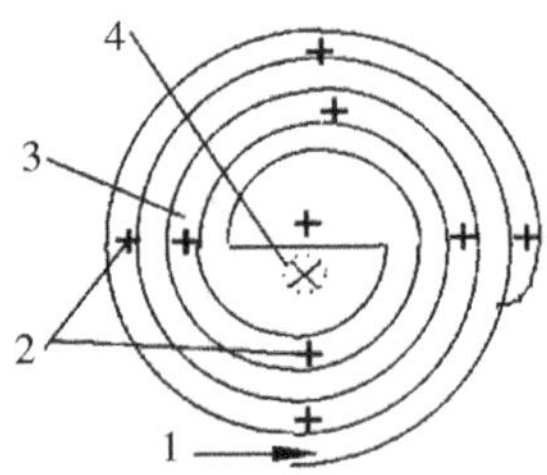

1 - Fluid outlet, 2 - Supported pole, 3 - Adsorbent, 4 - Fluid inlet

Figure 7.5 The diagram for the structure of spiral plate heat exchanger [24, 25]

of spiral plate is 2 m^2 and spiral space is 18 mm the filled activated carbon is 6 kg. A number of support poles between the gaps can be used to enhance the strength of the spiral plate. The wire mesh pipe in the spiral plate gap is used as the mass transfer channels of the adsorbent.

The main advantages in using the spiral plate heat exchanger in the adsorption refrigeration system are listed below:

1. The structure is compact, so the heat transfer temperature difference is smaller, which makes the temperature distribution more uniform.
2. The support pole not only can reinforce the strength but can also improve the heat conduction of the adsorption bed.
3. It has higher heat flux density.
4. With the increment of the flat area, the volume of spiral plate increases more slowly.
5. The manufacture is convenient, and the price is low.

Another special design of the adsorption bed is to use the tubesheet type heat exchanger as the adsorption bed (Figure 7.6). Similar to spiral plate type adsorption bed, the splint of the bending plate is the access of heat medium and cool medium, while the gap between the plates

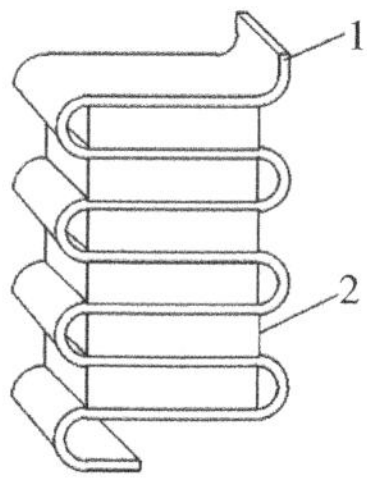

1 - Fluid channel 2 - Filled adsorbent

Figure 7.6 The structure of tubesheet type heat exchanger [26, 27]

is filled with the adsorbent. This kind of adsorption bed can be used for those occasions where the volume of adsorption bed is strictly limited.

7.2 The Influence of the Heat Capacity of the Metal Materials and Heat Transfer Medium on the Performance of the System

7.2.1 The Metal Heat Capacity Ratio vs. the Performance of the System

The heat capacity ratio between the metal materials of the adsorption bed and the filled adsorbent is called the metal heat capacity ratio. In the adsorption refrigeration system, the adsorption bed is heated to the desorption state and cooled to the adsorption state alternatively, which makes the metal heat capacity ratio have a significant impact on the actual performance of the system. The influence is reflected in the sensible heat loss of the metal material for the adsorption bed in the continuous heating and cooling process. The heat loss depends on the design of the adsorption bed. Besides, it is influenced by the system's working conditions, such as the desorption temperature, heat recovery process, and so on. The high metal heat capacity ratio can also influence the cycle time.

The metal heat capacity ratio of the adsorption bed will increase under the following conditions:

1. The small adsorbent density will lead to the small filled density of the adsorption bed. In order to meet certain refrigeration power the designed adsorption bed is often very large, so the metal heat capacity will significantly increase.
2. If the extended heat exchange area is used for the adsorption bed to improve the heat transfer, the metal quality for heat transfer will increase. Consequently the metal heat capacity ratio of the adsorption bed will also increase.
3. If the adsorption bed using the refrigerant has a positive pressure (such as ammonia, etc.), it is necessary to increase the metal wall thickness to bear the high pressure. This will lead to the increment of the metal heat capacity for the adsorption bed.

For the continuous heat recovery adsorption air conditioning/heat pump system [28] with activated carbon–methanol as the working pair, assuming that the heat exchange performance is independent from the metal heat capacity, the relationship between the metal heat capacity ratio and the system performance is shown in Figure 7.7. It indicates that the increment

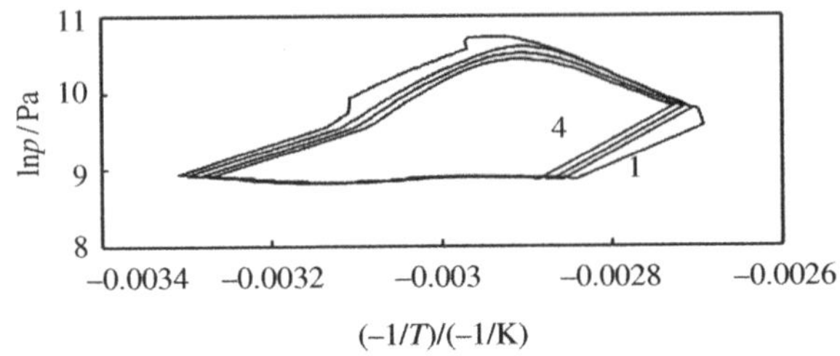

Metal heat capacity ratio: 1 - 0.8, 2 - 1.6, 3 - 2.5, 4 - 3.3 ($T_{heat,s}$=100 °C, T_{cool}=21 °C, t_{cycle}= 40 min, T_e = 9.5 °C, t_{reg} = 2 min)

Figure 7.7 The influence of metal heat capacity ratio on the system performance [28]

of the metal heat capacity ratio will cause the decrement of the system performance on the *p-T-x* chart. That is to say, more heat is transferred from heating medium to the adsorption bed metal in a cycle. The heat removed by the cooling medium mainly includes the sensible heat of the adsorption bed metal. This will influence the desorption performance of the adsorption bed, and thereby will reduce the refrigerating capacity of the system. At the same time, the operation efficiency of the system will reduce.

7.2.2 The Residual Heat Transfer Medium (Heating Fluid) in the Adsorption Bed and the Performance of the System

When the heating and cooling medium of the adsorption bed are different in the heating and cooling process, two kinds of heat transfer medium pipeline are required inside the adsorption bed. One kind is heating medium pipes, and another kind is cooling medium pipes. For the activated carbon–ammonia system, the desorption temperature of the adsorption bed can reach above 150 °C. For the choice of the heat transfer fluid water is generally used to cool the adsorption bed. However, water isn't used to heat the adsorption bed when the heating source temperature is higher than 100 °C. The adsorption bed can be heated by the heat conduction oil for this occasion. The designed adsorption bed is composed of two kinds of heat transfer medium pipelines. One is the water pipelines, and another is heat conduction oil pipelines. The adsorption bed is cooled by the water, and thereby completes the adsorption process. Some excess water will be held in the adsorption bed when the adsorption process is terminated and the adsorption bed is switched to be heated by the heat conduction oil. For this occasion the residual water in the pipes of the cooling medium will still consume a certain quantity of heat due to the heating, boiling, and evaporating process.

Even if employing the same heat transfer medium, the residual heat transfer medium still also influences the performance of the system. When the desorption process terminates, the stranded fluid won't return to the heating device, and instead will leave the system, and such a process will lead to heat loss. In the adsorption bed, the more stranded the medium is, the greater the heat is lost in the system.

Employing two adsorption beds can achieve a continuous refrigeration effect. Assuming that for such a system the heat recovery process is adopted, and assuming that the heating medium is water and the stranded water is 10 kg, then the heat loss will reach 1.25×10^3 kJ in the adsorption bed during a cycle when the heat recovery temperature is 65 °C and the final cooling temperature is 35 °C. The heat capacity of the heat transfer medium varied with the change of the heat transfer area in the adsorption bed. Generally speaking, the greater the heat transfer area and the more stranded water is in the adsorption bed, the more heat loss there will be.

As for the choice of the heat transfer medium for the adsorption bed, water, air, or heat conduction oil are commonly used. In the actual application the exhaust gas is also used frequently for the heating process but it generally can produce corrosion for the adsorption bed or scale formation inside the heat exchange pipelines. Such drawbacks might cause a leakage in the adsorption bed or the deterioration of heat conduction performance, and therefore will influence the performance of the system. Thus the method for heating the adsorption bed with the exhaust gas isn't directly recommended for the reasonable design of the system. For the utilization of the waste heat from the exhaust gas, generally we can employ the waste heat recovery boiler to retrieve the waste heat, and then heat the adsorption bed through the secondary heat exchange of water or heat conduction oil in the boiler.

As far as the selection of the heat medium and cool medium in the adsorption bed, one of the principles is to choose the same type of working medium for heating and cooling processes when the requirements for the processes can be satisfied. By such a method only one kind of heat transfer pipeline is designed, which can greatly reduce the heat capacity of the metal inside the adsorption bed. For example, for the working pair of silica gel–water the water is commonly used for the heating and cooling processes. For the working pair of activated carbon–ammonia the heat conduction oil is always used for the heating process after it exchanged the heat with the high temperature exhaust gas, and is used for the cooling process after it exchanged the heat with a low temperature cooling medium such as the cooling water. The second principle for choosing the heat transfer medium is to select the working medium with a good thermal physical property. Compared to the heat conduction oil, if the water is used as the heat transfer medium the heat loss is very large due to the larger heat capacity. But water also has the advantages of high thermal conductivity and the small viscosity, thus the flow speed of the water is higher than that of the heat conduction oil under the same restricted heat transfer conditions, and consequently the change rate of the water is faster when the water is selected as the heat transfer medium.

7.2.3 The Influence of the Ratio Between the Metal Heat Capacity and the Fluid Heat Capacity on the COP and SCP

As we mentioned before, there are two main parameters evaluating the performance of an adsorption refrigeration system, one is *COP*, and another is the *SCP*. The calculation formula of the *COP* and *SCP* is defined as follows:

$$Q_{ref} = M_a \times \Delta x \times L \tag{7.3}$$

$$SCP = \left(\frac{Q_{ref}}{t_{cycle} \times M_a} \right) \tag{7.4}$$

$$COP = \frac{Q_{ref}}{Q_h} \tag{7.5}$$

where Q_{ref} is the refrigerating capacity (kJ) of the adsorption bed in a cycle, t_{cycle} is the cycle time, Q_h is the required heat (kJ) for the desorption process, M_a is the adsorbent mass, Δx is the adsorption quantity of the cycle, L represents the latent heat of vaporization (kJ/(kg·°C)).

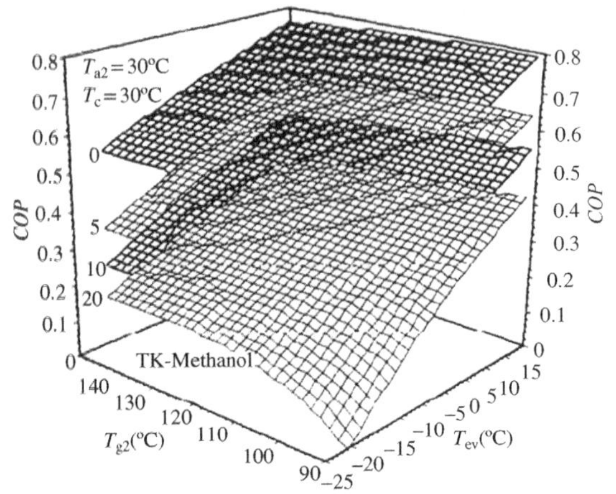

Figure 7.8 The heat capacity of the metal and fluid in the adsorption bed vs. COP [29]

The heat capacity of the metal and the fluid in the adsorption refrigeration system will influence these two parameters mentioned above, and the influences will be reflected by two aspects. Firstly, the heat capacity will influence the cycle time. From the calculation formula of Q_{ref}, it can be seen that the refrigeration capacity Q_{ref} is a constant value when the same adsorbent is used and the adsorbent mass is the same for different adsorption beds. If the heat capacity of the metal and the fluid inside the adsorption refrigeration system increases, the heating and cooling time of the system will be prolonged, namely the overall cycle time t_{cycle} will increase. However, the SCP will be reduced. As for the COP, the influence of the heat capacity of the metal and fluid is mainly reflected by the Q_h. Provided that the Q_{ref} is a constant, the consumed sensible heat will increase when the heat capacity of metal and the fluid increases. The Q_h will increase in Equation 7.5, and consequently the *COP* will reduce.

Take the adsorption working pair of coconut shell activated carbon–methanol as the example, the influence of the heat capacity of metal and the fluid on the COP is shown in Figure 7.8. The ratio between the heat capacity of metal and fluid and the heat capacity of adsorbent is investigated as the main parameter. Generally speaking, the ratio depends on the design of the system, that is, the value of the ratio mainly depends on the mass and the heat capacity of the adsorption bed metal materials. The ratio is 0, 5, 20, 50 [29], respectively, and is shown in Figure 7.8. From Figure 7.8 it can be seen that the *COP* of the system is rapidly reduced with the increment of the heat capacity of the metal and the fluid. It shows that the heat capacity of the metal and the fluid will greatly influence the performance of the system, which should not be overlooked.

In order to reduce the influence of the heat capacity of the metal and the fluid on the *COP* of the system, the following measures can be adopted:

1. Employing the consolidated adsorbent to improve the heat transfer performance by increasing the heat transfer coefficient. The heat transfer intensification can be reflected in two aspects. On the one hand, Δx of the system will increase when the cycle time is the same. Then, according to Equation 7.3 the Q_{ref} will increase; on the other hand, if the Δx is the same, the cycle time t_{cycle} of the system will decrease. No matter the increment of the Q_{ref} or the decrease of t_{cycle}, the *SCP* of the system will be effectively improved according to Equation 7.4. The relationship between the metal heat capacity ratio and the *SCP* of system

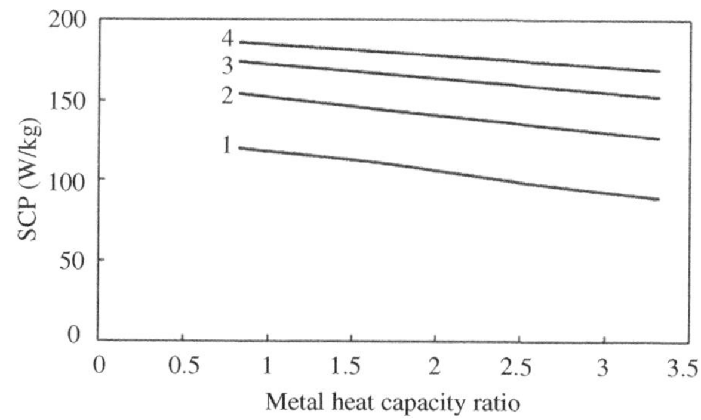

The heat transfer coefficient of adsorption Bed: 1- 60W/(m^2·°C); 2 - 90W/(m^2·°C); 3 - 120W/(m^2·°C); 4 -150W/(m^2·°C). ($T_{heat,s}$=100 °C, T_{cool}=21 °C, t_{cycle}=40 min, T_e=9.5 °C, t_{reg}=2 min)

Figure 7.9 The metal heat capacity ratio of adsorption bed vs. *SCP* [28]

is shown in Figure 7.9 when the activated carbon–methanol is used as the adsorption working pair and the heat transfer coefficients are different. Figure 7.9 showed that the SCP of the system will decrease when the metal heat capacity ratio increases. However, the *SCP* of the system will increase when the heat transfer coefficient increases. The SCP of the system is 150 W/kg when the metal heat capacity ratio is 3 and the heat transfer coefficient is 120 W/(m^2·°C), and the *SCP* increases by 40% if compared with that for the heat transfer coefficient of 60 W/(m^2·°C) and the metal heat capacity ratio of 1.

2. Using a porous medium as the matrix and increasing the mass transfer channel, and so on can enhance the mass transfer performance of the system. If both mass transfer performance and the heat transfer performance can be intensified, then the Q_{ref} can be improved when the cycle time is the same, or the cycle time t_{cycle} can be reduced when the cycle adsorption quantity is the same. Both methods can effectively improve the SCP.
3. Using the heat recovery process at the switch time can partly recover the sensible heat of the fluid and metal in the adsorption bed. According to Equation 7.5, such a process will reduce the Q_h, and thereby will improve the *COP*. For the activated carbon–methanol adsorption refrigeration system, the contrast between the simple cycle and continuous heat recovery cycle is made when the cycle time is 50 minutes, heating time is 2 minutes, cooling temperature is 25 °C, and the heating temperature is 100 °C. The results are shown in Figure 7.10. From Figure 7.10 we can see that COP will increase by 20–30% by using the heat recovery process if compared the results of the simple cycle.
4. Choosing the heat transfer media properly can reduce the disadvantages influenced by the heat capacity of the metal and the fluid. Taking the different heat transfer media into account, Figure 7.11a,b provides an overall understanding regarding the influence on the performance by different heat transfer media. It illustrates the performance of the shell and tube type adsorption refrigeration unit that uses the water and oil as the heat conduction medium. The density of heat conduction oil is 8.4 kg/m^3, and the specific heat capacity is 2 kJ/(kg·°C). Because the heat conduction performance of the oil is inferior relative to water, the adsorption quantity of the cycle with the heat transfer medium of oil will decrease compared with the cycle with the heat transfer medium of water for the same cycle time, and thus lead to the decline of the Q_{ref} that is calculated by Equation 7.4. As a result, the SCP will decrease by about 2.4% in Figure 7.11a. The *COP* will decrease with the decrement of Q_L in Figure 7.11b. But because the heat capacity of the retention oil in the adsorption

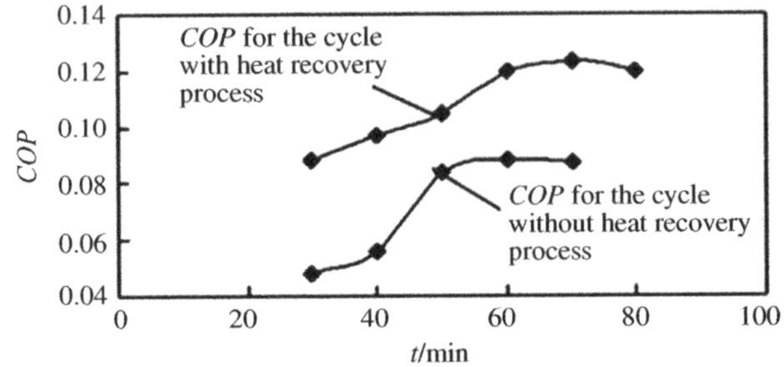

Figure 7.10 The comparison of the COP for the cycles with and without heat recovery process [30]

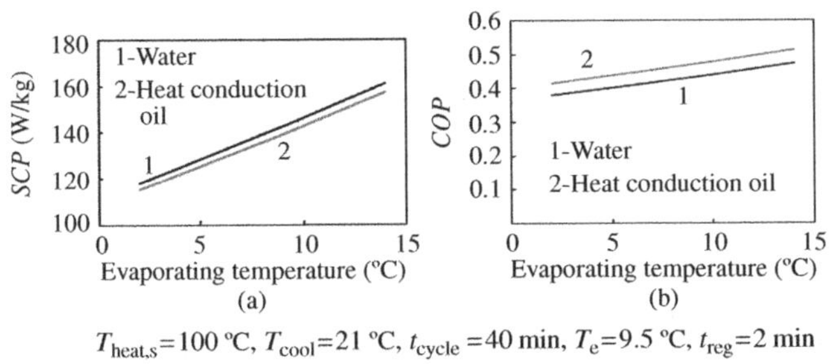

Figure 7.11 *SCP* and *COP* of the system using the oil and water as the heat transfer media [28]. (a) *SCP* vs. evaporating temperature and (b) *COP* vs. evaporating temperature

bed is smaller than that of water, Q_h calculated by Equation 7.5 will decrease, and consequently the *COP* of the system with the heat transfer medium of oil increases significantly. Compared to water, the *COP* of the system increases by 8.2–9.6%. If the heat source supplied is sufficient and the *SCP* is the main evaluation index, water can be chosen as the heating medium. If the heat source supplied is insufficient and then *COP* will be the main evaluation index, the system should choose the oil as the working medium. But on this occasion a high-power pump will be required owing to the large viscosity of the oil. Otherwise, the flow velocity will decrease, which will deteriorate the heat transfer performance of the adsorption bed.

7.3 Other Components of the Adsorption System

An adsorption refrigeration system can be classified into two categories: vacuum system and pressure system according to the working fluid. In the vacuum systems, methanol and water are generally used as the refrigerant, whereas the pressure systems use ammonia as the refrigerant. The refrigeration output of an adsorption refrigeration system isn't homogeneous, and therefore the indirect way for the refrigeration output is usually utilized. For pressure systems, the design of evaporator must firstly ensure that it is pressure-resistant, for which the shell and tube type heat exchangers always have a good performance under the conditions of high pressure. Structure and the size of evaporator are determined according to the refrigeration capacity of the system.

7.3.1 Design of Evaporator, Condenser, and Cooler of Low Pressure System

7.3.1.1 Evaporator

Vacuum systems are different from the pressure systems, and for the design of evaporator in a vacuum system the problem of mass transfer should be considered firstly. Because the evaporation pressure of the refrigerant, such as methanol and water, in a vacuum system is low especially for low temperatures (for example, when the evaporating temperature is 5 °C, the evaporation pressure of methanol and water are 5.7 and 0.95 kPa, respectively), the resistance for the flowing process of the refrigerant should be specially considered for the design of the evaporator. If the resistance for the flowing process of the refrigerant in the evaporator is quite high, the refrigerant needs to overcome the resistance firstly before it gets into the adsorption bed. Such a process will lead to the low adsorption pressure in the adsorption process of refrigerant, and it will be equivalent to the condition of a lower evaporating pressure, consequently it will affect the refrigeration performance of system. For example, for a methanol system if the resistance from the evaporator to adsorption bed is 0.2 kPa, the pressure inside the adsorber will be lowered and that is equivalent to the condition for the evaporation temperature drop of 1 °C. Thus the resistance of the refrigerant side should be reasonably calculated for the design of the evaporator, and the reasonable structure should be chosen. Secondly, heat transfer between refrigerant of evaporator and heat transfer medium should be considered to ensure that the evaporator could take away the cooling capacity of the system in time. Therefore, taking the above factor into account, if tube and shell or plate heat exchangers are utilized, the resistance for a flowing process will be high. If the flooded evaporator is utilized, the problem occurs where the static pressure of the liquid column for a refrigerant exists in the evaporator, which will make the evaporation temperature increase. The static pressure of the liquid column in particular has a great influence on the evaporation temperature. Taking methanol as an example, if the height of liquid column is 400 mm, when the temperature for the interface of vapor and liquid in evaporator is 10 °C, the temperature for the bottom of the evaporator is 13.3 °C. This will also affect the performance of the refrigeration system, and generally is only applied to the system with a low evaporation quantity. Design of an evaporator for a vacuum system can refer to the design for the evaporator of an absorption refrigeration system by using the structure of a spray evaporator to ensure the performance of the system.

Figure 7.12 is the working principle of a spray evaporator [31]. The output of the cooling capacity of the evaporator is 5 kW. The refrigerant in the evaporator is pumped into the liquid tray by the magnetic pump. There are many orifices for the liquid droplets at the bottom of the pan. Corresponding to the design of refrigerant pipe, heat transfer happens between the dripping refrigerant and the refrigerant pipe, which forms the evaporative cooling. Flow resistance of the adsorption bed in the adsorption process is mainly produced in the pipes between the adsorption bed and the evaporator, and therefore the flow resistance is small. The loop of the cooling medium for the refrigeration process is composed of eight flows, and each flow includes five light tubes.

Referring to the evaporation heat transfer process on the surface of the tubes in the LiBr absorption system, the heat transfer performance of the spray evaporator is estimated.

Firstly, for all the thermal resistance between cooling water and refrigerant, it should include thermal resistance on the water side, coefficient of the dirt on the surface of the pipes, thermal resistance of the tube, and thermal resistance of the refrigerant. The overall heat transfer

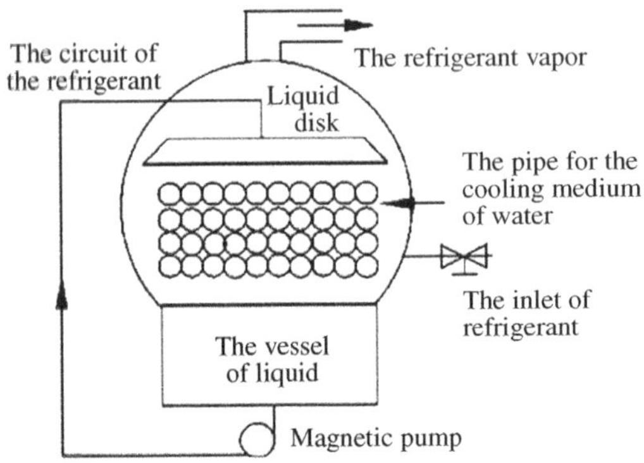

Figure 7.12 Diagram for the principle of the spray evaporator [31]

coefficient α is:

$$\frac{1}{\alpha} = \frac{1}{\alpha_{water}} + R_i + R_0 + \frac{1}{\alpha_{ref} \cdot S_A} \tag{7.6}$$

where α_{water} is the heat transfer coefficient of the water side; α_{ref} is the heat transfer coefficient of the refrigerant side; R_i, R_0 are coefficients of dirt and thermal resistance of tube, respectively; S_A is the surface area ratio (the ratio between the outside area of pipe and the inside area of the pipe).

The heat transfer coefficient of the water side inside the tube is:

$$\alpha_{water} = 0.023 \cdot \frac{\lambda_{water}}{d_{pi}} \cdot \mathrm{Re}^{0.8} \left(\frac{C_{pwater} \cdot \mu_{water}}{\lambda_{water}} \right)^{0.4} \tag{7.7}$$

Qualitative water temperature can be chosen as 9.5 °C, flow rate u_f of the water in the tube is 1.2 m/s. because:

$$\mathrm{Re} = \frac{u_f \cdot d_{pi}}{\nu} = 9264 > 2100,$$

The calculated α_{water} is 4800 W/(m^2 °C).

Liquid spraying on the tube forms a film dripping down from the tube. Evaporation only happens on the surface of the film around the tube. This film can be considered as a type of thermal resistance. The heat transfer mechanism is similar to a falling film evaporator. The empirical formula for evaporation on the surface is:

$$\frac{\alpha \cdot \delta}{\lambda_f} = C_{ra} \cdot \mathrm{Re} \cdot \mathrm{Pr}^{1/3} \tag{7.8}$$

where C_{ra} is the proportional coefficient determined by evaporator type; δ is thickness of the falling film; λ_f is thermal conductivity of the liquid; Pr is the Prandtl number.

If the flow rate from a row of pipe is m_f, length of tube is L_{pi}, and because of that the film is thin, and the flow rate is u_f:

$$u_f = \frac{m_f}{2 \cdot L_{pi} \cdot \delta}, \quad \text{correspondingly : } \mathrm{Re} = \frac{u_f \cdot \delta}{\nu} = \frac{m_f}{2 \cdot \nu \cdot L_{pi}}$$

$$\alpha = C_{ra} \cdot \mathrm{Pr}^{1/3} \left(\frac{\lambda_f}{\delta} \right) \left(\frac{m_f}{2 \cdot L_{pi} \cdot \nu} \right) \tag{7.9}$$

According to the equations above, the heat transfer coefficient of the refrigerant side is about 4000–5000 W/(m^2 °C). If assuming a coefficient of dirtiness is 0, thermal resistance of the tube is 0.001 m^2K/W as well the overall heat transfer coefficient of the evaporator is 2000 W/(m^2 °C). Therefore, for the evaporator 40 tubes of ϕ16 mm need to be chosen and the total heat transfer area is 2.11 m^2.

When considering how to lower the refrigerant flow resistance in an evaporator for a vacuum adsorption refrigeration system, Shanghai Jiao Tong University obtained optimal results by means of connecting adsorption bed, condenser, and evaporator. The specific design and performance analysis will be described in detail in Chapter 8.

7.3.1.2 Design of Condenser and Cooler

The condenser of the adsorption refrigeration system consists primarily of two types: air-cooling and water-cooling. Its design is similar to that of the condenser of a compressed refrigeration system. The design of the condenser should fit the capacity of the adsorption bed by considering the load of the system and the condensing pressure. This design should in particular consider that the load in the condenser will change when the desorption amount of the refrigerant from the adsorber changes. Due to the temperature change of the adsorption bed in the desorption process and non-equilibrium adsorption process, the desorption amount of the adsorption bed always changes. Usually in the initial period of desorption time of the adsorption bed, the desorption amount is the largest, simultaneously, the condensing load of the condenser is also the largest. Therefore, when determining the condensing load of the system, the required maximum condensing load must be considered according to the maximum desorption amount.

The structure of a condenser commonly uses a tube and shell heat exchanger, and its heat transfer coefficient can reach 1400–2900 W/(m^2 K). A plate heat exchanger can also be used for compact structure because it has a high heat transfer performance. The typical merit of the plate heat exchanger is the large heat transfer area, especially the large specific area, which is as high as 0.2 m^2/kg. Corrugated heat transfer surface is also used for promoting the heat transfer performance of the fluid, and the heat transfer coefficient can reach 2000–6000 W/(m^2 K).

Taking continuous heat recovery adsorption air conditioner/heat pump units for example, a plate heat exchanger should be used as a condenser and calculated as such. The corrugated heat exchange surface of a plate heat exchanger is the key part of the heat exchanging process. The fluid flow and corrugated surface will form into a certain inclination angle. By such a process when fluid flows through the corrugated plate it will form a tortuous flow path, and therefore the secondary flow is generated due to the flow change, which will increase the turbulence of the fluid, and consequently will enhance the heat transfer. Experiments show it will change into a turbulent flow when $Re > 200$.

The heat transfer characteristics, corrugated shape of the plate, and the combination of the size and the plate are closely related to each other according to the Maslov empirical formula:

$$\frac{\alpha \cdot d_e}{\lambda_f} = M_{Re} \cdot \mathrm{Pr}^{0.42}\left(\frac{\mathrm{Pr}}{\mathrm{Pr}_w}\right)^{0.25} \tag{7.10}$$

where M_{Re} is the function of the Reynolds number. According to the corrugated feature of the plate, select the correlation formula of $M_{Re} = 0.1815\mathrm{Re}^{0.65}$, the heat transfer coefficient of both sides can be calculated. After that the structure and size of heat exchanger can be determined by the appropriate trial method.

$$\mathrm{Re} = m_{uA} \cdot d_e / \mu_f$$

where m_{uA} is the mass flow rate per unit area (kg/(m^{2}s)); d_e is the equivalent diameter, which is twice the amount of spacing between plates; and Pr is the Prandtl number of media at the average temperature. Pr_w is the Prandtl number of the media under the plate surface temperature of the heat exchanger.

The resistance for the flowing process should be considered. Generally the flowing resistance of the plate type heat exchanger is higher than that of the common tube and pipe type heat exchanger. For the condenser, the flowing resistance of the methanol vapor shouldn't be too large. The flowing resistance of the pressure drop can be calculated by Smith and Troupe equations.

$$\Delta p = (38.96 n_s + 121.22) \cdot \left(\frac{u_f^{\,2} \rho_f}{g} \right) \cdot \mathrm{Re}^{-0.13/(n_s - 0.565)} \tag{7.11}$$

where u_f is the flow rate of media and n_s is the number of flow channels.

Assuming a condenser is used for an activated carbon–methanol refrigeration system, the desorption temperature of the system is 100 °C, the condenser temperature is 30 °C, and the condensed methanol is approximately 5.2 kg in one cycle, then the load of the condenser is:

$$Q_{cond,load} = M_{me} \cdot (L + C_{pme} \cdot \Delta T) = 5.2 \times (1180 + 0.78 \times 70) = 6420\ \mathrm{kJ} \tag{7.12}$$

where $Q_{cond,load}$ is the load of the condenser, M_{me} is the mass of methanol desorbed from the adsorber, L is the vaporization latent heat of methanol (L is 1180 kJ/(kg °C) for the temperature of 30 °C), C_{pme} is the specific heat of methanol gas, and ΔT is the temperature difference between 100 and 30 °C.

The cooler of an adsorption refrigeration system is usually used as the intermediate cooling equipment of the adsorption bed. Its design is similar to that of condenser and needs to be determined by the cooling load. If the adsorption temperature of the system is 30 °C, the cycle adsorption quantity in one cycle is 5.2 kg, and the temperature of the system after heat recovery is $T_{reg} = 60$ °C, the cooling load is the sensible heat and the adsorption heat from T_{reg} to the zone of the adsorption temperature:

$$Q_{cool,load} = \int_{30}^{T_{reg}} C_{pa} M_a dT + \int_{30}^{T_{reg}} C_{pme} M_a x dT + \int_{30}^{T_{reg}} C_{pm} M_{madb} dT + \int_{30}^{T_{reg}} H_a M_a dx < 10^4\ \mathrm{kJ} \tag{7.13}$$

According to the equations of the condenser and the cooler, and taking into consideration the types of the SWEP plate heat exchangers, the selected condenser and cooler are shown in Table 7.1.

Table 7.1 shows that the total heat transfer coefficient α is 2350 W/(m^2 °C), and the heat transfer area is 1.47 m^2. When the temperature difference between the refrigerant and the cooling water $\Delta T' = 2$ °C, the condenser heat $Q_{cond,heat}$ can be calculated by Equation 7.14

$$Q_{cond,heat} = \alpha \cdot A \cdot \Delta T' = 2350 \times 1.47 \times 2 = 6909\ \mathrm{kJ} \tag{7.14}$$

Therefore the selected condenser has a sufficient cooling capacity.

Table 7.1 Characteristic parameters of condenser and cooler

	Condenser	Cooler
Type	B10	B10
Number of heat exchange plates	48	60
Heat transfer area (m^2)	1.47	1.83
Overall heat transfer coefficient W/(m^2 °C)	2350	2730
Pressure loss (kPa)	0.2	52.7
Size (mm)	144×117×289	134×117×289
Weight (kg)	7.8	9.5

According to the data in Table 7.1, for calculating the heat transfer capacity of a cooler, when the temperature difference $\Delta T'' = 3\,°C$, cooling capacity is:

$$Q_{cool,heat} = \alpha \cdot A \cdot \Delta T'' = 2730 \times 1.83 \times 3 = 1.5 \times 10^4 \text{ kJ} \tag{7.15}$$

Obviously, the cooling capacity of the cooler is sufficient.

7.3.2 *Heat Exchanger for Ammonia*

For the adsorption system with ammonia as refrigerant, one key point for the design of a heat exchanger is to consider the compatibility between metal materials of the heat exchanger and the ammonia. An ammonia heat exchanger generally uses steel, and cannot use cooper because corrosion will occur between ammonia and copper. Another problem for the design of the heat exchanger for ammonia is the high pressure. For example, according to the rules for the management of the boiler and pressure vessel in China, if one of the following conditions exists the heat exchange vessel will be defined as a pressure vessel:

1. Maximum working pressure is higher than or equal to 0.1 MPa (gauge pressure), and the product of pressure and volume is higher than or equal to 2.5 MPa l.
2. Nominal working pressure is higher than or equal to 0.2 MPa (gauge pressure), and the product of the pressure and volume is higher than or equal to 1.0 MPa l.

The saturated vapor pressure of ammonia for the temperature of 40 °C in a heat exchanger is about 1.5 MPa. If the volume of the evaporator is more than 0.67 l, then it has to be manufactured according to the standards of pressure vessel. The manufacture process of the heat exchanger must be directed by rules for the design of boiler and pressure vessel manufacturing supervision, which involves the manufacturing license of manufacturer, license management, product performance, supervision, inspection, and so on. For small ammonia adsorption refrigeration equipment in the laboratory, the standard limit of pressure vessel needs to be considered in the design of the system for improvement of the security for the system.

7.3.2.1 Several Typical Types of Heat Exchanger for Ammonia

The ammonia evaporator and condenser used in the adsorption refrigeration system are shell and tube type. Take the shell and tube type flooded evaporator as an example, its structure is

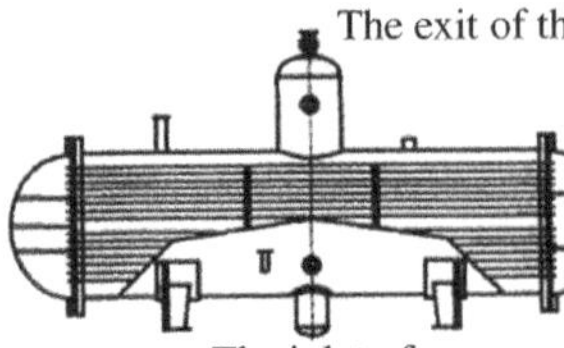

Figure 7.13 The shell and tube type evaporator for ammonia [32]

shown in Figure 7.13 [32]. When the heat exchanger acts as a condenser, it takes the same structure as above. The difference is when the heat exchanger is used as the condenser the upper ammonia outlet is the ammonia inlet of the desorption bed; the lower liquid ammonia inlet is the inlet of the condensed liquid ammonia.

When a tube and shell heat exchanger is used as a condenser, fluid at the tube side generally will not fall below 0 °C, so there is no danger of tube burst caused by freezing or expansion of cooling water. But when a heat exchanger is used as an evaporator, the column pipe burst might happen because of the reasons listed as follows [32]:

1. The influence of the system that provides ammonia. Physical and chemical properties of evaporator have a direct impact on the selection of the material, structure, and processing technique. Generally the liquid ammonia entering into the ammonia evaporator is provided by condensed liquid desorbed from the adsorption bed. The ammonia vapor evaporated from the evaporator is adsorbed by the adsorption bed. The pressure of ammonia vapor is low, and the unsteady adsorption process will lead to unstable ammonia vapor pressure, which is usually below 0.2 MPa. Occasionally evaporation pressure of ammonia is only 0.1 MPa. Under this pressure the evaporation temperature of liquid ammonia is about −20 °C. Except austenitic steels and related standards, minimum temperature of the pressure vessel and piping steels commonly needs to be higher than −20 °C. If a type of material often works at the lowest limit of the temperature then the material may break.
2. Influence of fluid resistance. In an ammonia evaporator the hot and cold fluids are mixed together, in which a part of the flow is forward flow, and another part of the flow is the reversed flow. Temperature unevenly distributes for the flowing process of the fluid. The flow at the shell side always changes, causing vibration. Consequently, the vibration will cause pressure fluctuation and shock, which will lead to additional stress on heat transfer tubes.
3. The influence of structure. Firstly, when the tube temperature is below freezing point, due to coagulation expansion of the water in the tubes the tubes might be broken. Secondly, the connecting tube of the liquid ammonia is arranged in the central part which makes it susceptible to corrosion because dead ends on both sides of the shell form turbulence there. At the junction between tube and support as well as inside tube and tube plate connections can exist gaps, which will generate stress concentration.
4. The corrosion caused by stress. The liquid ammonia exists at the shell side of the ammonia evaporator, which won't cause serious corrosion of carbon steel and low alloy steel. But when the passivating film on some parts of the material is incomplete or damaged, the corrosion will happen for these places. Due to the residual stress after manufacturing and

transformation processes, as well as the stress concentration points at the structure, such as stress concentration exists in tubes and baffles junction, the stress corrosion will happen by the function of the joint action of tensile stress and corrosive media, which will cause tiny cracks or damage on the heat exchanger.

5. The influence of the operation process. When the adsorption refrigeration system stops work we need to let the liquid ammonia at the shell side evaporate . Usually the valve of ammonia vapor is open, but after the end of the operation of the device, there is still some water in the bottom tube. Then, if the evaporation temperature is lower than 0 °C, the freezing process of the remaining water in the evaporator will form an additional force that might cause the cracks on tubes to burst.

Considering the influences above on the shell and tube type evaporator for ammonia, the design of the evaporator for an adsorption refrigeration system needs to pay attention to the following matters:

1. The pressure regulating valve needs to be installed to ensure the ammonia pressure within a certain range.
2. Sway the liquid ammonia interface to the side of the tube sheet to improve the properties of the flowing process.
3. Increase drainage tubes to prevent the accumulated remaining water that will freeze into the ice due to the occasional low-temperature when turned down in the system.

Another type of ammonia condenser is the heat exchanger with double plates. The double-plate heat exchanger is manufactured by laser welding two boards into a completely sealed plate group, then combining all of the groups together into a multi-plate group. The interface between every two board groups should be sealed. Such a double plate heat exchanger is a combination of the flow channels formed by welding and sealing methods alternately. Fluid with high pressure and strong permeability such as ammonia should flow through the welding path while fluid with low pressure and weak permeability such as water should flow through the sealing channels by the sheet gasket, which is shown in Figure 7.14.

The characteristics of a double plate heat exchanger are as follows:

1. It has high heat transfer performance. In the ammonia system, heat transfer coefficient K of a double plate ammonia condenser can be as high as 2326–4652 W/(m^2 °C). Due to the

The liquid flowing in the welding channels
The liquid flowing in the sealing channels by the sheet gasket

Figure 7.14 Double plate heat exchanger [33]

high heat transfer coefficient the temperature difference between the ammonia and water is also very small.

2. Due to the compact structure, the filling refrigerant inside the heat exchanger is less if compared with the tube and shell type heat exchanger. Consequently the weight of a double heat exchanger can decrease by 25–30%, the installation space of the heat exchanger can decrease by 40–50%, and the amount of the filling refrigerant is only 25–40% of that in the tube and shell type heat exchanger.
3. The double plate heat exchanger is composed of a number of plates with a bolt which is convenient to adjust the area for heat transfer. When working conditions change it is also very easy to remove the bolt to change the number of plates to adjust the heat transfer area.
4. The dirt inside the heat exchanger is less and it is easy to clean. For example, for in the fishing ice maker the cold sea water is used to cool the condenser directly for a double plate heat exchanger, and for such a process, due to the strong turbulence of the water between the plates, the dirt generated will clearly be hindered. The dirt also can be cleaned by simply removing the connecting bolts rather than removing the connecting pipes, which is much more convenient if compared with the tube and shell type heat exchanger.
5. The heat exchanger also is highly resistant to corrosion conditions. A double plate heat exchanger utilizes stainless steel as the material of the plate which has a strong resistance to corrosion. An adsorption marine ice maker for which the condenser is directly cooled by the cold seawater, generally uses titanium as the material to ensure a greater corrosion resistance.
6. When a double plate heat exchanger is used as evaporator, since the double plate heat exchanger has a good heat transfer performance the temperature difference between water and ammonia is small. Such a process could greatly improve the evaporation temperature of the ammonia as well as effectively reduce the subcooling degree of water outlet, consequently will eliminate the risk of freezing water at the outlet.

The comparison between the double plate heat exchangers and shell and tube heat exchangers is shown in Table 7.2. Table 7.2 shows that the weight of the ammonia double plate heat exchanger is reduced by 70% when compared with the shell and tube type heat exchanger. The amount of liquid inside the heat exchanger is only 25–40% of that in the shell and tube type heat exchanger, which is very safe for the operation process, and could avoid the shocks caused by the heat and pressure. But the maximum pressure of a double plate heat exchanger is limited, and it isn't suitable for high condensation temperature and pressure.

7.3.2.2 Shell and Tube Evaporator for Ammonia

For an adsorption refrigeration system testing rig the shell and tube heat exchanger is often used because its structure is relatively simple. In actual applications, there are mainly three types of shell and tube evaporators referring to the design of the evaporator in the compressed ammonia refrigeration system, that is, flooded evaporator, forced-circulation evaporator, and dry-type evaporator [33]. Three types of evaporator are shown in Figure 7.15.

For a flooded evaporator in Figure 7.15a, the working process is to store the mixture of liquid and vapor with a certain level of refrigerant liquid in the cylinder reservoir, and use a float valve to provide the liquid. The liquid refrigerant absorbs the heat of the cooling medium

Table 7.2 Comparison between double plate heat exchanger and tube and shell type heat exchanger for ammonia [33]

	Double plate	Tube and shell type		Double plate	Tube and shell type
Weight ratio	0.25–0.35	1.0	Area ratio	0.5	1.0
Filled liquid ratio	0.25–0.40	1.0	Heat and mass transfer ratio	1.1–1.7	1.0
Dirt factor ratio	0.1–0.5	1.0	Designed pressure	2.45 MPa	Not limited
Designed temperature	150 °C	No more than 200 °C	Standard material	Stainless steel, titanium	Carbon steel, stainless steel
Space expansion	Possible	Impossible	Water side cleaning	Possible	Possible at tube side
Heat and pressure shocks	Not dangerous	Dangerous	Vibration damage	Not exist	Exist in tube plate and baffled plate
Litter seal	Exist	Not exist	Risk for freezing	Low	High (shell side)
Freezing damage	Not exist	Exist			

in the evaporation pipe and evaporates. Its advantage is the high heat transfer coefficient at the refrigerant side, so when the large cooling capacity is required such a type of evaporator will be used, such as ammonia horizontal evaporator and helical cold water tank. For such a type of evaporator the refrigerant will be full of the cylinder, and consequently the filled amount of the refrigerant is large; its disadvantage is that due to the low temperature of ammonia for the evaporation, if the cooled medium is frozen in the tube the heat transfer tubes will face the risk of breaking up.

For the forced-circulation evaporator shown in Figure 7.15b, the working process is to pump the liquid refrigerant that is stored in the low-pressure accumulator to the evaporator, while the level of the refrigerant remains in the evaporation pipes at a certain level. For such a heat exchanger the coefficient of evaporator is higher than that of the flooded evaporator, but its drawback is its complex structure. The filling amount of the refrigerant in the system is large, thus it is dangerous in case the ammonia leakage happens by using the ammonia evaporator. Simultaneously, a forced-circulation pump will increase the electricity consumption of the entire system.

A dry type evaporator is shown in Figure 7.15c. The advantage of such an evaporator is that the refrigerant evaporated in the tubes as well as the cooling medium is cooled at the casing side with small flowing resistance, thus even if the cooling medium is frozen at the casing side the heat transfer tube will not split up. Since the refrigerant is completely vaporized in the dry type evaporator the filling amount of refrigerant can be greatly reduced.

The previous dry-type evaporator is mostly used in the Freon system because of better miscibility between Freon and mineral lubricants. Lubricating oil can flow back into the compressor

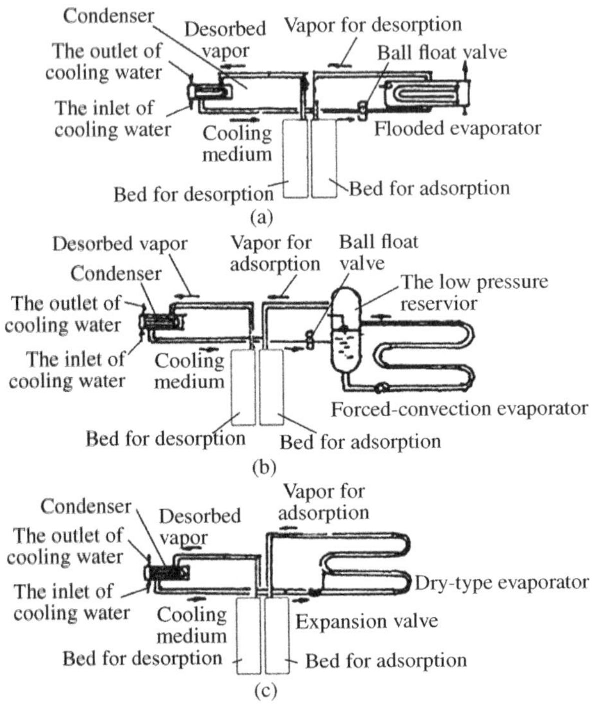

Figure 7.15 Three types of evaporator for ammonia. (a) Flooded evaporator; (b) forced-convection evaporator; and (c) dry-type evaporator

in the dry type evaporator. However, ammonia and mineral oil have poor miscibility, thus the lubricating oil will deposit on the inner wall of the evaporator in the evaporation process, and consequently will cause heat transfer deterioration. Thus, attention needs to be paid to three issues when applying ammonia into a dry type evaporator:

1. Use the machine oil which can be miscible with ammonia and can be used for the freezing condition.
2. Adopt the efficient heat transfer tube, for example the metal tube plated copper outside.
3. An indirect cooling system should be adopted to prevent the influence of the ammonia leakage of the system.

For the oil used for the freezing condition, Maekawa Company in Japan produced PAG (polyalkylene glycol) miscible with ammonia as the machine oil, which is utilized in the dry type ammonia evaporator. The condition of the oil is good even after 3000 hours of operation.

7.3.2.3 Aluminum Air Cooler for Ammonia

For an evaporator using air as a cooling medium the design of the adsorption refrigeration unit can also refer to the compressed refrigeration units, that is, utilizes the traditional ammonia evaporator. For example, the evaporator used the row of steel pipes with a diameter of

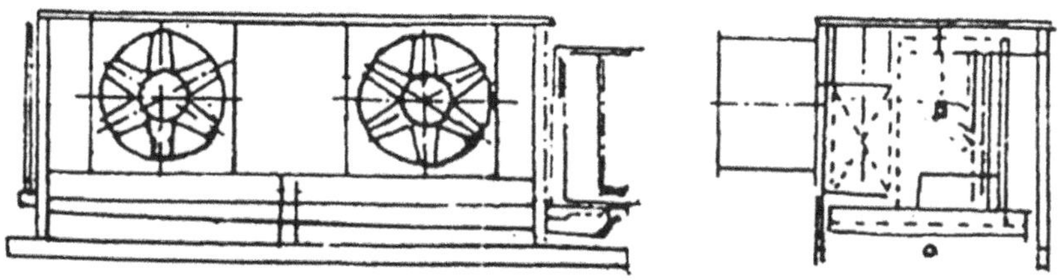

Figure 7.16 Shape of ammonia chiller [33]

φ51–57 mm and is cooled by a natural convection, and inside the tubes the gravity is used for the tube to provide the ammonia liquid. The problem for such a type evaporator is the slow cooling rate for cold storage, uneven storage temperature, difficulties of defrosting process, and a big workload during installation. Generally, the forced air cooling coil is used, that is, the cooling fan. A traditional ammonia cooling fan always has the steel plate casting outside steel pipe and is galvanized, commonly the steel pipe diameter is 16–25 mm, and the pitch of fins is 6–12 mm because for the temperature drop of 0 °C frost might form on the fin.

In Figure 7.16 the structure of an ammonia chiller is almost the same as the chiller with copper tubes and aluminum. Ammonia goes into the cooling coil through the dispenser side. Ammonia liquid flows from upwards to downwards, and gradually evaporats, thus preventing retention of lubricating oil in the tube. Usually the diameter of the aluminum tube is 16–19 mm as well as the space between two tubes being 50–60 mm. Aluminum is pressed and forms a spiral groove tube to improve the heat transfer coefficient at the ammonia side. Connection of the elbow may be manufactured by argon arc welding or brazing processes. Such a type of air cooler commonly will need the hot air to melt the frost, and for such a process the hot air should be blocked outside the circuit for the cold storage vessel by adding an air damper before the fan. Due to the poor miscibility between mineral oil and ammonia, such coolers generally use polyethylene glycol oil. The cooling fan is always the type with axial flow. Through stability test results it is shown that the corrosion of tubes for such a type of cooler is very small when ammonia, polyethylene glycol oil, water, and air coexist.

7.3.3 The Elements for the Control of the Flow

In the adsorption refrigeration system, the flow regulating valve is located in the middle of the condenser and the evaporator, which plays two roles. Firstly, it separates the high pressure part of the refrigerator with the low pressure part to prevent the high-pressure steam flowing into the evaporator. Secondly, it regulates the flowing process of the refrigerant liquid from the condenser to the evaporator, which is similar to the valve in the compressed refrigeration system. It is a main component for operating the system continuously. The throttle valve of the compressed refrigeration system can regulate the refrigerant flow according to the change of temperature at the inlet or outlet of the evaporator or variation of superheat degree that reflects the changes of the external thermal load. Unlike the compressed refrigeration system, the flow regulating valve of the adsorption refrigeration system is also needed to change the regulating process with the desorption process. This is mainly because the desorption quantity of the adsorption bed is influenced by the heat source temperature in the working process, and

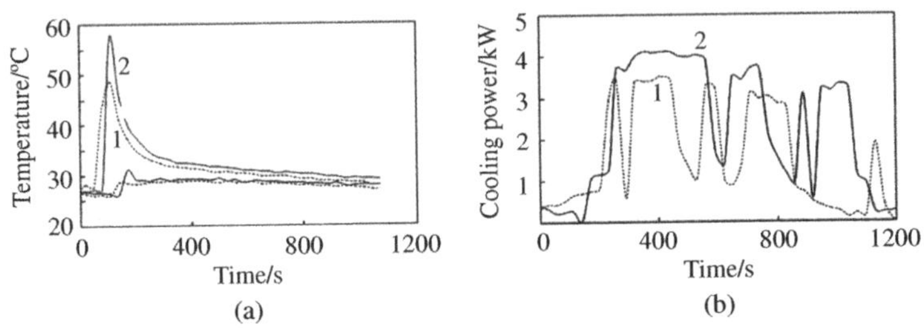

Opening stroke of flow adjustment valve, 1 - 2 mm; 2 - 5 mm

Figure 7.17 Comparison of the temperature for cooling water at the inlet and outlet of condenser [28]. (a) Temperature difference between inlet and outlet of condenser and (b) transient cooling power of working system

it will change with the changing temperature of the adsorption bed, thus the flow through the controlling valve is also subject to the influence of the temperature of the external heat source.

For the impact of the flow control valve on regulating the performance of the system, J Y Wu [28] tests a continuous adsorption air conditioning/heat pump system with heat recovery process. The flow regulating valve in the system utilizes a 2 mm needle valve which can adjust the stroke of 10 mm. The flow regulating valve adjusts the stroke of 2 mm (condition 1) and 5 mm (condition 2), respectively. The change of the condenser cooling water inlet temperature and the cooling capacity curve for different conditions are shown in Figure 7.17a,b. Results showed that the cooling water of condenser under condition 2 takes away more cooling power. The result also showed that for the working condition of the valve with a large opening degree, the desorption amount of system is relatively large. The reasons are mainly because opening the flow regulating valve drives the flow of the refrigerant, which evacuates the condenser space and consequently promotes the desorption rate of the adsorption bed. Thus, it makes a good foundation for the next adsorption process. From Figure 7.17 we can see that the transient cooling power of condition 2 is greater than that of condition 1.

Generally the performance of the adsorption refrigeration system is effectively improved by the degree of opening. But degree of opening of the flow regulating valve cannot be too large otherwise it will connect the condenser and the evaporator and consequently will influence the pressure in the condenser and evaporator. Such a process will maintain the evaporation temperature at a high level.

For an ammonia adsorption refrigeration system when using flooded evaporator and forced-circulation evaporator, the ball float valve is used between condenser and evaporator to control the liquid level of the evaporator to guarantee the normal operation of the evaporator. If using a dry-type evaporator, due to the great difference between condensing pressure and evaporation pressure, the throttle valve has to be adjusted according to a large pressure difference to ensure that the liquid ammonia is the mixture of cryogenic gas and liquid after throttling. For such a process the condenser requires a reservoir to be added between the condenser and the evaporator to ensure the liquid ammonia is throttled, otherwise the throttling valve won't fulfill the throttling process effectively and the cooling effect cannot be sufficiently obtained.

Taking the alternating heat pipe type compound adsorption refrigerator as an example, the application of the electronic expansion valve in the system is shown in Figure 7.18, for which the refrigerant is ammonia and the adsorbent is the composite adsorbent of the calcium chloride. An adsorption refrigeration system is composed of two adsorption beds, a heating

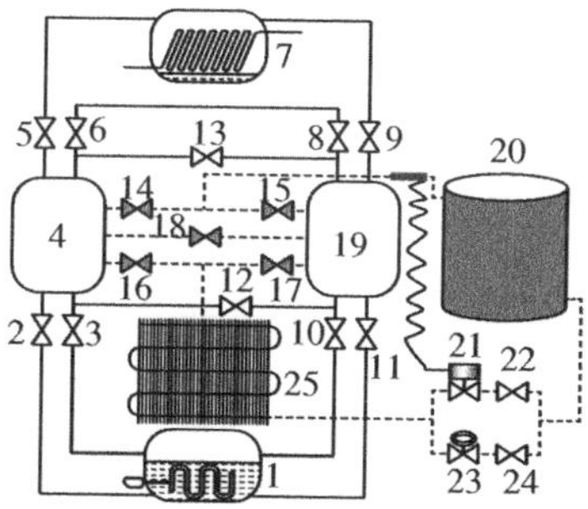

1 - Heat boiler; 2, 3, 5, 6, 8, 9, 10, 11, 12,13 - The valves for the working fluid in the heat pipe;
4, 19 - Adsorption bed; 14, 15, 16, 17, 18 - Valves for refrigerant; 20 - Ice maker;
21 - Thermal expansion valve; 22, 24 - Shut-off valves; 23 - Manual needle valve; 25 - condenser

Figure 7.18 Composite adsorbent-ammonia adsorption refrigeration system

boiler, an ice making machine, a condenser, and a cooler. The adsorption bed is heated by the steam in a boiler by the condensation process inside the pipes in an adsorption bed with the heat pipe principle. The adsorption bed is cooled by evaporation of the liquid in the pipes in the adsorption bed, and the evaporated vapor is condensed in the cooler and then flows back to the adsorption bed. Design of the heat pipe type adsorption refrigeration equipment will be described in detail in Chapter 8. The ice maker in the system is a dry-type evaporator. In order to compare the application effects between electronic expansion valve and the needle valve in a dry-type ammonia evaporator, two circuits are connected between condenser and ammonia ice maker. One is an electronic expansion valve control loop, and the other is a needle valve control loop. When using the electronic expansion valve to control the cooling performance of the system, close the shut-off valve in the needle valve loop. When using the needle valve to control the cooling performance of the system, close the electronic expansion valve in shut-off valve loop.

The typical diameter of needle valve used in the system is 15 mm. Repeated experiments were performed on the needle valve and the results showed that the optimum degree of opening is 1.5 mm. Under the condition of 29 °C cooling water temperature, −10 °C evaporation temperature, and 150 °C maximum heating temperature, the change of the absolute vapor pressure is shown in Figure 7.19 for the adsorption refrigeration system. Figure 7.19 showed that when the needle valve is used for the adjustment the absolute evaporation pressure drops sharply at first, and then rises slightly. This is mainly because at the beginning the adsorption bed

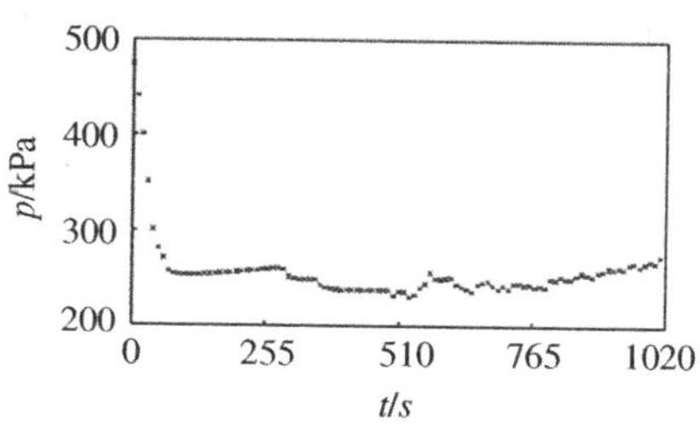

Figure 7.19 Evaporator pressure of ammonia under the condition of that the system is adjusted by the needle valve

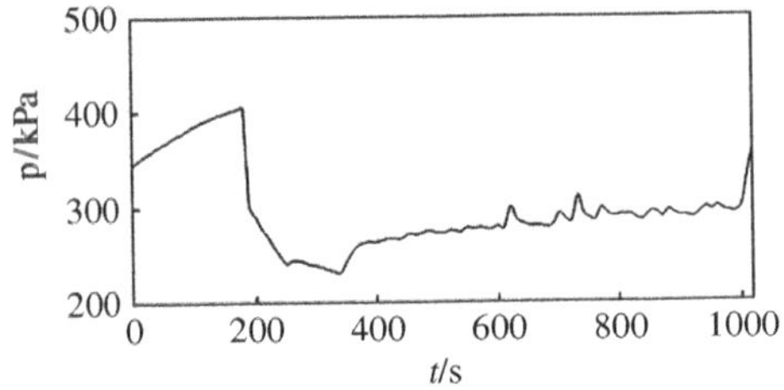

Figure 7.20 The pressure of the ammonia evaporator under the condition of that the thermal expansion valve is used as the regulating valve

has a strong ability to adsorb because it just completes desorption. With the development of the adsorption process, the adsorption capacity of adsorption bed is gradually weakened, and consequently the absolute evaporation pressure is increased slightly with the adsorption time.

By using the thermal expansion valve and setting the evaporating temperature at −10 °C, the flow of the expansion valve is adjusted according to the difference between the set evaporating temperature and the outlet temperature of the ice maker (ammonia evaporator). Under the conditions of 29 °C cooling water temperature and maximum 150 °C heating temperature, the trend of the evaporation pressure is shown in Figure 7.20. In the operation process the opening degree of thermal expansion valve changes frequently because the performance of the adsorption bed is at the unsteady state, consequently, as shown in Figure 7.20 the pressure of the evaporator fluctuates. By such a process the performance of the adsorption bed is reduced, and the evaporating temperature rises, which leads to a large difference between the set evaporating temperature and the outlet temperature of the evaporator, and consequently increases the opening degree of the thermal expansion valve. In this case, the amount of ammonia in the evaporator increases, but due to the decline of the adsorption capacity of the adsorption bed at this time, the increased ammonia in the evaporator cannot be adsorbed by the adsorption bed in time, which leads to the rising of the outlet temperature of the evaporator again, and consequently the opening degree of thermal expansion valve increases once more. In Figure 7.20 the evaporation pressure rises continually. Such a result indicates that the thermal expansion valve of compressed systems cannot be used for the adsorption refrigeration system.

Through the above analysis it can be found that compared with the thermal expansion valve used in the compression system, the needle valve is more suitable for an ammonia adsorption refrigeration system. But the shortcoming of needle valve is that it cannot adapt the adjustment to the change of the non-equilibrium parameters of the adsorption system. It can be used for the laboratory prototypes, but it will be difficult for the optimization of the performance if the adsorption refrigeration system is commercialized. This indicated that a special kind of expansion valve needs to be developed for an ammonia adsorption refrigeration system. Since the pressure of adsorption bed in the system has been at a non-equilibrium and non-steady state, and the pressure of the adsorption bed directly reflects the adsorption capacity of the adsorption bed, the ammonia expansion valve is preferably controlled by the pressure difference other than the temperature difference. Then when the capacity of the adsorption bed declines, pressure of the adsorption bed increases, which will result in a small pressure difference between evaporator and condenser, and consequently will reduce the opening degree of the expansion valve.

7.4 Operation Control of Adsorption Refrigeration System

The operation of an adsorption refrigeration system is mainly related to the program control system, energy regulation system, security system, and computer control system. The program control system includes the normal start up program, normal and improper shutdown procedure, and so on, which are necessary for keeping the system to work properly. The energy regulation system guarantees a normal output of the cooling capacity, as well as regulating the cooling capacity to fit with the external heat load. A security system ensures timely operation under the improper conditions. A computer control system monitors the state parameters of the units, ensures security, regulates the energy, and controls the program functions. The computer control system is a command center of the entire system.

7.4.1 Brief Introduction on Adsorption Refrigeration System and Its Energy Regulation System

Here we will take the operation control of a two-bed continuous adsorption refrigeration system with heat recovery process as an example. A diagram of the system is shown in Figure 7.21. The system comprises two adsorption beds, a condenser, an evaporator, a cooler, and a heat source. Heat source and related valves are used to heat the adsorption bed, and the cooling process can be achieved through the cooler and relating valves.

The control process of the energy regulation system is used to describe the continuous adsorption system with heat recovery process as shown in Figure 7.21. The processes for the control can be divided into two parts. One part is for the control of heating and cooling processes, and the other part is to balance the cooling capacity of the system and the demand of external cooling capacity.

The first part of the control mainly includes control of heating and cooling medium for adsorption bed as well as control of adsorption, desorption, and associated valves. For the operation of the system, in order to cool and heat two adsorption beds, respectively, as well as switch the system for continuous work, a number of shut-off valves must be installed on the

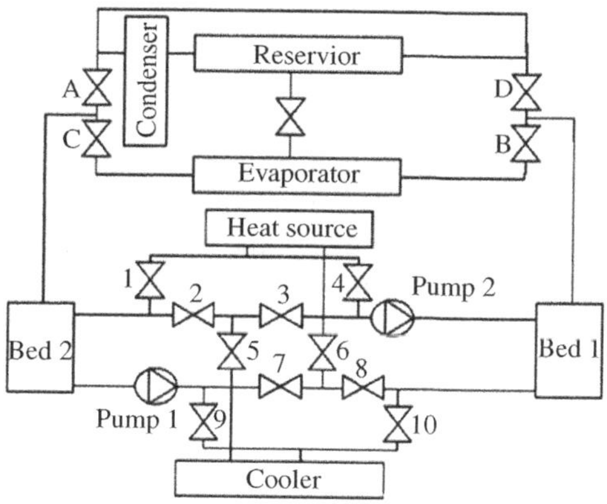

Figure 7.21 Diagram of adsorption refrigeration system [28]

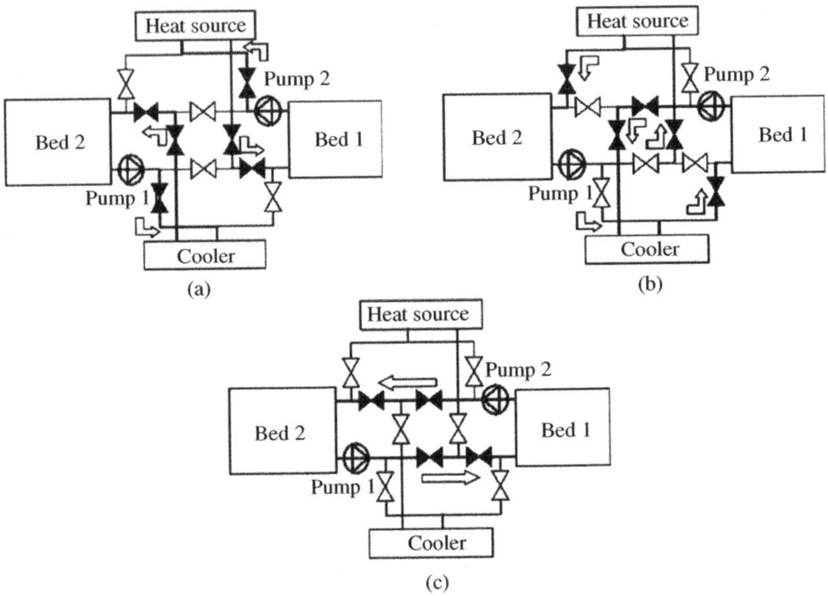

Figure 7.22 Working statuses of 10 control valves and two beds [28]. (a) Bed 1 is heated and bed 2 is cooled; (b) bed 2 is heated and bed 1 is cooled; and (c) heat recovery process between two beds

pipes. Through the coordination of these shut off valves, heating, cooling, and heat recovery processes can proceed. There are ten valves for heating, cooling, and heat recovery processes, and the installation positions of them are shown in Figure 7.21. Three working states will be achieved by these valves. Firstly there is the process shown in Figure 7.22a, and in the figure bed 1 is heated and bed 2 is cooled. For second process bed 2 is heated, and bed 1 is cooled, which is shown in Figure 7.22b. The last process is the heat recovery process between two beds that is shown in Figure 7.22c.

The control of the system depends on the control of the heating, cooling, and heat recovery time, as well as the adsorption and desorption processes, which could ensure the normal cooling capacity output. The adsorption bed needs to be connected with the condenser when it is heated, and it needs to be connected with the evaporator when it is cooled. But at the beginning when the adsorption bed begins to be heated we cannot connect the bed and the condenser until the pressure in the bed reaches the condensing pressure. Similarly we cannot connect the bed with the evaporator at the beginning of the cooling process until the pressure in the bed decreases to the evaporation pressure. For the two-bed system four valves are required for the control of heating, cooling, desorption, and adsorption processes, respectively. The installing positions of four valves are shown in Figure 7.21. At the switch time valves A, B, C, D are all closed. When bed 1 is heated and desorbed, bed 2 is cooled and adsorbed; valve states are shown in Figure 7.23a. When bed 2 is heated and desorbed, bed 1 is cooled and adsorbed; the valve states are shown in Figure 7.23b. The control system can control the adsorption bed to connect or disconnect with the condenser or evaporator according to the working status of the adsorption bed, the pressure within the bed, the condensing pressure, and the evaporation pressure.

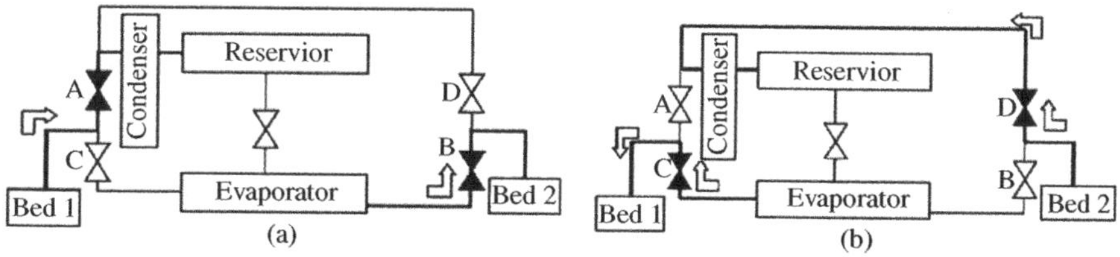

Figure 7.23 Working status of valves A, B, C, D and adsorption bed [28]. (a) Bed 1 desorbs and bed 2 adsorbs and (b) bed 2 desorbs and bed 1 adsorbs

The second part is the energy adjustment of the unit, that is, to ensure that the required external heat load and the cooling capacity of the system are matched. Take the continuous adsorption chiller with heat recovery process as an example, to control the energy the temperature of cold water needs to be controlled. For example, the flow rate of the heat source needs to be reduced (for example by reducing the flow of steam or hot water) when the exit temperature of cold water is below the required value. Whereas the flow rate of the heat source needs to be increased if the cold water temperature is greater than the required value. It needs to be emphasized that for a single bed adsorption refrigeration system the cooling capacity is not generated continuously, therefore, the effect cannot be obtained immediately when the heat quantity for the adsorption bed is adjusted, and generally the effect can be obtained by the next cycle.

7.4.2 Security System

The system security includes the security of the various components that can be summarized as follows.

1. **Adsorption beds.** Adsorption beds need to be protected from the high temperature in the heating process because each working pair has its working temperature range. Design for the adsorption bed is determined by the required temperature range of the working pairs. Furthermore, for either the vacuum system or the pressure system, ultra high-pressure in the adsorption bed must be avoided. The main reasons for ultra high-pressure phenomena includes the over large opening for the heating medium of heat source, non-condensable gases in the system, higher cooling water temperature, and so on. The methods used to protect the high pressure include the detection of the temperature and pressure of the adsorption bed in the heated process, and start the alarm signal when they exceed a predetermined value as well as turning off the heat source at the same time. For the adsorption bed in the cooling process the temperature of cooling water needs to be tested and protection needs to be carried out if the parameters exceed the preset values, which will be further described in the section on the security of the condenser.
2. **Evaporator.** The evaporator unit is the equipment of refrigeration production and cooling capacity output. For adsorption chillers they use water as a refrigerant, so the harm caused by the freezing process of water needs to be considered. Usually when the unit operates under normal conditions, the amount of the cold taken away by the secondary refrigerant should match the cooling capacity of the units to maintain the stable temperature of the

evaporator and cold water temperature. Once the cooling quantity is taken away by the second refrigerant is less than the cooling capacity of the unit, the temperature of the cold water will gradually reduce, and the freezing phenomenon will occur when it reduces to the freeze point, and consequently will lead to the frost crack on the evaporator tubes. Such a process will cause a major accident. Generally, two reasons can cause such phenomena. Firstly, if the external heat load is far less than the cooling capacity of the unit and the heat source isn't sufficient to adjust the temperature it will lead to the decline of the cold water temperature. Secondly, the equipment failure will decrease the temperature of the cold water, such as the sudden failure of the cold water pump, the inability to open the valves on the pipes of the chilled water system, too many impurities in the pipeline; as a result the flow of cold water will decline to a rate of 50% or less. Therefore, protection measures need to be adopted for the above two cases. Firstly, to collect the data of cold water temperature with a sensor, cut off the valve between the adsorption bed and the evaporator to stop the adsorption refrigeration when the temperature is below a certain value. Then reconnect the adsorption bed with the evaporator and continue the adsorption refrigeration process after the temperature rises. Secondly, install the flow controller on the cold water pipes, the flow rate controller will act and the alarm signal will start to stop the cooling operation of the unit when the water flow rate is less than 50% of the rated flow. The unit will be restarted when the flow is restored to more than 65% of the rated flow and the failure is settled.

3. **Condenser.** Refrigerant vapor desorbed by the adsorption bed is condensed into a liquid by the condenser, and cooling water plays an important role in this process. Once the temperature of cooling water is too high or is dried up, the unit will stop working. Therefore a couple of methods are used to control the cooling water temperature and flow channels. The first method is to install a flow controller on the cooling water pipes, when the cooling water flow rate is decreased to a certain value (e.g., reduced to less than 70% of the rated value), the alarm signal is activated at the same time as the heat transfer medium is cut off to stop the unit. The second method is to detect the import or export temperature of the cooling water, and start the alarm signal as well as to cut off the heat transfer medium if the temperature is found to be too high.
4. **Pump.** Over current protection of various circulating pump, shield pump, and fan needs to be adopted in the system.

7.4.3 Program Control System

The program control system consists of the normal starting up procedures, recirculation starting up procedures, shutdown procedures, normal shutdown procedures, and fault shutdown procedures. Take the continuous adsorption refrigeration system with heat recovery process, for example, to describe various procedures for controlling the processes as follows:

1. **Normal starting up procedures.** Press the button for normal start up, then the state of the bed at the start of the operation is determined by the state at the end of the last operation. If at the end of the last operation the state of bed 1 is desorption, then at the start the bed state is adsorption. The state of bed 2 will be the opposite. Turn on the associated valves of the

heating medium of the beds so that each bed gets into the heating or cooling processes. In this process the bed at the desorption state will communicate with the heat source, and the bed at the desorption state will communicate with the cooling source. Start the power device on the pipes of the heating or cooling media (pump or fan). (Note: Prior to starting various power equipment we need to complete the detection on the power equipment failure), then open the cooling water pump and open the pump for chilling medium. When the adsorption bed reaches a certain pressure open the valve between the bed and the condenser to start the desorption process. When the adsorption bed is cooled down to a certain pressure open the valve between the evaporator and the bed to begin the adsorption refrigeration process. Subsequently, the unit generally will output the refrigeration quantity normally following the second or third cycle.

2. **Recirculation start up procedures.** The unit re-starts up from the suspended state when the cold water outlet temperature is higher than a certain value. The initial state of the bed and maintaining cycle time can be derived from the suspended states of the unit. For example, if the system suspends the work due to the lower cold water temperature last time, and the desorption and adsorption processes of the adsorption bed are not completed according to a predetermined cycle time, the recirculation work must continue on the last cycle so as to ensure the unit gets into a stable state quickly. Firstly, start the power device for the heating transfer media, then open the valves between the adsorption bed and the evaporator. When the temperature of the desorbing bed increases to a certain value open the valve between the condenser and the adsorption bed, and the whole unit gets back to the normal operation of the cycle.
3. **Normal shutdown procedures.** For this process press the shutdown button to cut off the heat source of the system. Then determine the states of the adsorption beds and maintain the time of the cycle, which is essential for the starting up process of the next time. For example, if bed 1 is at the desorption state in the cycle, and the duration for desorption process is shorter if compared with the cycle time, then the bed will be determined at the adsorption state. If the duration time for the desorption process is longer if compared with the whole cycle time, then we can take the desorption state as the final state of the adsorber. Cut off the power equipment on the heating medium pipes, and then close off the valves between the adsorption bed and condenser or adsorption bed and evaporator. Cut off the power equipment on the cooling medium pipes, and switch the valves of the heating and cooling medium pipe to the state of heat recovery. When the heat recovery process finishes, close all the valves on the heating and cooling medium pipes, and turn off the cooling water pump and the refrigerant pumps and stop the operation on the unit.
4. **Recirculation shutdown procedures.** The unit will get to the suspend state when the cold water outlet temperature falls below a certain value. It is the same with the normal shutdown procedures of the unit, the duration of the cycle and the states of adsorption bed are recorded, and the turn down power equipment on the pipeline for heating the adsorption bed. Close valves between the adsorption bed and condenser and between the adsorption bed and evaporator, and get into the standby state.
5. **The fault shutdown procedures.** When the system fails, the security system is active and the system will stop working. Cut the heat source, start up the sound and light alarms, and display the reasons for the failure in the control system. The following procedures will be the same as the normal shutdown procedures.

7.4.4 *The Computer Control System*

The computer control system is the control center of the system. According to the detection and control of the system requirements, the functions of the computer control system can be divided into detection function, memory function, forecasting function, and executive function. A block diagram of computer control system is shown in Figure 7.24.

1. **Detection function.** The unit can fulfill the monitoring process of the working conditions, the control of the parameters, the diagnosis of the fault, and the security protection. The computer control system can also detect and display the main parameters for various parts of the unit. The main parameters which need to be detected are temperature, pressure, flow, and so on. The operating states of the unit can also be monitored, including the running state of adsorption beds, open state of valves, dynamic flow chart of parameters, cold water pump operation, chilling water pump operation, and failure monitoring process, and so on.
2. **Memory function.** The computer control system sets the data storage unit for storing important operating data, which facilitates the management of the unit, summary of the operating experience and trend analysis of the unit operation, and so on. The information stored in the computer unit includes the working principle of the unit, the basic method of operation, and maintenance methods. Users can refer to them at any time. The data record includes the total running time, operating parameters, the numbers of system failure, the details of the failure, and the specific parameters of the failure. In addition, the record of the data could provide the trend of the unit operation, such as the trend for the heating and cooling process of the adsorption bed, and the degree of the adsorption and desorption processes.
3. **Forecasting function.** The computer control system adds a prediction function to the failure of the system that is known as a fault management system in order to let the unit operate in a more safe and reliable way. The fault management system can predict the fault position, analyze the reasons of the fault, and suggest the methods for handling the faults; such a system could process the failure more efficiently as well as improve the efficiency and operational reliability of the unit. Computer fault diagnosis is divided into two types. The first type is the direct diagnosis that could give the appropriate conclusions according to the test of the main parameters of the system and compare the detected values with the set values. For example, when the temperature of the adsorption bed is too high, or the temperature of cold water is too low, or the cold water needs to be shut off, or the chilling water needs to be shut off. The second type is the indirect diagnosis that achieves the fault prediction function by acquiring parameters under several typical conditions, and calculating

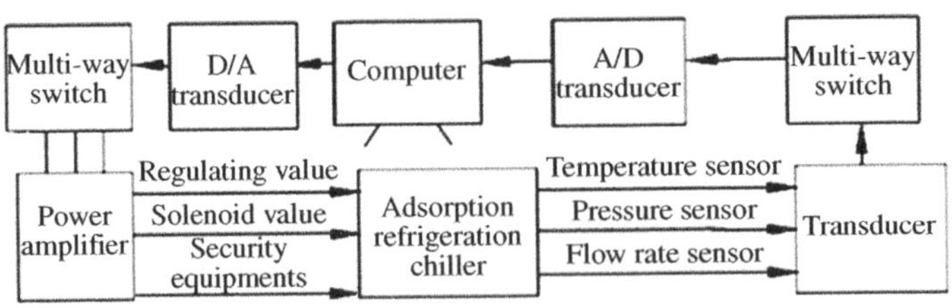

Figure 7.24 Diagram of computer control system

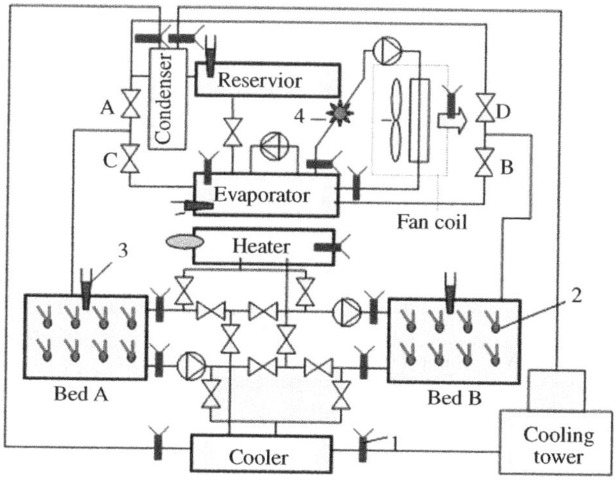

Figure 7.25 Distribution of sensors in the system [28]

the system by those data and the historical records. This approach can comprehensively analyze the system, and comprehensively evaluate a number of components to ensure that various components of the unit are in the best state, so as to prevent accidents. To fulfill such a type of indirect diagnostic function we must have a deep understanding of the coupling between the various parameters and the components of systems.

4. **Executive function.** An executive function of a computer control system includes control of the implementation of various components of the unit as well as the implementation of the safety control of various components. Both energy control systems and security systems are realized through the computer control system by the executive function.

A continuous adsorption air conditioning/heat pump unit with heat recovery process is taken as an example to describe the realization of the computer control system. The flowchart is shown in Figure 7.25. The main equipment of the system comprises two adsorption beds, a condenser, an evaporator, a heater, a cooler, and a circulating pump. The system is powered by the hot water in the heater. The cooler and the condenser are cooled by the cooling water circuit. The refrigeration quantity is transported by the circulation of the coolant water, and output by the two-fan coil.

According to the working media the system can be divided into four main loops.

1. The heating circuit of the adsorption bed that provides the driving heat to the system by the external heat source.
2. The cooling water circulation loop to ensure rapid heat releasing process to the environment.
3. Refrigerant circulating circuit (in the system the refrigerant is methanol) for which the refrigerant circulates at the different states and evaporates in the evaporator to provide the refrigeration output.
4. The chilling water circuit to ensure the cooling capacity of evaporator timely delivered and the stability of the evaporation temperature.

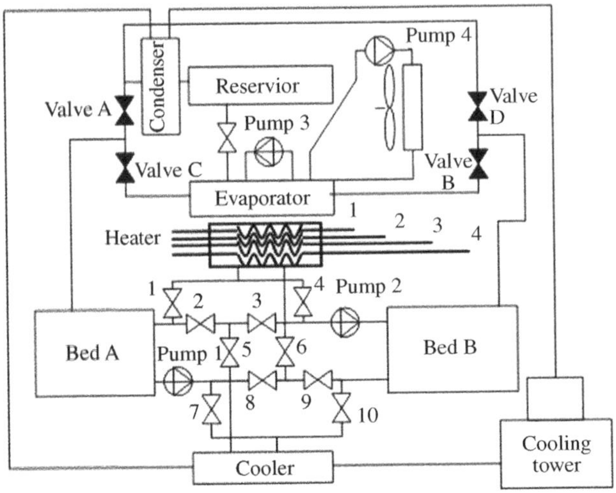

Figure 7.26 Distribution of control equipments in the system [28]

For the design of the detecting function of the system various components of the detection system and the working status of working medium are considered. These parameters are the basis for fulfilling the control, fault diagnosis, and security protection. Figure 7.25 shows the sensor arrangement of the system, which includes the temperature sensors, pressure sensors, and the flow sensors of the unit. The parameters monitored include the inlet and outlet temperatures of the heat source, inlet and outlet temperatures of the chilling water, the inlet temperature of the cooling water, the temperature of the adsorption bed, and the pressure of the bed, condenser, and evaporator.

The distribution of the system control devices is shown in Figure 7.26. The multiple control devices that could fulfill the automatic control and manual control is adopted in order to make the debug and operation process easier. The control equipments include pump 1 for heating and cooling process of adsorption bed A, pump 2 for heating and cooling processes of adsorption bed B, evaporator spray pump 3, driving pump 4 for the chilling water circuit, valve A between adsorption bed A and the condenser, valve C between adsorption bed A and evaporator, valve D between adsorption B and the condenser, valve B between adsorption bed B and the evaporator, valve 1–10 for the heating and cooling processes of the adsorption beds. On the control panel, 18 equipments can be operated by manual control, and the manual control can be switched into the automatic control. The lights for the operation conditions and the operating current are distributed on the panel, which makes the operation process of the boiler easier.

As mentioned above, the part of the control can be divided into five units.

1. Automatic control of the heat source (including four groups of heating wires in the electric boiler).

   The automatic control of the heat source aimed at controlling the outlet temperature of hot water, which is able to provide the heat source with steady or variable temperature. The heating desorption process is a non-steady process, thus it requires the different heat quantity at different desorption stages. Therefore the quantity of the heat needs to be changed to match the operation process.

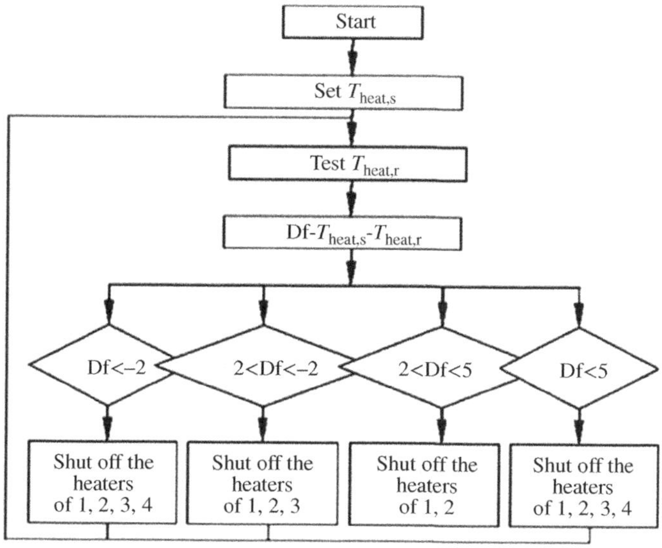

Figure 7.27 Computer control block diagram of isothermal heat source [28]

The control strategy for a total heating unit divides the whole system into four control units, respectively. The corresponding small unit is started according to the control requirements needed to start up all the heating wires together in the heating process to improve accuracy for the control.

Figure 7.27 shows the computer control block diagram of a steady temperature heat source. The control program can supply the required desorption heat of the adsorption bed as soon as possible. $T_{heat,s}$ in the diagram is the setting temperature of the heat source, and the $T_{heat,r}$ is the measured temperature of the heat source.

2. The automatic control of the fan flow (including four states: high speed, medium speed, low speed, and stop).

   Since the cooling capacity of the adsorption refrigeration system changes all the time, it is necessary to control the output of the fan coil to have a stable evaporation temperature of the system, that is, the cooling capacity of the fan coil needed to fit with the cooling capacity of the evaporator. The fan speed of the fan coil is controlled at four levels, and they are high speed, medium speed, low speed, and stop. The fan speed is controlled according to the level of evaporation temperature.

3. The automatic control of the valves on the heating and cooling pipes of adsorption bed.

   Heating and cooling processes of the adsorption bed can be realized, respectively, by the valve control of heating and cooling pipe. Generally there are three processes. The first process is to let bed A be heated for desorption, bed B to be cooled for adsorption (valve 1 opens and valve 2 closes). The second process is to let bed B be heated for desorption and bed A be cooled for adsorption (valve 1 opens and valve 2 closes). The third process is to connect two beds for heat recovery (valve 1 opens and valve 2 closes). Control of 10 valves can be determined according to the degree of desorption as well as adsorption of adsorption beds. There are two methods for judging the degree of adsorption or desorption.

The first method is by the observation of the actual operation temperature and pressure of the adsorption bed and the second method is by the cycle time based on the understanding of heat and mass transfer performance of adsorption beds.

4. Automatic/manual control of the solenoid valve for vacuum systems (including four valves separately for connecting the adsorption beds with the condenser and evaporator).

   The valves are controlled according to the values of the pressure in the adsorption bed to determine if it reaches the condensing/evaporating pressure or lower/higher than the condensing/evaporating pressure, and then open the valves automatically. In addition, control of the valves must be implemented on the basis of the above valve control for heating and cooling processes.

5. Automatic and manual control of the pump (including two heating pumps, one circulation pump for cooling process, a refrigerant circulating magnetic pump, and a pump for the chilling water).

   Control of water pump 1 and pump 2 is periodic and is determined by the operating state of the adsorption beds. The spray pump is normally open for the whole cycle. The control of the pump for chilling water can be determined by the operating state of the adsorption bed as well as the external heat load.

   Based on the operating characteristics of the system, the following security features are added in the computer control system: the protection for the sensors under the abnormal states; the protection on the heat source for high temperature and high pressure conditions, thermal relay protection of various pumps, the protection on the adsorption beds under the condition of high pressure; the protection on the condenser under the condition of high pressure, the protection on the chilling water pipes at the low temperature, and the shut-off procedures for the cooling water, and so on.

References

[1] Wang, R.Z., Wu, J.Y., Xu, Y.X. and Wang, W. (2000) Performance researches and improvements on heat regenerative adsorption refrigerator and heat pump. *Energy Convention and Management*, **2**, 233–249.

[2] Wang, L.W., Wang, R.Z., Wu, J.Y. and Wang, K. (2004) Adsorption performances and refrigeration application of adsorption working pair of $CaCl_2$-NH_3. *Science in China, Series E*, **47**(2), 173–185.

[3] Wang, L.W., Wang, R.Z., Wu, J.Y. and Wang, K. (2004) Compound adsorbent for adsorption ice maker on fishing boats. *International Journal of Refrigeration*, **27**(4), 401–408.

[4] Groll, M. (1992) Reaction beds for dry sorption machines. Proceedings of Solid Sorption Refrigeration Symposium, Paris, France, pp. 225–232.

[5] Sahnoune, H. and Grenier, P. (1989) Mesure de la conductivite thermique d'une zeolithe. *The Chemical Engineering Journal*, **40**(1), 45–54.

[6] Gurgel, J.M. and Grenier, P. (1990) Mesure de la conductivite thermique du charbon actif AC-35 en presence de gaz. *The Chemical Engineering Journal*, **44**, 43–50.

[7] Mauran, S., Prades, P. and Haridon, F.L. (1993) Heat and mass transfer in consolidated reacting beds for thermochemical systems. *Heat Recovery Systems and CHP*, **13**, 315–319.

[8] Pons, M. and Dantzer, P. (1994) Heat transfer in hydride packed beds, II. A new experimental technique and results on $LaNi_5$ powder. *Zeitschrift für Physikalische Chemie*, **183**, 213–223.

[9] Meunier, F. (1998) Solid sorption heat powered cycles for cooling and heat pumping applications. *Applied Thermal Engineering*, **18**, 715–729.

[10] Miles, D. and Shelton, S. (1996) Design and testing of a solid-sorption heat-pump system. *Applied Thermal Engineering*, **16**, 389–394.

[11] Wang, W., Wang, R.Z., Xu, Y.X. *et al.* (1998) Investigation on adsorption refrigeration with single adsorption bed. *International Journal of Energy Research*, **22**(13), 1157–1163.

[12] Wang, R.Z., Xu, Y.X., Wu, J.Y. and Wang, W. (1998) Experiments on heat regenerative adsorption refrigerator and heat pump. *International Journal of Energy Research*, **22**(10), 935–941.
[13] Qi, C.H., Tang, G.F., Li, D.Y. and Huang, D.K. (2002) System analysis and experimental study on triple effect cooling cycle adsorption system based on low grade heat recovery. *Fluid Machinery*, **30**(5), 45–48, ISSN: 1005–0329 (in Chinese).
[14] Coste, C., Mauran, S., and Crozat, G. (1983) Procede de mise en oeuvre de reaction gaz-solide. US Patent 4 595 774.
[15] Wang, S.G., Wang, R.Z. and Li, X.R. (2005) Research and development of consolidated adsorbent for adsorption system. *Renewable Energy*, **30**, 1425–1441.
[16] Bauer, J., Herrmann, R., Mittelbach, W. and Schwieger, W. (2009) Zeolite/aluminum composite adsorbents for application in adsorption refrigeration. *International Journal of Energy Research*, **33**(13), 1233–1249.
[17] Jakob, U. and Mittelbach, W. (2008) Development and investigation of a compact silica gel/water adsorption chiller integrated in solar cooling systems. International Seminar of Heat, Pumps, Refrigerators, and Power Sources, Minsk, Russia.
[18] Dawoud, B., Vedder, U., Amer, E.H. and Dunne, S. (2007) Non-isothermal adsorption kinetics of water vapour into a consolidated zeolite layer. *International Journal of Heat and Mass Transfer*, **50**(11), 2190–2199.
[19] Dunne, S. (1996) Carousel heat exchanger for sorption cooling process. US Patent 5 503 222.
[20] Vasiliev, L.L. (2003) Sorption refrigerators with heat pipe thermal control. Proceedings of The International Conference on Cryogenics and Refrigeration, Hangzhou, China, pp. 405–415.
[21] Xia, Z.Z., Wang, R.Z., Wu, J.Y., and Wang, L.W. (2004) Composite alternating heat pipe generator driven by low grade heat. Invention Patent of China 00410018291.3.
[22] Wang, L.W., Wang, R.Z., and Wu, J.Y. (2003) Marine adsorption ice maker similar to separated heat pump. Invention Patent of China 200310108924.5.
[23] Xia, Z.Z., Wang, R.Z., Wu, J.Y., and Wang, D.C. (2004) New efficient and reliable adsorption refrigerator by separated heat pipe. Invention patent of China 200410025398.0, 2003.
[24] Wang, R.Z., Wu, J.Y., Xu, Y.X. *et al.* (1998) Experiment on a continuous heat regenerative adsorption refrigerator using spiral plate heat exchanger as adsorbers. *Applied Thermal Engineering*, **18**(1–2), 13–23.
[25] Wang, R.Z., Wu, J.Y. and Xu, Y.X. (1999) A continuous heat regenerative adsorption refrigerator using spiral plate heat exchanger as adsorbers: improvements. *ASME Journal of Solar Energy Engineering*, **120**(1), 14–19.
[26] Restuccia, G., Recupero, V., Cacciola, G. and Rothmeyer, M. (1988) Zeolite heat pump for domestic heating. *Energy*, **13**, 333–342.
[27] Tchernev, D.I. and Emerson, D.T. (1998) High efficiency regenerative zeolite heat pump. *ASHRAE Transactions*, **94**, 2024–2032.
[28] Wu, J.Y. (2000) Cycle characteristics and experimental research on adsorption air conditioning/heat pump units with continuous heat recovery. PhD Thesis. Shanghai Jiao Tong University, Shanghai, China (in Chinese).
[29] Teng, Y., Wang, R.Z. and Wu, J.Y. (1997) Study of the fundamentals of adsorption systems. *Applied Thermal Engineering*, **17**(4), 327–338.
[30] Wang, L.W., Wu, J.Y., Wang, R.Z. and Xu, Y.X. (2004) Cycle characteristics analysis of t activated carbon – methanol adsorption ice maker. *Journal of Engineering Thermophysics*, **25**(2), 208–210, ISSN: 0253-231X (in Chinese).
[31] Yang, L.M. (1997) Study on solid adsorption refrigeration system with heat recovery. Master Thesis. Shanghai Jiao Tong University, Shanghai, China (in Chinese).
[32] Shen, P.W. (2003) Analysis for the problems of column pipe bursting in ammonia evaporator. *Chemical Production and Technology*, **10**(3), 36–37, ISSN: 1006–6829 (in Chinese).
[33] Zhou, Q.J. (1998) New development of ammonia refrigerating heat exchange. *Journal of Refrigeration*, (1), 36–39, ISSN: 0253-4339 (in Chinese).

8

Design and Performance of the Adsorption Refrigeration System

The performance of an adsorption refrigeration system is closely related to the design of it. Different design will have different heat and mass transfer performance, as well as different refrigeration performance. The design of the adsorption bed concerning the heat transfer has already been introduced in Chapter 7, and the technologies for the intensification of the heat transfer are mainly as follows: to extend the heat exchange area, to intensify the heat transfer coefficient, and to use the technology of the heat pipe to intensify the heat transfer performance of the fluid. In this chapter, the research done by Shanghai Jiao Tong University (SJTU) in the past 15 years are introduced. Several different adsorption refrigeration systems are analyzed in this chapter, as well as the heat transfer performance, mass transfer performance, and refrigeration performance being discussed.

8.1 Adsorption Chiller Driven by Low-Temperature Heat Source

For the design of the adsorption bed, the heat transfer area is extended for the beds of adsorption chiller driven by the low-temperature heat source to enhance the heat transfer performance [1]. The design of the system has the following merits if compared with the traditional adsorption chillers:

1. There are few valves for the refrigerants of water in the system. Such a design improved the reliability of the system as well as simplifying the maintenance process if compared with the traditional adsorption refrigeration system.
2. The novel design of the evaporator simplified the chilling water circuit for exchanging the refrigeration power from the evaporator, and again improved the reliability and minimized the loss for the cooling capacity in the evaporator.
3. Had a good performance when it was driven by the low temperature heat source, and could work effectively when the temperature of the heat source ranged between 65 and 85 °C.

Adsorption Refrigeration Technology: Theory and Application, First Edition. Ruzhu Wang, Liwei Wang and Jingyi Wu.

Companion Website: www.wiley.com/go/wang/refrigeration

Experiments on the system also showed that the heat and mass transfer performance of the system could meet the actual needs and the performance of the refrigerator could meet the actual requirements.

There are two suitable occasions for the application of the adsorption chiller driven by low-temperature heat source. One is for the waste heat, and the other is for the recovery of the solar energy. When the adsorption refrigeration system is driven by the waste heat the system will require less coefficient of performance (*COP*) because the heat amount is sufficient, but it will require higher specific cooling power (*SCP*) somehow because the volume of the system needs to be decreased effectively. However, when the system is driven by the solar energy or other limited heat energy it will require higher *COP* as well as higher SCP. Considering both occasions, the developed adsorption chiller has the cooling power of 10 kW for air conditioning condition. The *COP* of the system is higher than 0.4, and the performance of the system can be adjusted to adapt the different requirements for *SCP* and *COP*.

8.1.1 Choice of Adsorbent

For the design of the adsorption bed firstly the suitable adsorbent needs to be selected according to the application conditions, and then the structure of the adsorbent bed and the system needs to be determined, which could enhance the heat transfer performance in accordance with the selected adsorbent.

The adsorption chillers driven by the low-temperature heat source are mainly used under the condition that the temperature of the heat source is below 100 °C. The ideal adsorption refrigeration working pair is silica gel–water under such a condition. In the chiller the particulate silica gel with the spherical pore is selected, and the grain diameter of the silica gel is 0.5–1 mm. The fine-pored silica gel, also known as A-type silica gel, includes fine-pored spherical silica gel and fine-pored massive silica gel, and they are transparent or translucent glassy. The physical properties of selected silica gel are shown in Table 8.1.

8.1.2 The Innovation Design of the System and Refrigeration Cycle

For the design of the adsorption chiller driven by the low-temperature heat source using the working pair of silica–water, considering that the evaporation pressure of the refrigerant, which is water, in the adsorption process is low, the most important thing for the design should be to improve the mass transfer performance of the system as well as to ensure the operational reliability of the system.

Table 8.1 The physical properties of selected silica gel

Average pore size (nm)	Specific surface area (m^2/g)	Pore volume (ml/g)	Specific heat (kJ/(kg·°C))	Thermal conductivity (W/(m·°C))	Bulk density (kg/m^3)	Actual density (kg/m^3)	Apparent density (kg/m^3)
2.0–3.0	650–800	0.35–0.45	0.92	0.175	790	2200	1170

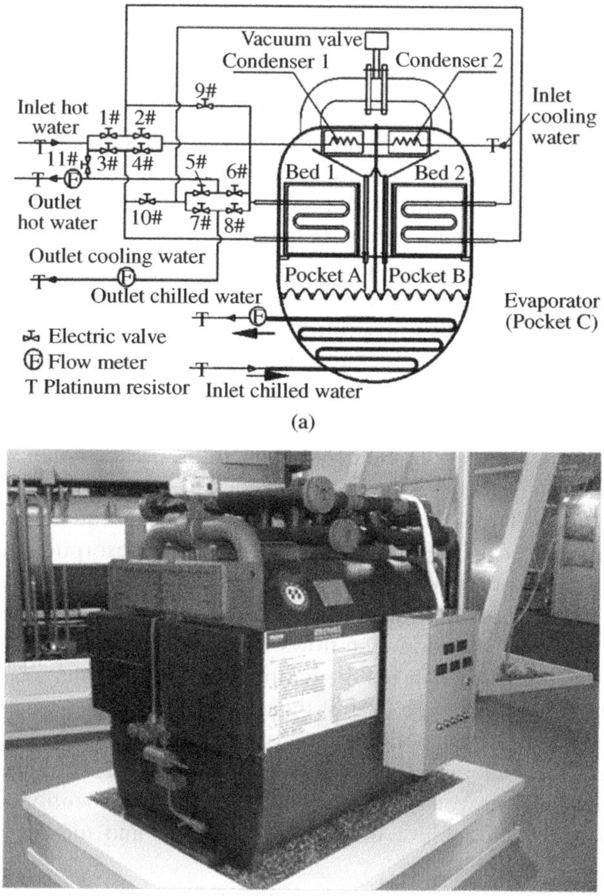

Figure 8.1 Silica gel–water adsorption chiller

For the operation process of the system reducing valves can effectively improve the reliability. For this reason and referring to the design of the literature [2] the developed silica-gel–water adsorption chiller (shown in Figure 8.1) includes two beds, two evaporators and two condensers. There are two adsorption/desorption vacuum chambers and every one had a condenser, an adsorption bed, and an evaporator. Every chamber is an independent adsorption refrigeration unit, and every adsorption refrigeration unit didn't have valves between the bed and the condenser as well as between the bed and the evaporator. The cross-sectional area between the evaporator and the adsorption bed and between the adsorption bed and the condenser is big, so it can ensure the mass transfer performance of the system. A vacuum valve for mass recovery is installed outside the adsorption/desorption chambers, by which the mass recovery process can proceed. In order to reduce the complexity of the chilling water circuit for the evaporator [2] and improve the reliability of the whole system, two adsorption/desorption vacuum chambers tightly integrated together through a gravity heat pipe type evaporator. The gravity heat pipe only can transfer the heat in one direction, by this property the evaporator that combined three evaporating heat exchangers is designed to convert the alternating refrigeration output of two beds into the continuous refrigeration output in one methanol evaporator. Three evaporating heat exchangers involve two symmetrical adsorption/desorption vacuum chambers at the top and a heat pipe heat exchanger at the bottom. The optimal working fluid for the heat pipe needs to have low latent heat of evaporation and high evaporation pressure, thus the methanol is chosen. For the design of the evaporator two water-evaporating heat exchangers were merged into a methanol evaporator, thus the chilling water circuit only needs to exchange the heat with the methanol evaporator, which makes the system extremely simple.

As mentioned above, the heat pipe evaporator actually is a combination of two water-evaporators and a heat pipe heat exchanger (methanol evaporator), and the working processes are as follows: When adsorption process proceeds in one vacuum chamber, water was evaporated on the surface of the water evaporator, which is also the condensing part for the heat pipe type evaporator of methanol. Then the temperature there will decrease because of the evaporation heat. When the temperature is lower than that of the working fluid in the evaporation section of the methanol evaporator, the methanol will evaporate from the evaporation section and then condense at the condensation section, and such a process will exchange the refrigeration quantity from the water evaporator to the methanol fluid circuit for outputting the refrigeration power. Meanwhile the desorption proceeds in another vacuum chamber. For this process the desorbed water vapor from the bed condenses in the condenser into the high temperature water. The condensate water is distributed through the liquid traps and liquid distributor and then flows to the water evaporator. In this process the water evaporator is heated, which makes the temperature of the water evaporator that is also the condensing part of the heat pipe higher than the evaporation section of the heat pipe, therefore, the methanol in the evaporator won't evaporate because there is no temperature potential for driving the evaporation, then the heat exchange between the water evaporator and the methanol evaporator won't happen. By the processes mentioned above the novel design of the evaporator can achieve the thermal isolation and meanwhile can output the refrigeration power effectively.

In order to make the arrangement of various components in the refrigerator reasonable and compact, the layout of a condenser, a bed, and an evaporator in a vacuum chamber is vertically from the upper side to the lower side, which satisfies the requirements of a compact structure, as well as making the level of the condensate liquid in the liquid distributor keep at a certain level, which will make the liquid uniformly distributed. The mass transfer channels in the vacuum chamber create the large cross-section; by such a structure the flow rate of the water is low, which will reduce the vapor pressure drop thereby will reduce the loss of the cooling capacity. Taking into account that the actual flow rate of the cooling water for the condenser cannot be as big as the condenser requires, and the cross section area that for mass transfer will be reduced by the installation of a condenser with a large condensing area, the connection way of the condenser is in series instead of in parallel.

The entire refrigeration system can be looked upon as composed by the cooling water system, hot water system, and chilled water system. The cooling water first flows into two condensers, and then flows through the switching valves of two beds to enter the adsorption bed. The hot water flows directly into the adsorption bed through switching valves. By the thermal isolation function of the heat pipe, the whole evaporator only has a chilled water inlet and a chilled water outlet, and the chilled water directly circuits between the methanol evaporator and the user side. Such a process won't need any switching valves. Therefore, compared with the system in the literature [2], it greatly simplifies the system and improves the reliability.

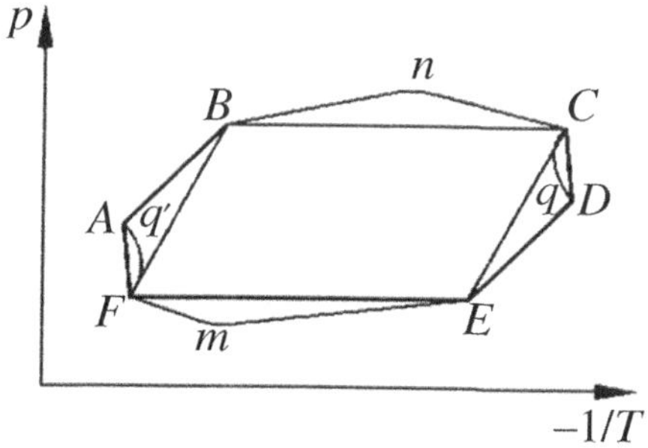

Figure 8.2 *p*-*T* diagram of the cycle

According to the characteristics of the structure of the adsorption chiller the heat and mass recovery cycle is chosen, but it is different from the traditional heat and mass recovery cycle. For one vacuum chamber the working processes are similar to the mode of basic intermittent cycle. But if we take two vacuum chambers as a whole, it is a kind of heat and mass recovery cycle. The pressure in adsorption/desorption chambers always changes during the whole cycle, as the process of A → B → n → C → q → D → E → m → F →q′ → A shown in Figure 8.2. Therefore, the cycle can be looked upon as double basic intermittent cycles combined with a heat and mass recovery process. In Figure 8.2, the process of A → B → C → D → E → F → A is the ideal process for the combination of double basic intermittent cycles and one heat and mass recovery process. The pressure change is linear. B → C → E → F → B is for the basic intermittent cycle. For completing one cycle the chiller needs six working processes.

1. The left bed is in desorption process and the right bed is in adsorption process (bed 1: B → n → C; bed 2: E → m → F). When the hot water is switched to the left bed through the valves, the temperature of the adsorbent rises in the left bed and the vapor pressure rises as well in the left chamber until the saturated vapor pressure reaches the condensing pressure, and then the left condenser begins to work. For the left evaporator, because its temperature rises to the condensation temperature which will be higher than the temperature of the working fluid in the methanol heat pipe evaporator, thus the heat exchanging process won't happen between the left water evaporator and the methanol evaporator. Meanwhile, the cooling water flows into the right bed, and it begins to be cooled and adsorb. The water vapor pressure of the right vacuum chamber decreases until the saturated vapor pressure reaches the evaporating temperature, and the water on the surface of the right evaporator begins to evaporate and the temperature quickly drops. When the temperature is below the temperature of the methanol in the methanol evaporator, the methanol at the evaporation part of the heat pipe type evaporator will evaporate and will condense on the condensing part of the heat pipe that is also the lower part of the water evaporator. By such a process the refrigeration power is output.
2. The mass recovery process from left bed to right bed (bed 1: C → q → D; bed 2: F → q′ → A). When the adsorption/desorption processes for both beds complete, the mass recovery vacuum valve is open, then the water vapor of the left chamber will rapidly flow to the right chamber under the condition of great pressure drop. By such a process the temperature of the left water evaporator will decrease while the temperature of the right water

evaporator will increase. The pressure of the left and right chambers will quickly reach the equilibrium state. Meanwhile, the water vapor desorbed from the left bed flows into the right bed through the mass recovery valve and will be adsorbed by the adsorbent there, which can be looked upon as the second adsorption and desorption processes. For both water evaporators, because their temperature is higher than that of the methanol in the methanol evaporator, they will be thermal isolated with the methanol evaporator.

3. The heat recovery process from left bed to right bed (bed 1: D → E; bed 2: A → B). When the pressure in left and right chambers reaches the equilibrium pressure, close the mass recovery vacuum valves. Then bypass the hot water and open the corresponding valves between two beds for fulfilling the heat recovery process. By this process the cooling water in the right bed flows into the left bed to exchange the heat with the hot water there. Then the temperature of the right bed will increase, and the temperature of the left bed will decrease.
4. Right bed desorbs and left bed adsorbs (bed 2: B → n → C; bed 1: E → m → F). Similar to the processes (1), the right bed desorbs by heating process, and the desorbed refrigerant vapor condenses in the condenser. The left bed adsorbs by the cooling process, and the refrigerant evaporates in the left evaporator.
5. The mass recovery process from right bed to left bed (bed 2: C → q → D; bed 1: F → q′ → A). Similar to the processes (2), the left bed adsorbs the water vapor desorbed from the right bed to achieve the second adsorption and desorption processes, and the temperature of the right evaporator rises while the temperature of the left evaporator decreases.
6. The heat recovery process from right bed to left bed (bed 2: D → E; bed 1: A → B). Similar to process (3), the remaining hot water of the right adsorbent bed flows into the left adsorbent bed until the temperatures of both beds are in equilibrium.

From the working processes of the adsorption chiller it can be seen that although the efficiency of the system will be lowered by two water evaporators and two condensers because they only work for half-cycle time, the reliability of the system is improved quite a lot because the switching valves are much less than the traditional adsorption refrigeration chillers. Similarly the gravity heat pipe type evaporator isolates the heat exchange between the high temperature water evaporator and the methanol evaporator, which reduces the number of valves and also reduces the loss of the refrigeration power, as well as improves the reliability of the system.

8.1.3 Design of the System Components

The main components of the adsorption chiller driven by low-temperature heat source are adsorbent beds, condenser, and evaporator. The condenser is a shell and tube type heat exchanger. The main innovations are in the design of the adsorption beds and the evaporator. The adsorption beds use the extended heat transfer area to enhance the heat transfer performance while the evaporator uses dual heat pipes to fulfill the thermal isolation among the water evaporators and the methanol evaporator.

8.1.3.1 Design of the Adsorbent Bed

The adsorption bed is a fin-tube heat exchanger. It uses the intensive fins to extend the heat transfer area and the space between two fins is only 2.5 mm. The thickness of the adsorbent

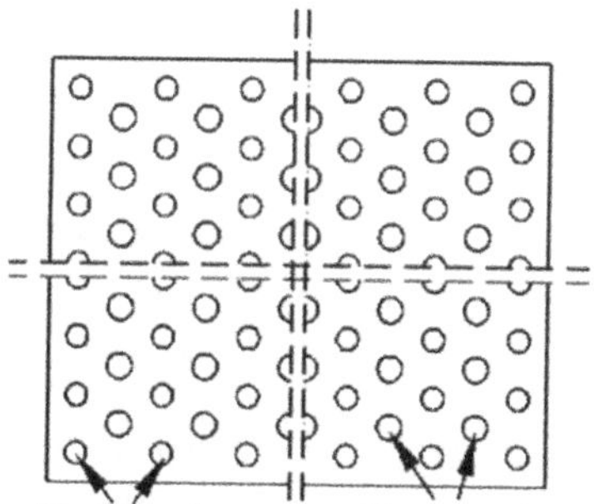

Figure 8.3 The diagram for the structure of the adsorbent bed

Table 8.2 The properties and parameters of the adsorbent bed

Parameters	Values			Unit
	Fin	Heat transfer tube	Adsorbent	
Mass	10	21.8	52	kg
Material	Pure aluminum	Red copper	Silica gel	°C
Size	δ0.12	Φ9.52 × 0.65	Φ0.5–1	mm
Thickness of the adsorption layer	2.5			mm
Mass transfer channel	Φ11			mm
Ratio of metal heat capacity	0.336			
Size of adsorbent bed	1050 × 320 × 300 (Length × Width × Height)			mm

on the fins is only 1.25 mm, so the thickness for heat transfer path is only 0.625 mm between the adsorbent and the fins. For mass transfer enhancement, the array of heat transfer tubes and mass transfer channels is alternatively arranged as shown in Figure 8.3. The largest thickness for mass transfer process, which is from the outside edge of the heat transfer tube to the outside edge of the mass transfer channel, is only 12.5 mm. The maximum average flow velocity of the refrigerant vapor in the adsorbent bed is 0.4 m/s, and it is 24.5 m/s in mass transfer channels. Under such conditions, the maximum pressure loss of the airflow caused by the mass transfer resistance inside the adsorbent bed is 2.3 Pa.

The parameters for the adsorbent bed are shown in Table 8.2. Calculation results showed that the overall heat transfer coefficient of the adsorbent bed is up to 1231 W/(m^2·°C) if the heat transfer resistance of the metal is neglected when the adsorbent bed is heated.

8.1.3.2 Design of the Evaporator

The evaporator is the heat pipe type evaporator under the condition of low pressure. Generally there are two kinds of evaporators for sorption refrigeration systems, and they are falling-film

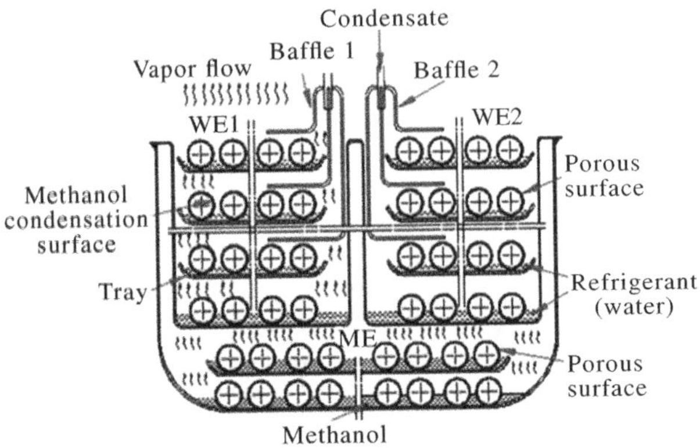

Figure 8.4 The schematic of heat pipe evaporator

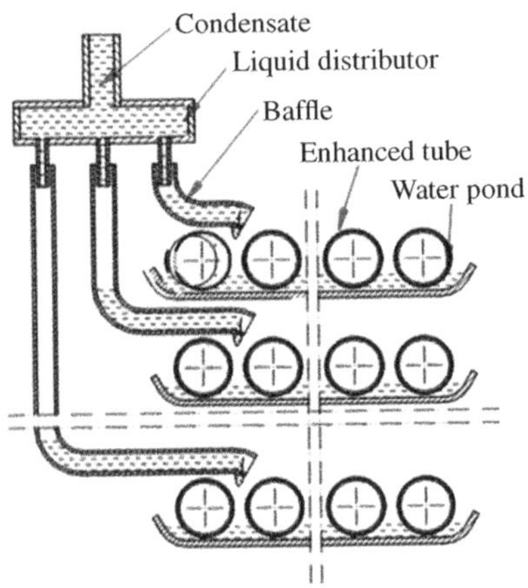

Figure 8.5 The principle of water evaporator

evaporator and climbing-film evaporator. Due to low power density of the adsorption refrigeration system, the cycle quantity of the refrigerant is small, so it is difficult to let the liquid uniformly distribute in a falling-film evaporator. Adding the refrigerant pump can make the situation better, but it will increase the complexity and the cost of the system, also the system will be less reliable. Therefore, the developed evaporator used a climbing-film evaporator.

The evaporator uses the outside rolling wire tubes as heat transfer tubes, and the capillary function of the evaporating surface is used for the enhancement of the heat transfer performance. The finned ratio of the tube is 7.49. The schematic of the evaporator is shown in Figure 8.4, and the principle of the water evaporator is shown in Figure 8.5.

In Figure 8.4 WE1 and WE2 are the refrigerant (water) evaporators which are also the condensation sections of the heat pipes, and ME is methanol evaporator which is also the evaporating section of the heat pipe. The water evaporator is composed by the semi-immersive trays, and the pipes are horizontally arranged in different layers of water tray. The numbers of water tray and heat transfer tubes are chosen by the cooling power. Condensate is uniformly arranged to water trays through the liquid distributor by gravity with the ratio of 1 : 1. About two-thirds below the centerline of pipes is immersed in the refrigerant liquid, and the capillary function of the wire rolling surface makes the refrigerant full of the evaporation surface, as shown in Figure 8.5. A side of the tubes is the inlet of methanol vapor, and the other side is used to collect methanol condensate. Methanol evaporator is the immersion evaporator, and the heat transfer tubes are made by the metal of brass, which is the same as the adsorbent bed and the condenser. The structure of the evaporator is multi-row and multi-slice. Chilled water flows in from the bottom of heat transfer tubes and flows out from the upside heat transfer tubes. Therefore, the temperature at the bottom of the heat transfer tubes is high, and the temperature of the upside heat transfer tubes is low. The liquid distribution at the top of the heat transfer tubes depends on the two rows of tubes between the heat transfer tubes of the upside and low side layers. The principle of the liquid distribution is similar with that of the thermosyphon heat pipe. The liquid distribution of the intermediate layer depends on the overflow at the top of the tray and the confluence of condensate. Due to the flow passages of the methanol vapor and the methanol condensate being separated, the cycle operates as a heat pipe loop.

Figure 8.6 is the schematic of a heat pipe loop in the evaporator.

From Figure 8.6 it can be seen that the working principle and the thermal isolation process of the evaporator are as follows: methanol absorbs the heat from the chilling water and begins to evaporate. Due to the downcomer having the function of the liquid sealing, the methanol at the bottom of the heat transfer tube boils and forms bubbles which are discharged by a riser pipe. A gas-liquid two-phase flow is generated in the discharging process of bubbles in the riser pipe, which pushes the methanol liquid flow to the tray. Meanwhile the methanol vapor flows to the water evaporator. A part of the methanol in the tray is evaporated outside of the heat transfer tubes, and another part of it overflows to interlayer heat transfer tubes and supplies the losses in the lower layer heat transfer tubes due to evaporation. Methanol vapor generated in a methanol evaporator will condenses at the surface of the tubes in lower-temperature water evaporator, thus transfers the evaporation heat from the water evaporator to the chilled water. The condensate collected will flow to the methanol evaporator tray through a U-tube. Meanwhile, another water evaporator doesn't work due to the desorption of the bed. The refrigerant liquid distributor disposes condensate uniformly into the water tray of the water evaporator.

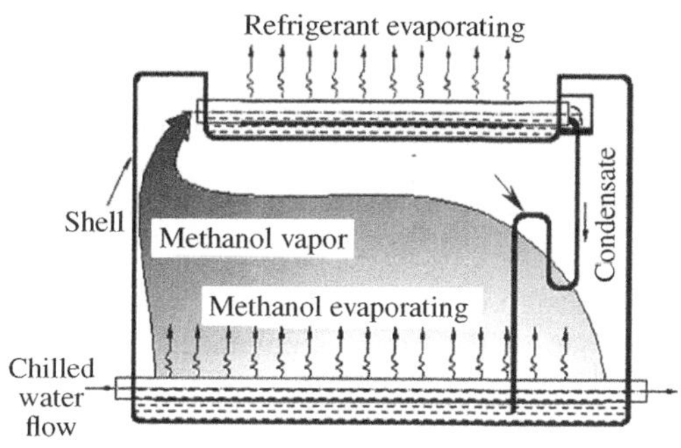

Figure 8.6 The schematic of heat pipe loop inside the evaporator

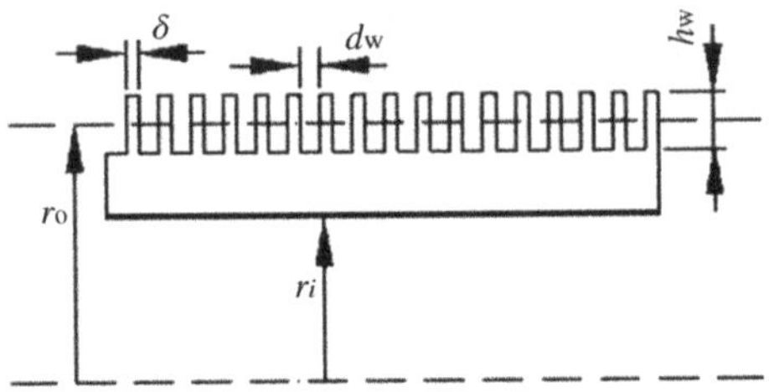

Figure 8.7 Physical model for heat exchanging process at the outside surface of the tube

During this period, the water evaporator is heated by the condensate and makes the temperature higher than that of another water evaporator as well as the methanol evaporator, therefore, the water evaporator neither exchanges heat with water evaporator or with methanol evaporator, that is, it is thermal isolated with the cold water evaporator and as well as the methanol evaporator. Hence, the water evaporator at this stage also can be known as a heat isolator.

Figure 8.7 is the physical model for heat exchanging process at the outside surface of the tubes, in which the rolling wire tubes are simplified into the rectangular low fin-tube type. The capillary pressure difference of the water in the capillary core of the rectangular channel is calculated by the following formula [3]:

$$\Delta p_{e,\max} = \frac{2\sigma}{d_w} \tag{8.1}$$

where σ is the tension force at the liquid surface (N/m); d_w is the channel width (m).

The maximum capillary height is:

$$H_{\max} = \frac{2\sigma}{d_w \rho_L g} \tag{8.2}$$

The evaporative thermal resistance of the liquid film is very small, and the main source for the evaporative thermal resistance is from the heat conductive resistance inside the liquid film. The entire outside surface of the wire rolling tube is normally covered by the liquid film, and the liquid film at the top of the fin is thinner than that between two fins. Postulate that the thickness of the liquid film at the top of the fin is the same as that between two fins, the equivalent thickness of the liquid film is:

$$\delta_{eff} = \frac{d_w h_w}{2h_w + d_w} = \frac{0.5 \times 0.3}{2 \times 0.5 + 0.3} = 0.115\ \text{mm} \tag{8.3}$$

Therefore, the evaporative heat transfer coefficient outside of tube is:

$$\alpha_{we,vap} = \frac{\lambda_L}{\delta_{eff}} = \frac{0.59}{0.115 \times 10^{-3}} = 5130\ \text{W}/(\text{m}^2 \cdot {}^\circ\text{C}) \tag{8.4}$$

Part of the parameters for the design of the evaporator are shown in Table 8.3.

The heat transfer performance of the tubes is enhanced by inserting the elements inside the tubes. There are mainly four methods for this technology [4]: To form a rotating flow, to produce the metathesis between the center fluid and the fluid near the tube wall, to partition the fluid and destruct the boundary layer, and to form a secondary flow. The common interpolation elements are ring, disc, ball, coil springs, spiral sheet, net, twist belt, drilling

Table 8.3 Part of the parameters for the design of the evaporator

Heat exchange components		Temperature difference for heat transfer (°C)	Heat transfer area (m^2)	Heat transfer coefficient ($W/(m^2{\cdot}°C)$)
Water evaporator	Water evaporation side	0.3	2.6	5130
	Methanol evaporation side	0.7	2.8	6890
Methanol evaporator	Methanol evaporation side	2	1.83	2586
	Chilled water side	1	3	3350

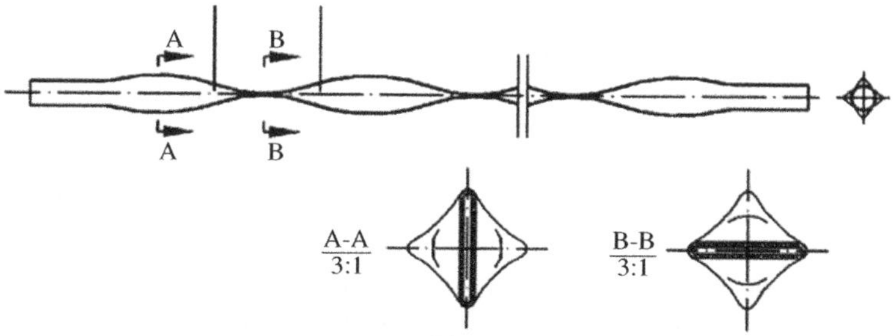

(a)

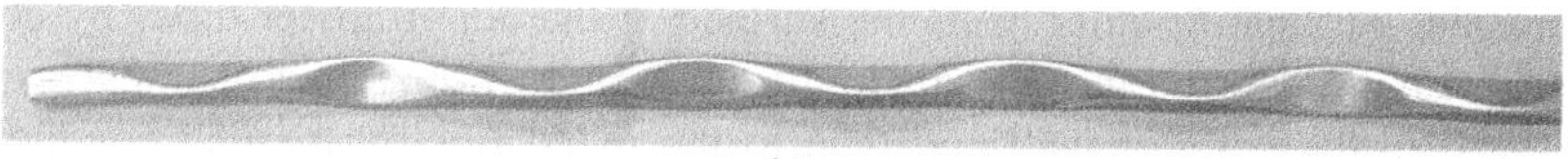

(b)

Figure 8.8 The developed turbulators for enhancing convective heat transfer performance. (a) Schematic and (b) photograph

distortion band, interval distortion band, capillament and cross serrated band, and so on. The developed turbulators which can enhance convective heat transfer performance at the chilled water side for chillers driven by low temperature heat source is shown in Figure 8.8. The reference [4] provided the test data of similar types of distortion effect on enhancing heat transfer performance, and it is predicted that the heat transfer coefficient could be improved more than 40% to reach up to 3000 $W/(m^2{\cdot}°C)$ in the range of Re = 6000–10000.

8.1.4 System Simulation

The mathematical model used the lumped parameter method. In order to simplify the calculation, we postulate that:

1. The temperature and vapor pressure inside the adsorbent bed are uniformly distributed.
2. The refrigerant is adsorbed uniformly by the adsorbent bed and it is liquid inside the bed.

3. The pressure drop between the adsorption bed and the condenser or evaporator can be ignored.
4. The thermal conductive process of the metal wall between the bed and the condenser as well as that between the bed and evaporator can be ignored. Two water evaporators are fully thermal isolated.
5. Inside the heat pipe evaporator the heat transfer is absolutely one-way, that is, it can only transfer the heat from the water evaporator to the chilling water.
6. In addition to heat exchanging processes between the hot water, cooling water, chilled water, respectively, with the environment, the heat loss from the system to the environment can be ignored.

8.1.4.1 Heat Transfer Processes of the Components

1. Heat transfer analysis of the adsorption bed.
For the calculation of the heat transfer process inside the adsorption bed, we only consider the convective heat transfer of one side fluid and the thermal conductive process of another side of adsorbent, which is porous material and is filled inside the adsorption bed. The heat transfer process of the adsorbent bed is a very complex thermal conductive process, which is accompanied by natural convection, forced convection, and radiation heat exchanging process, and so on. In the calculation these processes need to be simplified. The thermal conductivity of the adsorbent is very small, so we can think it is similar to the heat-insulating material, and the heat transfer resistance of the adsorbent bed is mainly at the adsorbent side. Take the heating process of the adsorption bed as an example, firstly, when the adsorbent bed is heated, the heating fluid heats the surface of the heat transfer tubes by convection and the heat is transferred to the wall fouling by conduction. Then, it transfers from the interior wall to the exterior wall and then to the metal fins by thermal conduction, and then goes through the metal wall and the adsorbent particles, between which the thermal resistance needs to be considered. Finally, the heat is delivered to the adsorbent bed mainly by conduction. Therefore, the heat transfer coefficient of the adsorbent bed can be obtained by Equation 8.5:

$$\frac{1}{\alpha_{adb}A_f} = \frac{1}{\alpha_f A_f} + R_f + \frac{1}{\alpha_m A_m} + R_t + \frac{1}{\alpha_{a,eff}A_{a,eff}} \tag{8.5}$$

The left side of the equation is the total thermal resistance of the adsorbent bed, and for the right side, the first part is the convective heat transfer thermal resistance between the heating/cooling fluid and the metal wall of the bed. The second part is the thermal resistance of the fouling between the fluid and the metal wall. The third part is the equivalent thermal resistance of the adsorbent bed metal, including the thermal resistance of inside wall and the fins. The fourth part is the thermal contact resistance between the metal wall and the adsorbent particles. The fifth part is the equivalent heat exchanging thermal resistance in the adsorbent. A_f, A_m, $A_{a,eff}$ are the heat transfer areas at the heating/cooling fluid side, the metal wall at the adsorbent side, and the effective heat transfer area for adsorbent, respectively. The effective heat transfer area in the adsorbent can be obtained by the area of fins in the adsorbent bed A_{fin} and the effective heat transfer area of the tubes. α_f, α_m, $\alpha_{a,eff}$ are the convective heat transfer coefficients at the fluid side, the metal wall of the adsorbent bed,

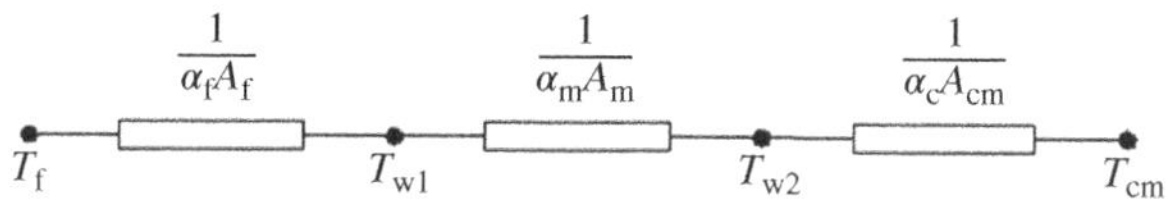

Figure 8.9 The network for the heat transfer process in condenser

and in the adsorbent, respectively. The heat transfer process between the fluid and the metal wall of the bed is turbulent convective heat transfer process.

2. Heat transfer analysis of condenser.
   As the normal shell-tube type condenser with fluorescent tubes for heat transfer, the network for heat transfer process of condenser is shown in Figure 8.9. α_f and A_f are the heat transfer coefficient and heat transfer area of cooling water. α_m and A_m are the heat transfer coefficient and heat transfer area of metal. α_c and A_c are the heat transfer coefficient and heat transfer area at the condensation side. α_f, α_m, and α_c can be obtained by the equations provided in the literature [5]. T_f, T_{w1}, T_{w2}, and T_{cm} are the temperature at the cooling water side, the metal side close to water, the metal side close to the condensation side, and the condensation side, respectively.
3. Heat transfer analysis of evaporator.
   Since the structure of the evaporator is gravity heat pipe type heat exchanger in the adsorption chiller, the location of evaporator is lower than the condenser, and it has an inclination of 90°. By such a structure the condensate can flow back to the evaporator by gravity. The gravity heat pipes don't need the die to supply the capillary force to drive the condensate flow back, so there isn't the phenomenon of drowning, liquid level fluctuations, and impact convection. The working process is only limited by the boiling process of the liquid.
   The heat transfer process inside the entire heat pipe evaporator can be represented by the physical model for the heat transfer process of the evaporator shown in Figure 8.10.
   If we neglect the heat transferred by the connection between the condensing part and the evaporating part of the heat pipe, seven heat transfer resistances need to be considered from the chilled water at the temperature of T_{chill} to the refrigerant vapor at the temperature T_{ew}. The equations and instructions are shown in Table 8.4. Therefore, the overall heat transfer coefficient of the heat pipe evaporator is:

$$\alpha_{eva} = \frac{1}{A_{evf}\sum_{i=1}^{7} R_i} \tag{8.6}$$

8.1.4.2 Mathematical Models

Adsorption equation in the simulation is the non-equilibrium adsorption Equation 3.75 of the working pair of silica gel–water.

The simulation of the chiller is different from that of the conventional adsorption refrigeration system. For the chiller the adsorbent bed, condenser, and water evaporator are in the same chamber, and the adsorbent bed is always connected with the evaporator and the condenser. Because the temperature of the water evaporator or condenser is constantly changing, the adsorption and desorption pressure of the bed in the chiller is also always changing. Especially

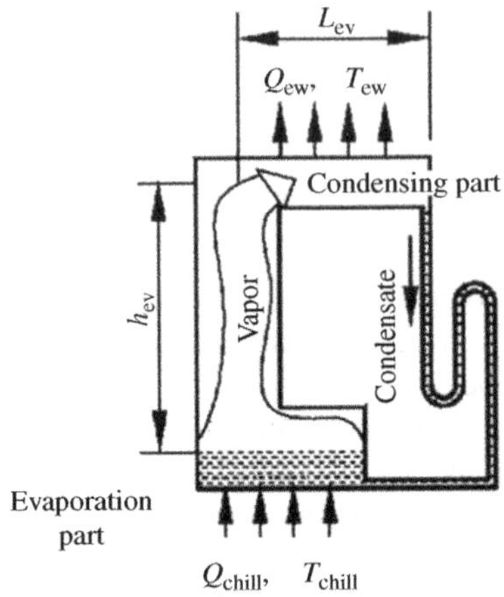

Figure 8.10 Physical model for heat transfer process of the heat pipe type evaporator

Table 8.4 Calculation of the thermal resistance for different parts of the heat pipe type evaporator

Thermal resistances	Symbol	Equations	Notes
Thermal resistance between chilled water and heat transfer tubes	R_1	$R_1 = 1/(\alpha_{f,in}A_{f,in})$	The convective heat transfer coefficient is obtained by literature [5], and the subscript of "in" represents the inside surface area
Metal thermal resistance between chilled water and methanol	R_2	$R_2 = \ln(d_{po}/d_{pi})/(2\pi\lambda_m l)$	l is the length, d_{po} and d_{pi} are the inner and outer diameter
Evaporating thermal resistance of methanol	R_3	$R_3 = 1/(\alpha_{pme}A_{f,out})$	"out" represents outside surface area
Thermal resistance for flowing process of methanol vapor	R_4	$R_4 = 128\mu_v T_v/(\pi d_v^4 \rho_v^2 g L_{gas})$	d_v is the equivalent diameter for the flowing process of the vapor [6]
Condensing thermal resistance of methanol	R_5	$R_5 = 1/(\alpha_{c,m}A_{fm,in})$	The condensation heat transfer coefficient is obtained by the literature [5], and A_{fm} is the surface area of condensation pipe
Metal thermal resistance between methanol and refrigerant	R_6	$R_6 = \ln(d_{po,m}/d_{pi,m})/(2\pi\lambda_{m,m} l)$	Subscript "m" represents methanol
Evaporating thermal resistance of refrigerant of water	R_7	$R_7 = 1/(\alpha_{pwater}A_{fm,out})$	α_{pwater} is the heat transfer coefficient of water

at the initial stage of the adsorption/desorption process after switching, the temperature of the evaporator in the desorption chamber is lower than the condensation temperature of the condenser, so the condensation process starts firstly in the evaporator until the temperature of the evaporator is the same as that of condenser. Then the condensing process starts in the condenser. In the desorption process, the desorption quantity of the refrigerant is gradually reduced and the condensation temperature is decreased. When the temperature of the condenser is lower than that of the evaporator, the condensation process occurs in the condenser. In the desorption process, the pressure in the desorption bed is influenced by the condenser and the evaporator, so the energy balance equations of the evaporator and the condenser are quite complex.

In simulation the lumped parameter method is used, and all the heat exchangers are looked at as a whole and the internal complex heat transfer process doesn't need to be considered. For each heat exchanger the energy and mass balance equations are as follows:

1. The energy balance equation for adsorption/desorption bed is:

$$\frac{\mathrm{d}}{\mathrm{d}t}\left\{\left[M_a\cdot\left(C_a+C_{Lc}\cdot x\right)+C_{mcu}\cdot M_{mcu}+C_{mal}\cdot M_{mal}\right]\cdot T_a\right\}=M_a\cdot H_{st}\cdot\frac{\mathrm{d}x}{\mathrm{d}t}$$
$$+(1-\delta_1)\cdot C_{Lv}\cdot M_a\cdot\frac{\mathrm{d}x}{\mathrm{d}t}\cdot(T_e-T_a)+m_w\cdot C_{water}\cdot(T_{ads,in}-T_{ads,out}) \qquad (8.7)$$

where M_a is the adsorbent mass in a single adsorbent bed, kg; C_a is the specific heat of adsorbent, kJ/(kg·K); C_{Lc} and C_{Lv} are the specific heat of water and water vapor, respectively, kJ/(kg·K); C_{mcu} and C_{mal} are the specific heat of copper for heat transfer tube and aluminum for fins, kJ/(kg·K); M_{mcu} and M_{mal} are the mass of the heat transfer tube and fins, kg; H_{st} isobaric adsorption/desorption heat, kJ/kg; T_e and T_a are the temperature of the evaporator and the adsorbent bed, K; $T_{ad,in}$ and $T_{ad,out}$ are the inlet and outlet temperature of the adsorbent bed, K; m_w is mass flow rate of the heating/cooling fluid, kg/s. δ_1 is determined by the following formula:

$$\delta_1=\begin{cases}1, & \text{Desorption process}\\ 0, & \text{Adsorption process}\end{cases} \qquad (8.8)$$

2. The energy balance equation of the condenser:

$$C_{mcu}\cdot M_{m,con}\cdot\frac{\mathrm{d}T_c}{\mathrm{d}t}=\delta_1\cdot\left[-L\cdot M_a\cdot\frac{\mathrm{d}x_{des}}{\mathrm{d}t}+C_{Lv}\cdot M_a\cdot\frac{\mathrm{d}x_{des}}{\mathrm{d}t}\cdot\left(T_c-T_a\right)\right.$$
$$\left.+\,m_{w,con}\cdot C_{water}\cdot\left(T_{cool,in}-T_{cool,out}\right)\right] \qquad (8.9)$$

In Equation 8.9 $M_{m,con}$ is the metal mass of condenser, kg; L is the latent heat of vaporization of water, kJ/kg; T_c is the condensing temperature, K; $T_{cool,in}$ and $T_{cool,out}$ are the water inlet and outlet temperature of condenser, K; $m_{w,con}$ is the mass flow rate of heating/cooling fluid of condenser, kg/s.

3. The energy balance equation of the evaporator.

For the simulation of the system, if the complex heat transfer processes in the evaporator is simplified as the heat transfer process for one heat pipe type evaporator, the energy balance equation is:

$$\begin{aligned}
&\frac{\mathrm{d}}{\mathrm{d}t}[(C_{Lc}\cdot M_{ew}+C_{mcu}\cdot M_{m,eva})\cdot T_e]\\
&=(1-\delta_1)\cdot\left[-L\cdot M_a\cdot\frac{\mathrm{d}x_{ads}}{\mathrm{d}t}+m_{w,eva}\cdot C_{water}\cdot\left(T_{chill,in}-T_{chill,out}\right)\right]\\
&+\delta_1\cdot\left[\delta_2\cdot C_{Lc}\cdot\left(T_e-T_c\right)\cdot M_a\cdot\frac{\mathrm{d}x_{des}}{\mathrm{d}t}-(1-\delta_2)\cdot L\cdot M_a\cdot\frac{\mathrm{d}x_{des}}{\mathrm{d}t}\right]
\end{aligned}\tag{8.10}$$

where $M_{m,eva}$ is the metal mass of the evaporator, kg; M_{ew} is the mass of the liquid refrigerant (water) in the evaporator, kg; $T_{chill,in}$ and $T_{chill,out}$ are the inlet and outlet temperature for the chilling water, K; $m_{w,eva}$ is the mass flow rate of heating/cooling fluid, kg/s. δ_2 is:

$$\delta_2=\begin{cases}1, & T_c\le T_e\\ 0, & T_c>T_e\end{cases}\tag{8.11}$$

The mass balance equation of the liquid refrigerant in evaporator:

$$\frac{\mathrm{d}M_{ew}}{\mathrm{d}\tau}=M_{e0}-M_a\cdot\frac{\mathrm{d}x}{\mathrm{d}t}\tag{8.12}$$

where M_{e0} is the mass of the refrigerant in the evaporator under equilibrium conditions.

4. The mass and energy balance equations in mass recovery process.
There are two forms of mass recovery processes, one is the mass recovery process coupled with heating/cooling process, and the other is the mass recovery process without heating and cooling process. According to the literature [7], the mass recovery with heating/cooling process is more effective, thus it is adopted in the running process of the chiller.

   The condenser doesn't work during the mass recovery process. The bed after adsorption and the bed after desorption will be connected for mass recovery process, and the "secondary adsorption" phenomenon driven by the pressure difference of both vacuum chambers will happen. The refrigerant will evaporate from the evaporator in the desorption chamber, and the refrigerant in the evaporator in the adsorption chamber begins to condense. In the mass recovery process, in addition to the condenser, the model equations of the adsorption/desorption bed and evaporator are different from ones of desorption by heating and adsorption by cooling processes. The adsorption process of the adsorbent bed is driven by the pressure drop between two chambers, and the desorption phase of the desorption bed is driven by the pressure difference between the desorption bed and the adsorption chamber. p_s in the adsorption equation is the pressure of the desorption chamber and the adsorption chamber, respectively, and in addition, it is the same as the heat transfer equations of the adsorption/desorption beds in the other processes. The biggest difference is the energy balance equation of the evaporator. It is

$$\begin{aligned}
&\frac{\mathrm{d}}{\mathrm{d}t}\left[\left(C_{Lc}\cdot M_{ew}+C_{mcu}\cdot M_{m,eva}\right)\cdot T_e\right]\\
&=-L\cdot\delta_3+\delta_4\cdot m_{w,eva}\cdot C_{water}\cdot(T_{chill,in}-T_{chill,out})
\end{aligned}\tag{8.13}$$

where

$$\delta_3=\begin{cases}m_{e,evap}, & \text{Desorption chamber}\\ m_{e,cond}, & \text{Adsorption chamber}\end{cases}\tag{8.14}$$

$$\delta_4 = \begin{cases} 1, & T_e \le T_{chill,in} \\ 0, & T_e > T_{chill,in} \end{cases} \tag{8.15}$$

The mass balance equation of the refrigerant is:

$$-M_a \frac{\mathrm{d}x_{des}}{\mathrm{d}t} + m_{e,evap} = m_{e,cond} + M_a \cdot \frac{\mathrm{d}x_{ads}}{\mathrm{d}t} = m_{mr} \tag{8.16}$$

In the above formulas, $m_{e,eva}$ and $m_{e,cond}$ are the flow rate of evaporating vapor in the water evaporator of the desorption chamber and that of condensing vapor in the water evaporator of the adsorption chamber during the mass recovery process, kg/s; m_{mr} is the mass flow rate of the vapor in mass recovery process, kg/s.

$$p_{des,wv} - p_{ads,wv} = \frac{v_{wv} m_{mr}^2}{2A_{mr}{}^2} \tag{8.17}$$

where v_{wv} is the specific volume of water vapor, m^3/kg; A_{mr} is the cross-sectional area of mass recovery channel, m^2.

In order to solve the models the state equation of gas needs to be given and it is the van der Waals equation for real gases.

$$\left(p_{wv} + \frac{a}{v_{wv}{}^2}\right)(v_{wv} - b) = RT_{wv} \tag{8.18}$$

where a and b are coefficients.

5. The energy balance equations for the heat recovery process.
During the heat recovery process, the heat transfer fluid circuits in two beds will be connected, and the cooling water in the bed after adsorption will exchange the heat with the hot water in the bed after desorption. In this process, the energy equations of adsorbent bed, the condenser and the evaporator are all the same as those in the conventional adsorption/desorption process, but the inlet water temperature of the bed after adsorption isn't the cooling water outlet temperature of the condenser, nor the inlet temperature of hot water, and it is the outlet water temperature from the bed after desorption. The relation is as follows:

$$T_{ads,in} = T_{des,out} \tag{8.19}$$

8.1.4.3 The Calculation of the Parameters for the System Performance

The cooling power and heating power of the system are calculated by the inlet and outlet temperature difference of the cooling water and heating water, respectively. The formulas are as follows:

$$\text{Cooling power: } Q_{ref} = \frac{\int_0^{t_{cycle}} C_{water} \cdot m_{w,eva} \cdot (T_{chill,in} - T_{chill,out})\mathrm{d}t}{t_{cycle}} \tag{8.20}$$

$$\text{Heating power: } Q_h = \frac{\int_0^{t_{cycle}} C_{water} \cdot m_{w,h} \cdot (T_{h,in} - T_{h,out})\mathrm{d}t}{t_{cycle}} \tag{8.21}$$

Table 8.5 Physical parameters and constants in the simulation

Symbol	Parameter	Value	Unit
a	Constant of the state equation for the vapor	1714.2	Pa·m^6/kg^2
b	Constant of the state equation for the vapor	1.7×10^{-3}	m^3
C_a	Specific heat of adsorbent (silica-gel)	0.924	kJ/(kg K)
H_{st}	Adsorption/desorption heat	2800	kJ/kg
M_a	Adsorbent mass in a single bed	52.0	kg
R	Rational gas constant of water vapor	461.5	J/(kg K)
R_p	Average particle radium of silica-gel	3.75×10^{-4}	m

Table 8.6 Simulation results

Hot water temperature (°C)	Cooling water temperature (°C)	Chilling water temperature	Refrigeration power (kW)	COP	Heating/cooling time (s)	Mass recovery time (s)
65	31	15	5.87	0.420	600	180
65	31	15	5.94	0.515	900	180
65	31	20	7.94	0.599	900	180
85	31	15	11.08	0.515	600	180
85	31	15	10.45	0.595	900	180
85	31	20	12.60	0.651	900	180

$$COP = \frac{Q_{ref}}{Q_h} \tag{8.22}$$

$$SCP = \frac{Q_{ref}}{2M_a} \tag{8.23}$$

The physical parameters and the performance of the heat exchangers for the simulation models are listed in Table 8.5.

8.1.4.4 The Simulation Results of the Performance

A part of the simulation results were listed in Table 8.6. Under the conditions of that the hot water temperature is 85 °C, the inlet temperature of the chilling water is 15 °C, and the cooling water temperature is 31 °C, the cooling power of the system is more than 10 kW and the COP is close to 0.6. If there is no working load for dehumidification, the inlet temperature of the chilling water is 20 °C, and the cooling power of the system is more than 12 kW which is more than 120% of full load and COP is up to 0.65. If the hot water temperature is 65 °C, the minimum cooling power of the refrigerator is 60% of the full load and COP is 0.4–0.6.

8.1.5 The Analysis on the Mass Transfer Performance of the Adsorbent Bed

The mass transfer process in the adsorbent bed includes three processes, and they are the mass transfer process inside the adsorbent particles, between the adsorbent particles, and in the channel outside of the adsorbent. Mass transfer modes can be divided into convective mode and molecular diffusion mode. For a fixed bed, the convective mass transfer process between the side of particles and the fluid generally is the diffusion mode, and molecular diffusion happens mainly inside the adsorbent particles. Taking the adsorption process, for example, firstly, the refrigerant vapor overcomes the resistance in the mass transfer channels of the adsorbent bed and spreads to the adsorption layer through the gap of the adsorbent particles due to convective diffusion; then, it overcomes the membrane resistance on the surface of adsorbent particles and flows to the micropores of the adsorbent particles due to molecular diffusion; finally, it distributes to the surface of the microspore due to the pore diffusion and surface diffusion. In this process the surface diffusion doesn't need to be considered when the concentration of the vapor isn't very high. The mass transfer inside the adsorbent particles is decided by the physical properties of the adsorbent particles, the geometric parameters of the refrigerant vapor, and the physical properties of the refrigerant vapor, but it has nothing to do with the design of the adsorbent bed. There is pressure resistance of the refrigerant vapor in the fixed adsorbent bed during the mass transfer process from the adsorbent bed to the surface of the adsorbent particles. The pressure drop for mass transfer is an important parameter to evaluate the mass transfer capacity of the adsorbent bed. The bigger the pressure drop is, the stronger the mass transfer capacity of the adsorbent bed is. At a constant evaporation temperature a bigger pressure drop will be equivalent to the increment of the saturation temperature of the adsorbent bed, which is good for the improvement of the adsorption capacity of the bed. The pressure drop for mass transfer in the mass transfer channel outside of the adsorbent is estimated by the Euler equation, and the pressure drop from the inside of adsorption layer to the adsorbent particles is calculated by the empirical Equation 8.24 of packed bed.

$$\Delta p = \frac{\mu \times u \times l}{k_p} \tag{8.24}$$

where μ is the dynamic viscosity, kg/(ms), u is the airflow velocity of the vapor in the gap of the adsorbent particles at the minimum section, m/s; l is the mass transfer scale, m; k_p (m^2) is the porous medium permeability which is from the (spherical particles in hexagonal array) Blake-Kozeny equation [8]:

$$k_p = \frac{d_a^2 {\varepsilon_a}^3}{150(1-\varepsilon_a)^2} \tag{8.25}$$

where ε_a is the porosity; d_a is the diameter of the adsorbent particles, m.

The mass transfer coefficient of packed bed is also an important parameter on mass transfer performance of the adsorption bed. The coefficient is commonly calculated by the Carberry formula [9]:

$$\frac{\varepsilon_a K_F}{u} \cdot Sc^{2/3} = 1.15\left(\frac{\mathrm{Re}}{\varepsilon_a}\right)^{-0.5}, 0.1 < \mathrm{Re} < 1000 \tag{8.26}$$

where Schmidt number $Sc = v/D_{ms}$; For the water vapor, $Sc = 0.6$; Reynolds number $\mathrm{Re} = ud_a/v$; d_a is the diameter of the adsorbent particles, m; D_{ms} is the mass diffusion coefficient of the fluid, m^2/s; K_F is the mass transfer coefficient of the fluid side, m/s.

According to the Equations 8.24–8.26 and the design parameters of the adsorbent bed, we can estimate the pressure drop for mass transfer and the mass transfer coefficient of the adsorbent bed. Table 8.7 lists the pressure drop for mass transfer and mass transfer coefficient of the adsorbent bed in the chiller driven by low-temperature heat source and in other literatures.

Table 8.7 Comparison of mass transfer parameters of adsorbent bed

Heat exchanger	Permeability (m^2)	Pressure drop for mass transfer (Pa)	Mass transfer coefficient (m/s)	Data source
Tube-fin type silica gel–water chiller	1.534×10^{-9}	14.9	1.213	Designed by SJTU
Plate-fin type silica gel–water chiller	6.667×10^{-10}	49	1.777	Literature [2]
Tube-fin type activated carbon–methanol refrigerator	4×10^{-11}	676	0.2	Literature [10]
Fin-tube type zeolite–water refrigerator	2.2674×10^{-9}	22.6	1.151	Literature [11]

It can be found from the data in the table that the comprehensive mass transfer performance of the adsorbent bed was not optimal, but the pressure drop for the mass transfer was much smaller compared with one in the literature [2, 10, 11], which makes the cooling capacity of adsorbent bed optimum.

8.1.6 Performance Analysis of the System

8.1.6.1 Introduction of the System

The system included a hot water loop, a cooling water loop, and a chilling water loop, and there is a hot water tank, a cooling water tank, and a chilling water tank, respectively, in the system to reduce the fluctuation of the temperature of the water, as shown in Figure 8.11. In the hot water loop, the hot water is pumped into the refrigerator, and when it flows out of the refrigerator it will be heated by a vapor steam of 4–6 kgf/m^2 in a countercurrent heat exchanger, then flows back to the hot water tank. The flow rate and temperature of the hot water are controlled by the valve V2 and V7, respectively. The chilling water from the chilling water tank is cooled in the evaporator and flows back, and the cooling capacity output from the chilled water tank is balanced by the part of the cooling return water, and the excess water in the chilling water tank flows to the cooling water tank via the overflow pipe to maintain the water level is the chilling water tank. Chilling water from the chilling water tank, majority of return cooling water, and the running water are mixed together in the cooling water tank, and then will be pumped by a cooling water pump into the refrigerator when it reaches the setting temperature, and then flows out from the refrigerator. A small portion of the cooling water that flows out of the refrigerator will flow into the chilling water tank by valve V4 to balance the cooling capacity that is produced by the refrigerator, and the other part of it will flow back to the cooling water tank through valve V5. The water-stop plates are installed between the cooling water return pipe and overflow pipe 2 in the water tank, as well as between the running water pipes. Overflow pipe 1 that maintains the water level in the cooling water tank is installed

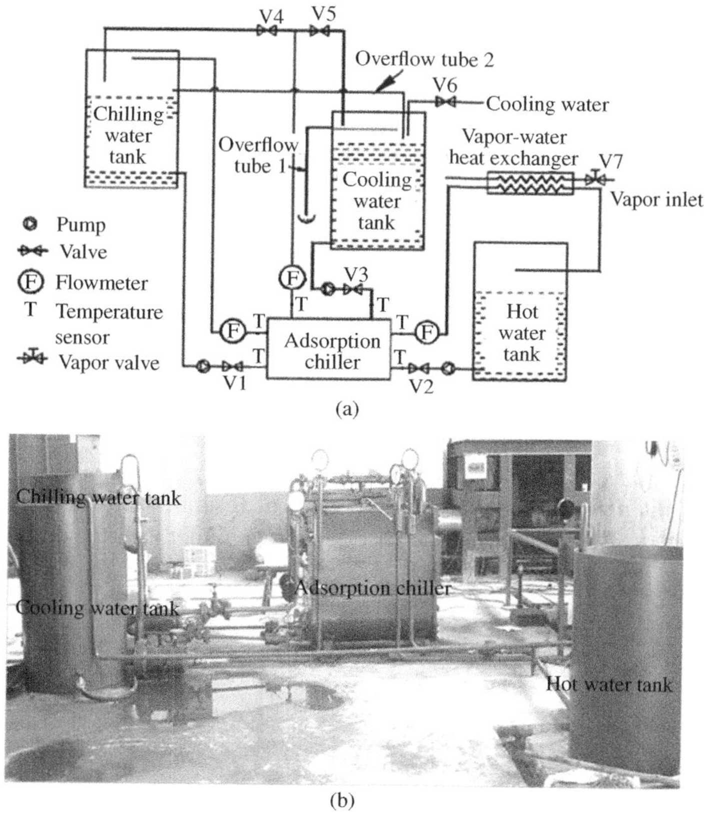

Figure 8.11 Experimental system. (a) System diagram and (b) photograph

at the side of the cooling water return pipe of the water stop plate. The flow rate of cooling water and chilling water are controlled by valve V3 and V1, respectively, and the temperature is controlled by valve V6 and V4, separately. Due to the heating/cooling power of each loop alternating it is difficult to control the inlet temperature of chilling water at a constant although there is a buffer tank. Therefore, the allowable fluctuation range of water temperature is set as ± 1.5, ± 0.5, and ± 0.5 °C for hot water, cooled water, and chilling water, respectively.

Not considering the measurement error of the adsorbent mass, according to the accuracy of the measuring instrument and performance parameters, the error of COP caused by the system is 8%, and the relative error of SCP is 1%. In addition, the time interval for data acquisition process cannot be unlimitedly short. Especially at the switching time, there will be a certain moment the measured inlet temperature of the hot water is the temperature of the hot water, while the measured outlet temperature of the cooling water is the cooling water temperature, which will cause the calculated value of the heating power to be too big. The deviation can be estimated through the temperatures of the hot water and the cooling water, the flow rate of hot water, the average heating power, and the cycle time. The influence on COP is estimated, and it is about 8%.

8.1.6.2 The Comparison for the Simulation and Experimental Results for the Performance

The simulation and experimental results are compared to analyze the performance of the system. Overall, the simulation results basically reflect the trends of the system performance under different conditions and are in accord with the experimental results. However, simulation results and experimental results have a gap in the values. The temperatures of hot water and the adsorbent bed are shown in Figures 8.12 and 8.13. From the diagram we can see that the calculation results are different from the experimental results for the heating mass recovery process and the heating desorption process, and the calculation results of the adsorbent bed temperature and the cooling water outlet temperature fit with experiment results during the cooling adsorption process. The results are analyzed, and the big differences between simulation and experiment results during the heating desorption process are mainly caused by the following aspects:

1. The simulation accuracy of lumped parameter model is not high enough.
2. In the simulation process we take the same value of reaction heat for the adsorption and desorption process, which fits with the adsorption heat rather than the desorption heat.

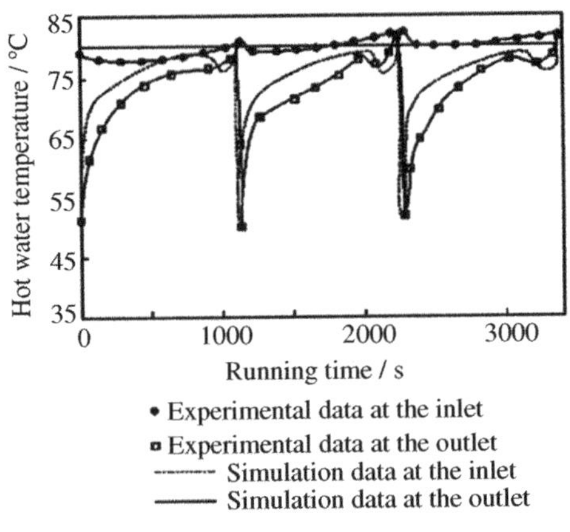

Figure 8.12 Comparison of the hot water temperature between the calculation and experimental results

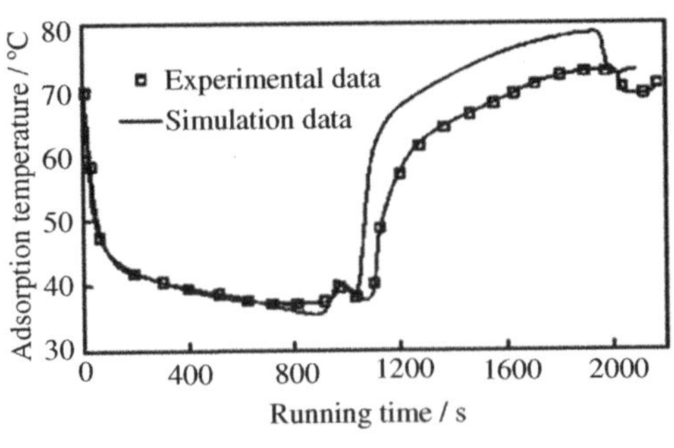

Figure 8.13 Comparison of the adsorbent bed temperature between the calculation and experimental results

3. There is a deviation between the heat transfer coefficient and the true value in the simulation. The values used in the simulation are constant, but in fact, in the actual experiments they are not constant. Especially the heat transfer coefficient of the adsorbent bed changes significantly in a cycle.
4. The correlation coefficient of the adsorption equations during the simulation are from the literature, not from the experimental data of silica gel that was used in the system.

Table 8.8 showed a comparison of performance parameters under two typical conditions between the calculation and experimental results. Under the same conditions of the hot water temperature and cooling water temperature, the higher the chilling water temperature is, the smaller the deviation is between the calculation and experimental results. The reason is that when the inlet chilling water temperature is higher, during the mass transfer and the switching processes, the temperature difference among the two water evaporators and the methanol evaporator is small, and the cooling losses caused by the heat transfer process of the heat pipe evaporator is small, that is, the condition is closer to the simulation assumptions.

Since under some conditions the simulation results are very different from the experimental results, the heat transfer coefficients of adsorbent bed, condenser, and evaporator are rectified according to the experimental results, and then the models are re-calculated and the simulation results are rectified based on cooling capacity loss that was estimated by the experimental results of the evaporator. Figure 8.14 shows the comparison of the bed temperature between calculation and experimental results after the heat transfer performance coefficients

Table 8.8 Comparison between calculation and experiment results

$T_h = 85\,°C$, $T_{cooling} = 31\,°C$		Calculation results	Experimental results	The percentage of deviation (%)
$T_{chilled,inlet} = 20\,°C$	COP	0.651	0.433	33.5
	Q_{ref}/kW	12.6	10.9	13.5
$T_{chilled,inlet} = 15\,°C$	COP	0.588	0.388	34
	Q_{ref}/kW	10.4	8.7	16.3

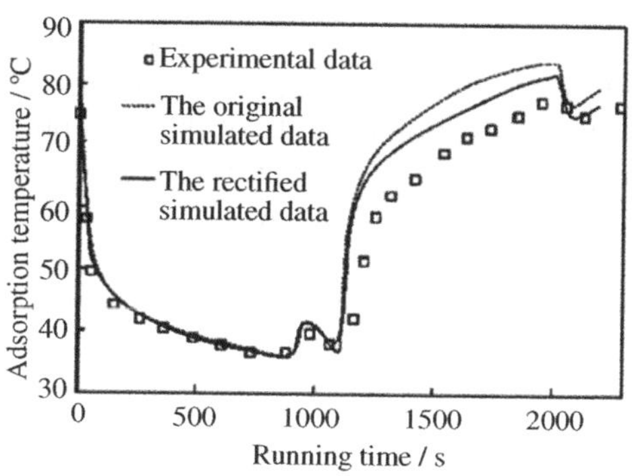

Figure 8.14 Comparison of the adsorbent temperature between the calculation and experiment results

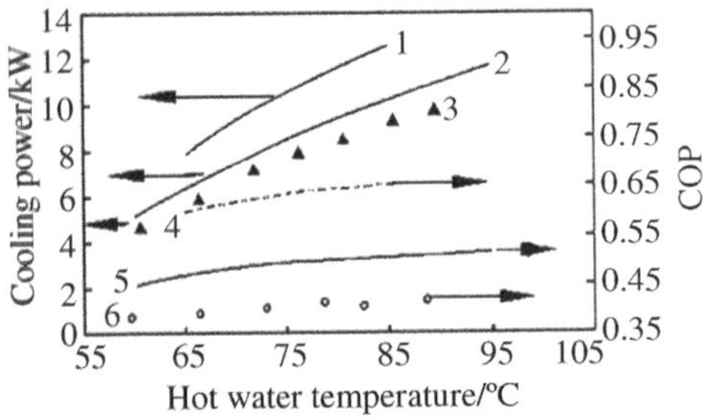

1,4 original simulation value; 2,5 the simulation value after the amendment; 3,6 experimental value

Figure 8.15 Comparison of the system performance between the calculation and experiment results

of heat exchangers are modified. Figure 8.15 shows the comparison of the system performance between the calculation and experimental results according to the above method under the conditions of the different hot water temperature. The diagram indicates that the simulated temperature is similar to that of the experimental results after a reasonable modification, as well as the simulated cooling capacity and COP, which are also similar to the experimental results.

Table 8.9 lists the experimental results of the adsorption chiller and the results in the literature. The chiller has the cooling power of 8.6 kW and COP of 0.383 under the condition of chilling water temperature of 12 °C. Compared with the experimental results in the literature [12], the chiller improved *SCP* by 95% and *COP* by more than two times when the hot water temperature is lowered by 2.5 °C and the chilling water temperature is lowered by 1 °C. If compared with the cooling power of the adsorption chillers in reference [7, 13, 14], *COP* and cooling capacity are close to the calculation results by Akahira *et al.* [7] and higher than the calculation results of Saha *et al.* [14], but *COP* increased by 14% if compared with the results by Boelman *et al.* [13] under the condition of the same chilling water temperature. Therefore, the performance of the adsorption chiller has been greatly improved. Meanwhile the number of vacuum valves is decreased significantly, which improved the reliability of the system.

8.1.6.3 The Thermodynamic Analysis of the Cycle

A-B-C-D-E-F-A in Figure 8.16 describes a complete thermodynamic cycle of the adsorbent bed, in which A-B and D-E are mass recovery processes. Figure 8.16 showed that the mass recovery process improves the adsorption quantity of the adsorption bed from x_{01} to x_1 and decreases the adsorption quantity of the desorption bed from x_{02} to x_2. Therefore, the cycle adsorption quantity increases from $(x_{01} - x_{02})$ to $(x_1 - x_{01})$.

In the region near to the point B of the Figure 8.16, there is a ring formed by the mass recovery and heat recovery process. Such a ring is caused by the adsorption heat released from the adsorbent bed during the mass recovery process, and this heat can't be released to the cooling water in time. Meanwhile there is a concave arc of D-E for the desorption bed, which is caused by the desorption heat needed by the bed during the mass recovery process, and it can't be supplied by the hot water in time. The ring and concave indicate that the adsorption and desorption during the mass recovery process are very quick, which reflects the unique thermodynamic properties of the chiller during the heat and mass recovery process. The area

Table 8.9 Comparison of the performance between the adsorption chiller developed by SJTU and the chillers in the references

Data sources	Hot water temperature (°C)	Cooling water temperature (°C)	Chilled water temperature (°C) Inlet	Outlet	Cooling power (kW)	COP	SCP (W/kg)	Cycle time (s)
Experiment data	78.8	31.3	20.5	16	8.32	0.31	80	1680
	81.8	31.3	20.7	16.3	9.33	0.34	89.7	1920
	86.8	30.9	21.1	16.3	10.62	0.398	102.1	1920
	59.7	30.4	20.5	18.2	4.80	0.385	46.2	2280
	69.1	30.3	19.6	16.2	7.57	0.378	72.8	2280
	84.4	30.5	21.5	16.5	10.88	0.432	104.6	2280
	80.3	30.2	15.8	12.1	8.26	0.382	79.4	2280
	82.5	30.4	15.8	11.9	8.69	0.388	83.6	2280
Literature [13] (experiment)	85	32	14	11.5	7.36	0.28	–	450
	85	30	14	10.5	10.3	0.34	–	450
Literature [7] (calculation)	80	30	14	–	8.6	0.39	–	1140
Literature [12] (experiment)	85	32	14	–	–	0.313[a]	–	1000
	85	30	14	–	–	0.361[a]	–	1000
	85	32	14	13	2.27	0.188[b]	43	1000
	85	30	14	12.8	2.79	0.212[b]	52.8	1000
Literature [14] (calculation)	85	32	14	–	–	0.28	–	–
	85	30	14	–	–	0.34	–	–

[a]The data for the second prototype is based on the literature [12].
[b]The data of the first prototype is based on the literature [12].

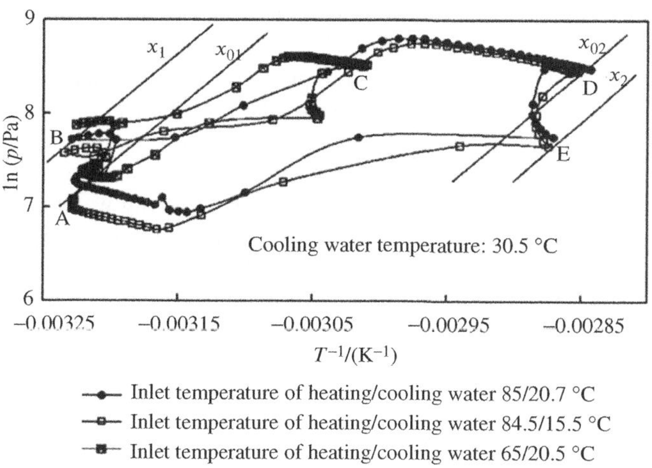

Figure 8.16 Experimental p-T diagram of the cycle

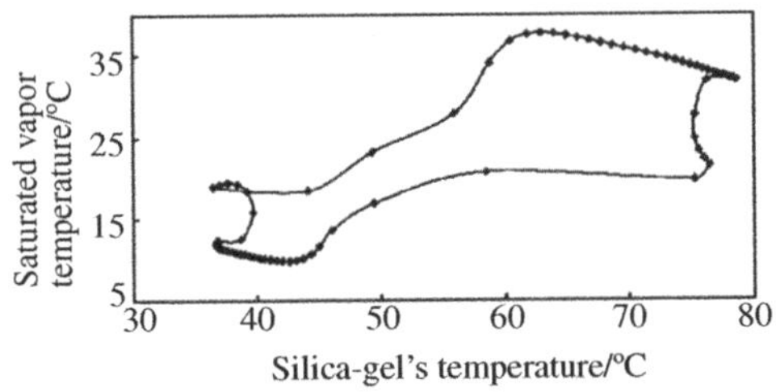

(Hot water temperature: 85 °C; Cooling water temperature: 31 °C; Chilling water inlet temperature: 20 °C)

Figure 8.17 Experimental Dühring diagram of system (T_s-T diagram)

of the thermodynamic cycle for the chiller in the *p*-*T* diagram is different when the temperature of the heat source is different, which is similar to the properties of traditional adsorption refrigeration systems. The higher the temperature of the heat source is, the larger the area of the cycle is. The temperature of the chilling water only has the influence on the adsorption process. From the *p*-*T* diagram, during the running process of the system, the pressure and temperature of the condenser and the evaporator always change with the heating and cooling processes of the adsorbent beds. Therefore, the thermodynamic cycle of the adsorption system developed can be defined as a heat and mass recovery cycle with the variable temperature and pressure.

From the cycle diagram of T_s-T in Figure 8.17 (the relationship between the saturation temperature of water vapor that is in accord with the saturation pressure p_s and the temperature of the adsorbent bed), the area of the adsorption stage is smaller, the temperature variation range is smaller. Whereas for the desorption stage, both the temperature range of the silica gel and the condensing temperature are larger. Under the experimental conditions shown in Figure 8.17, the temperature of the silica gel is in the range of 45–37 °C, so the adsorption function of the adsorbent bed is strong. When the temperature of the silica gel is at 60 °C, the desorption function of the bed is strong, so the corresponding condensation temperature is highest. With the progress of desorption process the desorption quantity is decreased and the condensation temperature dropped from 37.7 to 32 °C, meanwhile the temperature of the bed increases to 78.6 °C. During the desorption process, before the temperature of the adsorbent bed decreases to 60 °C, the heating power of the hot water at the side of the adsorbent bed is used to increase the temperature of the adsorbent bed. When the temperature is higher than 60 °C desorption will happen. At the end of the adsorption and desorption processes, the temperature difference between the adsorbent bed and the cooling water or hot water is about 7 °C. In addition, during the mass recovery process, the temperature change of the adsorption bed and the desorption bed is the same at 3.5 °C, but the saturation temperature change of the desorption chamber is bigger than that of the adsorption chamber, so it reduces the saturation temperature of the desorption chamber effectively, which is good for a quick cooling power output from the evaporator.

8.1.6.4 Non-equilibrium Performance of the System

For the adsorption chiller the evaporation, condensation, and heating/cooling conditions of two water evaporator, two condensers, and two adsorbent beds could not be completely symmetrical, which very easily leads to the different mass recovery amount for the different

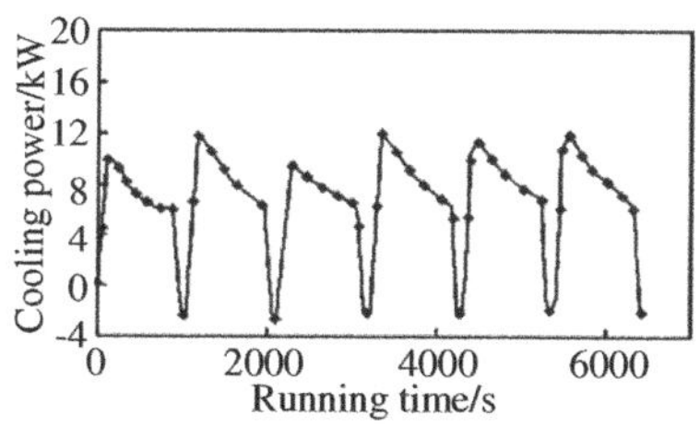

Figure 8.18 Non-equilibrium cooling power output at the mass recovery time of 180 seconds

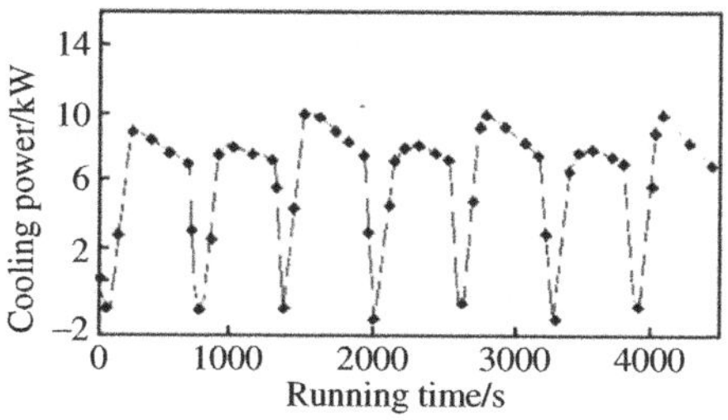

Figure 8.19 Non-equilibrium cooling power output at the mass recovery time of 60 seconds

mass recovery directions, and consequently will lead to a different refrigerant mass in both adsorption/ desorption chambers. The special structure of the evaporator requires the reasonable refrigerant mass that cannot be too much or too less. The excessive mass of the refrigerant will make the evaporation pipes submerge in the refrigerant, so the evaporation mode will be changed from the surface evaporation into a boiling evaporation, which requires a larger superheat degree. The influence will be equivalent to decreasing the cooling power by the decrement of the evaporation temperature. Conversely, if the mass of the refrigerant in the evaporator is too small, the cycle amount is small, and the refrigerant in the evaporator will be easily dried up, which will lead to the decline in the cooling capacity of the system. Therefore, the non-equilibrium of refrigerant in two adsorption/desorption chambers causes different cooling power output of two adsorbent beds in the adsorption process. The experimental results are shown in Figures 8.18 and 8.19. The non-equilibrium cooling capacity between the two adsorption refrigeration chambers is different when the mass recovery time is different. The non-equilibrium phenomenon is small when the mass recovery time is long, whereas for the short mass recovery time it is the opposite. In addition, with the long mass recovery time (180 seconds), the system will recover to the equilibrium cooling output easily, but with the short mass recovery time (60 seconds), the system cannot recover to the equilibrium state. Therefore, if the conditions permit the mass recovery time should be as long as possible.

8.1.6.5 Experiment Research on the System under the Condition of Unstable Heat Source

The temperature of heat source for some conditions cannot be kept stable, such as the waste heat from the exhaust gases of the locomotives, cars, and fishing boats, and so on, for which the engine is constantly start-stop or switch off or changes the rotation speed, so the heat source always changes. The unstable heat source also includes the solar energy applied to the adsorption refrigeration system. The unstable power or temperature, which is the thermodynamic driving force of the adsorption refrigeration system, has an important impact on the performance of the system.

Test Method

The novel silica gel–water adsorption chiller developed by SJTU has promising application foreground in the field of solar air conditioning because it can be driven by the low temperature heat source. Solar energy is green energy, but it's also an unstable energy, which will have a big influence on the performance of the chiller. The unstable heat source is simulated by the control of the steam valve V7 in Figure 8.11, and the degree of opening determines the heating power of the hot water, which will determine the hot water temperature. In the experiments, the steam valve V7 adjusts slowly step by step, and generally it is adjusted every 3–5 minutes. Due to the cushioning effect of the hot water tank, the hot water temperature changes smoothly under the control of the valve V7.

For two-bed continuous adsorption refrigeration system, the impact of the heat source on system performance generally will come up after a half cycle. Therefore, the data used for the calculation of the system performance shall be postponed to a half cycle. As the running time isn't a cycle or an integer multiple of one cycle in the experiments, the ratio between adsorption time and the cycle time is different under different conditions. Therefore, the ratio between adsorption time and desorption time, which is shown in Table 8.10, is an important parameter for the evaluation of the performance.

Experimental Results and Analysis

The change of the hot water temperature for the operating conditions 1–3 is shown in Figure 8.20, and they are in the range of 70–77 °C. The heating rate for the conditions of

Table 8.10 The ratio between the adsorption and the desorption time under different conditions

Condition number	1	2	3	4	5	6	7	8	9
Ratio between the adsorption and desorption time	0.891	0.802	0.865	1.107	1.107	0.964	0.579	1.791	0.812
Cycle time (s)	3100	4020	5400	4800	4800	3440	3440	3440	3440

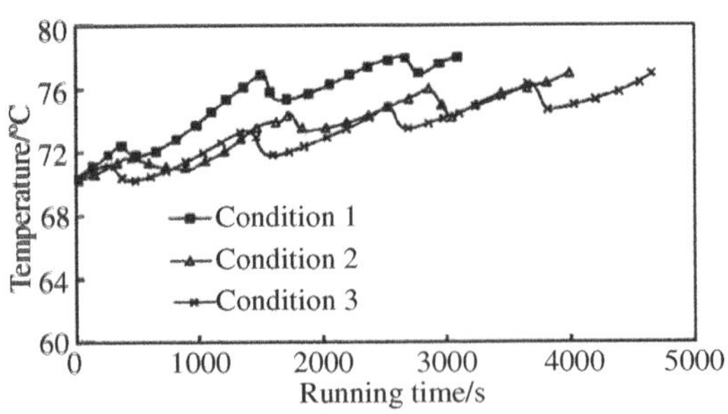

Figure 8.20 The temperature change of the hot water

Table 8.11 Experimental results under the conditions of different heat sources (the inlet temperature of chilling water is 20 °C)

Number	Average temperature of hot water (°C)	Temperature range of hot water (°C)	Change rate of temperature for hot water (°C/min)	Cooling water temperature (°C)	Refrigerating power (kW)	COP
1	75	70–77	0.135	30	7.34	0.359
2	73.6	70–77	0.105	30	7.3	0.355
3	73.8	70–77	0.091	30	7.56	0.379
4	66.6	77 to 56.7	−0.253	30	6.49	0.417
5	68.5	56.7–77	0.253	30	6.17	0.343
6	70	69.2–71.6	0.042	31	6.53	0.36
7	70	67–72	0.087	31	6.42	0.324
8	70	59.7–76.4	0.297	31	5.6	0.289
9	70	75 to 65.6	−0.164	31	6.27	0.382
10	75.7	60.1–85.8	0.12	31	7.35	0.345
		85.8 to 56.7	−0.211	31	7.53	0.435

Note: a negative value indicates the cooling rate.

1–3 are 0.135, 0.105, and 0.091 °C/min, respectively. The average temperature of hot water and the performance of the adsorbents are listed in Table 8.11. In theory, the more stable the heat source is, the higher the efficiency of the system is. When the temperature range and the average temperature of the hot water are fixed, the longer the time for the heating process is, the smaller the temperature change of the heat source in unit time is, and consequently the better the performance is. Such a trend can be seen from the data under the conditions of C2 and C3. Generally, when the temperature range of the heat source is the same for different conditions, the duration of the temperature change of the hot water has little impact on the performance of the refrigerator. Under the conditions as shown in Figure 8.20 the cooling power and COP don't change much, and the maximum discrepancy is only 6.8 and 3%, respectively. Especially under the conditions 1 and 2, the performance is almost the same although the average temperature of the hot water under the condition 1 is 1.5 °C higher than that under the condition 2.

Figure 8.21 shows the hot water inlet/outlet temperature change under the conditions of the same temperature range and same duration time. In the diagram, the temperature range of hot water is from 56.7 to 77 °C. Condition 4 is for the temperature decreasing process, and condition 5 is for the temperature increasing process. The temperature change rate of the hot water for both conditions is 0.253 °C/min. From Table 8.11, the cooling power and *COP* under condition 4 are higher than those under condition 5, and *COP* is relatively higher. Due to that the heat transfer from the hot water to the desorption bed will need time, the temperature change of the desorption bed is lagged compared with the temperature change of hot water, and the hot water outlet temperature also is lagged if compared with the inlet temperature. The temperature decreasing process of the hot water reduces the temperature difference between the hot water inlet and outlet of the desorption bed, while the temperature increasing process of the hot water is the opposite. In the temperature decreasing process of the hot water (condition 4), the temperature difference between the hot water and desorption

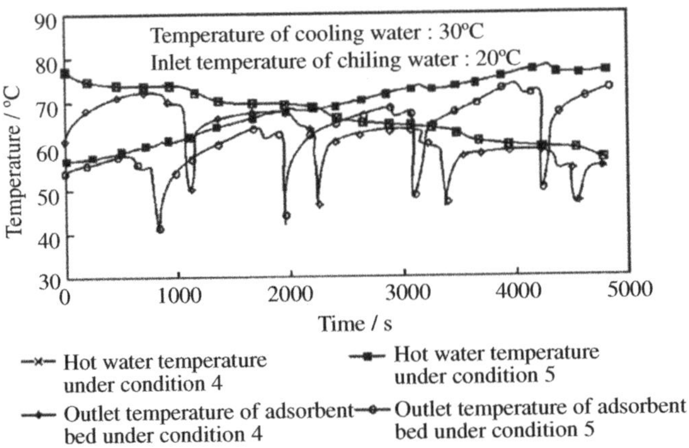

Figure 8.21 Temperature change of hot water

bed decreases, also the heating power of the hot water for the desorption bed decreases, but the temperature of the desorption bed is kept at a high level that corresponds to the high hot water temperature, and consequently the desorption quantity of the adsorbent is large. The temperature increasing process of the hot water is just the opposite. Therefore, with the same temperature range of the hot water and the same running time, the temperature decreasing process of the hot water is better than the temperature increasing process of the hot water on the system.

Experimental results under the conditions of different heat sources are shown in Table 8.11. Table 8.11 showed that except for condition 8, the cooling capacity of all other conditions of the system is almost the same. If the results were analyzed in more detail, the greater the temperature range of heat source is, the lower the cooling capacity is. In addition to the cooling process under the condition 9, the greater the temperature range of the heat source is, the lower the *COP* of the system is. The reason for the above results is that the change rate of the temperature for the hot water is too large.

The thermodynamic cycles for stable heat source and unstable heat source (condition 4 and 8) are shown in Figures 8.22 and 8.23. The process of A-B-C-D-E-F-A is the thermodynamic cycle driven by the stable heat source. F-A-B is desorption process by heating, C-D-E is adsorption process by cooling. B-C and E-F are the mass recovery processes. Due to the short time of the process the heat recovery cannot be represented in the diagrams.

Figure 8.22 is a cooling process. The process from point 1 to 7 is the first cycle of condition 4, and the second cycle is from point 7 to 13. Figure 8.22 showed the thermodynamic cycle for which the temperature of the heat source decreases. The process from point *a* to *k* in Figure 8.23 showed the thermodynamic cycle for which the temperature of the heat source temperature increases. It can be found from two diagrams that the cycle area of adsorbent bed with stable heat source is located inside the cycle area of the adsorbent bed with unstable heat sources. Compared with the performance of the cycle with a stable heat source, the cycle adsorption quantity of the adsorbent bed with unstable heat source is larger under the condition of high temperature (1-2-3-4-5-6-7), while it is smaller at the low temperature (7-8-9-10-11-12-13). Overall, under the same conditions of cooling and adsorption, the adsorption amount of the desorption bed at the end of the mass recovery process (such as the points of C, 3, 6 in

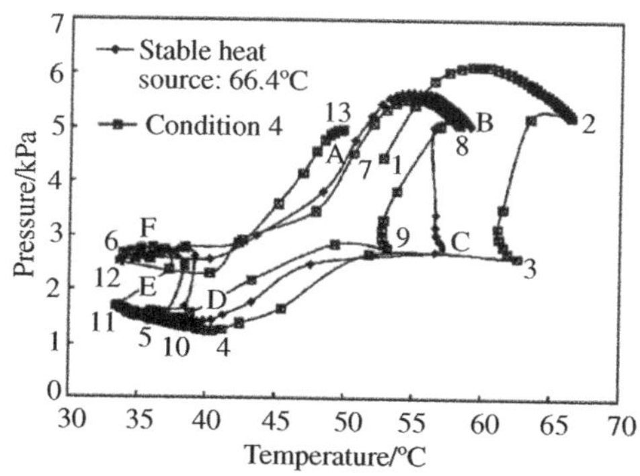

Figure 8.22 Comparison between condition 4 and the conditions with stable heat source

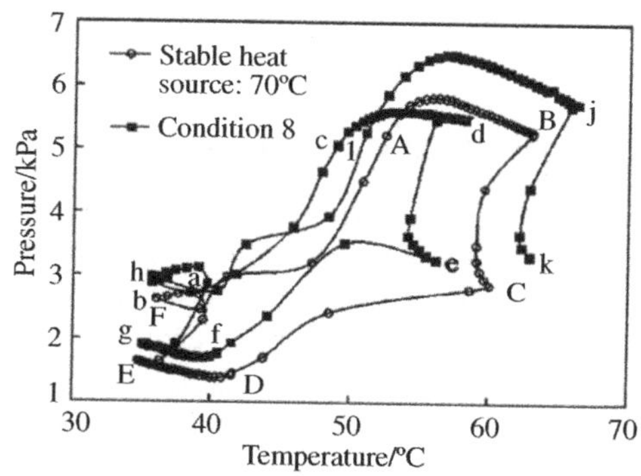

Figure 8.23 Comparison between condition 8 and the conditions with stable heat source

Figure 8.22 and the points of e, k, C in Figure 8.23) determines the cooling capacity of the adsorbent bed.

The Strategies for Decreasing the Influence of the Unstable Heat Source on the Performance of the Chiller

Research results of the system with unstable heat source showed that the performance of the system will be influenced. One method for solving this problem is to use the heat storage to improve the operating performance of the system. Such a method is tested for the application of the system as a solar energy powered air conditioner, and the results showed that it is conditional for the heat storage for improving the performance. The heat storage method is effective for improving the performance when the power consumption of the system isn't increased for keeping the heat source stable. Literature [15] discussed the influence of different heat storage processes on the performance when the heat source is intermittent, and it's based on the assumption that the temperature of the heat storage water tank was the same as the heat source temperature. Under the conditions of insufficient or small surplus heat source, the heat storage won't improve the performance of the system. Therefore, the choice of the heat storage device or the heat storage capacity is determined by different conditions, the power, and the actual application of the unstable heat source.

In addition, the experimental results indicate that the longer heating/cooling time is good for the improvement of the performance when the heat source temperature is low. In contrast, when the heat source temperature is high we need to choose the shorter heating/cooling time. Therefore, the system uses the variable heating/cooling running time in the variable temperature mode which will improve the performance. When the heat source temperature is low it uses a longer heating/cooling time, while the heat source temperature is high it uses a shorter heating/cooling time, which can improve the adaptability of the system. It needs a judgment criterion to determine the relationship between the unstable heat source and the heating/cooling time. The simplest method is to measure the inlet temperature of the hot water, and according to the duration for a certain variable temperature range to adjust the running time. Further, the pressure in the desorption bed can reflect the desorption state in it. Therefore the pressure change in the desorption bed can be a criterion for the heating time. However, due to the uncertainty of the unstable heat source, it requires a lot of experiments to obtain these judgment criterions. Moreover, such a method requires the relating complex control methods, which will increase the complexity and sacrifice the reliability and stability of the system.

8.2 Silica Gel–Water Adsorption Cooler with Chilled Water Tank

Based on the research on the silica–gel adsorption chiller driven by the low temperature heat source, a compact silica gel–water adsorption chiller without vacuum valves was manufactured and experimentally studied. This chiller contains two-adsorption/ desorption chambers and one chilled water tank. Each adsorption/desorption chamber consists of one adsorber, one condenser, and one evaporator. The chilled water tank is adopted to mitigate the variation of the chilled water outlet temperature. Again a mass recovery-like process, which is a heat recovery process between the two evaporators, was carried out in this chiller [16, 17].

8.2.1 Description of the Prototype

Figure 8.24 shows the schematic diagram of the studied adsorption chiller. The chiller contains two-adsorption/desorption chambers and one chilled water tank. Each adsorption/desorption chamber encloses one adsorber, one condenser, and one evaporator. The adsorber is arranged near to the condenser. Both absorber and condenser are placed over the evaporator. Five three-way valves (V1–V5) and three solenoid valves (V6–V8) are adopted in this chiller. All the valves used are water valves while vacuum valves are not adopted to enhance the reliability. A chilled water tank is integrated into this chiller in order to mitigate the chilled water outlet temperature variation. The compact silica gel–water adsorption chiller (shown in Figure 8.25) has a size of 1.2 m length, 0.7 m width, and 2.0 m height. All the water valves used are controlled by a SIEMENS LOGO controller.

Fin-tube heat exchangers are adopted as the adsorbers. Each adsorber contains 27 adsorber units (shown in Figure 8.26). In the adsorber unit, eight fin tubes with fin space of 2.5 mm are covered by wire mesh and then shaped by a porous metal plate. Twenty-seven adsorber units are stacked vertically to form the adsorber heat exchanger (AHE). Concave parts of the porous metal plate in each adsorber unit are used as the mass transfer channels of the adsorber. Microporous silica gel with diameters from 0.5 to 1.5 mm is filled in the fin space. Each adsorber contains 48.5 kg silica gel. The total heat transfer area in the adsorbent side is 59.33 m^2.

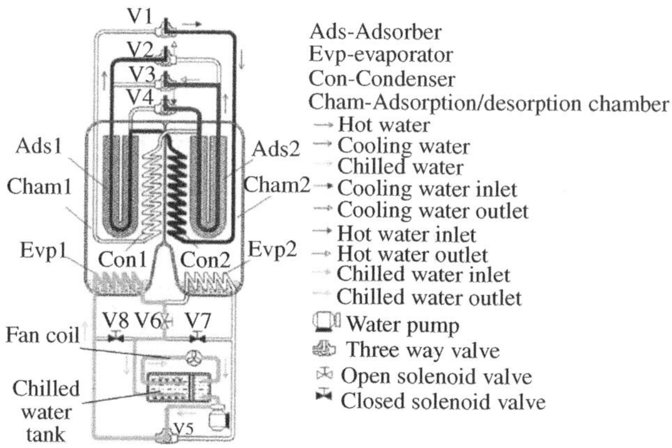

Figure 8.24 Schematic diagram of the compact adsorption chiller

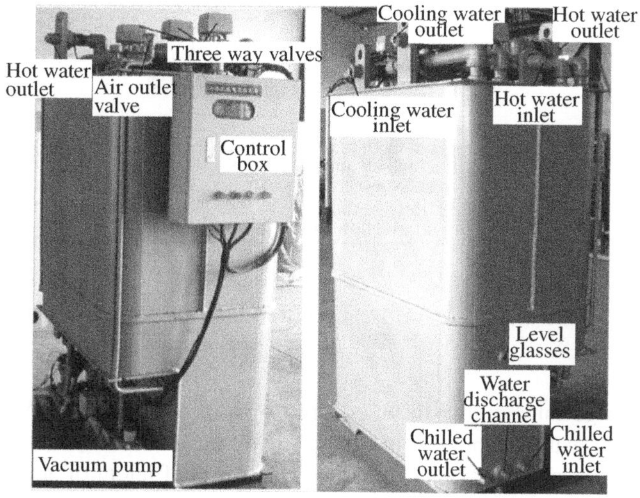

Figure 8.25 Photos of the compact adsorption chiller

The condensers are shell-tube heat exchangers with condensing areas in each condenser of 2.53 m^2, and the shell is also the wall of the adsorption/desorption chamber. The condenser is located over the evaporator and collects the refrigerant condensate; and then the condensate flows downwards into the evaporator. The evaporators are shell-tube heat exchangers too, and the shell is also the wall of the adsorption/desorption chamber. As shown in Figure 8.27 five trays are adopted in one water evaporator. Each tray contains nine copper tubes with outside micro-grooves, which were used as heat transfer tubes in the water evaporator.

Capillary-assisted evaporation is employed in these evaporators. The copper tubes are partly immersed in the refrigerant liquid (water). Some water flows upwards to the top of the copper tubes through the micro-grooves by capillary force. Thus the upper parts of the copper

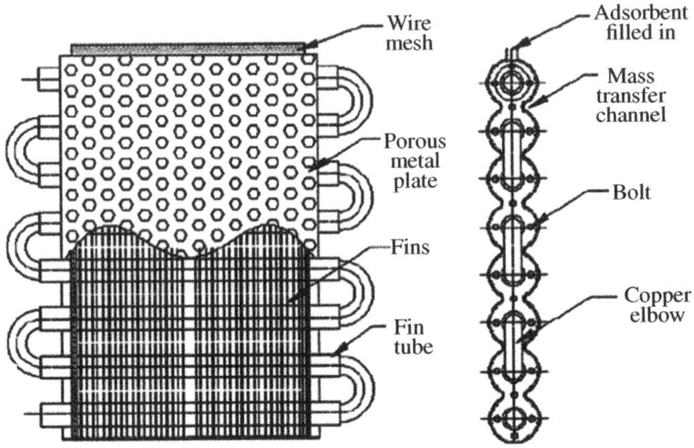

Figure 8.26 Schematic of adsorber unit (top view and left view)

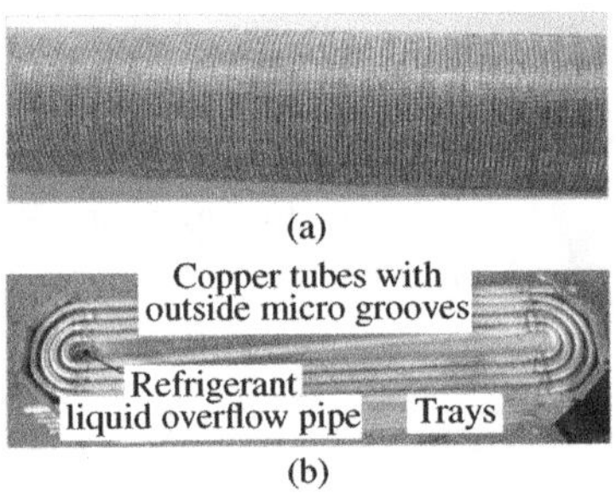

Figure 8.27 Photos of the evaporator. (a) Heat transfer tubes in evaporator and (b) structure of the evaporator

tubes above the refrigerant liquid will be covered by a thin liquid film. As a result, the heat transfer coefficient of water evaporator is enhanced. We have named this evaporator a rise film evaporator, which may have an evaporation heat transfer coefficient of about 5000 W/(m^{2}K).

The chilled water tank is divided into two parts (shown in Figure 8.28). The volumes for the left part and the right part are 88.2 and 18.6 l, respectively. Chilled water from the fan coil flows

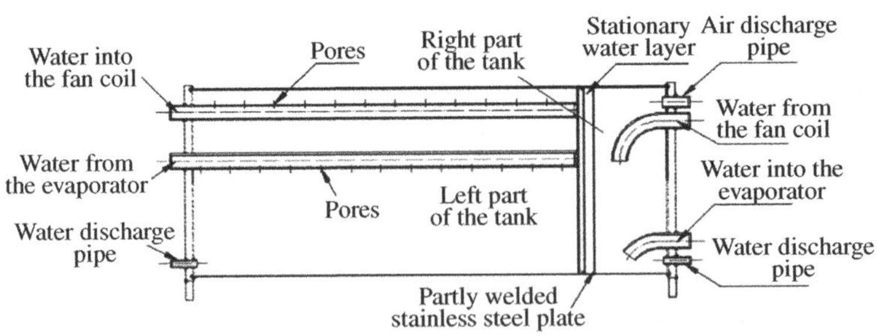

Figure 8.28 Schematic diagram of the chilled water tank

into the right part of the tank and mixes with the residual water, then the chilled water passes through the lower pipe in the right part of the tank into the evaporator and is cooled in the evaporator. In the left part of the tank, two long stainless steel pipes with pores are arranged. The chilled water rushes downwards into the left part of the tank. The chilled water mixes uniformly with the residual water in the left part of the tank, and then flows into the fan coil through the upper stainless steel pipe. A stationary water layer is formed by welding partly a stainless steel plate in the right part of the tank in order to reduce the heat exchange between the two parts of the chilled water tank.

As mentioned above, the studied adsorption chiller had compact adsorbers and highly thermal conductive evaporators. Vacuum valves are not adopted and the number of water valves is reduced by using three-way valves, which can greatly improve the reliability of the chiller. Meanwhile the special chilled water tank is adopted to reduce the temperature variation of chilled water during the heat and mass recovery process.

8.2.2 Working Principle

The process of the compact adsorption chiller consists of adsorption/desorption processes, mass recovery-like processes, and heat recovery processes. Schematic diagram of the working processes (six steps) of the compact adsorption chiller is shown in Figure 8.29. The six steps can be described consecutively as follows:

1. Adsorber 1 works in adsorption process and adsorber 2 is in desorption process (Figure 8.29a). Hot water flows into the adsorber 2 via V4, then it goes back to the hot water tank through V3. Thus the adsorber 2 desorbs water vapor. The desorbed water vapor firstly condenses in evaporator 2 since temperature of the evaporator 2 is much lower than that of condenser 2. When pressure of chamber 2 is higher than the saturation pressure under the temperature of the condenser 2, condensation will occur in the condenser 2, and then the condensate flows into the evaporator 2. Simultaneously, adsorber 1 is cooled by cooling water coming from the condenser 2 and adsorbs water (refrigerant) from evaporator 1. Meanwhile, chilled water rushes through the evaporator 1 via V5 and is cooled due to the evaporation of the refrigerant in the evaporator 1. Finally the chilled water flows back to the chilled water tank through V6. In this process, both condenser 1 and evaporator 2 are idle.
2. Mass recovery-like process (Heat recovery from evaporator 2 to evaporator 1, Figure 8.29b). A mass recovery-like process begins by the switch of V5, V6, and V8. Chilled water from the chilled water tank firstly flows into the evaporator 2 and pushes the residual chilled water inside the evaporator 2 into the evaporator 1. Meanwhile the residual chilled water inside the evaporator 1 is forced back to the chilled water tank through V8. Since the temperature of the residual chilled water from the evaporator 2 is higher than that of the evaporator 1, the evaporator 1 is heated. Therefore the adsorber 1 keeps on adsorbing refrigerant due to the increase of pressure in chamber 1. Simultaneously, the evaporator 2 is cooled by the chilled water. Since the pressure drops in the chamber 2, the adsorber 2 continues desorbing refrigerant. During this process, the adsorber 2 is still heated by hot water while the adsorber 1 is cooled by cooling water.
3. Heat recovery from adsorber 2 (Figure 8.29c). The valves V1, V4, V6, and V8 are switched. Chilled water flows through the evaporator 2 via V5 and flows back to the chilled water

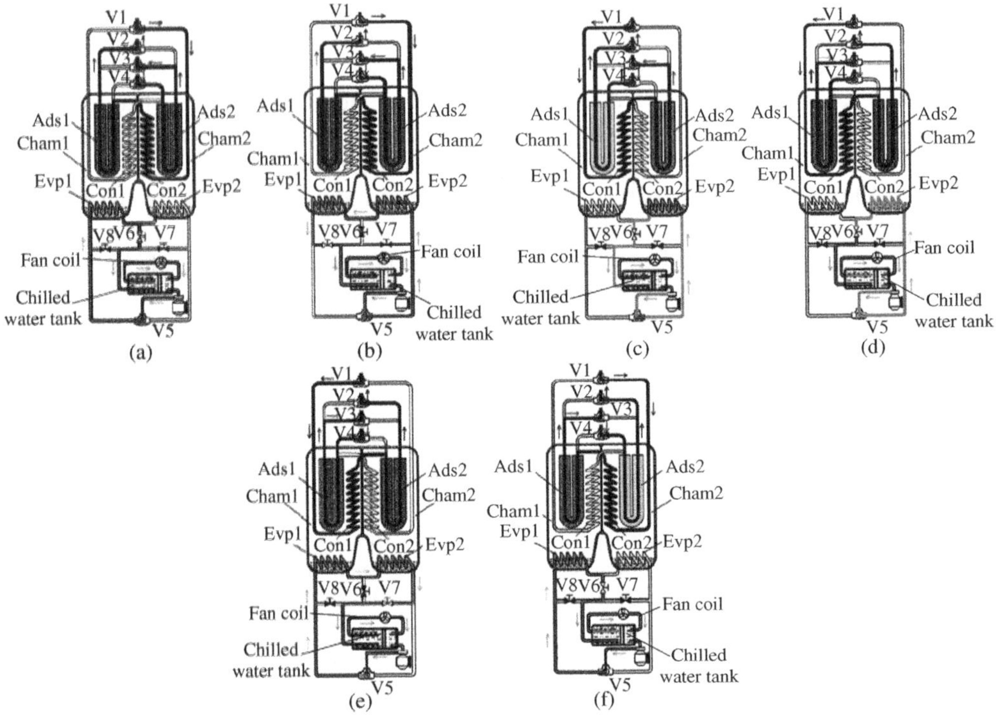

Figure 8.29 Six working steps of the studied adsorption chiller. (a) The first step; (b) the second step; (c) the third step; (d) the fourth step; (e) the fifth step; and (f) the sixth step

tank through V6. Simultaneously, hot water firstly flows into the adsorber 1 via V4, and the temperature of the hot water decreases sharply due to the low temperature of the adsorber 1. Then the hot water flows through a heat exchanger via V2 and is cooled by the cooling tower. As a result, the temperature of hot water approaches that of cooling water if proper heat recovery time is adopted, which means that the hot water turns into the cooling water during this process. Afterwards the cooling water is heated by the condensation of water vapor in the chamber 1 while flowing through the condenser 1. Finally the cooling water is further heated in the adsorber 2 and flows into the hot water tank via V3. Therefore the temperature of the adsorber 2 decreases. In conclusion, heat is recovered from the adsorber 2 by the circulation of hot water.

4. Adsorber 1 operates in desorption process and adsorber 2 works in adsorption process (Figure 8.29d). The adsorber 1 is heated by hot water while the adsorber 2 is cooled by the cooling water from the condenser 1. In the chamber 1, water vapor desorbed is condensed in the condenser 1 and the condensate flows downwards into the evaporator 1. Simultaneously, water in the evaporator 2 is adsorbed by the silica gel in the adsorber 2 so that the chilled water passing through the evaporator 2 is cooled. In this process, the condenser 2 and the evaporator 1 become idle.
5. Mass recovery-like process (Heat recovery from evaporator 1 to evaporator 2, Figure 8.29e). This process is similar to the second one. Chilled water flows into the evaporator 1 and the

evaporator 2 in sequence, and then rushes back to the chilled water tank. During this step, the evaporator 1 is cooled while the evaporator 2 is heated. Thus the adsorber 1 and adsorber 2 keep on desorbing and adsorbing refrigerant, respectively. In this process, the adsorber 1 and adsorber 2 were continuously heated and cooled, separately.

6. Heat recovery from adsorber 1 (Figure 8.29f). During this process, hot water flows firstly into the adsorber 2 and is sharply cooled. Then the hot water is further cooled while it flows through the heat exchanger of cooling water supply system. The hot water turns into cooling water if proper heat recovery time is adopted. Afterwards the cooling water flowing into the condenser 2 is heated by the condensation of the water vapor desorbed by the adsorber 2. Finally the cooling water is further heated in the adsorber 1 and runs into the hot water tank. When the sixth step is finished, the adsorption chiller switches to the first step and a new cycle starts.

8.2.3 Performance Test

8.2.3.1 Optimal Adsorption/Desorption Time

To determine the optimal adsorption/desorption time, the average hot water, cooling water and chilled water outlet temperatures are controlled at 78.5, 31.5, and 10.5 °C, respectively. In this section, the mass recovery-like time and heat recovery time is 40 and 30 seconds, respectively. As is shown in Figure 8.30, with the increase of adsorption/desorption time, the cooling power is increased sharply at first, and then it is decreased. When the adsorption/desorption time is 720 seconds, a maximum cooling power is obtained. The COP is also raised greatly at first, and then the increasing rate becomes small. When the adsorption/desorption time is 720 seconds, the adsorbers are insufficiently cooled or heated during the adsorption and desorption process, respectively. As a result, both the cooling power and COP are low. When the adsorption/desorption time is over 720 seconds, the cooling power is decreased since the water adsorbing rate of silica gel becomes smaller. However, the COP rises slowly when the adsorption/desorption time exceeds 720 seconds. The reason is that the heating power decreases due to the decrease of hot water inlet and outlet temperature difference when the desorption process of silica gel in desorption chamber nearly finishes.

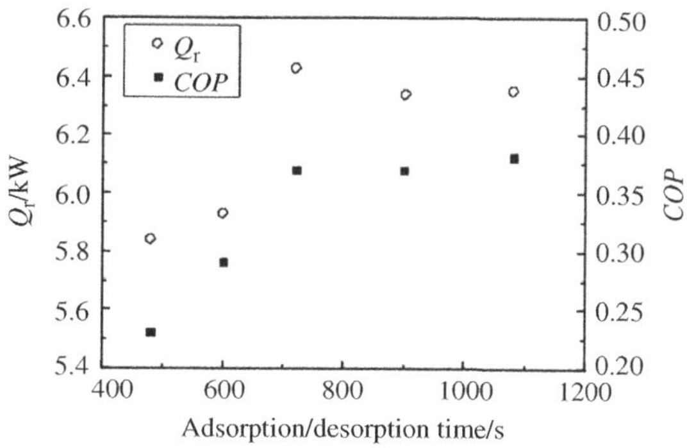

Figure 8.30 Optimal adsorption/desorption time

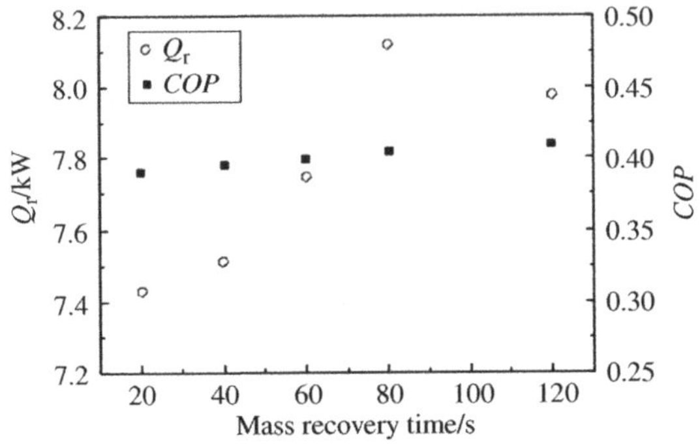

Figure 8.31 Influence of mass recovery-like process on the performance of the adsorption chiller

8.2.3.2 Effect of Mass Recovery-Like Time

The operating conditions are that the average hot water, cooling water, and chilled water outlet temperatures are 80.3, 31.3, and 11.2 °C, respectively. In this section, the adsorption/desorption time and heat recovery time are 720 and 30 seconds, respectively. As is shown in Figure 8.31 the cooling power increases sharply when the mass recovery time is smaller than 80 seconds. The cooling power decreases when the mass recovery time is over 80 seconds. Since the adsorber desorbs more refrigerant in a mass recovery cycle compared to a cycle without mass recovery process, the adsorber will adsorb more refrigerant from the evaporator in the next adsorption process, which contributes to a larger cooling power output. With the increase of mass recovery time, the cycled refrigerant mass will increase sharply at first, and then the increasing rate becomes smaller. With the increase of mass recovery time, the average cooling power increases firstly and then decreases due to the drop of the increasing rate of the cycle refrigerant mass during the mass recovery process. As a result, an optimized mass recovery time, which is about 80 seconds in this chiller, is adopted in order to get a maximum cooling power output. As is shown in Figure 8.31 the COP increases very less with the increase of mass recovery time. The reason is that the adsorber in desorption chamber is still heated by hot water during the mass recovery process. The more refrigerant desorbed by the silica gel in the desorption chamber is, the larger heating power input is needed. Therefore it is not always effective to improve COP by extending mass recovery time.

8.2.3.3 Influence of Heat Recovery Time

The heat recovery process is studied under the condition that the average hot water, cooling water, and chilled water outlet temperatures are 73.0, 30.9, and 11.4 °C, respectively. The adsorption/desorption time and mass recovery-like time are 720 and 80 seconds, separately. During the heat recovery process, hot water flow in the hot adsorber (for example, adsorber 1) will be switched off. The hot water temperature dropped due to the low temperature of the adsorber 1, and then it is further cooled in the heat exchanger by the cooling tower. Thus the temperature of the hot water flowing out of the adsorber 1 is equal to the cooling water outlet temperature. At the same time, cooling water flow in the cold adsorber, for example, in adsorber 2, will be heated. Then the cooling water with a high temperature flows back to the

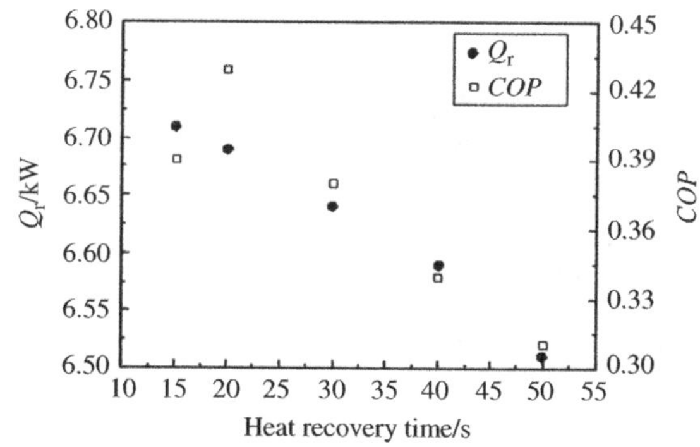

Figure 8.32 Cooling power and *COP* vs. heat recovery time

combustion heater. Therefore temperature of the cooling water coming from the adsorber 2 is considered to be the hot water outlet temperature. As a result, heat is recovered from the adsorber 2. The maximum recovered heat was obtained when the hot water outlet temperature is equal to the cooling water outlet temperature. As is shown in Figure 8.32 the cooling power decreases with the increase of heat recovery time, so mainly because of that there are no cooling power output during heat recovery process. The COP rises when the heat recovery time is shorter than 20 seconds. The reason is that the heat recovered is much larger than the cooling power decreased. When the heat recovery time is over 20 seconds, COP decreases sharply. As mentioned above, the heat was completely recycled with a heat recovery time of 40 seconds. However, the decrement of the cooling power is larger than the heat recycled, which will cause a smaller COP.

8.3 Adsorption Chiller Employing LiCl/Silica Gel–Methanol Working Pair

A novel composite adsorbent-methanol adsorption chiller was proposed and manufactured. It was filled by the adsorbent composed by Lithium Chloride and silica gel. Methanol was used as adsorbate and refrigerant. Experimental results showed that compared with silica gel–water chiller, SCP and COP of this novel chiller were improved [18, 19].

8.3.1 System Description

The schematic diagram of the composite adsorbent–methanol chiller is shown in Figure 8.33. The chiller is composed of three chambers. Both chamber 1 and chamber 2 contain one adsorber, one condenser, one evaporator. Chamber 3 is a heat pipe evaporator. By using the loop heat pipe in the evaporator, the vacuum valves are not used inside the chiller, so that the reliability of the chiller could be enhanced. Its working principle will be explained in the following section. Six three-way valves (V1–V6), one mass recovery valve (V7), and one check valve (V8) are adopted outside the chiller in this system. The photographs of the adsorber, condenser, evaporators, and photo of the chiller are shown respectively in Figure 8.33a–c.

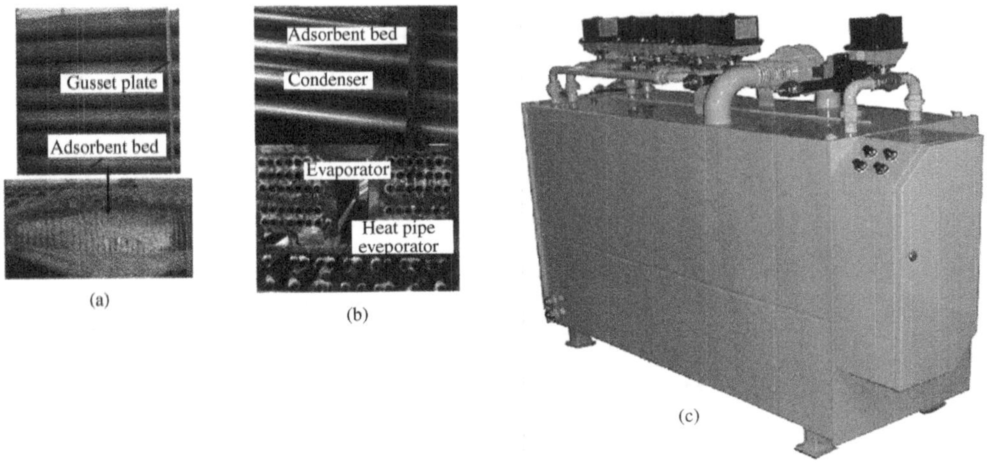

Figure 8.33 Schematic diagrams and photographs of the chiller. (a) Photo of adsorber; (b) photo of condenser and evaporator; and (c) picture of the chiller

8.3.2 Performance Test

8.3.2.1 Effect of Heating/Cooling Time

Figure 8.34 shows that the cooling capacity decreases firstly and then increases. When t_1 (heating/cooling time) is 680 seconds the value reaches the maximum, and after that it decreases again while *COP* keeps increasing until t_1 is 720 seconds. In Figure 8.34 the variation trend of cooling capacity is very complicated because there is a delay for the cooling capacity transfer from the evaporator to heat-pipe evaporator and the heat-pipe evaporator to the heat-transfer medium. Generally, when the cycle time is longer, the cooling capacity is larger, as Figure 8.35 shows. Sometimes cooling capacity may decrease as cooling capacity is relatively small. For the real application, the adsorption/desorption and condensation/evaporation processes cannot be proceeded completely. The desorption time is longer, the methanol desorbed will be more adequate, which means that in the next cycle, the adsorbent bed has a larger adsorption capacity

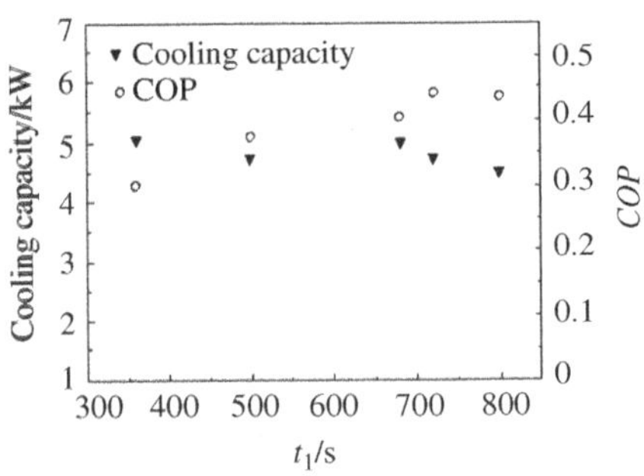

Figure 8.34 The cooling capacity and COP vs. the heating/cooling time (the temperature of the hot water inlet, the chilled water outlet, and the cooling water inlet is 85, 15, and 30 °C, respectively. The mass recovery time is 60 seconds and the heat recovery time is 20 seconds)

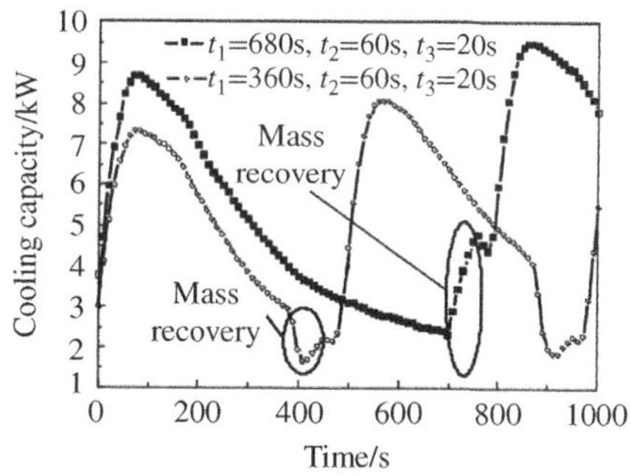

Figure 8.35 Cooling capacity vs. heating/cooling time. (The temperature of the hot water inlet, chilled water outlet, and cooling water inlet is 85, 15, and 30 °C, respectively. In the figure t_1 is heating/cooling time, t_2 is mass recovery time, and t_3 is heat recovery time)

because the concentration difference of the methanol inside and outside of the adsorber is larger. In the experiments, because the adsorption/desorption process nearly finished when the cycle time is 720 seconds, the highest cooling capacity can be obtained when heating/cooling time is between 680 and 720 seconds. t_1 should be longer if a higher *COP* is required. The largest *COP* can be obtained when t_1 is 720 and 800 seconds. The optimal cycle time is dependent on the requirements of *COP* or the cooling capacity.

8.3.2.2 Effect of Mass Recovery Times

The influences of the mass recovery process on the performance of the chiller are as follows:

1. The adsorption bed adsorbs the vapor from the desorption bed instead of the evaporator, so the evaporation in the heat pipe evaporator is weakened and thus the cooling capacity is seriously decreased.
2. The hot vapor flows from adsorber 2 to adsorber 1 because of the pressure difference in mass recovery process. Most of the hot vapor will be condensed in the condenser. However, when vapor velocity is too fast and according to the connection of the condenser 1 and evaporator 1, it is entirely possible that a small part of the vapor will be condensed in evaporator 1, and the temperature of the evaporator 1 will increase. The mass recovery time is longer, the desorption will be more completed, and the more refrigeration capacity will be obtained. Figure 8.36 shows that the mass recovery improves the cooling capacity and *COP*. However, when mass recovery time is larger than 60 seconds, the cooling capacity and *COP* almost remain constantly. Thus in the experiments 60 seconds is enough for the mass recovery process.

8.3.2.3 Operation Performance with Different Hot Water Inlet

Figure 8.37 shows the variations of cooling capacity and COP with the increasing hot water temperature. It demonstrates that the performance of the chiller is affected by hot water inlet temperature. The suitable temperature is about 80 °C. The machine can work even when the heat source is below 60 °C.

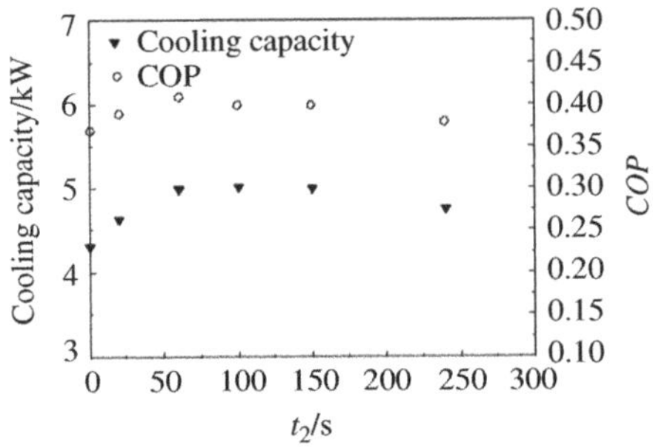

Figure 8.36 The cooling capacity and COP vs. mass recovery time. (The temperature of hot water inlet, chilled water outlet, and cooling water inlet are 85, 15, and 30 °C, respectively. The heating/cooling time is 680 seconds and the heat recovery time is 20 seconds.)

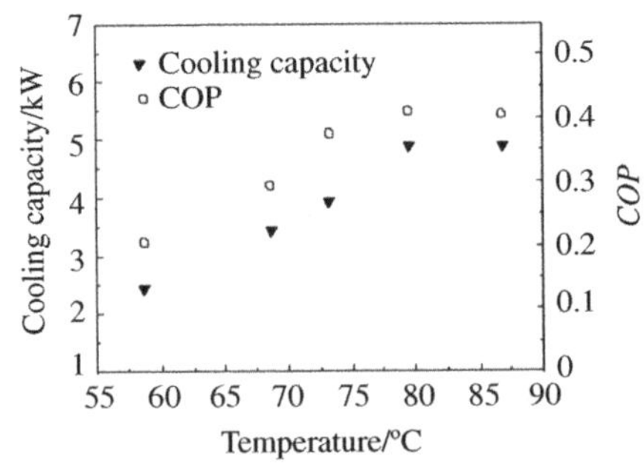

Figure 8.37 The cooling capacity and COP vs. the temperature of the hot water inlet. (The temperature of the chilled water outlet and the cooling water inlet are 20 and 30 °C, respectively. The heating/cooling time is 680 seconds, the mass recovery time 60 seconds, and the heat recovery time is 20 seconds.)

8.3.2.4 Performance Test under the Conditions of Different Evaporation Temperature

Figure 8.38 shows the temperature variation of the chilled water inlet and outlet when the evaporating temperature is 0 °C. Results showed that the chilled water outlet temperature rises sharply when the half cycle is ended. The reason is that at the low evaporating temperature the adsorption capacity is small. As a consequence a major part of the hot methanol vapor stays in the dead volume of the chamber. When the process of adsorber switches from the adsorption to desorption, some of the refrigerant vapor would be condensed and the temperature of the evaporator will increase. Such a process will influence the cooling capacity of the system. Such an influence will be more serious when the cooling capacity is small under the condition of the low evaporating temperature.

8.3.2.5 Performance Comparison with Silica Gel–Water Chiller

Different temperatures of hot water inlet, cooling water inlet, and chilling water inlet will result in different *COP* and cooling capacity. The *COP* and cooling capacity variations with

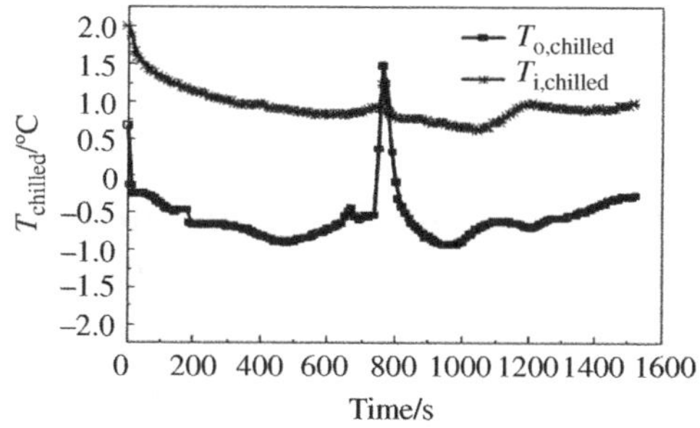

Figure 8.38 The temperature for the heat transfer fluid at the inlet and outlet of the evaporator. (The temperature of the hot water inlet, the chilled water outlet, and the cooling water inlet are 85, −3, and 29 °C, respectively. The heating/cooling time is 680 seconds, mass recovery time is 60 seconds, and heat recovery time is 20 seconds.)

inlet temperatures are shown in Figure 8.39a–c. The *COP* will increase with the increasing temperature of the hot water inlet and the chilled water inlet, and will decrease with increasing temperature of the cooling water. The cooling capacity has the same temperature trends as the *COP*. Compare the performance of the chillers with different adsorbents of pure silica gel and composite adsorbent, the *COP* of the chiller using composite adsorbent is slightly higher than the conventional chiller using the pure silica gel except conditions for the temperature of

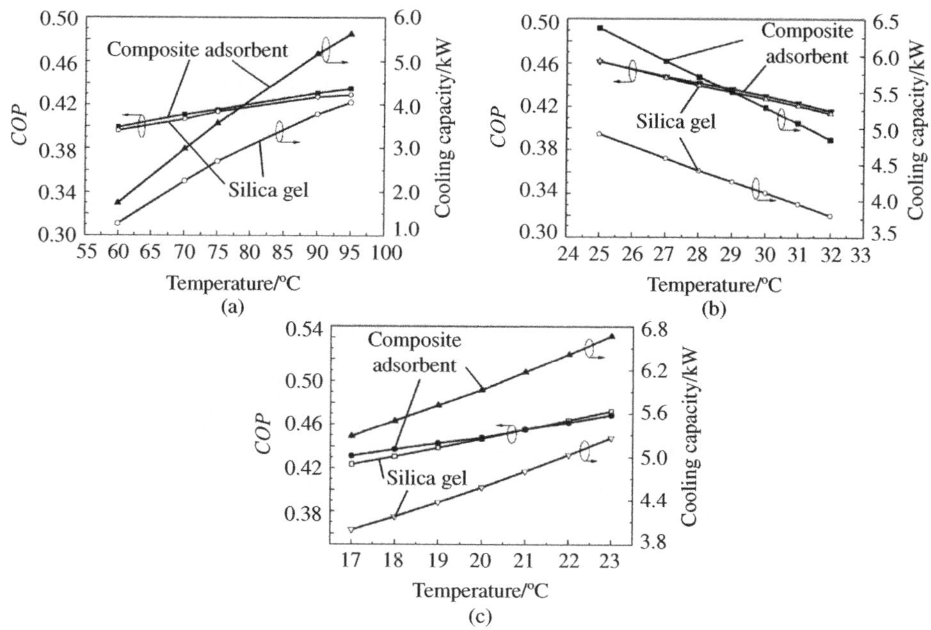

Figure 8.39 COP and cooling capacity of the chillers with different adsorbents. (a) The performance vs. the temperature of hot water inlet; (b) the performance vs. the temperature of cooling water inlet; and (c) the performance vs. the temperature of chilling water inlet

the chilling water inlet higher than 21 °C. The cooling capacity of the chiller using composite adsorbent is obviously higher than that of the conventional chiller.

8.4 Adsorption Ice Maker Adopted Consolidated Activated Carbon–Methanol Working Pair and Used for a Fishing Boat

Adsorption ice maker for a fishing boat adopted consolidated activated carbon–methanol as the working pair, for which the consolidated adsorbent could enhance the heat transfer performance of the adsorbent bed. Such an ice maker is very feasible for the fishing boats. If compared with the adsorption refrigeration system with water as the refrigerant, the advantage of the system is that it can be applied under the freezing condition for which the evaporation temperature is lower than 0 °C. Such a system can be driven by the waste heat of the diesel engine, and the cooling power is mainly used for the storage of the fish.

Because the large fishing boats are always equipped with the compression type refrigeration equipments, the adsorption ice maker developed here is mainly for application on the small and medium-sized fishing boats. For such type of boats the waste heat of the diesel engine is always directly emitted into the atmosphere with a temperature up to 300–400 °C. Meanwhile the boats always brought the ice for keeping the fish fresh and don't have a refrigerator onboard. Taking the 804 type boat that could carriage 100 tons fish as an example, there are two factors for the economic losses. Firstly, the annual cost is about 40 000 RMB Yuan for the cost of the ice, and secondly, under the condition that the time for the ship out at the sea is prolonged, the poor storage ability for the fish using just ice will lead to an economy loss amounting to 200 000 RMB Yuan per year. There are two advantages for the utilization of the adsorption ice makers on the fishing boats. Firstly, the waste heat is used to drive the system and can reduce the thermal pollution on the environment; secondly, it can save the cost of ice and reduce the economic losses caused by poor preservation conditions for the fish [20].

8.4.1 The Heat Transfer Intensification Technologies for the Adsorbent Bed

For adsorption ice makers on fishing boats with activated carbon–methanol as the working pair, due to the limited space on the fishing boats, it is necessary to enhance the heat transfer performance of the adsorbent beds, as well as improving the bulk density of the adsorbent, which can effectively make the system compact. So the consolidated adsorbent is developed. The activated carbon is the Hainan coconut shell activated carbon with 14–28 mesh and the block structure is shown in Figure 8.40. The cross-section size of the block is 116 × 272 mm, the thickness of the block is 10 mm, and the diameter of the hole on the block is 14 mm. The parameters of the granular activated carbon and the consolidated activated carbon are compared and the results are shown in Table 8.12. The bulk density of the consolidated activated carbon is far greater than that of the granular activated carbon, and the heat transfer coefficient of the consolidated activated carbon is also higher than that of the granular one. Thus the consolidated adsorbent could improve the heat transfer performance of the adsorbent bed as well as the volume filling density of the adsorbent [21].

In the adsorbent bed the numbers of mass transfer channels are added for the enhancement of mass transfer performance. The cross section of the bed is shown in Figure 8.41.

Figure 8.40 Structure for the cross-section of the consolidated activated carbon

Table 8.12 Parameters of adsorbent

Parameter	Granular activated carbon	Consolidated activated carbon
Density (kg/m^3)	360	600
Heat transfer coefficient (W/(m °C))	0.017	0.27–0.34
Specific heat (kJ/(kg °C))	0.93	0.93

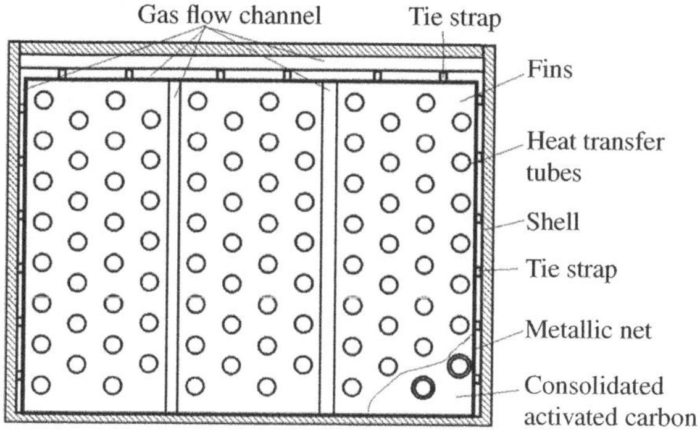

Figure 8.41 The diagram for the cross-section of the adsorbent bed

The adsorbent bed is the shell-tube type heat exchanger as shown in Figure 8.41. The heat transfer medium is the water. There are five mass transfer channels on the section of the bed and the minimum path is only 58 mm. The mass of the adsorbent in each bed is 60 kg and the refrigerant is methanol. The characteristics of the adsorbent bed are as follows:

1. The heat transfer area inside the bed is extended and the adsorbent is consolidated for the improvement of the heat transfer performance in the bed. The fin thickness of the fins is only 0.15 mm, and the mass of it is very small if compared with the mass of the whole bed thus it has little influence on the total metal heat capacity of the bed. Figure 8.41 showed that the distance between the fins (i.e., the thickness for the block of the activated carbon) in the bed is only 10 mm, which is very small and could enhance the heat transfer performance effectively.
2. The mass transfer path is reduced for the improvement of the mass transfer performance in the adsorbent bed. Because the consolidation process of the adsorbent will increase the complexity as well as the cost for the manufacture process of the bed, such a technology is only valuable when the volume filling quantity is increased greatly. Because the increment of the mass transfer channels will decrease the filling density of the adsorbent inside the bed, the density of the consolidated adsorbent is controlled at a level that could improve the filling density of the adsorbent effectively. For example, the packing density of the granular activated carbon in the adsorbent bed is 360 kg/m^3, and the consolidated adsorbent is chosen by the criterion that the volume filling density of the adsorbent is improved by more than 50% the density of the granular activated carbon. The mass transfer channel volume is 15.6% of the total volume in the adsorbent bed.

The heat capacity and the mass of the adsorbent, the refrigerant, and the metal of the bed are shown in Table 8.13. The material for the walls and the ribs of the adsorbent bed is steel, the material of the mesh is stainless steel, and the heat transfer pipes are made of copper. Due to copper being compatible with seawater, the adsorbent bed can be directly cooled by the cold seawater, which can reduce the electricity power of the water pump and enhance the heat transfer performance by the direct heat exchanging process.

8.4.2 Design of Activated Carbon–Methanol Adsorption Ice Maker

The structure diagram of activated carbon–methanol adsorption ice maker is shown in Figure 8.42 [21]. The system comprises two adsorbent beds, a condenser and an evaporator.

Table 8.13 Parameters of the adsorbent bed

Component	Material	Mass (kg)	Specific heat (kJ/°C)
Refrigerant	Methanol	60×0.13 (0.13 is the cycle adsorption quantity)	21.45
Metal	–	158.51	66.384
Adsorbent	Coconut shell activated carbon	60	55.8

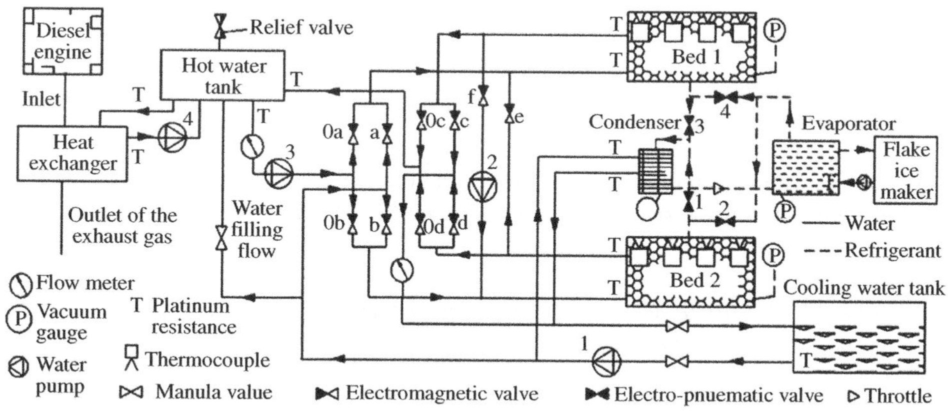

Figure 8.42 Diagram for the activated carbon–methanol adsorption ice maker

The system is driven by the waste heat of the diesel engine directly. In order to use the waste heat effectively, a heat storage tank is installed at the outlet of the gas-water heat exchanger. It has two functions: one is to recover the waste heat from the diesel engine during the heat and mass recovery processes between two adsorption beds, and the other is that to ensure there is enough water in the system that could prevent the vapor-liquid two-phase flow in the heat transfer pipes. The adsorption bed is directly cooled by the cooling source. Two adsorption beds are used to achieve a continuous cooling output. The heat recovery between two beds is controlled by pump 2 and valves f, c. Mass recovery is controlled by valves 3 and 1. The refrigerant evaporates in the evaporator by the function of the adsorption in the bed, and the cooling power is directly input to the flake ice maker to produce ice. The designed ice output of the system is about 350–500 kg ice per day, corresponding to the cooling capacity of 1.5–3 kW for the system.

In the working process, the pumps 1 and 4 and the magnetic pump between the evaporator and flake ice machine are normally open, and the rest pumps and valves are as follows:

1. For the heating process of bed 1 and cooling process of bed 2, open water valves 0b, 0d, a, c, and pump 3.
2. For the desorption process of bed 1 and cooling process of bed 2, open water valves 0b, 0d, a, c, pump 3, and valve 1.
3. For the heating process of bed 1 and adsorption process of bed 2, open water valves 0b, 0d, a, c, pump 3, and valve 4.
4. For the desorption process of bed 1 and adsorption process of bed 2, open water valves 0b, 0d, a, c, pump 3, valves 1 and 4.
5. For mass recovery process, open water valves e, f, pump 2, valves 3 and 1.
6. For heat recovery process, open water valves e, f, and pump 2.
7. For cooling process of bed 1 and heating process of bed 2, open water valves 0a, 0c, b, d, and pump 3;
8. For the cooling process of bed 1 and desorption process of bed 2, open water valves 0a, 0c, b, d, pump 3, and valve 3.

9. For the adsorption process of bed 1 and heating process of bed 2, open water valves 0a, 0C, b, d, pump 3, and valve 2.
10. For the adsorption process of bed 1 and desorption process of bed 2, open water valves 0a, 0c, b, d, pump 3, valves 2 and 3.

The condenser is a plate-fin type heat exchanger and the condensing area is 1.4 m^2. The evaporator is designed as a reservoir and the storage quantity of the refrigerant is 30 kg.

8.4.3 The Mathematic Model for the Activated Carbon–Methanol Adsorption Ice Maker

An activated carbon–methanol adsorption ice maker has the heat and mass recovery process between two beds, and the models mainly refer to the thermodynamic calculation of the basic cycle, two-bed heat recovery cycle and two-bed mass recovery cycle in Chapter 6 [22]. The contents introduced here are mainly the energy conservation models of the system [23].

For the heat and mass transfer process of the refrigerant inside the solid adsorbent bed, the parameters concerned are the vapor temperature T_{gas}, vapor density ρ_{gas}, vapor flow rate u_{gas}, vapor pressure p_{gas}, vapor heat flux J_{gas}, and the heat flux for the solid adsorbent skeleton J_{so}. These variables usually use the average measures to oriented micro-scale to the macro-scale (within the void space), which can obtain the macroscopic mathematical description for the heat and mass transfer process in porous medium. Each variable of the fluid can be looked as an average value, which is:

$$u_{f,aver} = \int_{(\Delta U_0)_V} u_{gas} \mathrm{d}U_V \bigg/ \int_{(\Delta U_0)_V} \mathrm{d}U_V \tag{8.27}$$

The variables of solid skeleton are defined as:

$$u_{s,aver} = \int_{(\Delta U_0)_s} u_{gas} \mathrm{d}U_s \bigg/ \int_{(\Delta U_0)_s} \mathrm{d}U_s \tag{8.28}$$

where $u_{f,aver}$ is the average value of the fluid inside the porous structure of the adsorbent; $u_{s,aver}$ is the generic variable of the skeleton of the porous adsorbent; $(\Delta U_0)_\text{v} = \beta \Delta U_0$ and $(\Delta U_0)_\text{s} = (1 - \beta)\Delta U_0$ are the void volume and the volume of the solid skeleton in the reference unit volume of ΔU_0, respectively.

There are six basic equations, and they are:

$$\frac{\partial \rho_{\text{gas}}}{\partial t} + \frac{\partial}{\partial L_{xi}}(\rho_{\text{gas}} u_{gas}) = -\frac{\beta_\text{s} \rho_{\text{so}}}{\beta} \frac{\partial x}{\partial t} \quad \text{Vapor continuity equations} \tag{8.29}$$

where t is the time (s); L_x is the coordinate of the plane coordinate system; ρ_{so} is the bulk density of the consolidated activated carbon (kg/m_3); x is the adsorption amount at the bed temperature of T and bed pressure of p (kg/kg); ρ_{gas} is the vapor density (kg/m^3). β_s and β are coefficients.

Neglecting inertial terms of kinematic equation, the vapor kinematic equation is:

$$u_{gas} = -\frac{k_{ij}}{\beta \mu_{\text{gas}}} \left(\frac{\partial p}{\partial L_{xj}} + \rho_{\text{gas}} g \frac{\partial L_y}{\partial L_{xj}} \right) \quad \text{Vapor kinematic equation} \tag{8.30}$$

where u_{gas} is the vapor flow rate (m/s); k_{ij} is the component of the permeability tensor for different medium (m^2); μ_{gas} is the vapor dynamic viscosity (Pa·s); g is the gravitational acceleration (m/s^2). L_y is the vertical upward direction.

Energy conservation equation of the vapor and the heat flux equation are:

$$\rho_{gas} C_{pgas}\left(\frac{\partial T_{gas}}{\partial t}+u_i\frac{\partial T_{gas}}{\partial L_{xi}}\right)=-\frac{\partial J_{gasi}}{\partial L_{xi}}-\frac{\partial}{\partial L_{xi}}\left(E_{ij}\frac{\partial T_{gas}}{\partial L_{xj}}\right)+\alpha_{gas}(T_{so}-T_{gas})+\varepsilon_{\mu} \quad (8.31)$$

$$J_{gasi}=-(\lambda_{gas})_{ij}\frac{\partial T_{gas}}{\partial L_{xj}} \quad (8.32)$$

In Equation 8.31, the first term on the right side of the equal sign is the heat obtained by conduction; J_{gasi} is the projection of the vapor conduction heat flux on the i axis; $(\lambda_{gas})_{ij}$ is the vapor thermal conductivity, W/(m·K). The second term is the heat obtained by the unit volume of vapor due to the thermal diffusion, for which the heat dispersion flux generated by the undulation of u and T. E_{ij} is the thermal dispersion coefficient. ε_{μ} is the energy consumption rate of fluid in the unit time and the unit volume caused by the fluid viscosity, W/m^3. α_{gas} is the heat transfer coefficient for the heat transfer process from the solid adsorbent skeleton to the refrigerant vapor, $W/(m^2 \cdot K)$. $\alpha_{gas}(T_{so} - T_{gas})$ is the heat transfer to the unit volume of vapor in the unit time due to the temperature difference between the vapor and the solid adsorbent skeleton. C_{pgas} is the gas average specific heat under constant pressure, J/(kg·K).

If $e_{so} = C_{so}$ is adopted as the formula for the calculation of the internal energy of the solid adsorbent skeleton, the average heat conservation equation of solid adsorbent porous skeleton is:

$$\frac{\rho_{so}}{1-\beta_p}C_{so}\frac{\partial T_{so}}{\partial t}=-\frac{\partial J_{soi}}{\partial L_{xi}}+\alpha_{so}(T_{gas}-T_{so}) \quad (8.33)$$

$$J_{soi}=-(\lambda_{so})_{ij}\frac{\partial T_{so}}{\partial L_{xj}} \quad (8.34)$$

where β_p is the porosity of the solid adsorbent; C_{so} is the specific heat of the solid adsorbent skeleton (assuming that it is independent of the time), J/(kg·K); $(\lambda_{so})_{ij}$ is the thermal conductivity coefficient of the solid adsorbent skeleton, W/(m·K); α_{so} is the heat transfer coefficient for the heat transfer process between refrigerant vapor and the solid adsorbent skeleton side, W/(m·K); J_{gasi} is the projection of the gas conduction heat flux per unit area on the i axis.

In order to make all independent unknown numbers equal to the number of equations, that is, equations closed, it is necessary to supplement the density of the real refrigerant vapor, the specific volume of the vapor at constant pressure, the thermal conductivity coefficient, and the viscosity of the vapor at the low pressure, and so on, and then simplify the energy equations. The simplified conditions are as follows:

1. Assuming that the skeleton of the adsorbent is incompressible (i.e. β does not change with the time τ), and a single block of solid adsorbent medium is isotropic (i.e. it does not vary with the spatial variation). $(\lambda_{gas})_{ij}$ and $(\lambda_{so})_{ij}$ can be simplified as the standard values of λ_{gas} and λ_{so}. k_{ij} can be instead by k.
2. Compared with the heat conduction flux, the thermal diffusion can usually be negligible, that is, $E_{ij} = 0$;

3. Compared to the other items, the dissipative term ε_μ in the energy equation is also very small, that is, $\varepsilon_\mu = 0$.

Substituting Equations 8.32 and 8.34 into Equations 8.31 and 8.33, Equations 8.31 and 8.33 can be written as follows:

$$\rho_{\text{gas}} C_{\text{pgas}} \left(\frac{\partial T_{\text{gas}}}{\partial t} + u_i \frac{\partial T_{\text{gas}}}{\partial L_{xi}} \right) = \frac{\partial}{\partial L_{xi}} \left(\lambda_{\text{gas}} \frac{\partial T_{\text{gas}}}{\partial L_{xi}} \right) + \alpha_{\text{gas}} (T_{\text{so}} - T_{\text{gas}}) \tag{8.35}$$

$$\frac{\rho_{\text{so}}}{1 - \beta_{\text{p}}} C_{\text{so}} \frac{\partial T_{\text{so}}}{\partial t} = \frac{\partial}{\partial L_{xi}} \left(\lambda_{so} \frac{\partial T_{\text{so}}}{\partial L_{xi}} \right) + \alpha_{\text{so}} (T_{\text{gas}} - T_{\text{so}}) \tag{8.36}$$

If the velocity of the fluid in the bed is very low, that is, the Reynolds number in the adsorption bed is very small, we can ignore the temperature difference between the vapor and the solid adsorbent skeleton, and take the local thermal equilibrium between the vapor and the solid, that is, $T_{\text{gas}} = T_{\text{so}} = T$. Then both sides of the energy Equation 8.35 multiply by β, both sides of the energy equation of the solid adsorbent skeleton (Equation 8.36) multiply $\beta_{\text{s}}(1 - \beta_{\text{p}})$, and after that we add two equations together. Because $\beta\alpha_{\text{gas}} = \beta_{\text{s}}(1 - \beta_{\text{p}})\alpha_{\text{so}}$, we can finally get a unified energy equation, and it is:

$$\left[\beta\rho_{\text{gas}} C_{\text{pgas}} + \beta_{\text{so}}\rho_{\text{so}} C_{\text{so}}\right] \frac{\partial T}{\partial t} + \beta\rho_{\text{gas}} C_{\text{pgas}} u_i \frac{\partial T}{\partial L_{xi}} = \frac{\partial}{\partial L_{xi}} \left\{ \left[\beta\lambda_{\text{gas}} + \beta_{\text{so}} \left(1 - \beta_{\text{p}}\right) \lambda_{so}\right] \frac{\partial T}{\partial L_{xi}} \right\} \tag{8.37}$$

Introduction of the effective thermal conductive coefficient λ_{eff} and the effective heat capacity C_{eff} of the adsorbent bed, and then the terms for the heat source and heat sink, such as desorption heat and adsorption heat, the simplified energy equation of the consolidated adsorbent bed is:

$$\frac{\partial}{\partial t}(C_{\text{eff}} T) + \frac{\partial}{\partial L_{xi}} \left(\beta\rho_{\text{gas}} C_{\text{pgas}} u_i T - \lambda_{\text{eff}} \frac{\partial T}{\partial L_{xi}} \right) = \beta_{\text{so}} S_{\text{h}} + \beta_{\text{so}} \rho_{\text{so}} q_{\text{st}} \frac{\partial x}{\partial t} \tag{8.38}$$

where S_h is the heat exchange rate in the unit volume by the solid adsorbent side (i.e., the heat exchange amount between the solid adsorbent and the fluid by the pipes and fins), W/m^3; q_{st} is the internal adsorption heat and desorption heat, J/kg.

The conventional control equation for the pure fluid is obtained by the unit volume of the fluid, but for the porous medium, only the β portion in the space of the porous medium per unit volume is the fluid, and the remaining portion is occupied by the solid skeleton (activated carbon and other solid objects). The object of the Equation 8.38 is the whole space of the unit volume adsorbent.

The adsorption and desorption equation of the adsorbent is the D-A equation:

$$x = x_0 \exp\left[-K \left(\frac{T}{T_s} - 1 \right)^n \right] \tag{8.39}$$

Experiments showed that x_0, K, and n are 0.367 kg/kg, 12.9, and 1.45, respectively.

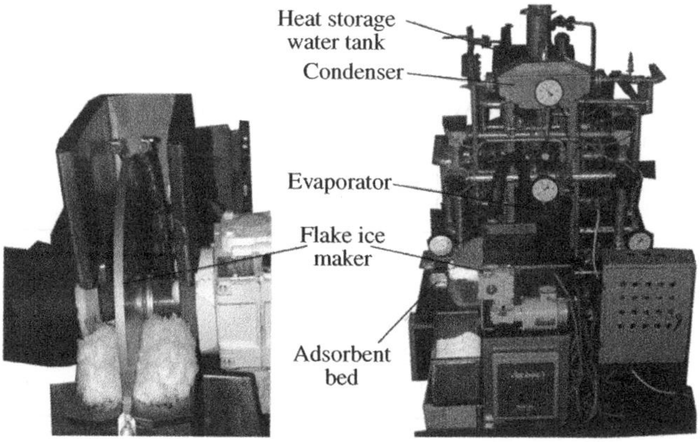

Figure 8.43 The prototype of adsorption ice maker

8.4.4 The Adsorption Refrigeration Performances of Activated Carbon–Methanol Adsorption Ice Maker

8.4.4.1 Experimental Prototype

The experimental prototype is shown in Figure 8.43. The main component of the flake ice maker is a stainless steel disk-shaped evaporator, and at the middle of the evaporator is the refrigerant channel. The disc-shaped evaporator has a diameter of 380 mm and a thickness of 13 mm. The refrigerant flows through the evaporator and takes away the heat on the surface, and water freezes outside the evaporator. The size of the adsorption prototype is 2.5 × 1.5 × 2 m (length × width × height). The heating process of the adsorbent bed is simulated by the waste heat of an oil burner.

8.4.4.2 The Heat and Mass Recovery Performance of an Adsorption Ice Maker

Under the conditions of the ambient temperature of 30 °C and cooling water temperature of 28 °C, the system was tested without heat and mass recovery process, and the arrangements as well as parts of performance parameters of the system are shown in Table 8.14. The heating power in Table 8.14 is calculated according to the water flow of the heat exchanger side as well as inlet and outlet temperature difference. The cooling capacity is calculated by the ice production, and the equation is:

$$W_{ref} = \frac{M_{ice} \times L_{water} + M_{ice} \times C_{water} \times T_{water} - m_{ice} \times c_{ice} \times T_{ice}}{t_c}$$

$$COP = \frac{W_{ref}}{W_h} \quad SCP = \frac{W_{ref}}{M_a} \tag{8.40}$$

Table 8.14 The arrangements of the experiments and the performance of the cycle without heat and mass recovery process

Number	Hot water temperature (°C)	Cycle time (min)	Adsorption /desorption time (min)	Average evaporation temperature (°C)	Average temperature of ice (°C)	Mass of ice (kg/h)	Cooling power (kW)	Average heating power (kW)	*COP*	*SCP* (W/kg)
1	62–107	30	15	−7.21	−4.5	11.3	1.34	27.92	0.048	22.4
2	66–108	40	20	−8.54	−5.5	12.3	1.58	28.00	0.056	26.4
3	68–109	50	25	−10.19	−5.70	13.5	1.73	20.69	0.084	28.8
4	67–112	60	30	−10.23	−6.31	12.8	1.56	17.73	0.088	26.0
5	68–118	70	35	−10.34	−6.92	11.5	1.48	16.93	0.087	24.6

where M_{ice} is ice making capacity in one cycle; L_{water} is the latent vaporization heat of water; C_{ice} is the specific heat of the ice; T_{ice} is the ice temperature; T_{water} is the water temperature; T_c is the cycle time; M_a is the mass of the activated carbon in a adsorbent bed; W_h is the heating power.

Table 8.14 showed that without heat recovery process, the temperature of the hot water is low in the heat storage tank. The cooling capacity of the adsorbent bed reached the maximum value when the cycle time is 50 minutes [24].

Under the same ambient temperature, the heat and mass recovery experiments are preceded. The mass recovery process lasts for 1.5 minutes. After mass recovery open the valves for the heat recovery and let the process lasts for 2 minutes. Such a process ensures that for the mass recovery the bed after adsorption is under the condition of low-temperature and low-pressure, and the bed after desorption under the condition of high-temperature and high-pressure, which will ensure a fully and sufficient mass recovery process. When the cycle time is 50 minutes, the temperature and pressure of the adsorbent bed with the heat and mass recovery process are shown in Figures 8.44 and 8.45 [25].

Assuming that the adsorbent bed 1 is just after the desorption and the adsorbent bed 2 is just after the adsorption, p_1 and T_1 in Figures 8.44 and 8.45 are the pressure and temperature of adsorbent bed 1, and p_2 and T_2 are pressure and temperature of the adsorbent bed 2, respectively. If the heat and mass recovery processes proceed synchronizedly, at the end of the heat and mass recovery, the pressure of the adsorbent bed 1 is almost equal to the pressure of the

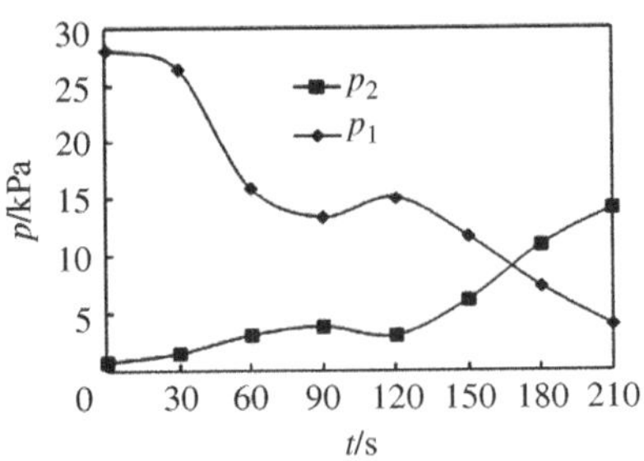

Figure 8.44 The pressure change of two beds for heat and mass recovery process

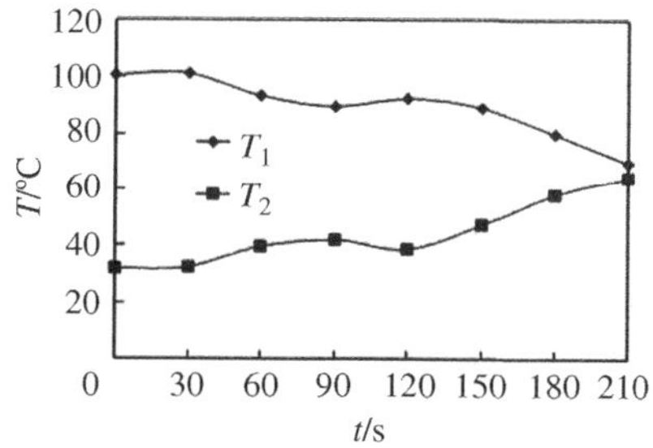

Figure 8.45 The temperature change of two beds for the heat and mass recovery process

Table 8.15 The arrangements and the performance of the experiments with heat and mass recovery process

Number	Hot water temperature (°C)	Cycle time (min)	Adsorption/desorption time (min)	Mass recovery time (min)	Heat recovery time (min)	Average evaporation temperature (°C)	Average temperature of ice (°C)	Mass of ice (kg/h)	Cooling power (kW)	Average heating power (kW)	*COP*	*SCP* (W/kg)
1	87–120	30	12	1.5	2	−5.41	−2.5	12.00	1.51	15.21	0.088	25.0
2	90–120	40	17	1.5	2	−9.47	−5.61	13.50	1.72	17.62	0.097	28.6
3	95–120	50	22	1.5	2	−10.23	−6.61	15.00	1.93	18.34	0.105	32.2
4	92–120	60	27	1.5	2	−10.31	−6.63	14.64	1.87	15.63	0.120	31.2
5	98–120	70	32	1.5	2	−10.97	−7.5	14.32	1.84	14.98	0.123	30.6
6	95–120	80	37	1.5	2	−10.53	−6.96	13.78	1.76	14.71	0.120	29.4

adsorbent bed 2. As shown in Figures 8.44 and 8.45, due to the fact that the experiments with heat recovery process is just after mass recovery process, at the end of the heat recovery of the adsorbent beds, the temperature difference between two adsorbent beds is small and the pressure of the adsorbent bed 1 is lower than that of the adsorbent bed 2.

The arrangements and the performance of the experiments are shown in Table 8.15.

Table 8.15 showed that the heat and mass recovery processes effectively improved the temperature of the hot water storage tank. The optimum value for the performance of the system is obtained when the cycle time is 50 minutes. The relationship between experimental and numerical simulation results of *COP* and *SCP* with and without heat and mass recovery processes are shown in Figure 8.46.

Since the simulation is carried out under the ideal conditions of no heat loss, the simulated *COP* is bigger than the experimental value. Heat and mass recovery can significantly improve the system *COP* by 20–30% and *SCP* by 7–11%. With the short cycle time such as 15 minutes, due to the insufficient heating and cooling time, the system *COP* and *SCP* are small. After that both *COP* and *SCP* increases when the cycle time increases. The *SCP* for the cycle with heat and mass recovery process gets to an optimum value when the cycle time is 50 minutes. Due to the fact that the heat and mass recovery process needs 7 minutes, when the cycle time is short, the cooling adsorption time was limited by a fixed cycle time and heat and mass recovery time, thus the simulated *SCP* with heat and mass transfer recovery is lower than that without heat and mass recovery.

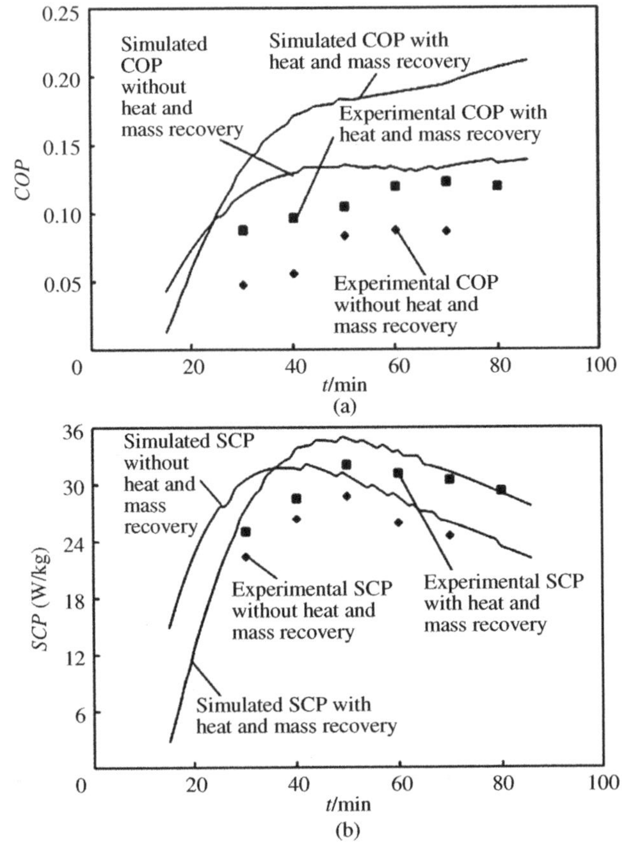

Figure 8.46 Comparisons of performance with and without heat and mass recovery. (a) Comparisons between experimental and simulated results of COP and (b) comparisons between experimental and simulated results of SCP

For a 6160A-type diesel engine the waste heat is about 34 kW, and experiments showed that the maximum heating power of the experimental prototype with the heat and mass recovery process is only about 19 kW, which means the waste heat from the diesel engine is sufficient for driving the ice maker. For such a condition the optimum cycle time is decided by *SCP*, that is, the optimum cycle time is 50 minutes, for which both the SCP with and without heat and mass recovery reaches the highest value.

8.4.4.3 p-T Diagram of the Refrigeration Cycle

Under the conditions of the ambient temperature of 30 °C and the cycle time of 50 minutes, the experiments with and without heat and mass recovery processes were tested. For heat and mass recovery process the heat recovery time is 2 minutes, which proceeds after 1.5 minutes mass recovery time. The relationship between pressure and temperature with and without heat and mass recovery processes (Clapeyron diagram) is shown in Figure 8.47.

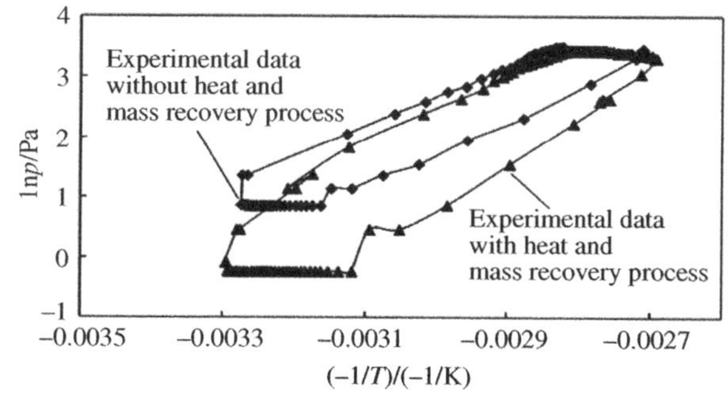

Figure 8.47 Clapeyron diagram

Desorption pressure for both cycles with and without heat and mass recovery processes is almost the same, but the adsorption pressure for the cycle with recovery process is much lower than that without the recovery process. Thus with heat and mass recovery process the driving force of the adsorption process will be greatly increased, that is, the heat and mass recovery processes improve the working conditions of the system [25] efficiently.

8.4.4.4 Heat and Mass Transfer Performance of the System

Heat and mass transfer performance of the adsorption bed is compared with the data of the adsorption bed with the granular adsorbent [24]. The heat transfer coefficient of the adsorbent bed is calculated as:

$$\alpha = \frac{W_h \times 1000}{A_{ads} \Delta T_m} \tag{8.41}$$

where W_h is the heating power (kW); A_{ads} is the heat transfer area of the adsorbent bed (m^2); ΔT_m is the logarithmic mean temperature difference (°C). The results are shown in Table 8.16.

Table 8.16 shows that the heat transfer coefficient of the adsorbent bed with the consolidated adsorbent is about four times that of the granular activated carbon adsorbent.

Table 8.16 Test arrangement and heat transfer coefficient

Adsorbent bed	Evaporation temperature (°C)	Cycle time (min)	Desorption and adsorption time (min)	Heat transfer coefficient (W/(m^2 °C))
Granular adsorbent bed	−10	30	15	25.44
Consolidated adsorbent bed	−9	30	15	86.32
Granular adsorbent bed	−10	40	20	19.13
Consolidated adsorbent bed	−10.5	40	20	90.5
Granular adsorbent bed	−10	50	25	20.12
Consolidated adsorbent bed	−10.5	50	25	81.2

The mass transfer performance of the adsorbent bed is mainly analyzed by the pressure of the adsorbent bed. The maximum difference between the evaporation pressure and the pressure in the bed with the granular adsorbent is about 1 kPa, whereas the maximum difference between the evaporation temperature and the pressure in the bed with the consolidated adsorbent is 0.6 kPa [24]. It shows that the consolidated adsorbent bed improves the mass transfer performance of the methanol vapor inside the bed effectively by reducing the distance of the mass transfer path.

8.4.4.5 The Analysis on the Refrigeration Cycle

The evaporation temperature of the adsorption ice maker is lower than the conventional air conditioning system, so the adsorption rate for the ice maker is slower. The relationship between the evaporation temperature and the working time as well as the relationship between the condensation pressure and the working time are shown in Figure 8.48 for the cycle with heat and mass recovery processes. When the desorption rate of the adsorbent bed began to decrease (condensing pressure began to decrease), the evaporation temperature still remains stable at about −15 °C, which shows the adsorption rate doesn't change very much. Hence, the optimum adsorption time of the adsorbent bed under the ice making condition is longer than the optimum desorption time.

In order to study the performance of the system further, the experiments were done under the conditions that the adsorption time of bed 1 is twice that of the desorption time, and the trends of the evaporation temperature and the condensing pressure are shown in Figure 8.49. Adsorption time was also longer than the desorption time, thus at the switch time the condensing pressure is still very high, and due to the longer adsorption time the evaporation temperature

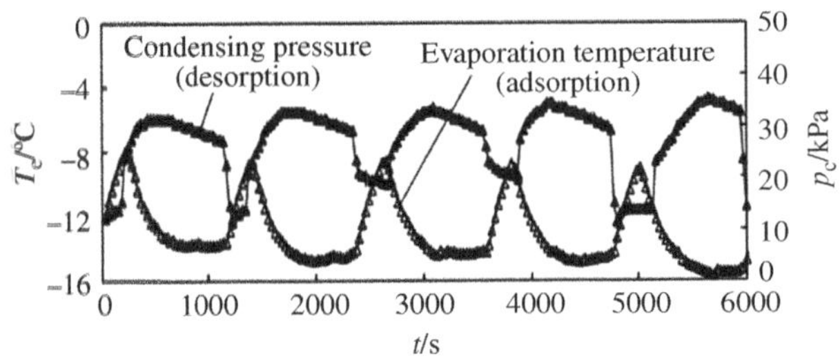

Figure 8.48 Evaporation temperature and condensation pressure vs. time

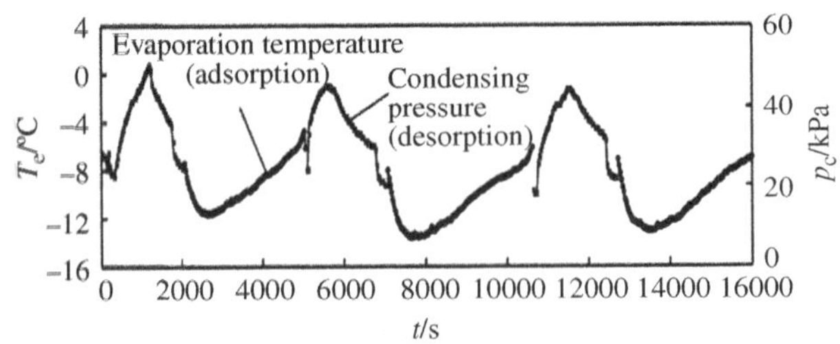

Figure 8.49 Evaporation temperature and condensation pressure vs. time under the condition of that the adsorption time is unequal with the desorption time

began to increase at the switch time. So the ratio of the optimum adsorption time and desorption time ranges from 1 to 2.

Usually two-bed continuous refrigeration cycle is adopted when the optimum adsorption time and desorption time are almost the same, and a multi-bed continuous refrigeration cycle is commonly used when the optimum adsorption time is different from the desorption time, which was introduced in Chapter 6. For this adsorption system the ratio between the optimum adsorption time and desorption time ranges from 1 to 2, so there are two choices, one is a two-bed continuous refrigeration cycle with heat and mass recovery process, and another is triple-bed continuous refrigeration cycle with heat and mass recovery process. The performances of the two-bed system have already been analyzed. The performance of a triple-bed system will be analyzed [21] as follows.

8.4.4.6 The Operation of the Triple-Bed System

The adsorption beds in a triple-bed system are defined as the bed 3, 4 and–5, respectively. When the optimum adsorption time and desorption time was 50 and 25 minutes, respectively, the working processes of the triple-bed system are shown in Table 8.17. Assuming at the beginning the adsorbent bed 3 was desorbed by heating and the adsorbent bed 5 adsorbed by cooling, then after 25 minutes bed 3 began to be cooled and the bed 4 began to be desorbed by heating, meanwhile the adsorbent bed 5 was continuously to be cooled; At the time of 50 minutes, adsorbent bed 3 was cooled continuously, bed 4 began to be cooled and bed 5 began to be heated; the first cycle was finished at the time of 75 minutes.

For the triple-bed system, the optimum adsorption time is twice that of the desorption time. The diagram for the operation of the triple-bed system is shown in Figure 8.50. Three adsorbent beds shared one evaporator and one condenser. The system was heated by the exhaust gas of the diesel engine, and was cooled by the cold water source. The storage tank is installed at the outlet of the heat exchanger for the exhaust gas and water, and it was used to collect the heat. The output cooling power was used to produce the ice in ice maker. The triple-bed system is more complex than the two-bed system. If the heat recovery process is adopted the system needs 20 solenoid valves, which is much more than the number of the valves for the two-bed system. In the figure f1, e1, f2, e2, f3, e3, e4, f4 are heat recovery electromagnetic valves.

When the system in Figure 8.50 starts to work, the pump 1 and 4 as well as the magnetic pump are normally open, and the control of other components is mainly as follows:

1. When bed 3 is heated, open f1, e1, 0g, h, and pump 3, and open valve 5 at the desorption stage. When bed 3 is cooled, open f1, e1, g, 0h, and pump 1, and open valve 6 at the adsorption stage.

Table 8.17 The operation processes for triple-bed system

Time (min)	Adsorbent bed 3	Adsorbent bed 4	Adsorbent bed 5
0	Start to be heated	–	Start to be cooled
25	Start to be cooled	Start to be heated	Continue to be cooled
50	Continue to be cooled	Start to be cooled	Start to be heated
75	Start to be heated	Continue to be cooled	Start to be cooled
100	Start to be cooled	Start to be heated	Continue to be cooled

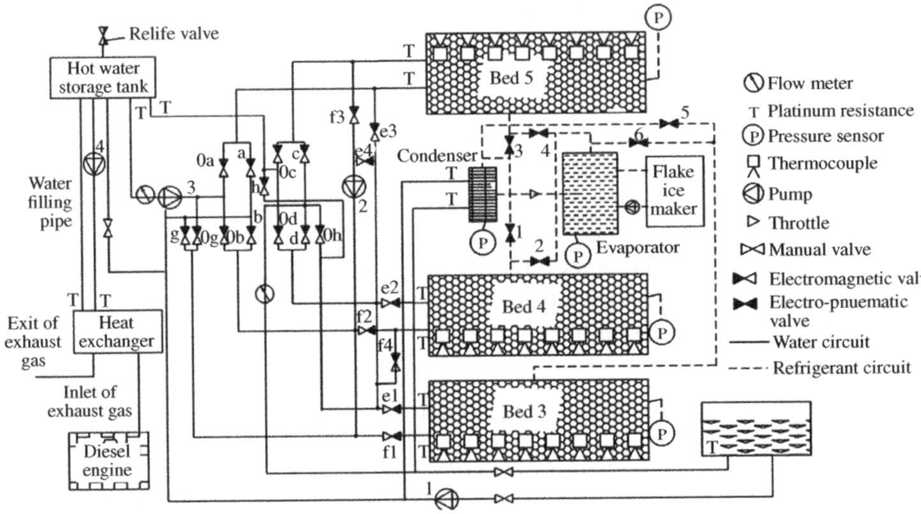

Figure 8.50 Triple-bed system

2. When bed 4 is heated, open 0b, 0d, e2, f2, and pump 3, and open valve 1 at the desorption stage. When bed 4 is cooled, open b, d, e2, f, and pump 1, and open valve 2 at the adsorption stage.
3. When bed 5 is heated, open 0a, 0c, and close e3, f3; open pump 3, and open valve 3 at the desorption stage. When bed 5 is cooled, open a, c, close e3, f3, and pump 1; open valve 4 at the adsorption stage.
4. For the heat recovery between bed 3 and 4, open e1, f1, e2, e4, f4, and pump 2. For the heat recovery between bed 4 and 5, open e2, f2, e3, f3, and pump 2. For the heat recovery between bed 3 and 5, open e1, f1, e3, f3, and pump 2.
5. For the mass recovery between bed 3 and 4 open valves 5 and 1. For the mass recovery between bed 3 and 5 open valves 5 and 3. For the mass recovery between bed 4 and 5 open valves 1 and 3.

The working processes of the system mainly include the following steps: firstly, the adsorbent bed 3 is heated, bed 4 and bed 5 are cooled; then bed 4 is heated, bed 3 and bed 5 are cooled; at last, bed 5 is heated, bed 4 and bed 3 are cooled. The heat and mass recovery processes proceed during the switching process.

Bed 1 of the system (Figure 8.42) was used to estimate the performance of the system, and the test arrangement and the results are shown in Table 8.18. It shows that the optimum cooling performance is obtained for the cycle time of 75 minutes, and the relationship between the cooling power and time is shown in Figure 8.51.

The mass of the adsorbent in the two-bed adsorption system is 120 kg (each bed has 6 0 kg adsorbent). If the total mass of the adsorbent is the same in the triple-bed system, then each bed should have 40 kg adsorbent. If the same experimental process is used and is as shown in Table 8.17 for the triple-bed system, then there are always two beds under the condition of cooling and adsorption and one bed under the condition of heat and desorption, that is, the

Table 8.18 Experimental arrangement and results

Number	Hot water temperature (°C)	Cycle time (min)	Adsorption /desorption time (min)	Average evaporation temperature (°C)	Average temperature of ice (°C)	Mass of ice (kg/h)	Cooling power (kW)	Average heating power (kW)
1	60	20	40	−7.38	−3.81	12	1.52	35.97
2	75	25	50	−7.14	−4.74	14.4	1.83	34.87
3	90	30	60	−7.08	−4.75	13.33	1.69	18.94

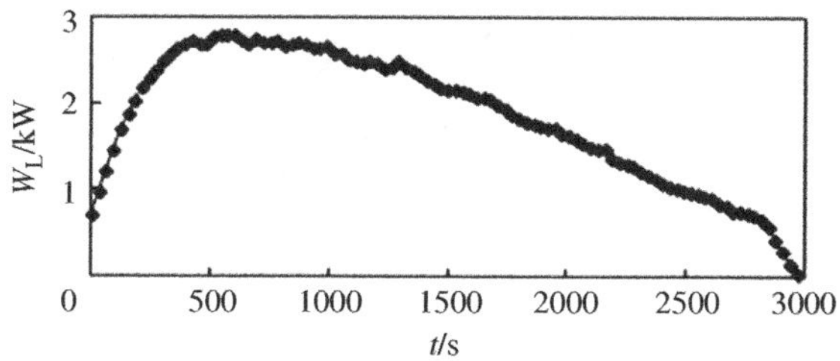

Figure 8.51 Cooling power vs. running time

adsorbent mass for adsorption is 80 kg and for desorption is 40 kg. Cooling power and heating power of the triple-bed system are:

$$W_{thref} = W_{ref} \times 8/6 \tag{8.42}$$

$$W_{heat} = W_h \times 4/6 \tag{8.43}$$

$$W_{sref} = W_{ref}/2 \tag{8.44}$$

$$COP = W_{thref}/W_{heat} \tag{8.45}$$

where W_{thref} is the cooling power of the triple-bed system; W_{ref} is the cooling power from the experiment (Figure 8.51 and Table 8.18). W_{heat} is the heating power of the triple-bed system; W_h is the heating power from the experiments (Table 8.18). W_{sref} is the cooling power of single bed.

If the experimental processes of the triple-bed system are the same as those shown in Table 8.18, according to the Equations 8.42–8.45 the performance of triple-bed system is shown in Table 8.19. When the cycle time is 75 minutes, the trends of the cooling power of each bed and the total cooling power of the whole system are shown in Figures 8.52 and 8.53.

Table 8.19 shows that for the same mass of adsorbent, the cooling power and *COP* of triple-bed system are greatly improved. For the experiments in Table 8.18 only the single bed is used for simulation, so there is no heat and mass recovery. If there are heat and mass recovery processes among the three beds, compared with the current system without mass recovery, the cooling power should be at least 7% increased and heating power should be at least reduced by 20%, and then the performance of the triple-bed system with heat and mass recovery processes should be as shown in Table 8.20.

Table 8.19 The performance of the triple-bed system

Number	Cycle time (min)	Adsorption time (min)	Desorption time (min)	Average cooling power (kW)	Average heating power (kW)	*COP*
1	60	20	40	2.03	23.98	0.085
2	75	25	50	2.44	23.25	0.105
3	90	30	60	2.26	12.62	0.179

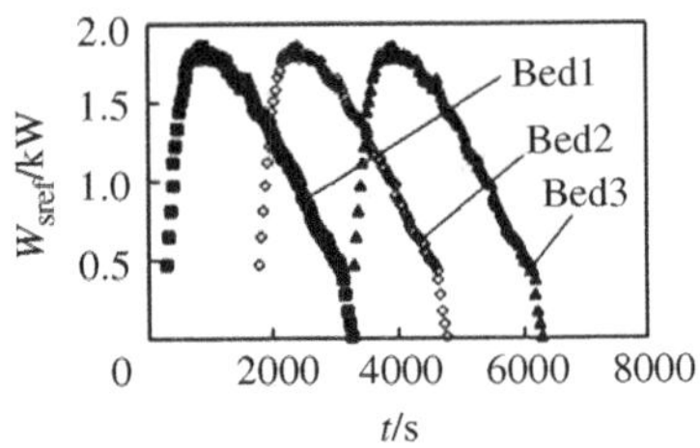

Figure 8.52 The cooling power of each bed vs. running time

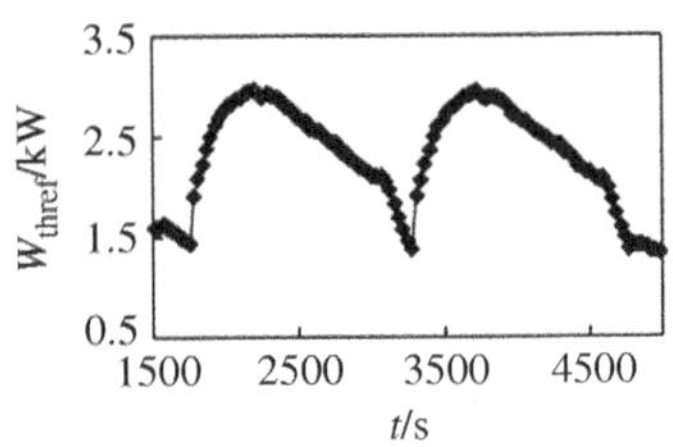

Figure 8.53 Cooling power of the triple-bed system vs. running time

Table 8.20 shows that under the condition of the same mass of the adsorbent in the system, the optimal cooling power of the triple-bed system with heat and mass recovery processes is 2.61 kW, which is 35.2% increased compared with that of the two-bed system. The best *COP* is 0.24, which is 95.1% increased if compared with that of a two-bed system. When the water temperature is 25 °C and the ice temperature is −7 °C, the triple-bed system can make ice of 21 kg/h and 504 kg per day, which can meet the requirements for a small fishing boat.

8.5 Heat Pipe Type Composite Adsorption Ice Maker for Fishing Boats

The heat transfer performance of the adsorption refrigeration system can be intensified by the heat pipe technology [20]. The heat pipe type composite adsorption ice maker for fishing boats adopts the composite adsorbent as well as the phase change heat transfer process for achieving high efficiency and refrigeration power.

Table 8.20 The performance of the triple-bed system with heat and mass recovery processes

Number	Cycle time (min)	Desorption time (min)	Adsorption time (min)	Cooling power (kW)	Average heating power (kW)	*COP*
1	60	20	40	2.17	19.18	0.11
2	75	25	50	2.61	18.60	0.14
3	90	30	60	2.42	10.10	0.24

One advantage of the heat pipe type composite adsorption ice maker is the composite adsorbent of activated carbon and calcium chloride (details of the adsorbent was shown in Chapter 5), which improved the adsorption performance of pure calcium chloride, and another advantage is that the system uses heat pipe technology to enhance the heat transfer performance of the heat transfer fluid, such a technology solves the corrosion problem caused by the seawater in the conventional ammonia adsorption refrigeration system.

The composite adsorption refrigeration system uses ammonia as the refrigerant, and the ammonia is not compatible with copper, so the metal material for the composite adsorbent bed is commonly steel. Thus the adsorption refrigeration system cannot be directly cooled by the seawater when it was used as the adsorption ice maker for fishing boats because the seawater will corrode the steel material. Meanwhile the conventional adsorption ice maker with ammonia as the refrigerant [26] generally uses the exhaust gas from the diesel engine to heat the adsorbent bed directly, and due to the sulfide in the smoke will form sulfate in the heat transfer tubes of the adsorbent bed, it will also cause the corrosion to the adsorbent bed. For the heat pipe type adsorption ice maker for the fishing boats, firstly, the phase change heat transfer process can improve the heat transfer performance of the fluid effectively, and secondly, the second heat exchange process is performed by the phase change heat transfer process, which can solve the corrosion problem both caused by the seawater and exhaust gas in the traditional adsorption ice maker for fishing boats.

Compared with the activated carbon–methanol adsorption ice maker for fishing boats, the advantage of heat pipe type composite adsorption ice maker for fishing boats is that the cycle adsorption quantity of the adsorbent is big, and consequently the cooling power is big because of the adsorbent as well as the enhanced heat transfer performance. The drawback of the system is that the ammonia is used as a refrigerant, which is not safe. But because the ammonia is lighter than air and it is easy to escape into the atmosphere, the safety problem can be resolved by installing the adsorption refrigeration unit at a reasonable place. Such as that we can install the part with the ammonia flow on the deck of the boat, and anti-freezing fluid is used as the heat transfer fluid to exchange the heat from the ammonia to the cabin for storing the fish there.

8.5.1 System Design of the Adsorption Refrigeration Test Prototype

The overall design of the heat pipe type composite adsorption refrigeration prototype adopts the separate heat pipe technology which consists of the heat pipe loop that includes the boiler and the adsorbent bed and the loop that is composed by the adsorbent bed and the cooler. For the experiments on the performance, the boiler is heated by the electric heater for simulating

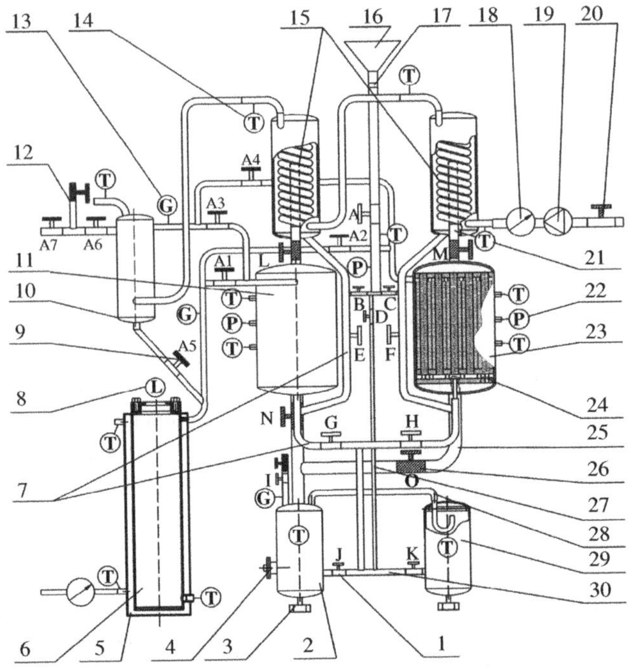

1. liquid valve for heat pipe; 2. heating boiler; 3. liquid tap interface;
4. electric heater; 5. glycol fluid jacket; 6. evaporator; 7. liquid tube for heat pipe;
8. megnetostrictive level sensor; 9. ammonia valve; 10. condenser; 11. adsorber 1;
12. relief valve; 13. pressure gauge; 14. temperature sensor; 15. coil pipe cooler;
16. filler; 17. screw interface; 18. flow sensor; 19. water pump; 20. water valve;
21. vapor channel for heat pipe for cooling phase; 22. pressure sensor; 23. adsorber 2;
24. vapor channel/liquid return pipe inside adsorber; 25. vapor channel for heat pipe for heating phase;
26. vapor valve for heat pipe; 27. pumping tube; 28. pressure balance pipe for boilers;
29. liquid pumping boiler; 30. liquid balance pipe for boilers

Figure 8.54 The heat pipe type composite adsorption refrigeration prototype

the heating process by the exhausted gas from the diesel engine, and the cooling power is balanced by the heat transfer fluid of the glycol. The adsorption and desorption quantities are tested by the magnetostrictive liquid level sensor with high accuracy. The design of the system is shown in Figure 8.54.

The system includes two adsorbent beds with the composite adsorbent, two coolers, a heating boiler, a liquid pumping boiler, a condenser, and an evaporator. The adsorbent bed is cooled by two coolers, and the rising section of the heat pipe is placed inside the cooler, which can greatly reduce the rising resistance of the steam in the heat pipe during the cooling process. The liquid filling process of the adsorbent bed is fulfilled by the liquid pumping boiler, the pressure balance tube of the boiler, and the liquid pumping tube. During the process, the heating boiler is under the condition of high temperature and high pressure, and there is pressure difference between the liquid pumping boiler and the adsorbent bed. With the control of the pressure balance pipe the liquid in the liquid pumping boiler can be pumped into the adsorbent bed. In order to reduce the convective heat transfer quantity between the liquid pumping boiler and heating boiler, the pressure balance pipe in the boiler is made as the elbow form, so the steam from the heating boiler won't directly exchange the heat with the liquid inside the liquid pumping boiler by a convection heat transfer process. The built-in electric heater is used in the heating boiler to reduce the heat loss in the heating process. During the heating process of the adsorbent bed, the rising section of the heat pipe is designed inside the adsorbent bed, which can reduce the heat loss and effectively reduce the dead space of the adsorbent bed. The rising section of the heat pipe can also be used as the liquid return pipe. The liquid return pipe can let the excessive fluid flow back to the heating boiler if too much liquid is pumped inside the adsorbent bed. Such a process could ensure the adsorbent bed is under the condition of flooded evaporation during the cooling process. In the liquid pumping and the liquid return processes, the working fluid of the heat pipe absorbs the adsorption heat and flows back to the heating boiler, which also constitutes a heat recovery process, that is, the adsorption heat is recovered. The condenser is the coil type heat exchanger, and the cooling power of the evaporator is exchanged by the heat transfer fluid of glycol. The adsorption and desorption quantity are tested by the liquid level sensor and the error is only 0.05%.

The working processes of the heat pipe type composite adsorption refrigeration prototype are controlled by the valves and pumps. For example, Figure 8.55 shows the working processes of the system, for which the initial working state of the adsorbent bed 2 is the liquid pumping process and the adsorbent bed 1 is under the condition of heating and desorption. The valves, pumps and electric heaters are normally open. The liquid pumping process of adsorbent bed 2 is shown in Figure 8.55a. In the process open the valves K, D, C, M, and F, the liquid in the boiler is pumped into the fin tube of the adsorbent bed under the pressure difference between the boiler and the adsorbent bed. Then open the valve O, and the excess liquid in the adsorbent bed will return to the heating boiler, which is also the heat recovery process of the adsorbent bed.

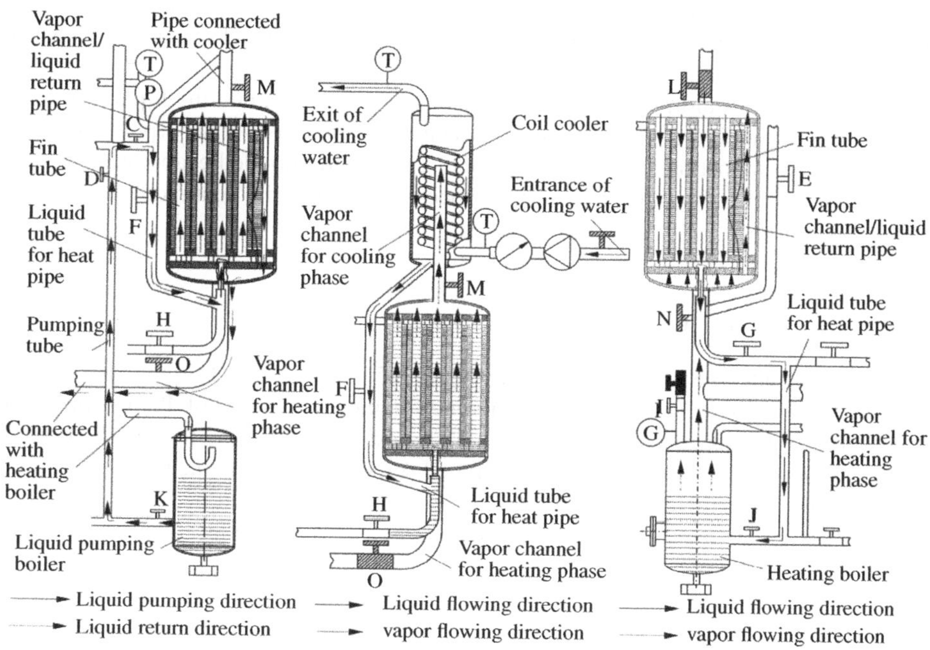

Figure 8.55 The working process of the compound adsorption system. (a) Liquid pumping process; (b) cooling process; and (c) heating process

After the end of the liquid return process, close the valves K, D, and O, so the adsorbent bed 2 is under the condition of the cooling and adsorption, and the working process is shown in Figure 8.55b. The liquid in the fin tube of the adsorbent bed evaporates along the direction of the vapor flow and condenses in the coil cooler, and the condensed liquid flows back to the adsorbent bed along the direction of the liquid flow in the diagram. At this time the adsorbent bed 1 is under the condition of heating and desorption, and the working process is shown in Figure 8.55c. Open the valves N, G, and J, and the liquid in the heating boiler evaporates along the direction of vapor and condenses inside the fin tube of the adsorbent bed, and then the condensed liquid flows back to the heating boiler along the liquid flow direction in the diagram. Valves J and K are used to balance the level between the heating boiler and the liquid pumping boiler.

A1 and A2 are the adsorption valves, and they control the adsorption of the adsorbent bed 1 and 2, respectively. A3 and A4 are the desorption valves, and they control the desorption of the adsorbent bed 1 and 2, separately. The mass recovery process is controlled by two desorption valves. Valve A5 is used to control the reflux of the condensate in the condenser which generally happens before mass recovery process.

The advantages of the heat pipe type composite adsorption refrigeration prototype are as follows:

1. It adopts the adsorbent bed filled with the composite adsorbent, as well as the heat and mass recovery processes for the performance improvement of the system.
2. The rising section of the heat pipe is inside the cooler for the cooling process, and such a design avoids the utilization of the elbows, which reduces the rising resistance of the heat pipe working fluid in the adsorbent bed during the cooling process.
3. The rising section of the heat pipe is designed inside the adsorbent bed during the heating process which can reduce the heat loss, as well as reducing the dead space of the adsorbent bed effectively.
4. The design of the liquid pumping boiler, liquid pumping tube, and the liquid return tube could complete the liquid filling process inside the heat pipes of the adsorption beds effectively in a very short time. Such a design also ensures the flooded evaporation in the fin tubes in the adsorbent bed for the cooling process, as well as the heat recovery process of the adsorbent beds.

8.5.2 Design of the Adsorbent Bed

The design of the adsorbent bed is shown in Figure 8.56. The selected fin tube has the inner diameter of 20 mm, outer diameter of 26 mm, outer diameter of the fin of 48 mm, fin pitch of 2.5 mm, and the fin thickness of 0.4 mm. The unit tubes are wrapped by fine mesh after it has been filled by adsorbent, then compresses the metal screen outside the fine mesh to ensure the strength for preventing the expansion of the adsorbent outward. There were 19 unit pipes in the adsorbent bed. The mass of $CaCl_2$ in each adsorbent bed is 1.88 kg. Assuming the cycle adsorption quantity is 0.6 kg/kg, the evaporation temperature is $-15\,^{\circ}C$, and such a design could ensure the cooling power of 1.3 kW if the cycle time is 40 minutes.

As shown in Figure 8.56, in order to reduce the heat loss of the adsorbent bed during the heating process, the rising section of the heat pipe is designed inside the adsorbent beds, which connects with the upper and lower steam chambers. In the heating process the steam from the boiler flows through the heat pipes to the lower steam chamber, and then flows into the

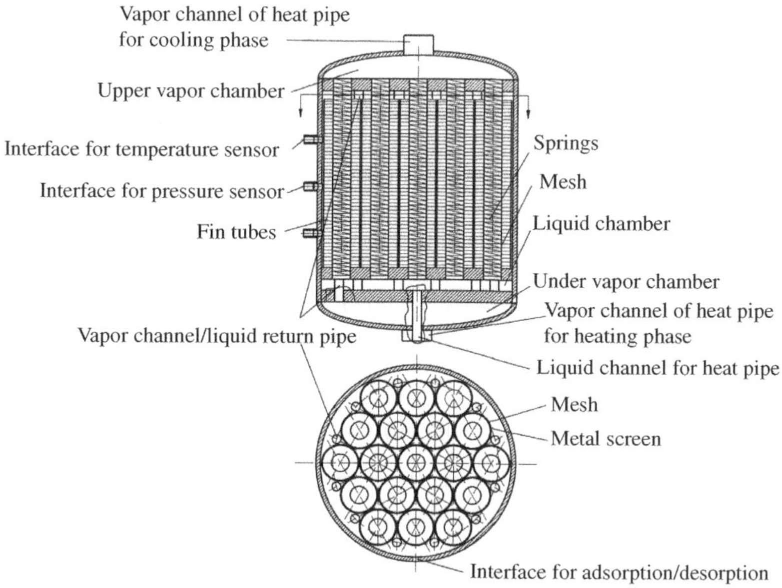

Figure 8.56 Design of the adsorbent bed with composite adsorbent

upper steam chamber, and finally condenses and releases heat in the inner fin tube. The condensed liquid flows through the heat pipe liquid pipes, and then flows back to the heating boiler. Another function for the rising section of the heat pipe inside the adsorbent bed serves as the liquid return pipe to control the liquid level in the liquid pumping process. After the liquid pumping boiler pumps fluid to the adsorbent bed, there will be excess liquid existing in the adsorbent bed as well as the cooler, then open the valve on the heat pipe between the adsorbent bed and the heating boiler, the remaining liquid in the adsorbent bed and the cooler will flow through the upper steam chamber, and then flow into the liquid return pipe. After that it will flow through the lower steam chamber and then flow back into the heating boiler. Such a process ensures that the amount of the working fluid in the fin tube is just sufficient for the flooded evaporation. The rising section of the heat pipe in the adsorbent bed or the liquid return pipe can also effectively reduce the dead space inside the adsorbent bed. The steel mesh with 80 mesh is put inside the inner pipe of the fin tube to increase the heat transfer area, and the steel mesh is fixed by springs. The spacing between the outer diameter of the fin tubes is only 1.5 mm, and the dead space of the adsorbent bed is about 21% of the total space. Under the condition of the condensation temperature of 25 °C and evaporation temperature of −15 °C, the difference between the minimum and maximum mass of ammonia vapor in the dead space of the adsorbent bed is 0.0239 kg. If the adsorption cycle quantity is 0.13 kg/kg, the mass of ammonia vapor is only 9.8% of the total adsorption quantity of the adsorbent, which won't influence the adsorption performance.

8.5.3 Simulation Model

The ice making condition is mainly simulated for an evaporation temperature lower than −15 °C. For physical adsorbent in the composite adsorbent, the cycle adsorption quantity was only 0.04 kg/kg under the condition where the evaporation temperature is −15 °C and the adsorption temperature is higher than 30 °C. While for $CaCl_2$ inside the composite adsorbent, under the same conditions the cycle adsorption quantity was 0.6 kg/kg. For the composite adsorbent the ratio between the chemical adsorbent and the physical adsorbent was 4 : 1, so the total adsorption quantity of the physical adsorbent was only 1.6% of that of the chemical adsorbent. Hence in the simulation, the adsorption quantity of the physical adsorbent is neglected.

During the working process of the composite adsorption system, due to the maximum ammonia mass in the dead space of the adsorbent bed being only 0.0239 kg, so in simulation heating and cooling power consumed by the ammonia in the dead space is also neglected. Meanwhile, in order to simplify the calculation, it was assumed that there was no heat loss between the adsorbent bed with the insulation material and the outside.

8.5.3.1 Calculation Model of the Adsorbent Bed

For the calculation model the energy conservation models of heating, cooling, desorption, and adsorption of the adsorbent beds has already been described in the Chapter 6.

Calcium chloride in the composite adsorbent is tested, and there is a serious adsorption hysteresis for adsorption and desorption process, so the adsorption and desorption equilibrium pressure equations are not the same. The pressure balance equations for the adsorption and desorption processes are as follows:

$$p_{eq} = \exp(-4.35 \times 10^3/T_a + 13.72) \quad \text{Adsorption process} \tag{8.46}$$

$$p_{eq} = \exp(-4.76 \times 10^3/T_a + 15.21) \quad \text{Desorption process} \tag{8.47}$$

where p_{eq} is the adsorption equilibrium pressure of the adsorbent bed ($\times 10^2$ kPa); T_a is the adsorption temperature (K).

The adsorption and desorption kinetic equations of calcium chloride are derived from the Equations 4.45 and 4.46, and the formulas are as follows:

$$\frac{dx_a(T)}{dT} = \frac{dN_g \times m_N}{dT \times m_C} \quad \frac{dx_d(T)}{dT} = \frac{dN_g \times m_N}{dT \times m_C} \tag{8.48}$$

where x_a is the adsorption quantity in the adsorption process (kg/kg); x_d is the adsorption quantity in the desorption process; N_g is the molar adsorption quantity (mol/mol); m_N and m_C are the molecular weight of the ammonia and calcium chloride, respectively.

The reaction heat of the chemical adsorbent is related with the number of complex ions [27], and for the adsorption process, ΔH_a is:

$$\Delta H_{\text{a}} = 2300 \times \int_{t_{a0}}^{t_{a1}} \left(\frac{dx_a(T)}{dT} \frac{dT}{dt} \right) dt \quad x_{\text{a}} < 0.613\ \text{kg/kg} \tag{8.49}$$

$$\Delta H_{\text{a}} = 2200 \times \int_{t_{a1}}^{t_{a2}} \left(\frac{dx_a(T)}{dT} \frac{dT}{dt} \right) dt \quad 0.613\ \text{kg/kg} < x_{\text{a}} < 1.225\ \text{kg/kg} \tag{8.50}$$

where t_{a0} is the initial adsorption time, t_{a1} and t_{a2} are the adsorption time. t is running time (s).
For desorption process, ΔH_d is:

$$\Delta H\mathrm{d} = -2200 \times \int_{t_{d0}}^{t_{d1}} \left(\frac{\mathrm{d}x_d(T)}{\mathrm{d}T} \times \frac{\mathrm{d}T}{\mathrm{d}t} \right) \mathrm{d}t \quad 0.613\ \mathrm{kg/kg} < x_\mathrm{d} < 1.225\ \mathrm{kg/kg} \tag{8.51}$$

$$\Delta H_\mathrm{d} = -2300 \times \int_{t_{d2}}^{t_{d3}} \left(\frac{\mathrm{d}x_d(T)}{\mathrm{d}T} \times \frac{\mathrm{d}T}{dt} \right) \mathrm{d}t \quad x_\mathrm{d} < 0.613\ \mathrm{kg/kg} \tag{8.52}$$

For the phase change heat transfer process, the heat transfer quantity of the adsorbent bed in the cooling process is:

$$W_\mathrm{c} = \alpha_\mathrm{e} A_{fa}(T_\mathrm{w1} - T_\mathrm{s1}) \tag{8.53}$$

where α_e is the evaporation heat transfer coefficient inside fin tube (W/(m^2 °C)), A_{fa} is the internal surface area of the fin tube (0.29 m^2), T_{s1} is the saturation temperature of the heat pipe working fluid inside the fin tube (°C), and T_{w1} is the wall temperature of the fin tube (°C). α_e can be calculated by the equation [28]:

$$\alpha_e = 7.915 q_e^{\ 0.662} p_v^{\ 0.0566} \tag{8.54}$$

where q_e is the heat flux density for the evaporation heat transfer process (W/m^2), p_v is the evaporation pressure (Pa).
The heat transfer equation for the heating and desorption process is:

$$W_\mathrm{h} = \alpha_\mathrm{fc} F_{fa}(T_\mathrm{s1} - T_\mathrm{w1}) \tag{8.55}$$

where α_{fc} is the heat transfer coefficient of inner fin tube in the adsorbent bed (W/(m^2 °C)). During the heating and desorption process, considering the influence of liquid film flowing down and vapor flow on the heat transfer performance, the heat transfer coefficient is [29]:

$$\frac{\alpha_\mathrm{fc} l_{ah}}{\lambda_l} = \frac{\left(\frac{g \rho_l^{\ 2} l_{ah}^{\ 3}}{\mu_l^{\ 2}} \right)^{1/3} \mathrm{Re}}{58 \mathrm{Pr}_s^{\ -0.5} \left(\frac{\mathrm{Pr}_w}{\mathrm{Pr}_s} \right)^{0.25} \left(\mathrm{Re}^{3/4} - 253 \right) + 9200} \qquad \mathrm{Re} = \frac{4 q_\mathrm{c} l_{ah}}{L_{hp} \mu_l} \tag{8.56}$$

where l_{ah} is the heat pipe height in the adsorbent bed (m), ρ_l is the density of the working fluid in the heat pipe (kg/m^3), λ_l is the thermal conductivity of the liquid (W/(m °C)), μ_l is the dynamic viscosity of the working fluid (kg/(ms)), Pr_s is the Prandtl number at the saturated temperature of the vapor, and Pr_w is the Prandtl number at the wall temperature. q_c is the heat flux for the condensing heat exchanging process (W/m^2), L_{hp} is the evaporation latent heat of the fluid inside the heat pipe (kJ). In simulation, water is taken as the working fluid of the heat pipe.
The mass recovery process is equivalent to that which the high-pressure bed desorbs at low pressure while the low-pressure bed adsorbs at high pressure, and the heat exchange

equations are:

$$(M_{madb}C_m + M_{ca}C_{ca} + M_aC_{\mathrm{am}}x_a)\frac{\mathrm{d}T}{\mathrm{d}t} = M_a\Delta H_{\mathrm{a}}\frac{\mathrm{d}x_a}{\mathrm{d}t} \text{ (Adsorption)} \tag{8.57}$$

$$(M_{madb}C_m + M_{ca}C_{ca} + M_aC_{\mathrm{am}}x_d)\frac{\mathrm{d}T}{\mathrm{d}t} = M_a\Delta H_{\mathrm{d}}\frac{\mathrm{d}x_d}{\mathrm{d}t} \text{ (Desorption)} \tag{8.58}$$

where M_{madb} is the metal mass for the adsorbent bed (kg), C_m is the specific heat of the metal (kJ/(kg °C)), M_{ca} is the mass of the composite adsorbent (kg), C_{ca} is the specific heat of the composite adsorbent (kJ/(kg °C)), M_a is the mass of $CaCl_2$ in the adsorbent bed, C_{am} is specific heat of the liquid ammonia refrigerant (kJ/(kg °C)), T is the temperature of the adsorbent bed (°C), and t is the running time (s).

At the end of the mass recovery process, the pressure of the two adsorbent beds is equal, and the formula is:

$$p_{ae} = p_{de} \tag{8.59}$$

where p_{ae} is the pressure of the adsorbent bed after adsorption; p_{de} is the pressure of the adsorbent bed after desorption.

The adsorption quantity of the adsorbent bed after the mass recovery is:

$$x_a = x_{am} - \Delta x_{md} = x_{am} + \int_{t_{m0}}^{t} \frac{\mathrm{d}x_d(T)}{\mathrm{d}t}\mathrm{d}t \tag{8.60}$$

where x_{am} is the adsorption quantity of the bed after desorption before the mass recovery; Δx_{md} is the desorption quantity during the mass recovery process; t_{m0} is the initial time.

During the mass recovery process, the adsorption quantity of the adsorbent bed in the adsorption state is:

$$x_a = x_{dm} + \Delta x_{ma} = x_{dm} + \int_{t_{m0}}^{t} \frac{\mathrm{d}x_a(T)}{\mathrm{d}t}\mathrm{d}t \tag{8.61}$$

where x_{dm} is the adsorption quantity of the adsorbent bed before the mass recovery. Δx_{ma} is the adsorption quantity during the mass recovery process.

8.5.3.2 Mathematic Models for Heating and Liquid Pumping Process

The electric heater inside the boiler is 4.5 kW. The following assumptions are adopted for the simulation process of the heating boiler:

1. Ignore the convective heat transfer through the vapor balance pipe between liquid pumping boiler and the heating boiler.
2. Ignore the heat loss between the boiler and the outside because the heating boiler is thermal insulated with the outside.
3. Assume that the volume of the heat pipe is always filled with the saturated working fluid.
4. The flowing rate of the liquid in the heat pipe of adsorbent bed is equal to the vaporization rate of the vapor inside the heating boiler for the heating process of the adsorbent bed.

5. The maximum temperature of the working fluid in the heating boiler is limited to 150 °C. The heating power absorbed by the working fluid in the heat pipe is:

$$W_h = M_{hb} \times C_{hb} \times \frac{dT_{hb}}{dt} + \frac{dM_{gas}}{dt} \times L_{hb} = W_H - M_{mh} \times C_{mh} \times \frac{dT_{hb}}{dt} \quad (8.62)$$

where M_{hb} is the total mass of the working fluid in the boiler (kg); T_{hb} is the temperature of the working fluid in the boiler (°C); M_{gas} is the mass of the working fluid that vaporized (kg); L_{hb} is the vaporization latent heat of the working fluid (kJ/kg); W_H is the heating power of the electric heater (4.5 kW); M_{mh} is the metal mass of the heating boiler (11.4 kg), and C_{mh} is the metal heat capacity of the heating boiler.

At the switching time of the adsorbent beds, the liquid filling process of the adsorption bed will proceed first. The first step is that the liquid level in the liquid pumping boiler is balanced with the liquid level in heating boiler, and then the working fluid in the heating pipe of the bed after adsorption flows into the liquid pumping boiler. The mass of the working fluid in liquid pumping boiler is:

$$M_{hp} = M_{eqh} + M_{ha} = \frac{M_z - M_{ha}}{2} + M_{ha} \quad (8.63)$$

where M_{hp} is the initial mass of the working fluid in the liquid pumping boiler; M_{eqh} is the mass of the working fluid in the liquid pumping boiler after its liquid level was balanced with that in the heating boiler; M_z is the total mass of the working fluid filled into the heat pipe system (10 kg); M_{ha} is the mass of the working fluid in the fin tube of the adsorbent bed and in the liquid chamber, which is 2.38 kg under the condition of the flooded evaporation.

The temperature of the working fluid in the liquid pumping boiler is calculated by the temperature of the working fluid in the heating boiler and working fluid in the adsorbent bed that returned to the liquid pumping boiler at the switch time. It can be calculated as follows:

$$T_{pb} = \frac{M_{eqh} \times T_{hb} + M_{ha} \times T_{sa}}{(M_{eqh} + M_{ha})} \quad (8.64)$$

where T_{pb} is the temperature of the working fluid in the liquid pumping boiler; T_{sa} is the saturation temperature of the working fluid in the fin tubes of the bed after adsorption.

The next step is the liquid pumping process. The liquid in the liquid pumping boiler flows through the liquid pumping tube into the adsorbent bed after desorption, and the excess liquid in the adsorbent bed flows back into the heating boiler through the liquid return pipe which can recover part of the sensible heat of the adsorbent bed. The energy balance equation is:

$$(M_{madb}C_m + M_{ca}C_{ca} + M_aC_{am}x_{a2})dT + (M_{eqh} + M_{ha} - M_{pbf}) \times C_{hb} \times dT_{pa} = 0 \quad (8.65)$$

where M_{pbf} is the mass of the liquid in the liquid pumping boiler that cannot be pumped into the adsorbent bed because the level of the liquid is below the liquid pumping tube (1.52 kg); T_{pa} is the temperature of the working fluid for the heat pipe working fluid after liquid pumping process and the liquid return process.

The water temperature in the heating boiler is:

$$T_{hb} = \frac{M_{eqh} \times T_{hb0} + (M_{eqh} - M_{pbf}) \times T_{pa}}{(2 \times M_{eqh} - M_{pbf})} \quad (8.66)$$

where T_{hb0} is the temperature of the boiler before the liquid returned from the heat pipes in the adsorbent bed to the heating boiler.

8.5.3.3 Performance Calculation

The cooling power is calculated by the following equation:

$$W_{ref} = M_a \frac{dx_a}{dt} \times L_{am} + (M_{ev} - M_a \times \Delta x_a + M_a \times \Delta x_d) \times \frac{dT_e}{dt} \tag{8.67}$$

where M_{ev} is the mass of the refrigerant in the evaporator(kg), T_e is the evaporation temperature (°C).

The total cooling power for the period of the half-cycle is:

$$Q_{ref} = \int_0^{t_{hc}} \left[M_a \frac{dx_a}{dt} \times L_{am} + \left(M_{ev} - M_a \times \Delta x_a + M_a \times \Delta x_d \right) \times \frac{dT_e}{dt} \right] dt \tag{8.68}$$

where t_{hc} is the time for the half-cycle.

SCP is:

$$SCP = \frac{Q_{ref} \times 1000}{t_{hc} \times M_a}, \quad \mathrm{W/kg} \tag{8.69}$$

COP is:

$$COP = \frac{Q_{ref}}{\int_0^{t_{hc}} W_h \times dt} \tag{8.70}$$

8.5.3.4 Simulation Results

Taking the adsorption refrigeration system for fishing boats that uses the working pair of metal chloride–ammonia as an example, the performance is simulated. In the calculation process it was found that the values of Δx_a and Δx_d are different in the mass recovery process. It has resulted in the different dynamic equations for adsorption and desorption processes. The kinetic equations in the adsorption process are influenced by the dynamic process among molecules, while the desorption process isn't affected by the dynamic process among the molecules. In the simulation, Δx_d for the mass recovery process of 30 seconds is about 0.47 kg/kg, while Δx_a is only 0.35 kg/kg. In order to simplify the calculation, the smaller recovered mass, that is, the adsorption quantity of Δx_a is taken as the recovery mass.

As the temperature of the sea water is generally about 25 °C, for the simulation the inlet temperature of the ice maker for the fishing boats is taken as 25 °C. The water is selected as the working fluid in the heat pipe. Under the conditions of the evaporation temperature of −15, −25, and −35 °C, and the heating power of 4.5 kW, the simulation results are shown in Figure 8.57.

In Figure 8.57a,b, for the heat pipe type composite adsorption ice maker for the fishing boats, under the conditions of heat and mass recovery processes and evaporation temperature of −15 °C, *SCP* was 798 W/kg and *COP* was about 0.47. The optimum cycle time was 46 minutes.

In Figure 8.57a,b, the performance decrement from −25 to −35 °C is larger than that from −15 to −25 °C. The main reason is that the adsorption ability is proportional to the exponential of the pressure difference (the difference between the evaporation pressure and the equilibrium pressure of the adsorbent bed) in Equation 4.46, thus the decrement of the evaporation pressure, that is, the evaporation temperature, had a big impact on the performance. In addition, if the cycle time was short, the adsorbent bed didn't get to the desorption temperature by a heating process, and the cycle adsorption quantity of the system was very small, so *SCP* and *COP*

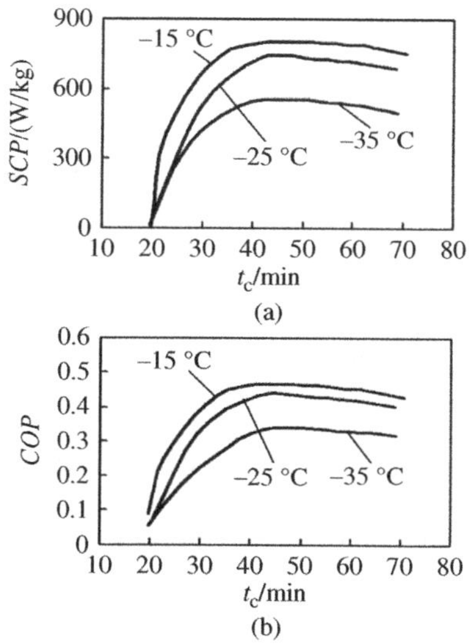

Figure 8.57 The performance of the system for the heating power of 4.5 kW. (a) SCP vs. cycle time and (b) COP vs. cycle time

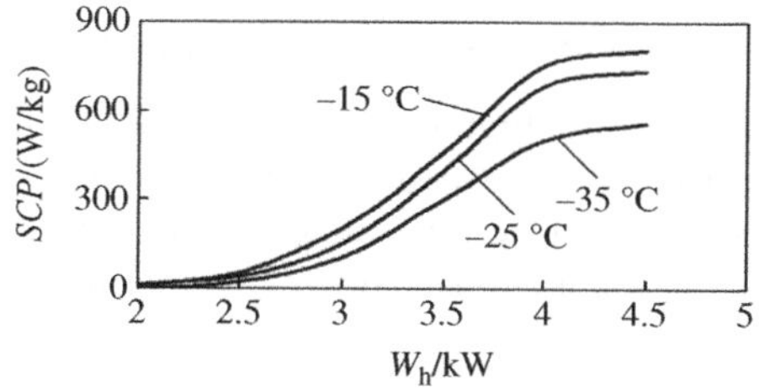

Figure 8.58 *SCP* vs. heating power at the cycle time of 46 minutes

were close to 0 when the cycle time was less than 20 minutes, as shown in the diagram. When the cycle time was 46 minutes, the simulation results of the performance under the conditions of different heating power were shown in Figure 8.58.

Figure 8.58 shows that when the heating power is less than 2.5 kW, the bed doesn't reach to the desorption temperature when the cycle time is less than 46 minutes, so the *SCP* of the system is close to 0. The cooling power output of the system increases when the heating power increases. The *SCP* reaches 750 W/kg when the heating power was 4 kW and the evaporating temperature was −15 °C. When the cycle time is 46 minutes the temperature change of the bed under the conditions of the different heating power is shown in Figure 8.59.

In Figure 8.58, when the heating power is less than 4 kW, the increment of *SCP* is large. When the heating power is higher than 4 kW, the increment of *SCP* is small. Figure 8.59 also shows that the highest desorption temperature of the adsorbent bed is 105 °C when the heating power is 4 kW, while the highest desorption temperature of the adsorbent bed is only about 80 °C when the heating power is 3 kW and the cycle time is 46 minutes. Such a result indicates that the

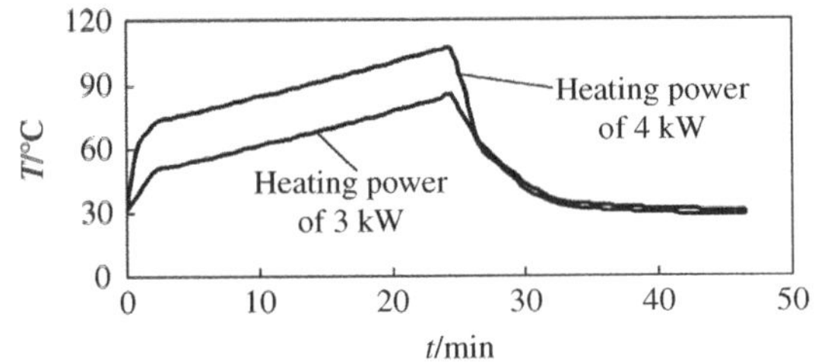

Figure 8.59 The temperature change of the adsorbent bed vs. cycle time

optimum cycle time is close to 46 minutes when the heating power is greater than 4 kW. When the heating power is less than 4 kW, the maximum desorption temperature of the adsorbent bed is lower when the cycle time is short, then according to Equation 8.47, the equilibrium pressure of the adsorbent bed is smaller, so the absolute pressure difference between the condensing pressure and the pressure of the adsorbent bed is small, and consequently the adsorption rate is small. In Equation 8.48, $dx_d(T)/dt$ was negative and the absolute pressure difference between the condensing pressure and the pressure of the adsorbent bed is large, so the desorption rate was great.

8.5.4 The Construction of the Adsorption Refrigeration System

The photo of the system was shown in Figure 8.60. The vapor and liquid valves for heat pipes are ball valves which can be used both for vacuum and high-pressure conditions. The interface for the filler can be used as the interface for leakage test and evacuation. The interfaces for the cooling medium on the evaporator connected with the thermostat, which is used to control

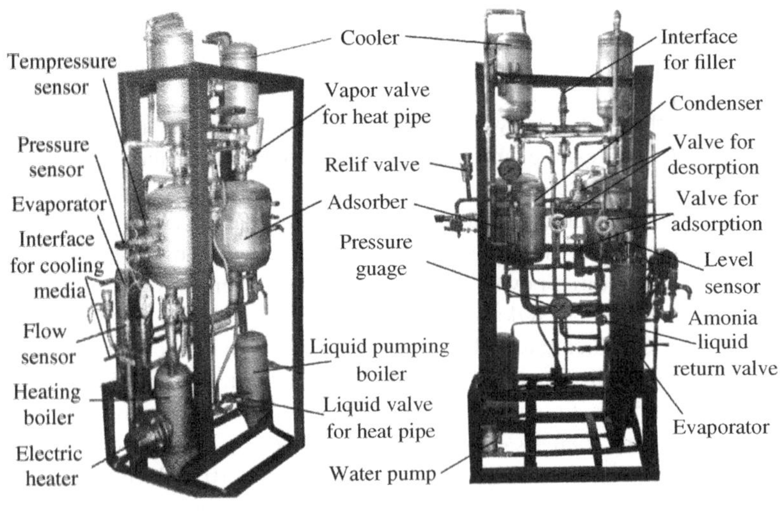

Figure 8.60 The photograph of the heat pipe type composite adsorption refrigeration prototype

the evaporating temperature. The size of the system has a height of 1.9 m, length of 1.2 m, and width of 0.6 m. Magnetostrictive level sensor, which only has the measuring error of 0.05%, is used to measure the adsorption quantity of the system. The cooling power of the system is calculated by the level change of the liquid inside the evaporator directly, such a process could avoid the testing error caused by the heat loss to the environment if compared with the heat transfer process by the cooling media circuit. Such a method is also feasible for studying both equilibrium and non-equilibrium adsorption and desorption properties. For traditional adsorption refrigeration systems due to that the adsorption and desorption amount couldn't be precisely calculated because the cooling power is generally tested by the cooling media circuit, it is very difficult to study the non-equilibrium adsorption and desorption characteristics. However, for the system designed here the non-equilibrium performance can be studied easily because the adsorption and desorption quantity can be tested easily.

The system used the thermostat to control the evaporation temperature, and the working fluid in the thermostat was ethanol. Voltage regulator was used to regulate the heating power output from 0 to 4.5 kW.

8.5.5 Studies on the Performances of the Adsorption Refrigeration Prototype

8.5.5.1 The Selection of the Working Fluid for Heat Pipe

The performances of methanol, acetone, and water are tested in the experiments, and experiments show that the methanol will decompose in the repeated experiments, which will generate non-condensable vapor and consequently will affect the performance of the system. The performances of acetone and water are stable.

The heat transfer coefficients of the heat pipe working fluid of acetone and water were studied, and they are calculated as follows:

$$\text{Heating stage} \quad \alpha A_{fin} \Delta T_{ah} = W_h \tag{8.71}$$

$$\text{Cooling stage} \quad \alpha A_{fin} \Delta T_{ah} = W_c = m_{water} C_{water} \Delta T_{wc} \tag{8.72}$$

where α is the heat transfer coefficient between the adsorbent and the vapor fluid in the heat pipe (W/(m^2 °C)). A_{fin} is heat exchange area of fin tube in adsorbent bed, and it is 4.26 m^2 by calculation. ΔT_{ah} is the temperature difference between the adsorbent and vapor inside the fin tube (°C), W_c is the cooling power (W) of the water in the coil cooler, m_{water} is the flow rate of water (0.0433 kg/s) in the coil cooler, and ΔT_{wc} is the temperature difference (°C) between the water inlet and outlet of the coil cooler.

The heat transfer coefficient of acetone is shown in Figure 8.61. The average heat transfer coefficient for heating stage is 90.1 W/(m^2 °C), and the average heat transfer coefficient for cooling phase is 84.8 W/(m^2 °C).

The heat transfer coefficient of the water fluid is shown in Figure 8.62, which is under the conditions that the filled volume of the liquid is 10 l and the heating power is 3 kW. The average heat transfer coefficient for heating stage is 109.3 W/(m^2 °C), and the average heat transfer coefficient for cooling stage is 105.1 W/(m^2 °C).

The water is chosen as the working fluid of the heat pipe because it has higher heat transfer coefficients during the heating and cooling phases if compared with the results of the acetone.

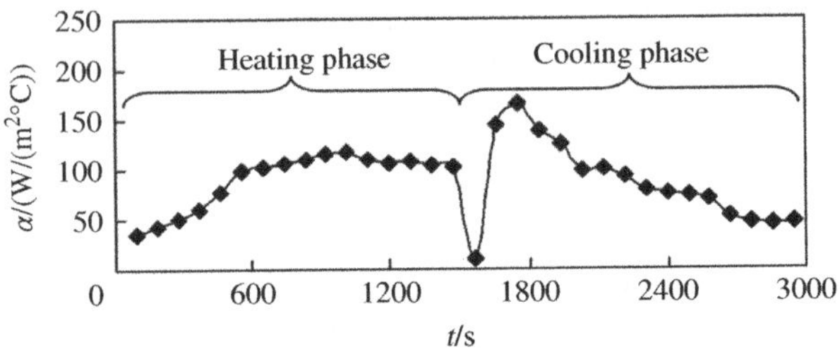

Figure 8.61 Heat transfer coefficient for the working fluid of acetone in the heat pipe

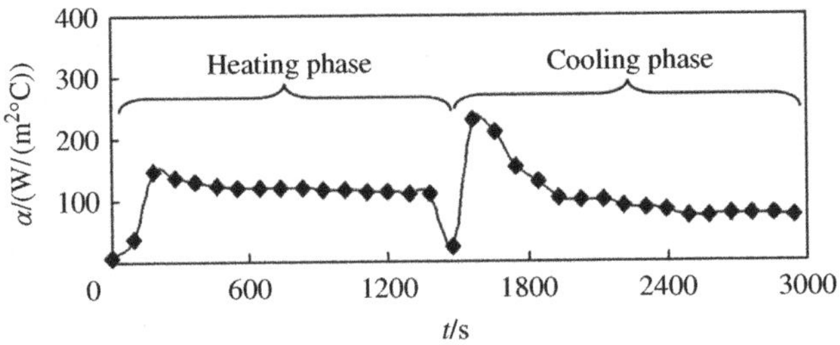

Figure 8.62 Heat transfer coefficient for the working fluid of water in the heat pipe

8.5.5.2 Experimental Study on Mass Recovery Process

Under the conditions of evaporating temperature of −25 °C, cooling water temperature of 25 °C, and the cycle time of 60 minutes, the performance of the mass recovery process between two adsorbent beds is measured, and the pressure change between two adsorbent beds are shown in Figure 8.63. Since ammonia is the refrigerant, the system is a system with higher pressure, so the pressure in the two adsorbent beds could quickly reach equilibrium during the mass recovery process. In Figure 8.63, the pressure balance time is only 47 seconds. Thus for the following experiments the mass recovery time is taken as 47 seconds at the switching process.

8.5.5.3 Non-equilibrium Adsorption and Desorption Performances

When the heat and mass recovery processes are adopted, the non-equilibrium adsorption and desorption performances of the adsorbent beds are tested under the conditions of evaporation

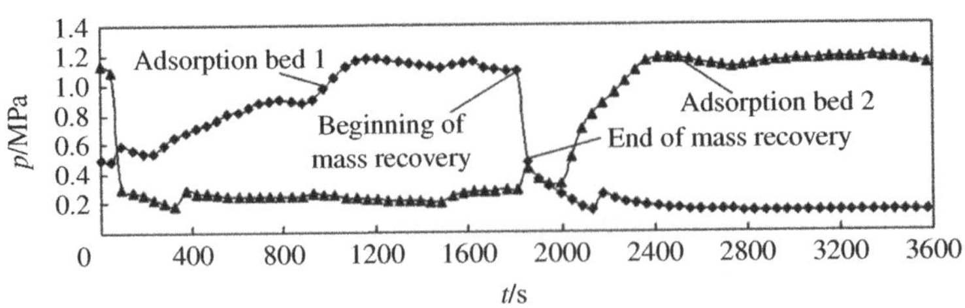

Figure 8.63 Pressure of two adsorbent beds

temperature of −15 °C, desorption temperature of 90, 96, and 105 °C, adsorption temperature of 43, 40, and 36 °C. The cycle adsorption capacity Δx is calculated as:

$$\Delta x = \frac{\Delta M_a}{M_a} \tag{8.73}$$

where ΔM_a is adsorption/desorption mass of ammonia (kg), M_a is the mass of $CaCl_2$ in adsorbent bed (kg).

Non-equilibrium adsorption curves are shown in Figure 8.64. As seen from Figure 8.64 the desorption temperature has a big influence on the desorption rate. When the desorption temperature is 90 °C, the average desorption rate is 0.040 (kg/kg)/min, while the desorption temperature increased to 96 °C, the desorption average rate is 0.066 (kg/kg)/min, which increases by 65%. While the desorption temperature rises to 105 °C the average desorption rate was 0.084 (kg/kg)/min, which is improved by 37% relative to the condition of desorption temperature of 96 °C. The adsorption rate at the temperature of 43, 40, and 36 °C is 0.028, 0.030, and 0.034 (kg/kg)/min, respectively. The adsorption rate is slower than the desorption rate.

Figure 8.64 also shows that the desorption can be completed in 7 minutes when the desorption temperature is 105 °C. But in the actual operation process, the temperature of the adsorbent bed is close to 105 °C when the adsorbent bed is heated for as long as 25 minutes. It means that the adsorbent bed cannot reach 105 °C quickly, so there will be a large deviation if Figure 8.64 is used to determine the desorption rate of the system. In order to study the actual working process more in detail, the relationship between the temperature change and the desorption rate of the adsorbent bed during heating process, as well as the adsorption rate during the cooling process were tested.

The adsorption and desorption rate are calculated as follows:

$$R_{\Delta x} = \frac{\Delta x}{\Delta t} \tag{8.74}$$

where $R_{\Delta x}$ is adsorption/desorption rate ((kg/kg)/min), Δx is the cycle adsorption quantity (kg/kg), and Δt is the adsorption/desorption time (minutes).

Under the conditions of different temperatures of the adsorbent bed, the relationship between $R_{\Delta x}$ with or without mass recovery process and the temperature change of the adsorbent bed

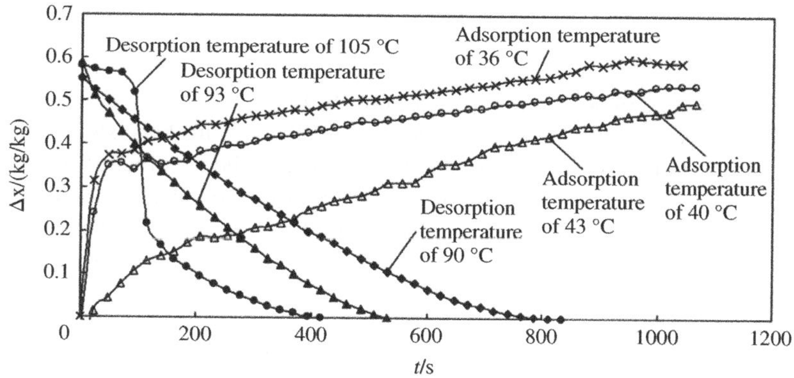

Figure 8.64 Non-equilibrium adsorption and desorption performances under the condition of stable adsorption and desorption temperature

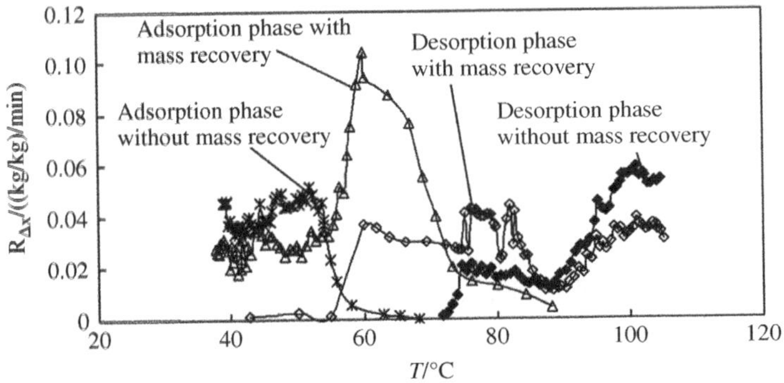

Figure 8.65 Non-equilibrium adsorption and desorption performance vs. adsorption temperature

is shown in Figure 8.65. Due to the adsorbent bed being saturated after the mass recovery process, at the beginning of the desorption phase (55–70 °C), its desorption rate is bigger than that without mass recovery process, and its adsorption rate is also much larger than that without mass recovery process at the beginning of the adsorption stage (the temperature decreased from 80 to 55 °C). With the increase of desorption and adsorption quantity, the desorption/adsorption rates for the cycles with and without mass recovery process are almost the same at the end of desorption (80–105 °C)/adsorption (lower than 55 °C) process.

In general, the desorption rate should increase when the temperature rises, while the adsorption rate decreases as the adsorption temperature decreases. However, from Figure 8.65, the adsorption and desorption rate are higher at the beginning of adsorption stage (high-temperature phase) and the beginning of the desorption stage (low temperature phase) for the cycle with mass recovery process. One reason for this phenomenon is that the adsorbent bed is saturated because of mass recovery, and the second reason is the heat transfer performance of the heat pipe type adsorption system. The phase change heat transfer process is used in the heat pipe technology, so the heat exchange power is large at the beginning of cooling and heating stage due to the large temperature difference, consequently the temperature change rate of the adsorption bed is also great. The change rate for adsorption/desorption temperature is as follows:

$$R_T = \frac{\Delta T_a}{\Delta t} \tag{8.75}$$

where ΔT_a is the change of adsorption temperature (°C).

The relationship between R_T and the temperature of the bed in the system was shown in Figure 8.66. At the initial cooling stage, the cooling rate of the adsorbent bed is large, thus the decrement rate of the pressure inside the adsorption bed also increases, and consequently the adsorption rate increases. At the beginning of the heating stage, the heating rate of the adsorbent bed is large, and also as the increment rate of the pressure in the adsorption bed increases, so the desorption rate increases.

Under the condition of variable temperature of the bed, if Figure 8.65 is used to analyze the performance, the non-equilibrium adsorption performance of the system is better than the non-equilibrium desorption performance. As shown in Figure 8.64, the desorption rate with and without mass recovery process are 0.029 and 0.018 (kg/kg)/min, respectively, and

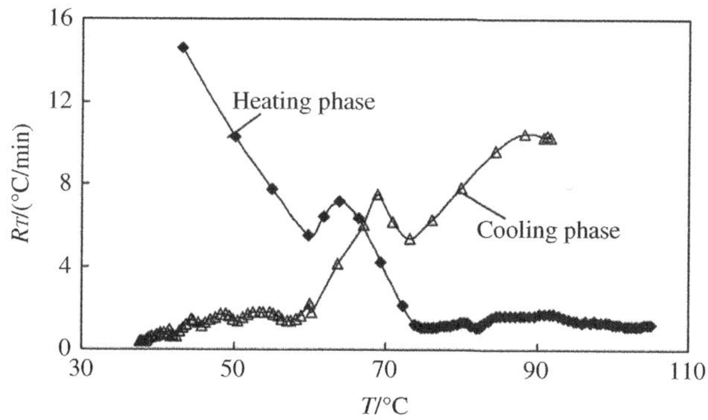

Figure 8.66 The temperature change rate in the adsorption bed vs. adsorption temperature

the adsorption rate with and without mass recovery process are 0.032 and 0.023 (kg/kg)/min, respectively. Taking temperature variation into consideration, that is, taking Figure 8.66 into consideration, the non-equilibrium adsorption/desorption rate $R_{\Delta xT}$, which is the adsorption and desorption rate with adsorption temperature change rate, that is:

$$R_{\Delta xT} = \frac{R_{\Delta x}}{R_T} \tag{8.76}$$

The non-equilibrium desorption performance is slightly higher than the performance of the non-equilibrium adsorption according to $R_{\Delta xT}$. For the cycle with the mass recovery process, average value of $R_{\Delta xT}$ is 0.014 (kg/kg)/°C in the desorption process, whereas it is 0.013 (kg/kg)/°C in the adsorption process.

As the values of $R_{\Delta xT}$ for desorption and adsorption process are similar, the cycle with identical adsorption and desorption time is chosen in the experiments.

8.5.5.4 Lowest Evaporation Temperature

The cooling water temperature is controlled at 25 °C to simulate the temperature of sea water. During the measuring process of minimum evaporation temperature, when the circuit of the ethanol in the jacket of the evaporator is disconnected, that is, the cooling power doesn't output, the evaporation temperature is tested and the curve is shown in Figure 8.67. When the system runs continuously for 3.6 hours, the lowest evaporation temperature in the evaporator is −42 °C.

8.5.5.5 Research on the Characteristics of the Cycle

Magnetostrictive level sensor in the evaporator is used to measure the adsorption and desorption capacity of the cycle. For example, assuming that the adsorbent bed 1 is in adsorption and the adsorbent bed 2 is in desorption before the switching time, in Figure 8.54 the valve A5 is off before the switch time, the desorption amount from the adsorbent bed 2 is collected in the condenser, and the measuring amount by the level sensor is the adsorption quantity of

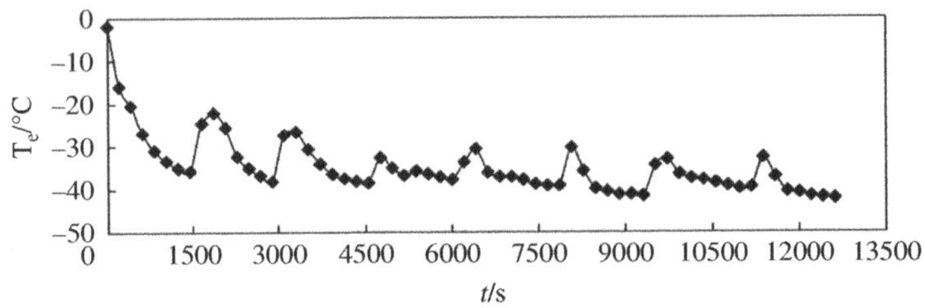

Figure 8.67 Refrigeration process without the cooling power output

the adsorbent bed. At the switch time, the adsorption and desorption valves close, and valve A5 opens, then the refrigerant desorbed from the adsorbent bed 2 in the half cycle time flows into the evaporator through valve A5, and the desorption quantity of adsorption bed 2 (Δm_d) is measured by the level sensor. At the end of the switch time, valve A5 closes again, and the adsorption quantity of adsorbent bed 2 (Δm_a) is measured by the level sensor.

The values measured by the level sensor are used to calculate the cooling power, and the equation is:

$$W_{ref} = W_{aref} - W_e - W_d = \frac{\Delta M_a}{\Delta t_a} \times L_{am} - (M_{ev} + \Delta M_d - \Delta M_a) \times C_{am} \times \frac{\Delta T_{ev}}{\Delta t_a} \quad (8.77)$$

where W_{ref} is the cooling power (kW), W_{aref} is the cooling power due to the adsorption of the adsorbent bed (kW), W_e and W_d are the cooling power consumed by the ammonia in the evaporator and the desorbed ammonia, respectively (kW), Δt_a is the cooling and adsorption time (s), M_{ev} is the mass of liquid ammonia in the evaporator measured by the level sensor before the condensate flows back to the evaporator (kg), ΔM_d is the desorption quantity of ammonia that flows to the evaporator through valve A5 (kg), ΔT_{ev} is the fluctuating value of evaporation temperature (°C).

The SCP and the COP of the system are calculated as follows:

$$SCP = \frac{W_{ref}}{M_a} \quad (8.78)$$

$$COP = \frac{W_{ref}}{W_h} \quad (8.79)$$

where M_a is the mass of $CaCl_2$ in the adsorbent bed (kg).

When the evaporating temperature is −25 °C, the performance of the adsorption system under the condition of different cycle times is shown in Table 8.21. The table showed that the optimum cycle time is 50 minutes, and corresponding COP and SCP are 0.36 and 627.7 W/kg, respectively.

8.5.5.6 Refrigeration Performance under the Condition of Optimum Cycle Time

When the cycle time is controlled at 50 minutes, the cooling water temperature is controlled at 23–27 °C, and the mass recovery time is controlled at 47 seconds, the system performance is

Table 8.21 The performance of the adsorption system under the condition of different cycle time

Cycle time t_c (min)	Maximum desorption temperature. T_{dmax} (°C)	Minimum adsorption temperature T_{amin} (°C)	Mass recovery time t_m (s)	Average cooling power W_{ref} (kW)	Average heating power W_h (°C)	*COP*	*SCP* (W/kg)
30	95.6	36.2	47	0.93	3.86	0.24	494.6
40	99.4	34.5	47	1.06	3.54	0.30	563.8
50	101.2	33.1	47	1.18	3.31	0.36	627.7
60	103.5	32.5	47	1.12	3.16	0.35	595.7

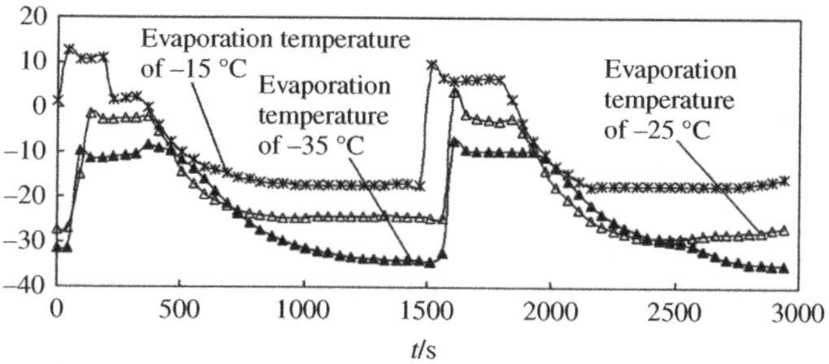

Figure 8.68 Evaporation temperature vs. running time

tested under the conditions of the minimum evaporation temperature of −15, −25, and −35 °C. The temperature fluctuation of the working fluid in the evaporator is shown in Figure 8.68. The temperature fluctuation is large, and it was mainly due to the refrigerant that desorbed from the adsorbent bed are accumulated in the condenser, and doesn't flow into the evaporator until the switch time.

The performances of the cycles with and without mass recovery process are compared when the evaporation temperature is −25 °C, and the results of the Clausius-Clapeyron diagram are shown in Figure 8.69. Figure 8.69 showed that the adsorption curve doesn't overlap on the

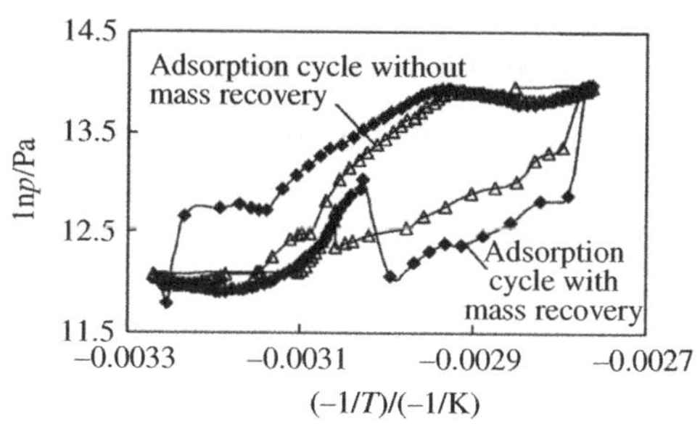

Figure 8.69 Clausius-Clapeyron diagram

Table 8.22 The refrigeration performance for the mass recovery cycle at different evaporation temperatures

Evaporation temperature T_e (°C)	Mass recovery time t_m (s)	Average cooling power W_L (kW)	Average heating power W_h (°C)	*COP*	*SCP* (W/kg)	*COP* increment (%)	*SCP* increment (%)
−15	47	1.37	3.34	0.41	731.0	41.3	39.8
	0	0.98	3.38	0.29	522.9		
−25	47	1.18	3.28	0.36	627.7	56.5	48.6
	0	0.79	3.43	0.23	422.3		
−35	47	0.89	3.30	0.27	473.4	79.8	67.9
	0	0.53	3.53	0.15	282.0		

desorption curve, which is mainly due to the hysteresis phenomenon between the adsorption and desorption process of the composite adsorbent.

As shown in Figure 8.69, for the cooling and adsorption process, the pressure for the cycle with mass recovery process is lower than that without mass recovery process. For heating and desorption process, the pressure for the cycle with the mass recovery process is higher than that without the mass recovery process. The area of the cycle on the *p*-*T* diagram with mass recovery process is bigger than that without mass recovery process, which means that the thermodynamic performance for mass recovery cycle is better than the cycle without mass recovery.

The refrigeration performances are tested for the cycles with and without mass recovery processes under the condition of different evaporation temperature, and the results are shown in Table 8.22. Table 8.22 shows that the lower the evaporation temperature is, the greater the performance is improved due to the mass recovery. It is mainly caused by the pressure difference between two adsorbent beds at the switch time. The lower the evaporation temperature is, the greater the pressure difference between two adsorbent beds is for the mass recovery process and the more positive the effects are. The average value of the *SCP* is 627.7 and 422.3 W/kg, respectively, for the cycles with and without mass recovery process under the condition of evaporating temperature of −25 °C, and the data for the mass recovery cycle increases by 48.6%. The average value of the *COP* for the cycles with and without mass recovery is 0.36 and 0.23, respectively, and the data for the mass recovery cycle increases by 56.5%.

8.5.5.7 Research on the Heat Transfer Performance

The temperature and the heat transfer coefficient of the adsorbent bed during the heating and cooling processes are analyzed at the evaporating temperature of −25 °C, which are shown in Figure 8.70. Compared with the results in Figure 8.62, it can be found that the heat transfer coefficients for both heating and cooling processes have been greatly improved due to the higher heat flux. In Figure 8.62, the heating power of the electric heater is 3 kW, while in Figure 8.70, the heating power of the electric heater is 4.2 kW. When the heating power increases, the heat flux of the adsorbent bed in the heating process will also increase, which makes the desorption quantity increase. Consequently the adsorption quantity of the adsorbent

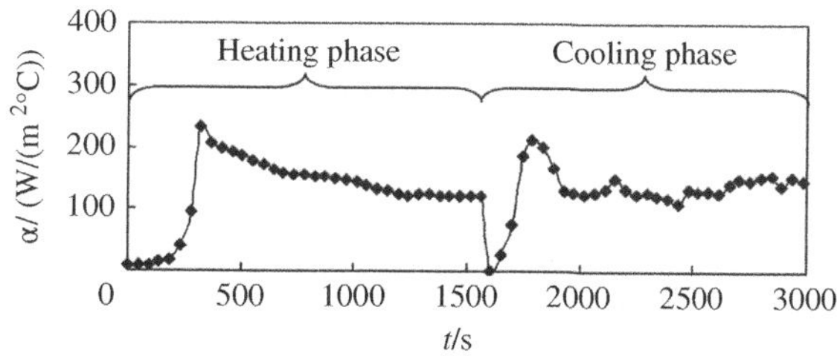

Figure 8.70 The heat transfer coefficient of the adsorption bed

bed as well as the adsorption heat will also increase, that is, the heat flux of the adsorbent bed in the cooling and adsorption processes increases. As a result, from Equations 8.54 and 8.56, the heat transfer coefficients of the adsorbent bed will increase too.

As shown in Figure 8.52, the heat transfer coefficient of the adsorbent bed in the heating and cooling process is 162.3 and 169.4 W/(m^2 °C), respectively. It is improved by about 88% if compared with the adsorbent bed of activated carbon–methanol adsorption ice maker [21].

The total volume of the system including two adsorption beds, one heating boiler, one liquid pumping boiler, and one cooler is only 0.064 m^3. The volumetric cooling capacity of the adsorption bed is 21.4 kW/m^3 at the evaporation temperature of −15 °C. If we further reduce the volume of heating boiler and cooler by a reasonable design, the value will be raised. Nowadays the cooling capacity is about 6 kW and the volume of the ice maker is about 0.587 m^3 for the comercialized adsorption ice making system for fishing boats with two adsorption beds, one waste heat recoverer and one water heat exchanger [26], and the volumetric cooling capacity is 10.2 kW/m^3. Compared with this data, the heat pipe type composite adsorption ice maker could improve the cooling capacity by 109.8%.

Comparing the data for the heat pipe type composite adsorption ice maker with the data in the references, the results are shown in Table 8.23. The best *SCP* of the system under the evaporation temperature of −15 °C is 1.2 times better than that for the air-conditioning condition (evaporation temperature is 1 °C) and is 2.2 times better than that for the ice-making conditions (evaporation temperature of −10 °C).

8.5.5.8 Refrigeration Performances for Unstable Conditions

The power of the diesel engine on fishing boats varies with the operating conditions, which will influence heating power of the exhausted gas and consequently will influence the heating power of the adsorption system. In order to study the performances of the heat pipe type adsorption system under unstable conditions, the performance of the adsorption system is tested under the conditions of different heating power.

Under the conditions of the heating power of 3.6 kW, the evaporation temperature of −15 °C, and the cooling water temperature around 23–27 °C, the performances of the system under different cycle times are tested and the results are shown in Table 8.24. As seen from Table 8.24, due to the heating power decreasing, the optimal cycle time is prolonged. When the cycle time is 60 minutes, the performance of the system is best and the cooling power is 0.92 kW, and the corresponding highest temperature of the bed is 94.1 °C.

Table 8.23 Comparison of the adsorption refrigeration performances

	Refrigeration temperature (°C)	Working pair	*COP*	*SCP* (W/kg)	Feature of the system	Data source
Data in the references	8	Activated carbon-NH_3	–	1000	Convective thermal wave cycle	Simulation [30]
	1	Activated carbon fiber/$CaCl_2$-NH_3	0.6	330	With thermal wave heating process	Experiments [31]
	3	Activated carbon-NH_3	0.67	557	Convective thermal wave system	Simulation [32]
	−10	$SrCl_2$-NH_3	0.32	230	Single effect system	Experiments [33]
	−25	($MnCl_2 + NiCl_2$)-NH_3	0.4	70 × 2	Double effect system	Simulation [33]
	−10	Metal hydride-ammonia	0.43	25 × 2	Thermal wave cycle	Experiments [34]
	3	Graphite/silica gel-water	–	35 × 2	Composite adsorbent	Experiments [35]
	5	Zeolite-water	0.9	125 × 2	Intermittent convection thermal wave cycle	Simulation [36]
Heat pipe type composite adsorption prototype	−15	Composite adsorbent-ammonia	0.41	731	Phase change heat transfer process	Experiments
	−25	Composite adsorbent-ammonia	0.36	627.7	Phase change heat transfer process	Experiments
	−35	Composite adsorbent-ammonia	0.27	473.4	Phase change heat transfer process	Experiments

Table 8.24 The adsorption performance of the system under different cycle time

Cycle time t_c (min)	Maximum desorption temperature T_{dmax} (°C)	Minimum adsorption temperature T_{amin} (°C)	Mass recovery time t_m (s)	Average cooling power W_{ref} (kW)	Average heating power W_h (°C)
40	88.3	34.9	47	0.62	3.27
50	91.2	33.1	47	0.81	3.12
60	94.1	31.5	47	0.92	2.90
70	96.7	31.1	47	0.89	2.87

Table 8.25 The adsorption performance of the system for different cycle time

Cycle time t_c (min)	Maximum desorption temperature T_{dmax} (°C)	Minimum adsorption temperature T_{amin} (°C)	Mass recovery time t_m (s)	Average cooling power W_{ref} (kW)	Average heating power W_h (°C)
50	80.9	33.8	47	0.31	2.58
60	83.4	32.3	47	0.49	2.46
70	85.2	31.4	47	0.67	2.39
80	87.1	31.2	47	0.64	2.34

Under the conditions of the heating power of about 3 kW and the minimum evaporation temperature of −15 °C, the performances of the system under different cycle times are shown in Table 8.25. Due to the heating power decreasing, the desorption cannot be completed when the cycle time is short, and the performance of the system will be influenced. The performance of the system is gradually improved when the cycle time is prolonged, and the cooling power of the system was 0.67 kW when the cycle time is increased to 70 minutes.

8.5.5.9 Study on the Performance Deterioration of the System

The heat pipe type composite adsorption prototype was established in October 2004. In December 2004 in order to test the heat pipe performance and verify the performance deterioration phenomenon of the adsorbent, the repeated heating and cooling processes were operated, and the maximum heating temperature of the adsorbent bed was about 145 °C. The performance stability of the system was tested during December 2004 to May 2005. Under the conditions of the maximum desorption temperature of the adsorbent bed being 105 °C, the cooling water temperature being 25 °C, minimum evaporation temperature being −15 °C, and cycle time being 50 minutes, the average values of *SCP* for different months are shown in Table 8.26. The cooling power per unit adsorbent at different months ranged between 720 and 750 W/kg, which indicated that the adsorbent bed doesn't have the performance deterioration phenomenon after it runs for more than eight months.

Table 8.26 Refrigeration performance per unit mass of adsorbent for different month

Month	Maximum desorption temperature T_{dmax} (°C)	Minimum adsorption temperature T_{amin} (°C)	Evaporation temperature T_e (°C)	Mass recovery time t_m (s)	Average *SCP* (W/kg)
2004-12	105.6	33.8	−15.2	47	728.3
2005-01	104.9	32.3	−17.8	47	731.0
2005-02	107.8	34.6	−14.8	47	746.8
2005-03	103.7	34.9	−16.4	47	722.9
2005-04	106.8	36.7	−15.8	47	736.7
2005-05	109.7	38.7	−16.7	47	741.4

8.5.6 *Comparison between the Experimental Results and the Simulation Results*

The heating power in the experimental processes is analyzed, due to the heat loss in the experiments the metal heat capacity consumed about 0.4 kW heat and the heat loss was about 0.38 kW when the adsorbent bed is heated from 30 to 110 °C in 25 minutes. Then when the heating power is 4.2 kW, the average heating power absorbed by the working fluid of the heat pipe and the metal heat capacity is actually about 3.8 kW. Similarly when the heating power is 3.6 kW, the data is about 3.2 kW, and with the heating power of 3.0 kW the data is about 2.6 kW. Due to the heat loss, the performance of the system should be affected and consequently it will have a large difference with the simulation results. But the comparison showed that the experimental results are similar with the simulation results. The experimental results for 4.2 kW heating power (the actual heating power was 3.8 kW due to heat loss) were better than the simulation results under the heating power of 3.8 kW shown in Figure 8.58 and almost the same as the simulation results of the heating power of 4.0 kW. The experimental results for the heating power of 3 kW (the actual heating power is 2.6 kW) were almost the same as the simulation results for the heating power of 3 kW.

The reason is analyzed, and the phenomenon was mainly caused by the adsorption kinetic models. In the simulation the chemical kinetic models in Section 8.5.3 are chosen because they have similar equilibrium adsorption performance with composite adsorbent. But actually the porous media in composite adsorbent made the adsorption process not able to be described by the chemical adsorption kinetic models accurately. In other words, the physical adsorbent that has porous structure in composite adsorption greatly improved the non-equilibrium adsorption properties, which results in the difference between the simulation and the experimental results. Consequently the experimental results are better than the simulation results.

8.6 Two Stage Adsorption Refrigerator

An innovative multifunction heat pipe type sorption refrigeration system is designed, in which a two stage sorption thermodynamic cycle based on two heat recovery processes is employed to reduce the driving heat source temperature, and the composite sorbent of $CaCl_2$ and activated carbon is used to improve the mass and heat transfer performances. For this test unit, the heating, cooling, and heat recovery processes between two reactive beds are performed by multifunction heat pipes [37–40].

8.6.1 *System Design*

The schematic diagram of a multifunction heat pipe type sorption refrigeration system is shown in Figure 8.71, and the corresponding photograph of a sorption refrigeration machine is illustrated in Figure 8.72. The experimental setup mainly consists of one waste heat source equipment, two reactive beds, a water cooler, a condenser, an evaporator (ice maker), and a refrigerant storage vessel, in which an electric-powered boiler was used to simulate the waste heat source equipment. The water in the boiler is heated by an electric heater to produce water vapor, and the pressure transducers and temperature sensors are installed to measure the relevant pressures and temperatures. The composite sorbent of activated carbon–$CaCl_2$, which has higher sorption performance than $CaCl_2$, was used to improve the mass and heat transfer

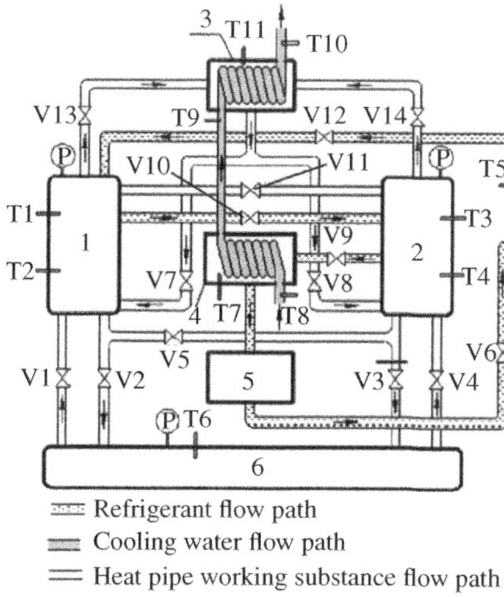

Figure 8.71 Schematic diagram of the multifunction heat pipe type sorption refrigeration system. 1, adsorber A; 2, adsorber B; 3, cooler; 4, condenser; 5, storage vessel; 6, waste heat equipment; 7, ice maker (evaporator); V1,V4, valves of heat pipes for water vapor path during heating phase; V2,V3, valves of heat pipes for water path during heating phase; V5, valve of heat pipe for water path during heat recovery phase; V6, valve of liquid refrigerant path; V7,V8, valves of heat pipes for water path during cooling phase; V9,V10,V12, valves of gas refrigerant path; V11, valve of heat pipe for water vapor path during heat recovery phase; V13,V14, valves of heat pipes for water vapor path during cooling phase

Figure 8.72 Photograph of the multifunction heat pipe type sorption refrigeration machine

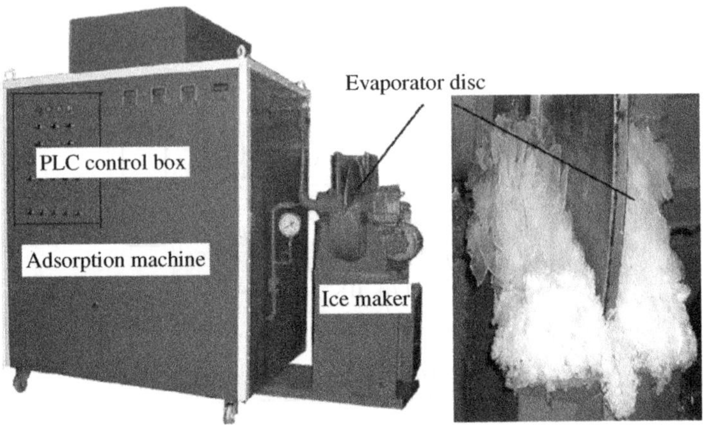

Figure 8.73 The photograph of two-stage sorption prototype

performance. Enhancement of heat transfer inside the reactive bed is one of the most important factors to improve the system performance. For the test unit, the heating, cooling, and heat recovery processes between two reactive beds are performed by multifunction heat pipes. For example, when the reactive bed is in desorption process, the bed is heated by a split heat pipe in which the heating boiler serves as an evaporator and the desorbing heat exchanger (DHE) serves as a condenser. On the contrary, if the bed is in the adsorption process, it would be cooled by the upper part split heat pipe in which the AHE (adsorption heat exchanger) serves as an evaporator and the cooler serves as a condenser. During the heat recovery process, low-temperature bed is pre-heated and high-temperature bed is pre-cooled by a split heat pipe in which the high-temperature DHE serves as an evaporator and low-temperature AHE serves as a condenser.

Figure 8.73 illustrates the photograph of the two-stage sorption prototype for ice making. The primary component of the ice maker is a stainless steel evaporator disc. The evaporator has a hollow circuitry and refrigerant ammonia flows through passages inside the evaporator disc removing heat from the freezing surface of the disc. Water is sprayed on both sides of the disc where it is frozen into a thin layer of ice. Two fixed horizontal blades harvest ice as the disc rotates in the vertical plane, and the harvested ice falls into a container. The sorption system operation is automatically controlled by a PLC control box.

8.6.2 Schematic Diagram of the Two-Stage Sorption Refrigeration Cycle

The two-stage sorption thermodynamic cycle with heat recovery process mainly includes two stages, and the experimental procedures are as follows:

1. During the first stage, reactive bed A serves as a desorber and reactive bed B serves as a pseudo-condenser, and the refrigerant desorbed from bed A flows to bed B directly. As the refrigerant concentrations in the adsorber and the desorber are at or near to equilibrium level, desorber A should be switched into adsorption mode while adsorber B would be changed into desorption state.

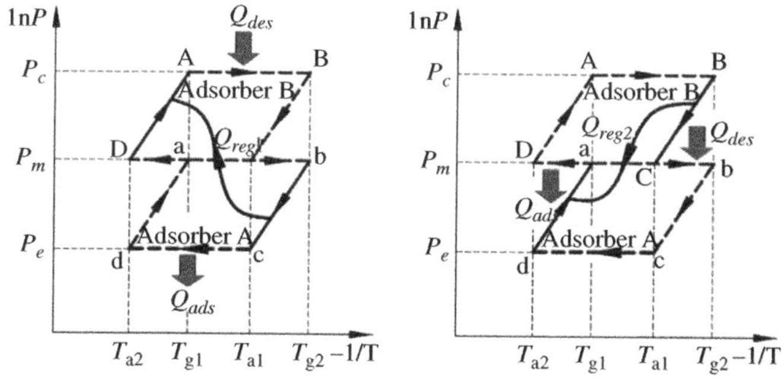

Figure 8.74 Schematic diagram of two-heat recovery processes of the two-stage sorption refrigeration cycle

2. During the switching process, the valves V5 and V11 between the DHE and AHE are opened, and the first heat recovery occurs to recover heat from the high-temperature bed A to the low-temperature bed B.
3. In the next stage, the reactive bed B is heated to desorb the refrigerant to the condenser while the reactive bed A is cooled to adsorb refrigerant from the evaporator, and the evaporation heat of ammonia inside the evaporator provides refrigeration effect, which is transported to the flake ice maker to make ice. This phase will end until the desorption process is finished inside the desorber B and the adsorption process is completed inside the adsorber A.
4. The second heat recovery is carried out to recover heat from the desorber to the adsorber. Thus, as shown in Figure 8.74, there are two heat recovery processes for the innovative two-stage sorption thermodynamic cycle to improve the heat recovery efficiency.

8.6.3 *Performance Test*

8.6.3.1 Temperature Evolution with Operating Time

The temperature evolution with operating time for the two-stage sorption cycle with heat recovery is shown in Figure 8.75. It was observed that two-heat recovery processes are employed to

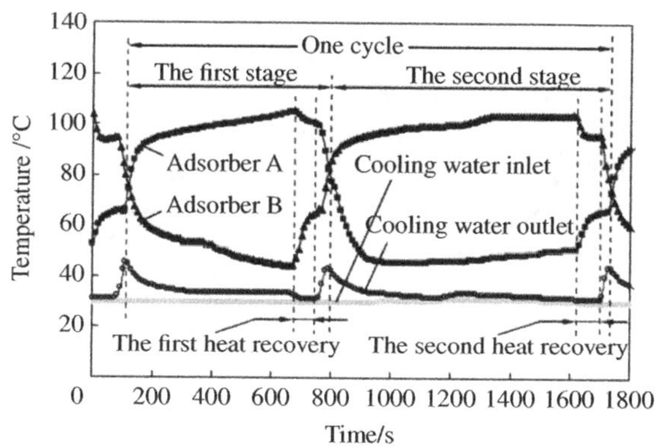

Figure 8.75 Temperature evolution of the two-stage sorption cycle with heat recovery

reclaim energy from hot bed to cold bed during one cycle, and the energy recovery strategy can reduce the heat consumption effectively. For example, the cold bed temperature is 43 °C and the hot bed temperature is 103 °C when adsorption and desorption are completed, in order to regenerate the sorbent, the cold bed temperature for a conventional two-stage cycle (without heat recovery) should be upgraded from 43 to 103 °C. However, for a two-stage cycle with heat recovery, the cold bed temperature would be upgraded from 66 to 103 °C due to the fact that the cold bed temperature has been lifted from 43 to 66 °C by the heat recovery process. At the same time, the hot bed temperature can be decreased from 103 to 92 °C due to rejecting heat to the cold bed, and the cooling load of cooler is also reduced.

8.6.3.2 Pressure Evolution with Operating Time

Figure 8.76 shows the pressure evolution with operating time for the two-stage sorption cycle with heat recovery. During the first stage, the reactive bed A is heated while the reactive bed B is cooled. It was observed that the pressures of two reactive beds are nearly equal during this phase because desorber A is connected to adsorber B directly, and the pressures decrease gradually as the reaction evolves, which means that the refrigerant desorbed from the hot bed cannot be entirely adsorbed by the cold bed. During the second stage, reactive bed B is heated to desorb the refrigerant to the condenser and reactive bed A is cooled to adsorb the refrigerant from the evaporator, and the desorption pressure is more stable when compared to the adsorption pressure. Combining temperature and pressure evolutions of two reactive beds, it was found that the pressure evolutions of desorber are completely different though the desorber temperatures are nearly equivalent during two operation stages, and the desorption pressure during the second stage is much higher than that in the first stage.

8.6.3.3 Evaporation Pressure Evolution with Operating Time

Figure 8.77 illustrates the evaporation pressure evolution with operating time for the two-stage sorption cycle with heat recovery. It was observed that the evaporation pressure increases during the first stage due to the sensible heat change of the liquid refrigerant inside the evaporator. During the second stage, the evaporation pressure reduces sharply under the adsorption function of adsorber A, and it rises slowly along with the adsorption process. When the

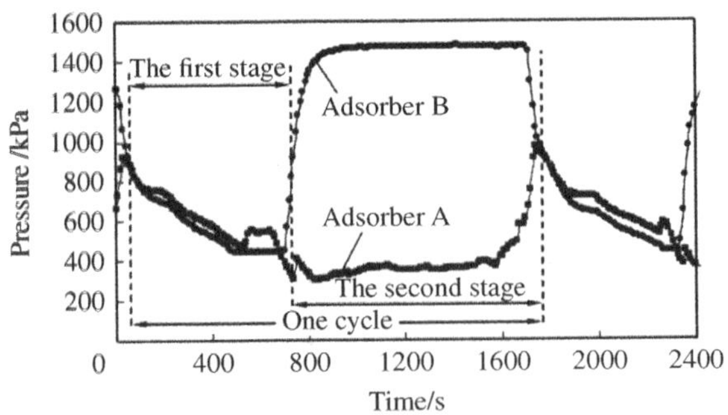

Figure 8.76 Pressure evolution of the two-stage sorption cycle with heat recovery

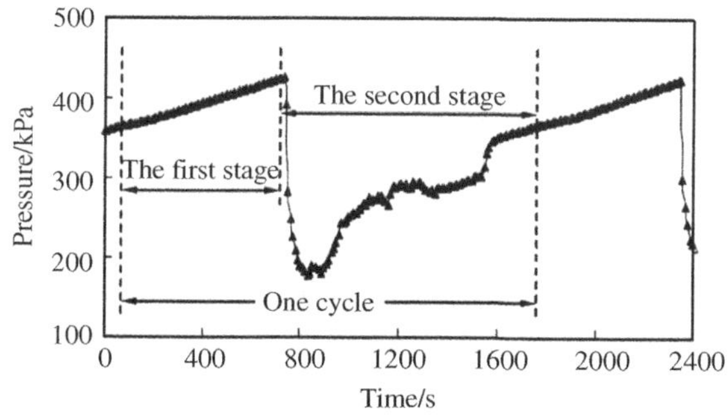

Figure 8.77 Evaporation pressure evolution of the two-stage sorption cycle with heat recovery

adsorber is almost saturated and the evaporation pressure is too high to make the flake ice, the system is switched by electromagnetic valves to the next sorption cycle.

8.6.3.4 Condensation Temperature Evolution with Operating Time

The condensation temperature evolution with operating time for the two-stage cycle with heat recovery is shown in Figure 8.78. During the first stage, the condenser is separated from the desorber, and its temperature decreases as it is cooled by the cooling water. When the system is switched into the second stage, a large amount of refrigerant is released from the reactive bed and flows into the condenser. Thereby, the condensation temperature increases promptly, and the temperature falls again as the desorption rate of refrigerant becomes low during the next process.

8.6.3.5 Experimental Clapeyron Diagram of the Two-Stage Sorption Cycle

Figure 8.79 shows the experimental Clapeyron diagram of the two-stage sorption thermodynamic cycle. It is obvious that a two-stage sorption cycle includes two sub-cycles: the upper one is a high pressure sub-cycle while the bottom one is a low pressure sub-cycle. The experimental

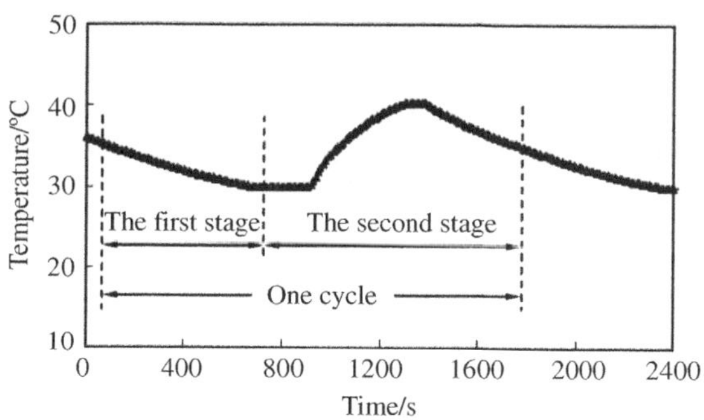

Figure 8.78 Condensation temperature evolution of the two-stage sorption cycle with heat recovery

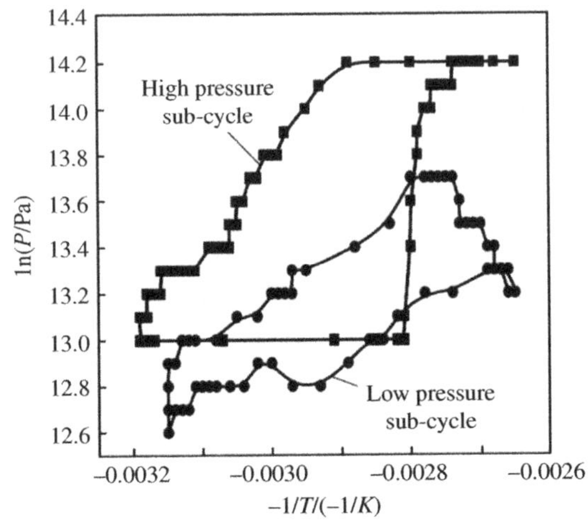

Figure 8.79 Experimental Clapeyron diagram of the two-stage sorption thermodynamic cycle

results are in accord with the theoretical results in terms of the temperature evolution, and the results show that the two sub-cycles have equivalent desorption and adsorption temperatures. In a theoretical cycle, desorption and adsorption processes operate at the constant pressures P_c and P_e, respectively, and they have the same pressures during the first stage. However, the pressures of desorption and adsorption fluctuate with the temperatures in the real sorption cycle. The reasons are as follows: first, the adsorption and desorption are assumed as ideal processes in a theoretical cycle and these assumptions are not compared to the practical results. For example, the adsorption capacity is inconstant during the isosteric pre-heating process, and the desorber is not connected to the condenser until the desorption pressure is higher than the condensation pressure in the real cycle. Secondly, the distributions of adsorption capacity and temperature inside beds are usually non-uniform due to the limitations of the heat and mass transfer, and the local refrigerant would be transferred inside beds. Moreover, the existence of the flow resistance and the variation of refrigerant flux also result in the fluctuation of pressure.

8.7 Adsorption Refrigerator Using $CaCl_2$/Expanded Graphite-NH_3

A novel $CaCl_2$/expanded graphite-NH_3 adsorption refrigerator is developed and tested. Only three valves are installed in this refrigerator and they all work at the same positive pressure condition, so the operating reliability of the refrigerator can be improved.

8.7.1 Structure of Adsorption Refrigerator

The refrigerator is shown in Figure 8.80. It includes two adsorbers, a cooler, a vapor generator, a pump, two condensers, two evaporators, a three-way vapor valve, a three-way water valve, and an ammonia valve. Each adsorber has 27 kg of $CaCl_2$ within the composite adsorbent. The composite adsorbent is filled in between the fins of the 37 adsorption unit tubes. It is agreed that the adsorbers must be heated and cooled discontinuously to realize the decomposition (corresponding to the desorption process) and combination (corresponding to the adsorption process)

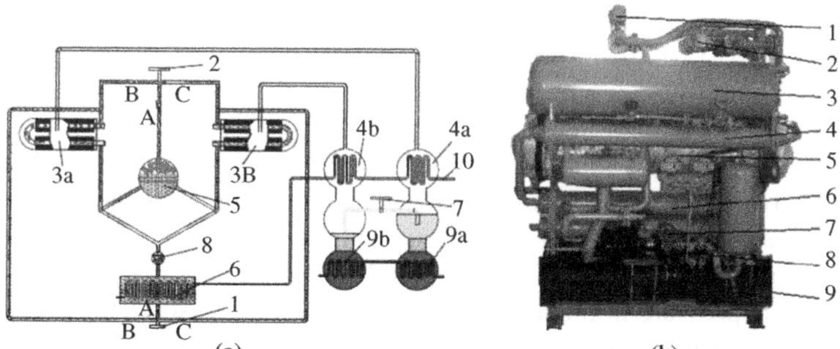

(a) (b)

1. Water valve; 2. Vapor valve; 3. Adsorbers (3a and 3b); 4. Condensers (4a and 4b);
5. Vapor generator; 6. Cooler; 7. Ammonia valve; 8. Pump; 9. Evaporators (9a and 9b);
10. Cooling water

Figure 8.80 Structure of refrigerator. (a) Schematic diagram and (b) photo of the refrigerator

processes between adsorbent ($CaCl_2$) and refrigerant of ammonia. So the adsorbers will be in desorption and adsorption processes alternately and discontinuously. In the desorption process, the adsorber is heated by the separate thermo-syphon heat pipe. The adsorber serves as the condenser and the vapor generator serves as the evaporator of the separate thermo-syphon heat pipe. The adsorber is connected with the vapor generator and is heated by the evaporated vapor of the heating fluid from the vapor generator via the separate thermo-syphon heat pipe. The heating fluid in the separate thermo-syphon heat pipe is heated to evaporation by the vapor generator, and then the evaporated vapor enters into the adsorber via the three-way vapor valve and is condensed into liquid when it touches the cold surface inside the adsorption unit tubes, releasing energy to heat the adsorber. Finally, the condensed liquid of the heating fluid returns to vapor generator to be heated to evaporate again [41, 42].

When the temperature of the adsorber reaches a sufficiently high value, the ammonia is desorbed from the adsorber. The evaporated vapor of the heating fluid is controlled by the valve ports that connect from A to B or A to C. When the A–B ports of the three-way vapor valve are connected, the evaporated vapor flows into the adsorber 3a to heat it. When the A–C ports are connected, the evaporated vapor flows into the adsorber 3b and the adsorber 3b is heated. In the adsorption process, the adsorber is connected to the cooler. The adsorber is cooled by cooling fluid from the cooler. The cold cooling fluid in the cooler is pumped into the adsorber via the three-way water valve, then the cooling fluid exchanges heat with the adsorber and becomes "hot" cooling fluid return to the cooler, then the "hot" cooling fluid exchanges heat with the cooling water in the cooler and becomes cold cooling fluid to cool the adsorber again. When the adsorber is cooled down, the refrigerant in the evaporator is adsorbed by the adsorber. The evaporator transfers the refrigerating capacity outside. The flowing process of the cooling fluid is controlled by switching the ports A–B or A–C in the three-way water valve. When the A–B ports of the three-way water valve are connected, the cooling fluid flows into the adsorber 3a to cool it. When the A–C ports are connected, the cooling fluid flows into the adsorber 3b or the adsorber 3b is cooled.

The refrigerator has three working processes. The first one is desorption of adsorber 3a and adsorption process of adsorber 3b, the second one is mass recovery process between two evaporators 9a and 9b and the third one is desorption phase of adsorber 3b and adsorption process of adsorber 3a. In the first working process (desorption process of adsorber 3a and adsorption process of adsorber 3b), A–B ports of the vapor valve and A–C ports of the water valve are connected, A–C ports of the vapor valve, A–B ports of the water valve are not connected and the ammonia valve is closed. At the end of the first working process, the ammonia mass and pressure in two evaporators are not equal, so the next process is the mass recovery process. In the second process (mass recovery process between evaporators 9a and 9b), the ammonia valve between two evaporators is open. The ammonia in the desorbing adsorber 3a quickly enters the adsorbing adsorber 3b, because the pressure in the desorbing adsorber 3a is much higher than that in the adsorbing adsorber 3b. The pressure in the desorbing adsorber 3a decreases and the pressure in the adsorbing adsorber 3b increases. This process increases the quantity desorbed from the desorbing adsorber 3a and the quantity adsorbed by adsorber 3b. When the pressures of the two adsorbers 3a and 3b are equal, the third process begins. In the third process (desorption process of adsorber 3b and adsorption process of adsorber 3a), A–C ports of the vapor valve and A–B ports of the water valve are connected, A–B ports of the vapor valve, A–C ports of the water valve are not connected and the ammonia valve is closed.

Figure 8.81a shows the structure of the adsorbers. There are 40 tubes in each adsorber, which are composed of 37 adsorption unit tubes and 3 plain tubes. The length of the fin tube is 1500 mm, the fin thickness is 0.3 mm, the external diameter of fin is 58 mm, the internal diameter of fin is 28 mm, and the fin pitch is 2.5 mm. The cross section of the adsorber is shown in Figure 8.81b. There are five groups of adsorption unit tubes in each adsorber, one group

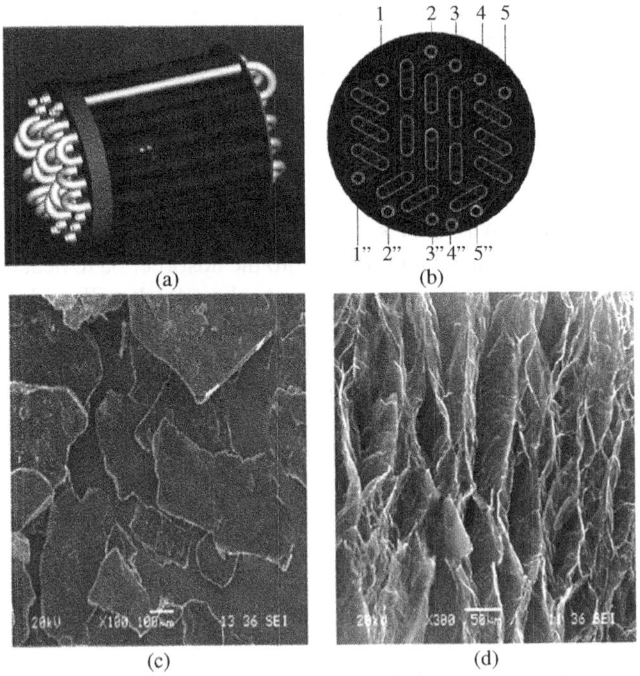

Figure 8.81 Structure of adsorbers and SEM image of graphite. (a) Structure of adsorbers; (b) cross section of the adsorbers; (c) SEM image of expandable graphite; and (d) SEM image of expanded graphite

(a) (b) (c)

Figure 8.82 Adsorption unit tubes. (a) Finned tube; (b) adsorption unit tube with the adsorbent inside; and (c) the final adsorption unit tube

is composed of eight adsorption unit tubes. SEM (scanning electron microscope) images of expandable graphite and the expanded graphite are shown in Figure 8.81c,d.

The adsorption unit tubes in each adsorber are shown in Figure 8.82. The adsorption unit tubes are filled with the compound adsorbent in the space between the fins of the finned tubes. A thin stainless wire mesh is utilized to wrap the fin tube to keep the shape of compound adsorbent, and it is fixed firmly by a stainless porous tube. The compound adsorbent contains calcium chloride and expanded graphite. The compound adsorbent is mixed with water to dissolve the calcium chloride before the adsorbent was filled in the unit tube. The water in compound adsorbent is driven out at a high temperature of 200 °C after it is filled into the adsorption unit tube. The mass ratio of the calcium chloride and expanded graphite is about 4 : 1. The mass of the compound adsorbent in each adsorption tube is 0.94 kg, which contains about 0.75 kg calcium chloride.

8.7.2 Performance Test

Figure 8.83 shows the test system of the adsorption refrigerator. It is composed of three systems. They are the boiler, cooling water, and chilled liquid systems. In the boiler system, the flow rate and temperature of the heating vapor flowing into the adsorption refrigerator from the boiler are not adjustable and controllable. So the heating power is calculated through the condensation water of the heating vapor flowing out of the adsorption refrigerator. In the cooling water system, the cooling water from the adsorption refrigerator flows into the cooling water tank. The retrieval cooling water mixes with the city water. The redundant water in the cooling water tank is drained out through the overflow pipe. The constant inlet temperature of the cooling water is controlled by a different opening of V1. In the chilled liquid system, a little of inlet cooling water flows into the coil pipe in the chilled liquid tank through valve V4 to balance the refrigerating power yielded by the adsorption refrigerator. The constant inlet temperature and flow rate of the chilled liquid are controlled by a different opening of valve V4 and V5. Finally, a constant evaporation temperature is achieved.

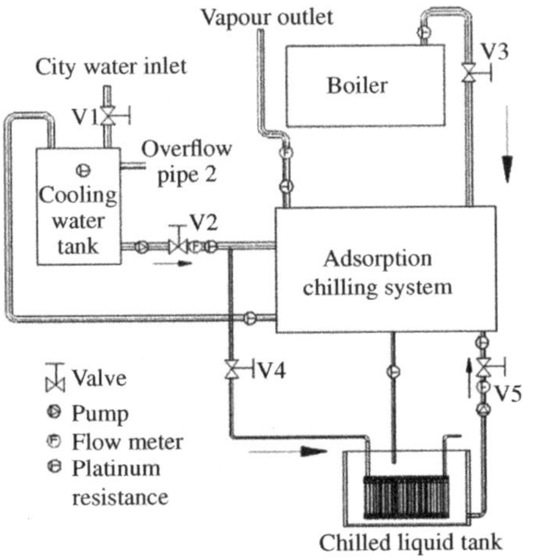

Figure 8.83 Test system of the adsorption refrigerator

8.7.2.1 Effects of Heating/Cooling Time

Heating/cooling time is a key parameter that influences the performance of the adsorption refrigerator. Theoretically, the shorter the cycle time for a given concentration/adsorption quantity change, the larger the refrigerating power, whatever the heat source and evaporation temperatures are. However, in practice, the adsorption/desorption process will not be completed and the refrigerating potential of the adsorber cannot be fully achieved if the heating/cooling time is too short. Meanwhile the heating sources coming from a boiler cannot output the heating media with constant temperature. Under such conditions the performances of the system are tested, and the results are shown in Table 8.27. In the table the performance of the adsorption refrigerator with different cycle time at the evaporation temperature of −15 °C, cooling water temperature of 25 °C, mass recovery time of 45 seconds are studied. When the cycle time is 25 minutes, the adsorption refrigerator has the highest refrigerating power of 11.4 kW and highest COP of 0.27, and the corresponding SCP, electricity COP, and cooling capacity are 422.2 W kg^{-1}, 32.6, and 57 kW m^{-3}.

8.7.2.2 Effects of Evaporation Temperature

Figure 8.84 shows the refrigerating power variations with the different evaporation temperature. Evaporation temperature increase by 1 °C can cause approximate 8.7 and 4.8%

Table 8.27 Cooling capacity variation with the heating/cooling time

Cycle time (min)	Highest heating temperature (°C)	Cooling capacity (kW)	COP	SCP (W/kg)
16	121.7	10.1	0.24	374.1
19	126.7	10.7	0.26	396.4
22	130.4	10.8	0.26	400.0
25	140.3	11.4	0.26	422.2
28	152.7	11.0	0.26	407.4

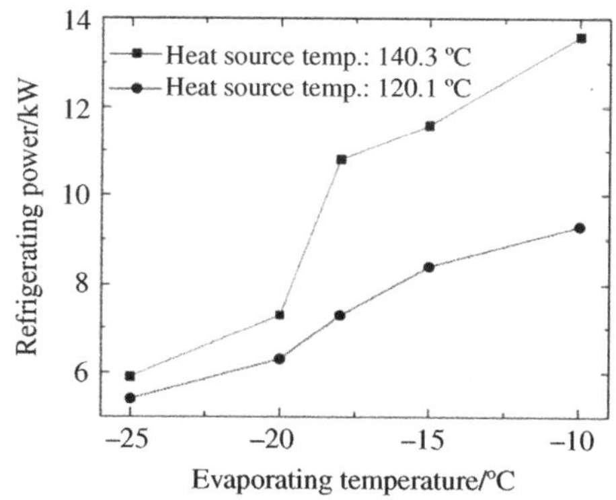

Figure 8.84 Refrigerating power vs. evaporation temperature

improvement of the refrigerating power at a higher heat source temperature of 140.3 °C and lower heat source temperature of 102.1 °C under the conditions of the same cooling water temperature, mass recovery time, and heating/cooling time.

8.7.2.3 Effects of Cooling Water Temperature

The cooling water temperature is another key parameter that influences the performance of the adsorption refrigerator. It influences not only the adsorption process but also the condensation process, because the cooling water firstly flows into the condenser to condensate the ammonia, then flows into the cooler to cool the adsorber. A higher cooling water temperature increases the condensation temperature and adsorption temperature. As a result, both desorption and adsorption processes are influenced by the cooling water temperature. Figure 8.85 shows the refrigerating power variations with the cooling water temperature. The lower the cooling water temperature is, the larger the refrigerating power. Decreasing cooling water temperature has a

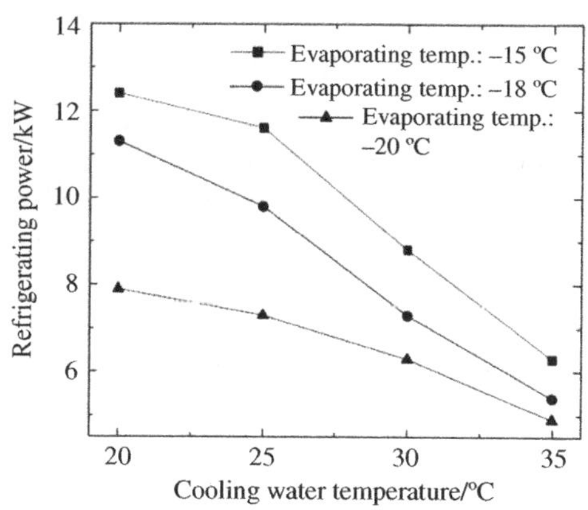

Figure 8.85 Refrigerating power vs. cooling water temperature

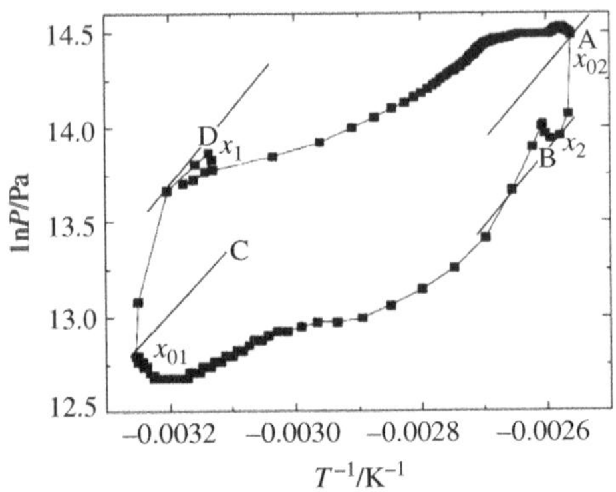

Figure 8.86 *P-T* diagram of the cyclic process obtained from the experimental results

similar effect on the refrigerator as increasing the evaporation temperature. When the cooling water temperature decreases by 1 °C, the refrigerating powers of the refrigerator can increase 6.5, 6.3, and 4.0% at evaporation temperatures of −15, 18, and 20 °C, respectively.

8.7.2.4 Clausius–Clapeyron Diagram

Figure 8.86 shows a full Clausius–Clapeyron cycle (A–B–C–D) obtained from the experimental results. A–B and C–D processes are the mass recovery processes. D/B process is the desorption process and B/D process is the adsorption process. The adsorption quantity of the adsorber decreases from x_{01} to x_1 and the desorption quantity of adsorber increases from x_{02} to x_2. The variation of the adsorption quantity with mass recovery process is $x_2 - x_1$, which is larger than that of $x_{02} - x_1$ without mass recovery process. The variation of the desorption quantity with mass recovery process is $x_2 - x_1$, obviously, this value is larger than that $x_2 - x_{01}$ without the mass recovery process. Therefore, the mass recovery processes can effectively increase desorption quantity and adsorption quantity.

8.8 Adsorption Refrigerator Using $CaCl_2$/Activated Carbon–NH_3

One adsorption refrigerator using two-phase heating with chemical/physical compound adsorbent of $CaCl_2$/activated carbon was designed and tested. Micro-porous activated carbon was mixed with salt to increase the mass transfer performance of the adsorbent and avoid the problem of agglomeration phenomenon. Because of the utilization of two-phase heating in the adsorption heating process, the heat transfer performance can be improved compared with single-phase heating process.

8.8.1 System Description

The chemical reactor is the most important part in the adsorption refrigerator since its heat and mass transfer performance will influence the system refrigeration performance directly.

Figure 8.87 Picture of adsorbent and adsorption reactor

Table 8.28 The physical parameters of 14–28 mesh activated carbon

Specific surface area (m^2/g)	Pore volume (cm^3/g)	Density (kg/m^3)	Specific heat (J/(kgK))	Porosity	Pore diameter A	Thermal conductivity (W/(mK))
1200	0.36–0.55	465–775	1003.2	0.5–0.6	20–50	0.14–0.35

In this system, finned tube is used as shown in Figure 8.87a. Then, a mesh and a metal screen were placed around the materials to avoid the leakage from fins, as shown in Figure 8.87b,c. Finally, the tubes were welded between the cover plates inside the chemical reactor, as shown in Figure 8.87d. $CaCl_2$ and activated carbon are used as the consolidated chemical/physical adsorbents. The mass ratio of $CaCl_2$ and activated carbon is 4 : 1, with 17.3 kg being the mass of $CaCl_2$, as shown in Figure 8.87e. The adsorbent was pressed inside the aluminum fins to enhance heat transfer performance. The SEM of activated carbon is shown in Figure 8.87f, which has high volumetric adsorption capacity and does not present agglomeration phenomena and performance attenuation. The physical parameters of activated carbon are shown in Table 8.28.

The adsorption refrigerator consists of boiler, chemical reactors, condensers, evaporators, cooling water tank, valves, and one water pump, as shown in Figure 8.88. The picture of the

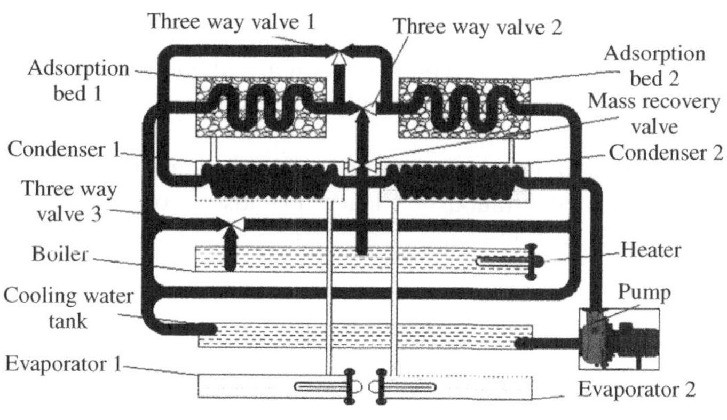

Figure 8.88 The schematic of the adsorption refrigerator

Figure 8.89 The picture of the adsorption refrigerator (insider)

adsorption refrigerator is shown in Figure 8.89. There are two innovations in this system. Firstly, there are only three water valves, which provide a higher reliability yet more convenient controlling process. Secondly, the chemical reactor is heated by a two-phase heat transfer that significantly improves the heat transfer performance.

There are three working processes:

1. Water is heated in the boiler and evaporates. The generated vapor enters the left chemical reactor through the heating vapor pipeline, and condenses when it touches the cold surface inside the finned tubes, providing heat for desorption. The condensed water flows back to the heating boiler. The ammonia is desorbed from the left hot reactor and the ammonia vapor enters the left condenser, wherein it is condensed into liquid and drops into the left evaporator.
2. The cooling water is pumped to the condensers and then to the right chemical reactor. The right reactor is cooled, and the adsorbent can adsorb ammonia from the right evaporator; thus the right evaporator can provide cooling capacity.
3. The final step is the mass recovery process. During this process, the ammonia valve between the hot reactor and the cold reactor is opened. The ammonia vapor in the hot reactor will enter the cold reactor quickly because of the pressure difference between the two beds. Therefore, the hot reactor can adsorb more ammonia, and both the adsorption capacity and cooling capacity can be improved significantly.

Then, the two reactors will be heated and cooled alternately and the adsorption refrigerator can provide cooling capacity continuously. Clearly, the adsorption system always keeps high pressure status both in the heating process and in the cooling process, which improves the system reliability.

8.8.2 Performance Test

In the adsorption experiment rig, all the valves and pump are controlled by the controller. The adsorption refrigerator test rig is shown in Figure 8.90.

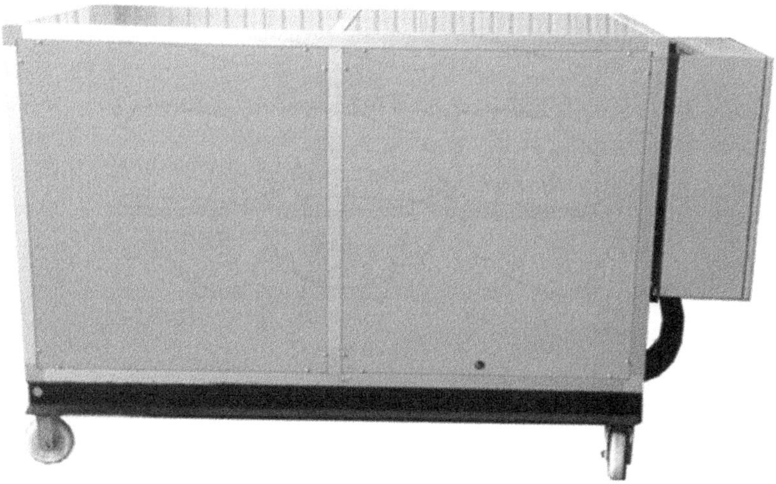

Figure 8.90 Adsorption refrigerator test rig

8.8.2.1 The Performance Relating with Mass Recovery

The adsorption performance variation with mass recovery is tested. The Clausius–Clapeyron diagram of the cycle with and without mass recovery is shown in Figure 8.91. For the cycle without mass recovery, the *P-T* area is small. However, for the cycle with mass recovery, the pressure of the hot reactor drops sharply during the mass recovery process, then declines slowly during the cooling process. Meanwhile, the pressure of the cold reactor rises sharply during the mass recovery process, and then rises slowly during the heating process. The total *P-T* area of the cycle with mass recovery is about 2.8 times that of the cycle without mass recovery, indicating that more moles of ammonia participate in the chemical reaction.

The adsorption characteristic of the adsorption cycle with mass recovery is shown in Figure 8.92. When the hot source inlet temperature is about 128 °C, and the cooling water

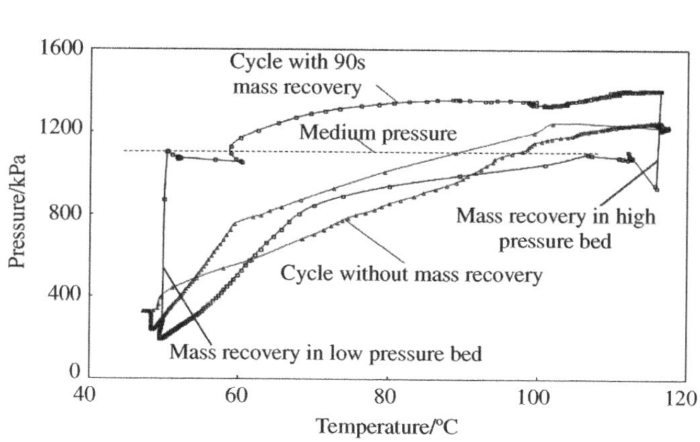

Figure 8.91 Clausius–Clapeyron diagram of the cycle with and without mass recovery

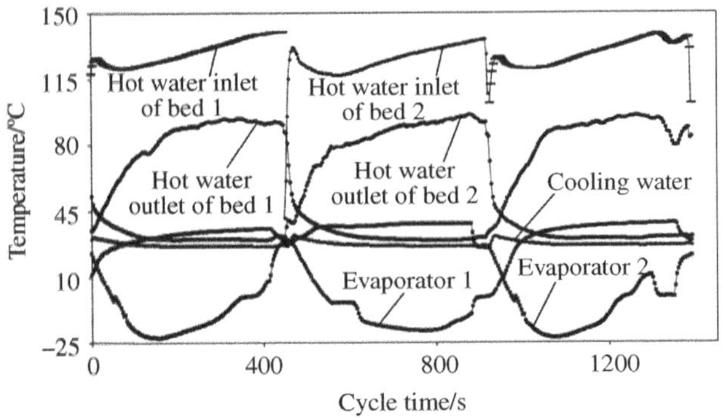

Figure 8.92 Adsorption characteristic of the adsorption cycle with mass recovery

inlet temperature is about 28 °C, the evaporating temperature can reach −17 °C. The two evaporators can provide cooling capacity continuously together. The cooling capacity, COP and SCP of the cycle without mass recovery are 2.0 kW, 0.13 and 229.4 W/kg, while they are 3.0 kW, 0.20 and 346.7 W/kg in the cycle with mass recovery. For the cycle with the mass recovery the cooling capacity, COP and SCP are increased by 33.3%, 53.8%, and 51.1%.

8.8.2.2 The Performance with Different Cycle Time

The relation between the adsorption performance and different cycle time is studied. The Clausius–Clapeyron diagram of the cycle with 440 and 880 seconds is shown in Figure 8.93. The cycle of 880 seconds expands the area of the cycle of 440 seconds. In the longer cycle, the chemical reactor can reach a higher temperature, and more ammonia participates in the chemical reaction; however, more heat is needed for the longer cycle time.

The hot water temperature of the adsorption cycle with a different cycle time is shown in Figure 8.94. When the average temperature of the hot water inlet is about 128 °C, the average temperature of the heat source for the cycle of 880 seconds is higher than that of the cycle of 440 seconds. When the hot source inlet temperature is about 128 °C, the cooling water is about 28 °C, and the evaporating temperature can reach −17 °C. The cooling capacity, COP

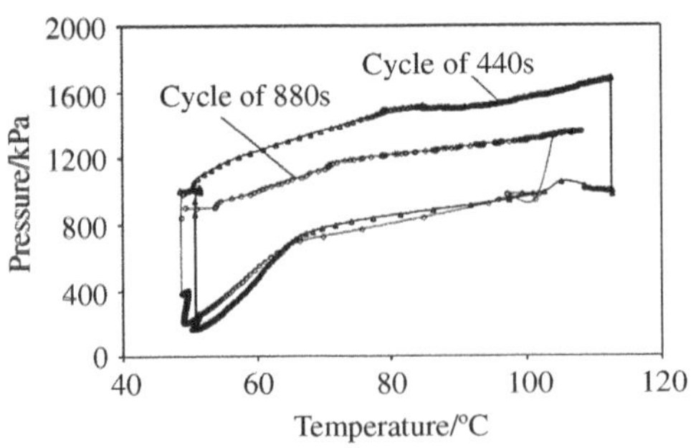

Figure 8.93 Clausius–Clapeyron diagram of the cycle with different cycle time

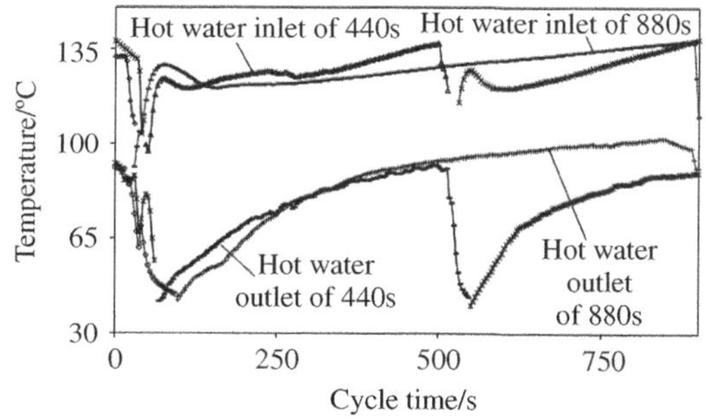

Figure 8.94 Temperature of the hot water for different cycle time

and SCP of the 440 seconds cycle are 3.0 kW, 0.20 and 346.7 W/kg, while they are 2.7 kW, 0.21 and 314.1 W/kg in the cycle of 880 seconds.

8.8.2.3 The Relation between the Adsorption Refrigeration Performance and the Heat Source Temperature

The relation between adsorption performance and different heat source temperature is studied, as shown in Figure 8.95. The adsorption cooling performance improves as the heat source temperature rises. When the cooling water inlet temperature is about 28 °C, and the evaporating temperature is about −17 °C, the heat source average inlet temperature ranges from 114 to 128 °C, the COP rises from 0.12 to 0.20 and the SCP rises from 154.6 to 346.7 W/kg, respectively.

8.9 System Design and Performance of an Adsorption Energy Storage Cycle

Due to the supplement and the demands of energy being strongly dependent on time and space for most occasions, people often need to store the energy that cannot be used up. Energy storage is also an important way to save energy, on the one hand, it can recover the low grade heat, and on the other hand it can adjust energy demand for rational and effective utilization.

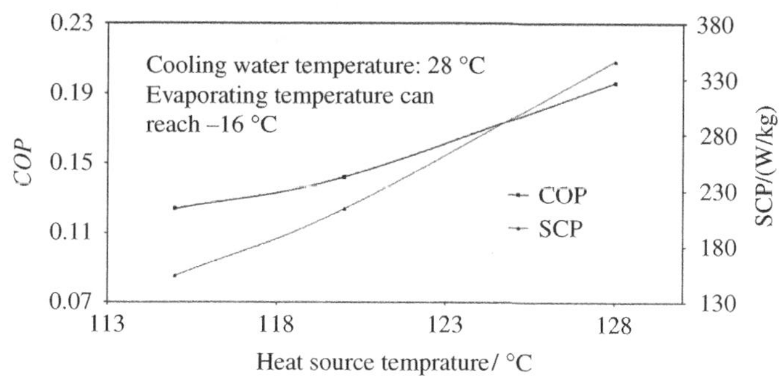

Figure 8.95 COP and SCP vs. heat source temperature

There are generally three types of energy storage. Firstly, storing the sensible heat of the solid or liquid medium. Secondly, storing the latent heat for the phase change process of some materials. The last way is to use the chemical energy in chemical reactions. The choice of the energy storage materials and the energy storage processes are commonly dependent on the application occasions and the economic requirements.

Sensible heat energy storage system has the advantages of a simple structure, convenient maintenance, and low technical requirements, and so on. Water or oil is usually used in the system with the liquid medium. A packed bed is generally used in the air system. Such a bed always consists of gravel or other solid materials that can be used for heat storage. Phase change energy storage density is high, and the common phase change materials are ice, eutectic salts, and so on. In recent years, there is the study of solid-solid phase change energy storage and the principle is that the molecular crystal polymorph of a certain special organic crystalline material can be changed with the temperature in a certain range, meanwhile the enthalpy also changes and it absorbs or releases a large amount of energy [43]. The chemical reaction energy storage uses the chemical reaction heat to store energy by a reversible chemical reaction process of chemical materials.

During the energy storage process, the phase change enthalpy of gas-liquid is greater than solid-liquid. For example, the vaporization latent heat of water is about 7.5 times that of the latent heat of melting heat of ice. But due to the large volume change of the materials during the liquid-vapor phase change process, it is difficult to use the vaporization process for energy storage, such as that the density of water and the saturated water vapor at 300 K are 1000 and 2.56×10^{-3} kg/m^3. If we can find a way to control the volume change during the phase change process, the liquid-vapor phase change process is undoubtedly an ideal method of energy storage. Such a process can be fulfilled by the solid adsorption technology because the volume does not change or changes very little when adsorbents adsorb vapor. In the adsorption system the adsorption heat can be used as for the heat storage process, while the evaporating process in the evaporator can be used as the cold storage process, and both processes are liquid-vapor phase change energy storage processes. The adsorption energy storage technology was first proposed by Close, and so on [44], and then Parrih *et al.* [45], Hisaki *et al.* [46], Tahat [47], and R.Z. Wang *et al.* [48] researched its application. To fulfill the energy storage we usually need to heat the adsorbent bed, and then let the refrigerant be desorbed from the adsorption bed and then condense in the condenser. By such a process the adsorption bed will recover the adsorption ability, which could produce the cooling power when it is at the environmental temperature. Then the valves between the adsorption bed and the evaporator/condenser will be closed. As long as the adsorption bed doesn't connect with the evaporator cooling/heating capacity won't be released. Such an energy storage process won't have much energy loss. When the energy is required, the adsorption bed will be connected with the evaporator, and the refrigerant evaporates in the evaporator and is adsorbed by the adsorption bed, which produces the cooling power in the evaporator and the heating power in the adsorption bed.

8.9.1 Thermodynamic Analysis of the Adsorption Energy Storage Cycle

8.9.1.1 The Basic Principle of the Cycle

The Clapeyron diagram (p-T-x) of the thermodynamic cycle for energy storage process is shown in Figure 8.96. It mainly includes three processes: heating and desorption process

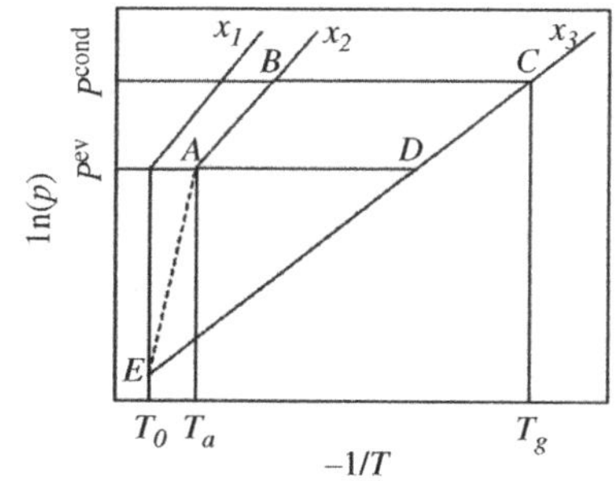

Figure 8.96 Clapeyron diagram of the thermodynamic cycle for adsorption energy storage

(*A-B-C*), cooling and energy storage process (*C-E*), and adsorption and refrigeration process (*E-A*). The heating and desorption process is the same as that in the general adsorption refrigeration cycle (*A-B-C-D-A*).

At the beginning of the heating process, the corresponding state is point *A*, and the adsorption bed isn't connected with the condenser or evaporator. Point *A* is also the end state of the bed after adsorption, thus its temperature is the adsorption temperature T_a, the pressure is the saturation pressure of the refrigerant that corresponds to the evaporating temperature, that is, $p_{ev} = p_s\ (T_{ev})$, and the equilibrium adsorption capacity x_2 is the function of the temperature T_a and the pressure $p_s\ (T_{ev})$, that is, $x_2 = x\ (T_a, T_{ev})$. When the bed is heated, the temperature and pressure rises, and the adsorption quantity doesn't change (*A-B*) until its pressure reaches the saturation pressure corresponding to the condensing temperature, that is $p_{cond} = p\ (T_{cond})$, then open the valve between the bed and the condenser. Then the refrigerant vapor is desorbed in the bed and condenses in the condenser, and finally flows into the evaporator. The desorption process (*B-C*) continually proceeds until the desorption temperature reaches the T_g, and the pressure keeps at p_{cond}. At the end of desorption, the valve between the bed and the condenser is closed, and the adsorption quantity is the minimum value, which is x_3.

The cooling storage process will start when the desorption process completes. For the cooling storage process (*C-E*) the adsorption bed can be cooled by the natural convection of the ambient air or the forced convection cooling process by the fluid through a heat exchanger. For such a process only the sensible heat of the adsorption bed is taken away. Because the bed isn't connected with the evaporator, the adsorption quantity inside the bed maintains the small value of x_3, and the pressure is also very low. Such a process can continue until the temperature of the bed is close or equal to the ambient temperature, which is related to the cooling mode and the time.

When the cold is needed, open the valve between the adsorbent bed and the evaporator, and the stored cold will be released (*E-A*), the liquid refrigerant evaporates within the evaporator, and then is adsorbed by the adsorbent in the bed. In this process the pressure of the adsorbent bed is close to the evaporation pressure. Due to the effect of adsorption heat, the temperature of the adsorbent bed will rise rapidly at the beginning, and subsequently the temperature will change, and the trends will be associated with the cooling mode. The adsorption amount of the adsorption bed is gradually increased as the pressure is maintained at p_{ev}. It must be noted that the *p-T-x* diagram for cold release process doesn't comply with the D-A equation, especially at the beginning stage, the pressure of the adsorbent bed rises rapidly, and the adsorption process is a non-equilibrium process. With the process going on, the non-equilibrium phenomenon will be gradually slowed down until it tends to equilibrium. Such a process cannot be accurately expressed in a *p-T-x* diagram, thus it is drawn by a dashed line from the starting point of the process (E) and the end point (A) (the two points can be considered as equilibrium adsorption). The end temperature of the adsorption bed is temperature T_a, and it's generally higher than the ambient temperature T_0. For some special circumstances, such as when the running time is long enough, the two values are equal, and consequently the adsorption quantity will be the maximum value of x_1.

8.9.1.2 Cooling Storage Ability

The cooling storage ability of the adsorption cycle can be calculated by the adsorption quantity of the adsorption bed at the beginning and the end of the cold releasing process. According to the D-A equation, the equilibrium adsorption capacity of the adsorbent bed is the function of the adsorption temperature T and the saturation temperature T_s (p):

$$x(T, T_s) = x_0 \exp\left[-K \cdot \left(\frac{T}{T_s} - 1\right)^n\right] \tag{8.80}$$

The theoretical released cold quantity is defined as the cooling power obtained from the system by a cold releasing process, and it is calculated as follows:

$$q_{c,st} = (x_1 - x_3)L_{ev} \tag{8.81}$$

where $q_{c,st}$ is the cold storage quantity per unit mass of adsorbent; L_{ev} is the latent evaporation heat of the refrigerant at the evaporating temperature; x_3 is the adsorption quantity in the cold storage process, $x_3 = x\ (T_g,\ T_{cond})$; x_1 is the maximum adsorption quantity that can be achieved at the end of the cold releasing process, that is, the adsorption quantity when the adsorption temperature is equal to the ambient temperature, $x_1 = x\ (T_0,\ T_{ev})$. For a certain adsorption cycle, the theoretical cooling storage capacity mainly depends on the adsorption quantity of the adsorption bed and the evaporation temperature. The maximum cooling storage capacity is:

$$q_{c,st,\max} = x_1 L_{ev} \tag{8.82}$$

It depends on the ambient temperature and the evaporating temperature.

The cooling storage capacity of unit mass of adsorbent and refrigerant inside the bed is:

$$q'_{c,st} = \frac{q_{c,st}}{1 + x_1} = \frac{x_1 - x_3}{1 + x_1} L_{ev} \tag{8.83}$$

8.9.1.3 Cold Releasing Process

For the cooling storage process due to the heat exchange process with the environment, the temperature of the bed is gradually close to the ambient temperature; therefore it can be thought to be equal to the ambient temperature. With the cold releasing process going on, the adsorbent bed is connected with the evaporator and adsorbs the refrigerant there, due to the adsorption heat, the temperature of the adsorbent bed rises rapidly, which is different to the cooling process of the general adsorption refrigeration cycle.

The theoretical cold output capacity of the adsorption energy storage system is:

$$q_c = (x_2 - x_3)L_{ev} \tag{8.84}$$

where x_2 is the adsorption quantity at the end of the cold releasing process, which depends on the adsorption temperature T_a and the evaporation temperature T_e.

The COP for the energy storage system is:

$$COP = \frac{q_c}{q_{in}} = \frac{\varepsilon \cdot q_{c,st}}{q_{in}} \tag{8.85}$$

where q_{in} is the heating power.

Cold releasing rate ε is defined as the ratio of the cold releasing capacity and the cold storage capacity:

$$\varepsilon = \frac{q_c}{q_{c,st}} = \frac{x_2 - x_3}{x_1 - x_3} \tag{8.86}$$

Different from the general cooling process of the adsorption refrigeration system, the cold releasing process of the energy storage system can be the forced convective cooling process or adiabatic. These two processes will be analyzed separately as follows.

1. Cold releasing phase with cooling process.

   The cooling process for a cold releasing phase is similar to the cooling process for the generally adsorption refrigeration system, and the main difference is that the initial adsorption temperature is low and the initial adsorption capacity is small. The adsorption temperature will rise initially and then fall down during the cold releasing process, and the process can't be simulated by the adsorption equilibrium equations.

   The cold output power in the cold releasing process is:

$$W_{ref} = M_a L_{ev} \frac{dx}{dt} \tag{8.87}$$

   The cooling process for the cold releasing phase can be carried out until the adsorption is close to the temperature of the cooling fluid.

2. Adiabatic cold releasing phase.

   The adiabatic cold releasing process is that the adsorption bed is thermal isolated from the surroundings in the cold releasing process. Actually the adsorbent bed will dissipate the heat to the environment, no matter how small it is. But considering that the natural convection cooling capacity is much smaller than the adsorption heat in the adsorption process, the heat exchanged by a natural convection process is always neglected. The process is taken as adiabatic if the bed doesn't exchange the heat with the environment by a forced convection process of the fluid. During the process, the adsorption heat is transformed into the sensible heat of the adsorbent and metal materials of the adsorber. The energy balance equation of the adsorption bed is:

$$M_a L_{ev} dx = M_a(C_a + xC_{ref})dT + M_{madb} C_m dT \tag{8.88}$$

   where M_a and M_{madb} are the mass of the adsorbent and metal materials, respectively. C_a, C_m, and C_{ref} are the specific heat capacity of the adsorbent, metal, and the refrigerant liquid, respectively.

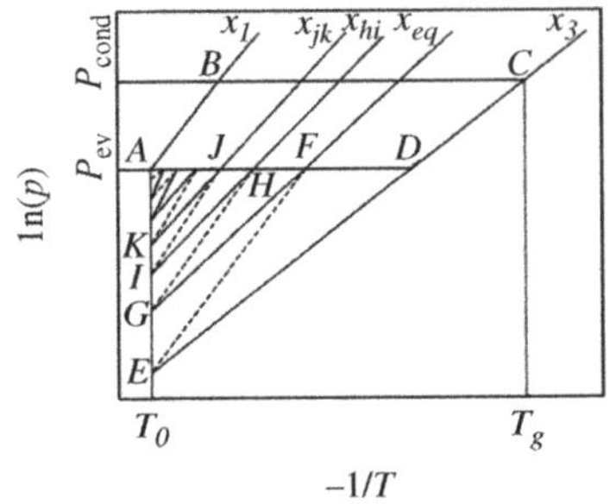

Figure 8.97 Clapeyron diagram of adiabatic adsorption process in the adsorption energy storage cycle

The adiabatic adsorption process is *E-F* in Figure 8.97. The maximum temperature of the bed at the end of the cold releasing phase has nothing to do with the adsorption rate, and can be obtained from the integral formula of Equation 8.88. The equilibrium adsorption quantity x_{eq} at the end of the cold releasing phase can be directly calculated by the D-A equation.

The total cold quantity released in the adiabatic adsorption process is:

$$Q_{ref,1} = M_a(x_{eq} - x_3) \cdot L_{ev} \tag{8.89}$$

The ratio of the cold released is:

$$\varepsilon_1 = \frac{Q_{ref,1}}{Q_{c,st}} = \frac{x_{eq} - x_3}{x_1 - x_3} \tag{8.90}$$

Because x_{eq} is smaller than x_1, the energy storage system in this case still has a considerable cold storage capacity. When the adsorption temperature reduces to *F-G*, the system can release the cold again, which is *G-H*. The multiple processes of cold releasing and storage processes, which is *E-F-G-H-I-J-K*, can run until all the cold storage capacity releases.

8.9.1.4 Thermal Analysis of Adsorption Energy Storage Cycle

The adsorption energy storage cycle adsorbed heat Q_{in} in the heating and desorption phase, which is used as the sensible heat Q_{sen} for the adsorption bed and the desorption heat Q_{des} (which is equal to adsorption heat Q_{ads}):

$$Q_{in} = Q_{sen} + Q_{des} \tag{8.91}$$

Q_{sen} is related with the running time of the energy storage process and the temperature difference between the system and the environment. It will gradually decrease to 0 during the energy storage process. Q_{des} can be stored in the long-term without heat loss. Hence, the short-term adsorption heat storage system actually is the adsorption heat pump without using the condensation heat, and the heat which can be obtained during the heat releasing process is $Q_{h,st}$ (i.e., heat storage capacity) and is the difference between the heat stored and the heat loss to the environment Q_{loss}:

$$Q_{h,st} = Q_{in} - Q_{loss} \tag{8.92}$$

where $Q_{loss} \leq Q_{sen}$.

For long-time energy storage system, the heat storage capacity is the adsorption heat,

$$Q_{h,st} = Q_{des} = M_a \Delta x \cdot L_{ev} = \frac{q_{ads}}{L_{ev}} Q_{c,st} \tag{8.93}$$

where q_{ads} is the average differential adsorption heat of the desorption process.

In this section the heat stored by the system refers to the heat for long-term storage but doesn't take into account the sensible heat storage.

From Equation 8.93, the ratio between the heat storage capacity and cooling storage capacity is just the ratio between the differential adsorption heat and evaporation latent heat of refrigerant. Thus, the research results of the adsorption cold storage cycle can also apply to the adsorption heat storage cycle if we change the evaporation latent heat of the refrigerant that corresponds to the cold storage capacity to the adsorption heat that corresponds to heat storage capacity. Taking a zeolite–water adsorption system, for example, the value for the differential adsorption heat is approximately 3600 kJ/kg, so the ratio between the heat storage capacity and cold storage capacity is:

$$\frac{q_{h,st}}{q_{c,st}} = \frac{q_{ads}}{L_{ev}} \approx 1.5 \tag{8.94}$$

8.9.2 *Adsorption Air-Conditioning Prototype with the Energy Storage Function*

Adsorption air-conditioning prototype with the energy storage function is designed mainly for the locomotive cab. Since the place for the application is the special place with serious vibration and limited installation space, it has high requirements for the safety, reliability, and maintainability. Such an occasion also requires that the system should have high adsorption refrigeration performance as well as a simple and compact structure.

8.9.2.1 Choice of the Adsorption Working Pairs

For the choice of the adsorption working pairs, the things we need to consider include the efficiency, safety, reliability, convenient application, and environmental protection. The outlet temperature of the exhaust gas from the diesel engine of the locomotive is generally 450 °C and even higher than 500 °C. The adsorption working pairs applicable for such high temperature heat source mainly include zeolite–water, activated carbon–ammonia, chemical adsorbent–ammonia, and composite adsorbent–ammonia. Because the water is safer than ammonia, the zeolite–water working pair is chosen as the adsorption working pair.

8.9.2.2 The Operation Mode of the Refrigeration Cycle

In order to achieve the continuous refrigeration output, except the system with two or more adsorption beds, we also can use an energy storage vessel to store the excess cooling capacity generated by the system. Comparing these two ways, the former one always has one adsorber heated and the other one cooled, which can ensure the waste heat is fully utilized. For the latter method the adsorption bed is alternately heated or cooled, which can only partially utilize the waste heat. In addition, in order to store the cold in the energy storage vessel, the actual cooling capacity of the system at the adsorption stage needs to be larger than the output cooling power.

Therefore, the previous research on solid adsorption refrigeration systems driven by waste heat always concentrates on the continuous refrigeration processes with two beds, but such a scheme has the following disadvantages if it is used as the locomotive air-conditioning system:

1. In order to effectively utilize the waste heat and control the back pressure, the two adsorbers must be installed closely to the chimney of the exhaust gas outlet, so it is difficult to arrange the position and space for two beds.
2. In the actual operation, the speed of the diesel engine on locomotive varies greatly, which is difficult for matching the cooling phase and heating phase for two adsorbers. Maybe one adsorber had already been desorbed completely, and the adsorption in another adsorber just begins.
3. More switching valves and moving parts will be needed in the refrigeration system, and then the system will be more likely to leak.
4. When the locomotive stops or the speed of the locomotive is low, the cooling power of the system is zero or very small, which can't satisfy the requirement.

Comparing various adsorption refrigeration cycles, and combining the requirements of the locomotive and the characteristics of the air conditioning, the adsorption air conditioner adopts a new operation mode with single adsorber and cold storage vessel, and it features:

1. Because the waste heat of the locomotive is very abundant, a small part of the waste heat is enough to drive the adsorption air-conditioner, so the exhaust gas will intermittently go through the adsorber, which can reduce the adverse impact that is caused by the back pressure increment of the exhaust gas outlet.
2. The structure is simple because of less valves, reliable operation, and convenient control.
3. For the special cases that the locomotive stops or slows down the speed, the air conditioner also can release the cooling capacity from the cold storage vessel to provide the refrigeration power.

A cold storage system also has other advantages, such as, the optimal ratio between the heating and cooling time can be used to optimize the performance of the system; the energy storage vessel can adjust the changes of the requirements on the actual cooling power; the energy storage vessel can match the cooling power output for air conditioning with different running modes of the locomotive, and so on.

8.9.2.3 The Experimental Prototype

The experimental prototype of the system is shown in Figure 8.98, and the photo of the system is shown in Figure 8.99. The main sub-systems of the prototype are as follows:

1. An adsorption refrigeration system that includes an adsorber, a condenser, an evaporator/cold storage vessel, and a reservoir.
2. A heating/cooling system that includes a burner, a blower, a fan, the exhaust gas/air switching valves, the exhaust gas/air channels, cooling water circuit, and the chilling water system.

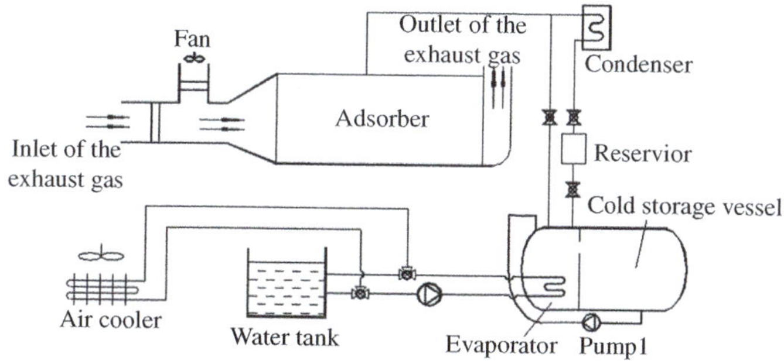

Figure 8.98 The system diagram of the adsorption air conditioning prototype

Figure 8.99 Photograph of the adsorption air conditioning prototype

The internal structure for the adsorber in the prototype shown in Figure 8.98 is also shown in Figure 8.100. The size of the adsorber is 1050 mm (length) × 1082 mm (width) × 458 mm (height). In order to enhance the heat exchange performance of the adsorber the fin-tube type heat exchanger is adopted for the design of the adsorber. The pipes are arranged in a rectangular pattern, and the number of pipes is 176. The heat exchange pipe is the stainless steel fin tube of $\Phi 32 \times 1.2$ mm thickness. The heat transfer fluid is inside the tube, and the adsorbent is the 13X zeolite mixtures with diameters of 1/8 in. (about 2/3) and 1/16 in. (about 1/3). The total mass of the adsorbent is 140 kg.

For the heat exchange process inside the adsorbent bed, the water is the refrigerant thus the pressure of the system is low. Under such a condition the consolidated adsorbent cannot be chosen because the mass transfer will be critical, so the granular adsorbent is used in the adsorber, and the heat transfer performance is improved by increasing the heat transfer area. To install the square copper fins outside the heat transfer pipes, the side-length of the square is 44 mm, the thickness of the fin is 1 mm, and the fin pitch is 11 mm. The granular adsorbent was filled among the fins. In order to enhance the mass transfer performance in the adsorber, the mass transfer channels that are formed by the stainless steel mesh of diameter $\Phi 8$ mm arrange every other column tubes (vertical direction).

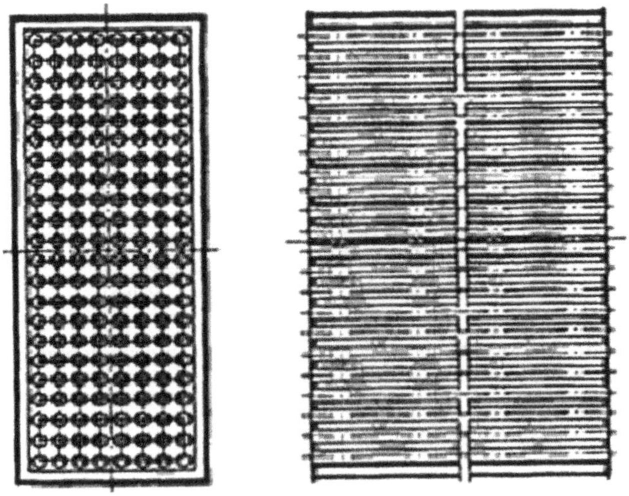

Figure 8.100 Internal structure diagram of the adsorber

For an experimental prototype, the condenser is cooled by the water, and comprises two parallel plate heat exchangers (CBE-B12 type plate heat exchanger produced by SWEEP International A.B. Company in Sweden). The reservoir is a cylinder-type container, mounting an endoscope at the side to observe and measure the condensed liquid in the desorption phase.

In Figure 8.98, the evaporator and cold storage vessel of the refrigeration system are designed as one component, which makes the system compact and will enhance the heat transfer performance for the cold storage and cold releasing processes. The left side of the component is an evaporator and the right side is a cold storage vessel. The heat exchanger in the evaporator is a fin tube type heat exchanger, which is immersed into the refrigerant water. The refrigeration power in the evaporator is transported outside by the heat exchange process between the chilling water inside the heat exchanger and refrigerant outside the heat exchanger. There is a metal partition plate between the evaporator and the cold storage vessel, and there is an overflow hole at the top of the plate. Pump 1 is a magnetic pump, and it opens in the cold storage process as well as the cold releasing process. When the pump is open the chilling water will be transported to the evaporator and flows back to the cold storage vessel by the overflow hole, thus the heat exchange between the cold storage vessel and the evaporator is fulfilled by the circuit of the chilling water. In addition, two endoscopes (Figure 8.99) were installed both at the upper side of the evaporator and the cold storage vessel, respectively, and the fluctuation of the water level can be observed. The reasonable mass ratio for the water inside the evaporator and the cold storage vessel should be 1 : 2 when there is no fluctuation. The total mass of the refrigerant inside the system is 185 kg, thus when the adsorption quantity is 0.2 kg/kg, the mass of water in the evaporator and the cold storage vessel is approximately 49 and 99 kg, respectively.

Because the cold storage medium and refrigerant are all water, the water can be mixed in both sides together to fulfill the heat exchange quickly, such a process will ensure the fast cold storage and cold releasing processes. Due to cold storage depending on the sensible heat of the water a large amount of water is needed, and consequently the volume of the cold

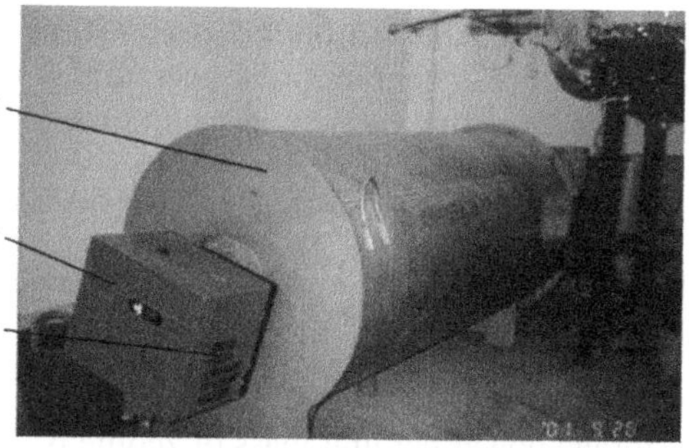

Figure 8.101 The photograph for the generator and the burner

storage vessel is large. If the phase change materials are used for cold storage the volume of the cold storage vessel will be decreased, but considering that the heat transfer performance between the water and the phase change material is small water is chosen instead of the phase change materials.

The heat source for the experimental prototype is the high temperature exhaust gas produced by burning diesel, which simulated the waste gas from the diesel engine in the locomotive. The prototype is cooled by the airflow from a blower. The generator for the high temperature exhaust gas is shown in Figure 8.101. The burner uses $0^{\#}$ light diesel oil, which can produce exhaust gas with a temperature of 1000 °C. In order to reduce the temperature of the exhaust gas to the required temperature for heating the adsorber, there are two air inlets on both sides of the generator. The fan will transport a part of cold air into the generator, which could mix with the high temperature exhaust gas to produce the exhaust gas with a temperature lower than 500 °C. According to different requirements of the experimental processes, the generator can be equipped with a burner that has the fuel consumption of 1–30 kg/h (12–363 kW). In the actual experiments the fuel consumption is 8–14 kg/h.

8.9.3 Experimental Study on Adsorption Cold Storage Cycle

In the experiments the valve between the condenser and the evaporator is open at the beginning, and the adsorber is heated by the high temperature exhaust gas that is produced by the burner, then the desorbed refrigerant vapor condenses in the condenser and flows into the evaporator. During the cold storage process the valve is closed, and the temperature of the adsorber decreases due to the natural heat convection. During the cold releasing process the valve is open and the adsorber adsorbs refrigerant adiabatically or with a forced convection cooling process by a blower. The cooling power produced in the evaporator is taken away by a chilling water circuit.

8.9.3.1 Storage Cooling Power and Heating Power

The adsorption working pair of zeolite–water is tested, and the D-A equation is:

$$x_{eq} = 0.261 \times \exp(-5.36 \times (T/T_s - 1)^{1.73}) \tag{8.95}$$

When the ambient temperature is 30 °C and the evaporation temperature is 5 °C, the maximum adsorption quantity of the system in actual operation is $x_1 = x\,(30, 5\,°\text{C}) = 0.240$, so the calculated maximum storage cooling capacity per unit mass of zeolite is:

$$q_{c,st} = x_1 L_{ev} = 0.2402 \times 2484 = 597\ (\text{kJ/kg}) \tag{8.96}$$

The maximum cold storage capacity per unit mass of working pair of zeolite–water is:

$$q'_{c,st} = q_{c,st}/(1 + x_1) = 481(\text{kJ/kg}) \tag{8.97}$$

The maximum heat storage power per unit mass of zeolite is:

$$q_{h,st} = \int_0^{x_1} q_{ads}dx = 885(\text{kJ/kg}) = 1.48 q_{c,st} \tag{8.98}$$

For the energy storage capacity per unit volume of the cold storage medium, the bulk density of granular zeolite is generally $\rho = 600–700\,\text{kg/m}^3$. In the prototype the adsorbent has the density of $\rho = 640\,\text{kg/m}^3$, then the maximum cold storage power and heat storage power per unit volume of zeolite are 382 and 566 MJ/m^3, respectively.

The cold storage quantity and heat storage quantity are related to the adsorption capacity in the energy storage process, that is, the adsorption quantity at the end of the desorption (x_3), which is determined by the desorption temperature T_g and the condensing temperature T_{cond}. Assuming that the condensing temperature is 35 °C, the trends of the cold storage quantity $q_{c,st}$ and heat storage quantity $q_{h,st}$ per unit mass of adsorbent vs. the desorption temperature (T_g) before the cold storage process are shown in Figure 8.102.

Regardless of the mass of the container for the energy storage vessel, the sensible cold storage capacity of the water is 4.2 kJ/(kg·K). The maximum temperature difference between the cold storage vessel and the environment is in the range of (25 – 5) 20 °C, and the corresponding cold storage capacity of water is 84 kJ/kg. If the ice is used for energy storage the energy storage capacity is 333 kJ/kg after taking into consideration the latent heat for the phase change process. Comparing these data with the data in Figure 8.102, we can see that the cold storage capacity of the zeolite–water adsorption cycle is larger than the method of using the sensible heat of water, as well as higher than the ordinary cold storage medium such as ice. The maximum cold storage capacity is 84 MJ and the total cold storage capacity is 124 MJ when the zeolite in the prototype is 140 kg.

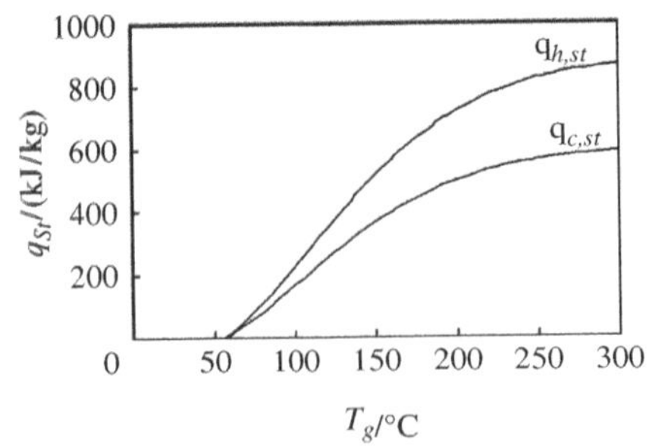

Figure 8.102 Cold storage quantity and heat storage quantity vs. desorption temperature

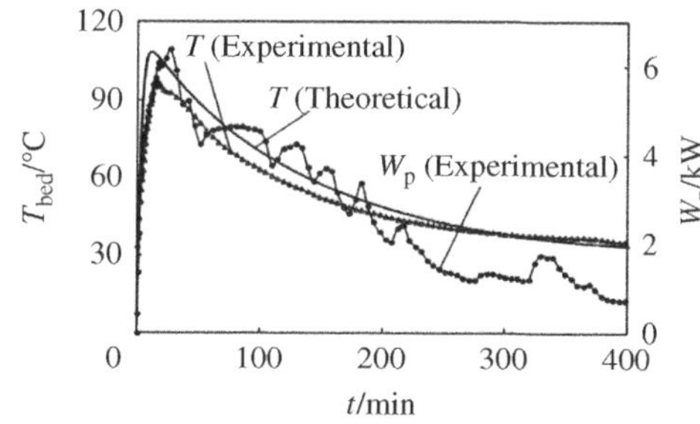

Figure 8.103 The adsorption temperature and refrigeration power during the cold releasing process

8.9.3.2 Cold Releasing Phase

Before the cold releasing phase, the desorption temperature of adsorber was 210 °C and the condensing temperature was 37 °C, thereby the adsorption quantity during the cold storage phase was $x_3 = 3.7\%$. The next day when the adsorption temperature decreased to the ambient temperature, the adsorber started to release the refrigeration power. In the experiments, the inlet temperature and flow rate of the cooling air are 33 °C and about 1000 m^3/h, respectively. The theoretical cold storage capacity of the adsorber is 78.4 MJ. Figure 8.103 showed the trends for the theoretical and experimental adsorption temperature.

As shown in Figure 8.103, the temperature of adsorber first rapidly increases due to the adsorption heat and then gradually decreases. The temperature change of the adsorber is slightly smaller if compared with the theoretical value. The cooling power is large at the beginning of the cold releasing phase, and then gradually reduces.

The cold releasing process lasts nearly 7 hours, and the final adsorption temperature is about 35 °C. The total output cooling capacity of the system is 66.9 MJ, therefore the cold releasing rate ε is 0.85. The cold storage capacity cannot be released completely because the system has the cold loss to the environment and the adsorption in the adsorber cannot reach the absolutely equilibrium. The COP for the adsorption cold storage cycle is calculated, and the value is 0.36. It is higher than that of the adsorption refrigeration cycle because of the lower final adsorption temperature obtained from a longer running time.

8.9.3.3 Adiabatic Cold Releasing Process

Assuming that the initial adsorption capacity of the adsorbent bed is zero, the ambient temperature is 30 °C, and the evaporation temperature is 5 °C, the theoretical cold storage capacity of the system is 84 MJ. The relationship between adsorption temperature and cold releasing rate in the adiabatic cold releasing process is analyzed, and the results are shown in Figure 8.104, where the numbers in the diagram indicate the times of the cold releasing process. In the cold releasing process with the increase of the adsorption quantity, the adsorption temperature and the cold releasing rate are gradually increased, and they are in an approximately linear relationship. At the end of the cold releasing process, the adsorption temperature will gradually decrease to the ambient temperature in the cold storage process, which makes the system prepared for the next cold releasing process. It can be seen that at the end of the cold releasing process, the adsorption temperature and the released refrigeration quantity will be gradually

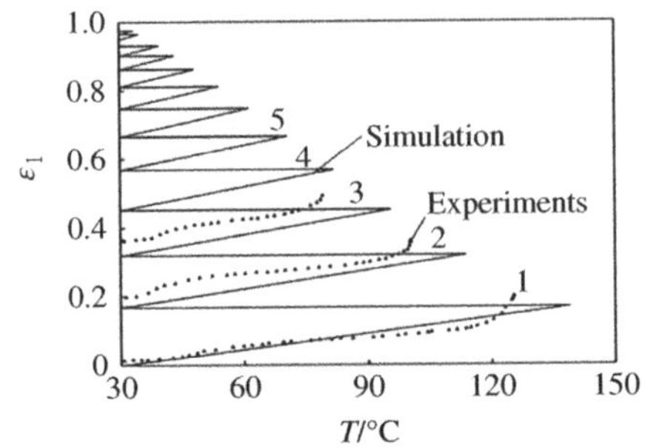

Figure 8.104 Cold releasing rate vs. adsorption temperature in adiabatic cold releasing process

reduced to the times for cold releasing processes, until the adsorption temperature is equal to the ambient temperature and the released cooling power tended to be zero.

Three cold releasing processes are studied, and the experiments are performed every other day. Before the first adiabatic cold releasing experiment, the adsorber is heated for desorption, and at the end of desorption the temperature of the bed is 267 °C when the condensing temperature is 43 °C. The calculated adsorption quantity in the adsorbent bed is $x_3 = 0.014$ kg/kg. When the ambient temperature is 30 °C, if T_{ev} is 7 °C x_1 is 0.243 kg/kg, the cold storage capacity is $q_{st} = 569$ kJ/kg per unit mass of adsorbent and the total cold storage capacity of the system is $Q_{st} = 80$ MJ.

Next day for the adiabatic cold releasing experiments on the adsorber, the trends for the temperature at the upper part of adsorber (T_{up}), the temperature at the middle part of adsorber (T_{mid}), the evaporation temperature (T_{ev}), and the cooling power (W_{ref}) are shown in Figures 8.105 and 8.106. At the beginning of the experiment, the adsorption temperature is 33 °C, the initial environment temperature is 30 °C, and the initial evaporation temperature is 12 °C. During the experiments the average evaporation temperature is 7.2 °C and the average ambient temperature is 28 °C.

Experiments show that the upper part of the adsorber firstly contacts with the refrigerant, which can be shown by the temperature change of the bed. As the valve between the adsorber and the evaporator opens, the upper part of the adsorbent bed is in contact with the refrigerant vapor firstly and adsorbs rapidly, this make the temperature at the upper part of adsorber increase quickly, while the temperature at the middle part of the adsorber almost doesn't

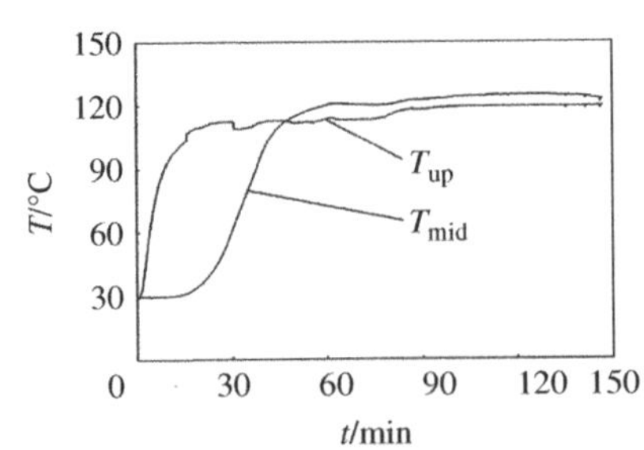

Figure 8.105 The adsorption temperature for the first adiabatic cold releasing process

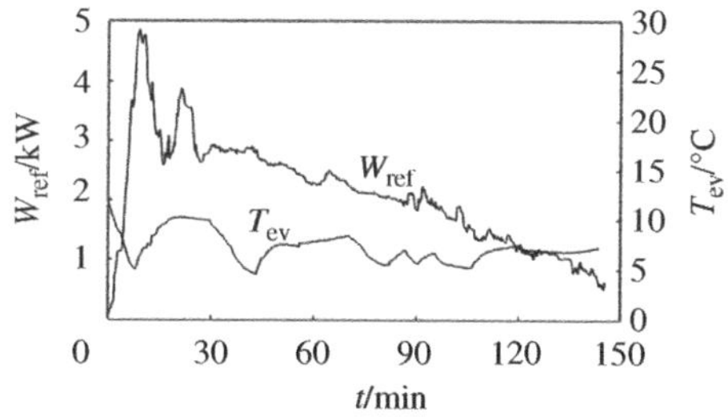

Figure 8.106 The cooling capacity and evaporation temperature at the first adiabatic cold releasing process

change for the first 15 minutes. After the temperature at the upper part of the adsorber is close to being in equilibrium, the middle part of the adsorber begins to adsorb. As there were only two thermometers at the longitudinal direction of the adsorber, the temperature at the lower part of the adsorber couldn't be tested directly, but we could deduce that then it will follow a similar trend, and it will take more time to reach equilibrium.

At the later stage of the adiabatic cold releasing process, the adsorption temperature reaches the maximum value and basically remains unchanged, while the cooling power output gradually approaches zero. As shown in the diagram, there is still a considerable amount of cooling power output when the adsorption temperature doesn't change, and the reason is mainly because the lower part of the adsorber is still in adsorption, and its temperature change isn't shown in the diagram. At the end of the cold releasing process, the average temperature of the adsorbent bed is 125 °C, the total cooling capacity output is 17.9 MJ, and the cold releasing rate is 0.21. As shown in Figure 8.104, at the end of the first cold releasing process, the theoretical average temperature of the adsorber is 139 °C, and the cold releasing rate is only 0.18. The actual temperature increment of the bed is smaller than the calculated value, and the difference is mainly that the bed in the experimental procedure exchanged the heat with the environment, and such a process makes its temperature decrease and the adsorption capacity increase, thus the actual cold releasing rate is bigger than its theoretical value.

At the end of the adiabatic cold releasing process, when the adsorption temperature decreases closely to the ambient temperature, the second and third time cold releasing processes are performed. The experimental results obtained are similar to the first process, except that the temperature increment of the bed and the cooling power are smaller. For these three adiabatic cold releasing processes the average adsorption temperature and the cold releasing rate are shown in Figure 8.104. Obviously, in the successive adiabatic cold releasing processes the temperature of the bed at the end of the process gradually decreases and so does the cooling capacity.

Figure 8.104 also shows that the error was large between the experimental results and the simulation results. The reasons are as followings:

1. The average temperature of the bed is only obtained from the upper and the middle part of the adsorber, and the temperature change along the longitudinal direction of the adsorber is very large during the actual adiabatic adsorption process. Especially when the temperature change at the lower part of the adsorber cannot be tested, so the average temperature obtained doesn't truly reflect the temperature of the bed.

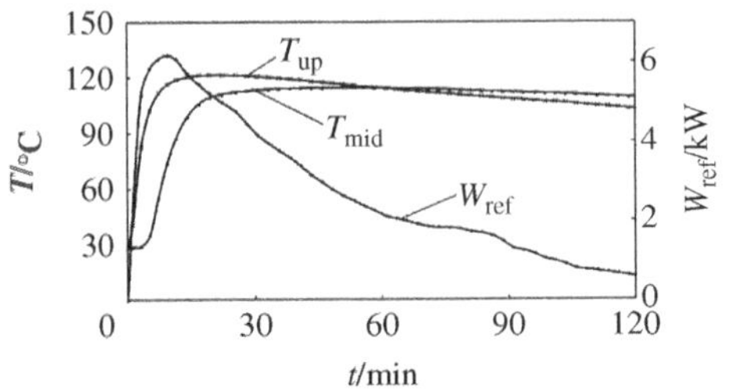

Figure 8.107 The first cold releasing process at lower ambient temperature

2. The adsorber transfers heat to the surrounding environment by a natural convection heat transfer process when the adsorption temperature rises, which makes the adsorption temperature at the end of the cold releasing process smaller than the calculated data. Consequently the cooling capacity will increase.
3. The simulation results are obtained under the condition that the adsorption in the adsorber is assumed uniformly (lumped heat capacity). In fact, in an adiabatic adsorption process, the adsorption inside the adsorbent bed isn't uniform.
4. The cooling power produced by adsorption is output through the heat exchange process between the water in the evaporator and the heat exchanger. Due to the mass of water inside the evaporator being big, the cold output tends to be delayed.

In addition, the experimental results are influenced by the ambient temperature, the initial adsorption capacity, and the adsorption time, and so on. The results are slightly different for different conditions. Figure 8.107 shows the experimental data that reflects the relationship between the adsorption temperature and cooling capacity at the first cold releasing process when the ambient temperature is 20 °C, and it indicates that the final temperature of the bed is 110 °C, the cooling capacity is 19.5 MJ, and the cold releasing rate is 0.24.

8.9.3.4 Analysis on the Optimum Refrigeration Performance of the Adsorption Air Conditioner

For the adsorption air-conditioning system using waste heat the cooling power is more important than COP because the waste heat is sufficient, so the maximum cooling power of the system is analyzed.

Figure 8.108 shows the cooling capacity and cooling power changes with the desorption temperature at different adsorption temperatures, and Figure 8.108 shows the cooling capacity and cooling power changes with the adsorption temperature at different desorption temperatures. For the experimental data in both figures the evaporation temperature and the condensation temperature are 7 and 40 °C, respectively.

In Figure 8.108, five groups of curves from left to right correspond to $T_a = 60$, 70, 80, 90, and 100 °C, respectively. In Figure 8.109, seven group of curves from left to right correspond to $T_g = 130$, 150, 170, 190, 210, 230, and 250 °C, respectively. The maximum cooling power is obtained when the adsorption and desorption temperatures are 80 and 230 °C, respectively, and the corresponding average cooling power output is 3.45 kW and the COP is about 0.25.

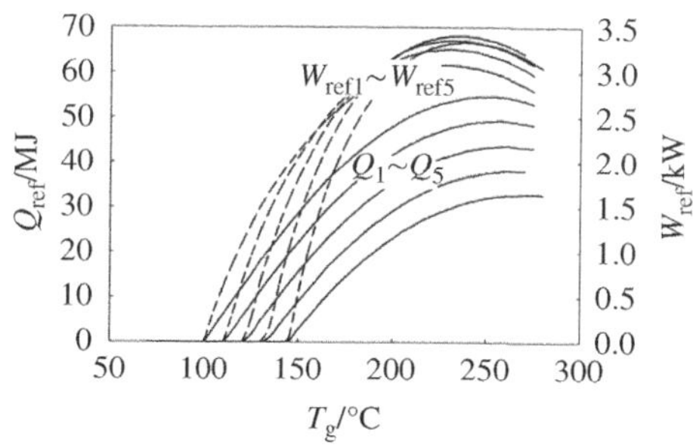

Figure 8.108 Cooling capacity and cooling power vs. desorption temperature at different adsorption temperature

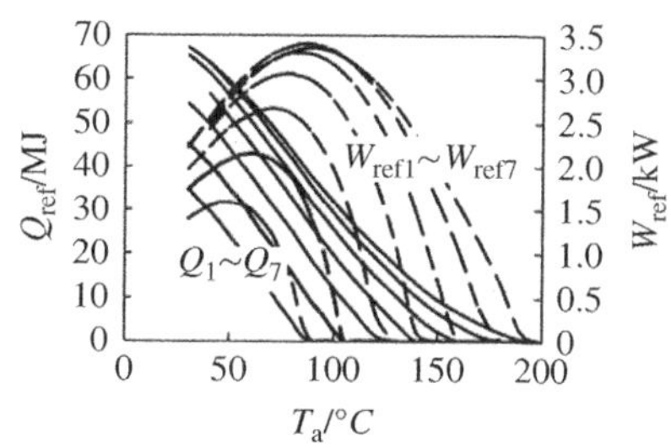

Figure 8.109 The cooling capacity and cooling power vs. adsorption temperature at different desorption temperature

8.9.4 Application of the Adsorption Energy Storage Cycle

The adsorption energy storage cycle can be used to recover and store the heat of solar energy, exhaust gas, waste water for air conditioning, or heat pump.

The energy stored in the adsorption system can be used for the first cycle or when the adsorption system stops running for a while. Then without external forced heating or cooling, the cooling or heating power output can be obtained by releasing the stored energy. For example, for the adsorption air conditioning chiller installed in the car or on the train driven by the exhaust gas of the diesel engine, when it stops, the stored energy can be used to provide the cooling power for the air conditioning process.

The adsorption air-conditioning system driven by the exhaust gas of the locomotive introduced here is also one application of the adsorption energy storage cycle. The air-conditioning system here also can be looked upon as an adsorption energy storage system.

In a journey the locomotive always needs to stop at several stations, and it sometimes even needs to stop for more than 10 hours in the train depot. As an energy storage system, the system can be heated for desorption in advance, and then the cold storage process will start if the locomotive needs to stop for a long time. Then the adsorption temperature decreases and the adsorption capacity will be kept unchanged. Compared with the general adsorption refrigeration cycle, the adsorption energy storage cycle has the following advantages:

1. The cooling process for the bed in which the adsorption quantity doesn't change proceeds when the locomotive stops, thus the operation time of the system for the next cold releasing process will be shortened and the performance coefficient will be increased.

2. The adsorbent bed is at a relatively low temperature at the beginning of the next cold releasing process, which can quickly output the cooling power.
3. From the above analysis we know that the cooling power at the beginning of the cold releasing process is large, with the adsorption process going on it gradually reduced, which just meets the requirements of the air conditioning load.
4. The adiabatic cold releasing process can also meet the demand for the short term parking of the locomotive.

In addition to the application in the refrigeration system, the adsorption energy storage cycle can also be used in other occasions. Because the bed has a cooling or heating capacity after desorption, the energy storage system after desorption can be moved to other places without energy loss for generating cold or heat. It can also be served as the cold storage or short-time air conditioning system.

References

[1] Wang, D.C. (2005) Development of a silica gel-water adsorption chiller and study on the characteristics of the adsorption systems driven by variable heat source. PhD Thesis. Shanghai Jiao Tong University, Shanghai, China (in Chinese).

[2] Liu, Y.L. (2005) Study on a solar driven silica gel-water adsorption refrigeration system. PhD Thesis. Shanghai Jiao Tong University, Shanghai, China (in Chinese).

[3] Ma, T.Z., Hou, Z.Q. and Wu, W.X. (1983) *Heat Pipe Technology*, Science Press, Beijing, ISBN: 7030020111 (in Chinese).

[4] Qian, S.W., Zhu, D.S., Li, Q.L. and Yang, L.M. (2004) *Heat Transfer Enhancement in Tube Type Heat Exchanger*, Chemical Industry Press, Beijing, ISBN: 9787502534455 (in Chinese).

[5] Yang, S.M. (2003) *Heat Transfer*, Higher Education Press, Beijing, ISBN: 9787502534455 (in Chinese).

[6] Cui, G.W. (1990) *Surface and Interface*, Tsinghua University Press, Beijing, ISBN: 730200675X (in Chinese).

[7] Akahira, A., Alam, K.C.S., Hamamoto, Y. *et al.* (2004) Mass recovery adsorption refrigeration cycle-improving cooling capacity. *International Journal of Refrigeration*, **27**, 225–234.

[8] Guilleminot, J.J., Chalfen, J.B., and Choisier, A. (1994) Heat and mass transfer characteristics of composites for adsorption heat pumps. Proceedings of International Absorption Heat Pump Conference, New Orleans, LA.

[9] Kitagawa, H. and Suzuki, K. (1983) *Foundation and Design for Adsorption*, Chemical Engineering Publishing Press, Beijing, ISBN: 150633488 (in Chinese).

[10] Lu, Y.Z. (2002) Adsorption air-conditioning system with zeolite-water working pair powered by waste heat of fuel gas exhausted from diesel locomotive. PhD Thesis. Shanghai Jiao Tong University, Shanghai, China (in Chinese).

[11] Wang, D.C., Wu, J.Y. and Xia, Z.Z. (2005) Design and heat and mass transfer performance research on adsorption unit pipe of adsorption refrigeration. *Journal of Engineering Thermophysics*, **27**(1), 139–141, ISSN: 0253-231X (in Chinese).

[12] Liu, Y.L., Wang, R.Z. and Xia, Z.Z. (2005) Experimental study on a continuous adsorption water chiller with novel design. *Journal of Refrigeration*, **28**(2), 218–230, ISSN: 0253-4339 (in Chinese).

[13] Boelman, E.C., Saha, B.B. and Kashiwagi, T. (1995) Experimental investigation of a silica gel-water adsorption refrigeration cycle-the influence of operating conditions on cooling output and COP. *ASHRAE Transactions*, **101**(2), 358–366.

[14] Saha, B.B., Boelman, E.C. and Kashiwagi, T. (1995) Computer simulation of a silica gel-water adsorption refrigeration cycle the influence of operating conditions on cooling output and COP. *ASHRAE Transactions*, **101**(2), 348–355.

[15] Wu, J.Y. and Wang, R.Z. (2003) The operation characteristic research of adsorption refrigeration system driven by intermittent heat source. *Journal of Engineering Thermophysics*, **24**(6), 926–928, ISSN: 0253-231X (in Chinese).

[16] Chen, C.J., Wang, R.Z., Xia, Z.Z. *et al.* (2010) Study on a compact silica gel–water adsorption chiller without vacuum valves: design and experimental study. *Applied Energy*, **87**, 2673–2681.

[17] Chen, C.J., Wang, R.Z., Xia, Z.Z. and Kiplagat, J.K. (2010) Study on a silica gel–water adsorption chiller integrated with a closed wet cooling tower. *International Journal of Thermal Sciences*, **49**, 611–620.

[18] Gong, L.X., Wang, R.Z., Xia, Z.Z. and Lu, Z.S. (2012) Experimental study on an adsorption chiller employing lithium chloride in silica gel and methanol. *International Journal of Refrigeration*, **35**, 1950–1957.

[19] Gong, L.X., Wang, R.Z., Xia, Z.Z. and Chen, C.J. (2011) Design and performance prediction of a new generation adsorption chiller using composite adsorbent. *Energy Conversion and Management*, **52**, 2345–2350.

[20] Wang, L.W. (2005) Performances, mechanisms and application of a new type compound adsorbent for efficient heat pipe type refrigeration driven by waste heat. PhD Thesis. Shanghai Jiao Tong University, Shanghai, China (in Chinese).

[21] Wang, L.W., Wu, J.Y., Wang, R.Z. *et al.* (2003) Experimental study of a solidified activated carbon-methanol adsorption ice maker. *Applied Thermal Engineering*, **23**, 1453–1462.

[22] Wang, L.W., Wang, R.Z., Wu, J.Y. *et al.* (2006) Design, simulation and performance of a waste heat driven adsorption ice maker for fishing boat. *Energy*, **31**, 244–259.

[23] Wang, S.G. (2003) Research of an adsorption refrigeration system driven by waste heat of diesel engine in vessels. PhD Thesis. Shanghai Jiao Tong University, Shanghai, China (in Chinese).

[24] Wang, L.W., Wu, J.Y., Wang, R.Z. *et al.* (2003) Study on the performance of activated carbon-methanol adsorption systems concerning heat and mass transfer. *Applied Thermal Engineering*, **23**, 1605–1617.

[25] Wang, L.W., Wu, J.Y., Wang, R.Z. and Xu, Y.X. (2004) Cycle characteristic analysis of adsorption ice maker with activated carbon-methanol as working pair. *Journal of Engineering Thermophysics*, **25**(2), 208–210, ISSN: 0253-231X (in Chinese).

[26] Li, D.Y. (2000) Adsorption refrigeration system with double generator. Invention Patent of China 20001260467.

[27] Iloeje, O.C., Ndili, A.N. and Enibe, S.O. (1995) Computer simulation of a $CaCl_2$ solid-adsorption solar refrigerator. *Energy*, **20**(11), 1141–1451.

[28] Chen, Y.G., Gao, M.C., and Xin, M.D. (1986) Experiments of heat transfer performance of separate type thermosyphon. International Heat Pipe Symposium, Osaka, Japan.

[29] Zhuang, J. and Zhang, H. (2011) *Heat Pipe Technology and Engineering Application*, Chemical Industry Press, Beijing, ISBN: 9787502527525 (in Chinese).

[30] Critoph, R.E., Tamainot-Telto, Z., and Davies, G.N.L. (1997) Design of an adsorption generator utilising a novel carbon-aluminium laminate. Proceedings of International Heat Pump Conference, Nottingham, UK, pp. 349–358.

[31] Vasiliev, L.L., Mishkinis, D.A., and Vasiliev, L.L. Jr. (1996) Multi-effect complex compound/ammonia sorption machines. Proceedings of International Sorption Heat Pump Conference, Montreal, Canada, pp. 3–8.

[32] Critoph, R.E. (1999) Forced convection adsorption cycle with packed bed heat regeneration. *International Journal of Refrigeration*, **22**(1), 38–46.

[33] Pons, M., Meunier, F., Cacciola, G. *et al.* (1999) Thermodynamic based comparison of sorption systems for cooling and heat pumping. *International Journal of Refrigeration*, **22**, 5–17.

[34] Willers, E. and Groll, M. (1999) Evaluation of metal hydride machines for heat pumping and cooling applications. *International Journal of Refrigeration*, **22**(1), 47–58.

[35] Eun, T.H., Song, H.K., Han, J.H. *et al.* (2000) Enhancement of heat and mass transfer in silica-expanded graphite composite blocks for adsorption heat pumps, Part II: Cooling system using the composite blocks. *International Journal of Refrigeration*, **23**, 74–81.

[36] Lai, H.M. (2000) An enhanced adsorption cycle operated by periodic reversal forced convection. *Applied Thermal Engineering*, **20**, 595–617.

[37] Li, T.X., Wang, R.Z., Wang, L.W. and Lu, Z.S. (2008) Experimental study on an innovative multifunction heat pipe type heat recovery two-stage sorption refrigeration system. *Energy Conversion and Management*, **49**, 2505–2512.

[38] Li, T.X., Wang, R.Z., Wang, L.W. *et al.* (2007) Performance study of a high efficient multifunction heat pipe type adsorption ice making system with novel mass and heat recovery processes. *International Journal of Thermal Sciences*, **46**, 1267–1274.

[39] Lu, Z.S., Wang, R.Z., Li, T.X. *et al.* (2007) Experimental investigation of a novel multifunction heat pipe solid sorption icemaker for fishing boats using $CaCl_2$/activated carbon compound-ammonia. *International Journal of Refrigeration*, **30**, 76–85.

[40] Lu, Z.S., Wang, R.Z., Wang, L.W. and Chen, C.J. (2006) Performance analysis of an adsorption refrigerator using activated carbon in a compound adsorbent. *Carbon*, **44**, 747–752.

[41] Li, S.L., Wu, J.Y., Xia, Z.Z. and Wang, R.Z. (2011) Study on the adsorption isosteres of the composite adsorbent $CaCl_2$ and expanded graphite. *Energy Conversion and Management*, **52**, 1501–1506.

[42] Li, S.L., Xia, Z.Z., Wu, J.Y. *et al.* (2010) Experimental study of a novel CaCl2/expanded graphite-NH_3 adsorption refrigerator. *International Journal of Refrigeration*, **33**, 61–69.
[43] Chen, C.F., Xi, F., Pan, Z.F. and Zhang, C. (1995) A new type energy storage material's development and its application prospects. *Chinese Space Science and Technology*, **10**(5), 31–36, ISSN: 1000-758X (in Chinese).
[44] Close, D.J. and Pryor, T.L. (1976) Behavior of adsorbent energy storage beds. *Solar Energy*, **18**(4), 287–292.
[45] Parrish, C.F., Scaringe, R.P., and Pratt, D.M. (1991) Development of an innovative spacecraft thermal storage device. Proceedings of the 26th Intersociety Energy Conversion Engineering Conference, Boston, MA, Vol. **4**, pp. 279–284.
[46] Hisaki, H., Kobayashi, N., Yonezawa, Y., and Morikawa, A. (1994) Development of ice-thermal storage system using an adsorption chiller. Proceedings of the International Absorption Heat Pump Conference, New Orleans, LA, pp. 439–444.
[47] Tahat, M.A. (2001) Heat-pump/energy-store using silica gel and water as a working pair. *Applied Energy*, **69**, 19–27.
[48] Wang, R.Z., Dai, Y.J., Xu, Y.X., and Wu, J.Y. (1999) Domestic air conditioning system driven by solar energy accumulator converts. Invention Patent of China 99124022.7.

9

Adsorption Refrigeration Driven by Solar Energy and Waste Heat

9.1 The Characteristics and Classification of Adsorption Refrigeration Systems Driven by Solar Energy

Under the dual pressure of the energy crisis and environmental pollution, as an inexhaustible and pollution-free natural energy, solar energy is considered to be the most promising energy source after the twenty-first century, and it has attracted great attention worldwide [1]. Nowadays, the development and utilization of solar energy have become the hot topics in the field of energy research, for instance, solar thermal utilization has been considered to be an essential technology for saving the energy used in buildings in Energy Conservation Policies for Buildings (1996–2010) made public by the Ministry of Construction in China. In general civilian buildings, air conditioning accounts for more than half of the total energy consumption. With the development of economic and living standards, the energy consumption on air conditioning keeps increasing every year, and it brings enormous pressure on energy, electricity, and the environment. In 1978, Dr D.I. Tehernev built the first intermittent solar adsorption refrigeration devices using zeolite-water. Since then, solar adsorption refrigeration has become an important branch of energy saving technology, and it has gradually become a major research topic [2–6]. Compared with other cooling systems, solar adsorption refrigeration system has the following characteristics:

1. The structure of the system is simple and the operation of the system is easy, and there are no solution pumps or rectifying devices required. Therefore, the running costs of the system are low. There are also no refrigerant contamination, crystallization, and corrosion problems. For a basic adsorption ice-making cycle driven by solar energy there are no moving parts or power consumption.
2. Different adsorption working pairs can be chosen for different heating and evaporation temperatures. For instance, a solar adsorption air-conditioning system with a silica gel-water working pair can be driven by the hot water of 65–85 °C for producing the chilling water at 7–20 °C. A solar adsorption ice maker with activated carbon–methanol working pair can be directly driven by solar radiation on the solar collectors.

Adsorption Refrigeration Technology: Theory and Application, First Edition. Ruzhu Wang, Liwei Wang and Jingyi Wu.

Companion Website: www.wiley.com/go/wang/refrigeration

3. The requirement of the cooling power of the system for air conditioning can match solar radiation. The stronger the solar radiation is, the hotter the weather is, and the greater the required cooling load is, consequently the larger the cooling power of the system.
4. Compared with the absorption and compression refrigeration systems, the cooling power of adsorption systems is relatively small. Because of the critical heat and mass transfer performance, if the cooling capacity increases, the mass of the adsorbent and heat exchanger will increase, and consequently the initial investment will increase. The machine will be large. In addition, due to the low energy density of solar radiation on the ground, a relatively large collector area to collect a certain amount of heating power is required. Because of the reasons above, it is difficult to develop the successful solar ice makers, refrigerators, or air conditioners.
5. Because of that the solar energy depends on the seasons and is provided periodically and intermittently, the solar driven adsorption refrigeration system usually needs an auxiliary thermal source when it is applied as an air conditioning or cold storage system.

Since the 1970s, much research in the world started to study the solar driven adsorption refrigeration systems. Adsorption refrigeration systems have been commercialized, firstly in the United States and Japan. In China, researchers in Beijing, Shanghai, Tianjin, Zhejiang, Hubei, Henan, and other provinces have started theoretical and experimental research on the solar adsorption refrigeration technology since the 1970s. Now there are different solar adsorption refrigeration systems with different structures. They can be classified by the application of the system, the working pair, and the adsorption refrigeration cycles. Some classifications are shown in Table 9.1.

9.2 Design and Application of Integrated Solar Adsorption Refrigeration Systems

The main feature of an integrated solar adsorption refrigeration system is the use of adsorption collector, which is an adsorbent bed and solar collector integrated into one component. During the day time it absorbs the solar energy and desorbs, and at night it produces the cooling capacity by adsorption. The adsorption refrigeration cycle is the basic intermittent cycle.

9.2.1 The Performance Index of Integrated Solar Adsorption Refrigeration System

The performance indexes of the integrated solar adsorption refrigeration systems are related to the performance of the collector and the refrigeration performance.

1. **Collector Performance**
   The performance of the solar adsorption collector can be estimated by the collector efficiency and collector temperature. The collector efficiency (η) is the ratio of the heat energy (Q_{seff}) transformed from the actual solar radiation (Q_{solar}).

$$\text{Collector efficiency: } \eta = \frac{Q_{seff}}{Q_{solar}} \tag{9.1}$$

Table 9.1 The classifications of solar driven adsorption refrigeration systems

Classifications		System	Characteristics
Application		Ice maker	For freezing condition, it uses basic adsorption refrigeration cycle and has simple structure
		Chiller/air conditioner	Supply the chilling water with the temperature of 7–20 °C, and the cycle is continuous
		Cold storage system	To store the food and other products in low temperature
		Dehumidification air conditioner	Dehumidifying the air by adsorption or used for an air conditioner with evaporative cooling technology
Cycles	Adsorption chiller	Intermittent cycle	Desorption at daytime, and adsorption at night time. The refrigeration output is intermittent
		Continuous cycle	Use two or more adsorbers that operated alternately for the continuous refrigeration output, and the performance can be improved by the heat and mass recovery process
	Dehumidification system	Rotary wheel dehumidifier or liquid dehumidifier	Desiccant directly absorbs moisture from the air, or being combined with the water evaporative cooling technology to deal with the air temperature and humidity
Adsorption working pair		Active carbon-methanol	Suitable for the ice maker
		Activated carbon-ammonia	Suitable for the ice maker that operated under positive pressure
		Strontium chloride-ammonia	Good performance for ice making condition, but the price for adsorbent is a little bit high
		Silica gel-water	Suitable for the solar air conditioner driven by the low temperature heat source
		Molecular sieve-water	Suitable for the condition with high desorption temperature

$$\text{Effective heating power: } Q_{seff} = \int_{T_1}^{T_2} (M_a(C_{pa} + xC_{pr}) + M_m C_{pm})\ \mathrm{d}T + L_{ref}\Delta x M_a \quad (9.2)$$

$$\text{Solar radiation: } Q_s = \int_0^t I(t) A_{seff}\,\mathrm{d}t \quad (9.3)$$

where M_a is the mass of adsorbent, C_{pa} is the specific heat of adsorbent, C_{pr} is the specific heat of the refrigerant, x is the adsorption quantity, $M_m C_{pm}$ is the heat capacity of the entire adsorption collector except the adsorbent and refrigerant, L_{ref} is the latent heat of refrigerant, I is the solar radiation intensity, A_{seff} is the effective collector area, and t is the duration time of sunshine.

Another important parameter that reflects the performance of the collector is the collector temperature. It is closely related to the degree of desorption of the adsorbent. But sometimes the higher collection efficiency cannot be obtained under the condition of a higher collector temperature if the structure of the adsorption collector isn't reasonably arranged.

2. **Refrigeration Performance**
The refrigeration performance of a solar adsorption refrigeration system can be indicated by the coefficient of performance (COP). Usually there are two coefficients. One is the coefficient of refrigeration performance of adsorption system. It can be expressed as the ratio of the refrigeration capacity to the effective heating power provided by the adsorption collector:

$$COP_{ref} = \frac{Q_{ref}}{Q_{seff}} \tag{9.4}$$

where Q_{ref} is the refrigeration capacity. It can be calculated by $Q_{ref} = Q_{eref} - Q_{cc}$, in which Q_{eref} is the evaporation refrigeration capacity in the evaporator, it can be calculated by the following equation:

$$Q_{eref} = \Delta x M_a L_{ref} \tag{9.5}$$

where $\Delta x M_a$ is the desorption mass of the refrigerant in the heating process of the adsorbent, it is also the cycle amount of the refrigerant in one cycle.

Q_{cc} is the sensible heat released by the refrigerant in the cooling processes from condensing temperature to the evaporation temperature.

$$Q_{cc} = \int_{T_e}^{T_c} M_a \Delta x C_{pr} \mathrm{d}T \tag{9.6}$$

Another important coefficient is the coefficient of solar refrigeration performance. It can be expressed as the ratio of the cooling capacity to the total solar radiation received by the adsorption collector:

$$COP = \frac{Q_{\mathrm{ref}}}{Q_{solar}} \tag{9.7}$$

9.2.2 The Design and Application of the Activated Carbon–Methanol Adsorption Ice Maker Driven by a Flat-Plate Type Solar Collector

9.2.2.1 Adsorption Refrigeration System

Since the desorption temperature of the activated carbon-methanol working pair is low (70–100 °C), it is an ideal working pair for a solar adsorption refrigeration system. The characteristic of this kind of system is that the adsorber is a flat-plate type collector, and the absorber is combined with the collector [7]. The structure of a typical flat-plate type solar adsorption ice-making machine is shown in Figure 9.1. The adsorption refrigeration system consists of four parts, the adsorber, condenser, evaporator, and throttle valve.

One of the main components of the system is the flat-plate type adsorption collector. It mainly consists of a flat adsorption bed, glass cover plate, insulation material, and throttle valve. The characteristic of the collector is that it cannot work under the condition of high pressure. As a

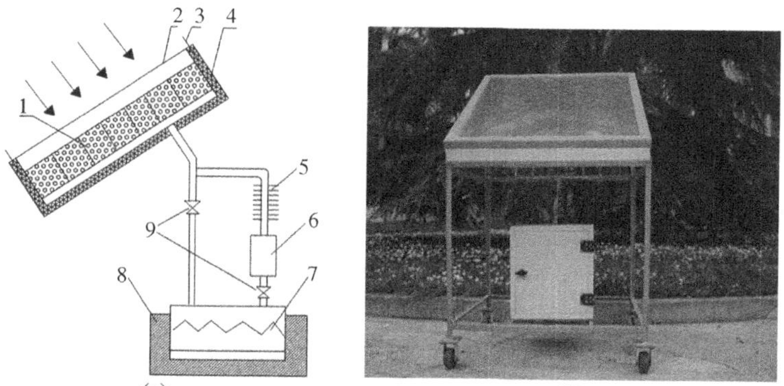

1 - Adsorber; 2 - Glass cover plate; 3 - Air throttle valve of adsorption bed;
4 - Thermal insulation material; 5 - Condenser; 6 - Liquid receiver;
7 - Evaporator; 8 - Refrigerator shell; 9 - Vacuum valve

Figure 9.1 The solar powered adsorption chiller. (a) Diagram of the working principle and (b) photograph of the system

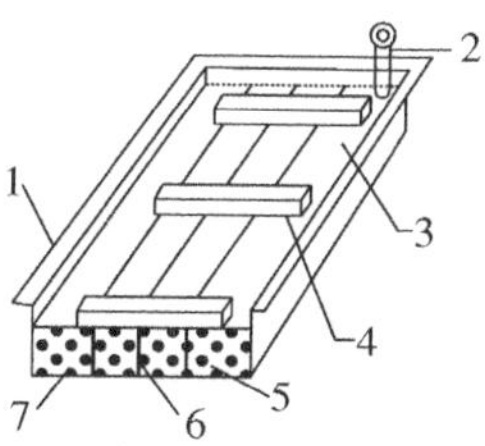

1 - The bottom surface of the adsorber after the back cover plate was dismantled;
2 - The inlet channel of the refrigerant;
3 - The chamber for collecting the desorbed refrigerant vapor;
4 - Strengthening rib; 5-Activated carbon; 6-Heat transfer fins;
7 - The former heat transfer plate of the adsorber

Figure 9.2 The schematic of flat-plate type solar adsorption bed

result the flat adsorption refrigeration system usually uses the working pair that can only work under the condition of vacuum state. The flat adsorption bed is usually filled with adsorbent such as activated carbon, zeolite molecular sieve. Its performance is mainly determined by the heat and mass transfer performance, which means that the heating power from the heating source should be transferred to the adsorbent in the adsorber as much as possible in the desorption process, so that the refrigerant can be desorbed. During the adsorption process, the latent heat of the adsorber and the adsorption heat should be released as soon as possible. So the performance improvement is closely related to the heat and mass transfer performance of the adsorber. Figure 9.2 shows the schematic of the flat adsorption bed.

Flat-plate type activated carbon-methanol adsorption refrigeration system works at a vacuum state, and the pressure in the system changes little. The adsorber that plays the role of compressor is driven by pressure difference of the system. So to ensure the refrigerant vapor flow through the sub-components of the system fluently, the flow channels of the sub-components

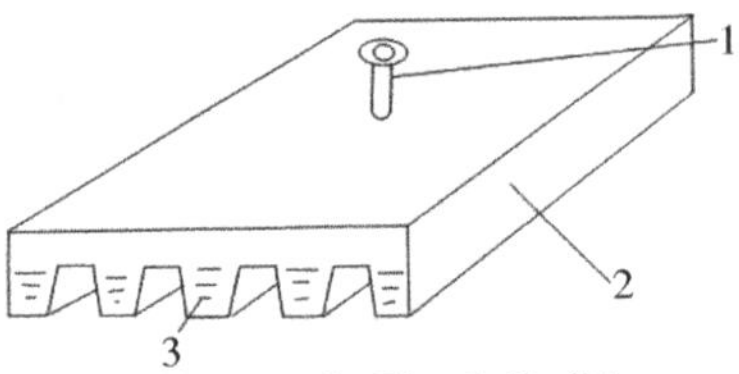

1 - The interface for the refrigerant; 2 - The shell of the evaporator; 3 - The liquid refrigerant

Figure 9.3 The schematic of the evaporator

of the system have to be smooth. Thus the condenser should use the finned tubes with relatively large diameter. It can enhance the cooling process for the desorbed refrigerant vapor, and it can ensure the vapor flows through the condenser smoothly.

For the design of evaporator, on the one hand, the latent heat of evaporation in the evaporator should be released to the environment quickly, so there should be a large heat transfer area; on the other hand, there should be enough space to accommodate the liquid refrigerant desorbed from the adsorber. Also, as an icemaker, the convenience for taking out the ice from the evaporator should be taken into account. For instance, the evaporator is designed in a rectangular shape with a saw tooth isosceles trapezoid. Figure 9.3 shows the schematic of the structure of the evaporator.

9.2.2.2 The Mathematical Models of Flat-Plate Type Solar Adsorption Collector

There are two ways to calculate the flat-plate type adsorption collector: the lumped parameter method and the distributed parameter method. Using the lumped parameter method, the adsorber is taken as a whole, and the performances were calculated by energy balance equations. The adsorption rate is simulated by D-A equation. Using the distributed parameter method, the governing differential equations describing the mass and heat transfer are built and they are solved using the numeric method. This method can calculate the temperature and concentration distribution within the adsorber, which has great significance for analyzing the heat and mass transfer processes, as well as for improving the performance of the adsorber.

The Lumped Parameter Method

A one-dimensional model depending on running time is used. The average temperature of the adsorption collector in one day can be calculated using this model. The adsorption process can be described by D-A equation. Because the cycle time is long, so it can be assumed that the adsorbent are in equilibrium with the refrigerant at any time.

For the isometric heating process, it can be concluded by energy balance equation:

$$A_s[S_{solar} - u_l(T - T_{me})] = (MC_p)_e \frac{\partial T}{\partial t} \tag{9.8}$$

where A_s is the area of solar collector (m^2), S_{solar} is the effective irradiation, u_l is the coefficient of heat loss, T is the temperature of adsorbent bed, T_{me} is the ambient temperature, and $(MC_p)_e$ can be calculated as follows:

$$(MC_p)_e = M_a C_{pa} + xM_a C_{pr} + M_m C_{pm}$$

For the isobaric heating process, it can be concluded by the energy balance equation:

$$A_s[S_{solar} - u_l(T - T_{me})] = -M_a H_{st}\frac{\partial x}{\partial t} + (MC_p)_m\frac{\partial T}{\partial t} \tag{9.9}$$

where at the right side, the first term is the desorption heat, the second term is the sensible heat relating to the temperature change, x is the adsorption quantity given by D-A equation:

$$x = x_0 \exp\left(-K\left(\frac{T}{T_s} - 1\right)^n\right) \tag{9.10}$$

Substitute Equation 9.10 into Equation 9.9, we can get:

$$A_s(S_{solar} + u_l T_{me}) = A_s u_l T + \left(-M_a H_{st}\left(-\frac{Kn}{T_s}\right)x_0\left(\frac{T}{T_s} - 1\right)^{n-1} \times \exp\left(-K\left(\frac{T}{T_s} - 1\right)^n\right) + (MC_p)_e\right)\frac{\partial T}{\partial t} \tag{9.11}$$

The thickness of the adsorption bed is 4.5 cm. When the adsorption working pair is activated carbon-methanol, the flat-plate solar adsorption collector shown in Figure 9.2 is calculated by the lumped parameter method. The relationships among the average temperature of adsorption bed, solar radiation intensity, and time are shown in Figure 9.4. The maximum temperature of the adsorber is up to 100 °C.

Distributed Parameter Method

The distributed parameter method analyzes the heat and mass transfer process in the adsorbent bed by considering the heterogeneity of the internal temperature field of the adsorption bed and the adsorption rate. Because the fins are uniformly distributed in the width direction of the bed, so only the activated carbon layer ABEF between two fins (shown in Figure 9.5) is analyzed. Taking into account the symmetry of the distribution of fins, the calculation can be further simplified so that half of the adsorbent bed between two fins is regarded as a calculation field, which is the area denoted by ABCD for the research on heat and mass transfer performance. Under this simplified condition, AB, BD, DC, and CA constitute the boundary of the model.

The adsorbent is usually the porous media with a large surface area (such as activated carbon and zeolite molecular sieves, etc.). The heat and mass transfer problems should be solved by the

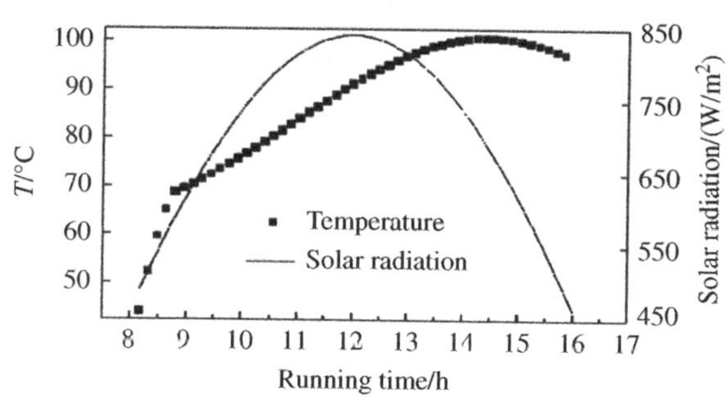

Figure 9.4 The temperature change of the adsorber vs. time

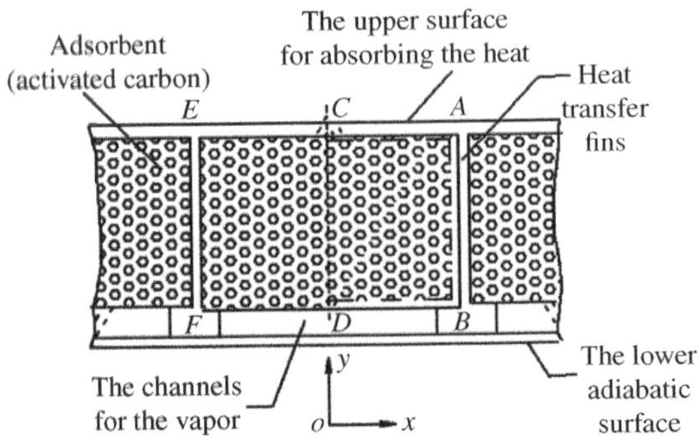

Figure 9.5 The calculation field for the cross-section of the adsorption bed

theory of porous media under the simplified conditions. The important parameters indicating the features of the porous media are porosity ε_a (the fraction of void pores in porous media (kg/m^3)), specific surface area Ω (the ratio of total volume of the porous media to the total surface area A_i of solid skeleton), and solid particle size d_p (the equivalent diameter when the solid particles are regarded as a sphere shape (m)).

In order to obtain the mathematic models for the internal heat and mass transfer processes in the adsorbent, the following assumptions are made:

1. The solid skeleton is incompressible (density is constant), and is stationary.
2. Liquid is adsorbed in the pores of the solid skeleton, and is also stationary.
3. Liquid is one-component and incompressible (density is constant), and the viscous dissipation of the liquid phase is negligible.
4. In each phase there is no chemical reaction. The gas compressibility is ignored. The viscous dissipation of gas is also taken as zero.
5. The temperature of the solid and fluid is the same [8], so: $T_{so} = T_l = T_g$.
6. The surface energy for the interfaces between the solid-liquid, gas-liquid, and gas-solid is negligible.

By the assumptions above the heat and mass transfer equations in the adsorbent can be obtained.

Mass Conservation Equation:

$$\text{Gas: } \rho_{gas}\frac{\partial \varepsilon_a}{\partial t} + \rho_{gas}\nabla(\psi_a \vec{W}_{gas}) + \rho_{gas}\frac{\mathrm{d}x}{\mathrm{d}t} = 0 \tag{9.12}$$

where ε_a is the adsorbent porosity, ρ_{gas} is the gas density, ψ_a is the ratio of airflow area to the cross-section area per unit mass of adsorbent. x is the adsorption quantity of adsorbent and $\rho_{gas}\frac{\mathrm{d}x}{\mathrm{d}t}$ indicates the desorption or adsorption rate of refrigerant in the heating or cooling process. For the solid and liquid phases, because the density of the materials is constant and the position is stationary, we don't need to list the equations.

Momentum conservation equation:

$$\text{Gas: } \rho_{gas}\frac{\mathrm{d}(\psi_a \vec{W}_{gas})}{\mathrm{d}t} = \nabla \mu_{gas}\nabla(\psi_a \vec{W}_{gas}) + \varepsilon_a \rho_{gas} g - \nabla(\psi_a p_{gas}) \tag{9.13}$$

where p_{gas} is corresponding to the gas pressure of flowing refrigerant (Pa); in the refrigerant gas adsorption process, the diffusion term $\nabla \mu_{gas} \nabla \vec{W}_{gas}$ in the right side of equation can be ignored. Again there are no momentum conservation equations for solid and liquid phases.

Energy conservation equation:

$$\rho C_p \frac{\partial T}{\partial t} + \varepsilon_a \rho_{gas} C_{pg} \vec{W}_{gas} \nabla T_{gas} = \nabla \lambda_{eff} \nabla T + \rho_{so} H_{st} \frac{\mathrm{d}x}{\mathrm{d}t} \tag{9.14}$$

where

$$\rho C_p = \varepsilon_a \rho_{gas} C_{pg} + (1 - \varepsilon_0) \rho_{so} C_{ps} + (\varepsilon_0 - \varepsilon_a) \rho_l C_{pl} \tag{9.15}$$

is the total heat capacity of the adsorbent bed (excluding the metal heat capacity). C_{ps}, C_{pg}, and C_{pl} are the isobaric specific heat of solid, gas, and liquid phases, respectively. λ_{eff} is the effective thermal conductivity, which is related to the types of adsorbent, pore structure of adsorbent, and the types of refrigerant. ε_0 is the proportion of the adsorbed refrigerant in gas and liquid phases. Usually it is measured by experiments. In Equation 9.14, at the left side of the equation, the first item is the internal energy change rate in the adsorbent bed (including the adsorbent and refrigerant); the second term is the heat exchange between the flowing refrigerant vapor and the solid adsorbent in the desorption or adsorption process, which is called convective heat transfer term. At the right side of the equation, the first term is the heat transferred by the thermal conductive process in the adsorbent bed, which is known as the diffusion term; the second term is the adsorption heat or desorption heat of the refrigerant, which is known as the term for the internal heat source.

Mass conservation equation of the refrigerant in adsorbent:

$$\rho_{gas} \frac{\mathrm{d}x}{\mathrm{d}t} = \nabla D_e \nabla x \tag{9.16}$$

where D_e is the effective diffusion coefficient of the refrigerant in the adsorbent, which is determined by three diffusion mechanisms, that is, the general diffusion, surface diffusion, and Knudsen diffusion. It is related to the types of the adsorbent, the pore structure of adsorbent, the adsorption rate, and the types of refrigerant [3, 9].

Solving the Equations

To solve the heat and mass transfer equations within the adsorber, we need to combine the mass conservation Equation 9.13, the momentum Equation 9.14, and the energy conservation Equation 9.15, as well as that we need to provide the data of desorption/adsorption rate and adsorption/desorption heat. For solar solid adsorption refrigeration cycle, the cycle is usually divided into four stages. The solving methods for heat and mass transfer processes can be divided into two types according to the characteristics of the cycle.

1. **The isosteric heating/cooling processes**
   At the first stage and third stage of the isosteric heating and cooling processes, there is no mass transfer between the adsorbent bed and other components of the refrigeration system, but only the heat transfer process in the adsorbent bed. As a result, there is no need to take into account the momentum conservation equation and mass conservation equations. The energy conservation Equation 9.14 is:

$$\rho C_p \frac{\partial T}{\partial t} = \nabla \lambda_{eff} \nabla T \tag{9.17}$$

This is a typical transient heat conduction problem. For the solar solid adsorption refrigeration cycle the solar radiation energy and the environmental parameters can be taken as boundary conditions, then the temperature change in the isosteric heating or isosteric cooling processes can be obtained.

2. **The heat and mass transfer at adsorption/desorption stage**
Solving the second stage corresponding to the desorption process and the fourth stage corresponding to the adsorption process, mass conservation Equations 9.12 and 9.16, momentum conservation Equation 9.13, and the energy conservation Equation 9.14 must be combined together. To get the adsorption rate in the adsorbent the formula [10] proposed by Sokoda and M. Suzuki is widely used as well as Equation 9.12.
To solve the energy Equation 9.14, the flow rate of the refrigerant gas in the desorption and adsorption processes in adsorbent bed must be calculated firstly, that is to solve the momentum Equation 9.13. As the flow rate of refrigerant gas in the adsorbent is low, then it is the Reynolds flow (Re < 10), so we can use Darcy's law to describe the gas flow in solid adsorbent during the desorption and adsorption process. For the flow of incompressible fluid in porous media, Darcy law is [11]:

$$\vec{W}_{gas} = -\frac{k_p}{\mu}(\nabla p_{gas} + \rho_{gas}\vec{l}_i) \tag{9.18}$$

where, k_p is the permeability of the adsorbent, μ is the viscosity of the refrigerant gas, and $\vec{l}_i$ is unit vector parallel to the direction of gravity. Usually the effect of gravity of gas is negligible, so the above equation can be written as:

$$\vec{W}_{gas} = -\frac{k_p}{\mu}\nabla p_{gas} \tag{9.19}$$

Equation 9.19 shows how the flow rate of the refrigerant vapor in the desorption and adsorption process changes with its pressure. The pressure distribution of the adsorbent bed can be obtained by the adsorption rate equation and some appropriate boundary conditions, and can be solved by the numerical heat transfer method. Then the pressure distribution is substituted into Equation 9.19, the velocity distribution of the gas field in the adsorber can be calculated. If the obtained velocity distribution field is substituted into the energy Equation 9.14, the temperature distribution within the adsorbent bed can be obtained.

The Initial Conditions and Boundary Conditions

The initial and boundary conditions must be provided in the models described above, so that the solution of the equations under specific conditions can be obtained.

The initial conditions are:

$$\text{When } t = 0, T(x, y) = T_{fin}(x, y) = T_0, p(x, y) = p_0.$$

Take the half distance between two fins in the adsorption bed as L_{fin}, the thickness (i.e., height) of the adsorbent bed as H_{adb}. The temperature boundary conditions at the four sides of the calculation field can be calculated as follows:

1. The CA side, which is the upper heat absorbing layer of the adsorbent bed, is under the boundary condition of:

$$\lambda_{eff}\left.\frac{\partial T}{\partial y}\right|_{y=H_{adb}} = \alpha(T_{mw} - T_i) \tag{9.20}$$

where λ_{eff} is the effective thermal conductivity between the adsorbent layers, α is the heat transfer coefficient between the metal shell and the absorbent, T_i is the temperature for the surface layer of the adsorbent, and T_{mw} is the temperature of the metal shell of the adsorption bed. T_i and T_{mw} can be calculated by the following equation:

$$M_m C_m \frac{dT_m}{dt} = W_{in} - A_{adb}\,\alpha(T_m - T_i) \tag{9.21}$$

where M_m is the mass of the metal shell of adsorption bed, C_m is the heat capacity of the metal shell, A_{adb} is the contact area between the metal shell of the bed and the absorbent, and W_{in} is the effective solar radiation absorbed by the metal shell of the adsorbent bed, which is calculated using the following equation:

$$(\tau_{solar}\varsigma_{solar})IA_{eff} - W_t - W_b = W_{in} \tag{9.22}$$

where I is the total solar radiation on the collector per unit surface area, it can be measured by solar radiation instrument; τ_{solar} is the sunshine transmittance through the glass cover; ς_{solar} is the absorbing rate of sunshine by the collector of adsorber, A_{eff} is the effective area of the metal shell for the adsorbent bed to absorb the sunshine; W_t is the facial heat loss; and W_b is the heat loss at the bottom which can be calculated as follows:

$$W_t = U_t A_{eff}(T_m - T_{me}) \tag{9.23}$$

$$W_b = U_b A_{eff}(T_m - T_{me}) \tag{9.24}$$

where T_m is the average temperature of the collector; T_{me} is the ambient temperature; U_t is the heat loss coefficient at the surface of the collector, which can be determined by the available empirical formula; U_b is the heat loss coefficient at the bottom of the collector; and W_b is usually smaller than 10% of the whole value of heat. The surrounding area of the collector is relatively small if compared to the area of the top and bottom of the collector. The heat loss around the collector can be ignored.

Because the collector is heat insulated with the surroundings except at the top of the collector, so the heat loss coefficient at the top U_t influences the performance of the adsorbent bed greatly. It is the function of the temperature for the heat absorbing plate T_p, environment temperature T_{me}, wind velocity u_w, the layer numbers of the glass cover N, the emissivity ε_g and ε_p of the glass cover and the heat adsorbing plate, and the angle of the collector β. Klein proposed a more accurate empirical formula to calculate U_t of the solar flat-plate type collector [12]. By using this formula, the tedious and repetitive iterative calculation can be avoided. Also it is suitable for the flat adsorption collector. It is as follows:

$$U_t = \left\{ \frac{N}{\frac{C}{T_p}\left[\frac{T_p - T_{me}}{N+f}\right]^e} + \frac{1}{h_w} \right\}^{-1} + \frac{\sigma_b(T_p + T_{me})(T_p^2 + T_{me}^2)}{[\varepsilon_p + 0.05N(1-\varepsilon_p)]^{-1}[(2N+f-1)/\varepsilon_g)] - N} \tag{9.25}$$

where σ_b is Boltzmann constant, C, f, h_w are calculated as follows:

$$C = 365.9(1 - 0.00883\beta + 0.00013\beta^2) \tag{9.26}$$

$$f = (1 - 0.04h_w + 0.0005h_w^2)(1 + 0.091N) \tag{9.27}$$

$$h_w = 5.7 + 3.8u_w \tag{9.28}$$

The boundary pressure is $\frac{\partial p}{\partial y}\Big|_{y=H_{adb}} = 0$ for CA side.

2. At the heat transfer side of AB, the temperature T_{fin} of the fin is calculated as follows:

$$T_{fin} - T|_{x=L_{fin}} = (T_p - T|_{x=L_{fin}})\frac{e^{my} + e^{-my}}{e^{mh} + e^{-mh}} \tag{9.29}$$

where T_{fin} is the temperature of the fin, T_p is the temperature of the heat absorbing plate. $m = \sqrt{\alpha l_{fin}/\lambda_{fin} A_{fin}}$, where A_{fin} is the area for the cross section of the fin, l_{fin} is the perimeter of the cross section, and λ_{fin} is the thermal conductivity of the fin. The temperature of the adsorbent in the direction of AB can be calculated as follows:

$$\lambda_{eff}\frac{\partial T}{\partial x}\Big|_{x=L_{fin}} = \alpha(T_{fin} - T|_{x=L_{fin}}) \tag{9.30}$$

For AB side $\frac{\partial p}{\partial x}\Big|_{x=L_{fin}} = 0$.

a. At the BD side, $\frac{\partial T}{\partial y}\Big|_{y=0} = 0$. When BD side is connected with the evaporator, $p|_{y=0} = p_e$ (evaporate pressure); when it is connected with the condenser, $p|_{y=0} = p_c$ (condenser pressure).
b. DC side is the symmetry plane of the computational field, and $\frac{\partial T}{\partial x}\Big|_{x=L_{fin}} = 0$, $\frac{\partial p}{\partial x}\Big|_{x=0} = 0$

9.2.2.3 The Performance of the Activated Carbon–Methanol Ice Maker

The characteristics of the adsorption bed are shown in Table 9.2, which lists the desorption time of the adsorbent bed, the corresponding highest desorption temperature, the time for the adsorbent bed to be cooled down to room temperature, and the highest temperature in adsorption process.

The experimental performance of a typical solar powered adsorption ice-making system can be expressed by the temperature variations of the adsorption collector, evaporator, and the ice bath [14].

The Regularity for the Temperature Change of the Adsorbent Bed

Figure 9.6 shows the temperature at the middle of the adsorbent bed (corresponding to the curve of T_2), which rises more quickly than that at the bottom of the adsorbent (corresponding to the curve of T_3) in the heating process. That means there is a relatively large temperature

Table 9.2 The working characteristics of the adsorption bed [13]

Area (m^2)	Solar radiation (MJ)	Time to heat the bed (environment temperature of 18–25 °C) (h)	Maximum desorption temperature of the adsorbent (°C)	Cooling process of the bed		Maximum adsorption temperature (environment temperature of 15–20 °C) (°C)
				Time (h)	Temperature (°C)	
0.75	13.6	8	91	5	24	59
1.5	29	7	90	5.5	23	65
1.5	27.93	6.5	103	3.5	23	61
0.75	14.36	8.5	94	3	21	47

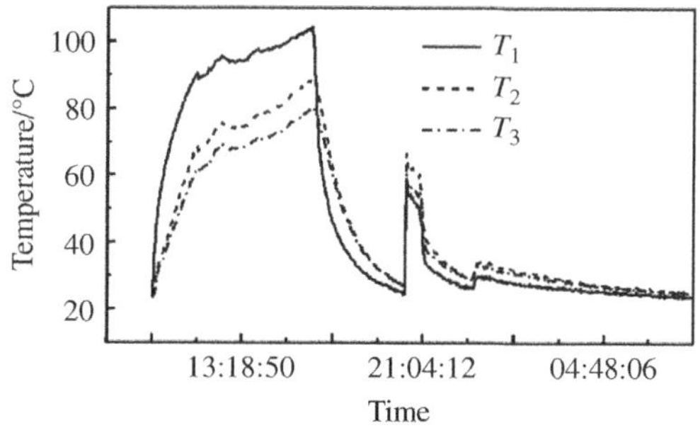

Figure 9.6 Temperature distribution of the adsorbent

gradient from the adsorbent (activated carbon) at the top to the adsorbent at the bottom. Although the length in the radial direction is only 4 cm, the temperature gradient between the top and the bottom is about 20 °C, which means that there is a large thermal resistance in the adsorbent. After 17:00 p.m. heating is stopped and the throttle of the adsorption bed is open, so that the adsorbent bed is cooled down by the air from outside. The temperature of the adsorbent is decreased to the ambient temperature. This process lasts about 4 hours until around 21:00 p.m. From the figure, it can be seen that the temperature of the upper surface of the adsorbent (curve of T_1) decreases faster than that at the lower level. The reason is that the upper surface is cooled directly by the air while at the lower level of the adsorbent the heat needs to be transferred to the upper surface firstly and then to the outside. At the night (around 21:00), when the temperature of the adsorption bed almost reaches the temperature of the ambient environment, open the valve between the adsorption bed and the evaporator, and the adsorbent in the adsorption bed adsorbs the refrigerant in the evaporator directly, resulting in the adsorption refrigeration effect. This process has been held until the next morning around 8:00. After the sun rises, the system begins a new adsorption refrigeration cycle. At the beginning of the adsorption, there is a very sharp increment of the adsorption temperature, which is about 40 °C. The reason is that at the beginning of the adsorption, the adsorption rate is the largest, and a large amount of the adsorption heat is generated in the adsorption process and cannot be released to the environment in time, which results in a significant temperature rising process. After that, with the cooling process continuing from the cold air from the environment, the temperature of the adsorbent decreases and it is close to the ambient temperature with the decreasing adsorption capacity. From the figure, it can also be concluded that the temperature of the adsorbent at the bottom is higher than that at the upper layer. This is because the adsorption channel is at the bottom of the adsorbent. Meanwhile the adsorption heat is released from the surface of the adsorbent. As the adsorption heat is released from the upper surface, it heats the adsorbent, so the difference for the temperature distribution in different places is not very great.

Figure 9.7 shows the temperature changes of the adsorbent between the fins. It should be noted that, desorption temperature (T_4) of the adsorbent closest to the heat transfer fins rises faster than that of the adsorbent farther away from the fins (temperature difference is about 8 °C), which indicates that the fins play an important role in the heat transfer process. Theoretically, the heat transfer will be better if there are more heat transfer fins, but more fins will also increase the metal heat capacity of the adsorbent bed. When the external input energy is constant it will affect the effective heat absorbed by the bed and therefore the number of the heat transfer fins should be optimized.

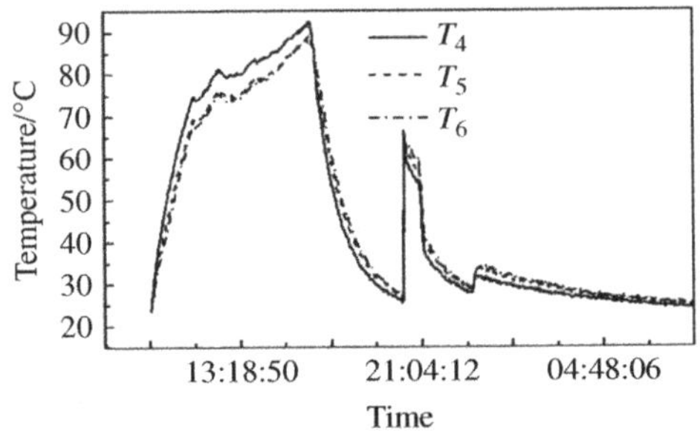

Figure 9.7 Temperature distribution of adsorbent between the fins

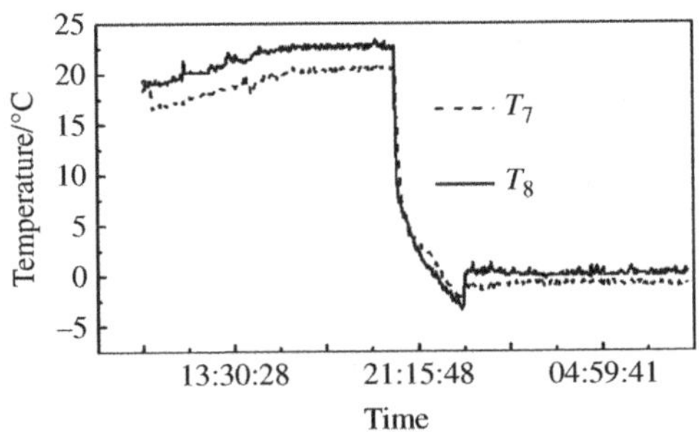

Figure 9.8 The temperatures of the refrigerant in evaporator and ice tank

Temperature Variation in the Evaporator and the Ice Tank

Figure 9.8 shows the temperature changes of the refrigerant in the evaporator (T_7 – dotted line) and the water in the ice tank (T_8 – solid lines) with time. In the heating process of the adsorbent bed, the desorbed refrigerant flows into the evaporator, and it results in the temperature increase of the refrigerant in evaporator. Because the evaporator is directly immersed in the ice box filled with water, the water temperature of the ice box rises. When the adsorption process begins, the liquid refrigerant in the evaporator evaporates quickly. The adsorption heat is released to the outside from the adsorbent bed. The refrigeration output is transferred to the water. From the figure we can see that at the beginning of adsorption the refrigeration capacity produced by refrigerant is large, and it takes away the sensible heat of the water quickly, thus the temperature of the water decreases sharply from ambient temperature to 0 °C. The refrigeration output is used for the frozen process of ice. This process lasts for 3 hours. It can also be concluded that the water temperature will rise a little bit when it freezes due to the releasing heat through the phase change process of the water. After that, the freezing process continues, and the water temperature is maintained at the freezing point temperature (0 °C). When the freezing process is completed, the water temperature will keep decreasing if the adsorbent still has a large adsorption capacity. In several experiments [14] the water can be frozen to −20 °C if the amount of water is little (less than 4 kg). If the amount of water is more than 12 kg there will be a mixture of ice and water in the ice tank. Also, from the experimental curves, the adsorption refrigeration effect is very significant. The temperature in the evaporator

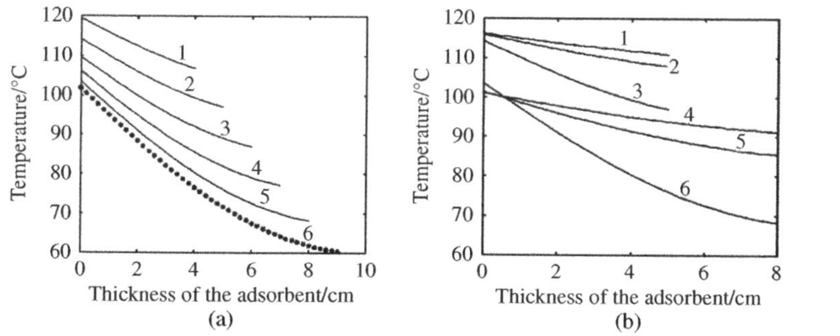

1,2,3,4,5,6 stand for the bed thickness of 4,5,6,7,8,9 cm, respectively.

1 and 4, 2 and 5, 3 and 6 stand for the thermal conductivity of 0.193,0.4,0.6W/(m °C)

Figure 9.9 The temperature distribution of the adsorbent without fins. (a) The adsorbent with different thickness and (b) the adsorbent with different thermal conductivity

is almost the same as that in the water tank. The reason for that is the evaporation area is very large, so that the cooling efficiency is high.

Analysis on the Heat Transfer Enhancement of Flat-Plate Solar Adsorption Bed

Usually, the thermal conductivity of the adsorbent is very small. It can be seen from Figure 9.9a. When the bed thickness is 4 cm the maximum temperature difference is about 10 °C; when its thickness is 8 cm, the maximum temperature at the heat absorbing plate is 103 °C, while the minimum temperature is 69 °C. The temperature difference reached 34 °C. The average temperature is only 82 °C. Therefore, to optimize the temperature distribution in the adsorbent bed the temperature gradient must be decreased and thermal conductivity should be strengthened.

Currently, the common methods to strengthen the thermal conductivity of the adsorbent bed are to embed the metal fin (extended heat transfer area) in the adsorbent bed and to add some metal particles in the adsorbent, such as copper, nickel foam to improve the thermal conductivity, and so on.

When the thermal conductivity of the adsorbent is improved from 0.4 to 0.6 W/(m°C) the temperature gradient in the bed will be significantly reduced. As Figure 9.9b shows, with the increase of the adsorbent thermal conductivity the slope for the temperature curve gradually decreases and finally it is nearly a straight line. The maximum temperature differences are 15 and 10 °C respectively when the thickness is about 8 cm.

If the thermal conductivity of the adsorbent is improved to 0.4 W/(m°C) with the fin space of 6.4 cm, the adsorption isotherm is shown in Figure 9.10a. The isotherms are distributed uniformly when the thickness of the adsorbent changes from 1 to 8 cm. It means that the temperature is approximately linear distributed along the direction of bed thickness. The maximum temperature difference is about 8 °C, the average temperature is about 97 °C.

The thermal conductivity will be greatly improved if the metal fin is embedded in the adsorption bed. However, the fin also absorbs heat in the heating process, and thus there are some negative impacts. Figure 9.10b shows the isotherms of the adsorbent with aluminum fins. Compared with Figure 9.9b, both the maximum temperature of the adsorption bed and the average temperature are improved by 10 °C. This is due to the usage of aluminum fin (density is 2660 kg/m^3, specific heat is 871 J/(kg °C), and the thermal conductivity is 162 W/(m °C)).

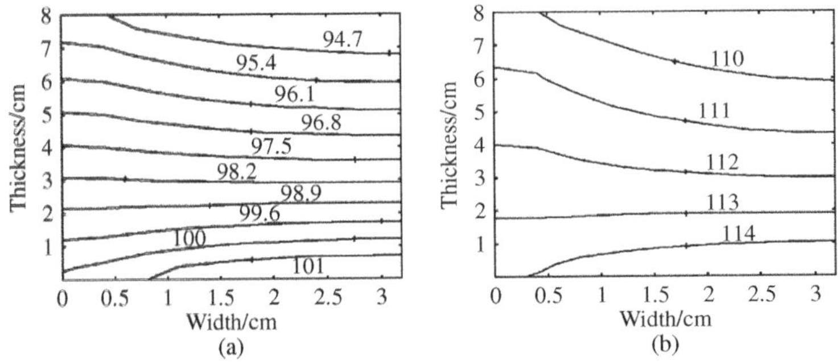

Figure 9.10 Temperature distribution (the space of the fins is 6.4 cm, the thickness of the bed $h = 8$ cm). (a) Thermal conductivity of the adsorbent is 0.4 W/(m°C) and (b) thermal conductivity of the adsorbent is 0.193 W/(m°C), with aluminum fins

Its heat capacity is only 63.5% of the carbon steel (density is 7840 kg/m^3, specific heat is 465 J/(kg°C), and the thermal conductivity is 49.8 W/(m °C)), but the thermal conductivity is 3.25 times higher than that of the carbon steel.

The Performance of the System

The experimental results of the adsorption ice maker are shown in Table 9.3, the COP of the system is from 0.12 to 0.147. In the experiments, the ice making capacity is from 4.7 to 6 kg per square meter of collector.

For the adsorption refrigerator shown in Figure 9.1b [8], when the area of the collector is 1 m^2 and the solar radiation is 18–22 MJ/m^2, we can get the ice production of 4–5 kg per day. The solar COP is as high as 0.12–0.14.

9.2.3 The Design Examples of the Activated Carbon–Methanol Ice Maker Driven by Evacuated Tube Collector

Heat loss of flat-plate solar adsorption collector is concentrated in the natural convection, radiation, and heat conduction between the absorber plate and the glass cover. In order to reduce the heat loss and improve the temperature of the collector, evacuated tube collectors have been proposed. The main feature of this kind of collector is that the space between the heat absorbing surface and the outer layer of glass is evacuated in order to effectively prevent the natural convection and heat conduction. The adsorbent layer is set in the center, and the space among

Table 9.3 The experiment results of the adsorption ice maker [14]

Area of collector (m^2)	Radiation energy received by the collector (MJ)	Mass of the ice (kg)	Desorbed methanol (kg)	COP
0.75	13.6	3.5	1.76	0.12
1.5	29	8.0	3.6	0.125
1.5	27.93	9.0	4.0	0.14
0.75	14.36	4.5	1.92	0.147

the adsorbent particles is regarded as mass transfer channels. The adsorption heat is taken away by the water pipes at the center of the adsorbent or heat exchanger.

The evacuated tube collector is developed based on the flat-plate solar collectors. According to the analysis of the instantaneous efficiency of solar collectors, for the air interlayer between the flat plat solar collector and the glass cover the heat loss caused by air convection is the main part of the total heat loss. To reduce this part of the heat loss the most effective method is to evacuate the space between the collector and cover. However it is very difficult to use this method because the cover should withstand the pressure of 1 ton/m^2 after the space is in vacuum state. So the evacuated tubes are developed. The space between inner tube and outer tubes is vacuumed, so that the heat loss caused by convection, radiation, and heat conduction are reduced. By connecting numbers of the evacuated tubes the tube collector is developed.

The vacuum tube is the core component of the collector; it is mainly composed of the internal heat absorbing body and outer glass tube. The surface of the absorber was deposited by selective absorbing spectrum coating by various means. As the mezzanine between the absorber and the glass tube maintains a high degree of vacuum, it can effectively inhibit heat loss by the air conduction and convection in the vacuum tube. Selective absorbing coating also has a low infrared emissivity which can significantly reduce the radiant heat loss of the endothermic board. These vacuum tube collectors can make maximize use of solar energy. It will have excellent thermal performance even under high operating temperatures and low ambient temperature conditions.

By heat absorbing materials the solar evacuated tube collectors can be classified into two kinds: glass evacuated tube collector (evacuated tubes are made only usung glass) and metal evacuated tube collector (evacuated tubes are made using both glass and metal).

The components of the glass evacuated tube collectors (Figure 9.11) are: the outer and inner glass tubes, selective absorption coating, spring bracket, air grievances, and so on. The shape is like a slender thermos bladder. One end of the tube is open and the inner and outer tube mouth is fused. The other end is sealed into the domed round, and there is a spring bracket support in the inner glass tube. Also it is retractable in order to buffer the pressure change caused by its thermal expansion and contraction. The mezzanine between internal and external glass tubes is highly vacuumed. The selective coating is painted outside the inner glass tube. The spring bracket is equipped with getters, which can collect the gas produced in the running process of the evacuated tubes and keep the tubes in a state of high vacuum.

There are many forms of metal evacuated tube collectors. But no matter what form the collector is, there are some advantages in common due to the metal absorber and the metal connector between the evacuated tube collector:

1. **High working temperature**. The maximum operating temperature exceeds 100 °C, and it is even up to 300–400 °C for some special collectors.

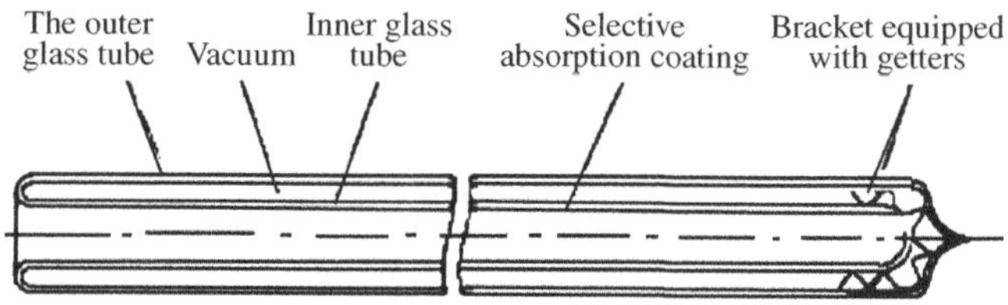

Figure 9.11 Glass evacuated tube type collector

2. **Can work under the condition of high pressure**. The vacuum tubes and the system can withstand the pressure from water and circulating pumps. Most collectors can also be used to generate the hot water whose pressure is higher than 10^6 Pa and high-pressure steam.
3. **Good resistance to thermal shock**. Even if the user accidentally operates the system the cold water will run into the heat collector immediately, and the tubes won't burst.

There are many advantages to the metal evacuated tube compared to other types of evacuated tubes, and various forms of evacuated tubes were developed to meet the needs of different occasions and expanding the scope for the application of solar energy.

9.2.3.1 Water-Cooled Evacuated Adsorption Collector in Adsorption Ice Maker

The vacuum adsorption collector of the adsorption ice maker has the shape of concentric cylinders. The outermost layer is a glass tube, the second layer can be a glass tube or the tube coated with selective coating. There are cooling water pipes in the middle of the tube, and the spaces between the cooling water tubes and inner tubes are filled with the adsorbent made by the sintering process. The pores of the adsorbent are by nature a mass transfer channel. The spaces between the glass tubes and inner tubes are evacuated. The glass tube allows visible light in and prevents the internal long-wave radiation, which guarantees good thermal insulation. The vacuum can prevent convective heat loss between the outer cylinder and the glass plate. Cooling water takes away the adsorption heat in the adsorption process and then turns into warm water for users. The whole adsorption cylinder is set above the surface which can reflect the light. Several adsorption cylinders are combined together to constitute an adsorber. The whole producing process of collectors is similar to the process of vacuum tube collectors. However, the adsorbent needs to be filled in the tubes. The higher the collection temperature is, the higher the desorption temperature can be if the structure parameters and the amount of the adsorbent are appropriate, and also the collection efficiency and the circulating mass of the refrigerant can be improved, and this will result in better performance of the system.

In Figure 9.12, the tubes across the center of the collector are the channels for water cooling or auxiliary heating. Both water inlets (1 and 2) and outlet (11) were processed and threaded in order to be easily connected with valves and other pipe connections. The water circuit is used for the cooling and adsorption process. If the solar radiation is insufficient, the hot water from

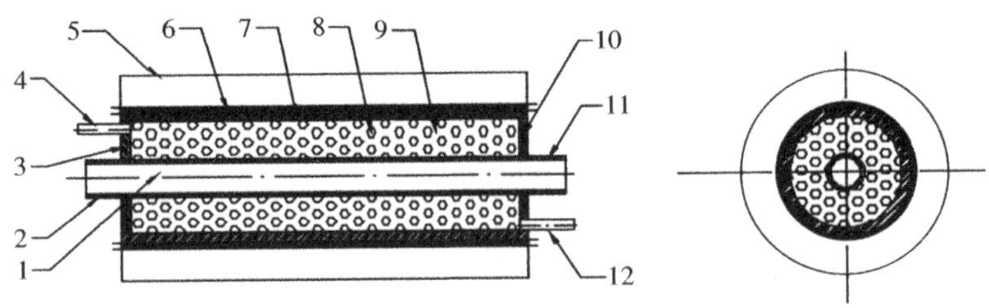

1,2 - Water inlet; 3 - The upper cover of the adsorption collector; 4 - Refrigerant inlet;
5 - External transparent glass tube; 6 - Selective coating; 7 - Inner cylinder filled with adsorbent;
8 - Adsorbent particles; 9 - Clearance among particles; 10 - Lower cover of the adsorption collector tube;
11 - Water outlet; 12 - refrigerant outlet

Figure 9.12 Evacuated glass adsorption collector

the auxiliary heat source is supplied into the pipe. Adsorption collector (with the cover plates of 3 and 10) installation requires a certain angle of inclination to make the flowing process of the refrigerant vapor (from 4 to 12) easier. In the figure 6 is the selective coating coated at the outer surface of 7, which can strengthen the adsorption rate.

In a practical application several evacuated tubes are connected by the water header and refrigerant header. Then it is combined with the evaporator, condenser, and other components to constitute a solar solid adsorption refrigeration/air conditioning system [9].

9.2.3.2 Mathematical Models of the Adsorption Collector

For adsorption collector the cylindrical coordinate system is taken as shown in Figure 9.13, where z direction is the length direction of adsorption collector, r is the radial direction, and θ is circumferential direction.

The energy balance equations for adsorption bed, the inside of the metal wall, the outer glass wall, and the cooling water pipe were established. Because the working period of the system is long, we assumed that the system reaches the adsorption equilibrium at a given temperature. By solving this equation, the temperature change of the adsorbent bed and the performance of the system can be simulated with their coupling relationships.

1. **Adsorbent layer**

   In order to facilitate the computational analysis, the models need to be simplified. The following assumptions are made: firstly we ignore the heat transfer between the adsorbate and its steam; secondly we assume that the adsorbent packing density and specific heat capacity are constant; thirdly we postulate all media are the homogeneous medium and the thermal conductivity λ is a constant; fourthly the refrigeration cycle of the corresponding adsorption bed is relatively long, and as a result, it is assumed that the adsorption bed is kept at equilibrium state and the adsorption rate of the adsorbent can be computed by D-A equation; lastly both upper and bottom surfaces of the adsorption collector meet the adiabatic condition.

   Based on these assumptions, the adsorption heat and desorption heat are regarded as the additional heat source for the energy equation. Then the basic heat transfer equation can be obtained.

   The energy balance equation in the adsorbent bed is as follows:

$$\rho C_p \frac{\partial T_a}{\partial t} = \lambda_a \frac{\partial^2 T_a}{\partial z^2} + \lambda_a \frac{1}{r\partial r}\left(r\frac{\partial T_a}{\partial r}\right) + \lambda_a \frac{1}{r^2}\frac{\partial^2 T_a}{\partial \theta^2} + \rho_a H_{st}\frac{\partial x}{\partial t} \tag{9.31}$$

   where H_{st} is adsorption/desorption heat. The adsorption heat is positive while the desorption heat is negative. ρC_p is the heat capacity of the adsorbent, $\rho C_p = \rho_a C_{pa} + \rho_a C_{pl} x$, where x

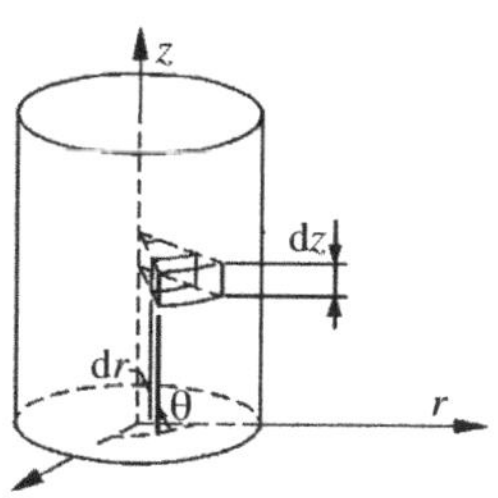

Figure 9.13 Coordinate diagram for adsorption bed

is the adsorption rate, T_a is the adsorbent temperature, and λ_a is the thermal conductivity of the adsorbent.

2. The energy balance equation for the outside metal tubes is:

$$\delta_m \rho_m C_m \frac{\partial T_m}{\partial t} = \delta_m \lambda_m \frac{\partial^2 T_m}{\partial z^2} + \delta_m \lambda_m \frac{1}{R_m^{\ 2}} \frac{\partial^2 T_m}{\partial \theta^2} + \alpha_{rg}(T_{go} - T_m) + \alpha_{gz}(T_z - T_m) + I\varsigma_m \tau_{go} \quad (9.32)$$

where T_m is the metal tube temperature, T_{go} is the temperature of the outer glass tube, δ_m is the tube thickness of the outer metal tube, τ_{go} is the sunlight transmittance of the glass tube, $\varsigma_{\rm m}$ is the absorption rate of the metal pipe, λ_m is the thermal conductivity of the metal pipe, and R_m is the radius of the metal tube.

3. The energy balance equation for the outside glass tubes is:

$$\delta_{go} \rho_{go} C_{go} \frac{\partial T_{go}}{\partial t} = \delta_{go} \lambda_{go} \frac{\partial^2 T_{go}}{\partial z^2} + \delta_{go} \lambda_{go} \frac{1}{R_{go}^{\ 2}} \frac{\partial^2 T_{go}}{\partial \theta^2} + \alpha_{rg}(T_m - T_{go}) + \alpha_{rs}(T_{sk} - T_{go})$$
$$+ \alpha_{am}(T_{me} - T_{go}) + q_r + I\varsigma_{go} \quad (9.33)$$

where δ_{go} is the thickness of the outer glass tube, ς_{go} is the sunlight absorption rate of outer glass tube, λ_{go} the thermal conductivity of outer glass tube, α_{rg} and α_{rs} are the radiation heat transfer coefficients of the outer glass tube to metal tube and to the sky, respectively, α_{am} is the heat transfer coefficient of the outer glass tube to the air, R_{go} is the radius of the outer glass tube, I is the total radiation intensity per unit area of the surface of the outer glass tube, including the direct sunlight intensity I_0 and the reflected sunlight intensity from back plate I_{ref}, q_r is the sum of the radiation between the two adjacent tubes q_{gg} and the radiation between the back plate and the tube q_{bg}, that is, $q_r = q_{gg} + q_{bg}$, T_{sk} is the sky temperature, and T_{me} is the ambient temperature.

4. **The channels for the cooling water**
Building the two-dimensional model of the cooling water channel the heat transfer in r direction is taken as the heat source. The energy equation is as follows:

Heating process: $$\delta_{mi} \rho_{mi} C_{mi} \frac{\partial T_{mi}}{\partial t} = \delta_{mi} \lambda_{mi} \frac{1}{R_{mi}^{\ 2}} \frac{\partial^2 T_{mi}}{\partial \theta^2} + \delta_{mi} \lambda_{mi} \frac{\partial^2 T_{mi}}{\partial z^2} + \alpha_{mi}(T_i - T_{mi}) \quad (9.34)$$

Cooling process: $$\delta_{mi} \rho_{mi} C_{mi} \frac{\partial T_{mi}}{\partial t} = \delta_{mi} \lambda_{mi} \frac{1}{R_{mi}^{\ 2}} \frac{\partial^2 T_{mi}}{\partial \theta^2} + \delta_{mi} \lambda_{mi} \frac{\partial^2 T_{mi}}{\partial z^2} + \alpha_{mi}(T_i - T_{mi})$$
$$- \alpha_{fi}(T_{mi} - T_f) \quad (9.35)$$

where δ_{mi} is the thickness of the cooling water tube, α_{mi} is the heat transfer coefficient between the adsorbent and cooling water tube, α_{fi} is the heat transfer coefficient of the cooling water and the surface of the tube, T_i, T_{mi}, T_f are the temperature of the adsorbent, tube wall and the cooling water, respectively.

5. **The energy balance equation for back plate**
As the metal back plate is usually thin and has a relatively large thermal conductivity, the heat equilibrium is analyzed with lumped parameter method.

$$M_{mb} C_{mb} \frac{\partial T_{mb}}{\partial t} = W_{rgb} + AI_0 \alpha \varsigma_{mb} + A_{0b} \alpha_{ab}(T_{me} - T_b) \quad (9.36)$$

$$A = \left(D - \frac{2R_{go}}{\sin\theta}\right) \times l \quad (9.37)$$

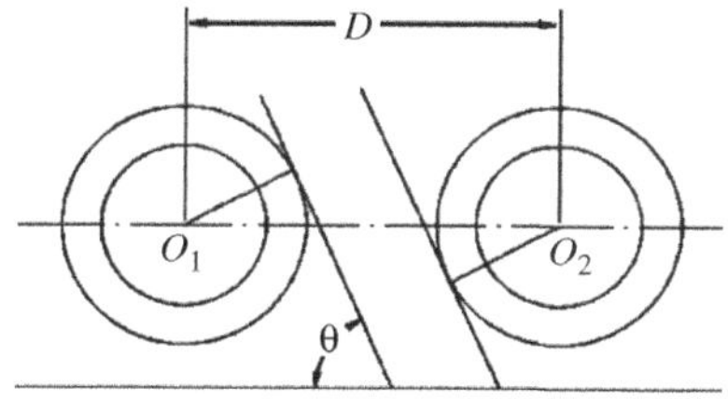

Figure 9.14 Schematic of the geometric relationship of two evacuated tubes

As Figure 9.14 shows, D is the distance between two adjacent evacuated tubes, A_{0b} is the area of two back plates, l is the length of evacuated tube, $A_{0b} = \mathrm{D} \times l$, A is the radiation area between two adjacent tubes, R_{go} is the radius of the outer glass tube, α_{ab} is the natural convection heat transfer coefficient, W_{rgb} is the radiation between the evacuated tubes collector and the back plate, and θ is the solar elevation angle, $\theta > \arcsin\left(\frac{2R_{go}}{D}\right)$.

6. **Initial conditions and boundary conditions**

For the models of the adsorber above, some initial conditions and boundary conditions should be added to solve the equations.

$$t = 0, T = T_{mi} = T_{mo} = T_{glass} \tag{9.38}$$

where T_{mo} is the temperature of the metal tube wall connected with the adsorbent and T_{glass} is the temperature of the glass tube.

Boundary conditions:

$$\text{The outer surface of the adsorbent } \lambda \frac{\partial T}{\partial r}\bigg|_{r=R_m} = \alpha_o(T_{mo} - T_o) \tag{9.39}$$

$$\text{The inner surface of the adsorbent } \lambda \frac{\partial T}{\partial r}\bigg|_{r=R_{mi}} = \alpha_i(T_i - T_{mi}) \tag{9.40}$$

$$\text{The top and bottom faces of the adsorbent layers } \frac{\partial T}{\partial z}\bigg|_{z=0 \text{ or } z=H} = 0 \tag{9.41}$$

where T_o is the temperature of the adsorbent at the tube wall.

Besides, the radiation heat transfer from the inside and outside of the tubes to the reflection tubes and between two adjacent glass tubes should be considered in detail. It is related to the specific construction of the rows. The evacuated tube collector with similar structure is analyzed in detail in reference [9].

9.2.3.3 Influence of the Construction Parameters on the Performance of the Adsorption Collector

According to the heating, cooling, adsorption, and desorption processes of the solar absorption evacuated tube collectors the energy conservation equations are built. The influence of the parameters of the collector on the performance is described as follows:

1. **Analysis of the evacuated tubes and the diameter of the inner channel**

The structure of the evacuated adsorption collector is analyzed. When the solar radiation intensity is fixed, the highest desorption temperature is constrained by the mass of the adsorbent, which means the desorption temperature is related to the diameter of the vacuum

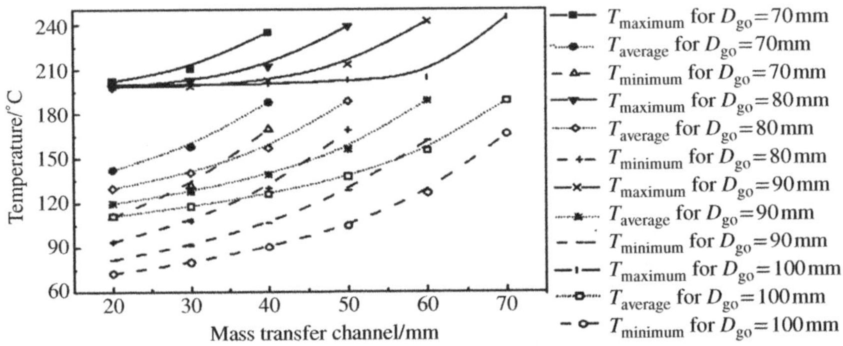

Figure 9.15 Temperatures variations of the adsorption collector with the diameters of the tubes and the inner channel

tube and diameter of the mass transfer channel. By the numerical calculation of adsorption collector with different diameters, the results are obtained as shown in Figure 9.15 (tube center distance $D=2D_g$, total radiation time $t=8$ hours, and the total radiation energy is 17 MJ/m^2).

The curves show that the maximum temperature, average temperature, and the change rate of minimum temperature of adsorption collector are increased with the increasing diameter of the mass transfer channel. When the diameter of the mass transfer channel is equal to the diameter of the evacuated tubes, temperatures of the evacuated tubes are almost constant. That means the collector temperature and the vacuum tube diameter have little relevance when the thickness of the adsorbent is the same (equal to the vacuum tube radius subtracts the radius of the mass transfer channel).

2. **Tube center distance**

When the evacuated tubes are applied as the adsorption collector, due to the limit of the mass of the adsorbent in a single tube, the tubes are usually combined to form the adsorber according to a required area. Meanwhile in order to increase the maximum temperature of the collector and reduce the cost, the distance between the tubes should be considered when the tubes are arranged.

Calculate the relationship of the center distance between two adjacent tubes and the temperature changes of an adsorption collector with a diameter of 70 mm, the results are shown in Figure 9.16. From the figure we can see that when the center distance is two times smaller than the tube diameter (i.e., four times of R_{go}), the average temperature of the adsorption collector is influenced by the tube center distance. When the center distance is equal or larger than twice that of the diameter of the tube, the average temperature reaches its highest and tends to be a constant.

3. **Collector efficiency**

Collector efficiency is usually an important indicator for the performance of the collector. The greater its value is the more radiant energy will be converted to heat. The calculation results of heat utilization efficiency of collectors with four kinds of evacuated tubes with different diameters and different diameters of mass transfer channels are shown in Figure 9.17. Obviously, when the diameter of the channel is constant, the thermal efficiency will increase with the increasing pipe diameter; when the diameter of the tube is constant, the thermal

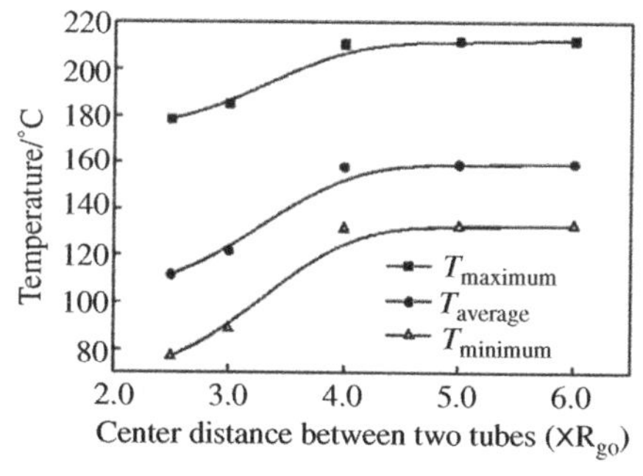

Figure 9.16 The relationship between the temperature of the evacuated tubes and the center distance of the adjacent tubes

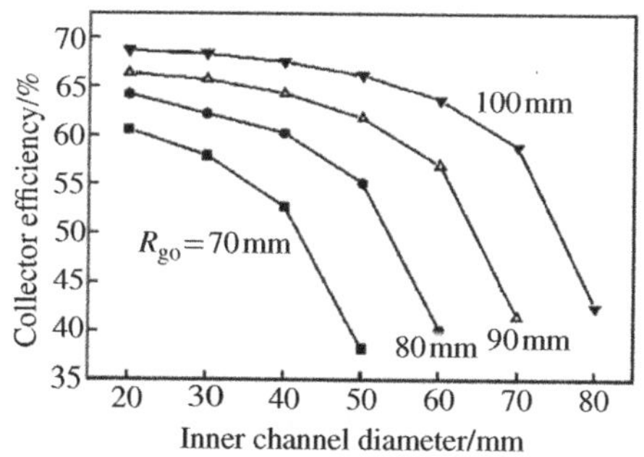

Figure 9.17 The collector efficiency vs. the diameter of the inner channel

efficiency will increase with the increasing channel diameter. When the difference between the diameters of the tubes and the channels is about 40 mm the decreasing rate is significant and the temperature increases rapidly. This is because of the small mass of the adsorbent and adsorbed refrigerant.

4. **The influence on system performance**
   When the solar radiation intensity is constant, the adsorbent temperature is related to the mass of the adsorbent, the smaller the diameter is, the smaller the adsorbent is, and the higher the temperature of adsorbent can be achieved, but the system cooling capacity is not always the highest. To obtain the best performance the optimal diameter should be chosen. Figure 9.18 shows the relationships between the temperature of the collector, COP and cooling capacity of the solar refrigeration system and the diameter of the tube when the diameter of the mass transfer channel is 20 mm and the center distance of adjacent tubes is twice that of the tube diameter. When the tube diameter is larger than 70 mm, the maximum temperature won't change. When the tube diameter is 70 mm the maximum COP and cooling capacity are obtained.
5. **Thickness of the adsorbent**
   Figure 9.19 shows the system COP changes with the diameter of the inner tube channel when the tube diameter is different. If the calculation results of four different tube diameters

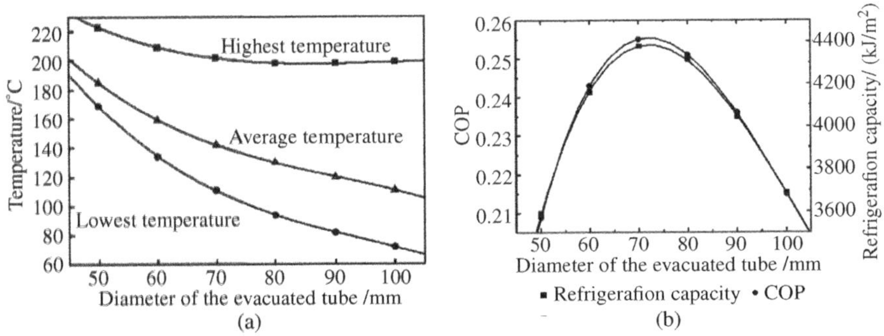

Figure 9.18 Influence of the tube diameter on the performance (tube center distance $D = 2D_g$, diameter of mass transfer channel = 20 mm). (a) Tube diameter vs. temperature and (b) tube diameter vs. COP

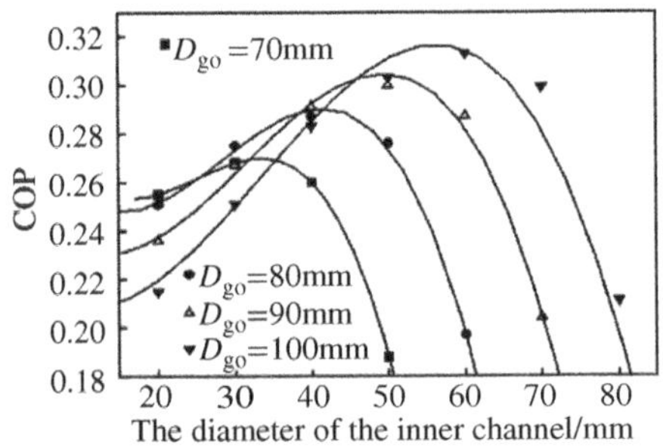

Figure 9.19 COP vs. thickness of adsorbent

are compared, it can be found that the best inner tube diameter is 40 mm smaller than the tube diameter. The thickness of the adsorbent is about 30 mm. COP will increase when the tube diameter increases.

9.3 The Introduction of the Typical Integrated Solar Adsorption System

9.3.1 *The Flat-Plate Solar Adsorption Ice Maker*

Based on research work on the adsorption refrigeration technology, Pons and Guilleminot thought it feasible to apply the adsorption refrigeration systems as the solar powered systems. In 1986 they developed a solar adsorption refrigeration system with activated carbon-methanol as the working pair. The structure of the system is shown in Figure 9.20 [3]. When the solar radiation is 20 MJ/(m^{2}day), the mass of ice produced is about 6 kg/m^2 of collector. Solar cooling coefficient is about 0.12.

The solar adsorption ice maker with activated carbon-methanol designed by Boubakri in France has been commercialized. It combined the adsorption collector and the condenser together. The schematic of the system is shown in Figure 9.21 [15]. The adsorption collector – condenser consists of two stainless steel shells with 90 mm thickness, 1 m^2 collector area and

Figure 9.20 The flat-plate solar adsorption ice maker

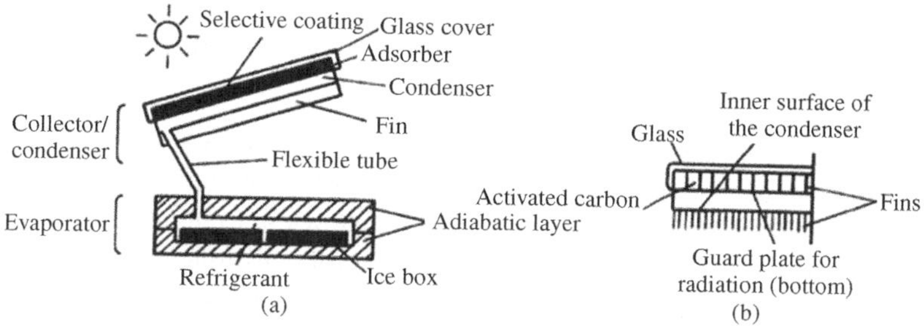

Figure 9.21 (a) Schematic of the ice maker and (b) cross-section of the collector-condenser

an angle of 20°. The upper shell plays a role as the adsorption collector coated with selective coating; the inner fins were used to improve the heat transfer performance. The bottom shell with fins is regarded as an air-cooled condenser. An evaporator (evaporation area is 0.3 m^2) is placed in an adiabatic ice tank (volume 5.2 l). This device in south Morocco (Agadir, 30°23′N, 9°34′WG, the average radiation is about 19.54 MJ/m^2, the average temperature at day and night is 24 and 15 °C, respectively) had run for more than two years. The ice making capacity is more than 4 kg/m^2 in 10% of the days in one year. The solar refrigeration coefficient is about 0.12.

Figure 9.22 shows a solar adsorption ice maker with activated carbon-methanol and throttle at the back. An adsorption collector is a collector with single glass cover. The condenser is cooled by natural air convection. Different to the traditional solar adsorption ice maker, the throttle is set at the back of the adsorption collector. The throttle is opened at night to intensify the heat dispersion of the collector at night. The testing results show that when the solar radiation is 22–25 MJ/(m^2day^2), the refrigeration coefficient of the solar ice maker is about

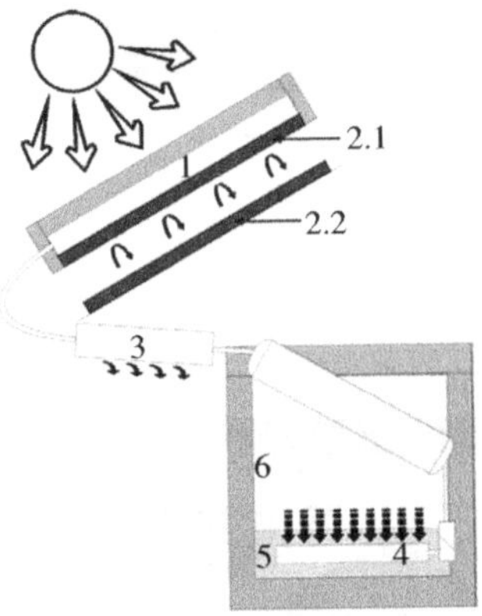

1 - Adsorber/collector; 2 - Throttle (position 1: close, and position 2: open);
3 - Condenser. 4 - Evaporator; 5 - Ice storage device; 6 - The shell of refrigerator

Figure 9.22 The flat-plate solar adsorption ice maker with throttle

0.09–0.1. The ice produced at night is stored in the refrigerator. The refrigerator temperature is maintained at 5 °C. The cooling performance is improved by 35% compared with the ice maker designed by Boubakri.

9.3.2 The Solar Adsorption Refrigeration System with Transparent Honeycomb Cover

Leite [12] proposed a flat-plate solar adsorption refrigeration system with a cover of transparent honeycomb material. Its working pair is activated carbon-methanol, and it is working at vacuum state. The main components of the device are shown in Figure 9.23.

The area of the adsorption collector is 1 m^2. The collector constitutes 13 copper tubes with a diameter of 76 mm and length of 1 m. Twenty kg activated carbon is packed inside the adsorber, but there are mass transfer channels in tubes. There is a transparent honeycomb covering at the surface of the adsorption collector and the bottom surface is insulated. The condenser is a metal tube (diameter of 50 mm, length of 600 mm, and 11 square fins with length of 150 mm)

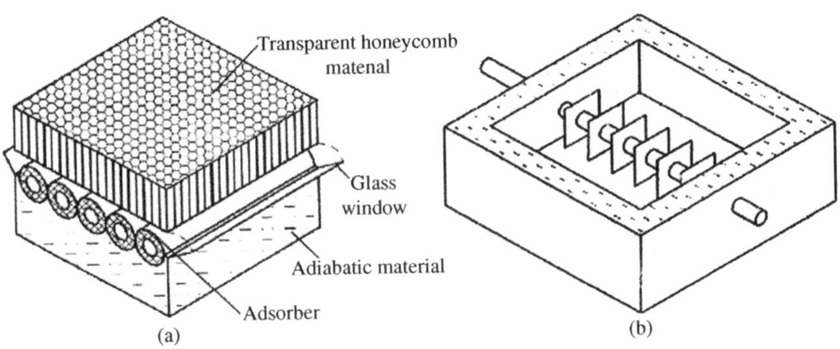

Figure 9.23 Solar adsorption collector with transparent honeycomb cover. (a) Condenser and (b) evaporator

placed in a 80 l thermal isolated tank. The heat transfer coefficient is 20–60 W/(m^2°C). The evaporator is made up of eight connecting horizontal tubes with a diameter of 50 mm and length of 500 mm. Studies have shown that the cooling capacity is about 7–10 kg/(m^{2}day) when the radiation is 20–23 MJ/m^2.

9.3.3 The Activated Carbon–Methanol Solar Adsorption Ice Maker with Reflective Plate

Figure 9.24 shows a novel activated carbon-methanol solar adsorption ice maker [13]. In this solar adsorption ice maker, the adsorber is placed in a glass container. A solar radiation

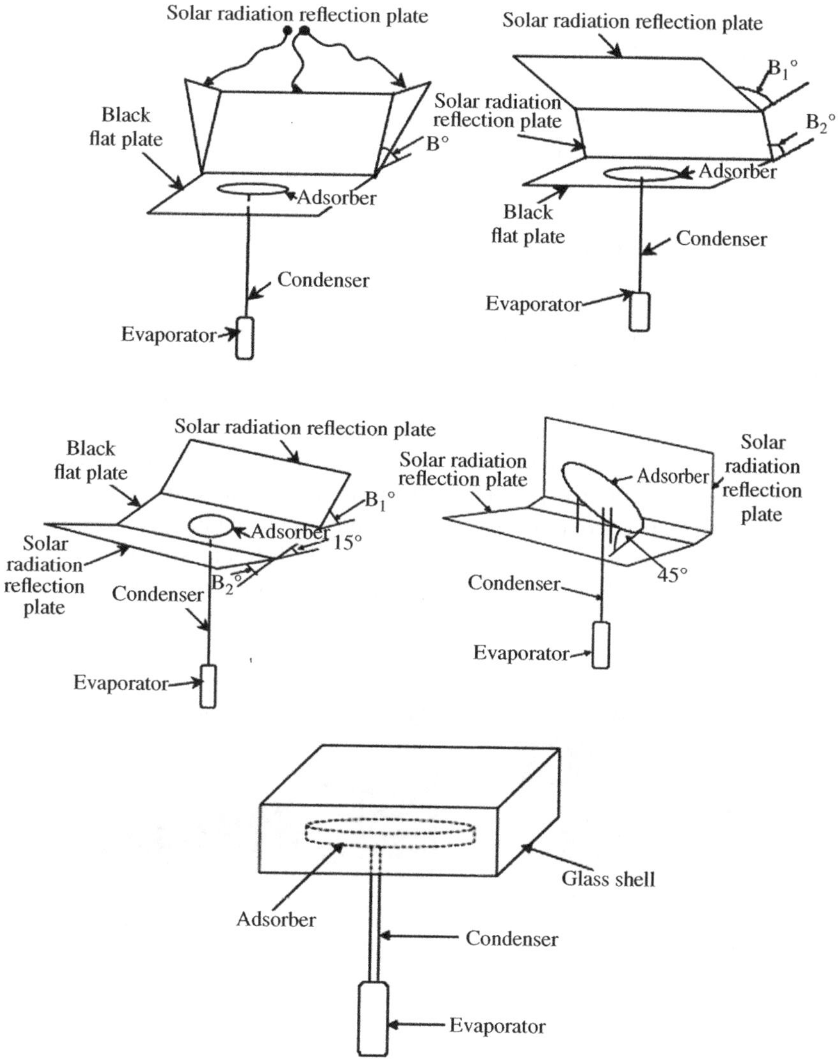

Figure 9.24 Five different positions for the solar adsorption ice maker with reflection plate

reflection plate is used to heat the adsorber. With this structure, on the one hand, the adsorbent can be uniformly heated through the entire outer surface; on the other hand, because the adsorber, condenser, and evaporator are placed in a glass container, it can reduce the possibility of any leakage. In the solar adsorption ice maker, the area of each solar radiation reflection plate is 0.04 m^2, the diameter of the cylindrical adsorber is 0.2 m and the thickness is 0.05 m. Khattab tested the highest desorption temperature for different angles of the reflection plate. In addition, Khattab proposed four other methods to improve the performance of the adsorber. The first is to wrap a black metal mesh at the outer surface of the adsorber, the second is to wrap a black metal plate at the outer surface of the adsorber, the third is to add the metal granular into the activated carbon adsorbents, and the last is to consolidate the granular activated carbon with metal granular particles.

After Khattab consolidated the granular activated carbon with metal granular particles, the ice maker is tested with the third kind of the device as shown in Figure 9.24. The test results show that when the solar radiation is 20 MJ/m^2, the mean temperature is 29 °C, solar refrigeration coefficient is higher than 0.16, the ice making capacity is about 9.4 kg/(m^2 day); when the solar radiation is 17 MJ/m^2, the mean temperature is 20 °C, the solar refrigeration coefficient is higher than 0.14, and the ice making capacity is about 6.9 kg/(m^2 day).

9.3.4 *The Adsorption Refrigeration System with the Working Pair of Activated Carbon–Ammonia*

Critoph in the United Kingdom studied the solar adsorption refrigeration device with activated carbon-ammonia as the working pair. Because the ammonia is used as the refrigerant, the system is operated at high pressure. Compared with the vacuum system, the lack of leakage with this the system is more reliable. In the 1990s, they developed a solar adsorption refrigerator for the storage of vaccine with activated carbon–ammonia, which is shown in Figure 9.25 [16]. By such a system the temperature of the refrigerator during the day can be maintained at 0 °C.

After that, Critoph carried out the theoretical and experimental [17, 18] study on the performance of the adsorption refrigeration device with different surface covers made by single-layer glass, double glazing, and transparency insulation materials. The cylindrical

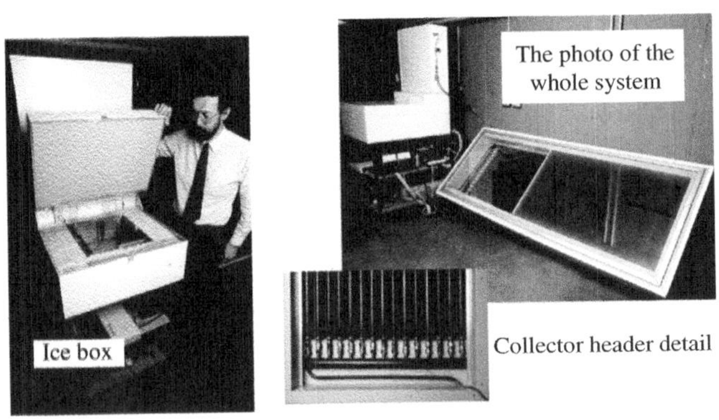

Figure 9.25 Activated carbon–ammonia solar adsorption device for the storage of vaccine

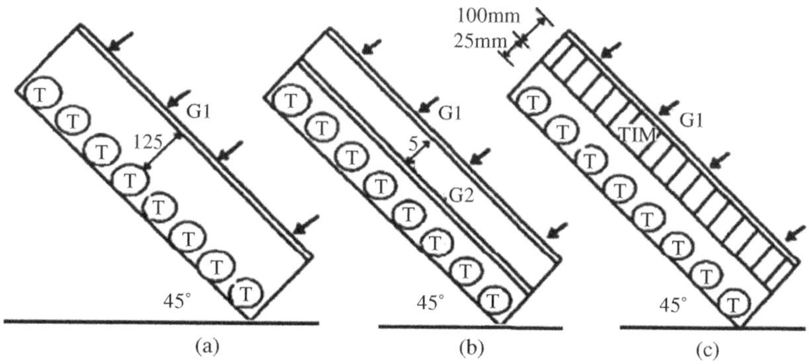

Figure 9.26 Adsorption collector with different structure (G: glass; TIM: Transparency insulation material; T: metal tubes). (a) Single–glass cover; (b) double-glass cover; and (c) single glass + transparency insulation material

Table 9.4 The comparison of cooling capacity and COP

Parameters	Single-glass cover	Double-glass cover	Single-glass + transparent honeycomb
Cycle time (h)	24	24	24
Condensation temperature (°C)	46	32	40
Evaporation temperature (°C)	0.9	0.7	0.9
The amount of the desorbed ammonia (g)	968	1109	723
Solar energy input (MJ)	19.8	19.8	13.8
Cooling capacity (MJ)	1.2	1.4	0.9
COP	0.061	0.071	0.065

adsorption collector has a good performance at high pressure, and it is shown in Figure 9.26. The adsorption collector is made up of a group of steel tubes with a diameter of 44 mm and length of 2 m. There is 17 kg 208C activated carbon filled inside the adsorber.

The experiment results are shown in Table 9.4. Compared with activated carbon–methanol, the COP is relatively low. It is lower than 0.1.

Critoph thought that in view of the cooling performance and economy, single-glass cover is the best choice. Although the transparency insulation material has some advantages, compared with the single-glass cover, the cost is too high, thus it is not yet competitive.

9.3.5 *Strontium Chloride–Ammonia Adsorption Refrigeration System*

The strontium chloride-ammonia working pair has good adsorption refrigeration performance. $SrCl_2$ adsorbs refrigerant when $T \leq 72\,°C$. It stops adsorption when $T > 72\,°C$. According to Li Chen *et al.* [19], there are two adsorption platforms for $SrCl_2$.When the adsorbent bed temperature is below 49 °C, the system begins to adsorb and output refrigeration power. When the bed temperature is below 25 °C optimal adsorption performance can be obtained. In the

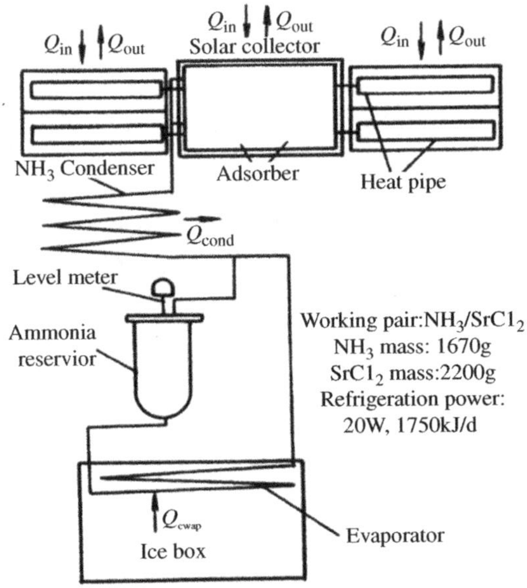

Figure 9.27 Solar adsorption ice maker with strontium chloride–ammonia as working pair

experiments, $SrCl_2$ can adsorb 8 mol ammonia per mole adsorbent at 35 °C. The adsorption capacity is zero when it is at 85 °C. Alfred Erhard and Erich Hahnet [20] used strontium chloride-ammonia in the solar adsorption heat pump, and results showed that under the condition of $T_{des} = 103\,°C$, $T_{ads} = 58\,°C$, $T_{cond} = 40\,°C$, and $T_{evap} = -10\,°C$, COP is in the range of 0.05–0.08 when 2.2 kg $SrCl_2$ and 1.9 kg NH_3 were used in the experiments. When $T_h = 375$ K, $T_{ads} = T_{cond} = 313$ K, $T_{ev} = 276$ K, and $Q_h = 2.53$ kJ the refrigeration quantity obtained from the experiments is $Q_{ev} = 1.0$ kJ.

A. Erhard, K. Spindler, and E. Hahne at Stuttgart University in Germany [21] studied a solar chemical adsorption refrigeration system with $SrCl_2$-NH_3 as the working pair. The device is shown in Figure 9.27. The adsorption bed is composed of two steel pipes. The heat pipe evaporator section taking away the adsorption heat is in the adsorption bed. More than 2000 experiments have been carried out, and Figure 9.28 shows a typical curve of the desorption process. The experimental results showed that the system COP was stable in the range of 0.05–0.08.

9.3.6 Silica Gel–Water Solar Adsorption Ice Maker

Hildbrand in Switzerland developed a solar adsorption ice maker with silica gel–water as the working pair. The collector area is 2 m^2. The system structure is shown in Figure 9.29 [22]. The system shown in Figure 9.29 had been tested for 68 days and the results showed that the environmental parameters such as solar radiation and outdoor temperature had a great influence on the system performance. When the solar radiation is higher than 20 MJ/(m^2 days), the outdoor temperature is in the range of 12–25 °C, solar COP is in the range of 0.12–0.23. When the outdoor temperature is below 20 °C solar COP is higher than 0.15. When the food

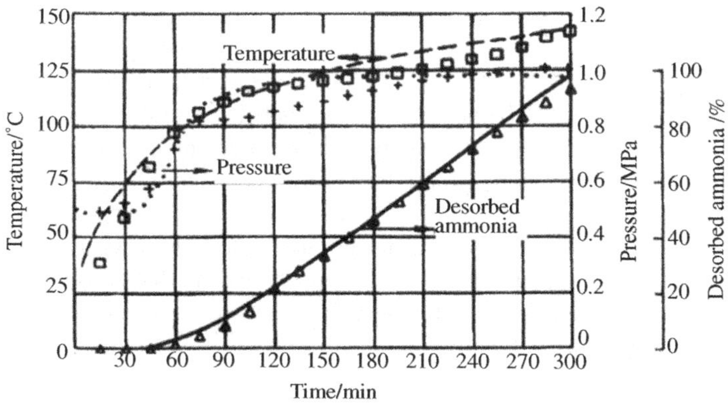

Figure 9.28 Experimental curves for the desorption process

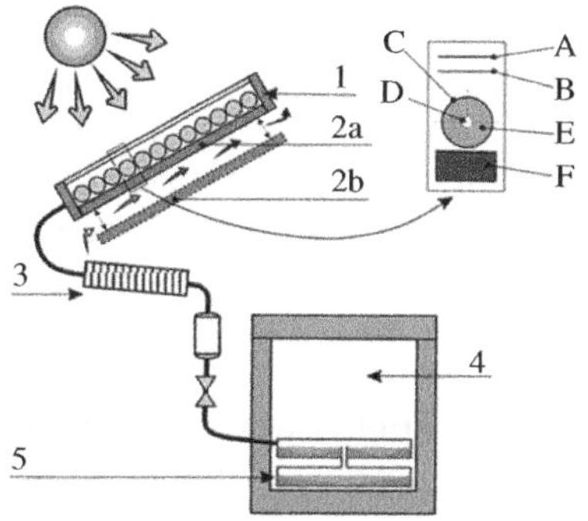

1 - Ads1 Adsorption collector (A - glass cover, B - special Teflon film, C - metal pipe coated with a selective coating, D - refrigerant pipe, E - silicone, F - insulation material); 2 - Throttle (a: turned off, b: turned on); 3 - Condenser; 4 - Storage tank; 5 - Evaporator and ice storage device

Figure 9.29 Silica gel-water solar adsorption ice maker

was in the refrigerator the results of the test in 30 days showed that the solar COP is 0.16. The ice making capacity is 4.7 kg/(m^2 days).

9.4 Design and Examples of Separated Solar Adsorption Refrigeration Systems

The characteristic of the separated solar adsorption refrigeration system is that the solar collector and the adsorption refrigeration system are independent. The solar collector recovered the energy to heat storage devices; the cycle time of the adsorption system is mainly determined by the working pair, the heat transfer performance, and the mass transfer performance of the bed.

For separated solar air conditioning systems, hot water for single stage absorption chillers is usually in the range of 88–90 °C, which generally is not very easily obtained from the common types of solar collectors. The effective time for radiation is only several hours in a day. Only when the solar radiation is very strong can the temperature get to a temperature higher than 85 °C. Meanwhile the thermal efficiency of the solar collectors will be reduced. Besides, the system investment for the solar system is high, and the capacity of the lithium bromide absorption chiller (minimum capacity of lithium bromide absorption chiller cooling capacity is usually larger then 70 kW) is too large for the household, thus it is difficult for the lithium bromide absorption air conditioning system to be applied for solar energy. In short, the high cost and the limitation in general usage affect the application of solar absorption air-conditioning systems. Solar adsorption air-conditioning can effectively solve the problems. The adsorption chillers driven by a low temperature heat source developed by Shanghai Jiao Tong University are a typical application in this area.

9.4.1 Design and Application Example of the Solar Air Conditioner for Green Building

The connotation of green building built in Shanghai City features energy saving and environmental protection. Renewable natural energy sources such as solar, wind and geothermal are used as much as possible in green building instead of coal, oil, and other fossil fuels. The basic idea of building integrated with solar energy is to use the most convenient, non-polluting energy for power generation, heating, and refrigeration, as well as to supply the hot water to meet living needs.

9.4.1.1 Solar Adsorption Air Conditioner in Green Building

The design for the solar adsorption air conditioner in green building is shown in Figure 9.30. Two adsorption chillers driven by low temperature heat source are used in the system. The design, installation, and the experiments of the chiller have been described in detail in Chapter 8.

The solar energy utilization system shown in Figure 9.30 consists of a solar collection system, a hot water supply system, a floor heating circuit, an air conditioning circuit, a natural ventilation strengthening system, a heat utilization system, and a control system. These systems are connected by tubes and heat storage tank, except for the control system which controls the solar collector system by the temperature difference measured by the temperature sensors.

The solar collection system in Figure 9.30 includes the solar collector array, the circulation pump of the collector and heat storage tank. The water at the bottom of the tank is pumped into the solar collector array by the circulation pump, and then flows back into the upper part of the heat storage tank. In order to increase the thermal efficiency, the stratified tank is used as the heat storage tank.

The heat utilization system includes the high efficient adsorption refrigerator, air-cooled cooling towers, floor heating coil, natural ventilation strengthened system by the fin-tube heat exchanger, and hot water heat exchanger. Efficient adsorption refrigerator, floor heating coil, natural ventilation strengthened system by the fin-tube heat exchanger, and domestic hot water heat exchanger connected in parallel by pipes. With the control of valves on the pipes, a domestic hot water heat exchanger runs throughout the year. While the other three terminal devices are operated according to the different requirements for different seasons.

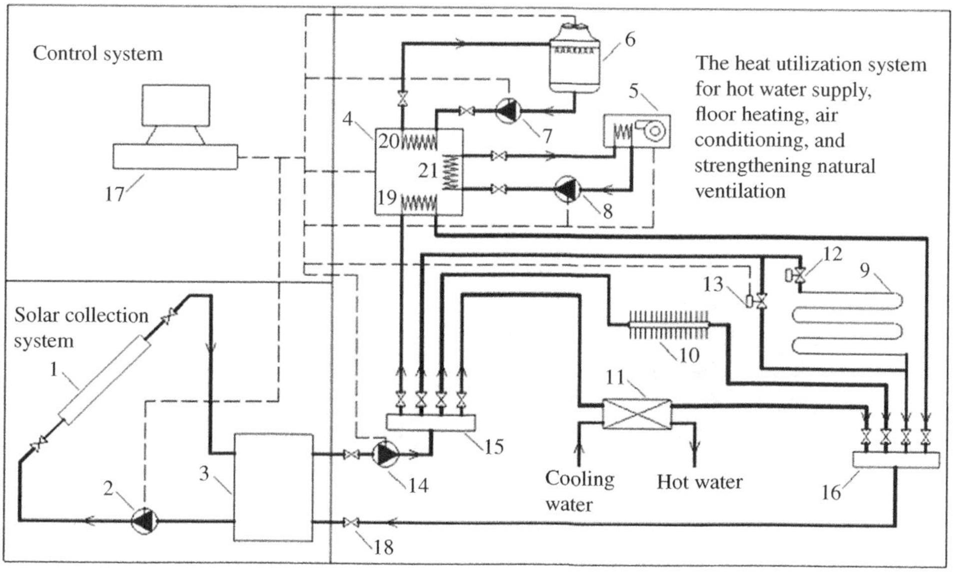

1 - The array of solar collectors, 2 - Circulation pump for collector, 3 - Heat storage water tank, 4 - Adsorption chiller,
5 - Fan coil, 6 - Cooling tower, 7 - Cooling water circulation pump, 8 - Chilling water circulation pump,
9 - Floor heating coil, 10 - Finned tube heat exchanger, 11 - Domestic hot water heat exchanger,
12 - Temperature control valve for floor heating, 13 - Bypass valve for floor heating, 14 - Hot water circulation pump,
15 - Bypass water tanks, 16 - Water collection tank, 17 - Industrial Personal Computer IPC, 18 - Flow control valve,
19 - Adsorbent bed, 20 - Condenser of adsorption chiller, 21 - Evaporator of adsorption chiller

Figure 9.30 Solar powered combined energy utilization system

Figure 9.30 shows the working process of the solar adsorption chiller. Firstly, the solar collector array heats the water at the bottom of the heat storage tank, then the water flows into the upper part of the tank, goes into the adsorption chiller, and heats the desorption bed. Meanwhile the other bed is cooled by the cooling tower, and it adsorbs the refrigerant vapor in the evaporator. The cooling capacity is produced in the evaporation process. The chilling water is supplied to the refrigerator for different application areas by the water pipes. In the working process of the refrigerator the water flow rate of the chiller is controlled by the indoor air temperature. When the temperature of the water in the upper part of the tank is lower than the set temperature of the system, the chiller, hot water circulation pump, cooling water circulation pump, chilling water circulation pump, and the terminal devices of air conditioning are all turned off.

9.4.1.2 The Installation of the Solar Adsorption Air Conditioners in Green Building

The Green Building, which is built by Institute of Building Science Research, Shanghai City, is shown in Figure 9.31. The building area is about 1984 m^2, the floor space is about 904 m^2, and the solar collector area is about 170 m^2.

The solar system has a solar collector of 170 m^2, in which the CPC (compound parabolic collector) produced by Linuo Paradigma Company was installed on the east side of the roof with an area of 90 m^2, while the CPC produced by the Wuxi High Tech Company was installed

Figure 9.31 The photo of the green building

on the east side of the roof with an area of 80 m^2. In architectural design, the integration of solar collectors and the construction of a sloping roof were fully considered to achieve the highest efficiency of solar energy. The inclination of the roof is set according to the local latitude. Shanghai is located at latitude 31°10′. Taking efficient utilization and the building elevation into consideration, the sloping roof inclination is designed as 40°. Figures 9.32 and 9.33 present a heat pipe solar collector array at the east side. There are 27 groups of heat pipe solar collectors, and each group contains 20 tubes at the east side of the roof. They are installed as three rows arranged in parallel. The collector area of each group is about 3 m^2, and it is divided into five branches arranged in parallel.

As the Green Building was built as the office for the Institute of Building Science Research, Shanghai City, and the demand for hot water is less, consequently the system is designed mainly for the needs of the heat pump and air conditioning. Besides, a fin tube heat exchanger is set as the natural ventilation duct. During the transitional seasons, water heated by solar energy heated the air in the duct, thus the natural ventilation is intensified. The energy utilization system combines solar water heating, air conditioning, floor heating, and natural ventilation systems. It contains the components like evacuated tube collector, adsorption chillers, cooling towers, floor heating coil, storage tank, fin-tube heat exchanger, and circulating pump. As the heat source driving the chiller and floor heating systems, the solar collector is the main part of this system. All the devices except the solar collector were put in the refrigerating station on the top floor shown in Figure 9.34. The area of the exhibition hall where the solar energy system is used is about 265 m^2, the sensible cold load for the air conditioning is about 15 kW, which is provided by the solar adsorption air conditioners.

The control system combines the data collecting and automatic controlling sub-systems with Industrial Personal Computer (IPC). The hardware includes: ADLINK NuPro-760 IPC, PCL-731 48-channel switching I/O card, PCLD-785B 24-channel I/O board, ADAM data acquisition card, Pt1000 temperature sensor, TBQ-2 standard radiometer, LXSGYR-15E~32E rotary type remote water meter, WSCB-24/16 multifunction data acquisition instrument. An

(a)

(b)

Figure 9.32 The photos of the collector arrays at east side. (a) The arrays on the east roof and (b) the steel support structure

IPC is used to monitor the system, collect the data, and control the system under different working modes. All the signals are sent to the control center. The entire system can be automatically operated. Automatic control procedure is composed of data acquisition, data storage, data analysis, automatic control, the running configuration, and troubleshooting modules. Five operating conditions should be considered for automatic control of the system. The first condition is to automatically start and stop the solar collection system and the anti-freezing control. Secondly, is to prevent the overheating phenomena for floor heating system. Thirdly, is to automatically start and stop the control for summer air-conditioner. Fourthly, is to control the numbers of chillers to be switched on. Lastly, is to automatically start and stop control of the natural ventilation strengthening system in the transition season.

(a)

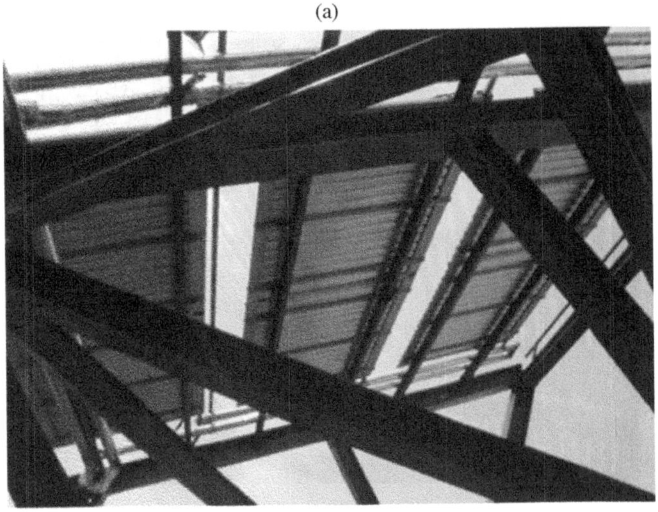

(b)

Figure 9.33 The photos of the collector arrays at the west side. (a) The arrays on the west side and (b) the steel support structure

9.4.1.3 Simulation on the Solar Powered Adsorption Chillers in Green Building

The models for the solar powered adsorption air conditioner is built by the energy balance equations of the collector and the refrigerant, as well as the phase equilibrium equation of adsorption and desorption. Also the whole modeling in software is built for the solar energy system and the operation characteristics are predicted for the full year. Use this software to simulate the performance for one day, and the comparison of the cooling capacity of the experiments and the simulation results are shown in Figure 9.35. $Q_{t,s}$ is the average cooling capacity of two chillers. The experimental result is about 12 kW. The simulation is in the

Figure 9.34 The photo for the solar powered adsorption chillers

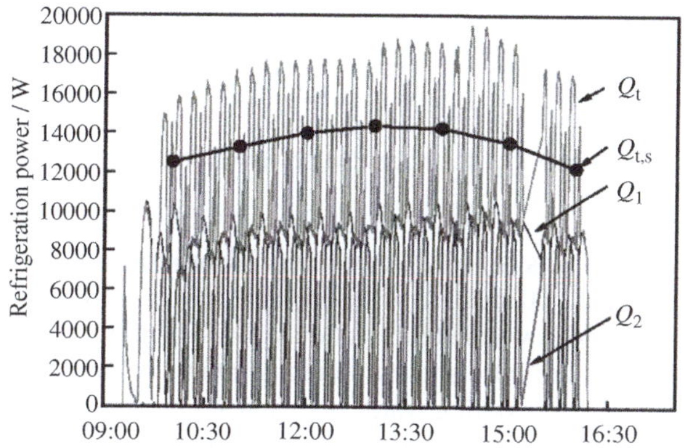

Figure 9.35 Comparison of the calculated cooling capacity and the experimental cooling capacity

range of 12.2–14.2 kW. It is basically consistent with the experimental values. Further comparison shows that the simulation results are basically consistent with the experimental results for different climates.

9.4.1.4 Experiments for the Solar Adsorption Chiller

Figures 9.36 and 9.37 show the experiments on a solar air-conditioner in one day. The temperature of water in the heat storage tank is heated to 65 °C by the solar collection system. The chiller is operated on for 8 hours (9:00–17:00) in one day. The difference of start time for two chillers is half a cycle. The daily solar radiation intensity is 19.2 MJ/m^2, and the average

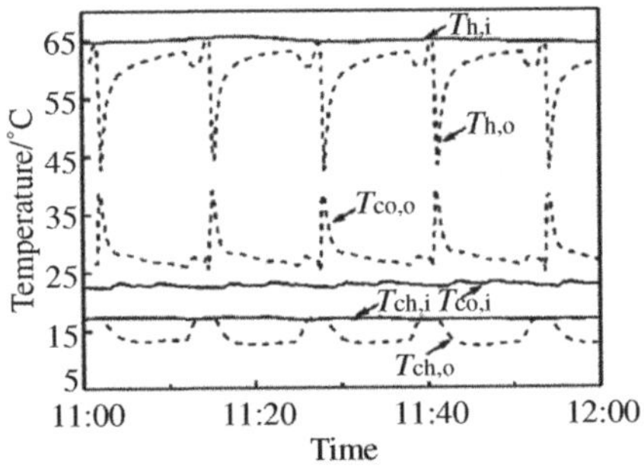

Figure 9.36 Temperature variation of the hot water, cooling water, and chilling water

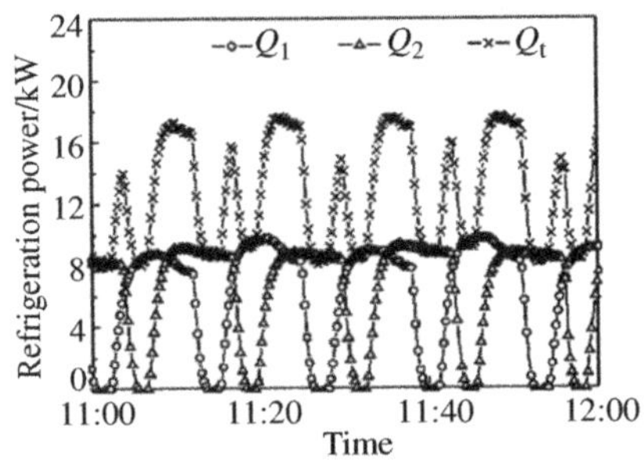

Figure 9.37 Refrigeration power vs. running time

efficiency of the solar collectors is 36%. Because the adsorption chillers run periodically, the performance from 11:00 to 12:00 are used to describe the refrigeration performance of the solar cooling system. Figure 9.36 shows the temperature variations of hot water, cooling water, and chilling water. $T_{h,i}$ and $T_{h,o}$ are inlet and outlet temperatures of hot water. $T_{co,i}$ and $T_{co,o}$ are inlet and outlet temperatures of cooling water, $T_{ch,i}$ and $T_{ch,o}$ are inlet and outlet temperatures of chilling water. The inlet hot water temperature is relatively constant, and the fluctuation is only about 1.4 °C/h. This is mainly due to the regulation function of the heat storage tank. The temperature difference between inlet chilling water and outlet chilling water is about 5 °C.

Figure 9.37 shows the trends of the refrigeration power, where Q_1 and Q_2 are the refrigeration power of two refrigerators. Q_t represents the total refrigeration power. The average refrigeration power in working hours is 12 kW, the average thermal COP of the adsorption refrigeration unit is 0.28, and the average electric COP is 4.0. During the operation process the average hot water temperature is 62.6 °C. Test results of the performance show that the key factor influencing the refrigeration power is the heat source temperature. Therefore, with the rise of the solar radiation intensity, the average hot water temperature will rise; thereby the refrigeration power of the chillers will be increased as shown in Figure 9.38.

The experiments of the solar air conditioner were carried out from June to August in 2005. The results show that the adsorption air conditioner could be driven by the hot water with a temperature higher than 55 °C from the solar collector. It ensured a long working period in summer. For example, on a sunny day it can work as long as 8 hours, the average refrigeration

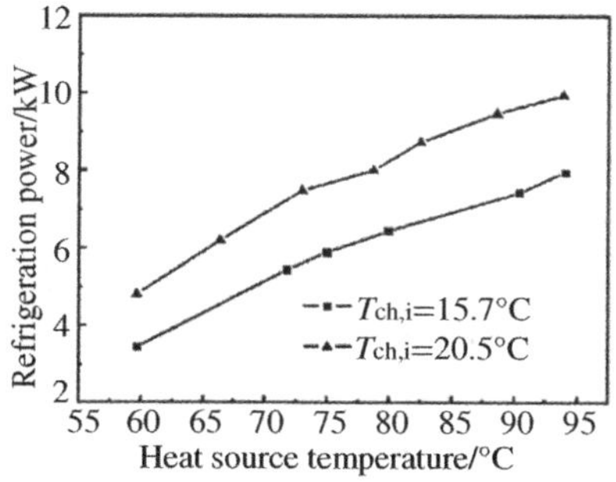

Figure 9.38 The refrigeration power of solar adsorption chiller vs. heat source temperature

Table 9.5 The experimental results of the solar adsorption chiller

Time	Average environment temperature	The total solar radiation	Average thermo-electric efficiency	Average refrigeration power	Average refrigeration COP	Average solar COP	Average hot water temperature	Average cooling water temperature	Average chilling water temperature
Month (mm/yy)	T_a (°C)	I (MJ/m^2)	η_{pj} (%)	Q_{chill} (kW)	COP_{chill}	$COP_{chill,solar}$	$T_{hw,in}$ (°C)	$T_{co,in}$ (°C)	$T_{chill,in}$ (°C)
06/05	29.86	17.71	35.64	10.29	0.29	0.11	62.42	24.31	20.12
07/05	33.77	18.96	37.41	11.23	0.36	0.12	65.44	26.68	22.33
08/05	31.89	17.24	38.02	10.77	0.32	0.11	63.80	26.75	21.52

power is 15 kW. The electrical COP of a solar air conditioner can reach 10 or more. The typical experimental results obtained from June to August are shown in Table 9.5. An average refrigeration power is about 10.76 kW, The thermal COP_{chill} is 0.32.The solar thermal $COP_{chil,solar}$ is 0.11. The dependency rate on solar energy is about 71.73%.

Compared with an energy system driven by the electricity, when the required heating power in winter, required refrigeration power in summer and required domestic hot water are the same the solar adsorption chiller can reduce the CO_2 emission by 57.02%, and can reduce the emission of other environment hazards of gaseous pollutants and solid pollutants by more than 70%, as well as reducing the SO_2 by 80%. Meanwhile the operating cost can be saved by 54% if compared with the energy system powered by the electricity.

9.4.2 Design and Application Example of the Solar Adsorption Chiller in Grain Storage System

9.4.2.1 Design and Construction of the System

The grain storage system with the refrigeration technology is attractive since the storage period is long and the grain can be kept well under the low temperature, and such a technology can also prevent the growing of insects and mold. At present the ventilation (mechanical ventilation

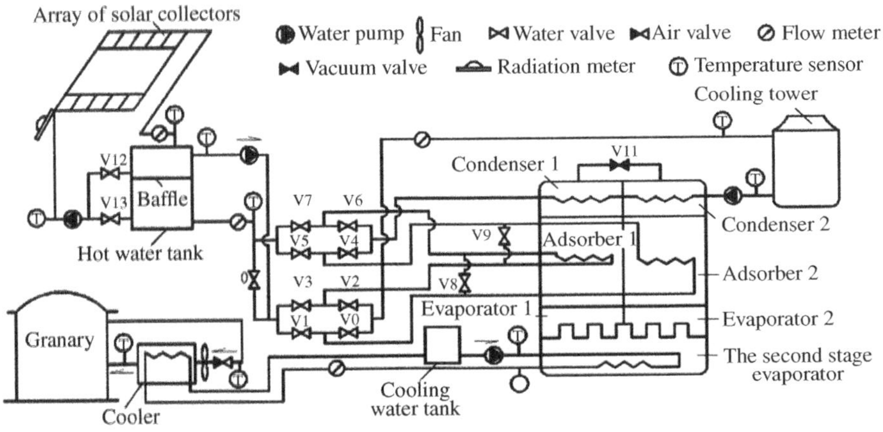

Figure 9.39 Diagram of the solar-powered adsorption refrigeration system for grain storage

and natural ventilation) and compression chillers are common used devices for the system. Ventilation is influenced by the environment and the compression chillers need to be driven by electricity, consequently have a high operation cost. Solar energy is clean and abundant in the environment. When it is used as the grain storage system, the biggest advantage is that when the weather gets hotter, the solar adsorption refrigeration system will have a greater refrigeration power, and the operating cost will be low.

The grain storage system with adsorption chiller includes the hot water system that collects the heat from the evacuated tube solar collector, the adsorption chiller, cooling tower, and fan coil. Figure 9.39 shows the schematic of the system, and the photograph of the system is shown in Figure 9.40.

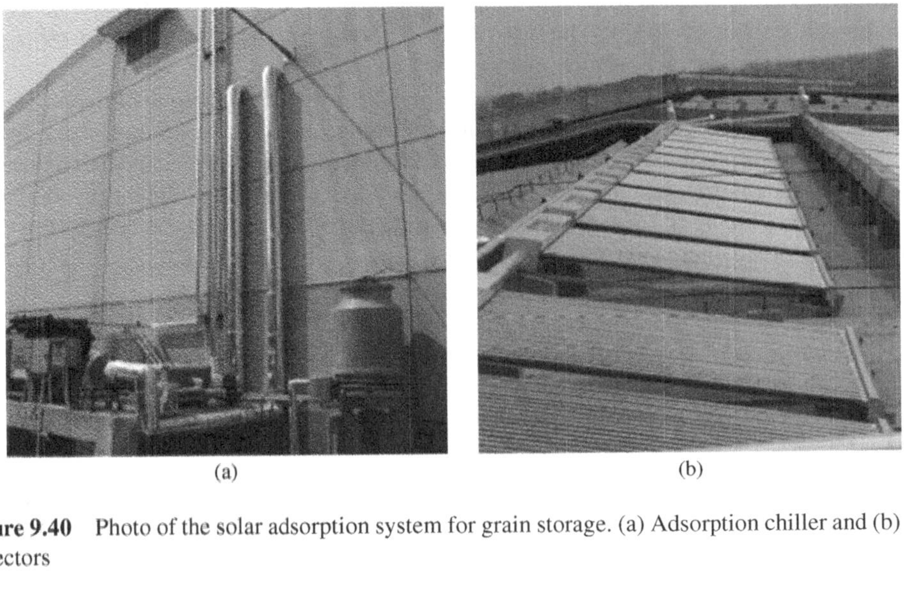

(a) (b)

Figure 9.40 Photo of the solar adsorption system for grain storage. (a) Adsorption chiller and (b) solar collectors

U-shaped all-glass evacuated tube collectors are used in the grain storage system, and the collector is 49.4 m^2. A circulating pump is controlled by the temperature difference. The total volume of stratified heat storage tank is 0.6 m^3. In the tank the volume of the part above the baffle is 0.24 m^3. In the morning before the refrigeration unit runs the valve 13 closes, valve 12 opens, and the water in the upper part of the hot water tank is heated rapidly. When the temperature of the water in the upper part of the hot water tank is above 65 °C, the refrigeration unit starts running. Thereafter the valve 13 opens, the water in the lower part of the hot water tank is gradually heated. During the afternoon, when the temperature of the water in the upper part of the hot water tank is below 65 °C, the refrigeration unit stops running.

The adsorption system contains two adsorption units and a secondary evaporator. The electric ball valve (V0–V10) and vacuum valve (V11) are controlled by the Programmable Logic Controller (PLC). The refrigeration unit can be automatically run with a heat and mass recovery cycle. Each adsorption unit contains an adsorber, a condenser, and an evaporator. Each adsorber is filled with about 50 kg of silica gel. Below the two evaporators of the adsorption unit a secondary evaporator is set up. The chilling water flows through the secondary evaporator, the refrigerant under the secondary evaporator is heated and evaporated, and it exchanges condensation heat with the evaporator of the adsorption unit. The refrigeration power is transmitted one-way, and the energy lost is reduced, which improves the system's performance.

The fan coil uses the fin type heat exchanger. The heat transfer area is 58.7 m^2, the rated power of fan is 750 W, the rated air volume is 1100 m^3/h, and the rated flow rate of the cooling tower is 8 m^3/h.

9.4.2.2 The Experiments on the Solar Powered Adsorption Chiller for Grain Storage

Table 9.6 shows hourly solar radiation amount, cooling power Q_{ref}, and cycle COP_{cycle} of a chiller. Its operating parameters during the experiments are shown in Table 9.7. Table 9.6 shows that when the system starts to work in the morning due to the low temperature of the water in heat storage tank, the cooling power and COP are low. Then, with the intensifying process of solar radiation, the hot water temperature is gradually increased, and consequently the cooling power and COP increases accordingly. For the solar powered adsorption system the suitable hot water temperature is in the range of 70–85 °C. In addition, the refrigeration capacity matches the requirement of the grain storage in its cooling capacity. For example, at about 13:00, solar radiation is strong and the temperature is high, the cooling demand is large. While at that time, the refrigeration capacity of the refrigeration system is also relatively large.

Table 9.6 Performance of the solar adsorption system for grain storage

Time	Solar radiation (MJ/m^2)	Q_{ref} (kW)	COP_{cycle}
9:50–10:50	2.667	2.96	0.247
10:50–11:50	3.005	4.13	0.279
11:50–12:50	3.103	4.77	0.307
12:50–13:50	2.987	4.82	0.313
13:50–14:50	2.491	4.69	0.305
14:50–15:50	2.094	4.53	0.294
15:50–16:50	1.338	4.15	0.283
16:50–17:50	0.609	3.21	0.258

Table 9.7 Operating parameters of the solar adsorption system for grain storage

Half cycle time (s)	Mass recovery time (s)	Heat recovery time (s)	Flow rate of hot water (m^3/h)	Flow rate of cooling water (m^3/h)	Flow rate of chilling water (m^3/h)	Temperature of inlet hot water (°C)	Temperature of inlet air (°C)
900	60	60	3.6	5.1	1.8	≥ 65	14–22

Table 9.8 Performance of the solar adsorption system for grain storage

Date	Solar radiation (MJ/m^2)	Running time (min)	Coefficient of the collector	COP_{solar}	$COP_{electrical}$	Q_{ref} (kW)
31/07/2004	19.6	474	0.45	0.123	2.62	4.19
06/08/2004	20.3	508	0.46	0.125	2.59	4.14
09/08/2004	17.4	423	0.42	0.096	2.03	3.25
15/08/2004	19.5	474	0.45	0.131	2.77	4.43
26/08/2004	18.7	457	0.44	0.124	2.63	4.21
19/09/2004	16.2	382	0.40	0.109	2.61	4.17

Performance experiments are carried out, and the operating parameters are listed in Table 9.7. The typical experimental results are shown in Table 9.8. It shows that when the solar radiation is 16–21 MJ/(m^2·day), the solar adsorption refrigeration unit can run for 6.5–8.5 hours daily. The daily average refrigeration power is about 3.3–4.4 kW, solar COP_{solar} is approximately 0.096–0.131, and electric $COP_{electical}$ is approximately 2.1–2.8. In addition, when the area of the collector and the volume of the hot water tank are increased, the running time of the refrigeration unit can be extended accordingly.

9.4.3 Examples for the Application of Separated Solar Powered Adsorption Refrigeration Systems

9.4.3.1 Silica Gel–Water Adsorption Chillers

Some developed countries in Europe have extensively researched on solar adsorption air-conditioning systems. France, Germany, Austria, Greece, Italy, Portugal, and Spain initiated "Climasol-Plan," and launched some demonstration projects of solar air-conditioners. The aim of the "Climasol Plan" is to promote a comprehensive approach to reduce the energy consumption of buildings and the development of passive cooling techniques. In 1999, in a Hospital of Freiburg University (Freiburg, Germany), solar powered silica gel-water adsorption air-conditioning system with a cooling power of 70 kW was installed, which is shown in Figure 9.41. The solar adsorption air conditioning system has a vacuum tube collector area of 230 m^2 to produce hot water. In summer hot water is used to drive adsorption chillers, and in winter the hot water is used as a heat pump. In summer the efficiency of the evacuated tube collector is about 32%, and the COP of the system is about 0.6. The total investment of the system is about 353 000 Euros, and the annual running cost is about 12 000 Euros.

In addition, there was one solar adsorption air-conditioning system installed in a cosmetics company in Sarantis S.A, Greece with a room area of 22 000 m^2 (130 000 m^3), as Figure 9.42

(a) (b)

Figure 9.41 The solar powered silica gel-water adsorption chiller installed in Freiburg in German. (a) Evacuated tube type collectors and (b) adsorption chiller

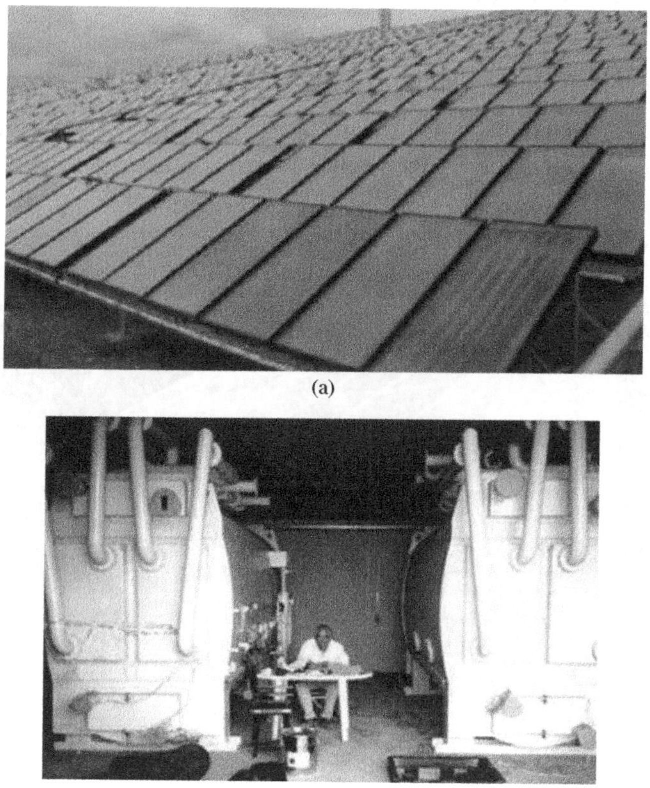

(a)

(b)

Figure 9.42 The silica gel–water adsorption system installed in Sarantis S.A, Greece. (a) Flat-plate solar collectors and (b) adsorption chiller shows. The solar adsorption air conditioning system with 2 700 m^2 of flat plate collectors can generate hot water of 70–75 °C. In summer the hot water is used to drive the silica gel-water adsorption chiller with the cooling power of 350 kW, and COP is about 0.6. In winter hot water is used for heating a room. The solar adsorption air-conditioning system is equipped with an oil-fired boiler as a supplementary heat source. The heat supplied by the solar energy is about 66% of the total heat throughout the year. The cost of a solar adsorption air conditioning system is about $1.3 million Euros. It can reduce annual CO_2 emissions by about 5100 ton.

9.4.3.2 Solar Adsorption Cold Storage System with Zeolite–Water Working Pair

Figure 9.43 shows a micro solar adsorption cold storage [23] house. The adsorption system uses zeolite-water as adsorption working pair, the area of the solar collector is about 20 m^2, and the space inside the house is 12 m^3. When the solar radiation is about 22 MJ/(m^2 day), the cold storage house can store 1000 kg vegetables. The solar refrigeration COP is 0.10. Such a system can be applied for the agricultural preservation in places with rich solar radiation resources.

9.5 Solar Powered Adsorption Refrigeration by Parabolic Trough Collector

9.5.1 The Research Work Done by Fadar

As a popular line-focus concentrator, a parabolic trough collector (PTC) focuses direct normal irradiation (DNI) onto a focal line on the collector axis. PTC has been used for many solar

Figure 9.43 Solar adsorption cold storage house with zeolite–water adsorption working pair [23]

(a) (b)

Figure 9.44 Photo views of LS-3 PTC collector for power plants [24]. (a) Front side and (b) back side

power plants and some of these commercial collectors have been tested under real operating conditions while the typical temperature ranges from 300 to 400 °C and the typical aperture width is about 6 m as shown in Figure 9.44 [24, 25]. However, the typical aperture width ranges from 1 to 3 m if the heat source is between 100 and 250 °C.

The solar collector is one of the most important elements of all solar powered adsorption refrigeration systems [26]. PTC seems to be a reasonable and promising alternative choice since they are the most developed type of solar collector [27]. However, many research works focused on the flat plate collector, evacuated tube collector, or compound parabolic concentrator, whereas little attention has been devoted to PTC, especially the experimental study of PTC.

Fadar *et al.* [28–30] had finished the study on a novel system in an attempt to overcome the intermittent characteristic of solar adsorption refrigeration systems, and test the applicability of PTC for these systems with the aim of improving their performance. In their work, a numerical investigation is performed to describe a two-bed continuous adsorption refrigeration cycle. As shown in Figure 9.45, the receiver is placed along the focal line of the concentrator. For producing cold continuously, the adsorbers have to be operated out-of-phase. A heat storage tank stored the solar energy as well as providing the heat source for the adsorption refrigeration system. In this work, the radial thickness of adsorbent bed is an important parameter for system optimization and Figure 9.46 depicted its influence on SCP (Specific Cooling Power) and COP. While in another paper, Fadar *et al.* gave the variation of COPs (solar COP) with both the aperture width of collector and external radius of adsorbent as shown in Figure 9.47. In their research, the conclusions had been made that in the ranges of numerical investigation, it was shown that the system has optimum COP of 0.14 when the adsorber external radius and aperture width of collector are of the order of 14.5 and 70 cm, respectively. The results showed a promising performance in comparison with other published data, which were obtained with other types of collectors.

9.5.2 *Introduction on the System Constructed by Shanghai Jiao Tong University*

In Shanghai Jiao Tong University, Li *et al.* [31] designed and built an adsorption ice making system driven by PTC as shown in Figure 9.48. The system mainly includes four parts,

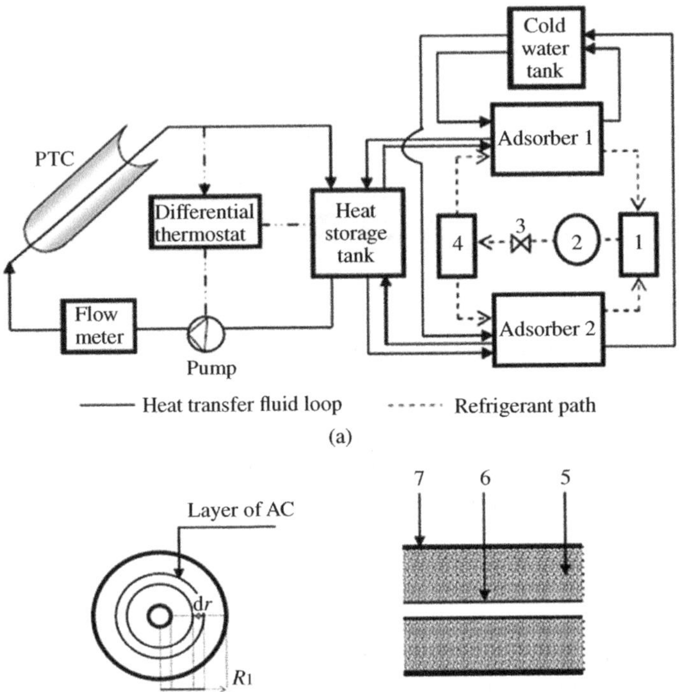

Figure 9.45 (a) Schematic diagram of the solar powered continuous adsorption refrigeration system and (b) longitudinal and cross sections of one adsorber [28]

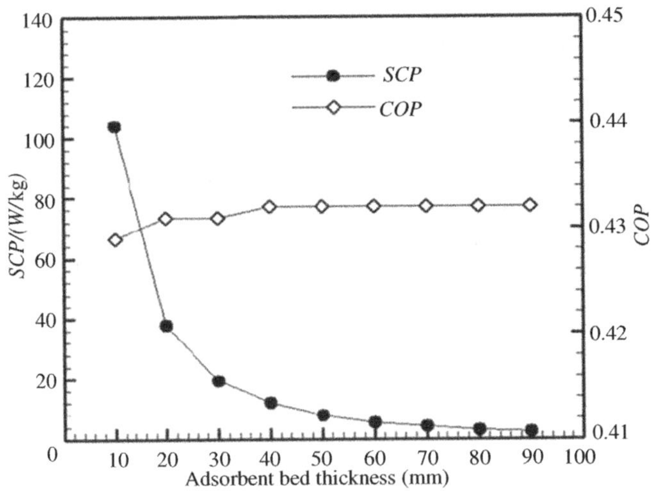

Figure 9.46 Influence of the adsorbent bed thickness on *SCP* and on *COP* ($T_{heat} = 100\,°C$) [28]

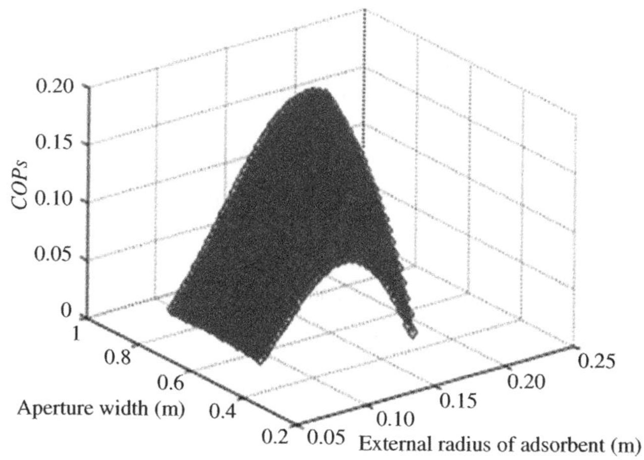

Figure 9.47 COPs variation vs. aperture width of collector and external radius of adsorbent [29]

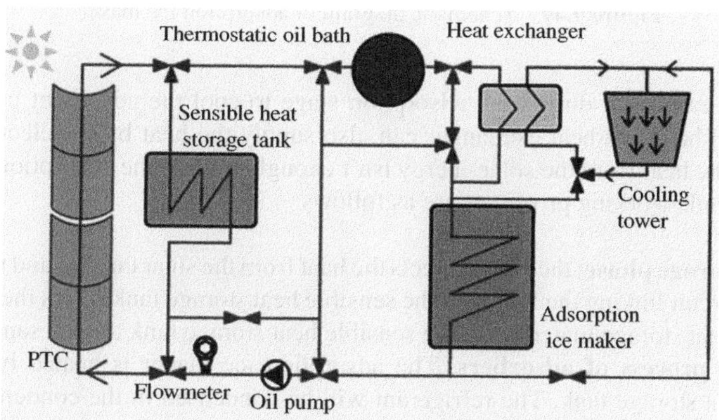

Figure 9.48 Scheme of solar adsorption ice making system with PTC [31]

adsorption ice maker, PTC, sensible thermal storage tank, and electrical thermostatic oil bath. By controlling the three-way valves, different cycles can be realized and the adsorption ice maker or the heat storage tank can be connected directly to the solar energy source-PTC or simulation heat source-electrical boiler. In the experiments, the adsorption ice- maker can be driven by PTC directly or preheating by sensible heat storage tank before the beginning of the desorption stage. The electrical oil boiler was connected in this system to simulate the stable solar heat source output and the performance of the ice maker was tested under different desorption temperature by setting the temperature of the thermostatic oil bath. The oil-water heat

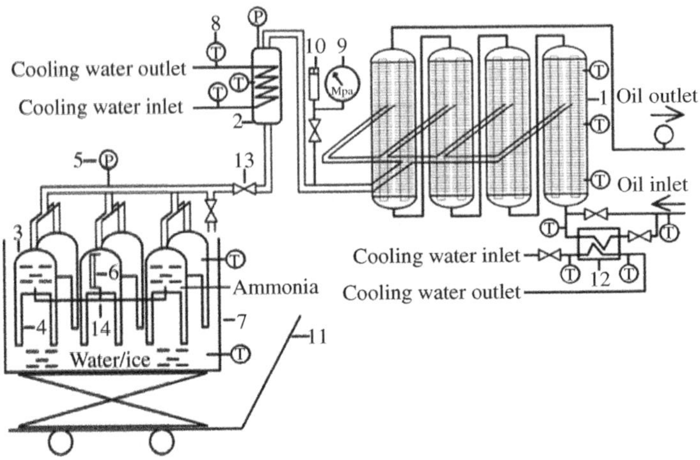

1 - Adsorbent bed, 2 - Condenser, 3 - Evaporator, 4 - Ice section of evaporator, 5 - Pressure transducer, 6 - Liquid ammonia level meter, 7 - Water tank, 8 - Temperature sensor, 9 - Pressure meter, 10 - Pressure relief valve, 11 - Hydraulic elevator, 12 - Oil-water heat exchanger, 13 - Manual valve, 14 - Communicating tube for evaporator units

Figure 9.49 Schematic diagram of adsorption ice maker

exchanger was employed during the adsorption stage to cool the adsorbent bed down. The thermostatic oil bath and heat exchanger can also supply the heat by the electricity heating process when the heat from the solar energy isn't enough to drive the adsorption refrigeration system. The whole working processes are as follows:

1. **The heat storage phase**: the PTC collects the heat from the solar energy, and the circulation by the oil circuit linking the PTC and the sensible heat storage tank stores the heat from the PTC. The heat storage materials in the sensible heat storage tank are the sand and stones.
2. **Desorption process of adsorbers**. The adsorption ice maker is heated by the PTC or sensible heat storage tank. The refrigerant will be condensed in the condenser (shown in Figure 9.49).
3. **Adsorption and refrigeration process**. For this process the system is cooled by the cooling tower, and the adsorbent will adsorb the refrigerant from the evaporator (shown in Figure 9.49). The evaporator transports the cooling power to the surroundings.

The schematic diagram of adsorption ice maker was shown in Figure 9.49, and the photo of the whole system is shown in Figure 9.50. The whole ice maker consists of adsorbent bed, condenser, and evaporator. Adsorption working pairs are the most important parts in the whole system. For ice making purposes, the commonly used refrigerant is ammonia or methanol, while the adsorbent can be metal chlorides, like $CaCl_2$ or activated carbon [32]. Compared with pure activated carbon and pure $CaCl_2$, the compound adsorbent by mixing $CaCl_2$ and activated carbon has a high volume adsorption capacity and has been used in many ice maker applications [33]. Thus it is chosen as the working pair in the system.

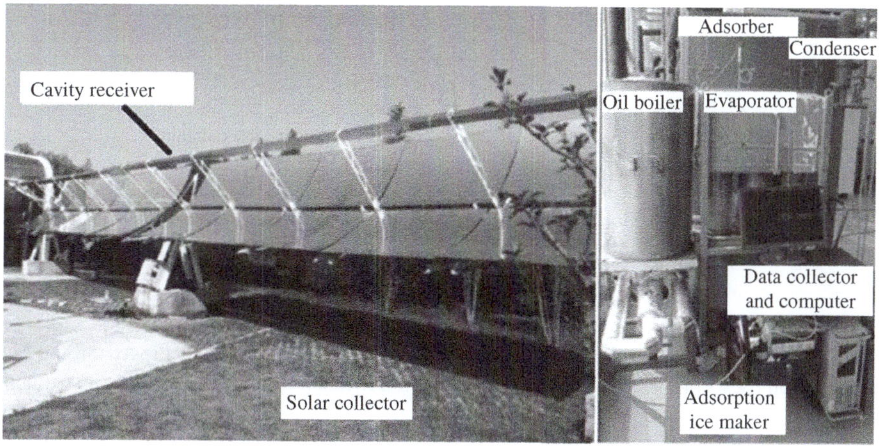

Figure 9.50 Photograph of the experimental setup

9.5.3 Experimental Results for the System Constructed by Shanghai Jiao Tong University

Firstly the sensible heat storage performance is studied, and the results are shown in Figure 9.51, two recharging processes and two discharging processes have been carried out. In the recharging processes, the oil inlet temperature is stable at 190 °C, while during the discharging processes, the inlet temperature is controlled at 120 °C. $T_{m,in}$ and $T_{m,out}$ indicated the inlet and outlet temperature of mixed sands in the storage tank. The recharging

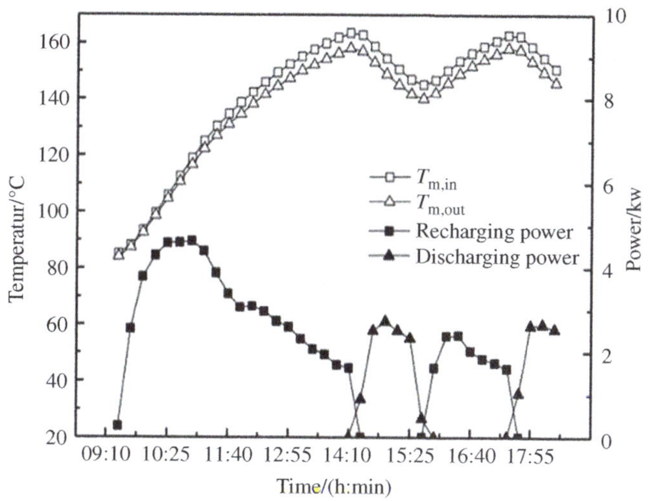

Figure 9.51 Temperature and power evolution of sensible heat storage

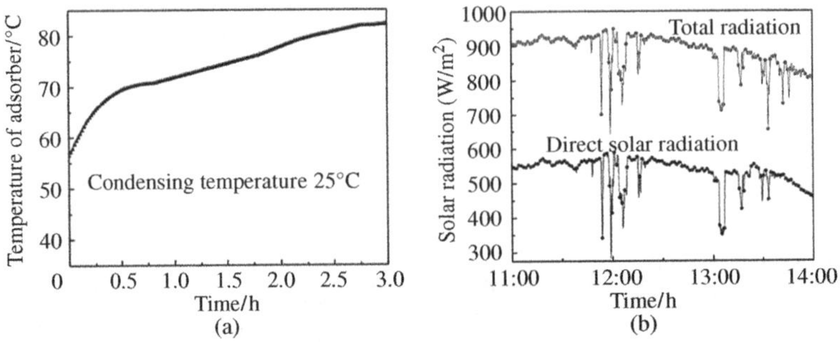

Figure 9.52 Adsorbent bed temperature evolution of cycle powered by PTC. (a) Adsorbent bed temperature evolution (desorption) and (b) radiation evolution measured by solar radiation instrument

heat, storing heat, and discharging heat are calculated, and they are 18.9, 17.6, and 7.2 MJ, respectively. The recharging power increases rapidly to 4.7 kW and then decreases to 1.7 kW with the temperature increment of the storage medium. The mean recharging power from 140 to 160 °C is 2.4 kW while the mean discharging power is 1.5 kW from 160 to 140 °C, individually. These experimental results verified the feasibility of the sensible thermal storage process for the cycle.

Figure 9.52a shows the adsorbent bed temperature evolution during the decomposition reaction process, powered by PTC, while the radiation data during the same time is shown in Figure 9.52b. In the first stage of desorption the adsorbent bed is preheated by PTC to about 57 °C in 3 hours. Then the adsorbent temperature increases slowly to 83 °C in 3 hours (Figure 9.52a) and the decomposition reaction was driven by PTC, where the condensation temperature was 25 °C. After the decomposition reaction is complete, the valve between condenser and evaporator is closed. In Figure 9.53a, the adsorbent temperature decreases from 81 to 43 °C in less than 3.5 hours. As shown in Figure 9.53b, both the evaporating temperature and pressure decreased synchronously due to the gradual increasing valve opening in 1 hour. Since the decomposition reaction is incomplete, the evaporating temperature increases in 1 hour after the valve is opened completely. The ice production in this case is only 18 kg.

As shown in Table 9.9, based on previous calculations, performance of different cases have been given including ice making capacity, COP, and COPs. In general, the higher the

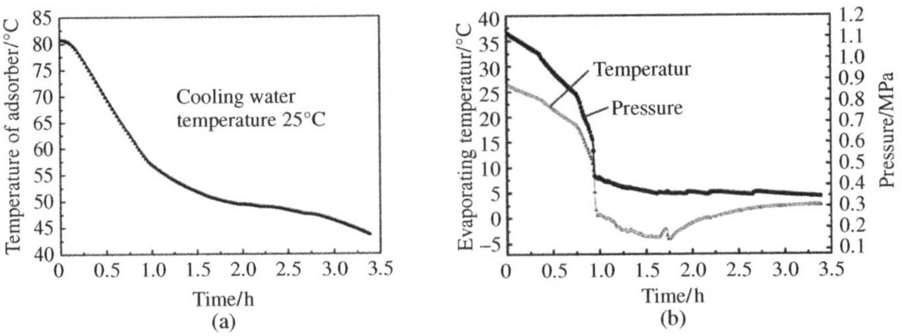

Figure 9.53 The temperature of adsorber, evaporating temperature, and evaporating pressure vs. time. (a) temperature of adsorber and (b) evaporating temperature and pressure

Table 9.9 Performance summary of different experimental cases

Case number	Desorption temperature (°C)	Heat input (MJ)	Ice making capacity (kg)	COP	COPs
1	75	50	0	0	0
2	83	82	18	0.08	0.04
3	85	83	20	0.08	0.05
4	89	90	27	0.10	0.05
5	94	95	28	0.10	0.05
6	95	100	30	0.10	0.05
7	100	155	52	0.12	0.06
8	105	115	50	0.15	0.08

desorption temperature is, the more ice can be produced. However, the COP and COPs are not only affected by the ice making capacity but also by the heat input. Case 8 shows that when the heat input and desorption temperature are 115 MJ and 105 °C respectively, the cooling COP is 0.15. While in case 1, since the desorption temperature is as low as 75 °C and the heat input is only 50 MJ, the ice making capacity is zero.

Research on the system shows that the system has the advantages of simple operation, low cost, and stable performance, but the low efficiency needs to be further investigated.

9.6 Other Types of Solar Adsorption Refrigeration Systems

9.6.1 Solar Cooling Tube

Figure 9.54a shows a special design for the solar adsorption cooling tube, which uses solid adsorption refrigeration technology in a tube, and outputs the cooling capacity periodically.

The outer wall of a solar adsorption cooling tube is made up of borosilicate glass which is heat-resistant and has a high penetration rate. The adsorbent is a molecular sieve which has a high adsorption rate. The external glass is black to fully absorb the solar energy. The cooling tube is divided into three sections: the adsorption bed, condenser, and evaporator. In the day time the adsorbent bed is heated and the water is desorbed and condensed in the condenser, while the condenser is cooled by the cooling water that flows to the bottom of the cooling tube. At night, due to the natural cooling function of the air the adsorbent bed is cooled, the molecular sieve recovers the adsorption capacity, and the water at the bottom of the cooling tube evaporates. All the heating, desorption, cooling, and adsorption processes were completed inside the tube. The research work by Zhen Liu has shown [34] that solar cooling efficiency of the solar adsorption cooling tube can reach 10–15%. Obviously such a tube has the advantages of simple structure and easy operation, and also the cost of the system is low. As the refrigerant works under negative pressure, the glass tube is easy to seal and keep at the vacuum state long term. According to the requirements of the refrigeration capacity, a number of cooling tubes can be easily assembled. The main disadvantage of such a type of system is that the adsorbent bed, condenser, and evaporator are all installed in one chamber, and the working processes are not composed of two constant-volume processes and two-isobaric processes, thus the COP and the refrigeration capacity will be reduced.

Figure 9.54b is the cold/hot water machine [34] composed of a solar cooling tube unit. The area of solar cooling tube unit is about 1 m^2. During the day, the desorption heat and condensing

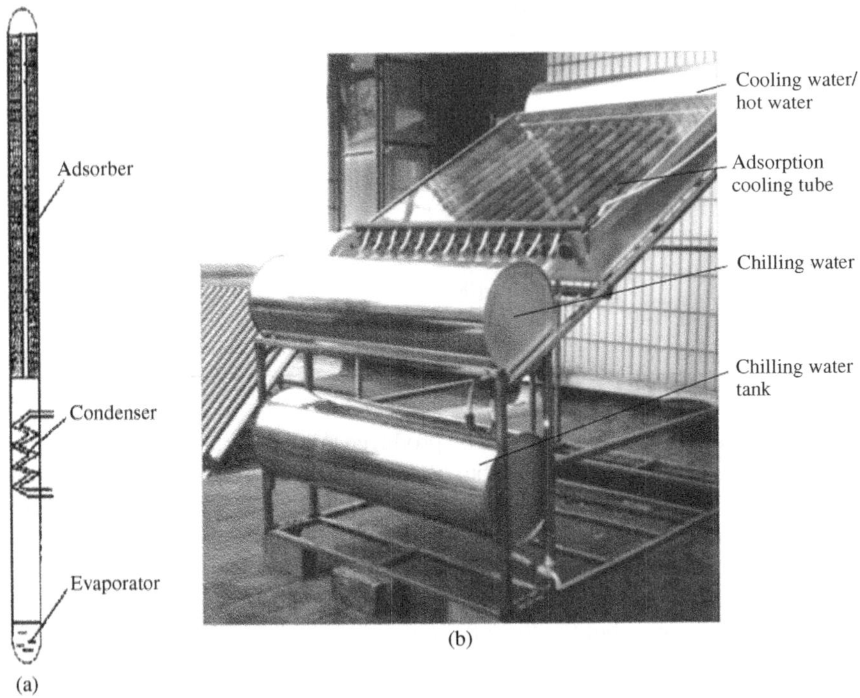

Figure 9.54 Solar cooling tube and its application. (a) solar adsorption cooling tube [34] and (b) cooling/heating water machine with solar cooling tubes

heat can be stored in the cooling water tank by natural circulation, therefore the system can produce low temperature hot water. Because the evaporator of the cooling tube is inserted into the tank, so at the night the chilling water can be produced. The chilling water can be transferred to the storage tank for refrigeration applications.

9.6.2 Solar Air Conditioner with Heat Storage Function

As the energy supply and demand are strongly dependent on the seasons and different locations, thus it is very promising if we can store the unused energy temporarily and release it later, and such a method is known as energy storage. The energy storage is also an important energy saving system. It can adjust the energy requirement to achieve the efficient and rational usage of energy. Solar air conditioner is running intermittently due to the variation of solar radiation. The simplest program for an air conditioner is the use of ice by a solar adsorption ice maker. The disadvantages of such a scheme are that firstly, it cannot continuously supply cooling power, and secondly, the efficiency is low because the evaporation temperature isn't high. If the evaporation temperature of the solar adsorption refrigeration system increases, energy storage devices can be taken as a supplemental method to overcome the disadvantages of the intermittent operation of the solar system, that is, the system can output the cooling power continuously.

Figure 9.55 is a continuous and stable solar air-conditioning system with energy storage function [35], which uses solid adsorption refrigeration and is driven by solar energy.

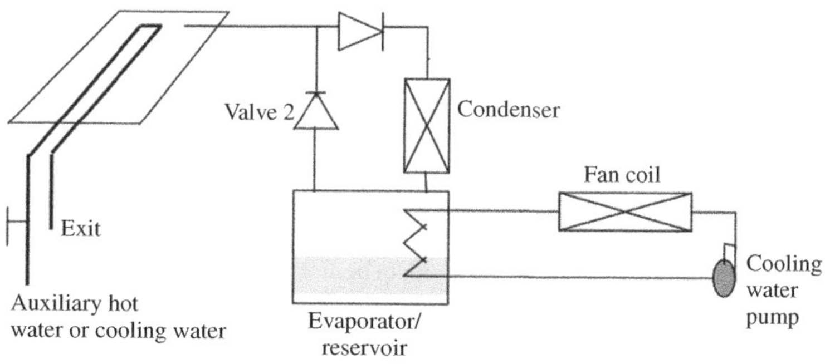

Figure 9.55 Schematic of the solar adsorption air conditioner with heat storage

It overcomes the disadvantages of the solar air condition system such as running intermittently and not being easy to adjust the refrigeration capacity with the combination of the adsorption potential and physical sensible heat storage, also it can produce hot water at a certain temperature using the adsorption heat.

The working pairs for solar adsorption air conditioner with heat storage are zeolite-water, silica gel-water, or activated carbon-methanol. The cooling principle is the same as the cooling device described previously, and the principle of the energy storage has been described in Chapter 8. The volume of refrigerant in the evaporator/reservoir is increased to store the cooling capacity. The aim of energy storage is to adjust the cooling capacity by combining it with the fan coil. The cooling capacity is stored as the physical sensible heat of the liquid in the reservoir. Adsorption potential energy is stored by the desorption process of the adsorption bed. After desorption, the adsorption bed has the adsorption capacity for refrigeration. The adsorption potential is reserved as long as we require the energy, and the energy can be released by connecting the adsorption bed with the evaporator. The advantage of such an energy storage system is that the adsorption potential energy can be stored long term without heat loss. When the adsorption potential energy is released it can produce both cooling capacity and heat to the outside. Figure 9.56 shows the variations of the cooling capacity per volume refrigerant

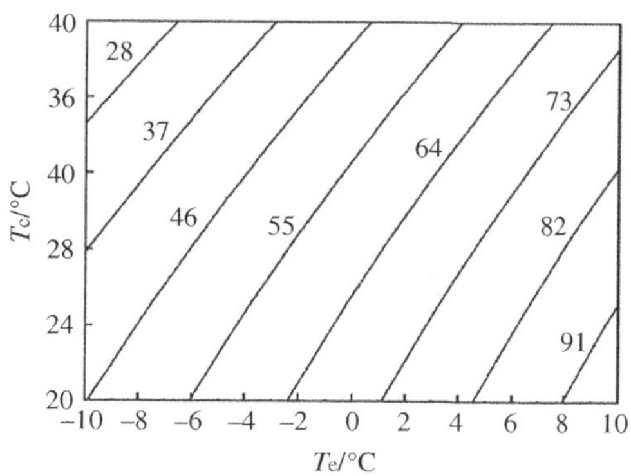

Figure 9.56 Cooling capacity per volume refrigerant vs. evaporation temperature and condensation temperature (unit is MJ/m^3)

with evaporation temperature and condensation temperature as the working pair is activated carbon-methanol.

This system is reliable and easier to be maintained. For example, to develop solar adsorption air conditioning for a bedroom with an area of $20\,m^2$, if the air conditioner works for 8 hours/day, and the cooling load is $100\,W/m^2$, then 57 600 kJ cooling capacity is needed for one day. If the COP of the system is between 0.2 and 0.3 and the daily irradiance is 1000 W, the adsorption collector can meet the demands of the refrigeration requirement if the area of the collector is in the range of $5-8\,m^2$, the adsorbent is 300–500 kg, and the refrigerant is 75–120 kg. Also a liquid receiver for evaporation (about 100–150 l) as well as a fan coil, vacuum valves, condenser, temperature controller, and flow rate controller, and so on, are needed.

A solar adsorption air conditioner with a heat storage system can also recover the adsorption heat for the production of hot water.

9.7 Adsorption Refrigeration Technology for the Utilization of Waste Heat

Development of human civilization requires the consumption of energy. Nowadays most of the energy required by industry is by fossil fuels. The burning of fossil fuels generates not only a lot of greenhouse gas in the form of CO_2, but also a large amount of waste heat is discharged into the atmosphere. Meanwhile with the gradually increment of living standards the requirement for air conditioners is growing. For example, air conditioners in cars, especially in taxis, usually use $CFC1_2$ or HFC134a as the refrigerant. For these situations, CFCs (chlorofluorocarbons), HFCs, HCFCs will eventually be emitted into the atmosphere. CFCs and HCFCs are stable in the stratosphere, but finally they will cause the destruction of the stratospheric ozone. From the point of view of global warming, almost all halogenated hydrocarbons produce infrared radiation like CO_2. As a result, refrigeration not only consumes a large amount of energy, but is also related to ozone depletion and the greenhouse effect of CFCS on the environment.

From the point of view of environment protection, the use of natural refrigerants or refrigerants with smaller environmental impact to replace CFCs have been paid more and more attention in recent years, as well as the development of the new type green refrigeration technologies. From the point of view of energy protection, fossil energy is limited, which has already led to the energy crisis which emerged in the 1970s in the West. So energy conservation and waste heat recovery have attracted the attention of researchers for a long time.

9.7.1 The Usage of Waste Heat from the Engine

Take the automobile, for example, when the car engine is running, the power output only accounts for 0–42% (diesel) or 25–30% (gasoline) of the total fuel combustion energy. Waste heat discharged to the outside of the car accounts for 58–70% (diesel) or 70–75% (gasoline) of the total fuel combustion energy, including the heat taken away by the circulating cooling water and exhaust heat taken away by the exhaust gas. Waste heat of the exhaust gas is characterized by high temperature, and it accounts for 25–45% (diesel) or 30–40% (gasoline) of the total combustion heat. The heat is roughly the same as the driving force. Engine exhaust temperature is about 450–550 °C. Taking into account the dew point corrosion of acidic oxides in the exhaust gas, the temperature of the exhausted gas should not be smaller than 180 °C. Generally

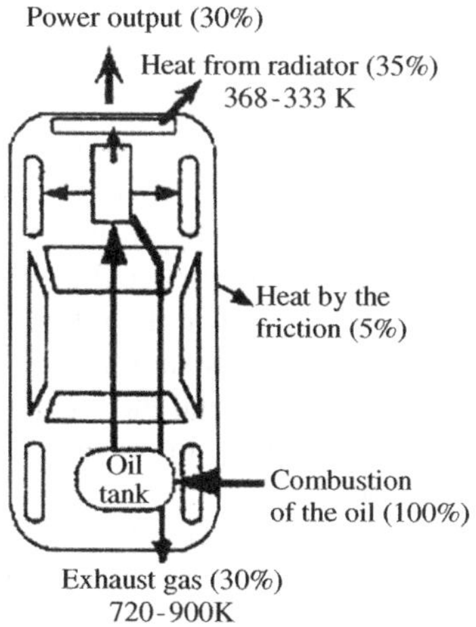

Figure 9.57 Diagram for the energy of the gasoline engine [36]

the utilizable waste heat accounts for 16% of the combustion energy. If this energy is used for the car refrigerator and the air conditioning system is the driving heat source, it can meet the refrigeration requirements and can reduce the running costs of the car.

Figure 9.57 is the diagram of the heat of the gasoline engine. The heat loss and the radiation through the heat sink accounts for 65–70% of the chemical energy. Cooling water temperature is usually controlled at 368 K by controlling the flow rate of cooling water in the radiator. The exhaust gas from the engine is 700–900 K at the outlet of the combustor, and then it exchanges the heat with the air and cools down when it flows through the exhaust pipe and muffler. The temperature of cooling water through the radiator is 333 K, and after that the cooling water returns to the engine. If the waste heat of the motor vehicle engine is combined with solid adsorption refrigeration technology, for a 2000 cc car the fuel consumption is 1.2 l/h (3×10^{-7} m^3/s) under ideal conditions of 4–5.5 l h ($1.1-1.5\times10^{-6}$ m^3/s) when the speed is 60 km/h. Taking into account that the gasoline combustion is 7770 kcal/l (3.25×10^{10} J/m^3), the fuel consumption is approximately 10 800 and 34 900–49 600 W. If the standard air-conditioner with 23 000–32 500 W cooling power is used, taking into account the dew point for corrosion problems of acidic oxides in the exhaust gas, the temperature of the final discharged exhaust gas should not be less than 180 °C, and 60% of consumption heat can be considered as the potential energy of the adsorption refrigeration system. If this energy is used as the driving heat source for car refrigerator/air conditioning systems, as long as the cooling system performance coefficient reaches 0.2, it can meet the refrigeration requirements of the corresponding vehicle.

9.7.2 *Waste Heat Recovery Methods*

People have considered how to use the engine waste heat for a long time, and they have obtained some of the following results:

1. The utilization of heat directly. Such as in cars in the winter, the cooling circulation water from the engine is used to regulate the temperature inside the car.

2. Use the heat for heating process of the boiler. The majority of passenger ships and cargo ships are equipped with an exhaust gas boiler, and the exhaust gas from the engine is used to produce the domestic water by heating boiler.
3. Use the heat to increase the specific power of the engine. The exhaust gas turbo charging system mechanically driven by the exhaust gas has been successfully used for many years. It greatly increased the power and the efficiency of the engine. M. Mostafavi (1996) [37, 38] proposed the combined system of the engine and absorption chillers which is driven by the high-temperature exhaust gas to produce the refrigeration power to cool down the inlet air of the engine. His calculation results showed that when the pressure ratio is 64 specific power will be increased by 29.96%, and the efficiency will be decreased by 12%. This indicates that this method is good for the improvement of the specific power, but not good for the improvement of the efficiency.
4. Use the waste heat as the heat source for the refrigerators. The refrigeration systems driven by the exhaust gas from the engine are mostly absorption or adsorption systems. COP of the absorption refrigeration system (compared with adsorption system) is high, but its structure is complex and cost is high. The absorbents for the absorption system are generally liquids, thus they commonly cannot adapt to the conditions of regular bumps and motion of the car. The adsorption system has a simple structure and low cost, and the system performance can be improved simply by heat and mass transfer processes. It is an ideal system for cars. The earliest research in this area was done by J.R. Akerman [39]. In the study of automotive air conditioning systems with the exhaust heat, he obtained a conclusion that it is feasible to use this system in the cars which run continuously under heavy load conditions. When the load of the car is low or idle, the refrigeration capacity of the chiller is very small. Heat storage devices or additional combustion chambers are needed to meet the requirement on the refrigeration capacity. Since then, Baoqi Han [40, 41] did research on a fishing vessel with an adsorption refrigeration system for fish storage. They used zeolite and water as a working pair. When the desorption temperature is 200 °C, the adsorption temperature equals to the atmospheric environment, and the cycle time is 3 hours, 1 kg of water can be cooled from 24 to 2 °C by a unit adsorption refrigeration tube.

9.7.2.1 Adsorption Working Pairs for Waste Heat Recovery

The choice of adsorption working pairs for waste heat recovery is based on the requirements on the heat source. As described in Chapter 2, the highest temperature of heat source for the working pairs of silica gel–water and activated carbon-methanol cannot be too high, thus can they only be used when the heat source temperature can be controlled at a reasonable level. Shanghai Jiao Tong University developed an adsorption refrigeration unit with activated carbon–methanol as the working pair for fishing boats. The unit uses a water tank to exchange the heat with the waste heat of the exhaust gas from the engine, while the water temperature is controlled below 120 °C by controlling the water temperature in the hot water tank. Also the silica gel–water refrigerator was used in the cooling, heating, and power cogeneration device. The heat source temperature is controlled at a suitable level for silica gel–water working pair.

Metal hydrides–hydrogen, zeolite–water, metal chloride–ammonia, and the composite metal chloride adsorbent-ammonia required a heat source with high temperature. They are suitable for the direct recovery of waste heat. If the adsorbent is metal hydride [42] the main drawback will be poor security because the hydrogen is flammable. Zeolite–water needs a heat source with a high temperature and it is often used in the waste heat recovery process.

Motoyuki Suzuki [36] calculated the possibility of using the adsorption refrigeration system in cars from the point of view of the heat balance equations, and the results showed that when the desorption/adsorption time is 60 seconds, 2 kg adsorbent (molecular sieve) is needed to meet the refrigeration capacity of 2300 W. Also, metal chloride–ammonia can be driven by a heat source with a high temperature. The calcium chloride–ammonia working pair has already been used on the fishing boats. Compared with metal chloride, the composite adsorbent of metal chlorides and porous media can effectively prevent the expansion and agglomeration phenomenon of the pure metal chloride, and it can greatly improve the volume cooling capacity. At present there is much research about its utilization for the waste heat recovery processes.

9.7.3 The Advantages of Adsorption Refrigeration Technology for the Waste Heat Recovery

The advantages of the adsorption refrigeration technology for the waste heat recovery process are as follows:

1. Various adsorption working pairs can be chosen for different conditions. If the adsorption working pair is different, then the required temperature of the heat source is different, so for different types of waste heat we can choose different working pairs. Besides, if the waste heat is recovered by the adsorption refrigeration technology it not only improves energy efficiency, but also will decrease the emissions of greenhouse gas.
2. The moving parts in a solid adsorption refrigeration system are few, or even none. The solid adsorbents are also more suitable for the vibration conditions if compared with the liquid absorption refrigeration technology, thus it can be applied to motor vehicles.
3. For a solid adsorption refrigeration system the condensing temperature and adsorption temperature can be increased according to the external heat source and the adsorption working pair. As a result it can be cooled by the air and we don't need to worry about the crystallization problem as happened in the absorption refrigeration systems.

However, its drawbacks are also obvious. Its cooling process is not continuous like the compression air conditioner. In order to ensure the adsorption and desorption processes, the adsorber (including the adsorption bed, the adsorbent, and adsorbate) is cooled and heated intermittently, so the COP is relatively low. For a simple cycle, the COP for an adsorption air conditioner is generally 0.2–0.3, and the COP for an adsorption ice maker is generally 0.05–0.15. In addition, compared to the compression refrigeration system, the volume of the adsorption refrigeration system is relatively large.

9.8 Application of Adsorption Refrigeration Systems Driven by Waste Heat

9.8.1 The Application of Zeolite–Water Adsorption System as Locomotive Air Conditioner

The locomotive air conditioner is the adsorption system introduced in Chapter 8, which uses the zeolite–water working pair and it has the heat storage function. The field experiments are divided into two steps: the first step is carried out from September to October 2000, the locomotive belongs to the Hangzhou section of the TRA, China, and the running section is from Hangzhou to Jinhua, Shanghai, and Ningbo; the second step is carried out from May to September 2002, the locomotive belongs to the Jinhua section of the TRA, Chiba, and its running section is from Jinhua to Hangzhou, Jiangshan, Qiandao Lake, and Lanxi.

9.8.1.1 The Structure of the Adsorption Air Conditioning System for a Locomotive

The air conditioning prototype was installed at the Dongfeng 4-2369 locomotive which belongs to the Hangzhou depot of Shanghai Railway Bureau (shown in Figure 9.58). The locomotive was designed and manufactured by the Dalian Locomotive Company in 1992. Its structure is shown in Figure 9.59. Its body is studio-style sidewall load-bearing structures, and there are driver's cages with the same equipments at both ends of the locomotive, thus the operating processes are the same. In the central area of the locomotive is the power house and there is a diesel generator and auxiliary equipment installed in this house. At the back of the house there is preheating boiler and electrical control equipment. The cooling room is in the engine room and the second driver's cage and there is a radiator arranged with a V-shape and axial cooling fan driven by a hydraulic motor.

The main components of an adsorption air conditioner are installed in the engine section of the locomotive. The system structure is shown in Figure 9.60.

1. **Adsorber**. The adsorber is located in the upper part of the engine section. Its shell is welded together with the roof of the engine section. A part of the adsorber is exposed outside the roof as shown in Figure 9.61. When the adsorber needs maintenance or the air conditioner won't be used long term, the roof can be removed and a standby ceiling can be put on instead.

Figure 9.58 Dongfeng 4B-2369 diesel locomotive

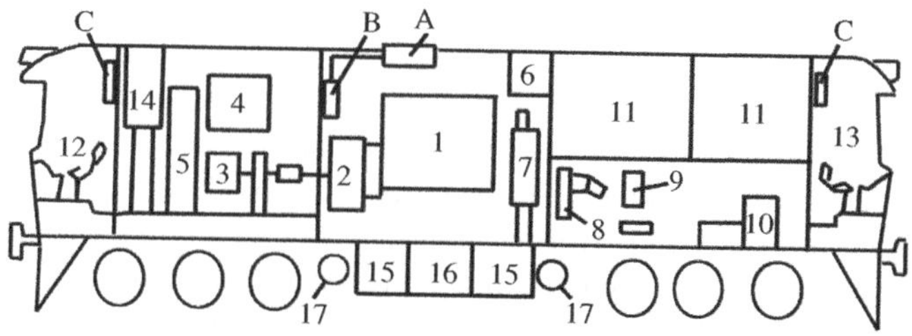

A - Adsorber; B - Evaporator/Energy storage tank; C - Fan coil; 1 - Diesel Engine: 2 - Traction generator; 3 - Start motor; 4 - Rectifier cabinet; 5 - Electricity cabinet; 6 - Water expansion tank; 7 - Preheating boiler; 8 - Hydraulic pump; 9 - Blower; 10 - Air compressor; 11 - Cooler; 12 - First cab of driver; 13 - Second cab of the driver; 14 - Braking resistor; 15 - Battery box; 16 - Fuel tank; 17 - Air cylinder

Figure 9.59 The schematic of Dongfeng diesel locomotive

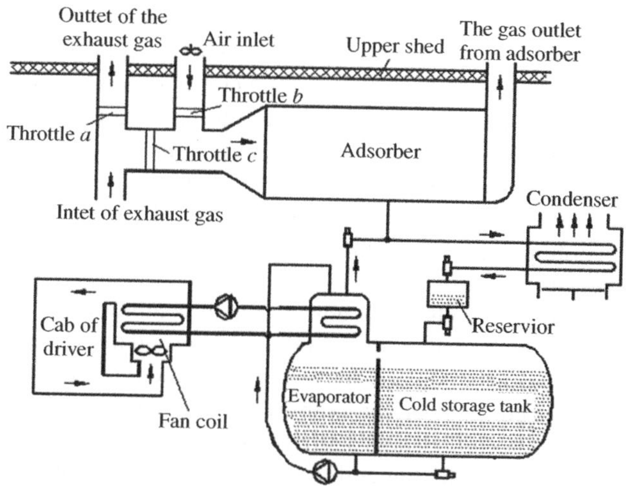

Figure 9.60 Schematic of the adsorption air conditioner installed in the locomotive

Figure 9.61 Adsorber of the air conditioner installed at the upper part of the locomotive

2. **Condenser**. Because of the particularity of the working environment of the locomotive, the condenser of the air conditioning system cannot be cooled by water because it isn't convenient. Therefore, the actual system is cooled by the air. There are four groups of air-cooled condensers at each side of the adsorber in the locomotive, and every group contains eight copper-aluminum composite pipes. The heat transfer tubes have the aluminum fins outside to enhance the heat exchange performance of air. The length of each fin-tube is 500 mm. The external diameter of the tube is 16 mm while the external diameter of the fin is 36 mm. The fin space is 3.2 mm. There a certain allowance was considered when the heat transfer area is calculated, and it meets the needs of the refrigeration system.
3. **Evaporator/regenerator**. The evaporator in the locomotive prototype is the spray type evaporator, and the heat exchangers are locate at the upper part of the evaporator. The evaporation and forced convection heat transfer processes happen outside of the heat exchanger, therefore the heat transfer coefficient between the refrigerant water and media water is intensified, and consequently the temperature difference between the chilling liquid and the refrigerant is reduced. There is still sufficient cooling capacity output when the evaporation temperature is relatively high. The heat exchange methods between the regenerator and evaporator are changed (as shown in Figure 9.60), the spray pump is turned on when the system is running. When the regenerator needs to produce the refrigeration, the electromagnetic pneumatic valve connected to the spray pump under the regenerator will be open, the outlet water from the regenerator and evaporator will be mixed and flow into the evaporator. When the water level of the evaporator rises the water will overflow into the regenerator. The cooling power of the fan coil in the driver's cab is set at 5 kW. The outlet air velocity is adjustable.
4. **Fan coil**. There are two fan coils in the adsorption air conditioner of the locomotive cab. They are hung parallel on the wall of the first driver's cab and the second, respectively. The chilling water flows through a fan coil by switching of the three-way valve. The refrigeration capacity is delivered to the driver's cab by the circulation of the water driven by the circulation pump. The fan coil releases a refrigeration capacity for the air conditioner. The designed cooling capacity of the fan coil in the driver's cab is about 5 kW, and the outlet velocity is adjustable.
5. **Exhaust gas/air switching device**. When the air conditioning system is operated, the exhaust gas of the engine and the cooling air alternately flow through the adsorber. The switch is operated by a special device, which is an electromagnetic cylinder and valve system. The exhaust gas at the outlet of turbocharger of the engine is exported into the atmosphere through two chimneys. The side of the chimney near to the first driver's cab is connected to the inlet of the adsorber for exhaust gas by a three-way pneumatic valve. There is an air inlet at both sides of the roof of the engine section besides the first driver's cab. The area of air inlet is about 250 mm × 140 mm, and including the head-on area of the baffle at both sides the total area is 395 mm × 140 mm. From the wind inlet the wind goes into the adsorber and cools down the adsorbent bed when the locomotive is running. There is a pneumatic valve between the air inlet and the gas inlet of the adsorber. In the heating/desorption process, the pneumatic valve cylinder opens and the valve is turned off. For other processes the cylinder is closed and the valve is open.
6. **Control and measurement system**. Electromagnetic – pneumatic valves, fan coil units, loop magnetic pump, and the switch of working conditions of the adsorption refrigeration system are controlled by the PLC control system. The control box is located at the locomotive electrical cab. The control system has manual and automatic operating modes,

and it can automatically start up and shut down, set off a fault alarm, collect the data, display, and store the data. It can also identify and deal with some failures when the system is running.

9.8.1.2 Some Parameters for the Diesel Engine of Dongfeng 4B Locomotive

The operation of the air conditioner is closely related to the conditions of a locomotive. The Dongfeng 4B locomotive uses four-stroke and a direct injection combustion chamber, exhaust turbocharger, locomotive diesel engine with the model type of 16V240ZJB. The main operating parameters of the diesel [43] engine are as follows:

1. **Speed**. The rated speed is 1000 rpm, maximum speed is 1100 rpm, and minimum idling speed is 400 rpm.
2. **Power**. The rated power is 3860 kW and the maximum power is 4246 kW.
3. **Fuel**. The type of fuel is GB252-87 top grade product or the first grade of light diesel oil. The standard fuel consumption under rated power is 208–215 g/(kWh), and the fuel consumption is 16 kg/h for the idling condition with an idling speed of 430 rpm.

For the total heat generated by the combustion of the fuel in a diesel engine, the heat injected into the atmosphere accounts for about 30–35% of the total energy, which is almost the same as the driving energy required by the adsorption air conditioner. Exhaust gas (500–650 °C)/(1.5–3 bar) enters the turbine and continues to expand and drive the compressor before it is discharged into the atmosphere. The exhaust gas from the turbine goes into the atmosphere through the chimney. There are two chimneys on the roof of the locomotive with an exit area of 576 cm^2 (240 mm × 240 mm). The matching parameters of the diesel engine and the turbocharger of Dongfeng 4B are shown in Table 9.10. When the temperature is 8 °C, the temperature of the exhaust gas of the diesel engine turbocharger is approximately 427 °C. As the air conditioner of the locomotive is generally used in the summer and autumn, the temperature of the exhaust gas is generally slightly higher than 450 °C.

The experiments on the adsorption air-conditioner contain three aspects as follows:

1. **The relations between the gas flow rate and the rotation speed.**
   The flow rate of the exhaust gas from the diesel engine turbocharger m_y and the fuel quantity m_{oil} are related to the air coefficient γ ($\gamma = m_{air}/m_{oil}$):

$$m_y = m_{oil}(1 + \gamma) \tag{9.42}$$

Table 9.10 Parameters of the diesel engine and the turbocharger

The rotate speed of diesel engine (revolutions/min)	The power of the diesel engine (kW)	Fuel consumption (g/(kW h))	Temperature of the outlet gas (°C)	Inlet temperature of the turbocharger (°C)	Air flow rate (kg/s)	Air coefficient	Environ. Temperature (°C)	Environ. pressure (kPa)
1008	3880	204.39	428	512	4.03	18.3	8.0	102
679	1288	222.90	427	533	1.30	16.3	8.3	102

The fuel consumption for the Dongfeng 4 locomotive is related to the excess air, engine rotation speed, the load of locomotive, and the speed of locomotive, and so on, but the change of the fuel consumption isn't big. For simplicity, assuming that the fuel consumption and excess air are constants, then the outlet exhaust gas of the diesel engine is essentially linear with the change of the diesel engine power or rotation speed. Fitting the data given previously we can get the relation between the rotation speed and the fuel consumption amount:

$$m_{oil} = 1.344\omega - 562.3 \tag{9.43}$$

where ω is the rotation speed (rpm), m_{oil} is the fuel consumption (kg/h). The air coefficient $\gamma = 16.7$, thus the relation between the exhaust gas flow rate and the rotation speed is:

$$m_y = 6.608 \times 10^{-3}\omega - 2.765 \ (430 < W < 1100) \tag{9.44}$$

where the unit for rotation speed is revolutions per minute , and the unit for gas flow rate is kilograms per second. The exhaust gas is discharged from two outlets of the turbocharger. So the exhaust gas going into the adsorber is only a half of the total exhaust gas.

2. **The experiments on the back pressure of the rack.**

   In order to improve the power of the internal combustion engine, the diesel locomotive used a turbocharger. If the back pressure of the turbocharger is too high, the power of the internal combustion engine will decrease or fuel consumption will increase. The Dongfeng 4B locomotive diesel engine turbocharger allows the back pressure to be around 200 mm H_2O (about 2000 Pa).

   When the absorber is in desorption phase, the outlet exhaust gas of the turbocharger will flow through the adsorber, and then go through the chimney, and finally be released into the atmosphere. The back pressure will be larger if there is an adsorber installed on the locomotive. To control the back pressure of the exhaust gas, the actual internal area of the cross-section of the exhaust pipe should be larger than that without the adsorber. When the locomotive stops, the system is idling or is in the adsorption phase, the exhaust gas is directly discharged into the atmosphere through the chimney, and the back pressure of the exhaust gas won't be influenced. The desorption time is about one-third of the total cycle time.

   When the system is designed, one-way resistance and local resistance are taken into consideration. However, whether the air conditioner will affect the operation performance of diesel engine needs to be verified by experiments. The experimental results on the back pressure of the rack in the adsorption chiller are shown in Figure 9.62. The abscissa in the figure is the engine rotation speed, and the vertical axis is the back pressure. It can be seen from Figure 9.62 that the back pressure is approximately exponential to the rotation speed. The back pressure is very small when the rotation speed is not high (< 800 rpm); if the rotation speed is high, the back pressure will increase with increasing rotation speed, but no more than the allowed range of back pressure. After being in operation for six months, it is concluded that the installation of an adsorption air-conditioning system haven't affected the performance of a diesel engine.

3. **The relationship between indoor temperature and ambient temperature.**

   Due to the heat generated by the staff in the cab, equipment and locomotive engine, the indoor temperature of the locomotive cab is higher than the outdoor ambient temperature. When the locomotive is running and the air conditioner isn't open (Jinhua–Hangzhou

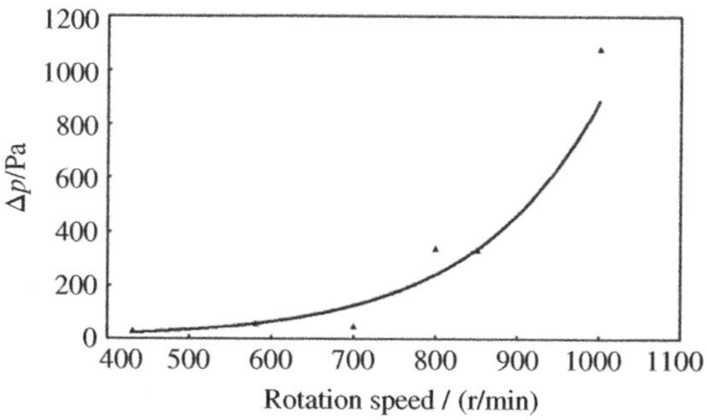

Figure 9.62 Back pressure of the rack for the locomotive with adsorption air conditioner

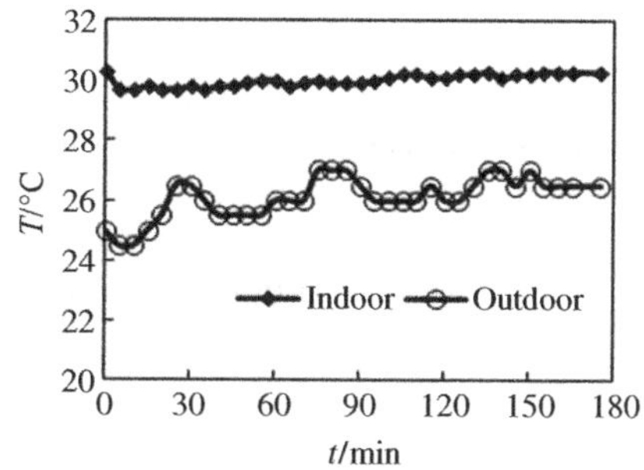

Figure 9.63 Trends of indoor and outside temperature when the windows of cab are closed

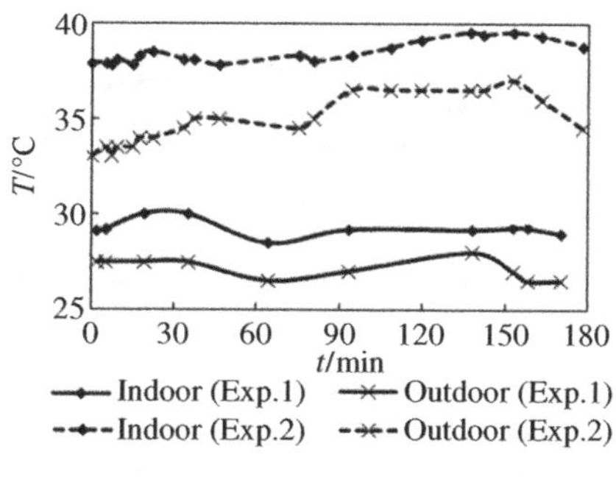

Figure 9.64 Trends of indoor and outside temperature when the windows of cab are open

scction), indoor and outdoor temperature variations with running time are shown in Figures 9.63 and 9.64. Figure 9.64 shows two groups of experimental results in different seasons when the windows are open, and there are three people in the driver's cab in the experiments. It can be seen from two figures that the indoor temperature is 4–5 and 2–4 °C higher than the outdoor temperature when the window is opened and closed, respectively. When the windows are closed, the indoor temperature is consistent and has little effect on the ambient temperature. When the windows are closed, the variation trend of the indoor temperature is similar to that of the ambient temperature. When the air conditioner is turned on, the windows are usually closed. Though the cab volume is very small (about 2.5 m × 2.0 m × 3.0 m), the experimental results show the air conditioning load is large, that is, the requirement for the cooling capacity is large. The cooling power of vapor compression air-conditioner in the existing diesel locomotive is about 4 kW.

9.8.1.3 Experiments on the Performance of the Adsorption Air Conditioner

The experiments are carried out on the section from Hangzhou to Jinhua [44]. The train number is 5103 from Hangzhou to Wenzhou, the entire distance is 190 km, and it takes about 3 hours with an average speed of 63 km/h for the journey. The average rotation speed is 640 rpm. The time for idle (at minimum speed) and parking is 63 and 20 minutes, respectively.

At the beginning of that the train departs from the start station, the adsorption heat storage will release at a rapid speed, which can be delivered to the driver's cab in time. In the experiments there are four people. At the beginning of the experiments, the adsorption bed is at the end of the desorption and the adsorption temperature is relatively low. At the beginning of the cycle, the temperature of the adsorption bed and the evaporator is 73 and 8.4 °C. During the operation, the adsorption temperature, saturation temperature corresponding to the adsorption pressure, chilling water temperature, cooling power, evaporator temperature, indoor temperature, and outdoor temperature vs. time are shown in Figures 9.65–9.67, respectively. The *p*-*T* diagram of the cycle is shown in Figure 9.68.

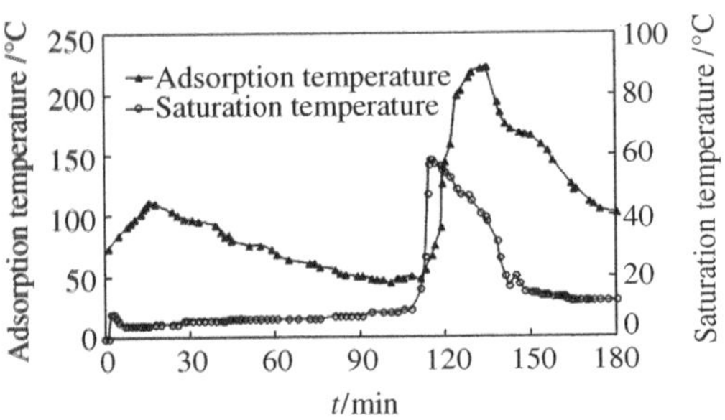

Figure 9.65 Variations of the adsorption temperature and saturation temperature vs. running time

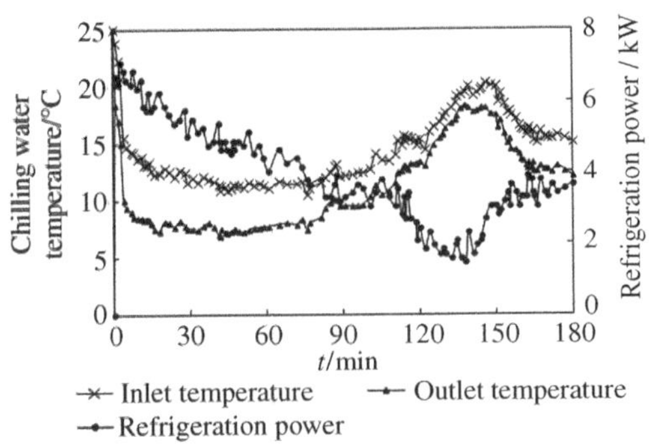

Figure 9.66 Variation of the chilling water temperature and the refrigeration power

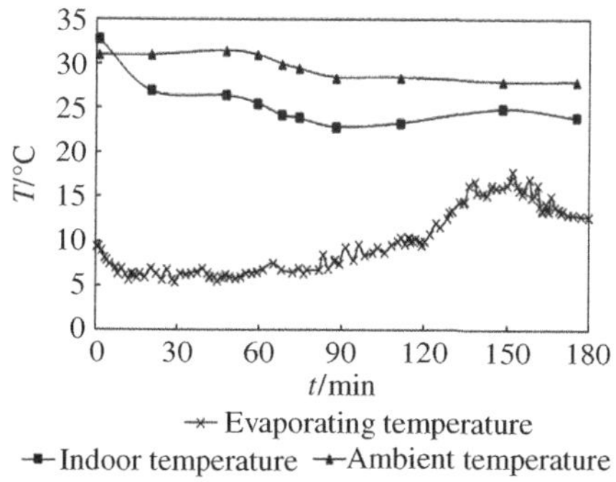

Figure 9.67 Variation of evaporation temperature, indoor temperature, and ambient temperature

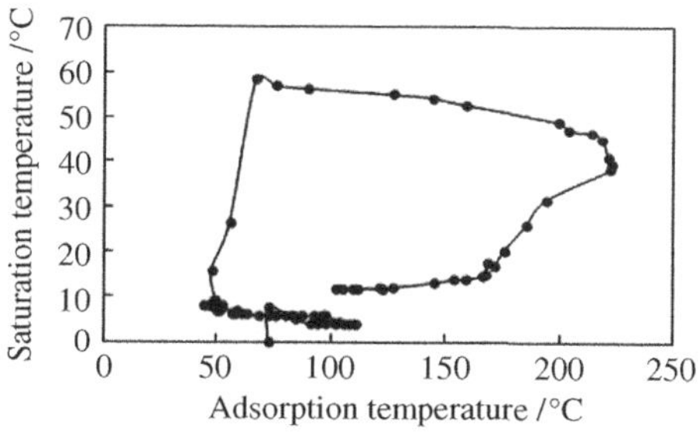

Figure 9.68 Adsorption temperature vs. saturation temperature

1. **Cooling and refrigeration phase.**

In this phase the valve between the adsorber and the evaporator is open after the locomotive departs from the start station, and then the adsorber is cooled down by the outside air and begins the adsorption and releases adsorption heat. From Figure 9.65 it can be seen that the heat releasing process is affected by the adsorption heat and the air. At the beginning, the influence of the adsorption heat is obvious. The temperature of the adsorber rises, and it reaches the maximum value $T_{max} = 110\,°C$ in the 18th minute. After that, the bed temperature decreases and the cooling/adsorption process lasts about 110 minutes. The final temperature of the adsorption bed is $T_a = 49\,°C$.

Before the adsorption, the pressure of the adsorbent bed is low. After the valve is open, the pressure rises to the evaporation pressure, and then decreases because of the strong adsorption function. After that the pressure gradually increased in the cooling/adsorption process due to the adsorption ability of the adsorber decreasing. The trend of the evaporation temperature is similar to that of the adsorption pressure. At the beginning and the end of the adsorption process, the evaporation temperature is 9.5 and 10.0 °C respectively.

Since at the beginning the evaporation temperature is quite low, there is refrigeration power output at the beginning of the adsorption phase. At that time the maximum output power is about 7.0 kW and the indoor temperature decreases rapidly. The output power also decreases gradually as the adsorption phase is going on. The output power is about 3.0 kW when the adsorption finishes. The average output refrigeration power is about 7.4 kW. The temperature difference of the interior and exterior of the driver's cab is basically maintained at 5 °C. The lowest indoor temperature reaches 23.0 °C when the ambient temperature is 28.5 °C.

As the system uses a spray type evaporator, and the heat exchangers are located at the upper part of the evaporator, so the evaporation of the liquid refrigerant at the surface of the heat exchanger is stronger than the liquid refrigerant at the surface of the evaporator. As a result, sometimes the temperature of the chilling water is lower than the refrigerant in the evaporator. When the refrigeration power is delivered out by the chilling water temperature gradually decreases, and it increases when the cooling capacity is smaller than 4.5 kW. The temperature of the chilling water actually reflects the difference of the refrigeration power and the load of the air conditioner. It shows that under this working condition (the ambient temperature is 30.5 °C, the room temperature is 24.3 °C, and four people in the cab), the load needed by the air conditioner in the driver's cab is 4.5 kW. The load will decrease when the temperature difference between interior and exterior decreases.

2. **Desorption and isosteric cooling phase.**

   After the refrigeration phase the adsorber is heated for 20 minutes. At the beginning desorption won't happen because at the later period of heating phase the locomotive is at idle state. Although the bed temperature rises slightly, the pressure of the bed decreases. The actual heating time is only 18 minutes. At the end of desorption, the desorption temperature reaches 220 °C. The saturation temperature corresponding to the adsorption pressure is 45 °C. After that in the process of the isosteric cooling process, the adsorption temperature decreases rapidly as there is only the sensible heat of adsorber exchanged. At that time, the pressure of the adsorber also decreases rapidly.

   In the desorption and isosteric cooling phase (32 minutes), the refrigeration power is delivered by releasing the sensible heat of the water in the regenerator and evaporator. The evaporation temperature and chilling water temperature are increased while the refrigeration power is smaller. The evaporation temperature increases from 10 to 17 °C, and the average output cooling power is 2.2 kW. The indoor temperature rises as the output power decreases.

3. **The second adsorption phase.**

   After the adsorber is cooled for 10 minutes open the valve between the adsorber and evaporator when the pressure of the adsorber is lower than that of the evaporator, the second adsorption phase happens. The adsorption temperature decreases slowly because of the adsorption heat. At the same time the pressure in the adsorber rises firstly and then drops. Compared with the first adsorption phase, the difference between the adsorption capacity and the equilibrium capacity is small for the second adsorption phase, so the output power is relatively small.

4. **Overall performance.**

   The total output refrigeration capacity in the above-mentioned 3 hours is 42.9 MJ. Taking into account that the temperature of water in the regenerator and the evaporator increases by 3.5 °C, the sensible heat is about 2.4 MJ, and as a result the refrigeration capacity is about 40.5 MJ. The cooling power is 3.75 kW.

9.8.1.4 The Optimization of the Operation and Simulation of the Air Conditioner

The non-equilibrium adsorption and desorption processes need to be analyzed for the locomotive prototype. There are two reasons. Firstly, the adsorption of water on the zeolite is non-equilibrium in the adsorption phase. Secondly, the flow rate of the exhaust gas in the desorption phase is large. The bed's temperature rises quickly and the desorption is also non-equilibrium.

The energy equation used in the simulation is the energy equation for a single-bed adsorption refrigeration system. In the equation when the cycle time is long and the mass transfer resistance is not large, it can be assumed that the adsorption and desorption reach equilibrium. The mass of the refrigerant in the adsorbent can be expressed as a function of adsorbent temperature and pressure:

$$x_{eq}(T_{bed}, T_s) = x_0 \cdot \exp\left[-K\left(\frac{T_{bed}}{T_s} - 1\right)^n\right] \tag{9.45}$$

where T_{bed} is the adsorbent temperature, T_s is the saturation temperature corresponding to the adsorbent pressure. In the adsorption process $T_s = T_{ev}$, and in the desorption process $T_s = T_{cond}$. x_0 is the largest adsorption capacity. According to the test results $x_0 = 0.261$, adsorption constant K and n are set as 5.36 and 1.73, respectively.

The equation of the adsorption rate in the running process of the locomotive prototype is:

$$\frac{dx}{dt} = k_1 \exp\left(-\frac{k_2}{T_{bed}}\right)(x_{eq} - x) \tag{9.46}$$

where the adsorption rates are set as $k_1 = 0.019\ (s^{-1})$, $k_2 = 906\ (K)$, respectively, according to the experimental results. It is clear that in the adsorption process $x < x_{eq}$ and $dx/dt > 0$; in the desorption process $x > x_{eq}$ and $dx/dt < 0$.

Equations 9.45 and 9.46 are suitable for the adsorption and desorption processes when the adsorption capacity changes. For heating or cooling processes, adsorption capacity doesn't change. For the isosteric heating process, adsorption capacity is the final adsorption quantity that is a function of adsorption temperature and evaporation pressure. Also, for the isosteric cooling process adsorption capacity is the final desorption quantity that is a function of desorption temperature and condensation pressure.

1. **The influence of the rotation speed and vehicle speed on the running process**

   The heating and desorption process are influenced by not only the initial state of the adsorbent bed (temperature and adsorption capacity), but also the rotation speed of the

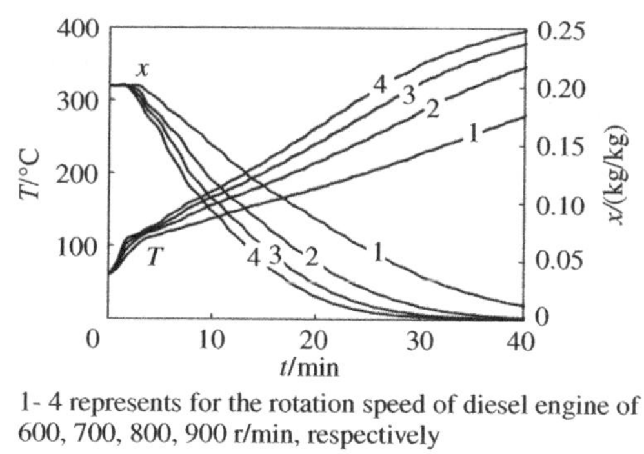

Figure 9.69 Influence of the rotation speed on the adsorption capacity and the bed temperature

diesel engine. The higher the rotation speed is, the larger the flow rate of the exhaust gas is, and the faster the desorption process is. Figure 9.69 shows the variation of the adsorption temperature and adsorption capacity in the desorption processed by simulation when the initial temperatures of the adsorbent bed, evaporator, and condenser are 60, 7, and 50°C, respectively, while the rotation speed of the diesel engine is 600, 700, 800, and 900 revolutions/min, separately.

It can be seen from Figure 9.69 that when the rotation speed is relatively low the impact of the rotation speed change on desorption process is significant. While the rotation speed is relatively high, the effect of improving the speed of the system on the desorption process is not obvious. The heating process of locomotive prototype is much faster than that of the laboratory prototype. The reason is that the exhaust gas of the locomotive prototype is much more than that of the laboratory prototype. The desorption process takes only 20 minutes when the engine speed is larger than 700 revolutions/min.

The airflow through the adsorber in the adsorption process has not been measured, but it can be calculated by the experiments of the cooling process with different speed and the energy balance equations. The results are shown in Figure 9.70. When the speed is low (< 25 km/h), the head-on air is difficult to force into the adsorber after overcoming the back pressure of adsorber, dampers, and pipes. The airflow is zero. From the figure we can see that the locomotive speed is not a linear relationship with the flow rate of air. Because the higher the speed is, the greater the influence of back pressure is. The locomotive speed is between 35 and 90 km/h, and the airflow can be fitted to a quadratic function of the speed:

$$m_{air} = -1.73 \times 10^{-4} u_{lo}^{2} + 4.83 \times 10^{-2} u_{lo} - 1.14 \tag{9.47}$$

where u_{lo} is the locomotive speed (km/h) and m_{air} is the airflow rate (kg/s).

Figure 9.71 shows the relationship between the adsorption temperature and the adsorption capacity in the cooling adsorption process when the desorption temperature, evaporating temperature, condensing temperature, and air inlet temperature are 250, 7, 50, and 40 °C, respectively, and airflow rate were 0.2, 0.5, 1.0, 1.5, and 2.0 kg/s, separately. It can be seen from the figure that like the influence of the rotation speed on desorption process, when the speed is not large the influence of the speed change on the cooling process is very obvious. This effect is getting smaller and smaller as the speed increases.

From the influence of the rotation speed and locomotive speed on the heating and cooling processes, it can be seen that with the increasing flow rate of the heat transfer fluid (exhaust gas or air), the desorption or adsorption rate is increased. With the increase of fluid flow, the influence of the flow change on the heating/cooling process is smaller. The reason is that the thermal resistance of heat transfer in the adsorption bed is basically a function of

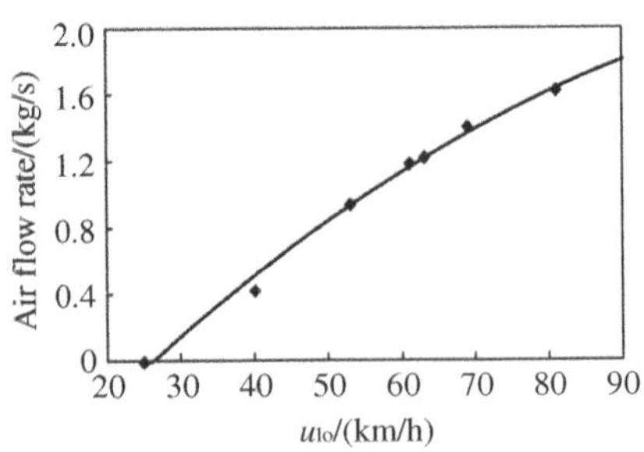

Figure 9.70 Airflow rate vs. locomotive speed

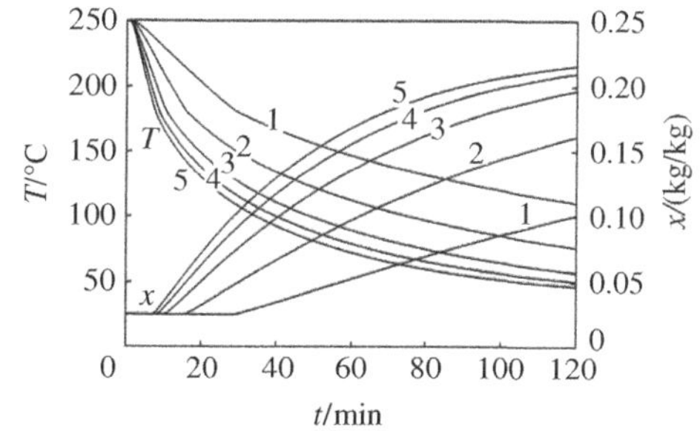

Figure 9.71 Influence of the airflow rate on the adsorption capacity and adsorption temperature

temperature and is relatively constant in the heat transfer process. The heat transfer coefficient at the fluid side is affected by the fluid flow. When the flow is small the heat transfer coefficient at the fluid side is small, which will be the largest thermal resistance in the heat transfer process. When the flow rate is large the heat transfer coefficient at the fluid side is large, then the heat transfer resistance at the adsorber will be the largest thermal resistance.

2. **Optimum operation when the rotation speed and vehicle speed are constants.**
The rotation speed and vehicle speed of the locomotive change with the sections and the train number. The flow conditions of heat transfer fluid (exhaust gas and air) and the ambient temperature cannot be changed arbitrarily. In the operation, set the condensing temperature $T_{cond} = 50\,°C$, evaporation temperature $T_{ev} = 7\,°C$, and inlet air temperature $T_{air,in} = 40\,°C$, the parameters which can be optimized are desorption temperature, adsorption temperature, start time for heating, and start time for cooling. In the following analysis, the most important target is the optimization of the average cooling capacity, and then the cycle time should be considered to be shortened, so that the cold storage demand can be decreased. The COP is not so important; however it can be regarded as the reference quantity for optimization.
In the following how to reach the optimum operation when the rotation speed is 700 revolutions/min and the vehicle speed is 70 km/h is researched. Under this condition among four parameters of desorption temperature T_g, adsorption temperature T_a, heating time t_{heat}, and cooling time t_{cool}, only two of them are independent. In the analysis take the desorption temperature and adsorption temperature as independent variables, the calculation results are shown in Figures 9.72 and 9.73.

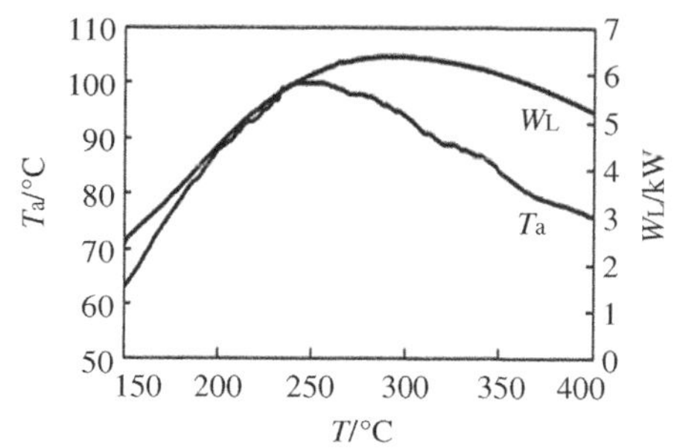

Figure 9.72 Adsorption temperature and refrigeration power vs. desorption temperature

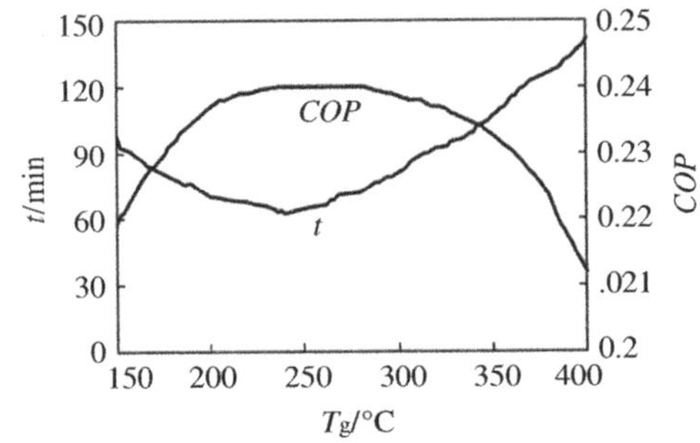

Figure 9.73 Cycle time and COP vs. desorption temperature

Generally when the desorption temperature increases the adsorption temperature should be increased correspondingly. However, the simulation results showed that the best adsorption temperature does not always increase with the increase of desorption temperature. When desorption temperature reaches a relatively high value, the adsorption temperature will decrease as the desorption temperature further increases. The reason for this phenomenon is that the bed temperature is very high. When the adsorption capacity is very small, with the increasing desorption temperature the change of the adsorption capacity is not obvious. The sensible heat of the bed accounts for a growing proportion of the total heat adsorbed by the system. To counteract this part of the useless sensible heat and increase the refrigeration power, the adsorption time should be extended for the smaller desorption temperature.

According to Figures 9.72 and 9.73, the maximum refrigeration power ($W_L = 6.4$ kW) occurs at $T_g = 290\,°C$ and $T_a = 95\,°C$, the corresponding COP = 0.24, and the cycle time $t_{cycle} = 76$ minutes. The minimum cycle time ($t_{cycle} = 62$ minutes) is obtained at $T_g = 240\,°C$ and $T_a = 100\,°C$, the corresponding COP = 0.24, and cooling capacity $W_L = 5.8$ kW. Therefore, under the above conditions, the suitable desorption temperature T_g and adsorption temperature T_a are in the range of 240–290 and 100–95 °C, respectively.

3. **The optimization of the actual working conditions.**

The variations of the adsorbent bed temperature and cooling capacity of the air conditioner in the cooling and heating processes are as shown in Figures 9.74 and 9.75. The figures also give the temperature variation of the adsorption bed in the experiments.

The above two figures show that the bed temperature obtained by simulation is similar to the experimental results. It indicates that the mathematical model is suitable for the actual operation. In the desorption process, the calculated desorption temperature is higher than the experimental value. In the adsorption process, the theoretical cooling power decreases. The theoretical cooling power sharply declines while the locomotive stops, while in the experiments, due to the cold storage function of the regenerator and evaporator, the change of cooling power is relatively smaller. When the cycle starts with the cooling process and heating process, separately, the average cooling power are 4.87 and 3.92 kW, respectively, and the cooling capacity is 52.6 and 42.0 MJ, which is 30 and 28% larger than the experimental values, separately. The main reason is that in the adsorbent bed, especially at the bottom of the adsorbent bed, the adsorption function isn't very strong. So the adsorption rate is smaller than that described in the formula. There are cold and heat losses to the

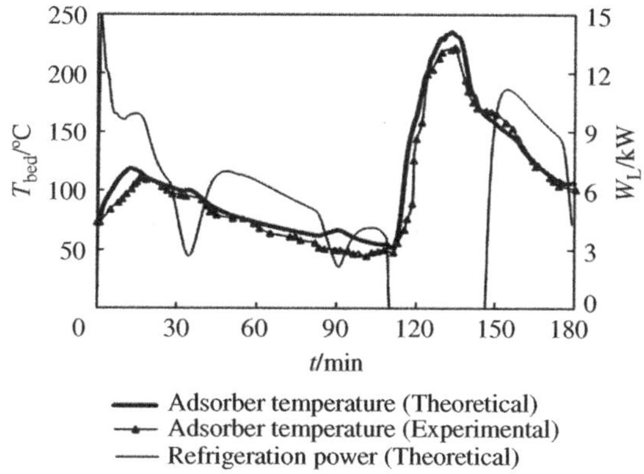

Figure 9.74 The performance under the condition of variable speed when the adsorber is cooled

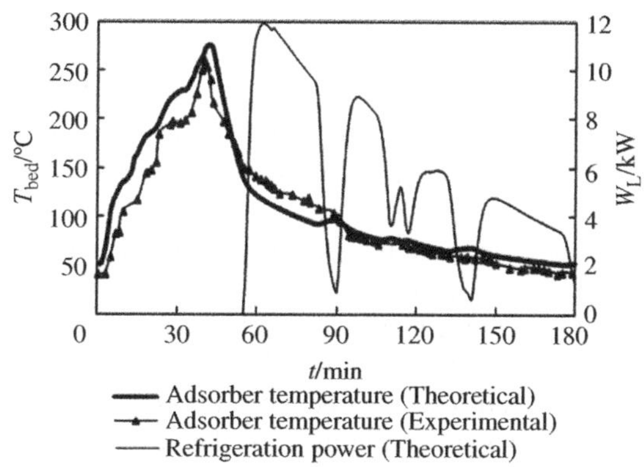

Figure 9.75 The performance under the condition of variable speed when the adsorber is heated

environment in the experimental system, and they are not considered in the calculation models.

4. **Analysis on the optimization of the operation.**

   There are many factors that affect the running of the air conditioner. Actual working conditions must be taken into account in optimization. By the previous analysis, several optimization methods of operation can be concluded:

   a. The adsorption storage is essential for the air conditioner, and similar systems should use this technology as far as possible. If the train stops for a long time, it is better to make the air conditioner at the end of the desorption just before the locomotive stops. Then when the locomotive starts, it will be in the adsorption process and could release the refrigeration capacity.

   b. The number of cycles is determined according to the road condition and running time. Under general conditions, a running time shorter than 2 hours per cycle should be chosen. If the running time of the locomotive is too long then the short cycle time can be

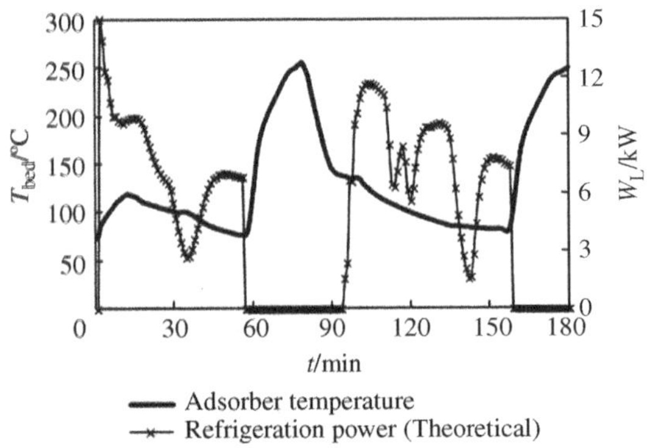

Figure 9.76 Optimized adsorber temperature and refrigeration power vs. running time

considered. The number of cycles can be two or more. For each cycle it starts with adsorption and ends up with desorption.

c. It is better to proceed the desorption process while the speed is high. Generally a switch from adsorption to desorption occurs when the locomotive starts. At this time the engine speed is high and the speed is slow, which is conducive to desorption but can be detrimental to adsorption.
d. When two or more cycles are used, it is better to set the desorption temperatures T_g and adsorption temperatures T_a in the range of 220–300 and 90–100 °C, also the change of locomotive speed and the engine speed should be taken into account.

According to the above optimization methods, the air conditioner in the locomotive from Hangzhou to Jinhua is optimized. The adsorbent bed temperature and refrigeration power of two cycles are simulated and shown in Figure 9.76. The cycle time for the two cycles is 77 and 103 minutes, respectively, and the cooling capacity is 24.7 and 27.9 MJ, separately, the total cooling capacity is 51.6 MJ, which equals to the one shown in Figure 9.74 (52.6 MJ). The optimized operation is chosen for the condition of that the cold storage capacity which is after the heating and desorption phase and for the stopping phase of locomotive is larger than that shown in Figure 9.74, and such a result is better for the next cycle.

9.8.2 *The Application of the Silica Gel–Water Adsorption Chiller in CCHP System*

For recovering the waste heat CCHP (cogeneration system for cooling, heat, and power) system is also regarded as the important energy technology development direction and has received more and more attention in recent years [45]. The CCHP system is an advanced energy utilization technology. It is a synthesized and distributed energy producing and utilization system with cascading utilization methods. Firstly, the primary energy (natural gas) is used to drive a variety of generator and then the waste heat is recovered by a variety of waste heat recovery equipment (absorption chillers, adsorption chillers, waste heat recovery boilers, dry dehumidification equipment, and heat exchangers, etc.). Such a system provides the electricity, refrigeration, heating, and sanitary hot water to the users.

9.8.2.1 The Advantages and Development of theCCHP System

Compared with the conventional distributed energy systems, the CCHP system has the following significant advantages:

1. **Energy conservation and high comprehensive utilization rate.**
   CCHP not only improves the utilization of low grade heat, it also improves the comprehensive utilization rate of energy. The form of energy in the conventional centralized power supply system is single. When the user requires power as well as other types of energy such as cold and heat supply, they can only be obtained by the conversion process of electricity. The distributed power supply is in small-scale and flexible and it can meet users' needs and achieve synthetic cascading utilization of energy. What's more, it overcomes the difficulties of the long-distance transmission of cold and heat.
   Take the heat utilization in the building as an example, we can compare the performance of aheating system driven by the electricity, boiler (coal, oil, or natural gas), and conventional vapor compression heat pump. The heat of the thermal energy of the primary energy fuel is regarded as the input energy, and the heat required by the terminal building is regarded as the output energy, the heating coefficient E is regarded as the ratio between the output and input. For the heat provided by the electricity, the thermal energy conversion of the electricity in power plants is only 33%. In this process the high efficient electrical energy is converted into the low grade thermal energy. For the heat provided by the boiler combustion, only 70% of the thermal energy is used due to the efficiency of the boiler and heat loss of the pipes in transmission. For the electrical heat pump, although there is only 33% of thermal energy in the power plant transformed into electricity at the user end, the heat pump can absorb heat from low temperature heat source (environmental environment) at a certain extent, and heating coefficient can reach 3, so $E = 0.33 \times 3 = 0.99$. There is lots of high-temperature waste heat in the electrical generation process, and it accounts for 70% of the whole energy input. If 55% of the waste energy is recovered the actual heating efficiency reaches up to $E = 0.99 + 0.55 = 1.45$. In fact, the complex energy system driven by the engine can generate power, heat, and cold simultaneously, and the energy utilization rate of such systems can reach more than 70% [46].
2. **Shift loads, alleviate the power shortage, and achieve the balance of energy consumption of different seasons.**
   The CCHP system uses lithium bromide absorption chillers and adsorption chillers as refrigeration devices. They are driven by low-grade heat. Compared with compression chillers, the advantage of absorption chillers and adsorption chillers is energy saving. For instance: absorption chillers with 3500 kW can save about 890 kW. So the installation of a lithium bromide absorption chiller is equivalent to the construction of a small power station [47]. Generally gas consumption is reduced and the electricity requirement increases in summer time, thus if the gas powered cooling unit can be used the peak of the power utilization can be reduced and the different energy consumption of the season can be balanced.
3. **Environmental protection.**
   Absorption refrigeration and adsorption refrigeration technology use the environmental benign and natural refrigerants, and the energy utilization efficiency of the CCHP system is high. The CCHP system has great potential to reduce carbon and pollution of air emissions. The experts estimate that if from 2010 onwards 50% of new buildings can use such

a technology, then up until 2020, carbon dioxide emissions will be reduced by 19%. If the proportion of existing buildings utilizing the CCHP technology increases from 4 to 8%, then carbon dioxide emissions will be reduced by 30% until 2020 [48–51].

4. **Energy security.**

   Distributed energy is a relatively independent energy supply system. It doesn't need to rely on external power system. So, how to effectively improve the energy efficiency of the distributed power supply system is one of the main obstacles for the development of distributed power supply technology. Compared with a simple power supply system, the CCHP system can significantly improve system energy efficiency while reducing environmental pollution. Therefore, CCHP technology is one of the main development directions of distributed power supply system.In 2000, <Regulations on the development of cogeneration>, that is, document [2000] No. 1268 is issued by the State Development Planning Commission, State Economy and Trade Commission, Ministry of Construction and the State Environmental Protection Administration of China. It states that China encourages the policy of developing CHP, and supports the projects of CCHP by gas turbine with natural gas, especially the small type of the units. At present, Beijing, Shanghai, and other cities plan to implement the CCHP technology. However, if this technology attains civilian and commercial development, it must be based on different conditions of different countries.

   From the current research, the CCHP system tends to be compact, integrated, intelligent, and have higher standards for environmental protection. The keys of making the system compact are the thermoelectric conversion devices and hot/cold switching devices. There are some typical systems, such as the micro-CCHP systems based on Stirling machine by SOLO company in German, it generates power in the range of 2–9 kW, the waste energy is in the range of 8–24 kW; a household CCHP system based on a Stirling machine manufactured by Whisper company in New Zealand, it has the power output of 730 W, and the waste heat is the 6 kW; a micro-CCHP system based on solid polymer fuel cell developed by EBARA Electric Manufacturing Company, the power generation efficiency is about 30%, and the combined heat efficiency is 70%; a small-scale waste heat solid adsorption chiller developed by Shanghai Jiao Tong University, its cooling power is in the range of 4–10 kW, and the cooling coefficient is in the range of 0.25–0.45.

9.8.2.2 The Choice of the Combustion Engine for the CCHP System

Shanghai Jiao Tong University developed a micro-CCHP system with silica gel–water adsorption chiller. The designed cooling capacity of the chiller is 10 kW.

The most important components of the CCHP system are thermoelectric conversion devices and hot/cold switching devices. Generally the thermoelectric conversion devices for commercialized CCHP systems are gas (oil) turbines, gas (petrol) engine-generator sets (internal combustion), and gas (oil) external combustion generator sets (Stirling). Fuel cells are still at the laboratory research stage. Table 9.11 shows three engines which are common in the market for the micro-CCHP system and performance parameters of the fuel cell which is still at the laboratory research stage.

Gas (fuel) drives the internal combustion engine work and then drives the turbine for power generation. In this process, the waste heat is generated in two parts. One part is from the high-temperature exhaust gas, whose temperature is in the range of 500–600 °C, and it can be

Table 9.11 Performance parameters of different engines for compact CCHP system [52, 53]

Parameters	Gas engine	Micro turbine	Stirling engine	Fuel cell (PEM cells)
Output power (kW)	10–200	25–250	2–50	2–200
Efficiency for power generation with full load (%)	25–45	25–30	15–35	40
Efficiency for power generation with 50% load (%)	23–40	20–25	=35	40
Thermoelectric efficiency (%)	75–85	75–85	75–85	75–85
Ratio between power and heat	0.5–1.1	0.5–0.6	0.3–0.7	0.9–1.1
Temperature range for output heat (°C)	85–100[a]	85–100[a]	60–80[a]	60–80
Maintenance period (h)	5 000–20 000	20 000–30 000	≈5000	–
Investment cost (US$/electric kW)	800–1 500	900–1 500	1300–2000	2500–3500
Maintenance cost (US$/kW)	0.012–0.02	0.005–0.015	0.01–0.018	0.01–0.03

[a]Hot water temperature.

recovered by the recovery devices (heat exchanger or waste heat boiler); another part is from the jacket cooling water of the internal combustion engine, whose temperature is in the range of 70–95 °C, and it can generate the hot water directly by heat exchanging process. Waste heat can be used for refrigeration and heat production. Figure 9.77 shows the distributed energy supply system, which uses an internal combustion engine as the engine. Figure 9.78 shows the energy utilization of a distributed energy supply system. It can be seen from Figure 9.78 that through the recovery of high-temperature exhaust gas and waste heat of jacket cooling water,

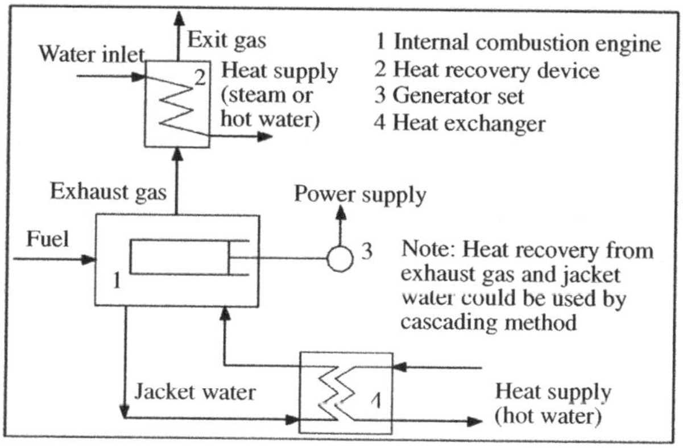

Figure 9.77 Schematic of the distributed energy supply system with the internal combustion engine

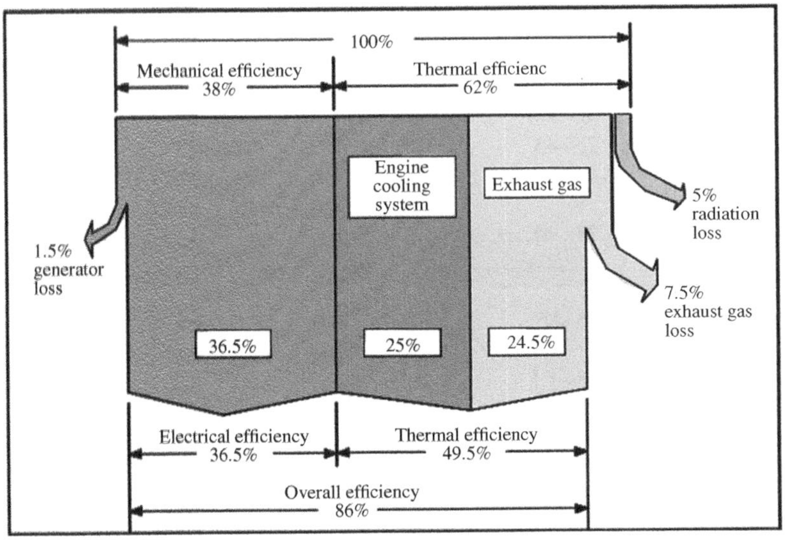

Figure 9.78 Schematic of the energy utilization of distributed energy supply system with the internal combustion engine

the effective energy utilization rate can be as high as 86%. The internal combustion engine in the CCHP system has the following advantages:

1. The specifications of the internal combustion engine available on the market are extensively from 4 to 5 kW. Users can choose the right type of the engine very easily. The investment of the internal combustion engine is lower than the gas turbines and Stirling engines.
2. An internal combustion engine can output hot water and low pressure steam according to the needs easily.
3. Internal combustion engine can start quickly, and this allows it to return to work quickly from stopping state. Under the peak or emergency circumstances the internal combustion engine can supply the power quickly according to the demand.
4. When the power of the engine is turned off suddenly little auxiliary power is needed to start the internal combustion engine, and the battery is enough for this work.
5. As the internal combustion engine still has high efficiency at partial load, it ensures that the engine can work economically under the conditions of different electricity loads. When the internal combustion engine runs at 50% load its efficiency is only 8–10% lower than that running at full load; while when the gas turbine runs at 50% load the efficiency is usually 15–25% lower than running at full load.
6. Practice has also proved that the operational reliability of the internal combustion engine is still quite high if it is properly maintained.
7. The proportion of NOx and micro particles in diesel emission is relatively high. However, the gas engine is fairly environmentally friendly. The main gas engine emission is NOx. Table 9.12 shows the comparison of NOx emission from different types of internal combustion engines. It can be seen that NOx emissions from the gas engine is low.

Table 9.12 Comparison of the emission of NOx with different fuels in combustion engine

Type of the combustion engine	Fuel	NOx (ppmv)	NOx (mg/kWh)
Diesel engine	Light oil	450–1350	7–18
Diesel engine	Heavy oil	900–1800	12–20
Gas engine	Nature gas	45–150	0.7–2.5

Compare absorption chillers and adsorption chillers, absorption chillers require the heat source with high temperature. The heat source needed by a single effect lithium bromide absorption chiller is generally the saturated steam at the pressure of 0.03–0.15 MPa or the hot water in the temperature range of 85–150 °C [52]. For small gas engines, the exhaust gas temperature is above 500 °C, the waste heat can be used to drive the double effect lithium bromide absorption chillers. But such a scheme cannot use the waste heat recycled from jacket water (about 80 °C). For a adsorption refrigeration unit it can recycle both types of the waste heat from the jacket water and exhaust gas, and the amount of the waste heat used is about two times that of only the waste heat from the exhaust gas. As the temperature of the jacket water for recovering waste heat is only 80 °C, so the adsorption working pair can be chosen for micro gas engine is only silica gel-water.

9.8.2.3 Design of a CCHP System

The micro type CCHP system includes a gas engine power generation subsystem, waste heat recovery subsystem of exhaust gas and jacket cooling water (waste heat utilization subsystem), and control subsystem, as shown in Figure 9.79 [54]. The gas engine power generation subsystem provides the electricity as well as the jacket cooling circulating water with a relatively high temperature as the heat source for the adsorption refrigeration system.

The gas engine power generation subsystem includes gas engine, three-phase AC induction generator sets and high-temperature exhaust gas heat exchanger. The gas engine is connected with the three-phase AC synchronous generator by coupling joint, and the three-phase AC induction generator sets output the power. The exhaust gas heat exchanger connects the exhaust port of the gas engine through the pipeline, and it recycles the waste heat of exhaust gas by the jacket cooling water. The waste heat utilization subsystem includes adsorption chiller, water heater for heating, water heater for living hot water, volumetric heat exchangers, cooling towers, terminal devices for air-conditioning, and terminal electrical equipments. Adsorption chiller and water heater for heating are paralleled installed, and they are controlled by the valves on the pipes. They connected with the water heater for living hot water and adsorption chiller in series by pipes. The cooling water tower is used for both the cooling process of the adsorption chiller and volumetric heat exchangers. The control subsystem includes protection components, the analyzing and controlling components for the temperature of the return jacket water of the gas engine. The protection components are mainly referred to the components for the overheated temperature protection of the return jacket water and shortage of circulation water in the system. For the return jacket water the parameters of temperature, flow rate and pressure in the system need to be analyzed. The micro-CCHP system consists of seven circuits,

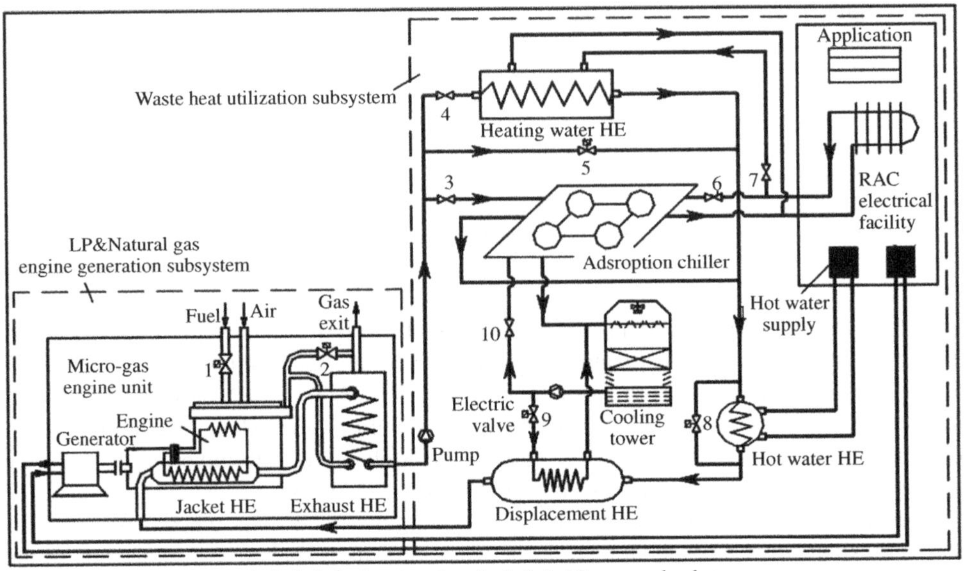

Figure 9.79 Schematic of the micro cooling, heating, and power generation system

and they are the jacket cooling water circuit, chilling water circuit, cooling water circuit, hot water circulation loop, the fuel supply circuit, the recovery loop of waste heat, and power supply circuit.

The working principles of the system are as follows: the natural gas drives the gas engine, which drives the generator to output the electricity. After the jacket water of the gas engine absorbs the waste heat of the cylinder liner and lubricants, it flows into the exhaust gas heat exchanger, and after that it goes into the heat using subsystem. Then the flow direction is determined by the working mode. If the system is at the refrigeration mode valves 3, 6, and 10 will open, the water from the high-temperature jacket will flow into the adsorption chiller. Adsorbent bed is heated and starts to desorb, while the desorption heat and condensation heat are taken away by the cooling water. The chilling water flows into the end devices (fan coil) of the air conditioner and the output has the cooling effect. If the system is at the heating mode valves 3, 5, 6, 10 close and valve 4, 7 open, the jacket water flows into the water heater, which supplies the hot water and achieves the heating effect. If the system works as the water heater for living hot water, valves 3, 4, 6,7,10 closes, and valve 5 opens, then the circulating jacket water flows into the water heater and supplies the living hot water for users. Finally the circulating jacket water at relatively low temperature flows into the jacket heat exchanger through the volume heat exchanger again. By different choices the system can satisfy the stable cooling, heating, and electricity supply for all seasons.

In the working process, the circulating water flow rate is kept constant. The inlet temperature of the cylinder liner heat exchanger of the gas engine is determined by controlling the cooling water inlet temperature of the combustion engine and the outlet cooling water temperature of the cylinder liner heat exchanger by the cogeneration system. The method of controlling the cooling water inlet temperature is as follows: by regulating electromagnetic valves 5, 8, 9, the cooling water inlet temperature of the jacket in the internal combustion engine is controlled within the allowable range; when the inlet water temperature is too high, the solenoid valves 5, 8 shut down, and the opening of the solenoid valve 9 is increased; when the inlet water temperature is too low, the opening of the solenoid valve 9 is decreased. For this process the solenoid valve 9 also can be closed for meeting the requirements, and correspondingly the opening of the solenoid valve 8 should be increased. If the solenoid valve 8 is opened fully and the temperature still cannot satisfy the requirement, the opening of the solenoid valve 5 should be increased. In addition, if the solenoid valve 9 opens, the solenoid valve 2 should open at the same time; if the solenoid valve 9 closes, the solenoid valve 2 also should be closed. The functional relationship of the two openings of the valves can be determined experimentally. The reason is that opening the solenoid valve 9 means the waste heat recovered from the heat recovery subsystem is higher than that of the requirement. This part of the waste heat should be consumed by the cooling tower, which increases the total energy consumption. So open solenoid valve 2, and a part of the exhaust gas could be bypassed to reduce the waste heat recovered, which could keep the waste heat amount provided balanced with that required. When the system is at refrigeration mode, cooling water tower and the adsorption chiller will be coordinately controlled. When the adsorption chiller starts to work, the cooling water pump and the fan will work at the same time. When the system is at heating or providing the living hot water modes, the cooling water tower and solenoid valve 9 will be coordinately controlled. When the solenoid valve 9 opens, the cooling water pump and the fan will work at the same time.

Compared with the conventional micro-CCHP system, this micro-CCHP system has three significant features:

1. Most conventional small (micro) CCHP systems use waste heat absorption chillers as hot/cold conversion devices, and some units use the waste heat adsorption chiller as hot/cold switching devices. The cooling capacity generally is more than 100 kW. In this research the CCHP system uses the small-scale waste heat adsorption chiller, and the cooling capacity is lower than 10 kW. It can be widely used as the cooling, heating, and power supply systems for family and small business premises (such as clubhouse, swimming pool, stadium, etc.).
2. Mostly conventional small (micro) CCHP systems use small gas turbines, micro gas turbines or diesel engines as a power generating devices, and most units recover the waste heat from high-temperature exhaust gas. The micro-CCHP system developed by SJTU (Shanghai Jiao Tong University) uses the small gas engine as the power generating device, and recovered the waste heat both from the exhaust gas and jacket cooling water. The jacket cooling water flows into the waste heat utilization system, thus a water-water heat exchanger is reduced, which saves the initial investment, and improves the system thermal efficiency.
3. The micro-CCHP system developed by SJTU has three operating modes, which makes sure that the system can be in normal operation under design conditions and un-design conditions for all seasons.

The parameters and performance of the developed micro-CCHP system are shown in Table 9.13.

Table 9.13 The parameters and performance of experimental system

Micro cooling, heating, and power generation system			
Maximum output electric power	Maximum output heating power	Maximum efficiency for generating electricity	The total system efficiency
>12 kW, 400 V/230 V, 50 Hz	>25 kW (>50 °C hot water)	>20%	>70%

Mini-type gas engine generator				
Maximum electric power output	Rotation speed	Maximum efficiency	Noise of bare machine	Noise outside the box
<12 kW	1500 revolutions/min	>20%	<95 dBA(1 m)	<75 dBA(1 m)

Adsorption chiller driven by waste heat	
Maximum cooling capacity	Maximum cooling COP
>10 kW	>0.3

The options of the various components in the system are described as follows:

1. **Gas engine generator sets.**
   If one only takes the system design and performance of the internal combustion engine generator set into account, the GEL17.5 gas generator produced by the U.S. Caterpillar Generator Company and GNAC12.5 gas generator produced by Cummins Generator Company meet the requirements and have good quality. But its investment is high (about $9-12 \times 10^4$ RMB Yuan/unit). From the point of view of economy, the high price of the equipment makes the payback period increase. Compare small gas engine generators produced by different manufacturers, the system uses TCS295-STC12 small gas engine generator produced by Shanxi Liquan Tianci Manufacturing Co., Ltd, China, and the price is only about $2.0-2.5 \times 10^4$ RMB Yuan/unit. Table 9.14 shows the main technical specifications and technical parameters of the unit.
2. **The selection of the small-scale adsorption chiller**
   The system uses the silica gel-water adsorption chiller as described in Chapter 8. Performance of the chiller is shown in Table 9.15.
3. **Design of heat exchangers**
   The micro-CCHP system has two heat exchangers. One is the heat exchanger of the exhaust gas and water, and the other is the heat exchanger of the waste heat of generator and the cooling tower.
   The exhaust gas-water heat exchanger is used to recycle the waste heat from the exhaust gas. The system uses a plate-fin heat exchanger as an exhaust gas heat exchanger. The working media are the exhaust gas and water. As there is a large temperature difference between the exhaust gas and water, flat type fins are selected. To increase the heat transfer coefficient, the cross-flow is used. The heat transfer coefficient based on the heat transfer area at the gas side is 63 W/(m^2 °C), and the heat transfer coefficient based on the heat transfer area at the water side is 96.7 W/(m^2 °C).

Table 9.14 Technical specifications and technical parameters of mini-type gas generator TCS295-STC12

Mini-type gas engine TCS295			
Fuel	Natural gas or LPG	Gas pressure (MPa)	0.03–0.05
Gas consumption (m^3/kWh)	0.248	Cooling water temperature (°C)	≤ 95
Rated power (kW)	13	Rated rotation speed (rpm)	1800
Exhausted gas temperature (°C)	580	Discharge pressure (Pa)	< 2000
Three-phase synchronous generators STC12			
Rated generator power (kW)	12	Rotation speed	1500
Line voltage (V)	400	Phase voltage (V)	230
Frequency (Hz)	50	Rated current (A)	21.7
Mini-type gas generator TCS295-STC12			
Peak generating capacity (kW)	12	Peak generating efficiency (%)	21
NO_x discharge (gm/kWh)	0.7	Noise (dBA)	< 75 (1 m)
The maintenance period (h)	48 000	Size (L × W × H, m^3)	1.26 × 0.61 × 1.13

Table 9.15 Technical specifications and performance of a mini-type adsorption air conditioner driven by waste heat

Parameter		Data
Rated cooling capacity (kW)	8.5	Wet condition (inlet/outlet temperature of chilling water is 15/11 °C)
	10.0	Dry condition (inlet/outlet temperature of chilling water is 20/15 °C)
Hot water temperature of driven source (°C)		60–95
Inlet/outlet temperature of cooling water (°C)		32/38
Flow rate of hot water (m^3/h)		4.0
Flow rate of cooling water (m^3/h)		5.0
Flow rate of chilling water (m^3/h)		1.8
COP		0.3–0.4 for wet condition and 0.35–0.5 for dry condition
The mass of the adsorbent (kg)		102
Size (L × W × H, m^3)		1.26 × 0.61 × 1.13
Price (×10^4 RMB Yuan)		5–6

The plate heat exchanger is used for the heat exchange between the internal combustion engine generator and cooling water tower. When the micro-CCHP system is in cogeneration mode, all the waste heat recovered must exchange the heat through the plate heat exchanger and cooling tower. A small stainless steel plate heat exchanger is used in the system, and the plate is in the form of corrugated type. It has the advantages of compact structure, high heat transfer efficiency, and it can also be operated and maintained easily.

9.8.2.4 Construction of the Micro-CCHP System

The actual installation of the experimental system is shown in Figure 9.80. Combined heat and power generation conditions and CCHP conditions are tested separately, and the environment has little impact on the experimental results.

In the performance testing process, the analog user of electricity, cooling, and heat are designed. The cooling water tower is the analog user of heat. The user of refrigeration is the end device of the air conditioning unit in the room-fan coil and the heat balance tank. Chilling water goes through the fan coil firstly, then goes into the heat balance tank with constant temperature, and finally flows back to the adsorption chiller. This design can precisely control the adsorption chiller and the inlet temperature of the chilling water. Power consumption of the heat balance tank is from the power generator sets. Taking into account that the rated power of the small internal combustion engine generator is12 kW, the electric analog users are made up of 60 bulbs whose power is 200 W and they are parallel installed in the circuit. These bulbs are divided into five groups, and the power of each group is 600, 1200, 2400, 3000, and 4800 W, respectively. Each group of bulbs can be controlled individually. So the electrical load can be increased gradually from 600 W to 12 kW (the interval is 600 W).

9.8.2.5 Economic Analysis of the Micro-CHP System

Economic analysis is based on the full load operation of the combined supply system in a building throughout the year. The waste heat recovered is used for providing refrigeration power in

Figure 9.80 Schematic of the installation of micro-CCHP system

Table 9.16 Running time and the output power of CCHP

	Working time (h)	Power generation (kW)	Refrigeration power (kW)	Heating load for heat pump (kW)	Heating load for hot water (kW)
Summer	3672	12	8.5 (wet condition)	0	0
Winter	2640	12	0	31	0
Transient seasons	2448	12	0	0	31

summer and the heat in winter, domestic hot water is provided in the transition quarter, and all the waste heat of the system is fully utilized. The application area is Shanghai city, China. The period for requiring the refrigeration in summer is from 15 May to 15 October, and the period for requiring heat in winter is from 20 November to 10 March. Operation period and energy output are shown in Table 9.16. The cost of the cold and heat is calculated based on the small heat pump type air conditioner, whose refrigeration coefficient and heating coefficient are 2.8 and 3.2, respectively. According to the electricity price of Shanghai hotels, commercial district, and residential district, the cooling/heating prices can be calculated, and the results are shown in Table 9.17. The price for domestic hot water is 16 RMB Yuan/ton [55]. It is increased from 25 to 55 °C in spring and autumn. Liquefied petroleum gas drives the micro-CCHP system. However, as the natural gas will be used to drive CCHP system in the future market, it was used as the input fuel for the economic analysis. The low heating value of the natural gas is 35 200 kJ/nm^3, the prices of natural gas are 1.4, 1.6, 1.7, and 1.9 RMB Yuan/Nm^3. The annual maintenance cost is 0.05 RMB Yuan/kWh. The initial investment of the system is shown in Table 9.18. The power consumption during system operation (including pumps, fan in cooling tower, and control cabinets, etc.) is 1.2 kW in the combined cooling and power generation mode and it is 0.37 kW in the combined heating and power generation mode. Table 9.19 shows the payback period of the CCHP system with different energy prices.

The investment payback period of the CCHP system is calculated as follows:

$$N = \frac{F}{A} = \frac{\text{Initial investment}}{\text{Annual net profit of the system}}\ \text{(Year)} \tag{9.48}$$

If the capital maintenance is considered, the investment payback period should meet this formula:

$$A\sum_{n=1}^{N}\frac{1}{(1+r)^n} = F \tag{9.49}$$

where A is the annual net profit of the system, r is the market discount rate (value of 5%), and F is the initial investment.

Table 9.17 The price for the energy of the CCHP system for different application occasions

Application	Price for electricity (RMB Yuan/kWh)	Price for refrigeration (RMB Yuan/kWh)	Price for heat (RMB Yuan/kWh)
Small hotel	0.824	0.294	0.258
Small commercial buildings	0.754	0.269	0.236
Residential building	0.610	0.218	0.191

Table 9.18 The initial investment of the CCHP system

Type	Mini-type gas generator	Adsorption chiller (including the cooling tower)	Heat exchangers, water tank, pump, and so on, as well as the fee for their installation	Total fee
Fee (RMB Yuan)	20 000	50 000	10 000	80 000

Table 9.19 The payback period of the CCHP system with different prices of energy

Application	The price of natural gas (RMB Yuan/Nm^3)	1.4	1.6	1.7	1.8	1.9
	The electricity price (RMB Yuan/kWh)	Payback period (Year)				
Small hotel	0.824	1.13	1.31	1.43	1.57	1.75
Small commercial building	0.754	1.30	1.56	1.73	1.94	2.20
Residential building	0.610	1.91	2.51	2.98	3.67	4.76

For residential buildings when the price of natural gas is 1.7 RMB Yuan/Nm^3. The investment payback period is about 3.3 years by calculation. From Table 9.19, it can be concluded that the price of the energy greatly impacts on the economy of the system. The higher the energy price is and the lower the price of the natural gas is, the better the economy of the system is. When the price of the natural gas is lower than 1.9 RMB Yuan/Nm^3, the user can recover the initial investment of the micro-CCHP system within five years. Therefore, it is feasible for the application.

9.8.2.6 Experimental Results and Analysis of the Micro CCHP System

The most important factor influencing the performance of the micro-CCHP system is to couple the generator sets and adsorption refrigeration unit reasonably. The technique of applying the waste heat of generators for the heat pump or hot water in the buildings had already been used extensively; however little research has been done on the adsorption refrigeration systems driven by the waste heat of the generator. For the CCHP system introduced here, the waste heat from the generator is supplied to the adsorption chiller. The system generates refrigeration and power at the same time.

1. **Power generation capacity of the system, the amount of waste heat recovered and the refrigeration capacity of the adsorption refrigeration unit.**
   In the experiments, when the adsorption chiller is driven by the waste heat of the generator, the variations of the power generation capacity of the micro-CCHP system and the heat required by the adsorption chiller are shown in Figure 9.81. The inlet water temperature of the jacket water is higher than 60 °C, so the minimum power output of the unit is

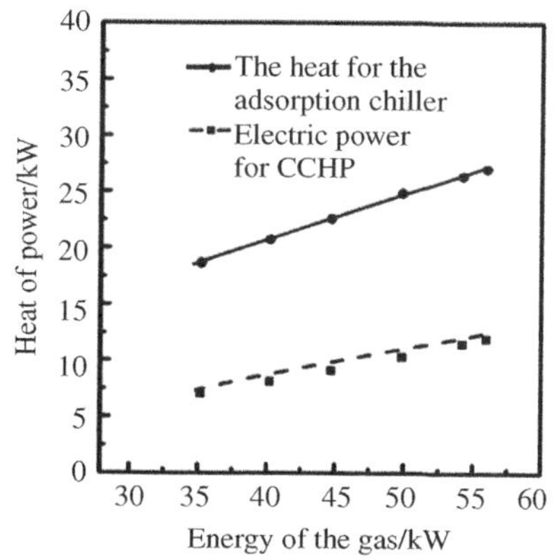

Figure 9.81 The power generation and the heat for the adsorption chiller

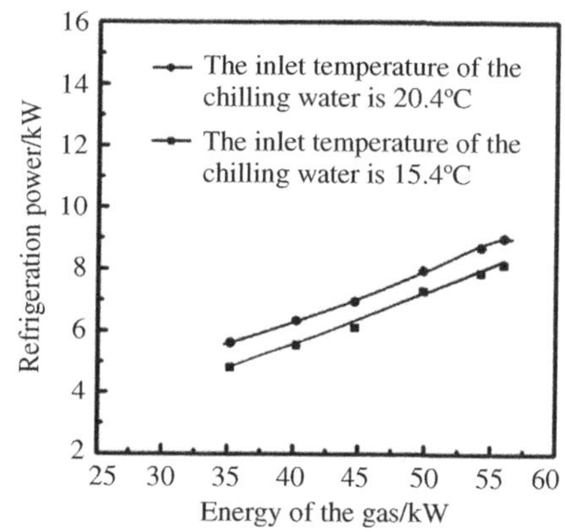

Figure 9.82 Refrigeration capacity of the adsorption chiller

7.02 kW as shown in Figure 9.81. When the unit runs at full load, whether it is under the dry condition or wet condition for the air conditioner, experimental results show that the waste heat recovered from the generator is 27 kW, and all the heat is supplied to the adsorption refrigeration units.

Figure 9.82 shows the variation of the cooling capacity of the adsorption chiller when the waste heat of the generator is preferentially supplied to adsorption refrigeration unit. From the figure it can be concluded that refrigeration capacity is in the range of 5.61–9.0 kW under the dry air conditioning condition while the refrigeration capacity is in the range of 4.83–8.14 kW under the wet air conditioning condition.

2. **The temperature measured in the experiments**

Figure 9.83 shows the variations of the hot water inlet temperature of the adsorption chiller when the CCHP works under the condition of cooling and power generation mode. It can be concluded that when the input energy increases the average hot water inlet temperature gradually increases, and consequently the cooling power and COP increases.

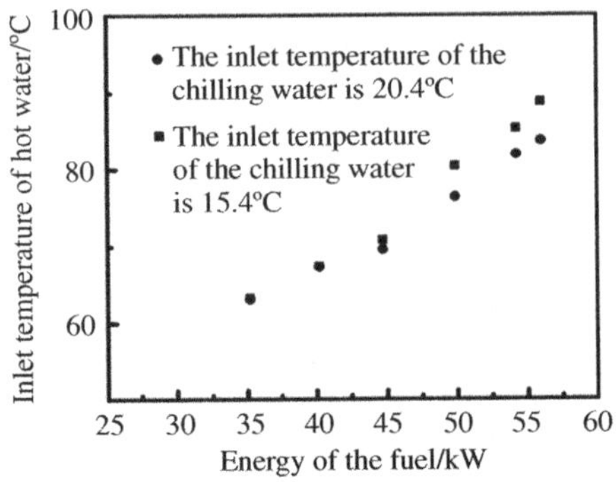

Figure 9.83 The average hot water inlet temperature of the adsorption chiller

During the experiments we keep the cooling capacity higher than 7.02 kW when the jacket water inlet temperature is higher than 60 °C. In this case, when the system works under dry air conditioning condition the minimum hot water inlet temperature is 63.1 °C and the maximum hot water inlet temperature is 83.7 °C. While under the wet air conditioning condition the minimum hot water inlet temperature is 63.2 °C and the maximum hot water inlet temperature is 88.8 °C.

The inlet and outlet temperature of the hot water and cooling water of other components of the CCHP in cooling and power generation mode are shown in Tables 9.20 and 9.21. Also the corresponding hot water flow rate, the chilled water flow rate and cooling water flow rate are listed in the table. Table 9.20 lists the experimental data under dry air conditioning condition, and Table 9.21 lists the experimental data under wet air conditioning condition. The amount of waste heat recovered from the generator, the cooling power of adsorption chiller, heating power, and the cooling power of the cooling towers can be calculated from the data in the tables. Theoretically the cooling power of the cooling tower should be equal

Table 9.20 Parameters under the dry condition

Input power of the CCHP (kW)	35.22	40.23	44.71	49.89	54.24	56.00
Power generation (kW)	7.02	8.12	9.10	10.30	11.49	12.00
Refrigeration capacity (kW)	5.61	6.32	6.95	7.95	8.70	9.00
Inlet water temperature of exhaust gas heat exchanger (°C)	61.7	65.6	67.7	74.2	79.3	81.0
Outlet water temperature of exhaust gas heat exchanger (°C)	63.2	67.4	69.7	76.5	82.0	83.8
Inlet hot water temperature of adsorption chiller (°C)	63.1	67.3	69.6	76.4	81.9	83.7
Outlet hot water temperature of adsorption chiller (°C)	59.1	62.8	64.8	71.1	76.3	78.0
Inlet hot water temperature of jacket water for cylinder of the generator (°C)	59.0	62.7	64.7	71.0	76.2	77.8
Outlet hot water temperature of jacket water for cylinder of the generator (°C)	61.7	65.6	67.7	74.2	79.4	81.0
Inlet chilling water temperature of adsorption chiller (°C)	20.1	20.4	20.8	20.7	20.4	20.3
Outlet chilling water temperature of adsorption chiller (°C)	16.5	16.3	16.3	15.4	14.7	14.4
Inlet cooling water temperature of adsorption chiller (°C)	30.4	30.1	30.1	30.1	30.6	30.6
Outlet cooling water temperature of adsorption chiller (°C)	36.0	36.0	36.9	37.2	38.1	38.5
Flow rate of the hot water (m^3/h)	4.01	4.01	4.02	4.03	4.01	4.02
Flow rate of the chilling water (m^3/h)	1.32	1.31	1.32	1.29	1.32	1.31
Flow rate of the cooling water (m^3/h)	3.90	3.84	3.85	3.89	3.87	3.86

Table 9.21 Parameters under the wet condition

Input power of the CCHP (kW)	35.22	40.23	44.71	49.89	54.24	56.00
Power generation (kW)	7.02	8.12	9.10	10.30	11.49	12.00
Refrigeration capacity (kW)	4.83	5.53	6.11	7.28	7.88	8.14
Inlet water temperature of exhaust gas heat exchanger (°C)	61.8	65.7	68.9	78.1	82.8	86.0
Outlet water temperature of exhaust gas heat exchanger (°C)	63.3	67.5	70.9	80.5	85.4	88.9
Inlet hot water temperature of adsorption chiller (°C)	63.2	67.4	70.8	80.4	85.4	88.8
Outlet hot water temperature of adsorption chiller (°C)	59.2	63.0	66.0	75.1	79.7	83.0
Inlet hot water temperature of jacket water for cylinder of the generator (°C)	59.1	62.9	65.9	75.0	79.6	82.9
Outlet hot water temperature of jacket water for cylinder of the generator (°C)	61.9	65.8	69.0	78.2	82.8	86.1
Inlet chilling water temperature of adsorption chiller (°C)	15.9	15.4	15.0	15.7	15.2	15.6
Outlet chilling water temperature of adsorption chiller (°C)	12.6	11.8	11.0	10.9	10.1	10.3
Inlet cooling water temperature of adsorption chiller (°C)	30.6	30.2	30.5	30.4	30.6	30.7
Outlet cooling water temperature of adsorption chiller (°C)	35.6	35.9	36.7	37.3	38.0	38.2
Flow rate of the hot water (m^3/h)	4.01	4.02	4.03	4.02	4.03	4.02
Flow rate of the chilling water (m^3/h)	1.33	1.30	1.32	1.31	1.32	1.32
Flow rate of the cooling water (m^3/h)	3.87	3.86	3.83	3.85	3.84	3.89

to the cooling capacity of the adsorption refrigeration plus heating power. The calculated results show that the actual cooling power of the cooling tower was slightly lower than the sum of cooling power and heating power, the error is less than 6%, which is mainly caused by the heat loss of the adsorption refrigeration unit.

9.8.2.7 Energy-Saving Analysis of the Micro-CCHP System

The energy-saving analysis of the micro-CCHP system is aimed at the optimization of the operation under different experimental conditions. The primary energy utilization of the system is analyzed. The conventional energy system which consists of electric air conditioner, centralized heating by boiler, and the electricity from electric network is compared with the CCHP system, and the cooling, heating, and power loads were calculated under the condition of full load.

The energy consumed by the different devices in the CCHP system has different grades. To compare the energy consumption they are all converted into primary energy consumption. The so-called primary energy ratio (PER) is the ratio of the output energy to the amount of primary energy consumption [56–60]. The higher the PER is, the better the energy saving performance is.

The PER of different combined system can be calculated as follows:

$$PER_{CHP} = \frac{Q_{recovered\_heat} + P_{el}}{Q_{LPG}} \tag{9.50}$$

$$PER_{CCP} = \frac{Q_{cooling\_load} + P_{el}}{Q_{LPG}} \tag{9.51}$$

$$PER_{CCHP} = \frac{Q_{heat\_load} + Q_{cooling\_load} + P_{el}}{Q_{LPG}} \tag{9.52}$$

where $Q_{recovered\_heat}$ is the heat recovered by the system, P_{el} is the power generating capacity of the system, $Q_{cooling\_load}$ is the cooling capacity obtained from the system, Q_{heat_load} is the heat supplied by the system, Q_{LPG} is the input energy of the system, PER_{CHP} is the PER for the cogeneration of the heat and power, PER_{CCP} is the PER for the cogeneration of the cooling and power; and PER_{CCHP} is the PER for the cogeneration of the heat, cooling, and power.

The primary energy utilization ratio of the conventional energy system, that is, distributed heating and power supply system, distributed cooling and power supply system, distributed heating, cooling, and power supply system can be calculated by the following formula:

$$PER_{HP\_conv} = \frac{Q_{recovered\_heat} + P_{el}}{Q_{recovered\_heat}/\eta_{boiler\_conv} + P_{el}/\eta_{el\_conv}} \tag{9.53}$$

$$PER_{CP\_conv} = \frac{Q_{cooling\_load} + P_{el}}{Q_{cooling\_load}/(COP_{el\_chiller\_conv}\eta_{el\_conv}) + P_{el}/\eta_{el\_conv}} \tag{9.54}$$

$$PER_{CHPS\_conv} = \frac{Q_{heat\_load} + Q_{cooling\_load} + P_{el}}{Q_{heat\_load}/\eta_{boiler\_conv} + Q_{cooling\_load}/(COP_{el\_chiller\_conv}\eta_{el\_conv}) + P_{el}/\eta_{el\_conv}} \tag{9.55}$$

where η_{boiler_conv} is the thermal efficiency of the boiler in the conventional distributed energy system, it is set as 85%; η_{el_conv} is the power generation efficiency of the conventional distributed energy system, and it is set as 30%; $COP_{el\_chiller\_conv}$ is the refrigeration coefficient for the air conditioner in the conventional distributed energy system, and it is set as 2.8; PER_{HP_conv} is the primary energy utilization ratio of the conventional distributed heating and power system; PER_{CP_conv} is the primary energy utilization ratio of the conventional distributed cooling and power system; and PER_{CHPS_conv} is the primary energy utilization ratio of the conventional distributed CCHP system.

The comparison of the PER of the micro-CCHP system and the conventional divided energy system is shown in Figure 9.84.

The calculated data in Figure 9.84 is based on the output energy corresponding to the different input energy of the micro-CCHP system. The horizontal axis represents the input energy. For combined heating and power system and combined cooling and power system they have different output electricity, output cooling capacity, and output heating capacity. Based on these parameters the amount of output energy can be obtained as well as the primary energy consumption of the distributed energy system, and then the primary energy consumption efficiency can be calculated. From the figure, when the CCHP works in the combined heating and power generation mode, the primary energy consumption efficiency of the combined system and the divided system will be decreased when the output energy of the system increases, and the decreasing rate becomes smaller with the increasing output energy. When the output energy of the combined system is larger than 300 W the primary energy consumption efficiency of the combined system is larger than that of the divided system, and with the increasing of the output energy the difference between these two systems will be larger. When the input energy of the combined system is larger than 35 kW the difference will be reached to a constant value. The reason is that the power generating efficiency won't change very much when the generating power increases to a certain value. The primary energy consumption efficiency of cogeneration system is 71.5%, which is 1.3 times higher than that of the conventional system. When the CCHP system works in the cooling and power generation mode the primary energy consumption efficiency of the combined system is lower than that of the conventional

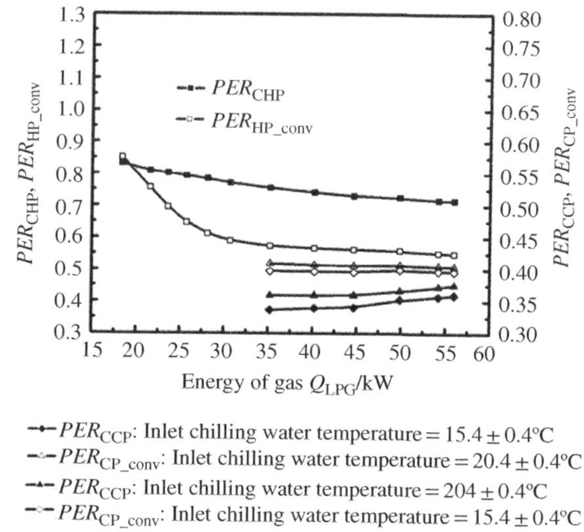

Figure 9.84 Primary energy ratio of the micro-CCHP system and the conventional energy system

system. The reason is that COP of the adsorption chiller driven by waste heat is low. With the increasing of the output energy COP increases, and it makes the difference of the primary energy consumption efficiency between two systems smaller. The primary energy consumption efficiency under the wet condition and dry condition for the air conditioner is 37.5 and 36.0%, respectively, when the output energy of the combined cooling and heating system reaches maximum value. It is 93 and 91% of the primary energy consumption efficiency of the conventional system.

When the energy input is the same as the primary energy consumption efficiency for heat, cooling, and power cogeneration is between that of the CHP system and the CCP system, so is the traditional divided energy system as well. From the view of the energy saving for the micro-CCHP system, because the primary energy consumption efficiency of the CHP system is larger than that of the CCP system, as a result, the proportion of the supplied heat for the user is larger the energy saving performance is better compared with that of the distributed energy system. A new performance parameter, the heat load friction, is proposed. It is the ratio of the heat for users to the recovered heat of the combined system. The larger it is the larger proportion of the heat for users is.

$$\theta = \frac{Q_{heat\_load}}{Q_{recovered\_heat}} \tag{9.56}$$

where θ is the heat load friction.

Figure 9.85 is the variation of the primary energy consumption efficiency and the corresponding energy input with the heat load friction when the system runs at a full load. When the heat load friction is 1.0 there is no cooling output, and it is for the combined heating and power generation. From the figure, the primary energy consumption efficiencies of these two systems increased with the increasing heat output. When the heat load friction is larger than 0.2, the energy conversion performance of the micro CCHP system is better than that of the divided

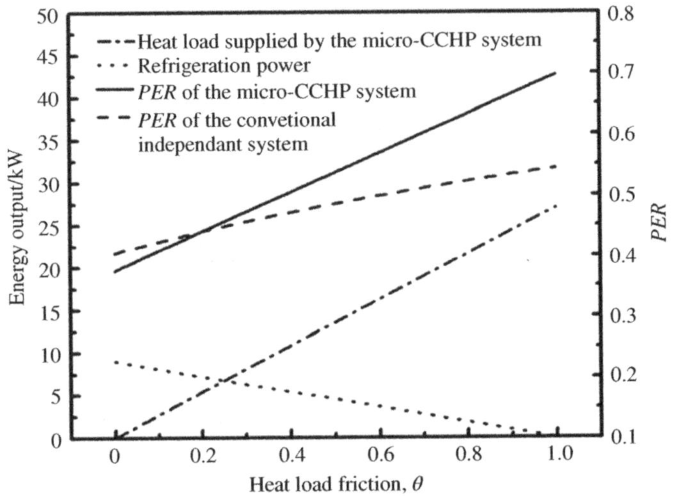

Figure 9.85 Primary energy consumption efficiency and the corresponding energy input vs. the heat load friction

system. With the increasing heat output, the energy conversion performance is better. When the heat load friction is larger than 0.5 the system supplies 14.1 kW heat and 4.7 kW cooling power. The primary energy consumption efficiency of the combined system and divided system is 55 and 50%, respectively. This means the combined system saves 9.1% of the primary energy. When the recovered heat by the combined system is used for the heat for users, it can save 23.1% of the primary energy if compared with the conventional divided system.

9.8.3 Other Examples of the Adsorption Refrigeration Systems for Waste Heat Utilization

9.8.3.1 Theoretical Research of the Tokyo University

M. Suzuki researched on the application of the solid adsorption air-conditioning systems in cars. The authors believed when the cooling load of the car is 2.3 kW, the molecular sieve – water is suitable for the working pair. When the desorption temperature is set as 473 K, the ambient temperature is 303 K, the adsorption temperature can reach 313 K, and the evaporation temperature is 283 K. If the adsorbent heat transfer distance is 5 mm, the thermal conductivity is 0.2 W/m, the overall heat transfer coefficient of adsorber (UA_0) is estimated can reach 990 W/(m^3K). After the modeling of the adsorbent and the optimization process, the overall heat transfer coefficient can be further improved. Figure 9.86 shows the adsorption capacity variation when $UA = 50UA_0$, the cycle time of adsorption – desorption is set as 60-60, 120-120, 180-180 seconds. The authors suggest that if the UA = 100 kW/(m^3K) the cycle time is 120 seconds (60-60 seconds) and the cooling capacity is 2.8 kW per kilogram adsorbent. If two adsorbent beds are used then 1 kg adsorbent in each adsorbent bed is required.

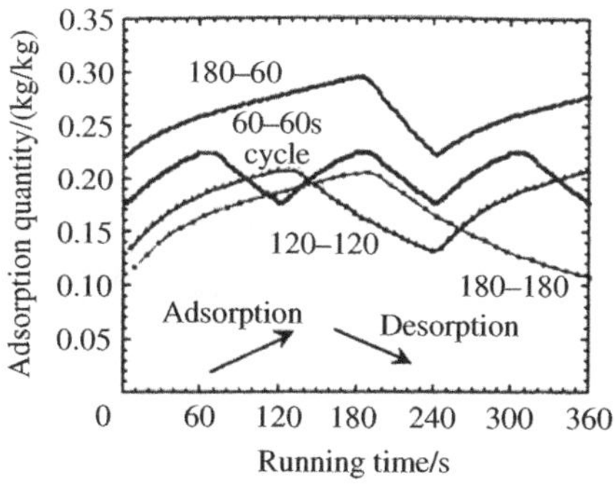

Figure 9.86 Variation of the adsorption capacity with cycle time [61]

9.8.3.2 Adsorption Chiller Developed by MYCOM Company, Japan (Mayekawa Mfg. Co.)

Maekawa (MYCOM) Company cooperated with Tohoku University, and they developed an adsorption air conditioner with silica gelwater adsorption working pair [62]. Two plate-fin type heat exchangers are used as adsorption reactor. Hot water at 55–100 °C (generally 75–95 °C) is used as the heat source for desorption, and the cooling water at 25–35 °C is used in the adsorbent bed (usually it is from the cooling tower and the temperature is at about 29 °C). Refrigeration unit output the chilling water at 9–14 °C, the cycle time is in the range of 5–7 minutes. Figure 9.87 shows the photos of the prototype. Figure 9.88 shows the arrangement diagram of the heat exchanger. Table 9.22 shows the operating parameters published by the Mayekawa Company.

Table 9.22 The parameters of the adsorption chiller

	Model	ADR-20	ADR-30	ADR-100
Hot water	Inlet and outlet water temperature (°C)	75/70	75/70	75/70
	Flow rate (m^3/h)	20	30	101
	Heating power (kW)	120	180	590
Cooling water	Inlet and outlet water temperature (°C)	29/33	29/33	29/33
	Flow rate (m^3/h)	41	62	205
	Cooling load (kW)	190	290	960
Chilling water	Inlet and outlet water temperature (°C)	14/9	14/9	14/9
	Flow rate (m^3/h)	12	18	61
	Cooling capacity (kW)	70	106	352
COP		0.6	0.6	0.6
The power of cooling water pump (kW)		3.7	5.5	18
The power of refrigerant pump (kW)		0.3	0.3	0.6
The power of vacuum pump (kW)		0.3	0.4	0.8
The weight of the chiller (ton)		7.5	11	25
Size (m × m × m)		2.4 × 2.1 × 2.8	3.1 × 2.2 × 2.8	6.3 × 3.1 × 3.5

Figure 9.87 Photo of the adsorption chiller developed by Mayekawa Company

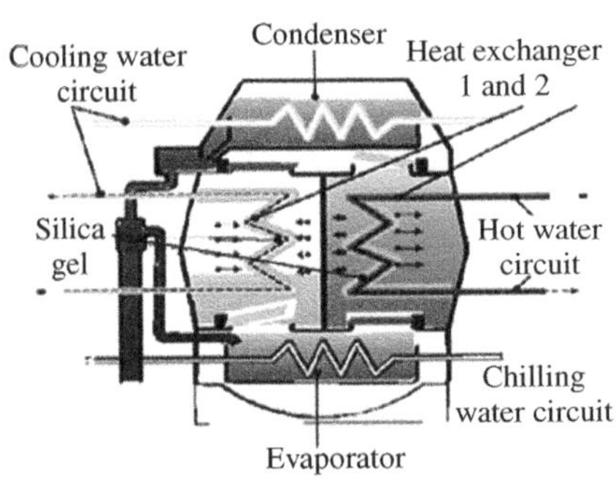

Figure 9.88 Schematic of the adsorption chiller developed by Mayekawa Company

9.8.3.3 Experiment on the Adsorption Refrigeration System Driven by the Waste Heat of the Engine by Dingyu Li

Dingyu Li *et al.* used waste heat of the engine as the heat source for desorption, and the halide salts–ammonia is used as the working pair in the car refrigerators and ice maker on fishing boats [63]. The refrigerators driven by the automobiles (Figure 9.89) has the volume of 60–120 l, and could keep the refrigerating temperature at −12 to 5 °C, when the desorption time is 0.5 hour. The refrigeration time can last 24 hours or maintain at the

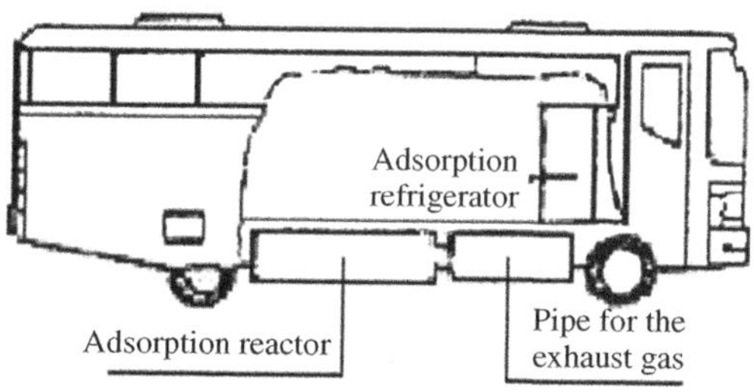

Figure 9.89 Adsorption refrigerator on bus

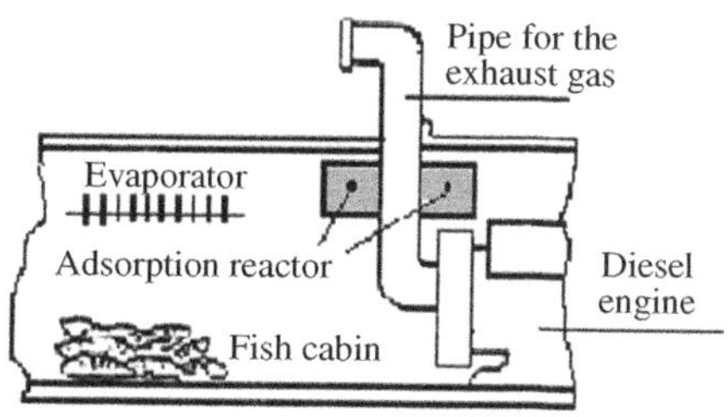

Figure 9.90 Adsorption ice maker on fishing boats

refrigeration temperature for 56 hours. The solid adsorption ice maker prototype on fishing boat (Figure 9.90) developed by them has the cooling power of 6 kW and ice making capacity of 38 kg/h. Generally the desorption process required the waste heat power of the diesel engine larger than 150 horsepower, and the cooling medium is water. The total volume of the adsorption ice maker is 0.59 m^3.

9.8.3.4 CCHP System in Kammenz of Germany and Nagoya of Japan

In Europe, a small CCHP system was installed in Malteser Hospital at Kammenz of Germany with an adsorption refrigerator. Also Nagoya of Japan, Tokai Optical Co., Ltd. installed a small CCHP system based on an adsorption chiller in April 2003. Both systems collected the waste heat from the fuel cell and the solar energy, and then supply heat and cooling by the adsorption chiller. The cooling capacity of the chiller is 105 kW. The system installed a compression chiller at the same time for cooling capacity adjustment. Tokai Optical Co., Ltd. in Nagoya, Japan used the CCHP system with the diesel engine system of 185 kW. The waste heat of this diesel engine can be used in CCHP system; also can be used for dehumidification and cooling. Such a method could reduce the energy consumption by 10% and CO_2 emissions by 12% every year.

References

[1] Research Group of Sustainable Development, Chinese Academy of Science (2000) *The Report for the Strategy of Sustainable Development 2000*, Science Press, Beijing, ISBN: 7-03-008310-5 (in Chinese).

[2] Tchernev, D.I. (1978) *Solar Energy Application of Natural Zeolites-Natural Zeolites: Occurrence, Properties and Use*, Pergamon Press, London.

[3] Pons, M. and Guilleminot, J.J. (1986) Design of an experimental solar-powered, solid-adsorption ice maker. *Transactions of the ASME Journal of Solar Energy Engineering*, **108**(4), 332–337.

[4] Pons, M. (1987) Experimental date on a solar-powered ice maker using activated carbon and methanol adsorption pair. *Transactions of the ASME Journal of Solar Energy Engineering*, **109**(4), 303–310.

[5] Iloeje, O.C. (1985) Design construction and test run of a solar powered solid absorption refrigerator. *Solar Energy*, **35**(5), 447–455.

[6] Hajji, A., Worek, W.M. and Lavan, Z. (1991) Dynamic analysis of a closed-cycle solar adsorption refrigerator using two adsorbent-adsorbate pairs. *Transactions of the ASME Journal of Solar Energy Engineering*, **113**(2), 73–79.

[7] Wang, R.Z., Dai, Y.J., Xu, Y.X., and Wu, J.Y. (1999) Flat-plate type adsorption solar collector. Invention patent of China 99240357.x.

[8] Li, M., Sun, C.J., Wang, R.Z. and Ca, W.D. (2004) Development of no valve solar ice maker. *Applied Thermal Engineering*, **2004**(24), 865–872.

[9] Li, C.H., Wang, R.Z. and Dai, Y.J. (2003) Simulation and economic analysis of a solar-powered adsorption refrigerator using an evacuated tube for thermal insulation. *Renewable Energy*, **28**(2), 249–269.

[10] Sukoda, A. and Suzuki, M. (1984) Fundamental study on solar-powered adsorption cooling system. *Journal of Chemical Engineering of Japan*, **17**(1), 52–57.

[11] Lin, R.T. (1995) *Introduction on the Heat and Mass Transfer of Porous Materials*, Science Press, Beijing, ISBN: 7030047087 (in Chinese).

[12] Leite, A.P.F. and Daguenet, M. (2000) Performance of a new solid adsorption ice maker with solar energy regeneration. *Energy Conversion and Management*, **41**, 1625–1647.

[13] Khattab, N.M. (2004) A novel solar-powered adsorption refrigeration module. *Applied Thermal Engineering*, **24**(17–18), 2747–2760.

[14] Li, M. (1999) The solid adsorption refrigeration cycle for the solar energy utilization. PhD Thesis. Shanghai Jiao Tong University, Shanghai, China (in Chinese).

[15] Boubakri, A., Guilleminot, J.J. and Meunier, F. (2000) Adsorptive solar powered ice maker: experiments and model. *Solar Energy*, **69**(3), 249–263.

[16] Critoph, R.E. (1993) Laboratory testing of an ammonia carbon solar refrigerator. Proceedings of Solar World Congress (ISES), Hungary.

[17] Critoph, R.E., Tamainot-Telto, Z.(1997) Solar sorption refrigerator. *Renewable Energy*, **12**(4), 409–417.

[18] Tamainot-Telto, Z. and Critoph, R.E. (1997) Adsorption refrigerator using monolithic carbon-ammonia pair. *International Journal of Refrigeration*, **20**, 146–155.

[19] Chen, L., Fang, L.G. and Tan, Y.K. (2002) Experimental research on the refrigeration performance of strontium chloride-ammonia working pair. *Acta Energiae Solaris Sinica*, **23**(4), 422–426, ISBN: 0254-0096 (in Chinese).

[20] Erhard, A. and Hahne, E. (1997) Test and simulation of a solar-powered absorption cooling machine. *Solar Energy*, **59**, 155–162.

[21] Erhard, A., Spindler, K. and Hahne, E. (1998) Test and simulation of a solar powered solid sorption cooling machine. *International Journal of Refrigeration*, **21**(2), 133–141.

[22] Hildbrand, C., Dind, P., Pons, M. and Buchter, F. (2004) A new solar powered adsorption refrigerator with high performance. *Solar Energy*, **77**(3), 311–318.

[23] Grenier, P., Guilleminot, J.J., Meunier, F. and Pons, M. (1988) Solar powered solid adsorption cold store. *Journal of Solar Energy – Transactions of ASME*, **110**(3), 192–197.

[24] Fernández-García, A., Zarza, E., Valenzuela, L. and Pérez, M. (2010) Parabolic-trough solar collectors and their applications. *Renewable and Sustainable Energy Reviews*, **14**, 1695–1721.

[25] Kalogirou, S.A. (2004) Solar thermal collectors and applications. *Progress in Energy and Combustion Science*, **30**, 231–295.

[26] Baker, D.K. and Kaftanoğlu, B. (2007) Predicted impact of collector and zeolite choice on the thermodynamic and economic performance of a solar-powered adsorption cooling system. *Experimental Heat Transfer*, **20**, 103–122.

[27] Badran, O. and Eck, M. (2006) The application of parabolic trough technology under Jordanian climate. *Renewable Energy*, **31**, 791–802.

[28] Fadar, A.E., Mimet, A. and Pérez-García, M. (2009) Modelling and performance study of a continuous adsorption refrigeration system driven by parabolic trough solar collector. *Solar Energy*, **83**, 850–861.

[29] El Fadar, A., Mimet, A., Azzabakh, A. *et al.* (2009) Study of a new solar adsorption refrigerator powered by a parabolic trough collector. *Applied Thermal Engineering*, **29**, 1267–1270.
[30] El Fadar, A., Mimet, A. and Pérez-García, M. (2009) Study of an adsorption refrigeration system powered by parabolic trough collector and coupled with a heat pipe. *Renewable Energy*, **34**, 2271–2279.
[31] Li, C., Wang, R.Z., Wang, L.W. *et al.* (2013) Experimental study on an adsorption icemaker driven by parabolic trough solar collector. *Renewable Energy*, **57**, 223–233.
[32] Wang, L.W., Wang, R.Z. and Oliveira, R.G. (2009) A review on adsorption working pairs for refrigeration. *Renewable and Sustainable Energy Reviews*, **13**, 518–534.
[33] Wang, L.W., Wang, R.Z., Wu, J.Y. and Wang, K. (2004) Compound adsorbent for adsorption ice maker on fishing boats. *International Journal of Refrigeration*, **27**, 401–408.
[34] Liu, Z.Y., Xu, H.F. and Lu, Y.Z. (2004) Research and development of the solar cooling pipes. *Acta Energiae Solaris Sinica*, **25**(1), 28–32, ISSN: 0254-0096 (in Chinese).
[35] Wang, R.Z., Dai, Y.J., Xu, Y.X., and Wu, J.Y. (1999) The household air conditioner with the solar energy storage function. Invention patent of China 99124022.7.
[36] Suzuki, M. (1993) Application of adsorption cooling systems to automobiles. *Heat Recovery Systems and CHP*, **13**(4), 335–340.
[37] Mostafavi, M. and Agnew, B. (1997) Thermodynamic analysis of combined diesel engine and absorption refrigeration unit-naturally aspirated engine with precooling. *Applied Thermal Engineering*, **17**(6), 593–599.
[38] Mostafavi, M. and Agnew, B. (1997) Thermodynamic analysis of combined diesel engine and absorption refrigeration unit-naturally aspirated diesel engine. *Applied Thermal Engineering*, **17**(5), 471–478.
[39] Akerman, J.R. (1969) Automotive Air-Conditioning System with Absorption Refrigeration. SAE paper 710037, pp. 821–829.
[40] Han, B.Q., Zhu, R.Q., and Wu, Y.J. (1986) A study on zeolite molecular sieve adsorption refrigeration using exhaust heat from diesel engine. Proceedings of International Conference of Energy Saving in Refrigeration, Xi'an Jiaotong University, China, pp. 104–109.
[41] Zhu, R.Q., Han, B.Q., Lin, M.Z. and Yu, Y.Z. (1992) Experimental investigation on an adsorption system for producing chilled water. *International Journal of Refrigeration*, **15**(1), 31–34.
[42] Ron, M.A. (1984) Hydrogen, heat pump as a bus air conditioner. Proceedings of International Symposium on the Properties and Applications of Metal Hydrogen IV, Eilat, Israel, pp. 9–13.
[43] Lv, C.S. (1994) *Operation and Maintenance of the Dngfeng 4B Locomotive with Internal Combustion Engine*, The Railway Press of China, Beijing, ISBN: 7-113-01683-9 (in Chinese).
[44] Lu, Y.Z. (2003) The zeolite-water adsorption air conditioner driven by the waste heat of the locomotive with the internal combustion engine. PhD thesis. Shanghai Jiao Tong University, Shanghai, China (in Chinese).
[45] Xu, J.Z. (2002) Prospective of the distributed CCHP system. *Energy Conservation and Environmental Protection*, **3**, 10–14ISSN: 1009-539X (in Chinese).
[46] Wang, R.Z. (2002) Some thoughts on the energy saving in the buildings and combined energy systems. *Acta Energiae Solaris Sinica*, **3**(23), 322–335, ISSN: 0254-0096 (in Chinese).
[47] Dai, Y.Q., Geng, H.B., Lu, Z. *et al.* (1996) *The Technology and Application of the Lithium Absorption Refrigeration*, China Machine Press, Beijing, ISBN: 7-111-05336-2 (in Chinese).
[48] Kaarsberg, T. and Elliott, N. (1998) *Combined Heat and Power: Saving Energy and the Environment*, The Northeast-Midwest Economic Review, Washington, DC.
[49] Maidment, G.G. and Tozer, R.M. (2002) Combined cooling heat and power in supermarkets. *Applied Thermal Engineering*, **4**(22), 653–665.
[50] Maidment, G.G. and Prosser, G. (2000) The use of CHP and absorption cooling in cold storage. *Applied Thermal Engineering*, **8**(20), 1059–1073.
[51] Maidment, G.G., Zhao, X., Riffat, S.B. and Prosser, G. (1999) Application of combined heat-and-power and absorption cooling in a supermarket. *Applied Energy*, **7**(63), 169–190.
[52] Vanhanen, J. and Loimaranta, O. (1999) Micro and Mini CHP, Technology and Market Review. Final report. Otaniemi Consulting Group Oy, Helsinki.
[53] Valkiainen, M., Klobut, K., Leppaniemi, S. *et al.* (2002) Micro CHP Systems Based on PEM Fuel Cell. Status report. VTT Tiedotteita-Meddelanden-Research Notes, VTT Chemical Technology, Espoo.
[54] Kong, X.Q. (2005) The micro-CCHP system based on the internal combustion engine and adsorption chiller. PhD thesis. Shanghai Jiao Tong University, Shanghai, China (in Chinese).
[55] Zhang, Y.M. (2000) The economic analysis for the district hot water supply. *Energy Research and Utilization*, **5**, 39–41, ISSN: 1001–5523 (in Chinese).

[56] Mone, C.D., Chau, D.S. and Phelan, P.E. (2001) Economic feasibility of combined heat and power and absorption refrigeration with commercially available gas turbines. *Energy Conversion and Management*, **42**, 1559–1573.

[57] Lai, Y.H. and Liu, Z.Q. (1999) The economic analysis of CCHP system. *Journal of Refrigeration*, **1**, 52–56, ISSN: 0253-4339 (in Chinese).

[58] Wang, R.Z. and Ding, G.L. (2002) *The Latest Refrigeration Technology*, Science Press, Beijing, ISBN: 7030106776 (in Chinese).

[59] Havelsky, V. (1999) Energetic efficiency of cogeneration systems for combined heat, cold and power production. *International Journal of Refrigeration*, **22**, 479–485.

[60] Ziegler, F. and Riesch, P. (1993) A review with regard to energetic efficiency: absorption cycles. *Heat Recovery System and CHP*, **13**, 147–159.

[61] Research Group of Sustainable Development, Chinese Academy of Science (2003) *The report for the strategy of sustainable development 2003*, Science Press, Beijing, ISBN: 703011085 (in Chinese).

[62] Ongiro, A., Isemt Ugursal, V., Tweed, A.L. *et al.* (1996) Thermodynamic simulation and evaluation of a steam CHP plant using ASPEN Plus. *Applied Thermal Engineering*, **16**(3), 263–271.

[63] Li, D.Y. (2000) The adsorption refrigeration systems with double generators. Invention patent of China 20001260467.

Index

Adsorption Refrigeration Technology: Theory and Application, First Edition. Ruzhu Wang, Liwei Wang and Jingyi Wu.

Companion Website: www.wiley.com/go/wang/refrigeration

Printed and bound by CPI Group (UK) Ltd, Croydon, CR0 4YY
16/09/2026
14963405-0002